Native Gold as a Specific Indicator Mineral for Gold Deposits

Native Gold as a Specific Indicator Mineral for Gold Deposits

Editor

Galina Palyanova

MDPI

Basel • Beijing • Wuhan • Barcelona • Belgrade • Novi Sad • Cluj • Manchester

Editor
Galina Palyanova
V.S. Sobolev Institute of
Geology and Mineralogy
Siberian Branch of the Russian
Academy of Sciences
Novosibirsk
Russia

Editorial Office
MDPI
St. Alban-Anlage 66
4052 Basel, Switzerland

This is a reprint of articles from the Special Issue published online in the open access journal *Minerals* (ISSN 2075-163X) (available at: www.mdpi.com/journal/minerals/special_issues/native_gold).

For citation purposes, cite each article independently as indicated on the article page online and as indicated below:

Lastname, A.A.; Lastname, B.B. Article Title. *Journal Name* **Year**, *Volume Number*, Page Range.

ISBN 978-3-0365-9381-4 (Hbk)
ISBN 978-3-0365-9380-7 (PDF)
doi.org/10.3390/books978-3-0365-9380-7

Cover image courtesy of Galina Palyanova

Contents

Preface

This Special Issue reprint includes one review and fifteen research articles. The review by Palyanova et al. (2023) summarized the available information on the palladian gold and proposed classifying the types of deposits based on the fineness, content, and set of impurities in palladian gold and minerals intergrown with it. Fifteen articles focused on the typomorphic features of native gold from gold deposits in Russia, India, Chile, Scotland, and Canada. This Special Issue aims to identify and generalize the reasons for variations in the compositions of native gold. I hope all the articles in this Special Issue will be helpful and valuable resources for anyone who is interested in native gold and gold deposits. I thank the Authors and Reviewers for their efforts and contributions. I am also grateful to the Editor-in-Chief, Editors and Assistant Editors of *Minerals* for their help.

Galina Palyanova
Editor

minerals MDPI

Editorial

Editorial for the Special Issue "Native Gold as a Specific Indicator Mineral for Gold Deposits"

Galina A. Palyanova

V.S. Sobolev Institute of Geology and Mineralogy, Siberian Branch of the Russian Academy of Sciences, Akademika Koptyuga pr., 3, Novosibirsk 630090, Russia; palyan@igm.nsc.ru

Citation: Palyanova, G.A. Editorial for the Special Issue "Native Gold as a Specific Indicator Mineral for Gold Deposits". *Minerals* **2023**, *13*, 1323. https://doi.org/10.3390/min13101323

Received: 31 August 2023
Revised: 6 October 2023
Accepted: 6 October 2023
Published: 13 October 2023

Native gold is the most common and significant industrial gold mineral. Pure native gold without impurities is rare. The main impurity in native gold is silver, which forms a continuous solid solution characterized by fineness from 0 to 1000‰. At some deposits, the fineness of native gold is high and varies over a narrow range, whereas at others, it is low and covers a wide range of values. Cu, Hg, and Pd concentrations vary from trace to high in native gold. These metals form solid solutions with gold (Au,Hg; Au,Pd) or intermetallic compounds (AuCu, Cu_3Au). Other elements, including Pt, Rh, Ir, Fe, As, Sb, S, Se, Te, Bi, Ti, Cr, Ni, Co, Mn, W, Sn, U, Th, and He, and rare earth, alkaline, and alkaline earth elements are less common and, when detectable, occur in minor amounts (between 1 and 0.01%) in native gold. The chemical composition of native gold is one of the most important typomorphic features, allowing experts to predict possible original sources of metal for placers. The quantity of impurities in native gold varies drastically between different types of deposits and primarily depends on the physicochemical conditions of formation and the metallogenic features of gold-bearing provinces. The concentration levels and ratios of various elements in gold grains can provide a geochemical history of ore-forming events. In recent years, special attention has been paid to studying mineral inclusions in native gold and matrix minerals since they are critical informative features. Minerals intergrown with native gold can be the best parameter to identify different styles of mineralization.

This Special Issue is a continuation of three previous Special Issues [1–3]. It includes 16 articles focused on the typomorphic features of native gold from gold deposits in Russia, India, Chile, Scotland, and Canada. This Special Issue aims to identify and generalize the reasons for variations in the compositions of native gold.

The paper by Savva et al. [4] contains the results of a detailed comparative description of the typomorphic features of native gold from 14 gold deposits in the northeast of Russia. These gold deposits are grouped as follows: gold–arsenic–sulfide disseminations in black shale strata (Natalka, Degdekan, Karalveem, Maldyak), gold–quartz veins in granitoids (Dorozhnoye, Butarnoye, Shkolnoye, Maltan), and epithermal gold–silver adularia in volcanogenic strata (Kupol, Olcha, Kubaka, Burgali, Primorskoe, Dalnee). The typomorphic features of native gold, such as the content of impurities, variations in fineness, internal structure, and mineral associations, are reliable indicators for identifying the three different settings of gold deposits.

In the article by Anand and coauthors [5], the petrography, mineral chemistry, fluid inclusion, and sulfur isotopic compositions of gold mineralization associated with magnetite and apatite were used to understand the possible source of ore-bearing fluids. The authors conclude that this deposit can be classified as an iron–copper–gold oxide (IOCG) and iron oxide–apatite (IOA) type of deposit.

The paper by Fridovsky et al. [6] studied five orogenic gold deposits (Malo-Taryn, Badran, Khangalas, V'yun, and Shumniy) in the central sector of the Yana-Kolyma metallogenic belt in northeast Russia. The native gold in quartz veins has a fineness of 800–900‰, and "invisible" gold is disseminated in arsenian pyrite-3 and arsenopyrite-1 in the ore zones of these deposits. They report new data on the microtextures, chemical composition, and stable sulfur isotopes of auriferous pyrite-3 and arsenopyrite-1 from proximal alterations in sediment-hosted (Malo-Taryn, Badran, Khangalas) and intrusion-hosted (V'yun, Shumniy) orogenic Au deposits. The analytical results indicate that subcrustal and metamorphic systems in the Late-Jurassic-to-Early Cretaceous Verkhoyansk-Kolyma orogeny were, probably, the primary source of S and Au and mineralizing fluids.

Mayorova et al. [7] investigated the typomorphic features of native gold from gold–sulfide–quartz and gold–arsenic–sulfide ore occurrences in the Manitanyrd region of the Polar Urals, Russia. These occurrences are localized in volcanogenic and volcanogenic-sedimentary Precambrian rocks. The isotope composition of sulfur in sulfides indicates a single deep magmatic source of sulfur in ore formation. The similarity of the geological–structural, mineralogical–geochemical, and isotope–geochemical features of the gold–sulfide–quartz and gold–arsenic–sulfide occurrences suggests their formation in a single hydrothermal system.

Kondratieva and coauthors [8] studied the typomorphism of native gold in the Aldan-Stanovoya gold-bearing province, Aldan Shield, Russia. Denudation processes and gold placer formations affected the morphological and geochemical properties of primary ores' native gold and eluvial deposits' supergene gold in this area.

Murzin and coauthors [9] investigated the mineralogy and sulfur isotope geochemistry in sulfides from one Au-PGE deposit, Ozernoe, at the Dzelyatyshor wehrlite–pyroxenite massif, Polar Urals, Russia. They concluded that the variations of sulfur isotopic composition in pyrite, chalcopyrite, and bornite indicated that a deep-seated magmatic basic melt was the source of the ore-forming fluid, ore components, and sulfur during Au-Pd ore formation.

Palyanova et al. [10] studied the composition and structure of uncommon Pd,Hg-rich placer gold from watercourses draining the Itchayvayam mafic-ultramafic complex, Kamchatka, Russia. Their paper contains a review and a considerable amount of data on the composition of minerals in the Au–Pd–Hg system and the formation conditions of the Pd,Hg-bearing gold and Au-bearing potarite in other regions across the world. These authors discussed the genesis of Pd,Hg-rich gold and concluded that meteoric waters or low-temperature hydrotherms rich in Pd and Hg could lead to replacing Pd,Hg-poor gold with Pd,Hg-rich gold.

The review by Palyanova et al. [11] summarized the available information on the palladian gold of several compositions and types of isomorphic impurities (Ag, Cu, Hg). The authors concluded that palladian gold more frequently corresponds to the Au–Pd–Ag, Au–Pd–Ag–Cu and Au–Pd–Ag–Cu–Hg systems and less often to Au–Pd, Au–Pd–Hg, Au–Pd–Cu, Au–Pd–Ag–Hg. They also concluded that palladian gold belongs with deposits where the main components of ores are PGE, Cr, Cu, Ni, V, and Ti. They propose classifying the types of deposits based on the fineness, content, and set of impurities in palladian gold and minerals intergrown with it.

Several articles focus on the typomorphic features and sources of native gold placer deposits [12–19]. Becerra et al. [12] studied the morphology, composition, and mineral inclusions in native gold from terraced placers in the Pureo Area (Southern Chile). The authors suggest two possible primary sources of the placer gold: Paleozoic-Triassic metamorphic rocks or hydrothermal deposits associated with Cenozoic intrusive activity. Their results have implications for exploring new placer deposits and gold-bearing hypogene deposits of the southern Chile Coastal Cordillera.

Nikiforova [13] studied the internal structures, chemical composition, and microinclusions of placer gold from the southeast Siberian platform in great detail. The author documented two stages of ore formation: the first in the Precambrian and the second in the Mesozoic. The internal structures typical of endogenous and exogenous conditions can be used for forecasting ore sources and types of gold deposits of the Siberian platform.

Lalomov and coauthors [14] studied the transformation of placer gold from the Sykhoi Log placer deposits in the Bodaibo gold-bearing district, Siberia, Russia. Weathering crusts developed along the zones of disjunctive dislocations near the Sukhoi Log gold deposit.

The morphology, chemical signatures, structure, and inclusions of placer gold from four different locations led to the conclusion that weathering rims develop mainly from chemical reactions with infiltrating fluid.

The article by Zmodik and coauthors [15] studied the morphology, composition, intergrowths, and microinclusions of native gold from the alluvial deposits of the Kamenny stream Ozerninsky area, Western Transbaikalia, Russia. The Kamenny placer is of particular interest due to finding gold–brannerite nuggets, unique both in size and in weight (from 1–2 g to 200 g), of various microtextures, and with mineralogical and geochemical features. The nuggets also contain hematite (±magnetite), W-bearing rutile, barite, less often muscovite, quartz, siderite, goethite, nano- and micro-inclusions of uraninite, native Pb or Pb oxide, petzite, tellurobismuthite, altaite, and single grains of chalcopyrite. Four types of placer gold were documented, and possible primary sources were postulated.

Okrugin and Gerasimov [16] studied palladian gold and platinum minerals in placers of diamonds and precious metals, widespread in the Anabar River basin on the northeastern part of the Siberian Platform, Russia. The composition of palladian gold, ferroan platinum, and other associated minerals and their microstructural relationships indicate a paragenetic connection between native gold and platinum group minerals. The authors conclude that the primary sources of the placers of the Anabar River basin are alkaline-ultrabasic massifs and carbonatites.

The paper by Gerasimov [17] focuses on the typomorphism of native gold from two modern placer occurrences, "Billyah" and "Nebaibyt", in the Anabar Region northeast of the Siberian platform, Russia. The results of this study indicate that the native gold in these modern placers comes from two different sources: paleo-placers and nearby "primary" ores.

The paper by Chapman et al. [18] is devoted to the alluvial gold from six localities south of Loch Tay in central Scotland. The compositional features and inclusions in native gold were interpreted to represent two broad geographical groupings. This paper demonstrates that research on gold composition and microinclusions can contribute to exploration campaigns.

Chapman et al. [19] reported the results of studying detrital gold from 160 localities of nine different deposit types in British Columbia. They documented that different deposits have specific compositional characteristics of native gold and specific suites of mineral inclusions. Their data demonstrate that mineral inclusion assemblages in gold particles provide far more information than the fineness of native gold. The authors developed compositional templates for identifying diagnostic signatures of native gold from different types of deposits, including orogenic, low-sulfidation epithermal, and alkalic porphyry settings.

A new Special Issue of *Minerals*, "Native Gold as a Specific Indicator Mineral for Gold Deposits, volume 2" [20], is planned. Three previous Special Issues [1–3] were printed as books [21–23]. This new Special Issue aims to further develop effective criteria for forecasting and searching for gold deposits beyond the scope of this Special Issue.

Funding: Work is conducted on the state assignment of the V.S. Sobolev Institute of Geology and Mineralogy, Siberian Branch, Russian Academy of Sciences (No. 122041400237-8).

Conflicts of Interest: The author declares no conflict of interest.

References

1. Special Issue "Experimental and Thermodynamical Modeling of Ore-Forming Processes in Magmatic and Hydrothermal Systems". Available online: https://www.mdpi.com/journal/minerals/special_issues/Experimental_Thermodynamical_Modeling (accessed on 13 December 2018).
2. Special Issue "Mineralogy of Noble Metals and "Invisible" Speciations of These Elements in Natural Systems". Available online: https://www.mdpi.com/journal/minerals/special_issues/noble_metals (accessed on 26 February 2020).
3. Special Issue "Mineralogy of Noble Metals and "Invisible" Speciations of These Elements in Natural Systems, Volume II". Available online: https://www.mdpi.com/journal/minerals/special_issues/noble_metals_volume_ii (accessed on 28 July 2021).
4. Savva, N.E.; Kravtsova, R.G.; Anisimova, G.S.; Palyanova, G.A. Typomorphism of Native Gold (Geological-Industrial Types of Gold Deposits in the North-East of Russia). *Minerals* **2022**, *12*, 561. [CrossRef]

5. Anand, A.; Singh, S.; Gantait, A.; Srivastava, A.; Mayachar, G.K.; Kumar, M. Geological Constraints on the Genesis of Jagpura Au-Cu Deposit NW India: Implications from Magnetite-Apatite Mineral Chemistry, Fluid Inclusion and Sulfur Isotope Study. *Minerals* **2022**, *12*, 1345. [CrossRef]
6. Fridovsky, V.Y.; Polufuntikova, L.I.; Kudrin, M.V. Origin of Disseminated Gold-Sulfide Mineralization from Proximal Alteration in Orogenic Gold Deposits in the Central Sector of the Yana–Kolyma Metallogenic Belt, NE Russia. *Minerals* **2023**, *13*, 394. [CrossRef]
7. Mayorova, T.P.; Kuznetsov, S.K.; Efanova, L.I.; Sokerina, N.V. Gold–Sulfide Mineralization in the Manitanyrd Region, Polar Urals, Russia. *Minerals* **2023**, *13*, 747. [CrossRef]
8. Kondratieva, L.A.; Anisimova, G.S.; Kardashevskaia, V.N. Ore Mineralogy and Typomorphism of Native Gold of the Spokoininsky Cluster of the Aldan–Stanovoy Gold Province. *Minerals* **2023**, *13*, 543. [CrossRef]
9. Murzin, V.; Palyanova, G.; Mayorova, T.; Beliaeva, T. The Gold–Palladium Ozernoe Occurrence (Polar Urals, Russia): Mineralogy, Conditions of Formation, Sources of Ore Matter and Fluid. *Minerals* **2022**, *12*, 765. [CrossRef]
10. Palyanova, G.; Kutyrev, A.; Beliaeva, T.; Shilovskikh, V.; Zhegunov, P.; Zhitova, E.; Seryotkin, Y. Pd,Hg-Rich Gold and Compounds of the Au-Pd-Hg System at the Itchayvayam Mafic-Ultramafic Complex (Kamchatka, Russia) and Other Localities. *Minerals* **2023**, *13*, 549. [CrossRef]
11. Palyanova, G.A.; Zhegunov, P.S.; Beliaeva, T.V.; Murzin, V.V.; Borovikov, A.A.; Goryachev, N.A. Palladian Gold: Chemical Composition, Minerals in Association, and Physicochemical Conditions of Formation at Different Types of Gold Deposits. *Minerals* **2023**, *13*, 1019. [CrossRef]
12. Becerra, P.; Sanchez-Alfaro, P.; Piquer, J.; Plissart, G.; Garroz, B.; Kunstmann, D. Gold Provenance in Placers from Pureo Area, Southern Chile Coastal Cordillera, and Their Relationship with Paleozoic Metamorphic Rocks. *Minerals* **2022**, *12*, 1147. [CrossRef]
13. Nikiforova, Z. Internal Structures of Placer Gold as an Indicator of Endogenous and Exogenous Processes. *Minerals* **2023**, *13*, 68. [CrossRef]
14. Lalomov, A.; Grigorieva, A.; Kotov, A.; Ivanova, L. Typomorphic Features and Source of Native Gold from the Sykhoi Log Area Placer Deposits, Bodaibo Gold-Bearing District, Siberia, Russia. *Minerals* **2023**, *13*, 707. [CrossRef]
15. Zhmodik, S.M.; Airiyants, E.V.; Belyanin, D.K.; Damdinov, B.B.; Karmanov, N.S.; Kiseleva, O.N.; Kozlov, A.V.; Mironov, A.A.; Moroz, T.N.; Ponomarchuk, V.A. Native Gold and Unique Gold–Brannerite Nuggets from the Placer of the Kamenny Stream, Ozerninsky Ore Cluster (Western Transbakalia, Russia) and Possible Sources. *Minerals* **2023**, *13*, 1149. [CrossRef]
16. Okrugin, A.; Gerasimov, B. Paragenetic Association of Platinum and Gold Minerals in Placers of the Anabar River in the Northeast of the Siberian Platform. *Minerals* **2023**, *13*, 96. [CrossRef]
17. Gerasimov, B. The Use of Typomorphic Features of Placer Gold of the Anabar Region for Determining Its Sources. *Minerals* **2023**, *13*, 480. [CrossRef]
18. Chapman, R.; Torvela, T.; Savastano, L. Insights into Regional Metallogeny from Detailed Compositional Studies of Alluvial Gold: An Example from the Loch Tay Area, Central Scotland. *Minerals* **2023**, *13*, 140. [CrossRef]
19. Chapman, R.; Mortensen, J.K.; Murphy, R. Compositional Signatures of Gold from Different Deposit Types in British Columbia, Canada. *Minerals* **2023**, *13*, 1072. [CrossRef]
20. Special Issue "Native Gold as a Specific Indicator Mineral for Gold Deposits, 2nd Edition". Available online: https://www.mdpi.com/journal/minerals/special_issues/A1D082156T (accessed on 24 August 2023).
21. Palyanova, G. *Experimental and Thermodynamical Modeling of Ore-Forming Processes in Magmatic and Hydrothermal Systems*; MDPI: Basel, Switzerland, 2019; Available online: https://www.mdpi.com/books/book/1081 (accessed on 30 August 2023).
22. Palyanova, G. *Mineralogy of Noble Metals and "Invisible" Speciations of These Elements in Natural Systems*; MDPI: Basel, Switzerland, 2020; Available online: https://www.mdpi.com/books/book/2273 (accessed on 30 August 2023).
23. Palyanova, G. *Mineralogy of Noble Metals and "Invisible" Speciations of These Elements in Natural Systems, Volume II*; MDPI: Basel, Switzerland, 2021. Available online: https://www.mdpi.com/books/book/4627 (accessed on 30 August 2023).

minerals

MDPI

Article

Typomorphism of Native Gold (Geological-Industrial Types of Gold Deposits in the North-East of Russia)

Natalia E. Savva [1], Raisa G. Kravtsova [2], Galina S. Anisimova [3,*] and Galina A. Palyanova [4]

1 Shilo North-East Interdisciplinary Scientific Research Institute, Far Eastern Branch, Russian Academy of Sciences, Portovaya Str. 16, 685000 Magadan, Russia; savva@neisri.ru
2 Vinogradov Institute of Geochemistry, Siberian Branch, Russian Academy of Sciences, Favorskiy Str. 1A, 664033 Irkutsk, Russia; krg@igc.irk.ru
3 Diamond and Precious Metal Geology Institute, Siberian Branch, Russian Academy of Sciences, Lenina Ave. 39, 677007 Yakutsk, Russia
4 Sobolev Institute of Geology and Mineralogy, Siberian Branch, Russian Academy of Sciences, Akademika Koptyuga pr. 3, 630090 Novosibirsk, Russia; palyan@igm.nsc.ru
* Correspondence: gsanisimova@diamond.ysn.ru

Abstract: This study presents the typomorphic features of native gold grains from three different geological-industrial types (GIT) of gold deposits in the North-East of Russia: (1) gold–arsenic-sulfide in black shale strata (Natalka, Degdekan, Karalveem, Maldyak deposits), (2) gold–quartz veins in granitoids (Dorozhnoye, Butarnoye, Shkolnoye, Maltan deposits), and (3) gold–silver adularia in volcanogenic strata (Kupol, Olcha, Kubaka, Burgali, Primorskoe, Dalnee deposits). The reliability of the geological interpretation is directly related to mineral associations, fineness variations, its internal structure and the content of microimpurities. Native gold is a reliable indicator for identifying various GIT of gold deposits at the early geological-prospecting stages of studying gold-bearing areas. Typomorphic features of native gold for each of the considered GIT are stable and do not depend on the age and scale of mineralization. It is shown that using an integrated approach obtains genetic information about a particular ore object, which makes it possible to predict the vertical range of mineralization and outline the technology for processing ores. The information obtained can also be effectively used in the search for placer deposits in nearby watercourses. Identification of typomorphic features of ore and placer native gold opens up wide opportunities for delineating the distribution areas of placer deposits.

Keywords: geological-industrial types; ore formations; typomorphic features of native gold; gold deposits in the north-east of Russia

check for updates

Citation: Savva, N.E.; Kravtsova, R.G.; Anisimova, G.S.; Palyanova, G.A. Typomorphism of Native Gold (Geological-Industrial Types of Gold Deposits in the North-East of Russia). *Minerals* **2022**, *12*, 561. https://doi.org/10.3390/min12050561

Academic Editor: Liqiang Yang

Received: 28 March 2022
Accepted: 26 April 2022
Published: 29 April 2022

1. Introduction

Successful exploration for gold is impossible without scientifically based forecasts, the reliability of which increases significantly due to the improvement of methods for integrated study of gold ores and the formation of their industrial grouping on this basis. Fundamentals of the geological and industrial grouping of mineral deposits were laid down by [1], who defined the geological-industrial type (GIT) as "a group of geologically similar deposits that have proven themselves in world and domestic practice as a real supplier of this type of mineral raw material". It was V.M. Kreiter who proposed to classify as industrial such natural geological and mineralogical types of deposits that provide more than 1% of the world production of a certain mineral type. It should be noted that, due to the fact that each deposit has some specific features, the unified technological schemes of concentrating plants at large mines are set up to process ores of one bulk geological and industrial type with all the range of its mineral varieties [2].

V.M. Kreiter's followers continued working in this direction [3–8]. They showed that the geological and industrial grouping of deposits is determined by their geological homogeneity and belonging to certain gold formations—groups of deposits with a similar material composition of ores and genesis. In Northeast Asia, gold deposits are most widely developed, which are grouped into the following GIT types: gold–arsenic-sulfide in black shale strata, gold–quartz veins in granitoids, gold–silver adularia in volcanogenic strata [3]. When studying a particular deposit and its geological and industrial typification, it often turns out that each mineral variety of ores may well be mistaken by researchers for a new or unconventional GIT. This often occurs on poorly studied fragments of gold deposits, ore fields and nodes. In this regard, at an early stage of prospecting and exploration, when predicting the industrial type, an important role belongs to the establishment of a set of typomorphic features of native gold obtained during its study. To be precise, these features give contrasting differences for each group of deposits. The methodical manual for the study of native gold in [9] allows researchers to develop a scientifically based forecast.

The description of the typomorphic features of native gold for each particular gold deposit makes it possible to obtain genetic information about a particular object. Native gold, due to chemical resistance, is capable of such information. To date, plenty of factual material has been obtained on the indicator properties of native gold in the study of ore and placer deposits for various regions of the world [10–23]. The beginning of a comprehensive study of this mineral was laid by outstanding works [24,25].

The concept of the typomorphism of minerals was formulated by A.E. Fersman [26] and developed in the form of the teaching of N.P. Yushkin [27,28]. It is that the mineral composition and properties contains information about its genetic nature, indicating conditions for the formation of the entire deposit. The authors of the doctrine emphasize that over the course of geological time, such information may be erased. In this regard, native gold can be attributed to one of the most stable minerals, preserving genetic information for a long time due to the ability to resist chemical weathering and the absence of brittle deformations. Thus, the complexity of constitutional features of native gold (composition, structure, mineral intergrowths) has excellent indicator qualities.

This study presents a solid dataset of using typomorphism of native gold grains from various types of gold deposits in northeastern Russia as examples to show the importance of applying such a concept. This will allow, with a limited amount of material, at an early prospecting and exploration stage of geological research, to determine the type of deposit, to predict the vertical range of mineralization and a possible method of processing ores.

2. Regional Geological Setting and GITs of Gold Deposits

2.1. Geological and Structural Position of Northeast Asia

In geological and structural terms, the north-east of Russia is a complex composition of the Kolyma Chukchi and Primorsky terranes of passive continental margins, island-arc and oceanic terranes, as well as middle cratons. According to A.I. Khanchuk [29], all layers of the earth's interior—from the Precambrian strata to the modern majestic volcanoes of Kamchatka—are exposed in this region and are available for the most detailed studies. Large gold deposits in the region are characterized by different tectonomagmatic stages of their development. The GITs of the deposits considered by us are confined to large structural elements, including the following: the passive margin of the Siberian craton (deposits of the Verkhoyansk complex); collisional structures in the zone of the Tenka deep fault with large zones of tectonomagmatic activation, Omolon cratonic terrane, Kedon—Late Paleozoic and Okhotsk-Chukotka—Mesozoic marginal-continental volcanic belts (Figure 1).

The article will consider three, contrastingly different, GITs of gold deposits in the north-east of Russia: (1) gold–arsenic-sulfide in black shale strata (Natalkinskoye, Degdekan, Maldyak, Karalveem); (2) gold–quartz veins in granitoids (Shkolnoye, Butarnoye, Dorozhnoye, Maltan), (3) gold–silver adularia in volcanogenic strata (Kubaka, Kupol, Olcha, Burgali, Primorskoe, Dalnee). Further in the text and captions, the abbreviated names of these GITs will be used—gold–arsenic-sulfide, gold–quartz veins and gold–silver adularia.

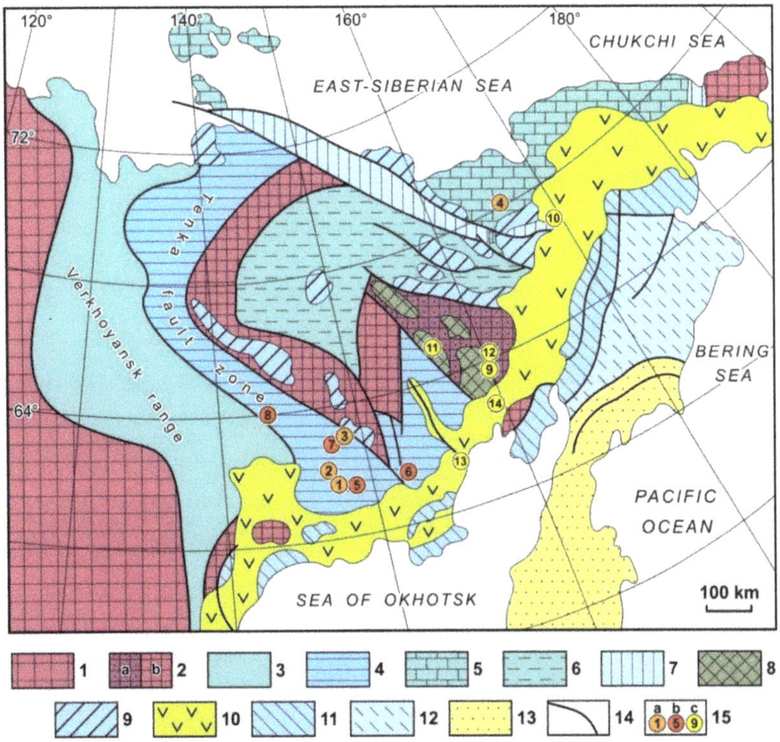

Figure 1. Location of research objects on a schematic geological and structural map of the north-east of Russia. The image was constructed by the authors of the present paper using data from [30]. The legend contains the following: 1—Siberian craton; 2—terranes with continental crust of the Siberian Craton (a—Omolon, b—others); 3, 4—deposits of the passive margin of the Siberian craton: 3—Paleozoic-Mesozoic deposits, 4—Mesozoic deposits; 5—folded cover of the Chukotka; 6—Paleozoic-Mesozoic deposits of the Verkhoyansk complex; 7, Late Mesozoic collisional sutures; 8—Kedon Late Paleozoic continental marginal volcanic belt; 9—Late Jurassic-Early Cretaceous volcanic belts; 10—Okhotsk-Chukotka Late Cretaceous volcanic belt; 11–13, Koryak–Kamchatka accretionary belt: 11—Late Jurassic–Early Cretaceous island-arc systems, 12—Late Mesozoic volcanic belts, 13—Cenozoic volcanic belts; 14—deep faults; 15—the main GITs of gold deposits: a—gold–arsenic-sulfide in black shale strata (1—Natalka, 2—Degdekan, 3—Maldyak, 4—Karalveem), b—gold–quartz veins in granitoids (5—Shkolnoye, 6—Butarnoye, 7—Dorozhnoye, 8—Maltan), c—gold–silver adularia in volcanogenic strata (9—Kubaka, 10—Kupol, 11—Olcha, 12—Burgali, 13—Primorskoe, 14—Dalnee).

2.2. *GITs of Gold Deposits*

2.2.1. Gold–Arsenic-Sulfide in Black Shale GIT

In genetic terms, these are hydrothermal-metamorphogenic deposits, localized mainly in the Permian-Triassic sedimentary strata. In the north-east of Russia, deposits of this type are mainly concentrated in the area of the Kularo-Nera terrane and are controlled by the Tenkinsky deep fault zone, which forms all gold deposits into a single Yano-Kolyma gold-bearing belt. In Chukotka, a similar type of deposits is also noted in the area of the Chukchi terrane. All the deposits of this GIT described the similarity of tectonomagmatic conditions of formation, paragenetic associations, as well as similar conditions for the development of the ore process, can be attributed to the gold–quartz geological formation of the medium-deep type.

Natalka Deposit

The Natalka deposit is one of the largest in Russia (the giant deposit) and is located in the Tenkinsky district of the Magadan region. Structurally speaking, the deposit is confined to the zone of the Tenkinsky deep fault and is associated with the collisional stage of development of the Yano-Kolyma gold-bearing belt as part of the Verkhoyansk-Chukotka folded system. The age of the host rocks is Late Paleozoic (P_3-T_1), and the ores are presumably Mesozoic (J_3-K_1). The NW bearing of the ore deposit (320–340°) in hydrothermally altered, predominantly silicified Permian-Triassic sedimentary strata is a mineralized zone penetrated by a network of quartz veins, lenses, brecciation areas, thin branching or parallel veinlets, with areas of massive silicification and arsenopyritization of varying intensity. The most common ore minerals are arsenopyrite and pyrite, less common are galena, chalcopyrite, sphalerite, pyrrhotite, rutile, and native gold. The size of gold grains is from 0.01 to 2 mm. A detailed description is given in [31].

Maldyak Deposit

The Maldyak deposit is located in the basin of the Berelekh river (upper reaches of the Kolyma river). It occurs in the Kularo-Nera terrane in the influence zone of the Tenka deep fault. The deposit was explored in 1938–1949, partially exploited in 1945–1948, and on a small scale in 1994–1998. It is estimated as large. At individual intersections defined by drilling, the concentration of Au in ores reaches 20.1 g/t [32]. Currently, prospecting and exploration work has been resumed at the Maldyak field. The ore field area of about 20 km^2 is composed of Middle Jurassic marine terrigenous deposits (J_2): shale, siltstone, sandstone. Intrusive rocks are represented by felsic and intermediate dikes belonging to the Ner-Bokhapcha complex ($\gamma\alpha J_3$ nr-bhr). The rocks form a complex linear folded structure with a general northwest bearing of 340°. Mineralization is developed both in sedimentary rocks and in dikes. Hydrothermal transformations of sedimentary rocks—silicification, in dikes—propylitization and beresitization. Ore bodies are represented by differently oriented veins, veinlets, and veined and veinlet-disseminated ore folds with a thickness of 0.5 to 3 m. Their mineral composition is dominated by quartz, accompanied by calcite, dolomite, sericite, chlorite, carbonaceous matter, and among ore minerals there is arsenopyrite, pyrite and native gold [10]. The size of gold grains is 0.01–3.50 mm.

Karalveem Deposit

The Karalveem deposit is located in the Bilibinsky district of the Chukotka Autonomous Okrug. Structurally, it is confined to the Anyui subterrane of the Chukchi terrane, which is considered as a fragment of the Late Paleozoic–Early Mesozoic passive margin of the continent [30]. The mineralization is located among Triassic gabbro-diabase sills, which, together with the enclosing sand-shale deposits of the same age, were turned into folds. The ore field, 15 km by 3 km in size, is elongated in a northwesterly direction along folded structures and is surrounded by outcrops of granitoids (K_1) that intersect diabases. Sedimentary rocks have intense contact changes and are transformed in a halo of about 1 km into cordierite-andalusite-biotite hornfelses. The mineralization is located among the Triassic gabbro-diabase sills, which, together with the enclosing sandy-shale deposits of the Triassic age, are crumpled into folds. The age of mineralization dated by K-Ar method is 130 Ma (SVKNII FEB RAS). The ores are partially transformed by Early Cretaceous granitoids (according to U-Pb determinations, 112 Ma). The deposit belongs to the vein type, where NW-bearing quartz vein bodies have a thickness of 0.5 to 2 m. The main vein minerals are quartz, ankerite, scheelite, ore minerals are arsenopyrite, galena, and native gold. The bonanza distribution of gold in veins with grades from traces to 190 g/t is typical [32].

Degdekan Deposit

The Degdekan deposit is located in the Tenkinsky district of the Magadan region (60 km north of the Natalka deposit). The ore-bearing area is confined to the Omchak-

Nelkoba metallogenic zone, which was revealed based on the interpretation of gravimetric data. Structural features of the Degdekan ore field are due to different scale manifestations of dynamometamorphism in Permian carbonaceous shales (P_{2-3}). Sedimentary rocks are intruded by Upper Jurassic (J_3) diorite porphyry dikes and Late Cretaceous (K_1) rhyodacites and dolerites of the Nera-Bokhapcha complex. Metasomatic transformations are represented by intense silicification, to a lesser extent by Fe-Mg carbonatization and albitization with cuts in carbonaceous matter. The ore bodies are represented by gold-bearing quartz-vein zones and quartz veins of sublatitudinal bearing (340–350°) up to 2 m thick. The main vein minerals are quartz, carbonates and ore minerals such as pyrite, pyrrhotite, marcasite, and chalcopyrite. The size of gold grains is from 0.1 to 1.5 mm. The age of gold mineralization dating is estimated at 133–137 Ma [33], i.e., the ore formation refers to the upper Jurassic—the beginning of the Early Cretaceous (J_3-K_1).

2.2.2. Gold–Quartz Veins in Granitoids GIT

Deposits of this type are predominantly concentrated in the Kularo-Nera terrane, along which the thick Tenkinskaya fault zone extends, cutting through the Permo-Triassic sedimentary sequences. A zone of tectonic-magmatic activation is oriented along this zone, marked by a series of Mesozoic granitoid intrusions (Figure 1). It is assumed that the main contribution of gold to intrusive magmatic systems was provided by gold-bearing Permo-Triassic sedimentary strata. Deposits of gold–quartz veins in granitoids or porphyry gold [5,7] of the GIT are associated with granitoid intrusions that by according to the level of formation, deep and medium–deep.

Shkolnoye Deposit

The Shkolnoye deposit is located in the Tenkinsky district of the Magadan region. Its structural position is determined by its confinement to the southeastern flank of the Yano-Kolyma gold-bearing belt, its Ayan-Yuryakh segment (Duskaninsky ore cluster). Fissure sublatitudinal fault tectonics had a great influence on the formation of ore bodies. The mineralization is confined to the Burgaginsky stock of granitoids with an area of about 2.6 km^2. The stock is composed of diorites, gabbro-diorites, tonalites, granodiorites, and biotite granites. The age of granitoids according to Rb/Sr determination by the North-East Common Use Center of the SVKNII FEB RAS is 127–152 Ma (J_3-K_1). Gold mineralization is localized in granodiorites and adamelites. Metasomatic changes are represented by silicification, muscovitization, carbonatization. The ore bodies are quartz veins up to 5 m thick and have a sublatitudinal bearing. They are confined to zones of cracking and increased fracturing of rocks. Gold is predominantly free in the ore. Gold mineralization is accompanied by the following: arsenopyrite, fahlore, silver lead sulfosalts; Au:Ag ratio—1:12 [34].

Butarnoye Deposit

The Butarnoye deposit is located in the Khasyn district of the Magadan region. It is confined to the central part of the Khurchan-Orotukan zone of tectonic-magmatic activation, traced in the submeridional direction for 150 km, with a width of 20–50 km. It is localized in a slightly eroded Late Jurassic-Early Cretaceous granitoid stock (J_3-K_1). The Butarny stockwork with an area of 4.6 km^2 is elongated in the submeridional direction for 2.9 km with a width of about 1.5 km. Granitoids of the Butarny stockwork belong to the Late Jurassic complex (J_3). Judging by the K–Ar dates obtained at the North-East Common Use Center of the SVKNII FEB RAS, the age of the granitoids is 142 ± 5 Ma. This age was also confirmed by U-Pb zircon dating at 150 ± 3 Ma [35]. The deposit is dominated by subvertical quartz veins and veinlets, NE-bearing (30–35°) with dip angles from 75° to vertical. The gold content in veins and veinlets reaches tens of grams per ton (up to 93.8 g/t). The ore bodies are represented by quartz veins up to 2 m thick and veinlets feathering them with poor nested-disseminated sulfide mineralization. Arsenopyrite predominates among ore minerals, rare finds of galena, galenobismuthite, Bi sulfotellurides, and maldonite are noted [36].

Dorozhnoye Deposit

The Dorozhnoye deposit is located on the right bank of the middle course of the Dorozhny brook in the Magadan region. It is localized in the Sylgytar granitoid intrusion. The territory is confined to the central part of the Burustakh synclinorium, which is part of the Inyali-Debinsk megasynclinorium of the Yano-Kolyma belt. It is composed of miogeosynclinal terrigenous sandy-silty flyschoid deposits of the Verkhoyansk complex (J_{1-2}). Ore bodies are gold-bearing quartz veins. They are gently dipping plates, 0.1–2.0 m thick, up to 800 m long, lying vertically inside the granitoid stock at intervals of 100–120 m. Presumably, these are contraction (concentric) and radial cracks that arose during the cooling of granitoids. Their bearing is as follows: northeast, dip northwest 10–15°. When leaving the granites in the hornfelses, the veins branch and wedge out. The average content of Au ranges from 8 to 17 g/t, and Ag from 10 to 350 g/t. The Au/Ag ratio varies from 1:1 to 1:20. At the Nadezhda site, the veins are steeply dipping (radial) cracks and veining zones, the prevailing thickness is 10–30 cm, the zones are up to 2 m, their bearing is as follows: sublatitudinal and northeast, dip to the northwest at an angle of 60–80°, length up to 125 m. Au content is from 6 to 25 g/t and Ag up to 300 g/t. The deposit is described in detail in a number of publications [37].

Maltan Deposit

The Maltan deposit is located in the Tenkinsky district of the Magadan region. The deposit is located on the northwestern flank of the Taryn ore-placer cluster. The ore field is limited by the adjacent Malo- and Bolshetarynskaya branches of the Adycha-Tarynsky fault, which separates the zone of the Verkhoyansk fold-thrust belt from the Kular-Nersky shale belt. It is localized in echeloned backstage (fractured bodies) of biotite gabbro, quartz diorite, granodiorite-porphyry and Cretaceous porphyritic granite cutting sedimentary rocks of the Middle and Upper Triassic (T_2–T_3). The length of the veins is 2.5–3.0 km. Mineralization is controlled by NE-bearing fractures, which are transverse with respect to the elongation of fractured intrusions. There are quartz, quartz-carbonate, and sulfide-quartz veins with a length of a few hundred meters, up to 50 cm thick. Together with gold, ore veins contain Bi (0.001–0.5%), As (0.1–1%) and W up to 1.0%; in rare cases, Mo is noted. The vein bodies are composed of quartz with large cluster-disseminated ore mineralization of arsenopyrite, native gold and bismuth, maldonite, bismuth tellurides, scheelite, and rarely molybdenite. The concentration of gold in ores is from 0.5 to 20 g/t. Gold mineralization is closely associated with bismuth and tellurium minerals.

2.3. *Gold–Silver Adularia GIT*

Gold–silver adularia GIT refers to the epithermal-volcanogenic, the most developed in the north-east of Russia. The deposits are confined to various volcanogenic belts. The considered objects are located in the Okhotsk-Chukotka and Kedonsky volcanic belts (Figure 1). According to the conditions of formation, the deposits are classified as shallow and near-surface. Most of them are composed of low-sulfidation, enriched ores.

2.3.1. Kubaka Deposit

The Kubaka deposit belongs to the class of large deposits. It is located in the basin of the Pravaya Aulandzha River in the North-Evensky district of the Magadan region. The ore field is composed of volcanics of the Kedon series (D_{2-3}-C_1 kd), represented by tuff sandstones, tuff siltstones, tuffs, and ignimbrites of intermediate and felsic composition. The structure of the ore field is a collapse caldera that underwent resurgent dome formation during intrusion of subvolcanic bodies and was influenced by Jurassic and Early Cretaceous magmatism (C_3-K_1). The ore bodies are composed veins and stockwork-type zones. The subscripts of the veins are sublatitudinal, and the thickness is unsteady along the strike with bulges up to 20 m and constrictions up to 10 cm, averaging 1–10 m. A series of ore veins stretches for 2 km [31,36]. Main minerals are quartz and adularia, and ore minerals in early veins are native gold, and in later veins include native gold, pyrite, arsenopyrite, acanthite, Sb-Ag and As-Ag sulfosalts and Ag selenides. The average grades of Au and Ag in the ore bodies of the Central zone are in the range 11–33 g/t and in the late ones the range is 12–23 g/t, with Au/Ag ratios of 1:1 and 1:100, respectively [31].

2.3.2. Kupol Deposit

The Kupol deposit is large in terms of reserves, located in the Anadyr region of the Chukotka Autonomous Okrug in the northwestern part of the Anadyr Highlands. The Kupolsky ore cluster is confined to the Upper Yablonsky metallogenic zone of the Central Chukotka sector of the Okhotsk-Chukotka volcanogenic belt. Currently, its first stage has already been put into operation. The ore field is composed of Late Cretaceous andesite lavas (K_1), less often basaltic andesites with interlayers of ash tuffs and Late Cretaceous tuffites. The ore body is a single thick (25–30 m) vein over 3 km long, presumably the mouth of a fissure volcano. Mineralization has been traced to a depth of more than 400 m. The mineral composition of the veins is quartz-adular, ore minerals are native gold, acanthite, Sb-Ag and As-Ag sulfosalts and Ag selenides. The concentration of Au in ores is from 0.1 to 230 g/t. Au/Ag ratio 1:10–1:100. Currently, Kinros LLC continues its operational work. The deposit is described in [31,38,39].

2.3.3. Olcha Deposit

Olcha deposit, average in reserves. It is located on the watershed part of the Khebikendzha mountains and the Yukagir plateau. The ore field is composed of sedimentary-volcanogenic strata attributed to the Kedon series of the Middle-Upper Devonian–Carboniferous (D_{2-3}-C_1), which are underlain by Archean gneisses, granite-gneisses, and other various Cambrian and Ordovician rocks. The recent relationships between Devonian volcanics and basement strata of different ages are tectonic. In the central part, the ore field is intersected by an extended and thick post-ore granodiorite-porphyry dike of northeastern orientation and rare thin submeridional dikes of andesitic porphyrites. The ore bodies of the deposit are multidirectional quartz, carbonate-quartz and adularia-quartz veins and vein zones of a simple lenticular and bead-like shape. The host rocks are predominantly rhyodacite tuffs. Ore bodies are a series of echelon-shaped quartz-adularia veins with a thickness of 0.4 to 8 m. More than 40 mineral species were identified in the ores of the Olcha deposit, but quartz, adularia, native gold, acanthite, selenides and Ag sulfoselenides predominate [40–42]. The ores are characterized by low sulfide content (no more than 2%, and in bonanza the amount of ore minerals reaches 70%). The distribution of useful components in ores is uneven—on average 10–15 g/t, Ag 70–120 g/t, and the Au content in bonanza is up to 14 kg/t, Ag is up to 10 kg/t; Au/Ag ratio in ores 1:1–1:100.

2.3.4. Burgali Deposit

The Burgali deposit is located in the Severo-Evensky district of the Magadan region. The deposit is present in the Kedon volcanic belt. The ore field is composed mainly of rhyodacite tuffs of the Kedon volcanic series (D_3-C_1 kd). The structure of the ore field is blocky. From the north and south, it is controlled by northeast-trending thrusts (20–35°), from the west and east—by steeply dipping northwest-trending faults (350°). Within these limits, a system of intersecting faults of a lower order (NE 40–45° and NE 60–65°) is developed. The gold-bearing stockwork is controlled by this fault system and consists of parallel enchelon zones, characterized by the most intense veining of quartz and carbonate-quartz composition. The stockwork is traced along the strike for 3700 m. The thickness of individual veins is 1.5–2 m, the length is 300–500 m [43]. Au in veins vary from 5 to 50 g/t, Ag—5–200 g/t, Au/Ag ratio—1:4. The vertical range of mineralization, according to exploration data, is about 250 m.

2.3.5. Primorskoye Deposit

The Primorskoye deposit is located in the south of the Omsukchansky district of the Magadan region. The territory is located at the junction of the Omsukchan (Balygychano-

Sugoi) tectonomagmatic zone with the Okhotsk volcanic zone of the Okhotsk-Chukotka volcanogenic belt, which developed on the folded base of the Sugoi marginal trough. The ore field is confined to an intrusive-dome structure and is localized in a gently sloping sequence of Late Cretaceous rhyolite ignimbrites (K_{1-2}) more than 700 m thick, which is intruded by dykes of intermediate and basic composition. The array of leucocratic granites (K_2), according to drilling data, is located under the deposit at a depth of 400–500 m. Metasomatic changes are largely due to the intruded intrusion and are represented by skarnoid associations of garnet–epidote–rhodonite composition. The veins and vein zones are 250–600 m long and 1–3 m thick, with a vertical span of 150 m [44]. The distribution of useful components is uneven. In the veins, Au content reaches 10–15 g/t, Ag reaches 100–150 g/t, and in ore columns values are 100 and 17,000 g/t, respectively. The Au/Ag ratio averages 1:10.

2.3.6. Dalnee Deposit

The Dalnee deposit, average reserves. It is located on the territory of the Severo-Evensky district of the Magadan region. The structure of the ore field is determined by its location in the southwestern part of the Gizhigin trough, with the structures of the Chukotka branch of the Okhotsk-Chukotka volcanic belt (Even volcanic zone) superimposed on it. In terms of Au reserves, the deposit belongs to the class of small deposits. A series of veins was found at the deposit, confined to faults of the northwestern and northeastern directions with a steep and vertical dip. These veins, together with subparallel smaller veins, form a veined and veinlet-disseminated system about 1300 m long and 150–300 m wide. The host rocks are represented by Late Cretaceous andesites (K_1). Isotopic determinations for host rocks by K-Ar are 92 ± 2 to 81 ± 2 Ma, and the age of Rb-Sr mineralization is 80 ± 5 Ma [45]. The ore bodies are represented by steeply dipping quartz, adularia-quartz veins and zones of explosive breccias with poor sulfide mineralization. Metasomatic alterations of host rocks are kaolinite-quartz-sericite. Vein minerals are mainly quartz and adularia, and ore minerals are predominantly pyrite, to a lesser extent native gold and silver, acanthite, pyrargyrite, and polybasite. Gold in ores is unevenly distributed and, on average, its content is 5.2–12.9 g/t, and in bonanza it is up to 500 g/t. The Au/Ag ratio ranges from 1:20 to 1:50. A detailed description of the deposit is given in the works [46,47].

3. Research Methods and Objects

3.1. Research Objects

The research objects for the study were ore samples and samples with visible gold from deposits of various GITs (see Sections 2.2.1 and 2.2.2), selected by the authors of this article in geological prospecting routes at a scale of 1:10,000 in different years of work, from 1978 to 2015. We also used technological samples provided to us for mineralogical study in different years by geologists of industrial organizations.

To date, the ore fields have been studied with varying degrees of details, since some of them are already in operation (Natalka, Butarnoye, Shkolnoye, Kubaka, Kupol, Primoskoye, Dalnee), while others (Maldyak, Degdekan, Maltan, Olcha, Burgali) are at the stage of detailed exploration.

3.2. Research Methods

From ore samples containing visible gold, silicates were dissolved with hydrofluoric acid and grains of native gold were isolated, which were then transferred to spectral quantitative analysis, and also placed in a compound for the preparation of polished sections.

The determination of microimpurities in native gold was carried out by a quantitative spectral method on a spectrograph from a micro-sample according to the method [48]. The advantage of this method is that before analysis, gold undergoes pre-treatment by rolling gold particles into the thinnest plate, followed by acid treatment to release it from mechanical mineral impurities.

The mineragraphic study of polished sections was carried out using optic microscopy (OM) on a Carl Zeiss AXIOPLAN Imagin reflected light microscope (Oberkochen, Germany). In these sections, minerals intergrown with native gold were studied and photographed using the MC-LCD visualization complex with the MMC software (LOMO, Petersburg, Russia).

The internal structure of native gold was revealed by etching the polished surfaces of the grains with standard reagents: for high-grade gold ($HCl + CrO_3$ of various concentrations) and for medium- and low-grade gold ($HCl + 4HNO_3$). Interpretation of the nature of internal structures was carried out in accordance with the recommendations from [9]. Photographing of mineral intergrowths and internal structures of native gold was carried out using a microphotographic attachment on a reflected light microscope.

The determination of the native gold fineness was carried out in polished sections on a modernized POOS-1 microspectrophotometer with an internal standard and a high degree of stabilization of the illuminator and PMT power source according to the method of L. N. Vialsova [49], fineness in each grain was measured by 3–5 points. In each ore sample, measurements were carried out by the authors on 30–50 grains of native gold. Processing of measurement results by methods of mathematical statistics was carried out in the GOLD program developed by S.V. Preis. A significant part of the gold was analyzed by X-ray spectral electron probe microanalysis (XSMA) on the CAMEBAX device, for four main elements—Au, Ag, Cu, Hg—by the operator E.M. Goryacheva (North-East Common Use Center of the SVKNII FEB RAS, Magadan); on the same device, equipped with the mounting attachment "OXFORD INSTRUMENT", photographing of native gold in reflected electrons was performed.

4. Results

4.1. Typomorphic Features of Native Gold

4.1.1. Gold–Arsenic-Sulfide in Black Shale Strata GIT

Fineness of Native Gold

This geological and industrial type is characterized by native gold of medium and high purity with maxima of this value in the areas of 750‰ and 950‰ (Figure 2). Trial histograms are unimodal with a relatively low dispersion of this indicator (from 600‰ to 1000‰).

Trace Elements

It has been established that, in addition to Ag, native gold grains constantly contain microimpurities of As, less often Sb, Pb, and Fe, and in rare cases Cu, Bi, Hg, and Sn. As concentrations range from 0.5 to 120 g/t, all other impurities do not exceed 50 g/t (Table 1). Sn microimpurities are usually associated with the influence of granitoids on the geochemistry of gold mineralization.

Table 1. Concentration of microimpurities in native gold from ores of gold–arsenic-sulfide GIT deposits.

Deposit (No. in Figure 1)	Element Concentrations, g/t							
	As	Sb	Cu	Bi	Fe	Hg	Pb	Sn
Natalka (1)	0.5–18.7	1.5–16.0	–	–	–	0.5–12.2	1.0–17.0	–
Degdekan (2)	1.0–41.0	0.1–11.3	10.2–35.3	–	1.5–9.6	–	–	–
Maldyak (3)	8.4–89.2	1.2–37.9	–	0.1–5.7	1.0–10.5	–	0.7–12.5	0.01–5.4
Karalveem (4)	5.0–120.1	–	1.0–45.5	–	–	–	0.5–10.4	–

Notes: —below detection limit.

Minerals in Intergrowth

In most cases, the study of mineral parageneses reveals an association of native gold with arsenopyrite, less often with galena and sphalerite. Gold mineralization is associated with late stages of mineralization evidenced by superimposing and filling cracks in previously deposited minerals and intergranular space (Figure 3a,b). A significant part of native gold is deposited directly in quartz. This type of deposit is favorable for enrichment of ores and is called "free gold".

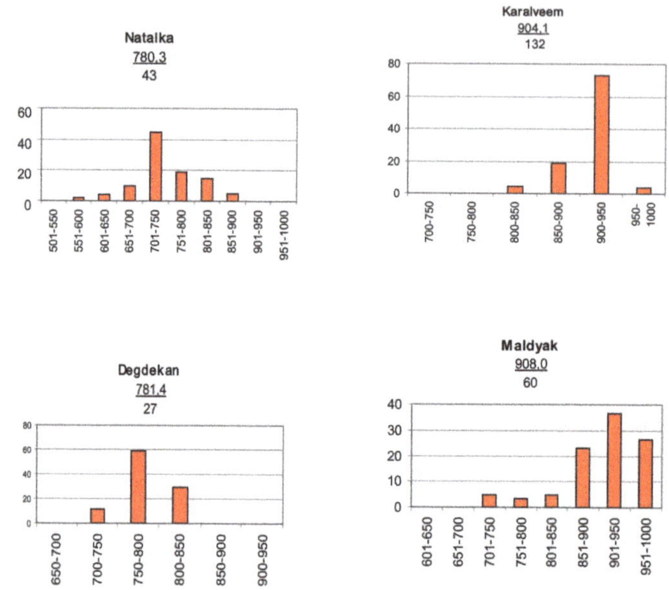

Figure 2. Histograms showing fineness of native gold for deposits of gold–arsenic-sulfide GIT on the abscissa axis—frequency of occurrence, %; along the *x*-axis, fineness intervals, ‰; in the numerator—fineness, ‰, in the denominator—the number of determinations.

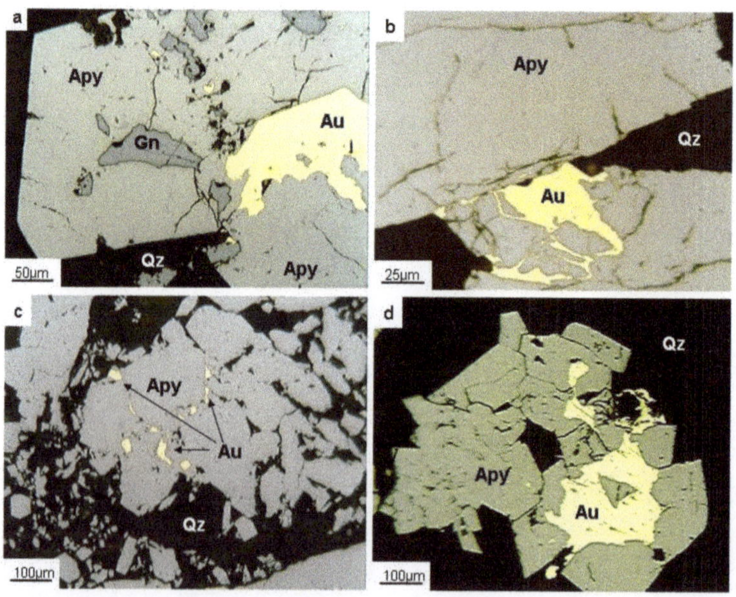

Figure 3. Typical mineral intergrowths of native gold in ores of gold–arsenic-sulfide GIT deposits: (**a**) intergrowth with arsenopyrite (Natalka); (**b**) development of native gold along cracks in arsenopyrite (Karalveem); (**c**) small inclusions of native gold in a single crystal of arsenopyrite (Degdekan); (**d**) overlay of native gold on an intergrowth of arsenopyrite crystals (Maldyak).

Internal Structures

Structural etching revealed a homogeneous polygonal-grain structure of native gold with simple twins (Figure 4), which indicates the medium-deep formation of this GIT and relatively stable conditions for gold crystallization. We associate the occurrence of twins with the formation of gold segregations in a turbulent tectonic setting, which is often evidenced by arsenopyrite cataclasis.

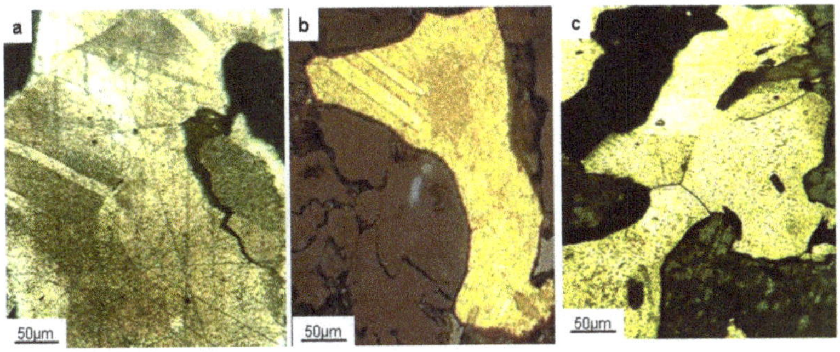

Figure 4. Polygonal-granular structure of native gold with simple twins in ores of various deposits of gold–arsenic-sulfide GIT: (**a**) Natalka, (**b**) Degdekan, (**c**) Maldyak.

4.1.2. Gold–Quartz Veins in Granitoids GIT

Fineness of Native Gold

Gold–quartz veins GIT is characterized by native gold of medium and high purity with maxima of this value in the regions of 850–950‰ (Figure 5). Fineness histograms are predominantly unimodal with a higher dispersion of this indicator (from 500‰ to 1000‰) compared to gold–arsenic-sulfide GIT. Often, in the frame of this type of deposits, there is a sharp decrease in the fineness of native gold and the appearance of silver mineralization (stephanite, freibergite, polybasite, acanthite, pyrargyrite).

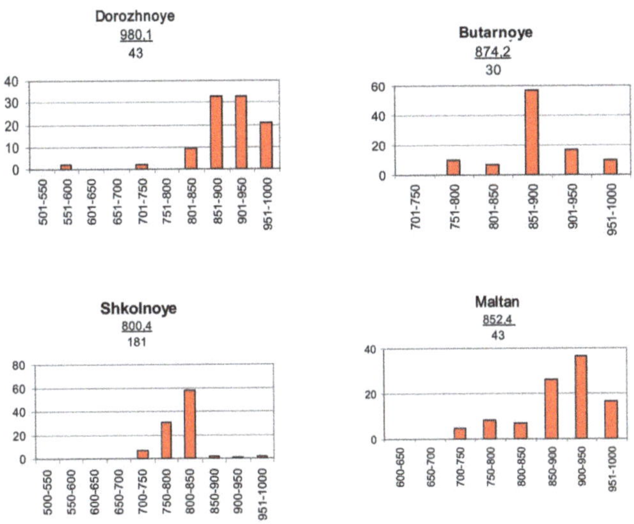

Figure 5. Fineness histograms of native gold for deposits of gold–quartz veins GIT along the abscissa axis—frequency of occurrence, %; along the x-axis, fineness intervals, ‰; in the numerator—fineness, ‰, in the denominator—the number of determinations.

Trace Elements

It has been established that native gold from deposits of this type constantly contains As in significant concentrations and Bi, Sb, Pb, Cu, and Fe are less common, while Hg, Sn are noted in rare cases (Table 2). As concentrations range from 0.8 to 300 g/t, all other impurities do not reach 50 g/t. Bi and Sn are classified by most researchers as granitogenic elements. Among the deposits of gold–quartz veins GIT, a gold-rare metal formation stands out [44–46], where Bi and Sn, and often W, form noticeable concentrations and are included, along with gold, in the list of useful components.

Table 2. Concentration of microimpurities in native gold from ores of gold–quartz veins GIT deposits.

Deposit (No. in Figure 1)	Element Concentrations, g/t							
	As	Sb	Cu	Bi	Fe	Hg	Pb	Sn
Shkolnoye (5)	100.0–300.0	1.0–30.0	1.0–7.5	1.5–12.4	30.0–100.4	30.1–99.6	–	1.7–31.1
Butarnoye (6)	0.8–100.0	–	–	1.5–22.3	1.0–12.7	–	–	–
Dorozhnoye (7)	1.0–15.3	–	–	1.5–21.0	–	–	1.0–82.0	–
Maltan (8)	10.0–150.0	2.5–16.3	–	1.0–50.0	–	–	1.5–12.3	–

Notes: —below detection limit.

Mineral Intergrowths

The widespread association of gold–quartz veins GIT is naturally quartz, which was the reason for its name. In addition, the diversity of mineral associations of native gold is determined by its intergrowth with Sb and Bi minerals, and often Te (fahlore, joseite, native bismuth, bismuthine, etc.), as well as with Fe-Sb sulfosalts (Figure 6). Intergrowths with vein minerals are characterized by the features of host rocks—granitoids. At the Dorozhnoye deposit, native gold often forms intergrowths with muscovite (Figure 6h). Figure 6a–f shows mineral intergrowths of native gold in the ores of the Shkolnoye deposit for various horizons. Changes in the average fineness value with depth change in waves, but the dispersion of this indicator naturally falls from the upper horizon to the lower one (Table 3).

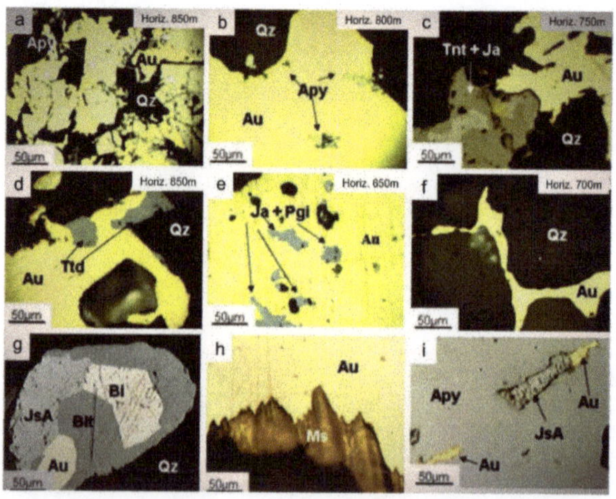

Figure 6. Intergrowths of native gold typical for gold–quartz veins GIT: (**a**–**f**) for different horizons of the Shkolnoye deposit: (**a**) native gold in the intergranular space of arsenopyrite, (**b**) in quartz with inclusions of small grains of arsenopyrite, (**c**) intergrowth with tennantite and jamesonite in quartz, (**d**) intergrowth with tetrahedrite in quartz, (**e**) inclusions in native gold of jamsonite and plagionite, (**f**) gold in the intergranular space of quartz; (**g**) (Maltan) inclusion of native gold in an intergrowth of bismuth, bismuthine, and joseite A; (**h**) (Dorozhnoye) intergrowth of native gold with muscovite; (**i**) (Butarnoye) small segregations of native gold in arsenopyrite intergrown with joseite A.

Table 3. Characteristics of native gold from ores of different horizons of the gold–quartz veins GIT Shkolnoye deposit.

Horizon, m	Scatter of Fineness Values Average, ‰	Fineness Dispersion	Mineral Paragenesis
850	250–850 775.4	11,375	Native gold + quartz + arsenopyrite + fahlore
800	700–850 782.7	1060	Native gold + quartz + arsenopyrite + tennantite
750	750–850 804.9	271.5	Native gold + quartz + arsenopyrite + tetrahedrite + jamsonite
700	700–800 753.5	223.5	Native gold + quartz + arsenopyrite + tetrahedrite
600	750–850 806.4	191.5	Native gold + quartz + arsenopyrite + jamsonite + plagonite

Internal Structures

For gold–quartz veins GIT deposits localized in granitoids, the structure of native gold is coarse-grained with rare simple and polysynthetic twins and patchy heterogeneity, presumably associated with the incorporation of microparticles of native Bi, which, unlike gold, undergoes intense etching even in air (Figure 7). A similar type of patchy heterogeneity for gold from deposits in granitoids was previously described in [8].

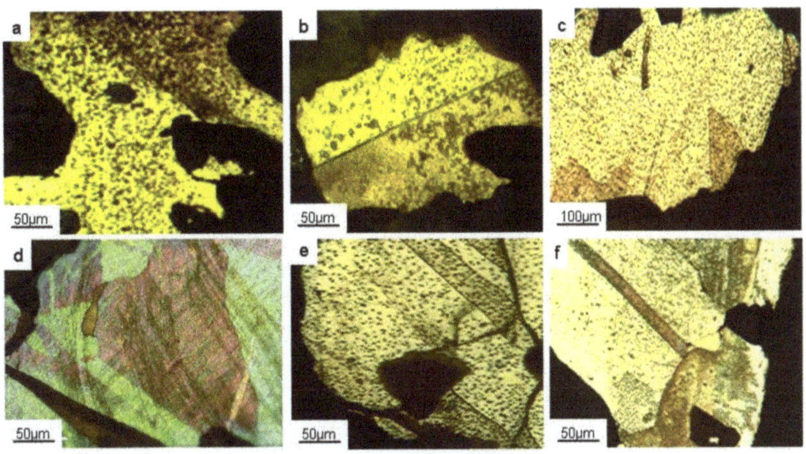

Figure 7. Polygonal-granular structure with simple polysynthetic twins and fine spotted heterogeneity in native gold from ores of gold–quartz veins GIT deposits: (**a**,**b**) Butarnoe; (**c**,**e**) Shkolnoye; (**d**) Maltan, (**f**) Dorozhnoye.

4.1.3. Gold–Silver Adularia in Volcanogenic Strata GIT

Fineness of Native Gold

The gold–silver adularia GIT is characterized by native gold of a relatively low fineness with a polymodal distribution on fineness histograms. The spread of the values of this indicator lies in the areas of 200–850‰ (Figure 8).

Deposits in genetic terms belong to the volcanogenic class, but often fall into areas of tectonomagmatic activation with late intrusions, which leads to the superposition of high-temperature processes on ores of early stages, to thermometamorphism of ores, differentiation and redistribution of matter, and staged mineral formation. It is these factors that are reflected in the histograms of gold in the form of polymodal graphs. Heterogeneity in fineness due to thermometamorphism can manifest itself within one grain of native gold (Figure 9).

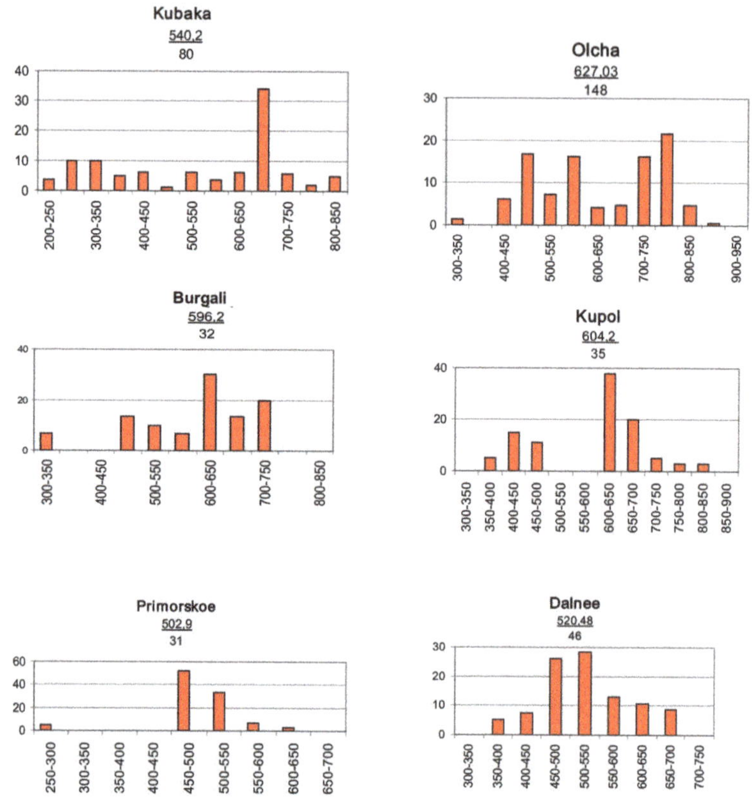

Figure 8. Fineness histograms of native gold for deposits of gold–silver adularia GIT: along the abscissa axis—frequency of occurrence, %; along the *x*-axis, fineness intervals, ‰; in the numerator is the fineness, ‰, in the denominator is the number of determinations.

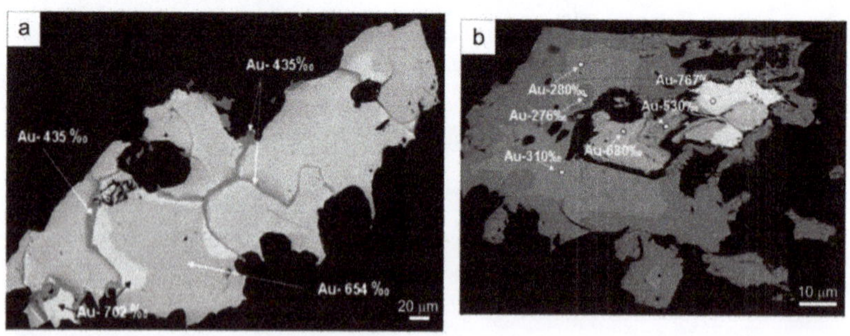

Figure 9. Heterogeneity of native gold within one grain: (**a**) Kupol deposit (from 435‰ to 702‰); (**b**) Dalnee deposit (from 276‰ to 767‰), taken in reflected electrons.

Trace Elements

It has been established that native gold from deposits of this type constantly contains impurities of Sb, Cu, Hg in noticeable concentrations (Table 4), and the highest concentrations reach Hg up to 1000 g/t in gold from the Dalnee deposit. Quite often, Fe is found, from 1.5 to 26.5 g/t (Kubaka, Burgali, Primorskoye), and also As, from 0.1 to 27.3 g/t (Kupol, Primorskoe, Dalnee).

Table 4. Concentration of microimpurities in native gold from ores of gold–silver adularia GIT deposits.

Deposit (No. in Figure 1)	Element Concentrations, g/t							
	As	Sb	Cu	Bi	Fe	Hg	Pb	Sn
Kubaka (9)	–	0.8–1.3	0.6–35.0	–	1.5–15.0	0.5–60.0		
Kupol (10)	0.1–27.3	0.1–25.5	–		–	0.1–1700.0		
Olcha (11)	–	1.3–16.4	0.8–45.3	–	–	0.7–45.5		
Burgali (12)	–	2.4–12.8	0.1–16.4	–	0.8–26.2	0.5–11.8		
Primorskoye (13)	0.1–0.9	1.0–12.3	2.0–68.4	1.0–24.4	0.3–14.7	0.1–65.2		
Dalnee (14)	2.7–12.1	1.5–21.1	0.1–7.3	–	–	0.1–1000.0		

Notes: —below detection limit.

Mineral Intergrowths

The ores of the gold–silver adularia GIT deposits are characterized by a wide variety of mineral intergrowths with Ag sulfides, selenides, and sulfosalts. The sharply gradient conditions for the formation of epithermal Au–Ag deposits predetermine the predominantly fine-grained character of native gold segregations and its intergrowth with a wide range of Cu, Pb, Zn, Fe sulfides and Ag sulfosalts and sulfoselenides. The deposits of the Omolon cratonic terrane show intergrowths of native gold with magnetite and hematite. This is due to the peculiarities of the rocks of the base of volcanic apparatuses, among which Archean-Proterozoic ferruginous quartzites are widely developed. At the same time, volcanic activity manifested itself much later—on the border of the Devonian and Carboniferous periods. Segregation forms are predominantly xenomorphic; in quartz they are interstitial. On the example of only one Olcha deposit, we have shown an exceptional variety of such intergrowths (Figure 10). For other deposits of this GIT, examples of the most common intergrowths of native gold in ores are given—with pyrite, tetrahedrite, sphalerite, galena, acanthite, polybasite (Figure 11).

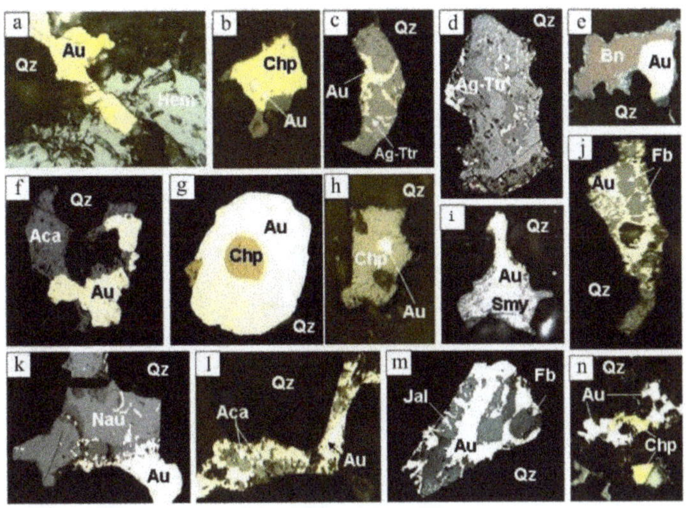

Figure 10. Types of intergrowths of native gold with ore minerals in quartz-sericite vein material at the Olcha deposit: (**a**) gold with hematite; (**b**) inclusion of kustelite in chalcopyrite; (**c**) close intergrowths of gold with argentotetrahedrite; (**d**) vein-like intergrowths of gold with argentotetrahedrite; (**e**) gold with bornite; (**f**) gold with acanthite; (**g**) inclusion of chalcopyrite in gold; (**h**) inclusion of gold in chalcopyrite; (**i**) intergrowth of gold with acanthite and stromeyerite; (**j**) gold with freibergite; (**k**) complex intergrowth of gold with naumannite and stromeyerite; (**l**) gold with acanthite; (**m**) gold-freibergite + jalpaite; (**n**) gold with chalcopyrite. (All taken at 100× magnification).

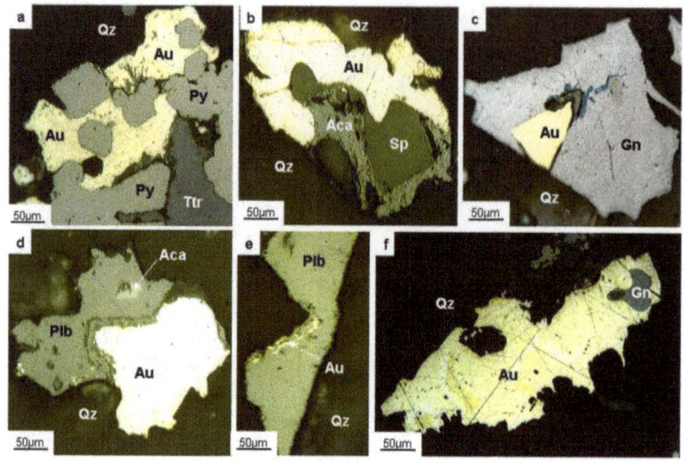

Figure 11. Mineral intergrowths of native gold from ores of gold–silver adularia GIT deposits in quartz: (**a**) with pyrite and tetrahedrite (Dalnee deposit); (**b**) with acanthite and sphalerite (Dalnee); (**c**) with galena (Primorskoe); (**d**) with polybasite and acanthite (Burgali); (**e**) with polybasite (Burgali); (**f**) heterogeneous in fineness of native gold with the inclusion of galena (Kupol deposit).

Internal Structures

The internal structures of native gold for gold–silver adularia GIT are predominantly zonal and are well identified by structural etching (Figure 12). In the case of thermometamorphism of ores, patchy heterogeneity often appears (Figure 13a) and there are granulation structures with the expansion of boundaries (Figures 12a and 13b,d) with expansion of grain boundaries. At the Olcha deposit, native gold has an unusual gulf-like shape with a zonal structure. This is due to the fact that it crystallized in quartz vacuoles filled with highly concentrated hydrothermal solutions (Figure 13).

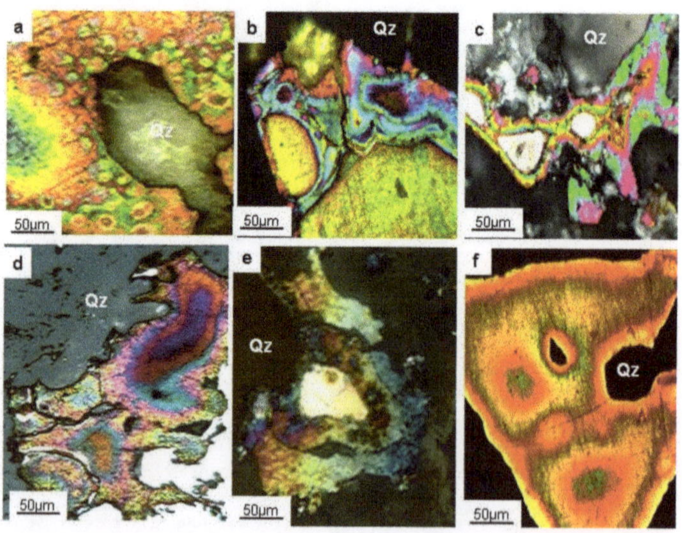

Figure 12. Internal structures of native gold from ores of gold–silver adularia GIT deposits: (**a**) spotted (Kubaka); (**b**) clear zonal (Dalnee); (**c**) zonal (Olcha); (**d**) zonal with a break in the boundaries (Kupol); (**e**) zonal (Burgali); (**f**) zonal (Primorskoe).

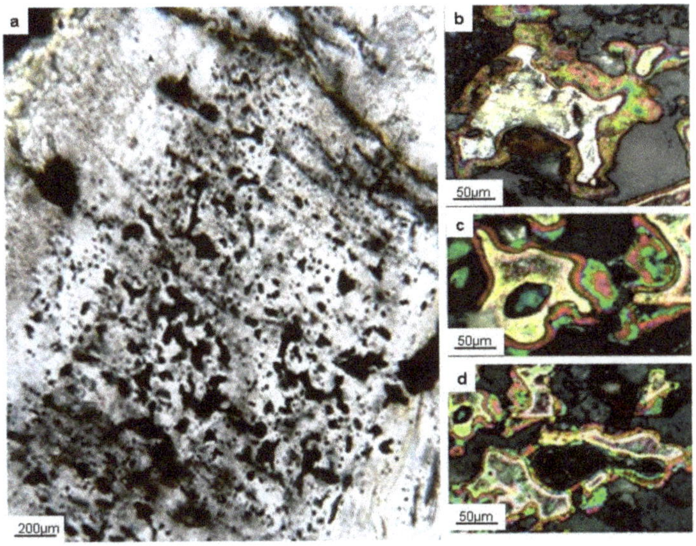

Figure 13. Filling of vacuoles with native gold in quartz at the Olcha deposit ((**a**) quartz in transmitted light without analyzer, black—vacuoles; (**b–d**) in reflected light, native gold in quartz, structural etching).

5. Discussion

All considered GIT deposits are hydrothermal and most of the features of native gold are associated with the depths of formation of these deposits. It is the depth that determines the duration, stability, and temperature zoning of mineral formation. To no lesser extent, the metallogeny of ore regions affects the formation of certain GIT deposits and their specific mineral types.

In genetic terms, the three most common GITs are generally accepted:

1. Orogenic [4]—hydrothermal-metamorphogenic in sedimentary strata (in most cases, these are black shale strata that have experienced metamorphism, mid-deep and deep (gold-sulfide quartz in sedimentary strata);
2. Porphyry ("intrusion-related") [7,9,50,51]—gold deposits associated with granitoid intrusions, deep and medium deep (gold–quartz veins in granitoids);
3. Volcanogenic (epithermal) [52]—in volcanic belts of any age associated with volcanic activity in basic volcanic rocks of the oxidized type, shallow (gold–silver adularia).

At medium-deep and deep deposits in GITs 1 and 2, the hydrothermal system is in a relatively stable state with a long cooling process. This creates conditions for the growth of large individuals during the crystallization of both native gold and any other mineral. A different picture in the shallow type of deposits is GIT type 3, where, during the outburst of volcanoes, sharply gradient conditions are created with high initial temperatures and a rapid cooling of the ore-forming system, which generally prevents the differentiation of natural gold-silver compounds (high content of silver in gold and, accordingly, its reduced fineness). In this unstable environment with a limited cooling time of the system, no conditions are created for the growth of large gold individuals—more often it is fine-grained. Therefore, the first two types are placer-forming (large gold, which quickly precipitates and accumulates in a water stream), and volcanic—although it carries fine and fine gold with a water stream during the destruction of ores, it precipitates more slowly, and in most cases, does not give industrial accumulations.

The scope of mineralization is determined mainly by the depth of deposit formation and the largest has an "orogenic" GIT (many hundreds of meters)—volcanogenic, 50–100 m—but in the case of localization of mineralization in the vent facies, it can reach 300–400 m.

At an early stage of geological prospecting at ore deposits, we do not consider the size of native gold as an indicator property, since the amount of gold particles found in specimens and lump ore samples is not representative enough for sieve analysis. However, in the study of placer gold, this typomorphic feature can be of great importance, in combination with roundness and flatness.

To determine the GIT at an early stage of prospecting and exploration at ore gold deposits, we considered the following typomorphic features of native gold: microimpurities, average fineness and fineness distribution of gold on histograms in combination with fineness dispersion value, as well as mineral intergrowths and internal structures of native gold. The given descriptions and examples of typomorphic features of native gold deposits for various GIT showed characteristic differences for each type (Table 5).

Table 5. Main typomorphic features of native gold for various GITs.

Geological and Industrial Type	Fineness Variations, ‰ / Dispersion	Microimpurities	Mineral Intergrowths	Internal Structures
Gold–arsenic-sulfide in black shale strata	780–908 / 408–1746	As (Pb)	Quartz + arsenopyrite + galena	Polygonal-grained with simple twins
Gold–quartz veins in granitoids	852–980 / 2127–5626 (191–11,375) *	As, Bi, Sn (W)	Quartz + arsenopyrite + minerals BiTe + Sb-Fe-(Ag) sulfosalts	Proligonal-granular with patchy heterogeneity
Gold–silver adularia	502–627 / 6320–17,287	Sb, Cu, Hg (As, Fe)	Quartz + hydromica + Pb, Zn Cu sulfides + pale ores + Ag sulfides and sulfosalts + Ag, Cu, Pb selenides	Clear and unclear zonal with border extension

Note: * for the Shkolnoye deposit, the dispersion values are determined horizontally (see Table 3).

For gold–arsenic-sulfide GIT in black shale strata there is relatively coarse Au and high fineness with low dispersion of this index, polygonal-grained structures with simple twinning and a constant microimpurity of As in the composition; for the gold–quartz-porphyry type in granitoids—coarse gold predominates, fineness decreases slightly and its dispersion increases, heterogeneities are noted in the structures, and granitogenic elements Bi, Te, W, Sn act as microimpurities; for the epithermal gold–silver adularia GIT is characterized by fine, relatively low-grade Au with a high dispersion of this indicator, up to native Ag, a constant increased admixture of Sb, Cu and Hg, zonal internal structures complicated by heterogeneities during thermometamorphism of ores.

The study of the typomorphic features of native gold, tied to the geological environment and named by N.P. Yushkin [28] with topomineralogy, is a rational way to display the features of native gold. It makes it possible to establish the spatial patterns of the distribution of these features in connection with the geological structure and morphostructure of the territory.

Here, we give an example of such a study for the Maldyak deposit (gold–arsenic-sulfide in black shale strata GIT). It was the indicator properties of native gold that made it possible to speak about the existence of a granitoid pluton under the deposit, as well as about the dome-ring structural position of the Maldyak ore field (Figure 14). The conclusion is justified by the concentric-zonal arrangement of typomorphic features of native gold associated with the radial and concentric orientation of ore bodies. The ubiquitous presence of Bi impurities and the appearance of Sn and W impurities suggest that they were introduced by an undiscovered intrusion that formed a dome-shaped structure. All this made it possible to predict the different positions of gold-bearing veins (ore-bearing cracks), both radial—steeply dipping, and concentric—consistent with the schistosity of sedimentary rocks, crumpled into folds. Thanks to this approach, one of the rich ore bodies, previously unknown, was identified during exploration work at the deposit in 2001 by geologist, who called it the "Sedlovidnya vein", due to its confinement to a radial crack and a sharp bend (fold) in the black shale sedimentary sequence.

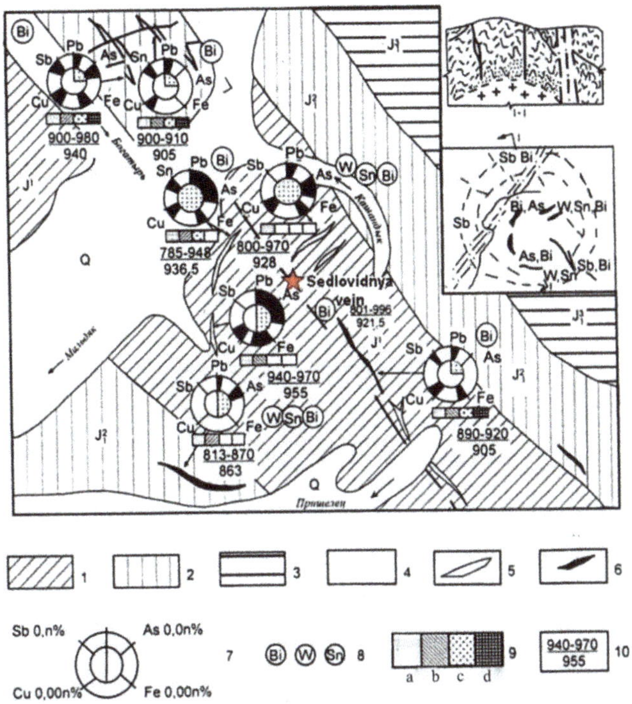

Figure 14. Topomineralogical map of typomorphic features of native gold in the Maldyak ore field. The image was constructed by the authors of the present paper using data from [10] with additions. 1–3—sedimentary deposits of the Jurassic age: 1—lower mudstone-siltstone sequence; 2—middle argillite sequence; 3—upper predominantly sandstone strata; 4—quaternary deposits; 5—residential zones; 6—mineralized dikes; 7—on the pie chart along the periphery shows the content of impurity elements Pb, As, Fe, Cu, Sb (the sector is 100%), in the center—the degree of hypogene transformations in native gold; 8—other microimpurities found in native gold; 9—mineral associations of native gold: a—gold-quartz, b—gold-arsenopyrite, c—gold-galena-sphalerite, d—gold-sulfoantimonite; 10—sample of native gold (in ppm): numerator—variations, denominator—average value.

For the Shkolnoye deposit (gold–quartz veins in granitoids GIT), we analyzed the average fineness value of native gold from well cores from six ore horizons. It turned out that changes in the average fineness value change with depth in waves, but the dispersion of this indicator naturally falls from the upper horizon to the lower one, indicating an increase in a more stable situation in this direction, as well as a consistent differentiation of gold-silver from the lower horizons to the upper ones (Figure 6a–f, Table 3).

It should be noted that the metallogeny of the region can influence the formation of various GITs. In this article, we considered the typomorphism of native gold for the North-eastern region—the Yano-Kolyma orogenic belt with its specific As, Sn, W metallogeny and block terrane composition with an extended covering complex of the Okhotsk-Chukotka volcanogenic belt, as well as the Kedon volcanic belt in a rigid block—Omolon craton terrane with Au–Ag specialization. The main axis of the orogenic gold mineralization here is the Ayan-Yuryakhsky anticlinorium with the Tenkinsky deep fault extending along it.

At the same time, the Verkhoyansk fold system adjacent to the Northeast from the west has an independent collisional terrane structure of the Verkhoyansk-Chukotka thrust belt and its own metallogenic Sb-Hg and Ag-Pb-Zn specialization. Here, along the Adycha-Taryn deep fault, gold deposits related to the gold–arsenic-sulfide (gold–antimony–mercury) GIT prevail—the Sarylakh, Kyuchus and other deposits. The study of the typomorphic features of native gold from this province showed the presence in its composition of high concentrations of As, Sb, especially Hg. Native gold may contain from 5 to 15 wt.% Hg (Figure 15). In intergrowths, along with arsenopyrite, aurostibite, antimonite, and berthierite are often found; a higher sulfide content of ores (up to 15%) is typical. For this GIT, the appearance of native gold is significantly different (Figure 15b), the most common spongy gold with a fineness of 940–970‰ intergrown with antimonite [53,54]. In the north-east of Russia, gold–antimony–mercury GIT deposits have not yet been found.

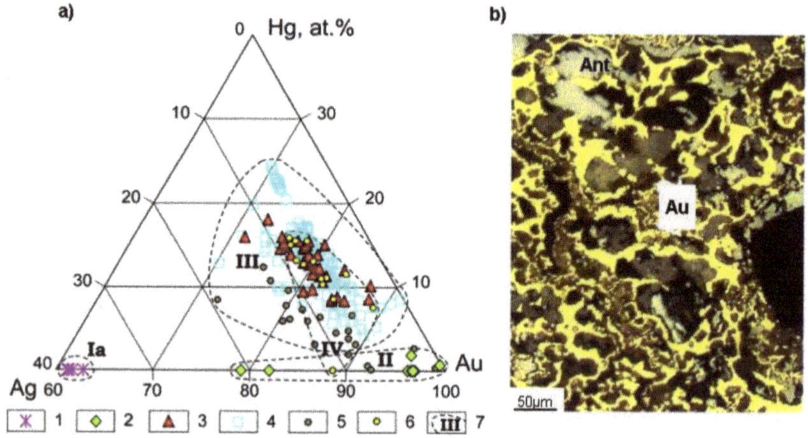

Figure 15. Typomorphic features of native gold (gold–antimony–mercury) GIT: (**a**) Ag-Hg-Au diagram of native gold of the Kyuchus deposit. The image was constructed by the authors of the present paper using data from [54]; (**b**) spongy gold intergrown with antimonite, Sarylakh deposit.

6. Conclusions

Summing up our study, we can state that native gold is an indicator mineral for various GITs of gold deposits. In a comprehensive study, it allows us to obtain the physical and chemical parameters of the main useful component of ores, to determine the deposit GIT, as well as genetic information about a particular object. All this makes it possible to preliminarily assess the scope of mineralization and outline schemes for processing ores, as well as to plan exploration work in order to identify placers.

On the example of different GITs and geological formational types of gold deposits in the north-east of Russia (gold–arsenic-sulfide—Natalka, Degdekan, Karalveem, Maldyak; gold–quartz veins—Dorozhnoye, Butarnoye, Shkolnoye, Maltan; gold–silver adularia—Kubaka, Olcha, Burgali, Kupol, Primorskoye, Dalnee) a comparative description of the typomorphism of native gold was carried out and it was shown that the reliability of geological interpretation is directly related to a comprehensive study of this mineral, including fineness variations, a spectrum of trace elements, internal structure and mineral intergrowths.

Author Contributions: Conceptualization, N.E.S., R.G.K. and G.A.P.; methodology, N.E.S., R.G.K., G.S.A. and G.A.P.; formal analysis, N.E.S., R.G.K. and G.S.A.; investigation, N.E.S., R.G.K., G.S.A. and G.A.P.; resources, G.S.A.; writing original draft preparation, N.E.S. and R.G.K.; writing—review and editing, R.G.K. and G.A.P.; project administration, N.E.S. All authors have read and agreed to the published version of the manuscript.

Funding: This research was funded with financial support from the Russian Foundation of Basic Research (project No 20-05-00393, 20-05-00142) and by within the framework of the state assignment of Shilo North-East Interdisciplinary Scientific Research Institute of Russian Academy of Sciences, Vinogradov Institute of Geochemistry of Russian Academy of Sciences, Diamond and Precious Metal Geology Institute of Russian Academy of Sciences, Sobolev Institute of Geology and Mineralogy of Russian Academy of Sciences.

Acknowledgments: The authors thank the analysts E.M. Goryacheva and N.S. Averchenko (North-East Common Use Center of the SVKNII FEB RAS, Magadan) for help in carrying out work on a scanning electron microscope, an X-ray spectral microanalyzer and a spectrograph.

Conflicts of Interest: The authors declare no conflict of interest.

References

1. Kreyter, V.M. Industrial types of deposits. In *Theoretical Foundations of Prospecting and Exploration of Solid Minerals*; Nedra: Moscow, Russia, 1968; pp. 9–27. (In Russian)
2. Shapovalov, V.S. About industrial types of own gold deposits of the Russian Federation. In *Collection of Abstracts of the IX International Scientific and Practical Conference, Proceedings of the Scientific and Methodological Foundations of Forecasting, Prospecting, Evaluation of Deposits of Diamonds, Precious and Non-Ferrous Metals, Moscow, Russia, 17–19 April 2019*; TsNIGRI Publ: Moscow, Russia, 2019; pp. 156–157. (In Russian)
3. Konstantinov, M.M.; Struzhkov, S.F.; Aristov, V.V. Geological and industrial grouping of gold deposits. Mineral Resources of Russia. *Econ. Manag.* **2007**, *4*, 15–18. (In Russian)
4. Groves, D.I.; Goldfarb, R.J.; Gebre-Mariam, M.; Hagemann, S.G.; Robert, F. Orogenic gold deposits: A proposed classification in the context of their crustal distribution and relationship to other gold deposit types. *Ore Geol. Rev.* **1998**, *13*, 7–27. [CrossRef]
5. Robert, F.; Poulsen, K.H.; Dubé, B. Gold Deposits and Their Geological Classification. In *Geophysics and Geochemistry at the Millennium: Proceedings of Exploration 97: Fourth Decennial International Conference on Mineral Exploration*; Gubins, A.G., Ed.; Prospectors and Developers Association of Canada: Toronto, ON, Canada, 1997; pp. 209–220.
6. Titley, S.T.; Zurcher, L. Regional geology and ore-deposit styles of the trans-border region, southwestern North America. In *Ores and Orogenesis: Circum-Pacific Tectonics, Geologic Evolution, and Ore Deposits*; Arizona Geological Society: Tucson, AZ, USA, 2008; pp. 75–294.
7. Sillitoe, R.H. Special Paper: Major Gold Deposits and Belts of the North and South American Cordillera: Distribution, Tectonomagmatic Settings, and Metallogenic Considerations. *Econ. Geol.* **2008**, *103*, 663–687. [CrossRef]
8. Nguimatsia, D.F.W.; Bolarinwa, A.T.; Yongue, R.F.; de Dieu Ndikumana, J.; Olajide-Kayode, J.O.; Olisa, O.G.; Abdu-Salam, M.O.; Kamga, M.A.; Djou, E.S. Diversity of Gold Deposits, Geodynamics and Conditions of Formation: A Perspective View. *Open J. Geol.* **2017**, *7*, 1690–1709. [CrossRef]
9. Nikolaeva, L.A.; Gavrilov, A.M.; Nekrasova, A.N.; Yablokova, S.V.; Pozdnyakova, N.N.; Shatilova, L.V. Typomorphism of native gold. In *Guidelines for Exploration Work*; Krivtsov, A.I., Vartanyan, S.S., Eds.; TsNIGRI: Moscow, Russia, 2021. (In Russian)
10. Savva, N.E.; Preis, V.K. *Atlas of Native Gold of the North-East of the USSR*; Nauka: Moscow, Russia, 1990. (In Russian)
11. Pal'yanova, G.A. Physicochemical modeling of the coupled behavior of gold and silver in hydrothermal processes: Gold fineness, Au/Ag ratios and their possible implications. *Chem. Geol.* **2008**, *255*, 399–413. [CrossRef]
12. Morrison, G.W.; Rose, W.J.; Jaireth, S. Geological and geochemical controls on the silver content (fineness) of gold in gold-silver deposits. *Ore Geol. Rev.* **1991**, *6*, 333–364. [CrossRef]
13. Minter, W.E.L.; Goedhart, M.; Knight, J.; Frimmel, H.E. Morphology of Witwatersrand gold grains from the Basal reef: Evidence for their detrital origin. *Econ. Geol.* **1993**, *88*, 237–248. [CrossRef]
14. Knight, J.B.; Mortensen, J.K.; Morison, S.R. Lode and placer gold composition in the Klondike district Yukon Territory, Canada, implications for the nature and genesis of Klondike placer and lode gold. *Econ. Geol.* **1999**, *94*, 649–664. [CrossRef]
15. Gammons, C.H.; Williams-Jones, A.E. Hydrothermal geochemistry of electrum; thermodynamic constraints. *Econ. Geol.* **1995**, *90*, 420–432. [CrossRef]
16. Olivo, R.; Gauthier, M.; Bardoux, M. Palladian gold from the Caue iron mine, Itabira District, Minas Gerais, Brazil. *Mineral. Mag.* **1994**, *58*, 579–587. [CrossRef]
17. Gas'kov, I.V. Major impurity elements in native gold and their association with gold mineralization settings in deposits of Asian folded areas. *Russ. Geol. Geophys.* **2017**, *58*, 1080–1092. [CrossRef]
18. Cabral, A.R.; Lehmann, B.; Kwitko, R.; Jones, R.D. Palladian gold and palladium arsenide–antimonide minerals from Gongo Soco iron ore mine, Quadrilátero Ferrífero, Minas Gerais, Brazil. *Appl. Earth Sci.* **2002**, *111*, 74–80. [CrossRef]
19. Chapman, R.J.; Leake, R.C.; Moles, N.R.; Earls, G.; Cooper, C.; Harrington, K.; Berzins, R. The application of microchemical analysis of gold grains to the understanding of complex local and regional gold mineralization: A case study in Ireland and Scotland. *Econ. Geol.* **2000**, *95*, 1753–1773.
20. Chapman, R.J.; Mortensen, J.K. Application of microchemical characterization of placer gold grains to exploration for epithermal gold mineralization in regions of poor exposure. *J. Geochem. Explor.* **2006**, *91*, 1–26. [CrossRef]
21. Chapman, R.J.; Leake, R.C.; Bond, D.P.G.; Stedra, V.; Fairgrieve, B. Chemical and Mineralogical Signatures of Gold Formed in Oxidizing Chloride Hydrothermal Systems and their Significance within Populations of Placer Gold Grains Collected during Reconnaissance. *Econ. Geol.* **2009**, *104*, 563–585. [CrossRef]
22. Palyanova, G.A.; Savva, N.E. Sulfides of gold and silver: Composition, mineral associations, formation conditions. *Chem. Technol.* **2007**, *8*, 421–442.
23. Palyanova, G.A. Gold and Silver Minerals in Sulfide Ore. *Geol. Ore Depos.* **2020**, *62*, 383–406. [CrossRef]
24. Petrovskaya, N.V. *Native Gold*; Nauka: Moscow, Russia, 1973. (In Russian)

25. Boyle, R.W. The geochemistry of silver and its deposits. *Bull. Geol. Surv. Can.* **1968**, *160*, 264.
26. Fersman, A.E. *Selected Works*; Academizdat: Moscow, Russia, 1953; Volume 2. (In Russian)
27. Yushkin, N.P. *Theory and Methods of Mineralogy*; Nauka: Leningrad, Russia, 1977. (In Russian)
28. Yushkin, N.P. Mineralogical mapping and predictive prospecting and evaluation practice. In *Mineralogical Mapping and Indicators of Mineralization*; Nauka: Leningrad, Russia, 1990; pp. 4–17. (In Russian)
29. Khanchuk, A.I. (Ed.) *Geodynamics, Magmatism and Metallogeny of the East of Russia: In 2 Books*; Dalnauka: Vladivostok, Russia, 2006. (In Russian)
30. Sokolov, S.D. Essay on the tectonics of the northeast of Asia. *Geotectonics* **2010**, *6*, 60–78.
31. Laverov, N.P.; Rundqvist, D.V. (Eds.) *Large and Super-Large Deposits of Ore Minerals*; IGEM RAN: Moscow, Russia, 2006; Volume 3, pp. 155–179, 179–203, 213–255. (In Russian)
32. Kondratiev, M.N.; Savva, N.E.; Gamyanin, G.N.; Kolova, E.E.; Semyshev, F.I.; Malinovsky, M.A.; Kondratiev, E.A. New data on the structure, mineralogy and geochemistry of the gold deposit Karalveem (Chukotka). *Domest. Geol.* **2017**, *3*, 26–44.
33. Akinin, V.V.; Voroshin, S.V.; Gelman, M.L. SHRIMP dating of metamorphic xenoliths from lamprophyre at the Degdekan gold deposit: On the history of transformations of the continental crust in the Ayan-Yuryakh anticlinorium (Yano-Kolyma fold system). *Geology* **2003**, *2*, 142–146.
34. Volkov, A.V.; Savva, N.E.; Sidorov, A.A.; Prokofiev, V.Y.; Goryachev, N.A.; Voznesensky, S.D.; Alshevsky, A.V.; Chernova, A.D. The Shkolnoe gold deposit in the North-East of Russia. *Geol. Ore Depos.* **2011**, *53*, 1–19. [CrossRef]
35. Akinin, V.V.; Prokopiev, A.V.; Toro, J.; Miller, E.L.; Wooden, J.; Goryachev, N.A.; Alshevsky, A.V.; Bakharev, A.G.; Trunilina, V.A. U–Pb shrimp ages of granitoides from the main batholith belt (North East Asia). *Dokl. Earth Sci.* **2009**, *426*, 216–221. [CrossRef]
36. Volkov, A.V.; Cherepanova, N.V.; Prokofiev, V.Y.; Savva, N.E.; Smilgin, S.V.; Trubkin, N.V.; Alekseev, V.Y. Gold deposit in the Butarny granite stock (North-East of Russia). *Geol. Ore Depos.* **2013**, *55*, 214–237. [CrossRef]
37. Savva, N.E.; Palyanova, G.A.; Kolova, E.E. Gold and silver minerals and condition of their formation at the Dorozhnoye deposit (Magadan region Rossia). *Nat. Resour.* **2014**, *5*, 478–495.
38. Volkov, A.V.; Prokofiev, V.Y.; Savva, N.E.; Sidorov, A.A.; Byankin, M.A.; Uyutnov, K.V.; Kolova, E.E.K. Conditions for the formation of the Kupol gold-silver deposit in the north-east of Russia based on the study of fluid inclusions. *Geol. Ore Depos.* **2012**, *54*, 293–301.
39. Savva, N.E.; Pal'yanova, G.A.; Byankin, M.A. The problem of genesis of gold and silver sulfides and selenides in the Kupol deposit (Chukot Peninsula, Russia). *Russ. Geol. Geophys.* **2012**, *53*, 457–466. [CrossRef]
40. Savva, N.E.; Shakhtyrov, V.G. Olcha gold-silver deposit: Tectonic position, structure, material characteristics. *Geol. Ore Depos.* **2011**, *53*, 460–484. [CrossRef]
41. Volkov, A.V.; Prokofiev, V.Y.; Savva, N.E.; Kolova, E.E.; Sidorov, A.A. Geochemical features of the ore-forming fluid of the Paleozoic Au-Ag epithermal Olcha deposit (North-East Russia). *Dokl. Earth Sci.* **2013**, *450*, 1–5. [CrossRef]
42. Savva, N.E. Silver Mineralogy of the North-East of Russia. In *Proceedings of the SVKNII FEB RAS*; Sidorov, A.A., Ed.; Triumph: Moscow, Russia, 2018; 544p. (In Russian)
43. Volkov, A.V.; Savva, N.E.; Ishkov, B.I.; Sidorov, A.A.; Kolova, E.E.; Murashov, K.Y. The epithermal Au-Ag deposit of Burgali in the Paleozoic Kedon volcanic belt (Northeast of Russia). *Geol. Ore Depos.* **2021**, *63*, 40–62. [CrossRef]
44. Savva, N.E.; Volkov, A.V.; Prokofiev, V.Y.; Kolova, E.E.; Murashov, K.Y. Epithermal Ag-Au deposit Primorskoye (North-East of Russia): Geological structure, mineralogical and geochemical features and conditions of ore formation. *Geol. Ore Depos.* **2019**, *1*, 52–74.
45. Gundobin, G.M.; Zakharov, M.N.; Smirnov, V.N.; Kravtsova, R.G. New data on the geochronology of igneous formations of the Even volcanic zone of the Okhotsk-Chukotka belt and associated ore mineralization. In *Geochronology of Granitoids of the Mongol-Okhotsk Belt*; Nauka: Moscow, Russia, 1980; pp. 145–156. (In Russian)
46. Kravtsova, R.G.; Dril, S.I.; Almaz, Y.A.; Tatarnikov, S.A.; Vladimirova, T.A. The first data on the Rb-Sr age and isotope composition of gold-silver ores of the Dal'nego deposit (Even ore district, North-East of Russia). *Dokl. Earth Sci.* **2009**, *428*, 240–243. [CrossRef]
47. Kostyrko, N.A.; Romanenko, I.N. Mineralogy of one of the near-surface gold-silver deposits in the North-East of the USSR. In *Mineralogy and Geochemistry of Ore Deposits of the North-East of the USSR*; SVKNII FEB RAS: Magadan, Russia, 1978; pp. 55–67. (In Russian)
48. Lantsev, I.P.; Nikolaeva, L.A.; Badalova, R.P.; Denisova, L.N. *On the Question of the Distribution of Impurity Elements in Native Gold from Various Deposits*; TsNIGRI: Moscow, Russia, 1971; Volume 1, pp. 130–137. (In Russian)
49. Vyalsov, L.N. Status and ways of development of research into the optical properties of ore minerals. In *Endogennye Ore Deposits*; Nauka: Moscow, Russia, 1980; pp. 230–265. (In Russian)
50. Stimac, J.; Goff, F.; Goff, C.J. Chapter 46: Intrusion-Related Geothermal Systems. In *The Encyclopedia of Volcanoes*; Academic Press: Cambridge, MA, USA, 2015; pp. 799–822.
51. Vikent'eva, O.V.; Prokofiev, V.Y.; Gamyanin, G.N.; Goryachev, N.A.; Bortnikov, N.S. Intrusion-related gold-bismuth deposits of North-East Russia: PTX parameters and sources of hydrothermal fluids. *Ore Geol. Rev.* **2018**, *102*, 240–259. [CrossRef]
52. Volkov, A.V.; Sidorov, A.A.; Savva, N.E.; Kolova, E.E.; Chizhova, I.A.; Murashov, K.Y. The Geochemistry of Volcanogenic Mineralization in the Northwestern Segment of the Pacific Ore Belt: Northeast Russia. *J. Volcanol. Seismol.* **2017**, *11*, 6389–6406. [CrossRef]

53. Gamyanin, G.N.; Nekrasov, I.; Zhdanov, J.J.; Leskova, I.V. Auroantimonat—A new natural compound of gold. *Dokl. Akad. Nauk SSSR Mineral.* **1988**, *301*, 947–950.
54. Chudnenko, K.V.; Palyanova, G.A.; Anisimova, G.S.; Moskvitin, S.G. Physicochemical modeling of formation of natural Ag-Au-Hg solid solutions: Kyuchyus deposit (Yakutia, Russia) as an example. *Appl. Geochem.* **2015**, *55*, 138–151. [CrossRef]

minerals MDPI

Article

Geological Constraints on the Genesis of Jagpura Au-Cu Deposit NW India: Implications from Magnetite-Apatite Mineral Chemistry, Fluid Inclusion and Sulfur Isotope Study

Abhishek Anand [1,2], Sahendra Singh [2,*], Arindam Gantait [1], Amit Srivastava [1], Girish Kumar Mayachar [3] and Manoj Kumar [4]

[1] State Unit Rajasthan, Geological Survey of India, Western Region, Jaipur 302004, India
[2] Department of Applied Geology, Indian Institute of Technology (ISM), Dhanbad 826004, India
[3] Geological Survey of India, NCEGR, Bengaluru 560078, India
[4] State Unit Madhya Pradesh, Geological Survey of India, Central Region, Bhopal 462016, India
[*] Correspondence: sahendra@iitism.ac.in

Abstract: The Jagpura Au-Cu deposit is situated within the Aravalli craton in the northwestern part of India. In the present work, petrography, mineral chemistry, fluid inclusion and sulfur isotopic compositions were used to study the Jagpura Au-Cu deposit. The ore mineral association of the deposit is arsenopyrite, loellingite, chalcopyrite, pyrrhotite and pyrite, along with native gold, magnetite and apatite. The gold fineness ranges from 914–937‰ (avg. 927‰). The presence of Au-Bi-Te phases, pyrite (>1 Co/Ni ratio), magnetite (≥1 Ni/Cr ratio, <1 Co/Ni ratio) and apatite (>1 F/Cl ratio) suggest the hydrothermal origin Au-Cu mineralization. A fluid inclusion study indicates the different episodes of fluid immiscibility with the homogenization temperatures varying between 120–258 °C and salinities range within the 8.86–28.15 wt% NaCl eq. The sulfur isotopic composition of sulfides varies from 8.98 to 14.58‰ (avg. 11.16‰). It is inferred that the variation in the sulfur isotopic compositions of sulfides is due to the cooling and dilution of the metalliferous fluid of mixed origin, derived from the basement meta-sedimentary rocks and the high saline basinal fluid. The iron oxide-copper-gold-apatite associations, structural control of mineralization, pervasive hydrothermal alteration, fluid salinity and sulfur isotope compositions indicate that the Jagpura Au-Cu deposit is similar to the iron oxide-copper-gold (IOCG)-iron oxide-apatite (IOA)types of deposits. Based on the ore geochemistry and the trace elements systematic of magnetite, the deposit is further classified as an IOCG-IOA type: IOCG-Co (reduced) subtype.

Keywords: Jagpura Au-Cu deposit; Aravalli Craton; high fineness gold; hydrothermal magnetite; fluid inclusions; sulfur isotopic composition; IOCG-IOA type: IOCG-Co (reduced) subtype mineralization

Citation: Anand, A.; Singh, S.; Gantait, A.; Srivastava, A.; Mayachar, G.K.; Kumar, M. Geological Constraints on the Genesis of Jagpura Au-Cu Deposit NW India: Implications from Magnetite-Apatite Mineral Chemistry, Fluid Inclusion and Sulfur Isotope Study. *Minerals* **2022**, *12*, 1345. https://doi.org/10.3390/min12111345

Academic Editor: Galina Palyanova

Received: 27 September 2022
Accepted: 19 October 2022
Published: 24 October 2022

1. Introduction

Gold occurs mainly in a native state and often contains Ag, Cu, Hg and other impurities [1,2]. Native gold is an indicator mineral of gold deposits [2,3]. On the basis of their genesis, the gold deposits are classified into different types, viz. Orogenic lode gold, Carlin-type gold deposit, Porphyry type gold and Iron oxide-copper-gold (IOCG) type gold deposit [4–10]. The majority of these gold deposits is of hydrothermal origin and constitute a significant portion of the world's gold resources [11].In these deposits, gold mineralization is commonly associated with pyrite and magnetite. Magnetite is stable across a wide variety of physicochemical circumstances and contains various trace elements, including Al, Ti, Mg, Mn, Zn, Cr, V, Ni, Co, and Ga [12–16]. These elements are useful petrogenetic tools for modern-day exploration [15,17–22]. Accordingly, magnetite's trace element composition is used to distinguish between IOCG, Volcanogenic Massive Sulfide (VMS), copper porphyry, Cu-Fe skarn, magmatic Fe-Ti-V-Cr, Ni-Cu-PGE, Kiruna-type iron oxide-apatite (IOA) and BIF deposits [14,16,18,20,22–28]. Apatite is also a vital pathfinder mineral for IOCG/IOCG-IOA type deposits and is effectively used as an indicator mineral for IOCG system [29,30].

The Jagpura Au-Cu deposit lies in the southern part of the Salumber-Ghatol Metallogenic Belt (SGMB), within the Paleoproterozoic Aravali Delhi Fold Belt (ADFB) Rajasthan, India. The Au-Cu mineralization of the SGMB, is hosted by carbonate rocks and albitite [27,31,32]. Previous workers advocated the magmatic-hydrothermal model for the origin of gold-sulfide mineralization within the Bhukia-Jagpura deposit [33]. However, recent investigations in the Jagpura deposit reveals that the Au-Cu lodes are hosted within the albitite and quartz-mica schist [34,35]. Since, the Jagpura Au-Cu deposit is a relatively recent finding; there are significant gaps in the understanding of the nature of the gold-sulfide mineralization, associated magnetite-apatite mineral chemistry, and source-transportation-precipitation mechanism of the ore bearing fluid to form the deposit. This requires a detailed investigation of the mineralization and hence, an integrated approach is attempted to constrain the genesis of gold-copper mineralization.

The present study was carried out with an aim to (i) characterize the gold mineralization associated with magnetite and apatite, (ii) understand the possible source of ore-bearing fluids and (iii) classify the Jagpura deposit in the light of the recent classification scheme for Cu-Au-Fe (CGI)/ IOCG deposits [36]. In this work, we present new data on the genetic aspects incorporating mineral chemistry of Au, Au-Bi, Bi-Te, Cu-sulfide phases, besides associated magnetite and apatite, the sulfur isotopic composition of major ore minerals, and fluid inclusion micro-thermometry of mineralized quartz veins of the Jagpura deposit. This study fills up the existing gaps in the understanding of the metallogeny of the Salumber-Ghatol Metallogenic Belt and has a broader exploration implication on this belt and allied areas of similar geological settings.

2. Regional Geological Setting

The Northwestern Indian shield is represented by 3.3–2.5 Ga Archean basement known as the Banded Gneissic Complex (BGC) and is overlain by 2.2–1.85 Ga Paleoproterozoic cover sequence of Aravalli-Delhi Fold Belt (ADFB), (Figure 1), [37–41]. The Archean basement consists of granite gneiss with meta-volcano-sedimentary rocks and intrusive rocks [38,39]. The BGC is in tectonic contact with or unconformably overlain by two Proterozoic supracrustal sequences, the Aravalli and Delhi supergroups [42]. The Aravalli Supergroup is widely distributed in the eastern and southeastern parts of the Aravalli-Delhi Fold Belt. The relationship between ADFB and BGC is unconformable along the entire Aravalli Fold Belt margin [42].

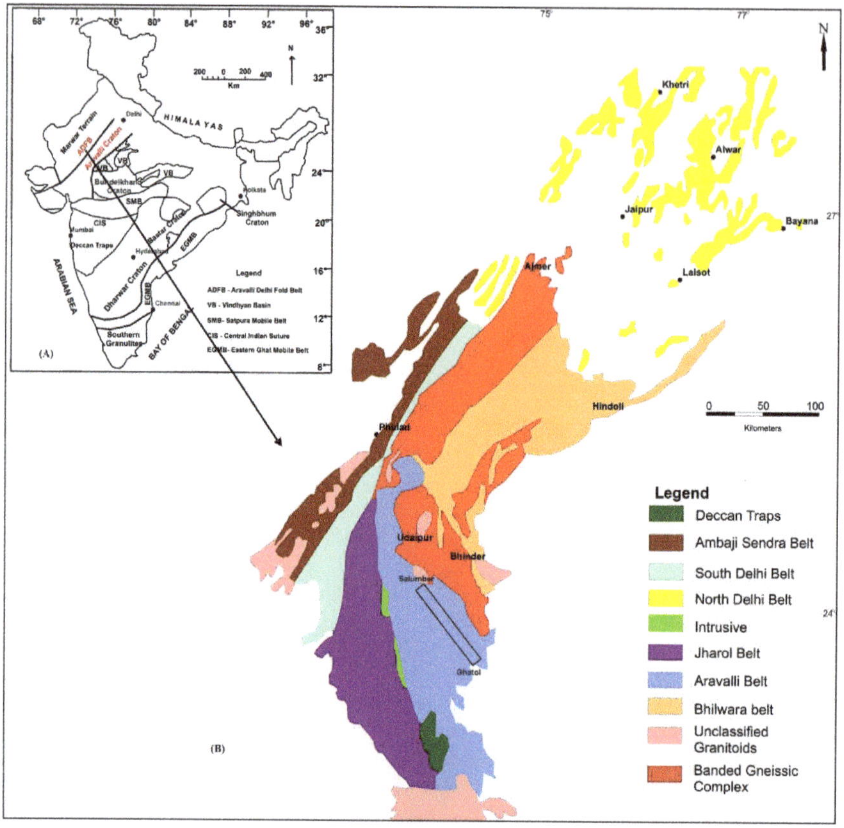

Figure 1. (**A**) Inset map showing the location of the Aravalli craton and Aravalli Delhi Fold Belt in Indian subcontinent, modified from reference [43]; (**B**) Geological map of the Aravalli–Delhi Fold Belt showing location of the basement Banded Gneissic Complex (BGC), Paleoproterozoic Aravalli Supergroup and Salumber-Ghatol metallogenic belt, modified from references [44].

The ADFB is composed of calcareous and argillaceous meta-sedimentary rocks, meta-volcano-sedimentary rocks, and intrusive rocks. The Aravalli sequence belongs to the continental rift basin settings [45,46]. Geochronological studies showed that the Aravalli sedimentation period ranges from ~2.3 to ~1.6 Ga [47,48]. The 4–6 km wide and 70 km long Salumber-Ghatol Metallogenic Belt (SGMB) is exposed in the extreme southeastern part of the ADFB and forms a part of the eastern margin of the Debari Group of the Aravalli Supergroup. It extends from Salumber in the northwest to Ghatol in the southeast, exposing a meta-volcano-sedimentary sequence unconformably overlying the BGC (Figure 2), [49,50]. Staurolite schist is a part of the basement rock [51]. The contact between the staurolite schist of the BGC and the meta-sediments of the Jagpura Formation of the Debari Group is sheared and referred to as Ghatol Shear Zone [31]. The lithostratigraphic sequence of the Debari Group, exposed in the SGMB is classified into five formations viz., the Gurali Formation (Basal quartzite and conglomerate), Delwara Formation (metabasalt with intercalations of conglomerate, marble, quartzite and feldspathic schist), Jaisamand Formation (conglomerate, feldspathic quartzite, mica schist and dolomitic marble), Mukandpura Formation (dolomite, phyllite and carbonaceous phyllite with intercalations of mica schist) and Jagpura Formation (quartzite, quartz-mica schist, garnet-biotite schist, dolomitic marble, calc-silicate rock and amphibolite).

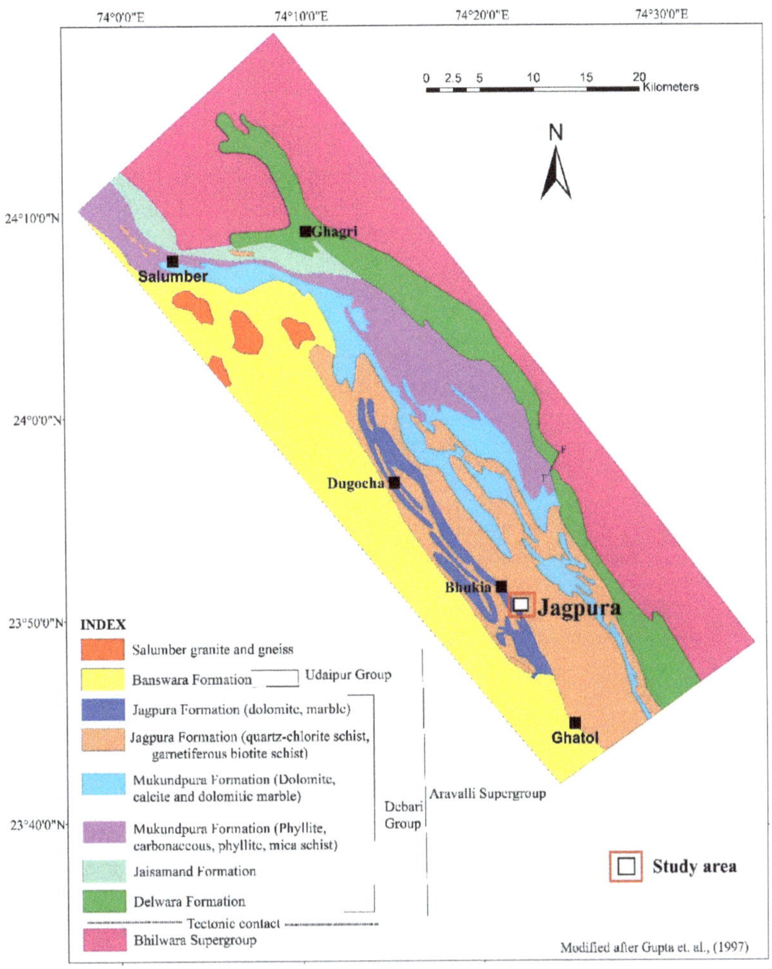

Figure 2. Geological map of Salumber-Ghatol metallogenic belt showing location of study area, modified from reference [51].

The general structural trend of the lithologic units of the SGMB is NNW-SSE with moderate to steep dips towards WSW. The rocks of the SGMB are affected by three phases of deformation events [33,52] and have undergone upper greenschist to middle amphibolite facies metamorphism [53]. The lower part of the Aravalli Supergroup i.e., Debari Group, lying close to the BGC, hosts several Au-Cu deposits/prospects in the SGMB viz. Bhukia, Jagpura, Delwara, Dagocha, Ghagri etc. [31]. The Geological Survey of India (GSI) has discovered gold-copper mineralization in the Bhukia area of SGMB in 1993 [49]. The subsequent exploration by the GSI has established 114.78 mt of gold resource with an average grade of 1.95 g/t gold, 0.15% associated copper, 93 g/t nickel and 130 g/t cobalt resource [54]. The interface of the BGC and the Aravalli Supergroup between Delwara and Ghatol in southeastern Rajasthan is mineralized (Cu-Au-iron oxide-graphite), [32]. In SGMB, the deposits/prospects situated in the southern part are more promising gold deposits in Western India [31,32,39,51,55].

The mineralization in this belt is represented by pyrrhotite, chalcopyrite, arsenopyrite, pyrite, loellingite along with magnetite, goethite and native gold. The host rock for auriferous mineralization is carbonate rocks and albitite. The U-Pb zircon ages of albite-rich rock from Bhukia deposit ranges from 1740 to 1820 Ma [56]. Gold-copper mineralization in this belt occurs as disseminations, massive ore, veins, stringers and smears along with shear fractures. The localization of ore is controlled by shears, genetically related to the D_2 phase of deformation. The ore is localized along with hinges of F_2 folds and F_2 axial plane which are parallel to D_2 shear planes. The associated alteration is characterized by pervasive Na-Ca-Mg-Fe-B-Ti alterations [31]. The present work is focused on Jagpura Au-Cu deposit, located in the southern part of the SGMB belongs to the Jagpura Formation of the Debari Group of the Aravalli Supergroup.

3. Deposit Geology

The lithologic units of the Jagpura deposit are classified into two different tectonostratigraphic domains: one is part of the Archean basement known as the Banded Gneissic Complex (BGC) and the other is part of Paleoproterozoic meta-sedimentary units of the Debari Group of the Aravalli Supergroup (Figure 3). The BGC is represented by medium grade staurolite schist, overlain by meta-sedimentary rock sequence comprising of dolomitic marble, amphibole quartzite, quartz-mica schist and albitite of the Jagpura Formation, Debari Group of the Aravalli Supergroup with a tectonic contact.

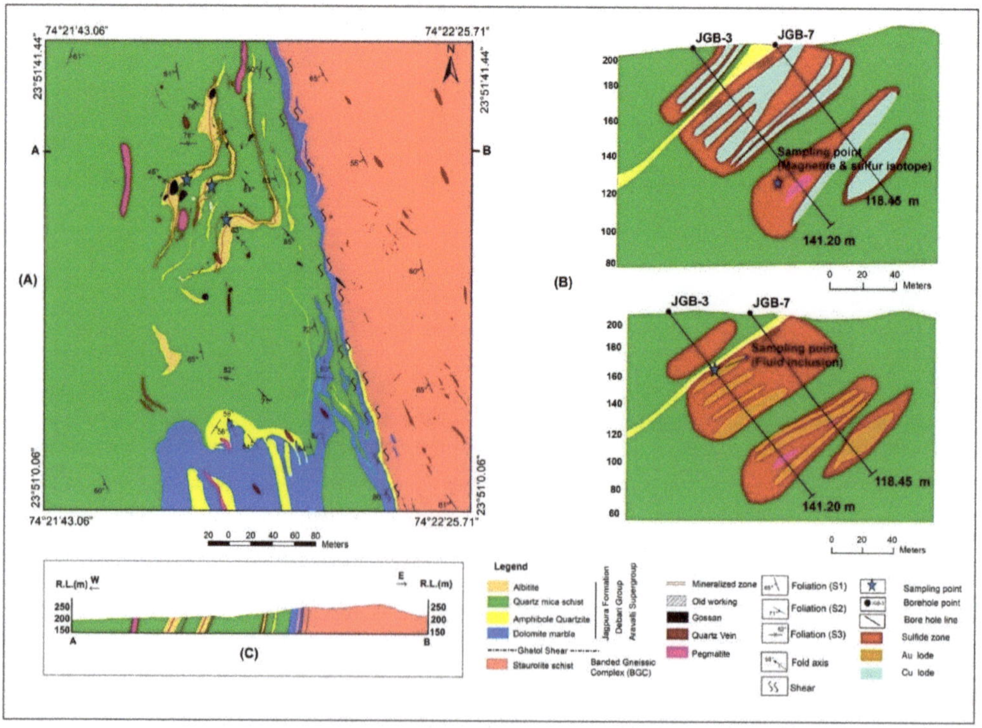

Figure 3. (**A**) Geological map of the Jagpura deposit showing different litho-units and mineralized zones, modified from reference [57], Albitite is from reference [27]; (**B**) longitudinal borehole cross-section showing the position of ore lodes; (**C**) Geological cross-section along line A–B in Figure 3A.

Staurolite schist is asilver-grey to dark grey colored fine to medium grained rock. The contact between staurolite schist and dolomitic marble is sheared, marked by various kinematic indicators in which sigma structure of quartz porphyroblast indicates a dextral sense of shearing. Dolomitic marble is medium to coarse-grained, fawn colored crystalline rock and exhibits well-developed elephant skin weathering and saccharoidal texture. Amphibole quartzite is exposed in the southern part of the deposit and is brown to pink colored, medium grained, hard and compact rock. It is composed mainly of quartz, actinolite and K-feldspar. Quartz-mica schist is the most dominant lithologic unit in the deposit. It is abuff-grey to greenish-grey colored, fine-grained rock with a silvery sheen at places with more muscovite concentration. The rock shows the development of quartz porphyroblasts, crenulation cleavage, well-developed schistosity marked by a strong preferred orientation of muscovite, chlorite and biotite (Figure 4A). The rock is traversed by foliation parallel magnetite veins and highly ferruginized at places (Figure 4B–D). Albitite is a brown to brick-red colored, fine grained, hard and compact rock (Figure 4E). It occurs as competent bands within quartz-mica schist, contains foliation parallel tourmaline rich bands and ferruginized at places.

Figure 4. Field photograph showing: (**A**) Well-developed foliation in quartz-mica schist; (**B**) Quartz-mica schist is characterized by the presence of foliation parallel magnetite veins; (**C**) Magnetite veins within quartz-mica schist; (**D**) Intense Fe alteration in quartz-mica schist; (**E**) Albitite occurring within quartz-mica schist as competent bands; (**F**) Ferruginized quartz vein within quartz-mica schist; (**G**) Well developed gossan zone within quartz-mica schist; (**H**) Old workings at contact of quartz-mica schist and albitite; (**I**) Malachite staining in wall of old workings; (**J**) Hydrothermal ferruginized breccias within quartz-mica schist indicate brecciation due to hydrothermal activity; (**K**) Na-B alteration marked by occurrence of albite and tourmaline veins; (**L**) Two generations of quartz veins occurring parallel to S_1 and S_2 foliation, depicting cross cutting relationship with each other.

Intrusive quartz veins are observed in all the lithologic units of the study area. Two generations of quartz veins are observed: the first-generation quartz veins are disposed of parallel to the S_1 foliation plane, the second-generation quartz veins are disposed parallel to the S_2 foliation plane and show a cross-cutting relationship with each other (Figure 4L). The second-generation quartz veins are prominent and associated with ore mineralization. The length of these quartz veins varies from 2 cm to 5 m and the width varies from 2 cm to 1 m. The quartz veins are ferruginized at places (Figure 4F). Quartz-mica schist and albitite are also traversed by pegmatite veins. The length of pegmatite veins varies from 5 cm to 4 m and the width varies from 2 to 15 cm. Calcite veins are present in quartz-mica schist, albitite and dolomitic marble lithologic unit and are variable in size ranging from 1 to 20 cm in length and 2 to 10 cm in width. Iron oxides in the deposit are represented by magnetite which occurs ubiquitously within quartz-mica schist and albitite as discontinuous veins parallel to the S_2 foliation plane (Figure 4B,C) and also as disseminations at places. The length of iron oxide veins varies from 15 cm to 2 m and the width varies from 1 to 10 cm.The mineral assemblages in different lithologic units of Jagpura deposit indicate upper greenschist to middle amphibolite facies of metamorphism.

3.1. Structure

The lithologic units of the Jagpura deposit show NNW-SSE to N-S structural trends with moderate (50°) to steep (82°) dip towards WSW and ESE. Three distinct phases of deformation have been identified in the area. F_1 type rootless, isoclinal folds represent D_1 phase deformation. The trends of the first-generation axial traces are NNW-SSE to NW-SE. D_2 phase deformation is marked by symmetrical to asymmetrical, tightly to openly inclined, moderately to steep plunging F_2 folds towards NW. F_2 folds control the outcrop pattern of the area and axial traces strike towards NW-SE. Broad warps with widely spaced fracture cleavage and chevron folds represent the D_3 phase of deformation. The trends of the third-generation axial traces are ENE-WSW to E-W. The third generation of folds is less pervasive and has least affected the outcrop pattern of the area. The contact between basement and cover rock is traversed through a ductile shear zone which is evident from several kinematic indicators along with mylonitic foliation. The sense of shear is dextral. The NNW-SSE trending shear zone is exposed in the eastern side of the study area and is part of the regional Ghatol shear zone. The shear zone is sympathetic to the regional D_2 deformation and shear plane is parallel with S_2 foliation. The ore microscopic studies show that the mineralization is localized along shear plane and fold hinges related to D_2 phase deformation event. In the study area, second-generation quartz veins occurring parallel to S_2 foliation are prominent and associated with mineralization. The fluid inclusion study of mineralized quartz vein suggest that the high saline ore fluids injected during the D_2 phase of deformation is responsible for the transportation of metals in the ore system as metals chloride complex.

3.2. Mineralization

In the Jagpura deposit, surface indications of mineralization are gossan zones, old workings, malachite-azurite stains, hydrothermal alteration, veins and disseminations of iron oxide (magnetite), as well as the presence of visible sulfides (Figure 4G–I). In the study area, 12 old workings are present which are semi-circular in shape and open-pit type. The length of these old workings varies from 08 to10 m and the width varies from 2 to 5 m. Gossan zones are yellow, deep brown to brick-red, massive, hard and compact, and parallel to sub-parallel discontinuous bands. At places, native gold is seen as fine flakes and specks within the gossan zone. Albitite and quartz-mica schist are the host rocks for gold-copper mineralization. Three parallel gold-copper mineralized zones are exposed in the deposit. The cumulative strike length of the mineralized zones is about 1050 m (400 m, 350 m and 300 m), with the width varying from 20 to 30 m.The ore mineral association of the Au-Cu mineralization is native gold, arsenopyrite, loellingite and chalcopyrite with associated pyrrhotite, pyrite and abundant magnetite. Alongside these mineral associations, maldonite (Au_2Bi) and hedleyite (Bi_7Te_3) are also observed as bismuth phases. Apatite, quartz, chlorite, biotite, albite, tourmaline and muscovite are common gangue minerals associated with the Au-Cu ore.

Subsurface mineralization is shallow in nature and all the three mineralized zones have been intersected in drillholes. Four different styles of mineralization are present in drillhole core samples i.e., (1) semi-massive to massive type, (2) vein and fracture fill type, (3) foliation parallel disseminations/smears and (4) patchy and stringer type (Figure 5A–F). The Geological Survey of India (GSI) has augmented a total resource of 6.07 mt gold with an average grade of 1.67 g/t Au and 8.05 mt copper with an average grade of 0.23% Cu [34,58]. Gold-copper mineralized zones occur along the hinges of F_2 fold and F_2 axial planes parallel to the D_2 shear planes. The mineralization is also observed in epigenetic quartz and pegmatite veins (Figure 5G,H).

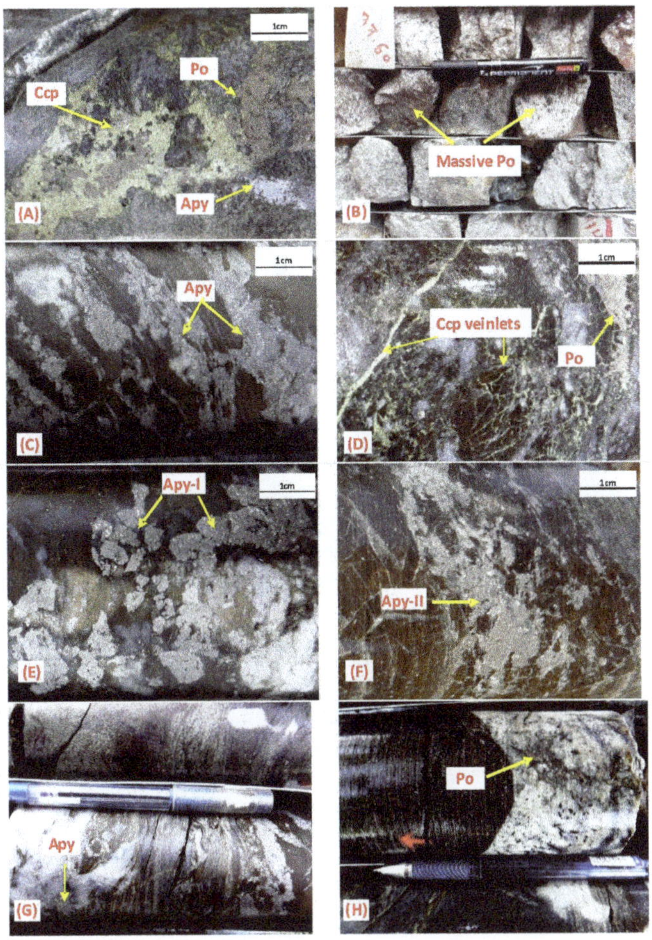

Figure 5. Photograph of drill core showing nature of sulfide mineralization in the Jagpura deposit. (**A**) Association of pyrrhotite, chalcopyrite and arsenopyrite in borehole core; (**B**) Massive to semi massive type pyrrhotite mineralization; (**C**) Foliation parallel semi massive arsenopyrite; (**D**) Vein and fracture filled chalcopyrite, pyrrhotite and arsenopyrite; (**E**) Euhedral coarse grained arsenopyrite specks of first generation; (**F**) Fine grained massive arsenopyrite of second generation; (**G**) Second

generation arsenopyrite in quartz vein; (**H**) Stringers of pyrrhotite and disseminations of arsenopyrite in pegmatite vein. (Apy = arsenopyrite, Po = pyrrhotite, Ccp = chalcopyrite).

3.3. Alteration Pattern

Hydrothermal alteration in the Jagpura deposit is represented by pervasive Na-B alteration besides chloritization, sericitization, silicification, ferruginization and hydrothermal iron oxide breccias present close to the gold-copper mineralized zones. In the Jagpura deposit, hydrothermal alteration is dominated by Na-B metasomatism (Figure 4K), marked by the ubiquitous presence of albite and tourmaline in all the lithologic units of the Jagpura Formation [34,35]. In the study area, B metasomatism postdates Na metasomatism. Within quartz-mica schist, Na metasomatism was so intense that it has formed albitite rock. Silicification is present along the S_1 and S_2 foliation. However, quartz veins occurring parallel to the S_2 foliation are dominant and close to the ore zones. Chloritization (of biotite and amphibole) near sulfide mineralization is mainly associated with the shear zone Calcic alteration is marked by the presence of calcite veins besides tremolite, actinolite (calcic amphibole) and oligoclase, andesine mineral assemblages. A lesser degree of K-alteration is also observed along the shear planes marked by biotite enrichment and albite replacement by K-feldspar. Alteration of feldspar to sericite is noted throughout the host rocks Intense ferruginization (magnetite, goethite) is present in the quartz-mica schist, albitite and in quartz veins. Hydrothermal iron oxide breccia occurs in patches within the host rock. It is composed of rounded to angular clasts made up of polycrystalline quartz set in a ferruginized matrix (Figure 4J). The occurrence of albite, tourmaline, magnetite, chlorite, actinolite, tremolite, calcite and sphene represents Na-B-Fe-Mg-Ca-K alteration within the host rock proximal to Au-Cu ore zones.

4. Materials and Methods

4.1. Sampling

For petrography, mineral chemistry, fluid inclusion, and sulfur isotope study, samples were collected from the Jagpura deposit of the SGMB. Both surface and drill core samples of host rocks and ore zones were collected for analysis (Figure 3). Samples from all stages of ore formation, representing different ore and gangue mineral assemblages, were collected. These samples were used for petrographic study and representative samples were used for electron probe micro analyzer (EPMA). Auriferous quartz vein (n = 10) occurring parallel to S_2 foliation was sampled for fluid inclusion study. A pure fraction of sulfides viz. pyrite, arsenopyrite, pyrrhotite and chalcopyrite were collected from the different ore zones for sulfur isotope analysis.

4.2. Petrography and EPMA

Thin polished and polished sections of host rocks (n = 25) were studied for ore mineralogical and petrological studies using Leica DMRX advanced polarizing petrological microscope at the Regional Petrology Laboratory, the Geological Survey of India, Jaipur, India. Representative sections (n = 10) were studied to determine the mineral chemistry and to understand the nature of the gold and sulfide mineral phases using Electron Probe Micro Analyzer (EPMA), CAMECA SX-100 with five wave dispersive spectrometers (WDS) at National Centre of Excellence in Geoscience Research (NCEGR), GSI, Faridabad, India. The polished thin sections were examined using an optical microscope to characterize the mineralogical and textural relationship of the ore and associated gangue minerals. EPMA analysis was carried out using the operating column conditions of 20 kv accelerating voltage and a probe current of 15 nA. Analyses were carried out using a beam diameter of 1 µm. Peak and background counting times for major elements were 20 s and 10 s, respectively, whereas those for trace elements were 40 s and 20 s, respectively. Natural standards were run before and after analysis to determine the analytical error. Two sections were studied to determine the mineral chemistry of magnetite using Electron Probe Micro Analyzer (EPMA), CAMECA SX-5 at Central Research Facilities in Indian Institute of Technology

(Indian School of Mines), Dhanbad, India. The analyses were carried out for Si, Ti, Al, Fe, Mn, Mg, Ca, Na, K, Ni, Co, V and Cr using the operating column conditions 15 kv accelerating voltage, probe current of 12 nA and beam diameter of 1 μm. Standards were run before and after analysis to determine the analytical error. The standards used for calibration of the instrument are as follows: spinel for Mg and Al, ilmenite for Ti, vanadium metal for V, chromite for Cr and Fe, and manganese oxide for Mn. Vanadium concentrations were corrected for the overlap between Ti K_β and V $K\alpha$ peaks by analyzing V-free and Ti bearing standards of rutile and metallic Ti. Matrix correction was carried outusing the ZAF correction. The analytical inaccuracies and uncertainties of the analysis are negligible and as follows: <0.1% for Cr; <1% for Al, Si, K, Ca, Fe, Mg, Ti, Co, V; <2% for Mn and Na.

4.3. Fluid Inclusion Micro Thermometry

At the Fluid Inclusion Lab of NCEGR, GSI, Bengaluru, India, the fluid inclusions were observed under a microscope at high magnification and the temperatures of thermally induced phase transitions in fluid inclusions were evaluated using a calibrated LINKAM THMSG 600 heating/freezing stage. The heating and freezing stage is equipped with an Olympus BX-50 transmitted light microscope, an LNP-95-LTS liquid nitrogen chilling unit, and a digital video capturing system for temperature monitoring. LINKSYS is the program used to study fluid inclusions. The system was calibrated, and the reproducibility of measurements at room temperature was tested using natural fluid inclusion standards from quartz crystals tested in other established laboratories, as well as LINKAM's standard CO_2 standards, with reproducibility achieved within the standard accepted error limits. The apparatus operates at temperatures ranging from 196 to 600 °C.

4.4. Sulfur Isotopic Composition

A total of 31 sulfide samples were selected for conducting the sulfur isotope studies and the respective data were generated from the Isotope Ratio Mass Spectrometry (IRMS) laboratory, NCEGR, GSI, Bengaluru, India. Sulfide grains were separated by hand from gangue minerals under a stereomicroscope and then powdered avoiding any contamination. Using the ANCA GSL (Automated Nitrogen and Carbon Analyzer for Gas Solids and Liquid) peripheral system; the separated powdered samples were analyzed for sulfur isotopes in continuous flow mode using an Isotope Ratio Mass Spectrometer (Make: SerCon, Model: Geo 2020). Around 3550 mg powder of each sulfide mineral was manually packed into tin capsules and placed into the auto-sampler unit above the ANCA, with the positions noted. Each tin capsule was put one by one into a furnace at 1050 °C with an additional oxygen pulse during the analysis. In the presence of oxygen, the tin ignites and burns exothermally, raising the temperature to around 1800 °C and oxidizing the sample. Water is removed using a Nafion dryer and anhydrous magnesium perchlorate trap. Helium is the carrier gas used (99.99%). The gas stream is routed through a gas chromatograph, where sulfur is separated and injected into a mass spectrometer for isotope analysis of sulfur. Each sample's total analysis time is 9 m 10 s. Analytical techniques and fractionation mechanism was followed in accordance with [59,60]. Each sample was analyzed three times in a batch, along with international reference (NBS) and internal laboratory standards. All sulfur-bearing samples are measured using the VCDT scale [61].

4.5. Geochemical Analysis

For geochemical analysis, drill core samples (n = 10) were collected from mineralized zone. The samples were crushed to −100 mesh size, sieved, coned and quartered, for major, trace, base metal and REE analysis. The geochemical analysis was carried out at NABL accredited (ISO/IEC17025:2005) chemical laboratory of Geological Survey of India, Western Region, Jaipur.The trace elements including REE were analyzed using Varian 820 Inductively Coupled Plasma Mass Spectrometry (ICP-MS instrument, following the closed digestion sample preparation technique). Standard reference materials (SRMs) were employed to estimate accuracy and precision of the instrument. Gold was analyzed using fire assay technique with ICPMS/GTAAS instrument. The base metal (Cu, Ni, Co, Pb, Zn) was analyzed using Flame Atomic Absorption Spectrophotometer (FAAS, Agilent Duo 280 AAS) instrument. The samples were pulverized to −200 mesh size using vibratory cup mill and taken in tightly packed zip plastic bag. For digestion, 0.2 g of the sample was weighed in a test tube and 5 mls of HNO3: HCl mixture in 1:2 ratios were added to it. The solution was heated over hotplate for 4 h and then made up with Type-I water and allowed to settle. The final solution was used for the determination of Cu, Ni, Co, Pb, Zn.

5. Results

5.1. Petrography and EPMA Study

5.1.1. Petrography of Host Rocks

Detailed petrographic studies of gangue and ore minerals of the host rocks were carried out to understand the relationship between gangue and ore minerals, nature of gold mineralization, microstructures, mineral paragenesis and hydrothermal alteration pattern. Quartz-mica schist is composed of muscovite, chlorite, sericite and quartz with an accessory amount of apatite, feldspar, tourmaline, biotite, ilmenite, rutile, hornblende, orthoclase and sphene (Figure 6A–F). A few sections also observe the alteration of biotite and amphibole to chlorite. The rock shows well-developed schistosity, marked by the parallel alignment of chlorite, muscovite, and sericite. S-C fabric, mica fish and pressure shadow are also observed within the rock. Orientation of mica fish and S-C fabric shows a dextral sense of shearing (Figure 6J) and alteration of feldspar to sericite. (Figure 6K). The rock shows well-developed crenulation cleavage and microfolds associated with the second phase of deformation. In chlorite, Fe/Fe+Mg ratio varies from 0.42 to 0.61 and Mg/Fe + Mg ratio ranges from 0.39 to 0.58. Most chlorites are ripidolite with few pynochlorite and diabantite (Figure 8E). Feldspar is mostly albite ($Ab_{98.6–91}An_{8.2–0.4}Or_{3–0.1}$) to oligoclase [($Ab_{84.8–70.9}An_{27.1–14.6}Or_{2–0.3}$), (Figure 8F) and tourmaline is dravite variety.

Albitite comprises albite ($Ab_{98.7–90.8}An_{9.1–0.7}$) and tourmaline as major minerals (Figure 6D) along with apatite, biotite, muscovite, actinolite, quartz, orthoclase, rutile and ilmenite as accessory minerals. Albite occurs as subhedral, polygonal grains and generally untwined in nature. In some sections, albite constitutes up to 90% volume of the rock. Tourmaline occurs as euhedral to subhedral grains and shows strong pleochroism in brown shades. The concentration of tourmaline ranges from 5 to 50% volume of the rocks. A higher concentration of tourmaline is observed in the mineralized zones. Albite and tourmaline grains in albitite are elongated and aligned, parallel to the shear foliation close to the shear zone (Figure 6G,H).

5.1.2. Ore Petrography and Ore Chemistry

The sulfide mineral assemblages of the Jagpura deposit are represented by pyrrhotite, chalcopyrite, arsenopyrite, loellingite and pyrite in the order of abundance (Figure 7A–O). Moreover, native gold, maldonite (Au_2Bi) and hedleyite (Bi_7Te_3) are present as gold and bismuth ore, respectively (Figure 8D). The sulfide minerals occur as semi-massive to massive type, vein and fracture fill type, foliation parallel type, patchy and stringers type. Magnetite is the dominant oxide phase and is closely associated with gold-sulfide mineralization (Figure 7G,J,M); it is euhedral to subhedral in shape and occurs as disseminations and veins (Figures 7E,I,M and 8A,B). Apatite is euhedral, hexagonal in shape and occurs as disseminations parallel to the foliation plane of host rocks (Figure 7L). Apatite is associated with magnetite and chalcopyrite at many places (Figures 7L and 8C). The ore minerals indicate replacement, remobilization and deformational textures. Major sulfide phases show replacement textures. Two types of arsenopyrite are observed in the ore zones: one is coarse grained, granular and intensely fractured with cataclastic texture (Figure 7A) and the other is idiomorphic, euhedral, rhombic in shape, fine to medium grained without deformation texture (Figure 7B). The healing of fractures within the arsenopyrite by chalcopyrite shows that the former is replaced by the later (Figure 7C). The convexity of chalcopyrite grain boundary within pyrrhotite shows that pyrrhotite has formed earlier (Figure 7D). Occurrence of sulfides along weak planes viz. fractures, veins and shear planes

(Figure 7H,K) and the hinge of F_2 folds represents ore remobilization (Figures 6E,F and 7F). In mineralized zones, sulfide veins cross-cutting the albite-tourmaline grains (Figure 6I), deformed albite twin lamellae and cataclastic fragments of sulfide minerals indicate ore deformational textures (Figure 7K). Sulfide minerals such as pyrite and arsenopyrite are associated with gold [62]. Arsenopyrite has a higher gold content than pyrite, and the gold content of arsenian pyrites increases with arsenic content [63].

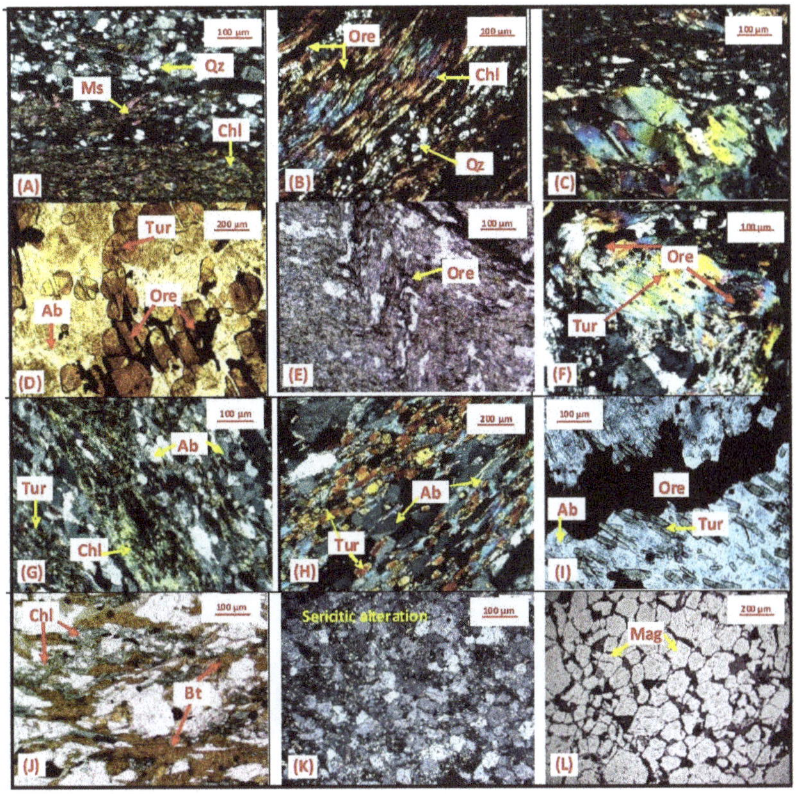

Figure 6. Photomicrograph showing mineral assemblages related to alteration zones, D_2 deformation and shear system. (**A**) Quartz-mica schist is mainly consists of muscovite, chlorite and quartz; (**B**) Alternate bandings of quartz and chlorite rich layers within quartz-mica schist; (**C**) Mica fish structure within quartz-mica schist indicating dextral nature of the shear zone; (**D**) Gold-sulfide mineralization is associated with extensive hydrothermal alterations viz. albitization and tourmalinization, also to note that mineralization is occurring parallel to the fractures of tourmaline grains; (**E**,**F**) F_2 fold in quartz-mica schist with disposition of opaque minerals at the hinge; (**G**,**H**) Elongated albite and tourmaline grains in albitite close to shear zone; (**I**) Sulfide vein cross cutting the foliation of albite-tourmaline grains in albitite; (**J**) Alteration of biotite to chlorite in proximity to sulfide disseminations; (**K**) Sericitic alteration in albitite (PPL); (**L**) Angular grains of magnetite within iron oxide breccia. (Mag = magnetite, Ms = muscovite, Qz = quartz, Chl = chlorite, Ab = albite, Tur = tourmaline, PPL = plane polarized light).

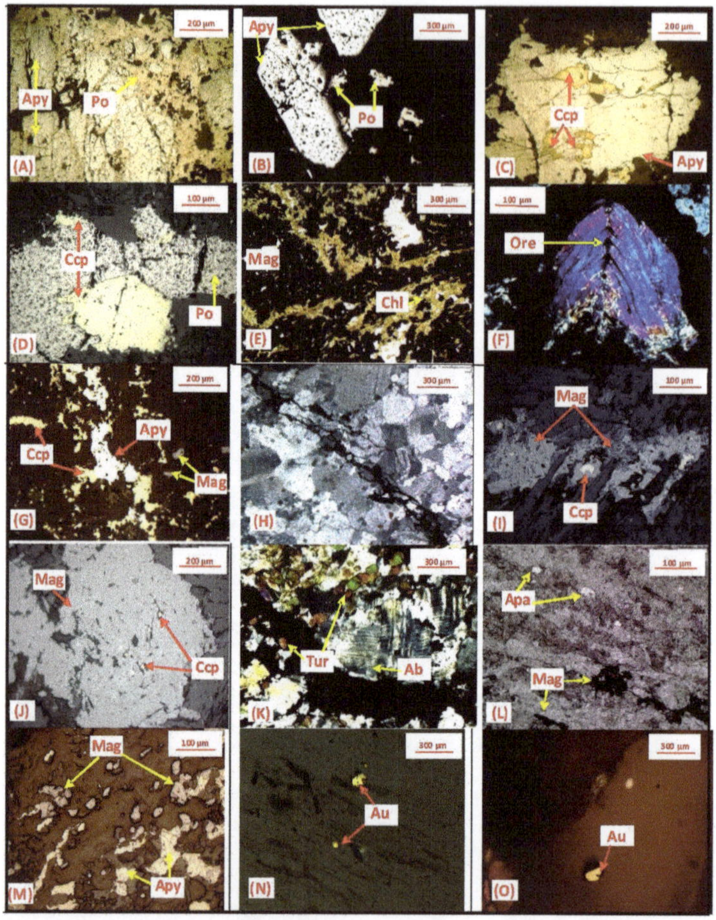

Figure 7. Photomicrograph showing various generations of mineral assemblages such as magnetite-apatite-sulfides-gold and their linkages to the shear zone characteristics and deformation patterns. (**A**) Catascatically deformed arsenopyrite of the first generation, where fractures are healed by pyrrhotite; (**B**) Undeformed isomorphic, euhedral, rhombic shaped second generation arsenopyrite; (**C**) Fractures of arsenopyrite are healed by later formed chalcopyrite; (**D**) Convexity of chalcopyrite grain in pyrrhotite indicate that pyrrhotite has formed earlier; (**E**) Occurrence of abundant subhedral magnetite grains within quartz-mica schist; (**F**) Saddle reef structure having sulfide mineralization at the hinge of microfold indicating ore remobilization; (**G**) Disseminations of arsenopyrite, pyrrhotite, chalcopyrite and magnetite of early stage mineralization; (**H**) Nature of mineralized shear system showing shear planes cross cutting the albite grains; (**I**) Magnetite vein of second stage mineralization, occurring parallel to shear fracture; (**J**) Coarse grained magnetite with associated chalcopyrite; (**K**) Deformed albite twin lamellae in albitite close to mineralized vein; (**L**) Occurrence of subhedral apatite grains associated with magnetite in quartz-mica schist; (**M**) Disseminated magnetite and arsenopyrite of early stage mineralization; (**N**) Native gold grains in quartz-mica schist; (**O**) Skeletal gold speck in second generation quartz vein intrusive within quartz-mica schist. (Apa = apatite, Mag = magnetite, Apy = arsenopyrite, Po = pyrrhotite, Ccp = chalcopyrite, Au = gold, Tur = tourmaline, Ab = albite, Chl = chlorite).

Ore petrography revealed that gold is mainly present within the arsenopyrite and loellingite. Within the arsenopyrite, gold occurs in association with loellingite, maldonite and hedleyite. Gold occurs as specks, fine flakes and films varying in size from a fraction of a micron to 40 microns (Figure 7N). Gold also occurs along the fractures within the arsenopyrite and loellingite. Epigenetic quartz vein within the host rocks also contains native gold (Figure 7O). Arsenopyrite and loellingite are intimately associated and form the main gold bearing minerals in the study area. EPMA analysis of native gold indicates that Au concentration ranges from 89.25 to 94.72 wt% and it contains silver as a minor impurity (6.15 to 8.46 wt%). Gold fineness ranges from 914–937‰, with average value of 927‰ (n = 6).

EPMA analysis of the sulfides also indicates the presence of invisible gold. Gold content ranges from 0.02 to 2.58 wt% in arsenopyrite, 0.02 to 0.20 wt% in chalcopyrite, 0.08 to 0.11wt% in pyrrhotite and up to 0.17 wt% in loellingite. The EPMA study reveals that loellingite and arsenopyrite occur as coexisting phase within the ore zones. However, loellingite is the preferred residence for gold mineralization compared to arsenopyrite (Figure 8D). Arsenic content is higher in loellingite (72.20 to 72.36 wt%) than arsenopyrite (47.07 to 51.05 wt%). Loellingite and arsenopyrite are also rich in nickel and cobalt wherein, loellingite shows higher values compared to arsenopyrite [Table 1]. The nickel and cobalt values in loellingite range from 2.10 to 2.45 wt% and 1.08 to 1.44 wt%, respectively. The nickel and cobalt values in arsenopyrite are 0.39 to 2.22 wt% and 0.18 to 0.40 wt%, respectively.

Table 1. EPMA data of Au, Au-Te, Au-Bi and sulfide phases (in Wt%) from the Jagpura gold-copper deposit, western India.

Lithology: Albitite, Sample No. JGB-10													
Mineral Name	**Gold (Au)**			**Arsenopyrite (FeAsS)**			**Pyrrhotite (FeS)**			**Chalcopyrite ($CuFeS_2$)**			
EPMA Analysis No.	1/1	1/2	1/3	1/4	1/5	1/6	1/7	1/8	1/9	1/10	1/11	1/12	1/13
S	0.09	0.02	0.02	17.74	17.70	17.33	38.27	39.04	39.17	34.14	34.18	34.03	33.88
Fe	0.68	0.12	0.12	31.10	31.60	34.68	59.85	58.75	59.22	30.08	29.35	30.02	29.92
Co	bdl	bdl	0.02	2.22	0.88	0.39	0.01	0.03	bdl	bdl	0.01	bdl	0.03
Ni	0.10	bdl	bdl	0.40	0.29	bdl	0.07	0.04	bdl	bdl	0.03	bdl	bdl
Cu	0.08	0.02	0.01	bdl	0.02	0.03	0.01	0.01	0.01	32.66	33.79	32.6	33.02
Zn	bdl	0.10	0.02	bdl	bdl	bdl	bdl	bdl	bdl	bdl	0.02	0.05	0.02
As	0.25	0.08	0.01	48.10	49.14	47.07	bdl	0.02	bdl	bdl	0.04	bdl	bdl
Mo	bdl	bdl	bdl	bdl	0.24	0.22	0.51	0.56	0.49	0.46	0.50	0.43	0.58
Ag	7.20	6.15	6.58	0.03	bdl	0.04	bdl	bdl	0.02	0.08	bdl	bdl	bdl
Cd	0.14	0.02	0.08	bdl	0.01	bdl	bdl	0.02	bdl	bdl	0.06	bdl	bdl
Te	bdl	bdl	bdl	bdl	bdl	bdl	bdl	0.02	bdl	2.44	bdl	2.42	2.19
Au	89.95	92.12	92.21	0.13	0.06	0.07	0.11	0.08	0.10	0.10	0.20	0.06	0.02
Pb	bdl	bdl	bdl	0.15	bdl	0.13	bdl	bdl	bdl	bdl	0.13	bdl	bdl
Bi	bdl	bdl	bdl	0.10	bdl	bdl	bdl	bdl	bdl	bdl	bdl	bdl	bdl
Total	98.49	98.63	99.07	99.97	99.94	99.96	98.83	98.57	99.01	99.96	98.31	99.61	99.66
Lithology: Quartz-mica schist, Sample No. JGB-7													
Mineral Name	**Gold (Au)**			**Loellingite ($FeAs_2$)**			**Maldonite (Au_2Bi)**		**Hedleyite (Bi_7Te_3)**		**Arsenopyrite (FeAsS)**		
EPMA Analysis No.	1/1	1/2	1/3	1/4	1/5	1/6	1/7	1/8	1/9	1/10	1/11	1/12	
S	0.05	0.07	0.01	1.29	1.17	1.30	0.07	bdl	0.04	0.02	16.65	18.25	

Table 1. *Cont.*

Fe	0.22	0.72	0.21	24.71	24.06	24.38	1.52	0.32	0.32	0.30	31.03	32.13
Co	bdl	bdl	bdl	1.08	1.44	1.41	0.06	0.04	0.07	0.09	0.69	0.67
Ni	0.01	0.11	0.05	2.45	2.50	2.10	0.11	0.06	0.14	0.15	0.29	0.18
Cu	0.03	0.07	0.04	bdl	bdl	bdl	bdl	bdl	bdl	bdl	bdl	bdl
Zn	bdl	bdl	bdl	bdl	bdl	bdl	bdl	bdl	bdl	bdl	bdl	0.02
As	0.08	0.21	0.03	70.36	70.20	70.34	0.32	0.22	0.49	0.42	47.55	48.32
Mo	bdl	bdl	bdl	0.02	bdl	0.04	bdl	bdl	bdl	bdl	0.19	0.28
Ag	8.46	7.60	7.12	bdl	0.08	bdl	0.12	0.09	bdl	bdl	0.02	bdl
Cd	0.03	0.12	bdl	bdl	bdl	bdl	0.19	0.12	bdl	bdl	bdl	0.07
Te	bdl	bdl	bdl	bdl	bdl	bdl	bdl	bdl	20.39	21.30	0.05	bdl
Au	89.93	89.25	91.12	bdl	bdl	0.17	64.71	66.47	bdl	bdl	2.58	0.02
Pb	bdl	bdl	bdl	bdl	bdl	0.11	bdl	bdl	0.10	0.08	bdl	bdl
Bi	bdl	bdl	bdl	bdl	bdl	bdl	32.68	31.52	78.11	77.18	0.80	bdl
Total	98.81	98.15	98.58	99.91	99.45	99.85	99.78	98.84	99.66	99.54	99.85	99.94
Lithology: Quartz-mica schist, Sample No. JGB-8												
Mineral Name	**Pyrite (FeS_2)**											
EPMA Analysis No.	1/1	1/2	1/3	1/4	1/5	1/6	1/7	1/8	1/9	1/10	1/11	
S	53.00	53.25	53.11	53.25	52.49	53.15	52.86	53.13	53.00	53.11	53.25	
Fe	45.88	45.62	46.25	45.62	46.33	46.36	45.80	46.34	45.88	45.25	45.62	
Co	0.06	0.15	0.10	0.15	0.04	0.03	0.19	0.07	0.06	0.10	0.15	
Ni	0.03	0.06	0.04	0.06	0.03	0.02	0.10	0.06	0.03	0.03	0.06	
Cu	bdl	bdl	0.02	bdl	bdl	0.02	0.02	0.01	bdl	0.02	bdl	
Zn	bdl	0.04	bdl	0.04	0.03	bdl	bdl	bdl	bdl	bdl	0.04	
As	bdl	0.01	bdl	0.01	bdl	0.01	0.01	0.02	bdl	bdl	0.01	
Ag	0.01	0.02	0.04	0.02	bdl	0.04	bdl	0.03	0.01	0.04	0.02	
Total	98.99	99.17	99.53	99.16	99.11	99.82	98.99	99.96	99.33	99.13	99.61	
Co/Ni	1.96	2.54	2.40	2.54	1.17	1.76	1.89	1.17	1.96	3.20	2.54	
S/Fe	1.16	1.17	1.15	1.17	1.13	1.15	1.15	1.15	1.16	1.17	1.17	
Fe/(S+As)	0.87	0.86	0.87	0.86	0.88	0.87	0.87	0.87	0.87	0.85	0.86	
bdl = below detection limit												

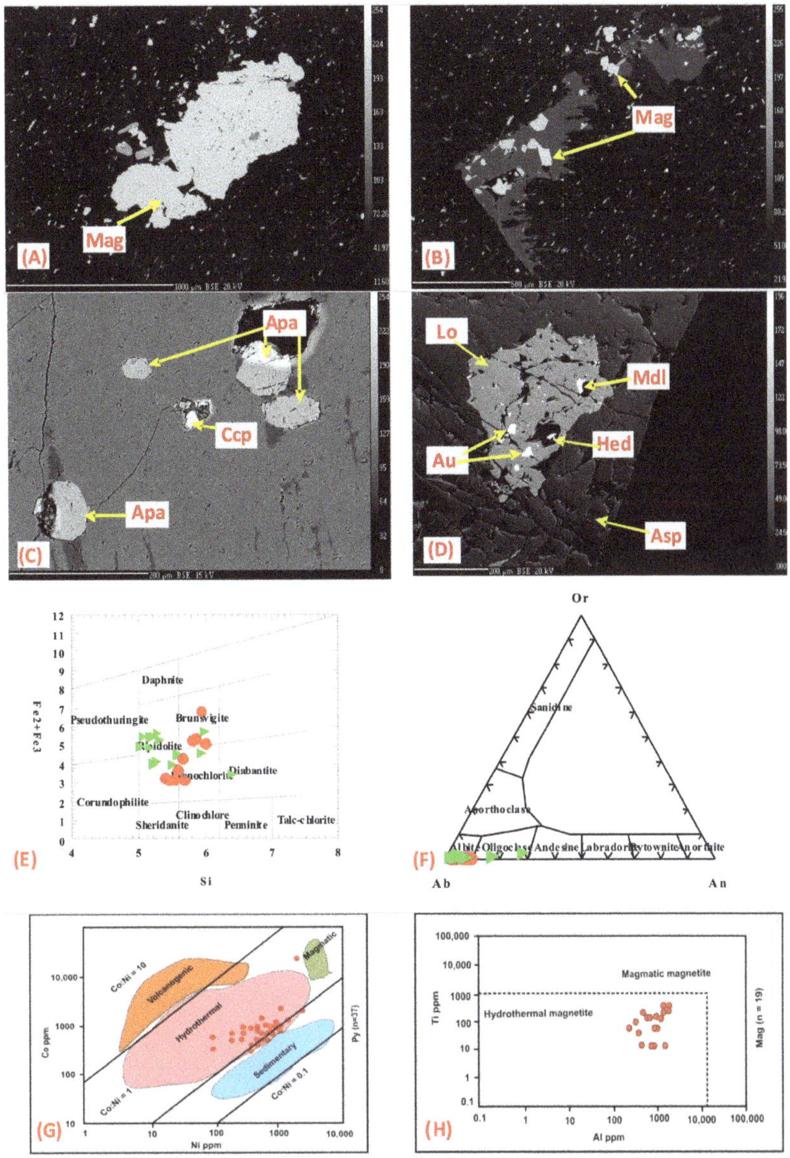

Figure 8. BSE images of magnetite-apatite-gold-sulfide mineral assemblages along with plots suggesting hydrothermal nature of pyrite and magnetite in the study area. (**A**) Magnetite megacryst in quartz-mica schist; (**B**) Disseminated magnetite within quartz mica-schist; (**C**) Subrounded to rounded apatite grains associated with chalcopyrite; (**D**) Association of gold, loellingite, arsenopyrite, maldonite and hedleyite; (**E**) Mineral composition classification diagram of chlorite, (Red = albitite, green = quartz-mica schist); (**F**) Or-Ab-An classification diagram of feldspar. (Red = albitite, green = quartz-mica schist); (**G**) Co vs. Ni plot of pyrite show hydrothermal signature, fields for pyrite composition from references [64]; (**H**) Ti vs. Al covariation plots of magnetite indicates magnetite is hydrothermal in nature, fields from reference [65]. (Apa = apatite, Mag = magnetite, Ccp = chalcopyrite, Au = gold, Mdl = maldonite, Lo = loellingite, Hed = hedleyite).

The EPMA study of magnetite from the study area reveals that the concentrations of TiO_2 ranges from(0.01 to 0.11 wt%), Al_2O_3 (0.06 to 0.28 wt%), MnO (up to 0.01 wt%), FeO (83.39 to 89.02 wt%), MgO (0.01 to 0.08 wt%), V_2O_3 (0.03 to 0.26 wt%), Cr_2O_3 (0.01 to 0.09 wt%), NiO (0.04 to 0.31 wt%), CoO (0.02 to 0.05 wt%) and CaO (0.01 to 0.08 wt%) [Table 2]. Magnetite shows low concentrations of TiO_2 (0.01 to 0.11 wt%) and Al_2O_3 (0.06 to 0.28 wt%). The Co/Ni ratio of magnetite is <1 (0.04 to 0.67) and Ni/Cr ratio is ≥1 (1 to 4). Co/Ni ratio of pyrite from the study area is >1 (1 to 2.54, n = 22). The EPMA analysis of apatite reveals that apatite is of fluorapatite variety with F content >1 wt% and >1F/Cl ratio [Table 3]. Apatite has a higher concentration of F (4.23 to 5.97 wt%, n = 10) and a lower concentration of Cl (0.08 to 0.73 wt%), FeO (0.01 to 0.28 wt%), and MnO (0.01 to 0.07 wt%).

Table 2. EPMA data of magnetite from the Jagpura gold-copper deposit, western India.

Lithology: Quartz-mica Schist, Sample No. JGB-8										
Mineral Name	**Magnetite**									
EPMA Analysis No.	1/1	1/2	1/3	1/4	1/5	1/6	1/7	1/8	1/9	1/10
Wt%										
SiO_2	0.02	0.17	0.06	0.10	0.02	bdl	0.05	0.04	0.10	0.06
TiO_2	0.05	0.01	0.05	0.07	0.01	0.10	0.05	0.09	0.01	0.05
Al_2O_3	0.12	0.14	0.13	0.11	0.10	0.28	0.15	0.27	0.26	0.21
Cr_2O_3	0.05	0.06	0.04	0.05	0.07	0.06	0.07	0.04	0.05	0.03
V_2O_3	0.20	0.17	0.18	0.07	0.17	0.15	0.08	0.15	0.19	0.17
FeO	89.02	87.69	88.49	87.88	87.82	83.39	83.82	88.57	86.63	88.98
MnO	bdl	0.06	bdl	0.06	bdl	0.01	0.03	bdl	bdl	bdl
MgO	0.05	0.08	bdl	bdl	0.03	bdl	0.01	bdl	bdl	bdl
CaO	bdl	0.03	bdl	bdl	bdl	0.01	0.02	bdl	0.02	bdl
Na_2O	0.02	0.01	0.05	0.03	bdl	0.08	0.01	0.01	0.01	0.05
K_2O	0.04	0.03	bdl	bdl	bdl	0.02	bdl	0.01	0.02	bdl
NiO	0.05	0.05	0.05	0.06	0.07	0.05	0.06	0.07	0.05	0.05
CoO	0.02	0.03	0.02	0.03	0.04	0.02	0.03	0.04	0.02	0.02
Total	89.64	88.53	89.07	88.46	88.33	84.17	84.38	89.29	87.36	89.62
apfu Number of ions on the basis of 4 (O)										
Si	0.01	0.08	0.03	0.05	0.01	bdl	0.02	0.02	0.05	0.03
Ti	0.03	0.01	0.03	0.04	0.01	0.06	0.03	0.05	0.01	0.03
Al	0.06	0.07	0.07	0.06	0.05	0.15	0.08	0.14	0.14	0.11
Cr	0.03	0.04	0.03	0.03	0.05	0.04	0.05	0.03	0.03	0.02
V	0.11	0.10	0.10	0.04	0.10	0.08	0.04	0.08	0.11	0.10
Fe	69.20	68.16	68.78	68.31	68.26	64.82	65.15	68.85	67.34	69.17
Mn	bdl	0.05	bdl	0.05	bdl	0.01	0.02	bdl	bdl	bdl
Mg	0.03	0.05	bdl	bdl	0.02	bdl	0.01	bdl	bdl	bdl
Ca	bdl	0.02	bdl	bdl	bdl	0.01	0.01	bdl	0.01	bdl
Na	0.01	0.01	0.04	0.02	bdl	0.06	0.01	0.01	0.01	0.04
K	0.03	0.02	bdl	bdl	bdl	0.02	bdl	0.01	0.02	bdl
Ni	0.04	0.04	0.04	0.05	0.06	0.04	0.05	0.06	0.04	0.04

Table 2. *Cont.*

Co	0.01	0.02	0.01	0.02	0.03	0.01	0.02	0.03	0.01	0.01
Co/Ni	0.25	0.50	0.25	0.40	0.50	0.25	0.40	0.50	0.25	0.25
Ni/Cr	1.18	1.00	1.48	1.67	1.20	1.00	1.06	2.00	1.18	1.95

Lithology: Albitite No. JGB-10									
Mineral name	**Magnetite**								
EPMA Analysis No.	1/1	1/2	1/3	1/4	1/5	1/6	1/7	1/8	1/9
Wt%									
SiO_2	0.10	0.10	0.02	0.04	0.01	0.03	0.03	0.09	0.03
SiO_2	0.10	0.10	0.02	0.04	0.01	0.03	0.03	0.09	0.03
TiO_2	0.03	0.03	0.06	0.04	0.11	0.02	0.03	0.07	0.01
Al_2O_3	0.17	0.19	0.19	0.08	0.25	0.09	0.06	0.25	0.17
Cr_2O_3	0.04	0.08	0.07	0.01	0.06	0.09	0.04	0.09	0.09
V_2O_3	0.15	0.12	0.26	0.20	0.13	0.13	0.16	0.03	0.08
FeO	86.27	88.91	85.43	88.23	87.69	87.28	88.19	84.06	88.37
MnO	bdl	bdl	bdl	bdl	0.10	0.06	0.03	bdl	0.09
MgO	0.04	bdl	0.01	bdl	bdl	bdl	bdl	0.01	bdl
CaO	0.04	bdl	0.04	bdl	bdl	0.01	bdl	0.05	bdl
Na_2O	bdl	0.08	0.03	0.03	0.02	bdl	0.01	0.02	0.01
K_2O	bdl	bdl	0.02	bdl	bdl	0.01	0.03	bdl	0.01
NiO	0.04	0.07	0.09	0.05	0.09	0.09	0.04	0.31	0.14
CoO	0.02	0.05	0.02	0.03	0.05	0.04	0.02	0.02	0.04
Total	86.90	89.63	86.24	88.71	88.51	87.85	88.64	85.00	89.04
apfu Number of ions on the basis of 4 (O)									
Si	0.05	0.05	0.01	0.02	bdl	0.01	0.01	0.04	0.01
Ti	0.02	0.02	0.04	0.02	0.07	0.01	0.02	0.04	0.01
Al	0.09	0.10	0.10	0.04	0.13	0.05	0.03	0.13	0.09
Cr	0.02	0.05	0.05	0.01	0.04	0.06	0.02	0.06	0.06
V	0.08	0.07	0.15	0.11	0.07	0.07	0.09	0.02	0.04
Fe	67.06	69.11	66.41	68.58	68.16	67.84	68.55	65.34	68.69
Mn	bdl	bdl	bdl	bdl	0.08	0.05	0.02	bdl	0.07
Mg	0.02	bdl	0.01	bdl	bdl	bdl	bdl	0.01	bdl
Ca	0.03	bdl	0.03	bdl	bdl	0.01	bdl	0.04	bdl
Na	bdl	0.06	0.02	0.02	0.01	bdl	0.01	0.01	0.01
K	bdl	bdl	0.02	bdl	bdl	0.01	0.02	bdl	0.01
Ni	0.03	0.06	0.07	0.04	0.07	0.07	0.03	0.24	0.11
Co	0.01	0.04	0.01	0.02	0.04	0.03	0.01	0.01	0.03
Co/Ni	0.33	0.67	0.14	0.50	0.57	0.43	0.33	0.04	0.27
Ni/Cr	1.50	1.20	1.40	4.00	1.75	1.17	1.50	4.00	1.83

Table 3. EPMA data of apatite (in Wt%) from the Jagpura gold-copper deposit, western India. Lithology: Quartz-mica schist, Sample No. JGB-11.

Mineral Name				**Apatite**			
EPMA Analysis No.	1/1	1/2	1/3	1/4	1/5	1/6	1/7
SiO_2	0.04	0.21	0.04	0.06	0.53	0.04	0.23
TiO_2	bdl	bdl	0.07	bdl	bdl	0.19	bdl
Al_2O_3	bdl	0.02	bdl	0.01	0.07	0.01	0.03
Cr_2O_3	bdl	0.12	0.03	bdl	bdl	bdl	bdl
FeO	0.15	0.01	0.04	0.04	0.28	0.13	bdl
MnO	bdl	bdl	0.07	0.05	0.01	bdl	bdl
MgO	bdl	bdl	bdl	bdl	bdl	bdl	bdl
CaO	54.42	53.55	54.34	54.10	53.45	54.46	53.23
Na_2O	0.54	0.07	0.03	0.09	0.08	0.03	0.07
K_2O	0.02	0.01	bdl	bdl	0.01	0.02	0.02
P_2O_5	38.10	39.91	39.92	41.05	41.14	38.99	41.12
F	5.89	5.20	4.74	4.32	4.23	5.97	5.10
Cl	0.73	0.14	0.12	0.29	0.16	0.08	0.16
Total	99.89	99.23	99.40	100.00	99.94	99.91	99.96
F/Cl	8.05	38.52	38.19	15.06	26.75	79.53	32.28

bdl = below detection limit

5.1.3. Paragenesis Study

In the Jagpura deposit, the ore minerals are semi massive to massive type, fracture fill type, vein type, disseminations/stringers type, patchy and foliation parallel types. Based on ore and gangue minerals assemblages, the texture of ore minerals and their grain boundary relationship, three different stages of ore formation are recognized (Figure 9).

Stage-I: This stage is called the dissemination/stringer stage and is characterized by fine disseminations/stringers of magnetite, arsenopyrite, pyrrhotite and chalcopyrite grains within the host rocks (Figure 7G). The common gangue mineral assemblages of this stage are quartz, biotite, apatite, albite, tourmaline, muscovite, biotite, chlorite and sericite.

Stage -II: This is the shear stage marked by the regional D_2 phase of deformation and it is the most dominant phase of mineralization in the Jagpura deposit. Most hydrothermal alterations were formed during this stage, viz. chloritization, sericitization, silicification, ferruginization and iron oxide breccia. The common gangue mineral assemblages of this stage are quartz, chlorite, sericite, biotite, apatite, albite, tourmaline and muscovite. Semi-massive to massive type ore, foliation parallel ore and vein filled ore were formed during this stage. Pyrrhotite, pyrite, chalcopyrite, magnetite, arsenopyrite, loellingite, gold, maldonite and hedleyite mineral assemblages represent the shear stage.

Stage-III: This stage is the final ore formation stage, marked by the second phase of hydrothermal fluid activity. It exhibits mineralization along fracture planes and replaces early formed ores with later ones, i.e., arsenopyrite is replaced by pyrrhotite and chalcopyrite. Native gold was precipitated in this stage, showing stringers and patchy textures.

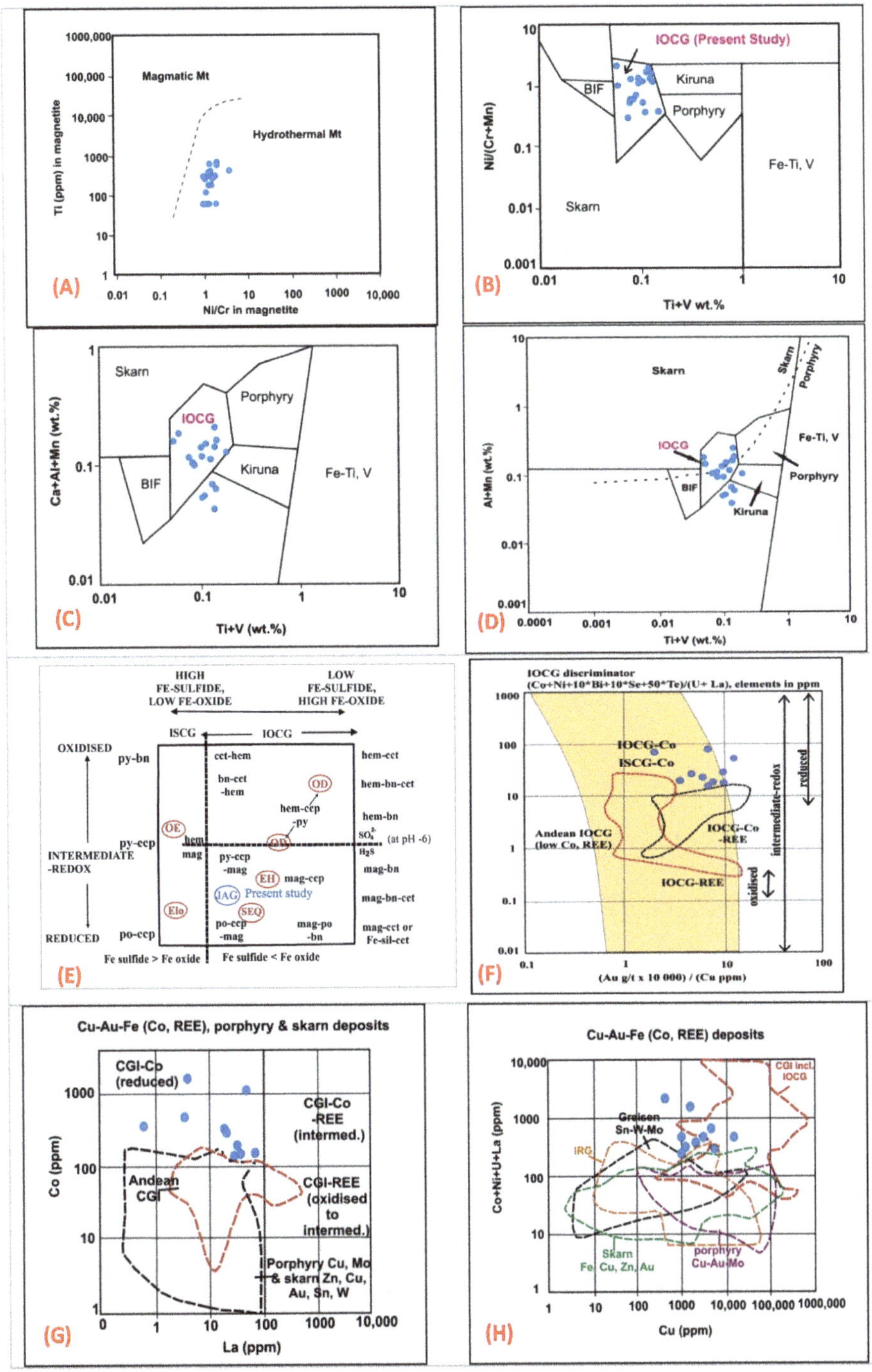

Figure 9. Trace elements plots of magnetite and geochemical discrimination diagrams suggesting IOCG type mineralization style in the study area. (**A**) Ti vs. Ni/Cr ratio of magnetite from reference [14]

suggest hydrothermal nature of magnetite; (**B**) Ni/(Cr+Mn) vs. Ti+V plot from reference [18] suggesting IOCG type deposit; (**C**) Ca+Al+Mn vs. Ti+V plot from reference [18] indicates IOCG type mineralization style; (**D**) Al+Mn vs. Ti+V plot from reference [26] also corroborate the IOCG affiliation based on magnetite EPMA data; (**E**) The IOCG-ISCG cube of the Jagpura deposit from reference [36] shows reduced mineralogical subtypes based on Fe-Cu-O-S mineral assemblages. (Key deposits are shown in red text: EH—Ernest Henry deposit, Elo—Eloise, OD—Olympic Dam, OE—Orlando East, Seq—Sequeirinho-Sossego); (**F**) The (Au g/t × 10,000)/ Cu ppm vs. (Co + Ni + 10*Bi + 10*Se + 50*Te)/(U + La) IOCG discriminator diagram for geochemical subtypes of IOCG deposit from reference [36] indicates that the Jagpura deposit is IOCG-Co (reduced) subtype; (**G**) The La vs. Co discrimination diagram from reference [36] also indicates that the Jagpura deposit is IOCG-Co (reduced) subtype and unrelated to porphyry and skarn type deposit; (**H**) The Cu vs. Co + Ni + U + La discrimination diagram from reference [36] further suggest that the Jagpura deposit is IOCG type and unrelated with porphyry Cu-Au-Mo, skarn Fe-Cu-Zn-Au, greisen Sn-W-Mo and Intrusion-Related Gold (IRG) deposits).

5.2. Fluid Inclusion Study

5.2.1. Fluid Inclusion Petrography

Fluid inclusions were investigated using fluid inclusion assemblage (FIA) methods [66]. The size of the inclusions varies from 2 to 30 μm. The size of water vapor/CO_2 bubbles varies from 0.50 to 6.85 μm. These inclusions are classified into four types based on the phases present: type-I primary monophase carbonic inclusions, type-II primary aqueous carbonic inclusions, type-III primary aqueous bi-phase inclusions, and type-IV secondary aqueous inclusions (Figure 10). Type-I inclusions are more common and abundant in mineralized quartz veins than any other type of inclusion. These inclusions have only one phase, but two phases have been observed at room temperature in some cases. The CO_2 vapor phase is perfectly circular, spherical, and occasionally oval in shape, and at room temperature, a few of the CO_2vapor bubbles have a dark rim on the periphery due to the presence of a thin film of liquid CO_2 over vapor CO_2 and are homogenized into the liquid phase. Type-II inclusions are less common and appear in isolated patterns containing CO_2 (liquid) + CO_2 (gas) + H_2O (liquid) + NaCl. Aqueous carbonic inclusions frequently have varying degrees of fill. In general, the vapor content ranges from 25% to 60% by volume. Type-III inclusions are more abundant than any other type of inclusion, and they occur as isolated inclusions and fluid inclusion assemblages (FIA). These are usually tiny, rounded, and irregular, with two phases: liquid (H_2O+NaCl) and a vapor bubble H_2O homogenized to liquid upon heating. The type-IV inclusions are secondary aqueous inclusions and are very similar to the aqueous inclusions of type-III.

5.2.2. Microthermometry

The melting temperatures (T_mCO_2) of type-I inclusions range from −56.6 to −60.2 °C, with the maximum depression in $TmCO_2$ (−60.2 °C) indicating an admixture of other gases, most likely CH_4 or N_2 [67], and the melting temperatures (Tm, CO_2) are graphically illustrated in histograms. The temperatures at which CO_2 was homogenized into the vapor phase ranged from 1.9 to 28.0 °C (density varies from 0.72 to 0.81 g/cm^3) and were graphically plotted in a histogram. The primary aqueous carbonic inclusions (type II) are less abundant and occur in isolated and cluster patterns. These are identified by the formation of vapor bubbles upon cooling, which corresponds to the composition CO_2 (liquid) + CO_2 (gas) + H_2O (liquid) + NaCl. These aqueous carbonic inclusions frequently have varying degrees of fill and coexist with aqueous-rich inclusions, inferring that fluid immiscibility occurred later in the crystallization process [66]. The total homogenization temperature (Th_{total}) ranges from 235 to 258 °C. The CO_2clathrate temperature (Tm_{clath}) ranges from 6 to 12 °C. The initial eutectic temperatures (T_e) of ice melting range from −26.5 °C to −32.0 °C. This implies that the fluid system's main component in the aqueous phase is NaCl ± $MgCl_2$. The melting temperatures of solid CO_2 range from −56.6 °C to 58.2 °C. The maximum melting temperature indicates a lesser amount of methane [68] and the homogenization temperature of CO_2 varies from 2.5 to 4.2 °C. The density of aqueous the carbonic phase varies from 0.87 to 0.98 g/cm^3. Type-III inclusions are aqueous inclusions that freeze at temperatures ranging from −52 to −72 °C. The homogenization temperature ranged from 146 to 252 °C, and the first melting (eutectic) temperatures (*Teu*) observed during the heating runs ranged from −51 to −30 °C, with an average of −41.65 °C, indicating that the major component in the fluid system is NaCl ± $FeCl_2$. The presence of $CaCl_2$ ± $FeCl_2$ with NaCl and H_2O may be indicated by the maximum eutectic temperature of −51 °C [69]. The final melting temperature of ice (*T*m, ice) ranges from −6.9 to −30.0 °C, with salinities ranging from 10.35 to 28.15 wt% NaCl equivalent. The density of the aqueous phase varies from 0.98 to 1.10 gms/cm^3. The type-IV, secondary bi-phase inclusions are predominantly aqueous-rich. The first melting temperature (eutectic) varies from −22.5 to −44 °C with an average value of −32.1°C. This suggests that the major component in the aqueous phase is $MgCl_2$ in the fluid system. The final ice melting temperatures range from −4.5 to −22.7 °C, corresponding to the wide range of salinity varying from 7.10 to 24.09 wt% NaCl equivalents. The total homogenization temperature (Th) varies from 120 to 238 °C. The density of inclusions varies from 0.93 to 1.11 gm/cm^3 (Figure 11), [Table 4].

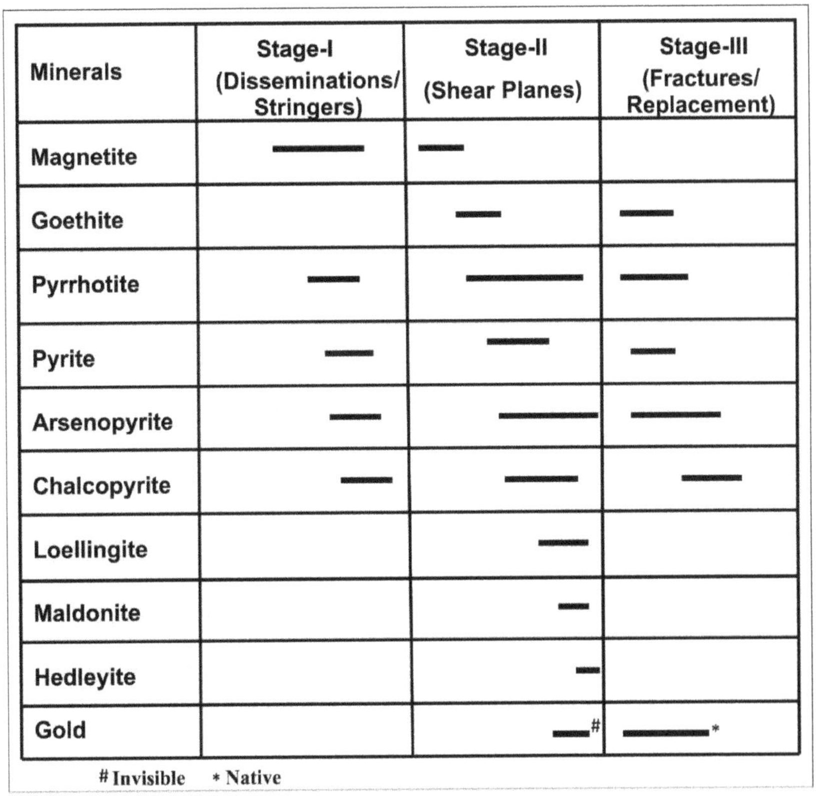

Figure 10. Paragenetic sequence showing evolution of gold mineralization with respect to various oxide and sulfide minerals based on relationships between host rock mineral assemblages, ore textural and micro-structural relationships.

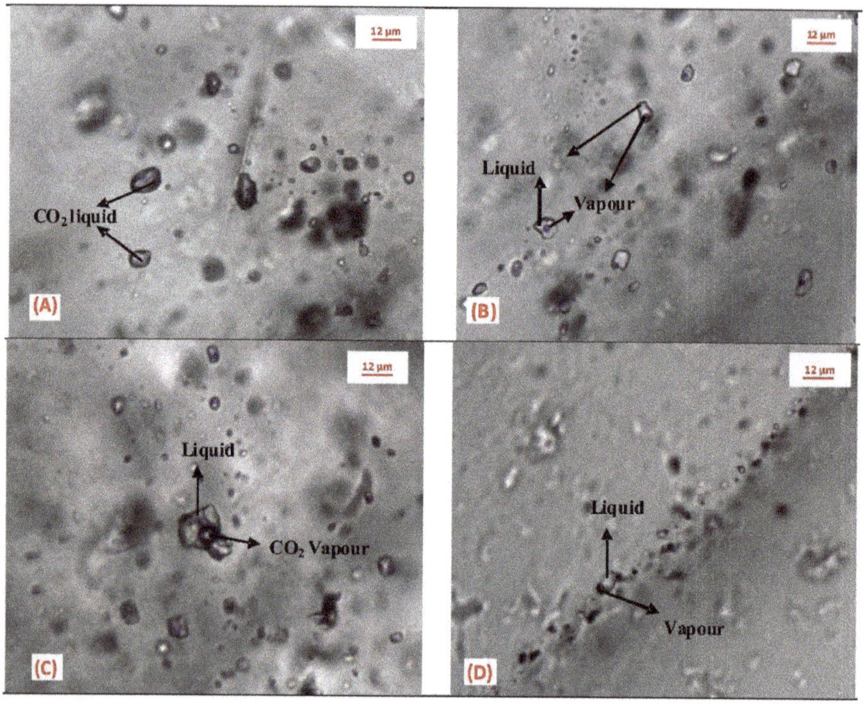

Figure 11. Types of fluid inclusions observed from study area. (**A**) Primary monophase carbonic (CO_2) inclusions; (**B**) Primary aqueous carbonic inclusions; (**C**) Primary aqueous bi-phase inclusions; (**D**) Secondary aqueous bi-phase inclusions.

Table 4. Fluid inclusion results of different inclusions from the Jagpura deposit.

Samples	Descriptions	Type	Origin	Size (μm)	$TmCO_2$	Te	Tmice	$ThCO_2$	Th Total	Salinity	Density
Mineralized Quartz veins	Mono phase carbonic inclusions	I	Primary	2–15	−56.6 to −60.2 °C	-	-	1.9 to 28 °C	-	-	0.66 to 0.92 g/cc
	Aqueous carbonic inclusions	II	Primary	6–30	−56.6 to 58.8 °C	−26.5 to −32 °C	−3 to −7 °C	2.5 to 4.2 °C	235 to 258 °C	4.86 to 10.48	0.87 to 0.98 g/cc
	Aqueous inclusion	III	Primary	2–28		−30 to −51 °C	−6.9 to −30.0 °C	-	146 to 252 °C	10.35 to 28.15	0.98 to 1.10 g/cc
	Secondary aqueous inclusions	IV	Secondary	1–15		−22.5 to −44 °C	−4.5 to −22.7 °C	-	120 to 238 °C	7.10 to 24.09	0.93 to 1.11 g/cc

5.3. Sulfur Isotopic Composition

The $\delta^{34}S_{VCDT}$ (‰) values obtained from major sulfide minerals of the Jagpura deposit show a narrow range from +08.89 to +14.58‰ with an average value of +11.16‰ [Table 5]. The $\delta^{34}S$ value of pyrite ranges from 10.28 to 11.05‰, pyrrhotite from 8.89to 12.34‰, chalcopyrite from 10.14to 12.14‰ and arsenopyrite from 10.98 to 14.58‰. The average $\delta^{34}S$ values of pyrite, pyrrhotite, chalcopyrite and arsenopyrite are +10.60‰, +10.65‰, +11.15‰ and +12.31‰, respectively, in the increasing order of heavier isotope enrichment of $\delta^{34}S$ values.

Table 5. $\delta^{34}S_{VCDT}$ (‰) values of various sulfide minerals of the Jagpura deposit.

Sl. No.	Sample No.	$\delta^{34}S_{VCDT}$‰	St. dev.	Sulfide Mineral Phase
1.	JPSI-1	8.89	0.20	Pyrrhotite
2.	JPSI-2	9.80	0.17	
3.	JPSI-3	10.87	0.37	
4.	JPSI-4	10.67	0.16	
5.	JPSI-5	10.52	0.27	
6.	JPSI-6	10.82	0.31	
7.	JPSI-7	11.36	0.14	
8.	JPSI-8	12.34	0.37	
9.	JPSI-9	12.81	0.11	Arsenopyrite
10.	JPSI-10	10.98	0.20	
11.	JPSI-11	11.00	0.07	
12.	JPSI-12	11.42	0.13	
13.	JPSI-13	11.73	0.16	
14.	JPSI-14	13.68	0.11	
15.	JPSI-15	14.58	0.08	
16.	JPSI-16	10.77	0.10	Chalcopyrite
17.	JPSI-17	10.90	0.05	
18.	JPSI-18	11.18	0.26	
19.	JPSI-19	10.99	0.03	
20.	JPSI-20	10.14	0.09	
21.	JPSI-21	10.67	0.33	
22.	JPSI-22	11.56	0.18	
23.	JPSI-23	12.03	0.28	
24.	JPSI-24	12.14	0.15	
25.	JPSI-25	10.28	0.06	Pyrite
26.	JPSI-26	10.39	0.16	
27.	JPSI-27	10.29	0.05	
28.	JPSI-28	10.71	0.46	
29.	JPSI-29	11.05	0.52	
30.	JPSI-30	10.62	0.30	
31.	JPSI-31	10.87	0.59	

5.4. Geochemical Results

The values of Fe in oxide range from 11.25 to 42.22 wt% and values of Fe in sulfide range from 4.27 to 22.37 wt%. The value of nickel ranges from 70 to 400 ppm, cobalt from 130 to 1600 ppm, Cu from 470 to 15,200 ppm, Cr from 155 to 348 ppm, Au from 0.20 to 18.60 ppm, Pb from 25 to 400 ppm, Zn from 10 to 100 ppm, La from 0.64 to 75.61 ppm and U from 2.42 to 3.84 ppm [Table 6]. The total REE varies from 76.64 to 346.38 ppm. The geochemical result shows enrichment of Co, Ni in Cu-Au ores and low values of REE and U.

Table 6. Geochemical results of samples from theJagpura deposit.

Lithology: Quartz-Mica Schist										
Trace Elements & REEs (ppm)										
Ga	38	31	28	54	69	46	53	46	51	109
Sc	10	11	10	23	20	09	09	9	13	11
V	78	85	92	122	90	104	87	88	92	88
Th	20	21	30	27	26	30	23	21	27	22
Pb	120	400	70	40	50	30	25	25	25	40
Ni	300	140	100	90	130	70	80	100	70	400
Co	1100	950	480	860	480	730	290	330	840	1600
Cr	155	176	179	306	348	223	234	215	218	228
Sr	69	70	72	78	64	56	63	68	68	53
Zr	142	134	121	97	76	119	102	105	114	60
Cu	4600	2400	1200	15200	4700	4600	3100	7100	1100	4700
Zn	20	100	10	40	20	20	10	10	10	10
Be	3.88	4.30	3.47	2.67	2.94	3.00	2.34	1.82	2.48	1.54
Ge	1.64	1.72	1.52	1.03	0.98	1.32	1.23	1.31	1.42	1.11
Y	34.86	28.68	19.68	47.61	96.01	23.40	19.14	19.99	18.95	38.82
Mo	1.54	1.74	2.10	2.38	4.23	2.28	3.29	3.00	2.88	3.00
Sn	3.76	3.76	3.73	3.81	2.74	3.98	2.75	3.21	3.76	5.06
La	50.29	75.60	34.52	10.63	32.64	31.74	21.73	19.90	16.53	24.16
Ce	89.24	133.66	59.79	0.06	4.93	56.72	39.25	35.17	65.10	6.87
Pr	10.41	15.44	6.99	0.27	0.86	6.79	4.96	4.37	7.83	1.02
Nd	40.40	59.37	28.11	1.72	4.35	27.60	19.83	17.54	31.42	4.74
Sm	6.88	9.26	4.73	0.86	1.67	4.31	3.53	2.78	5.05	1.21
Eu	1.24	1.62	0.84	0.34	0.49	0.89	0.75	0.66	0.95	0.41
Gd	6.32	7.62	4.04	2.71	5.40	3.85	3.25	2.84	4.17	2.34
Tb	1.18	1.27	0.73	1.00	1.98	0.74	0.62	0.58	0.69	0.77
Dy	7.05	6.15	4.07	8.34	17.44	4.61	3.60	3.63	3.78	6.74
Ho	1.31	1.11	0.74	1.76	3.60	0.89	0.72	0.76	0.68	1.39
Er	3.72	3.09	2.23	4.84	9.88	2.61	2.01	2.17	2.00	3.91
Tm	0.58	0.45	0.31	0.71	1.40	0.37	0.28	0.31	0.27	0.56
Yb	3.39	2.65	2.04	3.93	7.50	2.42	1.74	2.02	1.82	3.26
Lu	0.46	0.38	0.30	0.52	0.95	0.31	0.25	0.29	0.26	0.45
Hf	4.36	4.20	4.15	3.27	2.83	4.61	3.80	3.88	3.89	2.82
Ta	1.27	1.14	1.16	1.46	0.66	1.26	1.14	1.11	1.06	0.88
W	27.00	6.42	6.47	6.76	4.76	14.72	11.26	11.41	11.70	10.11
U	3.15	3.31	2.42	3.62	2.98	3.83	3.48	3.39	3.16	2.89
Bi	0.16	0.27	0.38	2.65	2.94	1.31	2.17	2.33	1.05	5.46
Te	1.26	1.09	1.02	0.20	0.27	1.21	0.97	0.92	1.02	1.01
Se	0.20	0.20	0.20	0.20	0.20	0.20	0.20	0.20	0.20	0.20
Au	4.62	1.70	1.20	3.10	1.70	2.20	2.40	4.30	1.40	3.30

6. Discussion

6.1. Nature and Control of Mineralization

The mode of occurrence of mineralization as (1) semi-massive to massive type, (2) vein and fracture fill type, (3) foliation parallel disseminations/smears and (4) patchy and stringer type indicates hydrothermal origin of the mineralizing fluid. Textural investigations indicate the mineralization occurring as a fracture and open space-filling. The mineral boundary relationship, exsolution and the replacement and deformation textures further substantiate the hydrothermal nature of mineralization. This is further corroborated by the presence of iron oxide breccia, epigenetic quartz and pegmatite veins within the deposit. The ore mineralization is associated with pervasive hydrothermal Na-B alteration besides Fe-Mg-Ca-K alteration. Disseminated mineralization occurs parallel to S_0 II S_1 planes, related to the first deformation episodes. The ore mineralization is remobilized and concentrated within the hinge zone of F_2 folds and sympathetic shears (Figures 6E,F and 7F,H) during second phase of deformation. The NW-SE trending shear planes are parallel to F_2 axial planes of the second stage deformation event. The ore localization along these structures suggests distinct structural control of mineralization in the study area.

6.2. Geochemistry of Native Gold, Pyrite, Magnetite, Apatite and Their Implications

EPMA analysis of gold indicates that Au concentration ranges from 89.25 to 94.72 wt% and it contains silver as minor impurity (6.15 to 8.46 wt%). Gold fineness ranges between 914–937‰ with an average value of 927‰ indicating high purity gold. The Co/Ni ratio of pyrite is >1 (1 to 2.54) suggests hydrothermal origin [64,69–71], [Table 1; Figure 8G].

The Co/Ni ratios [15], Ti and V concentrations [14,72,73] are important parameters to differentiate hydrothermal magnetite to magmatic magnetite. The hydrothermal magnetite is depleted in Ti (<2 wt%) and Al (<1 wt%), [14,15,18,20,26,74]. Low TiO_2 (0.01 to 0.11 wt%) and Al_2O_3 (0.06 to 0.28 wt%) concentrations in magnetite indicates the presence of hydrothermal magnetite in the study area [Table 2]. The Ti vs. Al and Ti vs. Ni/Cr bivariate plot of magnetite from study area also indicates magnetite is hydrothermal in origin (Figures 8H and 12A). The Ni/(Cr + Mn) vs. Ti + V plot, Ca + Al + Mn vs. Ti + V and Al + Mn vs. Ti + V plot of magnetite suggest IOCG deposit style (Figure 12B–D). Apatite from study area is of fluorapatite variety with F content >1 wt% (4.23 to 5.97 wt%) and >1 F/Cl ratio [Table 3]. Apatite has a higher concentration of F and a lower concentration of Cl, FeO, and MnO, which indicate that it is hydrothermal in nature and is sourced from metavolcano-sedimentary units [75].

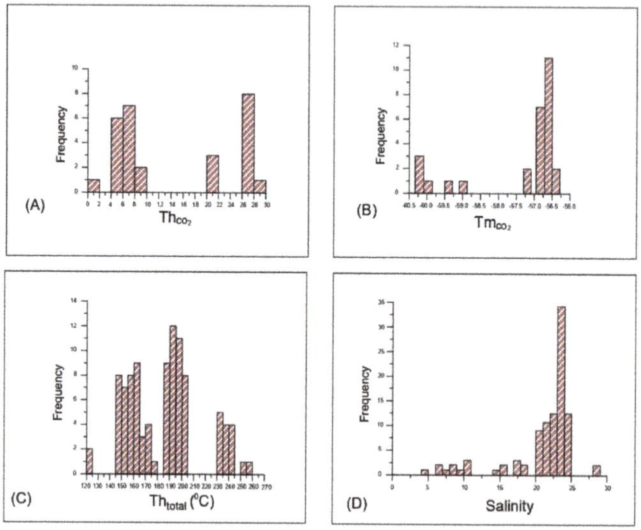

Figure 12. Histograms showing the relationship between: (**A**) Melting temperature of CO_2 ($Tmco_2$) vs. frequency; (**B**) Homogenization temperature of CO_2 ($Thco_2$) vs. frequency; (**C**) Homogenization temperature (Th_{total}) vs. Frequency; (**D**) Salinity (wt% NaCl equivalent) of the inclusion vs. frequency.

6.3. Evolution of Ore Fluids

The characteristics of the mineralized quartz vein hosted fluid inclusions are listed in Table 4 and Figures 9 and 10. The homogenization temperature versus salinity diagram indicates that the mineralization within the Jagpura Au-Cu deposit is the result of the isothermal mixing of ore fluids with boiling/effervescence (Figure 13A), [76]. Most hydrothermal deposits have coexisting liquid rich, highly saline, and vapur rich low saline inclusions, indicating that the different phases are generated by immiscibility [77].

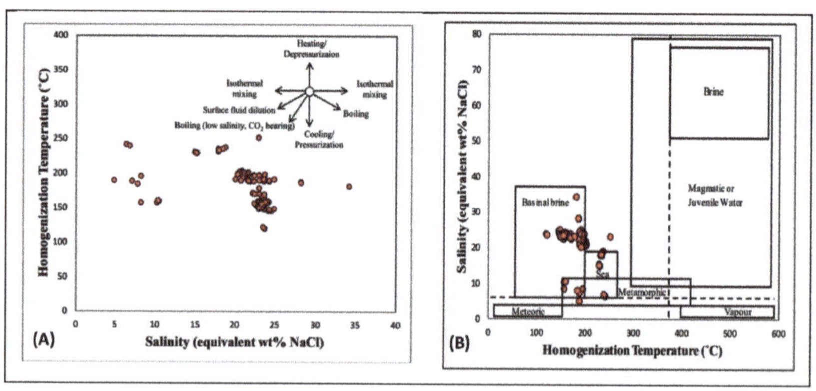

Figure 13. Homogenization temperature vs. salinity diagram of fluid inclusions from study area: (**A**) showing typical trends in homogenization temperature-salinity space due to various fluid evolution processes, from reference [78]; (**B**) Salinity vs. Homogenization temperature diagram from reference [79] indicates mixing of basinal brine, sea water and metamorphic fluids in ore fluid.

In the study area, type-III inclusions are liquid rich [(H_2O + NaCl (l)] and highly saline (10.35–28.15 wt% NaCl equivalent); however, type-II inclusions are vapor rich and low saline (4.8–10.48 wt% NaCl equivalent). These data indicate that there is an immiscibility of ore fluids. Generally, salinity of metamorphic fluids ranges between 3–10 wt% NaCl, with homogenization temperature varying between 150–425 °C and salinity of the basinal brine ranges between 5–40 wt% NaCl, with low homogenization temperature varying between 75–200 °C. In the Jagpura area, type-II inclusions match with the metamorphic fluids whereas type-III inclusions show similar character with basinal brine. Salinity versus homogenization temperature diagram indicates that the ore fluid is the result of the mixing of basinal brine, sea water and metamorphic fluids (Figure 13B), [80]. The high saline ore fluid transported metals in the system as metals chloride complex.

6.4. Source of Metals, Sulfur and Ore-Forming Fluids

The $\delta^{34}S_{VCDT}$ values of sulfides from the Jagpura Au-Cu deposit are within the range of 8.98- to 14.58‰, with an average value of 11.16‰ (Figure 14A). This narrow range of sulfur isotopes indicates that (i) sulfur has originated from one uniform source under stable physicochemical condition or (ii) local disequilibrium between two phases of mineralization [81].

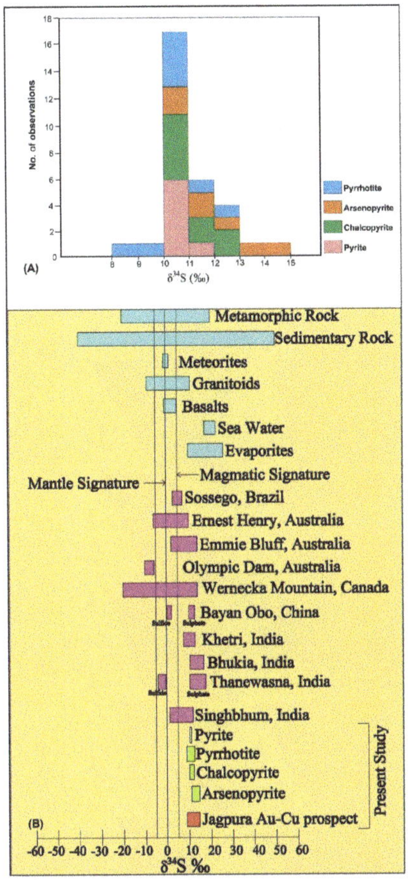

Figure 14. Sulfur isotope compositions of different sulfide minerals from the Jagpura deposit: (**A**) Histogram showing values of various sulfide minerals; (**B**) Range of $\delta^{34}S$ for magmatic, sedimentary, metamorphic, seawater and evaporite systems from references [59,82]. Plot showing the variation in $\delta^{34}S$ values of sulfides from the Jagpura IOCG deposit, and comparison of the $\delta^{34}S$ values of sulfides from the Jagpura deposit with major geological sulfur reservoir and with various IOCG deposit worldwide from reference [24,83] and India from references [27,28,31,39,44,84].

The major gold bearing sulfide mineral within the Jagpura Au-Cu deposit is arsenopyrite and it contains a higher isotopic ratio than the other sulfide phases. The higher isotopic range in the sulfur isotope compositions imply that sulfide minerals are enriched in heavy isotopes (Figure 14B). In various deposits around the world, isotopically heavy sulfur ($\delta^{34}S > 5$) suggests the presence of non-magmatic sulfur sources. Higher positive values of the sulfur isotope rule out magmatic ($\delta^{34}S = 05$), [85,86], mantle ($\delta^{34}S = 01$), [87], and seawater/biogenic sulfur sources ($\delta^{34}S = > 20$), [83]. The presence of relatively higher $\delta^{34}S$ rich sulfides indicates that the possible source of heavy sulfurs is (i) seawater, (ii) evaporitic water and (iii) oxidized meta-sedimentary fluids [88,89]. Therefore, the moderately high $\delta^{34}S$ values of the Jagpura deposit values suggest a meta-sedimentary source for sulfur.

The lack of sulfate minerals in the Jagpura deposit indicates that the sulfur was present in the hydrothermal fluids as reduced sulfur (H_2S). The potential source of reduced H_2S was the basement rocks. The sulfur isotope compositions of sulfides from the Jagpura deposit are similar to the major geological sulfur reservoirs and other IOCG type deposits in India and the world (Figure 14B). External fluids, primarily basinal brines and modified seawater with high $\delta^{34}S$values (>+10), play an important role in IOCG ore-forming systems [90]. The sulfur isotope data indicate the mixed source of sulfur derived from the non-magmatic hydrothermal fluids, and the high saline brine (marine/evaporitic). Variation in the sulfur isotopic compositions of sulfides is the result of dilution and cooling of the metalliferous fluid derived from the basement meta-sedimentary rocks, which along with highly saline brine triggered the precipitation of Au along structural weak planes.

6.5. Geochemical Characterization of the Jagpura Deposit

There are three mineralogical subtypes of IOCG deposits based on dominant ore assemblages i.e., (i) oxidized (ii) intermediate-redox and (iii) reduced mineralogical subtypes [36]. The reduced mineralogical subtypes are represented by pyrrhotite and/or Fe-silicates and variable modal proportions of magnetite, pyrite, and chalcopyrite. The reduced IOCG-ISCG deposit group commonly includes phases hosting Co, Ni, As and Bi; however, REE, U, fluorite, barite or anhydrite phases are rare [36]. The Jagpura deposit is represented by the dominant ore assemblages of pyrrhotite-chalcopyrite-arsenopyrite-pyrite-magnetite. The deposit includes phases hosting Co, Ni, As and Bi; however, REE, U, fluorite and barite are not present. The IOCG-ISCG cube diagram based on Fe-Cu-O-S mineral assemblages (Figure 9E) suggests that the Jagpura deposit is characterized by reduced mineralogical subtypes. The geochemical result shows enrichment of Co, Ni in Cu-Au ores and low values of REE and U [Table 6]. The occurrence of >15% Fe in oxide, (Au g/t× 10,000)/ Cu ppm vs. (Co + Ni + 10*Bi + 10*Se + 50*Te)/(U + La), La vs. Co and Cu vs. Co+Ni+U+La geochemical discrimination diagrams (Figure 12F–H) for geochemical subtypes of Cu-Au-Fe (± Co, REE) deposits including IOCG deposits indicates that the Jagpura is IOCG-Co (reduced) subtype deposit and unrelated to porphyry Cu-Au-Mo, skarn Fe-Cu-Zn-Au, greisen Sn-W-Mo, and intrusion-related gold (IRG) deposits.

6.6. Genesis of the Jagpura Au-Cu Deposit

The genesis of an ore deposits is closely related to the geological environments, i.e., pressure -temperature conditions, and the nature/source of the hydrothermal fluids [66,73,78,79]. Fluid inclusion study is a vital tool to understand the genetic aspects, whereas the sulfur isotope signatures help in the understanding of the nature and source of the mineralizing fluids of an ore deposit. [90–92]. Sulfur isotope data also provides the information about the source of metals in a deposit [59,93]. The sulfur isotope variations in an ore mineralizing systems are very complex and, hence, best understood in the context of the total geological framework of a deposit in conjunction with the fluid inclusion study.

Au-Cu sulfide mineralization within the Jagpura deposit is hosted by quartz-mica schist and albitite. The mineralization is localized along NW-SE trending D_2 shear planes parallel to F_2 axial planes and F_2 fold hinges and correlated to the second deformation phase (Figure 15). The ore mineral association is represented by the iron oxide (magnetite), copper, gold and apatite. The mineralization includes arsenopyrite, loellingite, chalcopyrite, pyrrhotite and pyrite, along with the native gold and abundant magnetite. Further the maldonite and hedleyite occur as Au-Bi-Te phases. The mode of occurrence of mineralization as open space-filling, vein and fracture fill type indicates hydrothermal nature. The mineralization is associated with pervasive hydrothermal Na-B alteration besides Fe-Mg-Ca-K alteration. The occurrence of maldonite, hedleyite associated with gold lode, epigenetic quartz-pegmatite veins in host rocks and pervasive hydrothermal alteration indicate hydrothermal origin of ore fluids.

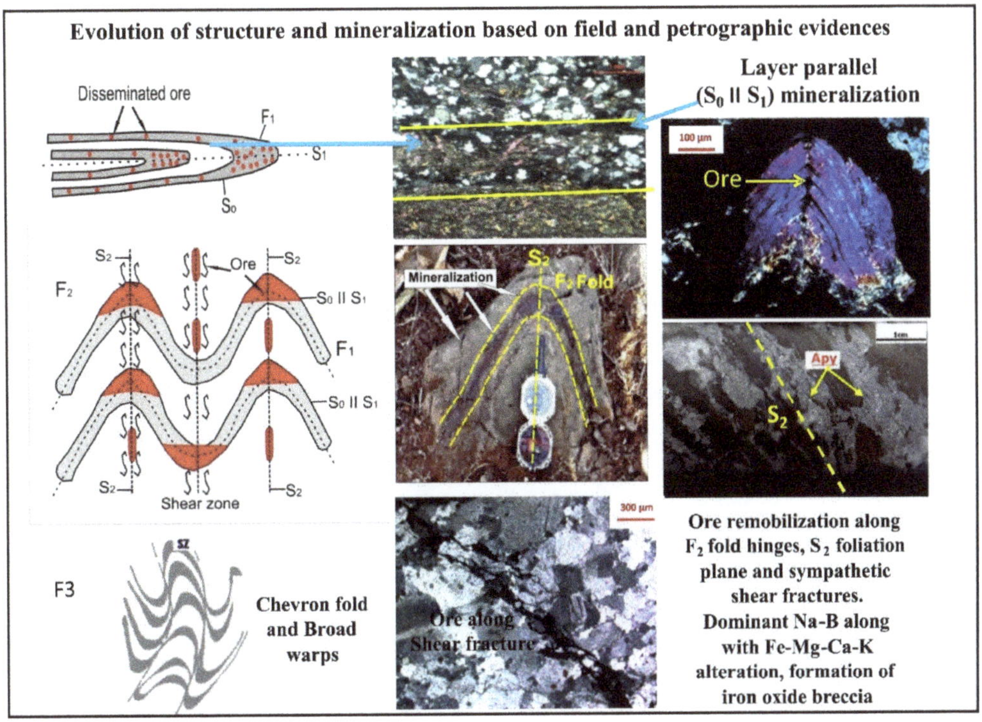

Figure 15. Schematic diagram showing the stage of evolution of the structure and polymetallic mineralization at theJagpura deposit based on field and petrographic studies.

The Co/Ni ratio of >1 in pyrite, gold-sulfide associated low TiO_2, Al_2O_3 hydrothermal magnetite (Co/Ni ratio < 1 and Ni/Cr $\geq$ 1) and fluorapatite (>1 F/Cl ratio) support the hydrothermal origin of mineralization. Apatite has a higher F concentration and a lower concentration of Cl, FeO, and MnO, indicating its source from the metavolcano-sedimentary units. Fluid inclusions micro-thermometry indicates the presence of aqueous rich, mixed carbonic-aqueous and minor vapor rich aqueous inclusions. The FI's show episode of fluid immiscibility with the homogenization temperatures and salinities varying between 120–258 °C and 8.86–28.15 wt% NaCl eq., respectively. This data indicates that the high saline ore fluids are injected during the D_2 phase of deformation. The high salinity ore fluid transport metals in the ore system as metals chloride complex.

The iron oxide-copper-gold (IOCG) type deposits are characterized by abundant (>10%) hydrothermal iron oxides (magnetite or hematite) and economic grade Cu and/or Au. It also contains Ag, U, Th, F, Co, Bi, W, rare earth elements (REE) and other metals. These deposits are diverse in age (Neo-Archaean to the Cenozoic), tectonic setting, P–T conditions, characteristic Na-Ca-K alterations, host-rock package and mineralization style [36,90,94–98].The field characteristics, trace element geochemistry of pyrite, apatite and magnetite, fluid inclusion and sulfur isotope compositions of the Jagpura Au-Cu deposit are compared with the Epithermal gold deposit of high sulfidation and low sulfidation, Carlin type gold deposit, Orogenic lode gold and IOCG type Cu-Au deposits [4–10] and this suggeststhat the Jagpura gold-copper ore system is similar to the IOCG type. The Jagpura Au-Cu deposit indicate the: (i) iron oxide-copper-gold-apatite association of hydrothermal origin, (ii) distinct structural control of mineralization, (iii) Na-B-Mg-Fe-Ca-K hydrothermal alteration(dominant Na-B alteration), (iv) presence of iron oxide breccia, (v) abundant hydrothermal Fe oxides (>15%), (vi) trace element geochemistry of magnetite (Ti, Al + Mn, Ti + V concentrations and Ni/(Cr + Mn) vs. Ti + V, Ca + Al + Mn vs. Ti + V, Al + Mn vs. Ti + V ratios of magnetite Figure 8J–L) suggests IOCG style of mineralization, (vii) presence of late-stage sulfides with economic gold-copper resources, (viii) high saline ore fluid and sulfur isotope compositions within the range of IOCG deposits worldwide (−30 to +26‰) and (ix) The (Au g/t × 10,000)/ Cu ppm vs. (Co + Ni + 10*Bi + 10*Se + 50*Te)/(U + La), La vs. Co and Cu vs. Co + Ni + U + La and discrimination diagrams (Figure 12F–H), [36] indicates a IOCG-Co (reduced) subtype deposit and is unrelated to porphyry Cu-Au-Mo, skarn Fe-Cu-Zn-Au, greisen Sn-W-Mo, and intrusion-related gold (IRG) deposits. These characters of the Jagpura deposit are well corroborated with the IOCG-IOA type and the IOCG-Co (reduced) subtype deposit.

7. Conclusions

The Jagpura Au-Cu deposit shows pervasive hydrothermal alteration zones associated with the iron-oxide and sulfide mineralization along the brittle-ductile shear zone. The ore petrography and EPMA study show the presence of Au-Bi-Te phases, Co/Ni ratio of >1 in pyrite, gold-sulfide associated low TiO_2, Al_2O_3 hydrothermal magnetite (≥1 Ni/Cr ratio, <1 Co/Ni ratio). The presence of apatite with higher F concentration (>1 F/Cl ratio) and a lower concentration of Cl, FeO and MnO is consistent with its hydrothermal origin. Fluid inclusion microthermometric studies indicate the presence of aqueous rich, mixed carbonic-aqueous and minor vapor rich aqueous inclusions. The FI's show episodes of fluid immiscibility with the homogenization temperatures and salinities range of 120–258 °C and 8.86–28.15 wt% NaCl eq., respectively. These data indicate a mixture of highly saline ore fluid of basinal brine and metamorphic origin, responsible for the transport of metals in the system as metals chloride complex. The $\delta^{34}S$ values of sulfides are within a narrow range of 8.98‰ to 14.58‰, with an average value of 11.16‰. This indicates the non-magmatic origin of sulfides. Variation in the sulfur isotopic compositions of sulfides resulted from the dilution and cooling of the metalliferous fluid derived from the basement rocks. The mixing of high saline brine triggered the gold precipitation along the structural discontinuities. The Ti, Al + Mn, Ti + V concentrations and Ni/(Cr + Mn) vs. Ti + V, Ca + Al + Mn vs. Ti + V, Al + Mn vs. Ti + Vratios of magnetite suggest an IOCG type of mineralization. The (Au g/t × 10,000)/ Cu ppm vs. (Co + Ni + 10*Bi + 10*Se + 50*Te)/(U + La), La vs. Co and Cu vs. Co + Ni + U + La geochemical discrimination diagrams indicates that the Jagpura Au-Cu deposit is an IOCG-Co (reduced) subtype deposit. Iron oxide-copper-gold-apatite association of hydrothermal origin, nature and distinct structural control of mineralization, pervasive hydrothermal alteration, trace element characteristics of magnetite, pyrite and apatite, high saline ore fluid and sulfur isotope compositions suggest that the Jagpura deposit is an IOCG-IOA type and is similar to those described for the shear fault-controlled-IOCG deposits found elsewhere in the world. The classification of the Jagpura deposit as an IOCG-IOA type: IOCG-Co (reduced) subtype has significant implications for the deeper level sub-surface exploration in the Salumber Ghatol Metallogenic Belt within the Aravalli Craton.

Author Contributions: Conceptualization, A.A., S.S.; methodology, A.A., S.S., A.G.; validation, S.S., A.A., A.G.; investigation, A.A., A.S., M.K.; writing original draft preparation, A.A., A.G., M.K., G.K.M.; writing, review and editing, S.S., A.S.; visualization, A.A., A.G.; Formal analysis, A.A., G.K.M., M.K.; supervision, S.S., A.S. All authors have read and agreed to the published version of the manuscript.

Funding: This study is supported by the Geological Survey of India, Western Region, Jaipur.

Institutional Review Board Statement: Not applicable.

Informed Consent Statement: Not applicable.

Data Availability Statement: The data presented in this study are mainly contained within the article and available in the references listed. To a minor degree, the data presented are not publicly available due to privacy and available on request from the first author.

Acknowledgments: The authors are extremely grateful to S. Raju, Director General, Geological Survey of India (GSI), Kolkata, Shri Jaya Lal, Addl. Director General &HoD, Western Region (WR), Jaipur and Sanjay Das, Deputy Director General, State Unit, Rajasthan, GSI, WR, Jaipur for their continuous encouragement and permission to publish this paper. AA extends sincere thanks to GSI for providing an opportunity to carry out research in the Jagpura area (project no.99) and extends sincere gratitude to Shri R. L. Jat, Shri Sanjay Singh, Shri Hemant Kumar Singh and Shri Lalit Mohan, GSI for providing help and guidance during field work. AA extends sincere thanks to M. L. Dora, GSI for their valuable suggestions and help to improve this manuscript. We also acknowledge and express sincere thanks to Shri Manish M. John, IRMS Lab, GSI, Bengaluru, Smt. Sonalika Joshi, EPMA Lab, GSI, Faridabad and personnel of EPMA Lab, IIT, ISM, Dhanbad for analyzing the samples. AA thanks Mohini S. Sathe, Geologist, GSI, WR for wholehearted support and encouragement during the writing of this article. The authors also acknowledge GSI's in house reviewers for their constructive and thoughtful comments, which helped in improving the quality of the manuscript. Authors extend their sincere thanks and gratitude to the Editor-in-Chief Paul Sylvester, Special Issue Editor Galina Palyanovaandanonymous reviewers for their continuous encouragement and kind suggestions for improving the quality of the manuscript.

Conflicts of Interest: The authors declare that they have no known competing financial interest or personal relationships that could have appeared to influence the work reported in this paper. There is no conflict of interest as this work is Ph.D work of first author, due acknowledgment has been given to pervious work. Author has intimated the competent authority of Geological Survey of India (GSI) to publish this work in the said journal.

References

1. Palyanova, G.; Murzin, V.; Borovikov, A.; Karmanov, N.; Kuznetsov, S. Native Gold in the Chudnoe Au-Pd-REE Deposit (Subpolar Urals, Russia): Composition, Minerals in Intergrowth and Genesis. *Minerals* **2021**, *11*, 451. [CrossRef]
2. Liu, H.; Beaudoin, G. Geochemical signatures in native gold derived from Au-bearing ore deposits. *Ore Geol. Rev.* **2021**, *132*, 104066. [CrossRef]
3. Savva, N.E.; Kravtsova, R.G.; Anisimova, G.S.; Palyanova, G.A. Typomorphism of Native Gold (Geological-Industrial Types of Gold Deposits in the North-East of Russia). *Minerals* **2022**, *12*, 561. [CrossRef]
4. Groves, D.I.; Goldfarb, R.J.; Gebre-Mariam, M.; Hagemann, S.G.; Robert, F. Orogenic gold deposits: A proposed classification in the context of their crustal distribution and relationship to other gold deposit types. *Ore Geol. Rev.* **1998**, *13*, 7–27. [CrossRef]
5. Goldfarb, R.J.; Groves, D.I.; Gardoll, I. Orogenic gold and geologic time: A global synthesis. *Ore Geol. Rev.* **2001**, *18*, 1–75. [CrossRef]
6. Cline, J.S.; Hofstra, A.H.; Muntean, J.L.; Tosdal, R.M.; Hickey, K.A. Carlin-type gold deposits in Nevada: Critical geologic characteristics and viable models. In *Economic Geology 100th Anniversary Volume 2005*; Society of Economic Geologists: Littleton, CO, USA, 2005; pp. 451–484.
7. Sinclair, W.D. Porphyry deposits. Mineral Deposits of Canada: A Synthesis of Major Deposit-Types, District Metallogeny, the Evolution of Geological Provinces, and Exploration Methods. In *Geological Association of Canada, Mineral Deposits Division, Special Publication*; Geological Association of Canada: St. John's, NL, Canada, 2007; Volume 5, pp. 223–243.
8. Sillitoe, R.H. Porphyry copper systems. *Econ. Geol.* **2010**, *105*, 3–41. [CrossRef]
9. Groves, D.I.; Bierlein, F.P.; Meinert, L.D.; Hitzman, M.W. Iron oxide copper-gold (IOCG) deposits through Earth history: Implications for origin, lithospheric setting, and distinction from other epigenetic iron oxide deposits. *Econ. Geol.* **2010**, *105*, 641–654. [CrossRef]
10. Goldfarb, R.J.; Groves, D.I. Orogenic gold: Common or evolving fluid and metal sources through time. *Lithos* **2015**. [CrossRef]
11. Frimmel, H.E. Earth's continental crustal gold endowment. *Earth Planet. Sci. Lett.* **2008**, *267*, 45–55. [CrossRef]
12. Klemm, D.D.; Henckel, J.; Dehm, R.M.; Von Gruenewaldt, G. The geochemistry of titanomagnetite in magnetite layers and their host rocks of the eastern Bushveld Complex. *Econ. Geol.* **1985**, *80*, 1075–1088. [CrossRef]
13. Nadoll, P.; Koenig, A.E. LA-ICP-MS of magnetite: Methods and reference materials. *J. Anal. At. Spectrom.* **2011**, *26*, 1872–1877. [CrossRef]
14. Dare, S.A.S.; Barnes, S.J.; Beaudoin, G. Variation in trace element content of magnetite crystallized from a fractionating sulfide liquid, Sudbury, Canada: Implications for provenance discrimination. *Geochim. Cosmochim. Acta.* **2012**, *88*, 27–50. [CrossRef]

15. Dare, S.A.S.; Bames, S.J.; Beaudoin, G.; Meric, J.; Boutroy, E.; Potvin-Doucet, C. Trace elements in magnetite as petrogenetic indicators. *Miner. Depos.* **2014**, *49*, 785–796. [CrossRef]
16. Huang, X.W.; Beaudoin, G. Textures and chemical composition of magnetite from iron oxide-copper-gold (IOCG) and Kiruna-type iron oxide-apatite (IOA) deposits and their implications for ore genesis and magnetite classification schemes. *Econ. Geol.* **2019**, *114*, 953–979. [CrossRef]
17. Carew, M.J.; Mark, G.; Oliver, N.H.S.; Pearson, N. Trace element geochemistry of magnetite and pyrite in Fe oxide (+/−Cu–Au) mineralised systems: Insights into thegeochemistry of ore-forming fluids. *Geochim. Cosmochim. Acta.* **2006**, *70*, 5. [CrossRef]
18. Dupuis, C.; Beaudoin, G. Discriminant diagrams for iron oxide trace element fingerprinting of mineral deposit types. *Miner. Depos.* **2011**, *46*, 319–335. [CrossRef]
19. Angerer, T.; Hagemann, S.G.; Danyushevsky, L. High-grade iron ore at Windarling, Yilgarn Craton: A product of syn-orogenic deformation, hypogene hydrothermal alteration and supergene modification in an Archean BIF-basalt lithostratigraphy. *Miner. Depos.* **2013**, *48*, 697–728. [CrossRef]
20. Nadoll, P.; Mauk, J.L.; Hayes, T.S.; Koenig, A.E.; Box, S.E. Geochemistry of magnetite from hydrothermal ore deposits and host rocks of Mesoproterozoic Belt Supergroup, United States. *Econ. Geol.* **2012**, *107*, 1275–1292. [CrossRef]
21. Boutroy, E.; Dare, S.A.S.; Beaudoin, G.; Barnes, S.J.; Lightfoot, P.C. Magnetite composition in Ni-Cu-PGE deposits worldwide and its application to mineral exploration. *J. Geochem. Explor.* **2014**, *145*, 64–81. [CrossRef]
22. Huang, X.W.; Qi, L.; Meng, Y. Trace element geochemistry of magnetite from the Fe (-Cu) deposits in the Hami region, Eastern Tianshan Orogenic Belt, NW China. *Acta. Geol. Sin.* **2014**, *88*, 176–195. [CrossRef]
23. Ciobanu, C.L.; Cook, N.J. Skarn textures and a case study: The Ocna de Fier-Dognecea ore field, Banat, Romania. *Ore Geol. Rev.* **2004**, *24*, 315–370. [CrossRef]
24. Monteiro, L.V.S.; Xavier, R.P.; de Carvalho, E.R.; Hitzman, M.W.; Johnson, C.A.; Filho, C.R.D.S.; Torresi, I. Spatial and temporal zoning of hydrothermal alteration and mineralization in the Sossego iron oxide–copper-gold deposit, Carajas Mineral Province, Brazil: Paragenesis and stable isotope constraints. *Miner. Depos.* **2008**, *43*, 129–159. [CrossRef]
25. Skirrow, R. "Hematite-group" IOCGÂ± U ore systems: Tectonic settings, hydrothermal characteristics, and Cu-Au and U mineralizing processes. In *Exploring for Iron Oxide Copper-Gold Deposits: Canada and Global Analogues*; Corriveau, L., Mumin, H., Eds.; Geological Association of Canada: St. John's, NL, Canada, 2010; Volume 20, pp. 39–58.
26. Nadoll, P.; Angerer, T.; Mauk, J.L.; French, D.; Walshe, J. The chemistry of hydrothermal magnetite: A review. *Ore Geol. Rev.* **2014**, *61*, 1–32. [CrossRef]
27. Mukherjee, R.; Venkatesh, A.S. Chemistry of magnetite-apatite from albitite and carbonate hosted Bhukia Gold Deposit, Rajasthan, western India–An IOCG-IOA analogue from Paleoproterozoic Aravalli Supergroup: Evidence from petrographic, LAICPMS and EPMA studies. *Ore Geol. Rev.* **2017**, *91*, 509–529. [CrossRef]
28. Dora, M.L.; Upadhyay, D.; Randive, K.R.; Shareef, M.; Baswani, S.R.; Ranjan, S. Trace element geochemistry of magnetite and pyrite and sulfur isotope geochemistry of pyrite and barite from the Thanewasna Cu-(Au) deposit, western Bastar Craton, central India: Implication for ore genesis. *Ore Geol. Rev.* **2020**, *117*, 103262. [CrossRef]
29. Belousova, E.A.; Griffin, W.L.; O'Reilly, S.Y.; Fisher, N.I. Apatite as an indicator mineral for mineral exploration: Trace-element composition and their relationship to host rock type. *J. Geochem. Explor.* **2002**, *76*, 45–69. [CrossRef]
30. Imai, A. Variation of Cl and SO3 contents of micropheno-crystic apatite in intermediate to silicic igneous rocks of Cenozoic Japanese island arcs: Implications for porphyry Cu metallogenesis in the Western Pacific Island arcs. *Res. Geol.* **2004**, *54*, 357–372. [CrossRef]
31. FareeduddinKirmani, I.R.; Chander, S. Petrology, geochemistry and fluid inclusion studies of Cu-Au mineralization in paleoproterozoicSalumber-Ghatol belt, Aravalli Supergroup, Rajasthan. *J. Geol. Soc. India* **2012**, *80*, 5–38. [CrossRef]
32. Mukherjee, R.; Venkatesh, A.S.; Fareeduddi. Albitite hosted gold-sulfide mineralization: An example from Paleoproterozoic Aravalli supracrustal sequence, Bhukia area, Western India. *Episodes* **2016**, *39*, 590–598. [CrossRef]
33. Deol, S.; Deb, M.; Large, R.R.; Gilbert, S. LA-ICPMS and EPMA studies of pyrite, arsenopyrite and loellingite from the Bhukia-Jagpura gold deposit, southern Rajasthan, India: Implications for ore genesis and gold remobilization. *Chem. Geol.* **2012**, *326–327*, 72–87. [CrossRef]
34. Kumar, M. *Report on Exploration for Gold-Copper Mineralization in Jagpura Area, Banswara District, Rajasthan*; Unpublished Report; Geological Survey of India: Kolkata, India, 2014.
35. Anand, A. *Report on Ore Microscopic, Petrographic and Metamorphic Studies of Prospect Sequence of Bhukia-Jagpura Area, Banswara District, Rajasthan*; Unpublished Report; Geological Survey of India: Kolkata, India, 2016.
36. Skirrow, R.G. Iron oxide copper-gold (IOCG) deposits—A review (part 1): Settings, mineralogy, ore geochemistry and classification. *Ore Geol. Rev.* **2021**, *140*, 104569. [CrossRef]
37. Gopalan, K.; Macdougall, J.; Roy, A.; Murali, A. Sm-Nd evidence for 3.3 Ga old rocks in Rajasthan, northwestern India. *Precambrian Res.* **1990**, *48*, 287–297. [CrossRef]
38. Roy, A.B.; Kröner, A. Single zircon evaporation ages constraining the growth of the Archaean Aravalli craton, northwestern Indian Shield. *Geol. Mag.* **1996**, *133*, 333–342. [CrossRef]
39. Golani, P.R.; Pandit, M.K.; Sial, A.; Fallick, A.E.; Ferreira, V.P.; Roy, A.B. B-Na rich Paleoproterozoic Aravalli metsediments of evaporitic association, NW India: A new repository of gold mineralization. *Precambr. Res.* **2002**, *116*, 183–198. [CrossRef]

40. Kaur, P.; Chaudhri, N.; Raczek, I.; Kröner, A.; Hofmann, A.W.; Okrusch, M. Zircon ages of late Palaeoproterozoic (ca. 1.72–1.70 Ga) extension-related granitoids in NE Rajasthan, India: Regional and tectonic significance. *Gondwana Res.* **2011**, *19*, 1040–1053. [CrossRef]
41. Banerjee, D.M. Aravalli Craton and its Mobile Belts: An Update. *Episodes* **2019**, *43*, 88–108. [CrossRef]
42. Roy, A.B.; Sharma, B.L.; Paliwal, B.S.; Chauhan, N.K.; Nagori, D.K.; Golani, P.R.; Bejarniya, B.R.; Bhu, H.; Ali Sabah, M. Lithostratigraphy and tectonic evolution of the Aravalli Supergroup: A proto geosynclinal sequence. In *Rift Basins and Aulacogens*; Casshyap, S.M., Ed.; GyanodayaPrakashan: Naini Tal, India, 1993; pp. 73–90.
43. Pradhan, V.R.; Meert, J.G.; Pandit, M.K.; Kamenov, G.; Gregory, L.C.; Malone, S.J. India's changing place in global Proterozoic reconstructions: A review of geochronologic constraints and Paleomagnetic poles from the Dharwar, Bundelkhand and Marwar cratons. *J. Geodyn.* **2009**, *50*, 224–242. [CrossRef]
44. Deb, M.; Sarkar, S.C. Proterozoic tectonic evolution and metallogenesis in the Aravalli Delhi Orogenic belt, northwest India. *Precambr. Res.* **1990**, *46*, 115–137. [CrossRef]
45. Singh, S.P. Stratigraphy and sedimentation pattern in the Proterozoic Delhi Supergroup, Northwestern India. *Mem. Geol. Soc. India* **1988**, *7*, 193–206.
46. Bhattacharya, H.N.; Bull, S. Tectono-sedimentary setting of the Paleoproterozoic Zawar Pb–Zn deposits, Rajasthan, India. *Precambr. Res.* **2010**, *177*, 323–338. [CrossRef]
47. Ahmad, T.; Dragusanu, C.; Tanaka, T. Provenance of Proterozoic Basal Aravalli mafic volcanic rocks from Rajasthan, Northwestern India: Nd isotopes evidence for enriched mantle reservoirs. *Precambr. Res.* **2008**, *162*, 150–159. [CrossRef]
48. McKenzie, N.R.; Hughes, N.C.; Myrow, P.M.; Banerjee, D.M.; Deb, M.; Planavsky, N.J. New age constraints for the Proterozoic Aravalli-Delhi successions of India and their implications. *Precambr. Res.* **2013**, *238*, 120–128. [CrossRef]
49. Grover, A.K.; Verma, R.G. Gold mineralization in the Precambrian (Bhukia Area) of southeastern Rajasthan—A new discovery. *J. Geol. Soc. India* **1993**, *42*, 281–288.
50. Chander, S.; Sisodia, C.P. Gold mineralization in the Palaeoproterozoic rocks of Sanjela-Manpur-Dugocha belt, Salumber area, Udaipur district, Rajasthan. *J. Geol. Soc. India* **2003**, *61*, 463–470.
51. Gupta, S.N.; Arora, Y.K.; Mathur, R.K.; Iqbaluddin Prasad, B.; Sahai, T.N.; Sharma, S.B. The Precambrian geology of the Aravalli region, southern Rajasthan and northeastern Gujarat. *Mem. Geol. Surv. India* **1997**, *123*, 262.
52. Sengupta, S. Structures and stratigraphic relations of Aravallis, southeastern Rajasthan. *J. Geol. Soc. India.* **1976**, *4*, 461–470.
53. Sharma, R.S. Patterns of metamorphism in the Precambrian rocks of the Aravalli Mountain Belt. *Mem. Geol. Soc. India* **1988**, *7*, 33–75.
54. Geological Survey of India. *Base Document on Precious Metals and Minerals: Gold, Diamond, PGE and Precious Stones for CGPB Committee–II*; Geological Survey of India: Kolkata, India, 2020; *Unpublished Report*.
55. Deb, M.; Goldfarb, R.J. Gold metallogeny: India and beyond. *Miner. Depos.* **2010**, *46*, 835–836.
56. Deb, M. Some key issues of gold metallogeny in India. In *International Workshop in Gold Metallogeny*; Abstract; Delhi University: Delhi, India, 2008; pp. 50–53.
57. Singh, S. *Report on Investigation for Gold and Associated Basemetal Mineralization in Jagpura Block, Banswara District, Rajasthan*; Geological Survey of India: Kolkata, India, 2012; *Unpublished Report*.
58. Kumar, M.; Mohan, L. *Exploration for Gold-Copper Mineralisation in Jagpura Area, Banswara District, Rajasthan*; Geological Survey of India: Kolkata, India, 2016; *Unpublished Report*.
59. Hoefs, J. *Stable Isotope Geochemistry*; Springer: Berlin/Heidelberg, Germany, 2009; p. 288.
60. De Groot, P.A. *Handbook of Stable Isotope Analytical Techniques*; Elsevier: Amsterdam, The Netherlands, 2009; Volume 2, p. 1372.
61. Zhang, Q.L.; Ding, T.P. Analysis of the reference material NBS-123 and the atomic weight of sulfur. *Chinese Sci. Bull.* **1989**, *34*, 1086–1089.
62. Boyle, R.W. *Gold History and Genesis of Deposits*; Van Nostrand-Reinhold Company: New York, NY, USA, 1987; pp. 583–585.
63. Fleet, M.E.; Chryssoulis, S.L.; MacLean, P.J.; Davidson, R.; Weisener, G. Arsenian pyrite from gold deposits: Au and As distribution investigated by SIMS and EP, and color staining and surface oxidation by XPS and LIMS. *Can. Mineral.* **1993**, *31*, 1–17.
64. Bajwah, Z.U.; Seccombe, P.K.; Offler, R. Trace element distribution, Co/Ni ratios and genesis of the Big Cadia iron-copper deposit, New South Wales, Australia. *Miner. Depos.* **1987**, *22*, 292–300. [CrossRef]
65. Canil, D.; Grondahl, C.; Lacourse, T.; Pisiak, L.K. Trace elements in magnetitefrom porphyry Cu-Mo-Audeposits in British Columbia, Canada. *Ore Geol. Rev.* **2016**, *72*, 1116–1128. [CrossRef]
66. Roedder, E. Fluid Inclusions. Reviews in mineralogy. *Min. Soc. Am.* **1984**, *12*, 644.
67. Brown, P.E.; Lamb, W.M. P-V-T properties of fluids in the system H2O-CO2-NaCl: New graphical presentation and implications for fluid inclusion studies. *Geochim. Cosmochim. Acta.* **1989**, *53*, 1209–1221. [CrossRef]
68. Van den kerkhof, A.M. The System CO_2–CH_4–N_2 in Fluid Inclusions: Theoretical Modelling and Geological Applications. Ph.D. Thesis, Vrije Universiteit Amsterdam, Amsterdam, The Netherland, 1988; p. 206.
69. Shepherd, T.J.; Rankin, A.H.; Alderton, D.H.M. *A Practical Guide to Fluid Inclusion Studies*; Blackie: Glasgow/London, UK, 1985; p. 239.
70. Cook, N.J.; Ciobanu, C.L.; Mao, J. Textural control on gold distribution in As-free pyrite from the Dongping, Huangtuliang and Hougou gold deposits, North China Craton (Hebei Province, China). *Chem. Geol.* **2009**, *264*, 101–121. [CrossRef]
71. Zhao, H.X.; Frimmel, H.E.; Jiang, S.Y.; Dai, B.Z. LA-ICP-MS trace element analysis of pyrite from the Xiaoqinling gold district, China: Implications for ore genesis. *Ore Geol. Rev.* **2011**, *43*, 142–153. [CrossRef]
72. Liu, W.; Zhang, J.; Sun, J.; Zhou, L.; Liu, A. Low-Ti iron oxide deposits in the Emeishan large igneous province related to low-Ti basalts and gabbroic intrusions. *Ore Geol. Rev.* **2015**, *65*, 180–197. [CrossRef]

73. Zhou, M.F.; Chen, W.T.; Wang, C.Y.; Prevec, S.A.; Liu, P.P.; Howarth, G.H. Two stages of immiscible liquid separation in the formation of Panzhihua-type Fe–Ti–Voxide deposits, SW China. *Geosci. Front.* **2013**, *4*, 481–502. [CrossRef]
74. Ray, G.; Webster, I. Geology and chemistry of the low Ti magnetite-bearing Heff Cu-Au skarn and its associated plutonic rocks, Heffley Lake, south-central British Columbia. *Explor. Min. Geol.* **2007**, *16*, 159–186. [CrossRef]
75. Blanc, P.P. Distinction between magmatic and hydrothermal signature in apatite from north Portugal Cathodoluminescence, SEM and electron microscopic study. *Bull. Soc. Geol. France.* **1994**, *165*, 329–339.
76. Gehrig, M.; Lentz, H.; Franck, E.U. Thermodynamic properties of water–carbon dioxide–sodium chloride mixtures at high temperatures and pressures. In *High Pressure Science and Technology*; Timmerhaus, K.D., Barber, M.S., Eds.; Physical Properties and Material Synthesis; Springer: Berlin/Heidelberg, Germany, 1979; Volume 1, pp. 539–542.
77. Vityk, M.O.; Bodnar, R.J.; Schmidt, C.S. Fluid inclusions as tectonothermobarometers: Relation between pressure–temperature history and re-equilibration morphology during crustal thickening. *Geology* **1994**, *22*, 731–734. [CrossRef]
78. Wilkinson, J.J. Fluid inclusions in hydrothermal ore deposits. *Lithos* **2001**, *55*, 229–272. [CrossRef]
79. Kesler, S.E.; Haynes, P.S.; Creech, M.Z.; Gorman, J.A. Application of fluid inclusion and rock gas analysis in mineral exploration. *J. Geochem. Explor.* **1986**, *25*, 201–215. [CrossRef]
80. Kesler, S.E. Ore forming fluids. *Elements* **2005**, *1*, 13–18. [CrossRef]
81. Seal, R.R. Sulfur Isotope Geochemistry of Sulfide Minerals. *Rev. Mineral. Geochem.* **2006**, *61*, 633–677. [CrossRef]
82. Holser, W.T. Catastrophic chemical events in the history of the ocean. *Nature* **1977**, *267*, 402–408. [CrossRef]
83. Zhao, L.; Chen, H.; Zhang, L.; Xiaoping, X.; Zhang, W.; Li, D.; Lu, W.; Liang, P.; Li, R.; Yang, J.; et al. Geology and ore genesis of the late PaleozoicHeijianshan Fe oxide–Cu (–Au) deposit in the Eastern Tianshan, NW China. *Ore Geol. Rev.* **2017**, *91*, 110–132. [CrossRef]
84. Pal, D.C.; Barton, M.D.; Sarangi, A.K. Deciphering a multistage history affecting U-Cu (–Fe) mineralization in the Singhbhum Shear Zone, eastern India using pyrite textures and compositions in the Turamdih U–Cu (–Fe) deposit. *Miner. Depos.* **2009**, *44*, 61–80. [CrossRef]
85. Neilsen, H. Sulfur isotopes. In *Lectures in Isotope Geology*; Jager, E., Hunziher, J.E., Eds.; Springer: Berlin/Heidelberg, Germany, 1979; pp. 283–312.
86. Ohmoto, H.; Goldhaber, M.B. Sulfur and carbon isotopes. In *Geochemistry of Hydrothermal Ore Deposits*; Barnes, H.L., Ed.; John Wiley and Sons: Hoboken, NJ, USA, 1997; pp. 517–611.
87. Eldridge, C.S.; Compston, W.; Williams, I.S.; Harris, J.W.; Bristow, J.W. Isotope evidence for the involvement of recycled sediments in the diamond formation. *Nature* **1991**, *353*, 649–653. [CrossRef]
88. Bastrakov, E.N.; Skirrow, R.G.; Davidson, G.J. Fluid Evolution and origins of iron oxide Cu-Au prospects in the Olympic Dam District, Gawler Craton, South Australia. *Econ. Geol.* **2007**, *102*, 1415–1440. [CrossRef]
89. De Haller, A.; Fontboté, L. The Raul-Condestable iron oxide copper-gold deposit, central coast of Peru: Ore and related hydrothermal alteration, sulfur isotopes, and thermodynamic constraints. *Econ. Geol.* **2009**, *104*, 365–384. [CrossRef]
90. Chen, H.Y. External sulfur in IOCG mineralization: Implications on the definition and classification of the IOCG clan. *Ore Geol. Rev.* **2013**, *51*, 74–78. [CrossRef]
91. Ault, K.M.; Williams-Jones, A.E. Sulfur and lead isotope study of the EL Mochito Zn-Pb-Ag deposit. *Econ. Geol.* **2004**, *99*, 1223–1231. [CrossRef]
92. Hutchison, W.; Finch, A.A.; Boyce, A.J. The sulfur isotope evolution of magmatic-hydrothermal fluids: Insights into ore-forming processes. *Geochim. Cosmochim. Acta.* **2020**, *288*, 176–198. [CrossRef]
93. Rollinson, H.R. *Using Geochemical Data: Evaluation, Presentation, Interpretation*; Longman: London, UK, 1993.
94. Hitzman, M.W. Iron oxide-Cu-Au deposits: What, where, when, and why. In *Hydrothermal Iron Oxide Copper-Gold and Related Deposits; A Global Perspective*; Porter, T.M., Ed.; Australian Mineral Foundation: Adelaide, Australia, 2000; Volume 1, pp. 9–25.
95. Williams, P.J.; Skirrow, R.G. Overview of iron oxide–copper–gold deposits in the Curnamona Province and Cloncurry District (Eastern Mount Isa Block). In *A Global Perspective*; Porter, T.M., Ed.; PGC Publishing: Brisbane, Australia, 2000; Volume 1, pp. 105–122.
96. Williams, P.J.; Barton, M.D.; Johnson, D.A.; Fontbote, L.; De Haller, A.; Mark, G.; Oliver, N.H.S.; Marschik, R. Iron oxide copper-gold deposits: Geology, space-time distribution and possible modes of origin. *Econ. Geol.* **2005**, *100*, 371–405.
97. Davidson, G.J.; Hamish, P.; Sebastien, M.; Ron, F.B. Characteristics and origin of the Oak Dam East breccia-hosted, iron oxide Cu–U–(Au) deposit: Olympic Dam region, Gawler Craton, South Australia. *Econ. Geol.* **2007**, *102*, 1471–1498. [CrossRef]
98. Barton, M.D. Iron oxide (-Cu-Au-REE-P-Ag-U-Co) systems. In *Treatise on Geochemistry*, 2nd ed.; Elsevier Inc.: Amsterdam, The Netherlands, 2014; pp. 515–541.

minerals

MDPI

Article

Origin of Disseminated Gold-Sulfide Mineralization from Proximal Alteration in Orogenic Gold Deposits in the Central Sector of the Yana–Kolyma Metallogenic Belt, NE Russia

Valery Yurievich Fridovsky [1,*], Lena Idenenovna Polufuntikova [1,2] and Maxim Vasilievich Kudrin [1]

1 Diamond and Precious Metal Geology Institute, SB RAS, 677000 Yakutsk, Russia
2 Faculty of Geology and Survey, M.K. Ammosov North-Eastern Federal University, 677000 Yakutsk, Russia
* Correspondence: fridovsky@diamond.ysn.ru; Tel.: +7-4112-33-58-72

Abstract: The Yana–Kolyma metallogenic belt, NE Russia, is a world-class gold belt with resources numbering ~8300 tons of gold. The belt is localized in the central part of the Verkhoyansk–Kolyma orogen, formed by a collage of diverse terranes. The Tithonian-to-Early-Cretaceous orogenic gold deposits are hosted in a sequence of Permian–Triassic and Jurassic clastic rocks and altered Late Jurassic andesite, dacite, granodiorite, trachyandesite, and trachybasalt dykes. High-fineness gold (800–900‰) in quartz veins and invisible gold in disseminated arsenian pyrite-3 (Py3) and arsenopyrite-1 (Apy1) are present in ores. Here, we present new data about microtextures; the chemical composition and stable sulfur isotopes of auriferous pyrite-3 and arsenopyrite-1 from proximal alterations in sediment-hosted (Malo–Taryn, Badran, Khangalas); and intrusion-hosted (V'yun, Shumniy) orogenic Au deposits in the central sector of the Yana–Kolyma metallogenic belt to better constrain the ore-forming process and tracking their evolution. Detailed petrography defined the following generations of pyrite: syn-sedimentary/diagenetic Py1, metamorphic Py2 and hydrothermal Py3, and Apy1. Hydrothermal Py3 and Apy1 are localized in the proximal pyrite–arsenopyrite–sericite–carbonate–quartz alteration in ore zones and make a major contribution to the economic value of the veinlet-disseminated mineralization with "invisible" gold in the orogenic deposits of the Yana–Kolyma metallogenic belt. Electron microprobe analysis (EMPA) of Py3 in both types of deposits shows concentrations of As (up to 3.16 wt%), Co, Ni, Cu, Sb, and Pb. Py3 in intrusion-hosted orogenic gold deposits reveals elevated concentrations of Co (up to 0.87 wt%), Ni (up to 3.52 wt%), and Cu (up to 2.31 wt%). The identified negative correlation between S and As indicates an isomorphic substitution of sulfur by As^{1-}. Py3 from igneous rocks is characterized by a high degree of correlation for the pairs $Fe^{2+} \rightarrow Co^{2+}$ and $Fe^{2+} \rightarrow Ni^{2+}$. For hydrothermal Apy1, Co (up to 0.27 wt%), Ni (up to 0.30 wt%), Cu (up to 0.04 wt%), and Sb (up to 0.76 wt%) are typomorphic. According to atomic absorption spectrometry, the concentration of Au in Py3 reaches 159.5 ppm; in Apy1, it reaches 168.5 ppm. The determination of the precise site of the invisible gold within Py3 and Apy1 showed the predominance of solid-solution Au^{+} in the crystal lattice. The values of $\delta^{34}S$ in Py3 and Apy1 (from −6.4 to +5.6‰, mean value of about +0.6‰), both from sediment-hosted and from intrusion-hosted deposits, display a relatively narrow range and are characteristic of the hydrothermal ore stage. Our analytical results showed no systematic differences between the chemical and stable sulfur isotope compositions of both auriferous pyrite-3 and arsenopyrite-1 from the proximal alteration in sediment-hosted (Malo–Taryn, Badran, Khangalas) and intrusion-hosted (V'yun, Shumniy) orogenic Au deposits, indicating that the primary source of sulfur, gold, and mineralizing fluids was likely from subcrustal and metamorphic systems in the Late-Jurassic-to-Early-Cretaceous Verkhoyansk–Kolyma orogen.

Keywords: disseminated Au-sulfide mineralization; orogenic gold deposits; Yana–Kolyma metallogenic belt; sulfide microtextures and mineralogy; stable sulfur isotopes; mineral chemistry; source of gold

check for updates

Citation: Fridovsky, V.Y.; Polufuntikova, L.I.; Kudrin, M.V. Origin of Disseminated Gold-Sulfide Mineralization from Proximal Alteration in Orogenic Gold Deposits in the Central Sector of the Yana–Kolyma Metallogenic Belt, NE Russia. *Minerals* **2023**, *13*, 394. https://doi.org/10.3390/min13030394

Academic Editor: David Banks

Received: 18 January 2023
Revised: 1 March 2023
Accepted: 7 March 2023
Published: 11 March 2023

1. Introduction

Orogenic gold deposits (OGDs) are an economically important source of gold in the world. According to estimates [1], the relative significance of OGDs (including intrusion-related OGDs) in past production and known reserves and resources is 32%. In addition, Goldfarb R.J. et al. [2] indicated that most (45%) of gold deposits worldwide containing more than 1 Moz of gold belong to orogenic gold deposits, much more than other types of deposits.

The Yana–Kolyma metallogenic belt (YKMB), NE Russia, is known for its large bulk-tonnage OGD at Natalka (1500 t Au), Degdekan (400 t Au), Pavlik (169 t Au), Drazhnoe (50 t Au), and other sites [3–6]. Similar Paleozoic-to-Mesozoic major deposits of this type are also known in Russia (Sukhoi Log), Uzbekistan (Muruntau), Kyrgyzstan (Kumtor), Australia (Bendigo, Ballarat), USA (Juneau, Treadwell), China (Jiaodong), and in other countries [5–8]. The endowment of the YKMB is ~8300 tons of gold [3,6]. However, the YKMB's main gold production came from placer deposits, sourced from OGDs.

Gold in the orogenic gold deposits of the Yana–Kolyma metallogenic belt is mainly found as "visible" gold grains in quartz veins/veinlets and rarely as inclusions in the pyrite and arsenopyrite of the proximal alteration envelopes in the host rock. In disseminated pyrite and arsenopyrite (up to 5–8%) in carbonaceous clastic rocks with pyrite–arsenopyrite–sericite–carbonate-quartz alterations, there is also "invisible" gold, which is in the form of a solid solution or nanoparticles [6,9–12]. Currently, the greatest economic value in the YKMB is Au-quartz veins and/or the Au-sulfide–quartz-veinlet-disseminated styles of mineralization. The reserves and resources of bulk YKMB deposits can be estimated in the Au-sulfide–quartz-veinlet-disseminated type of mineralization [4,13]. The geological, mineralogical, geochemical, and geochronological characteristics of the vein and veinlet mineralization in the YKMB are described in detail elsewhere (see [14–24] and references therein), but the study of disseminated mineralization and its contribution to the overall gold content of orogenic deposits in the central sector of the YKMB is in its infancy [11,12,25,26]. In addition, there are not many detailed investigations of the mineralization processes.

Fridovsky V.Y. et al. [11,26] and Kudrin M.V. et al. [12] showed that pyrite grew in clastic rocks in the central sector of the YKMB from sedimentation due to late metamorphism and finally formed together with arsenopyrite from hydrothermal fluids. Recent geochemical studies [11,12,26] revealed different concentrations of trace elements (As, Co, Ni, Cu, Pb, and Sb) in pyrite generations. However, the origin and sources of ore-forming fluids and metals in gold-bearing, disseminated mineralization from proximal alterations in sediment-hosted and intrusion-hosted OGDs in the YKMB remain unclear. Several genetic models of gold-bearing quartz veins and disseminated ores in the YKMB have been discussed: (a) magmatic–hydrothermal [17]; (b) formation from metamorphic fluids and boiling during uplift and decompression [22]; (c) and juvenile and metamorphic sources [23,27]. This creates serious difficulties in generating representative models and identifying potential targets.

The microtextures and mineral chemistry of pyrite and arsenopyrite are very useful for studying the processes of Au mineralization (e.g., [11,12,28–38]). These minerals can incorporate various metals and semi-metals (As, Co, Ni, Cu, Pb, Sb, etc.), showing the chemical evolution of gold ores [11,15,22,39–48] and many others. It is equally important that gold is either in solid solutions or micro- to nano-sized inclusions (e.g., [11,12,48,49]). Our understanding of the sources of ore-forming metals and fluids and the processes that form OGDs are also constrained by sulfur isotope analyses on sulfide minerals that are co-genetic with gold [11,15,20,21,23,26,39,43,50–56]. Additionally, sulfur isotope analyses allow us to assess the role of intrusions adjacent to hosting orogenic gold deposits in the providing metals (e.g., [43]).

Here, we present new data about microtextures and analyses of the chemical composition and stable sulfur isotopes of auriferous pyrite and arsenopyrite from proximal alterations in sediment-hosted and intrusion-hosted orogenic gold deposits in the central sector of the YKMB. New characteristics and information about the origin and gold grade in disseminated mineralization from proximal alterations can contribute to the further development of prospecting programs and provide important data for the reassessment of known deposits and the discovery of new deposits in the central sector of the YKMB. Observations of microtextures using scanning electron microscopy (SEM) were combined with major and trace-element analyses performed using electron probe microanalysis (EPMA) and the sulfur isotope data of pyrite and arsenopyrite. This information was used to understand ore-forming events and to reinterpret the source of gold and the hydrothermal fluid-flow processes that formed the sediment-hosted and intrusion-hosted OGD orogenic Au systems in the YKMB.

2. Geological and Metallogenic Background

2.1. Regional Geological Setting

The Yana–Kolyma metallogenic belt is localized in the central part of the Verkhoyansk–Kolyma orogen, formed by a collage of diverse terranes [57]. Parfenov L.M. and Kuzmin M.I. [57], Parfenov L.M. [58], Parfenov L.M. et al. [59], Toro J. et al. [60], and Nokleberg W.J. et al. [61] provided detailed descriptions of terranes. The majority of OGDs in the YKMB are located in Upper Permian, Triassic, and Lower Jurassic clastic rocks and, less often, in Late Jurassic dykes of mafic, intermediate, and felsic composition, as well as granitoids of the small intrusions complexes in the Kular–Nera terrane, the adjacent Verkhoyansk fold-and-thrust belt, and the Polousny-Debin terrane [6,14,23,57,62] (Figure 1). Clastic rocks are represented by proximal and distal sediments in the eastern margin of the Siberian craton, metamorphosed under conditions no higher than the greenschist facies [57]. The rocks are deformed into the linear folds of the northwestern strike, fold-parallel faults, and transverse/oblique faults. The Adycha–Taryn and Charky–Indigirka faults are the longest faults, having an NW strike and limiting the terranes of the eastern margin of the Siberian craton. There are also cross-cutting faults in the NE and WE strikes.

Clastic rocks are intruded by granitoids from the main (Kolyma) batholitic belt (154–144 Ma, zircon, U-Pb SHRIMP-RG [63], SHRIMP-II [64]), volcanic rocks and intrusive rocks with intermediate and felsic compositions from the Tas–Kystabyt belt (151–148 Ma, zircon, U-Pb SHRIMP-II [64,65]), and Late-Jurassic-to-Early-Cretaceous granitoids from small massifs and dykes with basic-to-felsic compositions (151–143 Ma, zircon, U-Pb SHRIMP II [62,66]). Based on the geodynamic setting, structural geometry and control, host rocks, alterations, ore mineralogy, geochemistry, isotopes, and geochronological characteristics, the studied deposits have been classified as orogenic [5,6,21,23,67,68]. The morphological styles of the ore bodies are veins, breccias, and veinlet-disseminated. The deposits are localized mainly near the Adycha–Taryn and Charky–Indigirka thrusts and subsidiary faults.

As reviewed by [13,18,19,21,23,69], the ore formation of OGDs in the central sector of the YKMB occurred at pressures ranging from 1.2 to 0.25 kbar and temperatures ranging from 300 to 200 °C. Ore-forming fluids contained a dilute water–salt solution (salt concentration 0.5–9.3 wt% NaCl-eq.), carbon dioxide (8.5–15 mol%), and methane (up to 1 mol%). The hydrothermal fluid contained the following major micro-components: As, B, Sb, Zn, Fe, and Cu. The parameters and composition of the ore-forming fluids in the studied deposits are typical of Phanerozoic OGDs [43].

The combined structural observations of the central sector of the Yana–Kolyma metallogenic belt [57,70], including deposit-scale observations [20,21,67–69,71], new Ar-Ar and Re-Os dating of minerals from ore veins [21,27], and the U-Pb dating of igneous rocks [62,66], allowed us to decipher the sequence of deformation events and their connection with OGDs [67,68]. A three-stage deformation history is proposed. D1 compression led to the formation of folds and faults in the northwest strike. When these progressive deformations were completed at the turn of the Late-Jurassic-to-Early-Cretaceous and at the beginning of the Early Cretaceous, the formation of OGDs in the YKMB took place during accretion–collision events on the eastern margin of the Siberian craton and in the Verkhoyansk–Kolyma orogen. D2 sinistral strike-slip and D3 dextral strike-slip movements activated faults in the NW strike. The youngest events are a reflection of superimposed subduction–accretion processes at the rear of the active Cretaceous East Asian continental margin.

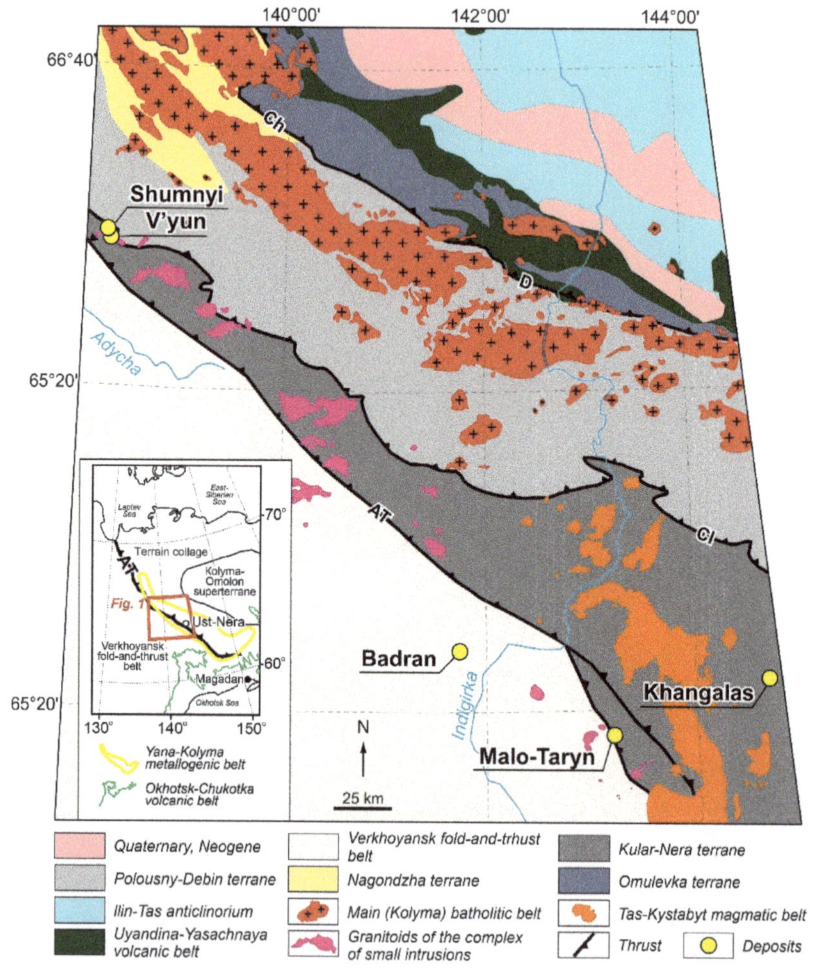

Figure 1. Geology of the central part of the Verkhoyansk–Kolyma orogen with main tectonic units, Jurassic–Cretaceous magmatism, and studied deposits. Faults: AT—Adycha–Taryn, CI—Charky–Indigirka, D—Darpir, Ch—Chibagalakh.

2.2. *Ore Deposit Geology*

The studied sediment-hosted (Malo–Taryn, Badran, Khangalas) and intrusion-hosted (V'yun, Shumnyi) gold deposits are typical OGDs with vein–veinlet quartz and disseminated mineralization with proximal pyrite–arsenopyrite–sericite–carbonate–quartz alterations controlled by faults [6,21,67,72,73]. Geological sketches of individual deposits are shown in Figures 2 and 3, and the common characteristics of the studied OGDs are presented in Table 1. Below is a brief geological description and information about the mineral composition of these ore deposits.

Table 1. General characteristics of the studied OGDs, central YKMB.

Characteristics	Malo–Taryn	Badran	Khangalas	V'yun	Shumnyi
Mineral type	Au-Py-Apy	Au-Py-Apy	Au-Py-Apy	Au-Py-Apy	Au-Py-Apy
Coordinates (N/E)	63°54′/143°11′	64°14′/141°31′	64°06′/144°55′	65°58′/138°16′	66°00′/138°08′
Metallogenic zone	Adycha–Taryn	Mugurdakh–Selerikan	Olchan–Nera	Olchan–Nera	Olchan–Nera
Ore cluster	Taryn	Selerikan	Khangalas	Burgandzha	Burgandzha
Magmatism/ composition	dyke/ trachybasalts	No	No	Dykes/andesite, dacite, granodiorite, trachyandesite, trachybasalt trachyandesite	Dykes/andesite, dacite, granodiorite, trachyandesite, trachybasalt, trachyandesite
Age, Ma: composition/ method	145–160?/trachybasalts/Rb-Sr	No	No	147/trachyandesite/ U-Pb SHRIMP II	151, 146/andesite, trachyandesite/ U-Pb SHRIMP II
Host rock	Clastic rocks (T_3)	Clastic rocks (T_3)	clastic rocks (P_2)	Dykes (J_3)/ clastic rocks (T_3)	Dykes (J_3)/clastic rocks (T_3)
Alteration	Pyrite–arsenopyrite–sericite–carbonate–quartz	Pyrite–arsenopyrite–sericite-carbonate–quartz	Pyrite–arsenopyrite-sericite–carbonate–quartz	Pyrite–arsenopyrite–sericite–carbonate–quartz	Pyrite–arsenopyrite–sericite–carbonate–quartz
Ore location	Reverse fault in the SW wing of the Malo–Taryn syncline	Thrust in the SW wing of the Mugurdakh syncline	Reverse fault in the SW wing of the Khangalas anticline	Transverse NE faults to the folding of the NW strike	Transverse NE faults to the folding of the NW strike
Ore bodies	Mineralized faults with Au-quartz veins/veinlets and Au-sulfide-disseminated			Au-quartz veins/veinlets and Au-sulfide-disseminated	
Mineral associations *	Au-Bi, Py-Apy-Qz metasomatic, Py-Apy-Qz vein, Au-polysulfide-Qz, Au-Sb, sulfosalt–carbonate, berthierite–antimony, Ag-Sb	Py-Apy-Qz metasomatic, Apy-Py-carbonate-Qz, Ccp-Sp-Ab-Qz, Ttr-Ser-Qz, Ag-Qz, Sbn-carbonate-Qz	Py-Apy-Qz metasomatic, Py-Apy-Qz vein, Au-polysulfide-Qz, sulfosalt-carbonate-Qz	Py-Apy-Qz metasomatic, Py-Apy-Qz vein, Au-polysulfide-Qz, fahlore-Qz	Py-Apy-Qz metasomatic, Py-Apy-Qz vein, Au-polysulfide-Qz, Sbn-Qz
Sulphide content, vol%	1–3	1–2	1–5	1–3	1–2
Au fineness, ‰	894–995, predominantly 901–925	462–998, predominantly 800–899	780–850, predominantly 820–830	700–920, predominantly 800–899	800–900
Au, C_{Au}, ppm	4.2	7.8	3.9	12.1	80.0
Au reserves/Au resources, t	~12.5/-	~9.3/-	~9.5/-	~2.5/~1.9	-/6.8

Compiled using [12,21,62,66,68,74–85]. Abbreviations [86]: Py—pyrite, Apy—arsenopyrite, Au—native gold, Qz—quartz, Ccp—chalcopyrite, Sp—sphalerite, Ab—albite, Ttr—tetrahedrite, Ser—sericite, Sbn—stibnite. * Mineral associations: Au-Bi—gold–bismuth, Py-Apy-Qz metasomatic—pyrite–arsenopyrite–quartz metasomatic, Py-Apy-Qz vein—pyrite–arsenopyrite–quartz vein, Au-polysulfide-Qz—gold–polysulfide–quartz, Au-Sb—gold–stibnite, sulfosalt-carbonate-Qz—sulfosalt–carbonate–quartz, Ag-Sb—silver–stibnite, Apy-Py-carbonate-Qz—arsenopyrite–pyrite–carbonate–quartz, Ccp-Sp-Ab-Qz—chalcopyrite–sphalerite–albite–quartz, Ttr-Ser-Qz—tetrahedrite–sericite–quartz, Ag-Qz—silver–quartz, Sbn-carbonate-Qz—stibnite–carbonate–quartz, fahlore-Qz—fahlore–quartz, Sbn-Qz—stibnite–quartz.

2.2.1. Geology and Mineralization of the Sediment-Hosted Orogenic Gold Deposits

Information about the geology, magmatism, and mineral composition of the Malo–Taryn deposit is provided in [27,62,76,77,84,85,87,88]. The deposit is located on the left bank of the Malo–Taryn River, 100 km southeast of the settlement of Ust-Nera. Tectonically, the deposit is localized along the border of the Verkhoyansk fold-and-thrust belt and the Kular-Nera terrane, in the axial part of the Malo–Taryn branch of the Adycha–Taryn fault. The host clastic Upper Triassic rocks (Figure 2A) are deformed into folds of several generations. The early ones associate with the NW-trending thrusts, and the later ones associate with sinistral and dextral movements along the reactivated NW thrusts. In the Malo–Taryn fault zone, the stratified rocks dip at 30°–55° to 60°–70°, including overturned beddings. The Early Cretaceous Samyr (144.5–143 Ma, zircon, U-Pb SHRIMP-II [62]), and Kurdat (141–137 Ma, biotite, $^{40}Ar/^{39}Ar$ [89]) granitoid massifs of the Tas–Kystabyt magmatic belt are located NW of the deposit at the intersection of the Malo–Taryn fault and the Kurdat transverse fault zone. Dykes of trachybasalts (145–160 Ma, Rb-Sr, rock [84,85]) are usually localized in the ore zones of the NW and NNW strikes. Their thickness is 0.1–2.0 m, and their lengths are up to the first tens of meters. The age of polychronous mineralization was constrained by different methods ($^{40}Ar/^{39}Ar$, K-Ar, Re-Os) at 148 to 126 Ma [21,27,87,90]. The ore bodies of the deposit occur along the faults of the NW and NS strikes with a dip to the SW and W at 30° to 85°. They are up to 40 m thick and can be traced 6 km along the strike.

The geological structure and mineral composition of the vein bodies and their localization at the Badran deposit have been studied in detail [74,91–94]. Underground gold mining at the deposit started in 1984. The deposit is located on the left bank of the Indigirka River, 130 km west of Ust-Nera in Upper Triassic clastic rocks (Figure 2B). The deposit is localized along the Badran–Egelyakh fault, up to 30 m wide, traceable for a distance of more than 5 km, and more than 1.2 km downdip. The northwestern and southeastern segments strike west–east and dip at 50°–60°. In the central segment, the fault dips at 24°–30° to the northeast. The quartz veins/veinlets and proximal pyrite–arsenopyrite–sericite–carbonate–quartz alteration are gold-bearing. The thickness of the quartz veins varies from tens of centimeters up to 4–5 m, with a length of up to 200 m. The gold content in altered rocks is at first grams per ton, reaching several kilograms per ton in the veins.

The structure and structural–tectonic setting of the formation and localization, as well as the mineralogical–geochemical and isotope–geochemical characteristics of the Khangalas deposit are described in [12,68,95–97]. The deposit is located on the left bank of the Nera River, 135 km southeast of the Ust-Nera settlement. Structurally, the deposit is confined to the Dvoinaya anticline (Figure 2C). The core of the anticline is composed mainly of Upper Permian sandstones; the limbs consist of Lower and Middle Triassic siltstones and mudstones. The Khangalas fault of the NW strike is the major ore-controlling fault. The ore bodies are represented by five extended (up to 1400 m) mineralized faults with quartz–carbonate veins/veinlets and pyrite–arsenopyrite-disseminated mineralization in the proximal alteration. The thickness of the faults is up to 32 m; their dip varies from SW to S and SE at 30°–50° to 70°–80° [67]. Re-Os data on the native gold from the quartz vein of the deposit showed that it was 137 million years old [27]. The gold content in the alteration is up to 5.29 ppm Au, with a mean of 0.73 ppm.

2.2.2. Geology and Mineralization of the Intrusion-Hosted Orogenic Gold Deposits

Detailed geological, mineralogical, and geochemical characteristics for the *V'yun and Shumnyi deposits* are provided in [66,75,78,98]. Underground gold mining at the V'yun deposit started in 2020. The V'yun and Shumnyi deposits are located in the upper reaches of the Adycha River. The tectonic structure of the deposit is controlled by longitudinal NW fold-and-thrust structures and transverse NE faults activated by late strike-slip faults (Figures 4 and 5). The Upper Triassic clastic rocks are crumpled into linear isoclinal and compressed asymmetric folds in the northwest strike. Mineralization is localized in the Late Jurassic dykes of trachybasalts, andesites, trachyandesites, dacites, and granodiorites

(151–145 Ma, zircon, U–Pb SHRIMP-II [66]) and faults in the NE strike (Figure 4). Dykes are 2 to 8 m thick, extending for more than 2 km. The style of ore bodies is vein–stockwork and disseminated. In the veins and stockworks, the main minerals are quartz and carbonate made up of, at most, 1–3% sulfides, represented mostly by arsenopyrite. Rarer sulfides and sulfosalts are pyrite, galena, sphalerite, chalcopyrite, tetrahedrite, and argentotetrahedrite. There are insignificant quantities of bournonite and antimony. Gold in the veins is free and characterized by high grades of up to several hundred grams per tonne. The disseminated gold–sulfide mineralization is localized both in the fault zones up to several tens of meters thick and in dykes. The main ore minerals in them are auriferous pyrite and arsenopyrite. The host alteration is sericite–chlorite–quartz in composition.

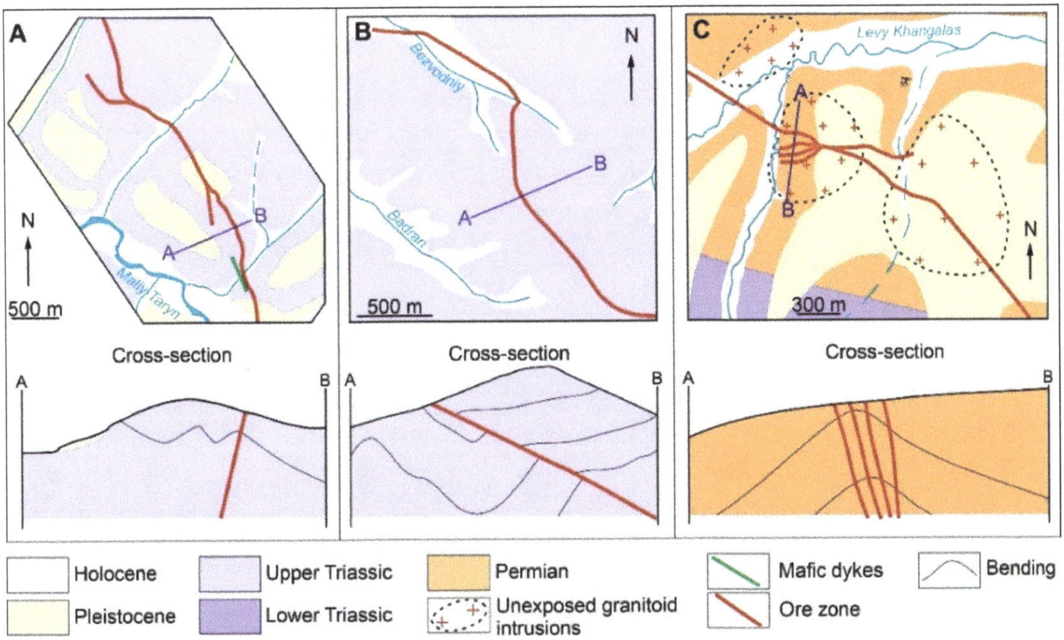

Figure 2. Geologic map and simplified cross-sections of the sediment-hosted OGDs, central YKMB: (**A**) Malo–Taryn; (**B**) Badran; (**C**) Khangalas.

2.3. *Mineral Composition of the Deposits and Mineralization Types*

There are several types of mineralization at the Malo–Taryn, Badran, Khangalas, V'yun, and Shumnyi deposits: gold–quartz/gold–sulfide–quartz, stibnite, and silver–stibnite (Table 2) [23]. Gold–quartz mineralization developed in quartz veins and veinlets; gold is native here. Disseminated mineralization with "invisible" gold is localized in the proximal arsenopyrite–pyrite–sericite–carbonate–quartz alteration. Berthierite–stibnite and silver–stibnite mineralization is superimposed on OGDs sporadically.

Gold–quartz/gold–sulfide–quartz-type mineralization is characterized by a number of successively alternating paragenetic associations: arsenopyrite–pyrite–quartz alteration, pyrite-arsenopyrite–quartz veins, Au-polysulfide-quartz, and sulfosalt–carbonate [18–21,23]. The number of ore minerals takes up 1–3% or, less often, up to 5%. A detailed mineralogical–geochemical and isotopic characterization of the disseminated Py-Apy-Qz paragenesis is provided in Section 4 of this article.

The pyrite–arsenopyrite–quartz vein association is composed of coarse- and medium-grained subhedral and anhedral milky-white quartz (85–95%), carbonate, chlorite, albite, and sericite. The texture of the veins is massive, banded (Figure 6A–C), and breccia-like

(Figure 6D,E); the structure is hypidiomorphoblastic (subhedral–blastic structure), and small veinlets are characterized by a comb structure. Pyrite4 and arsenopyrite2 occur in the form of individual crystals up to 1–2 mm in size; veinlets up to 1 mm in thickness; and nests up to 3–4 cm in size (Figure 6I) or, less often, 50–70 cm or more.

Pyrite4 and arsenopyrite2 form intergrowths and mutual inclusions (Figure 7A). The primary euhedral grains of early sulfides are cataclazed and corroded. Native gold, quartz, and sulfides from late mineral paragenesis occur in microcracks in arsenopyrite2 (Figure 7B,C) (Au-Qz-polysulfide and sulfosalt–carbonate).

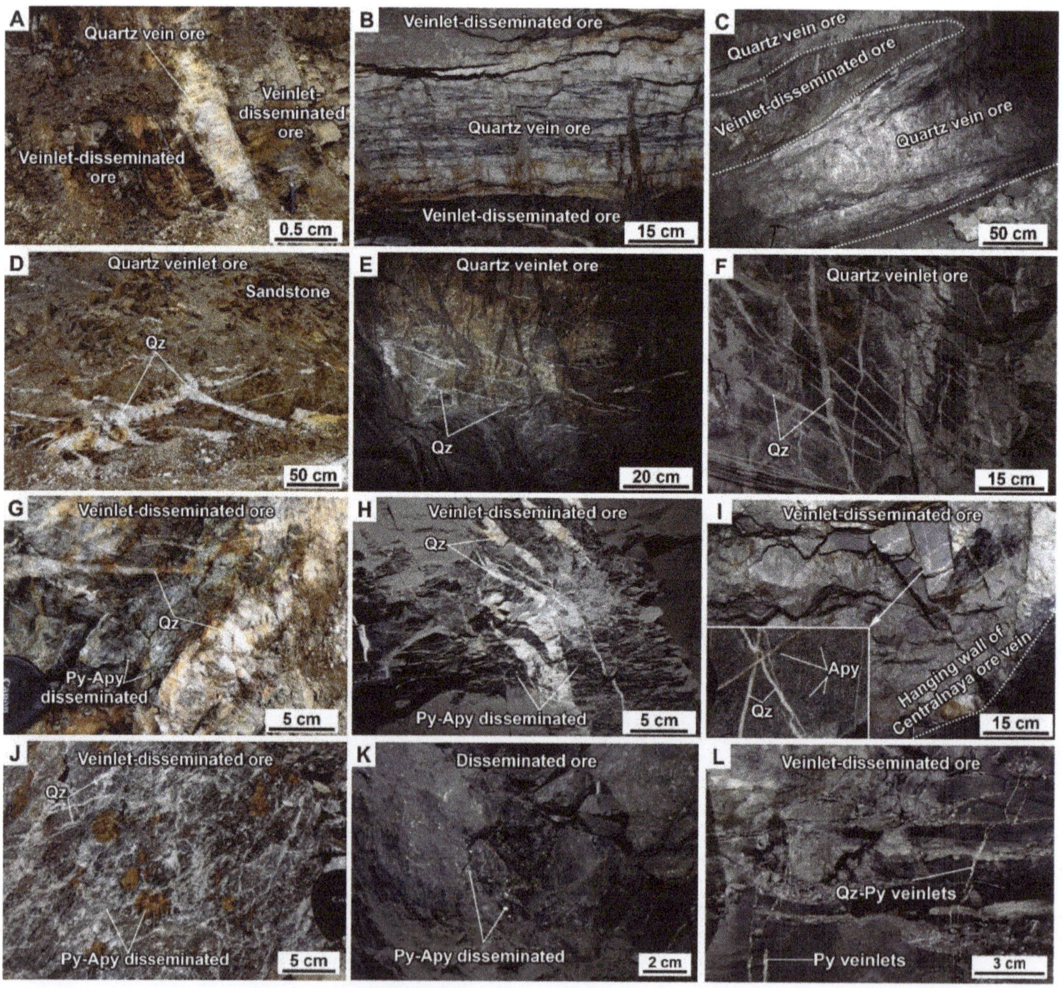

Figure 3. Ore bodies of (**A**,**D**,**G**,**J**) Malo–Taryn; (**B**,**E**,**H**,**K**) Badran; and (**C**,**F**,**I**,**L**) Khangalas orogenic gold deposits: (**A**–**C**) veins; (**D**–**F**) veinlets; (**G**–**J**,**I**,**L**) veinlet-disseminated; (**K**) disseminated. Abbreviations: Qz—quartz, Py—pyrite, Apy—arsenopyrite.

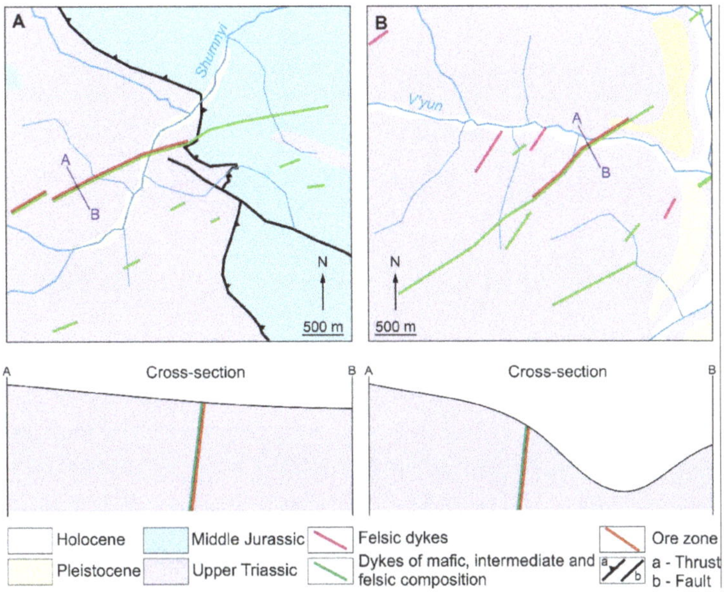

Figure 4. Geologic map and simplified cross-sections of the intrusion-hosted OGDs, central YKMB: (**A**) V'yun and (**B**) Shumnyi.

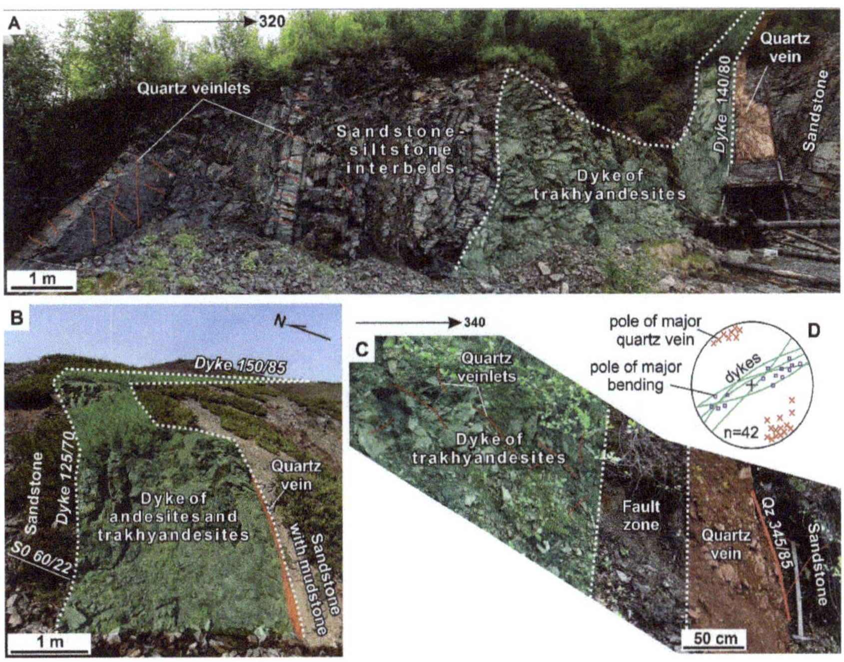

Figure 5. Photographs of the ore bodies in the intrusion-hosted OGDs, central YKMB: (**A**,**C**) V'yun; (**B**) Shumnyi; (**D**) stereogram of the veins, bedding, and dykes. Abbreviations: S0—bedding, Qz—quartz vein, D—dyke. Structural data were plotted on the upper hemisphere of the Wulff stereographic net.

Table 2. Mineral composition of ores and mineralization types of the studied OGDs, central YKMB.

Mineralization	Paragenesis	Minerals	Malo-Taryn	Badran	Khangalas	V'yun/ Shumnyi
Au-Qz/Au-sulfide-Qz	Py-Apy-Qz metasomatic	Pyrite	———	———	———	———
		Arsenopyrite	———	———	———	———
		Danaite	········			
		Fe-gersdorfite	········		········	
	Py-Apy-Qz vein	Quartz	━━━	━━━	———	━━━
		Sericite	- - - -	- - - -	- - - -	- - - -
		Chlorite		- - - -		- - - -
		Albite		———		- - - -
		Pyrite	———	———	———	———
		Arsenopyrite	———	———	———	———
	Au-polysulfid	Gold	- - - -	- - - -	- - - -	- - - -
		Galena	- - - -	- - - -	- - - -	- - - -
		Chalcopyrite	- - - -	- - - -	- - - -	- - - -
		Sphalerite	- - - -	- - - -	- - - -	- - - -
	Sulfosalt-Carbonate-	Carbonate	━━━	━━━	━━━	- - - -
		Tetrahedrite	········	········	········	········
		Boulangerite	········	········	········	
		Jamesonite	········	········		········
		Bournonite		········		········
Stibnite	Berthierite-Stibnite	Berthierite	········			
		Stibnite	- - - -	········		········
		Quartz	········	········		········
Ag-Stibnite	Ag-Stibnite	Quartz	- - - -		- - - -	
		Pyrite	········			
		Arsenopyrite	········			
		Tetrahedrite	········	········		
		Argentite		········		
		Acanthite			········	

The width of the lines represents the relative abundance of minerals.

The Au-polysulfide-quartz association is composed of sphalerite, galena, and chalcopyrite, which, in the form of small aggregates and micro-veinlets, are confined to the minerals in earlier paragenetic associations (Py-Apy-Qz metasomatic and Py-Apy-Qz vein). The intergrowths of the minerals in the Au-polysulfide-Qz association are up to the first few millimeters. Anhedral–blastic structures and solid solutions (emulsions) developed (Figure 7D). When interacting with earlier sulfides, microtextures due to corrosion (Figure 7E,F) and crack-filling are formed. Native gold formed simultaneously with the sulfides in this association, as evidenced by their joint presence and close intergrowths (Figure 7E). Common native gold developed in interstitial quartz grains or filled micro-cracks in earlier minerals. The shape of the gold particles is mainly elongated–lumpy (Figure 7G), lumpy–branched, and dendritic (Figure 7H), fractions of a millimeter to the first millimeter. The fineness of native gold is 800–900‰.

In the sulfosalt–carbonate association, the leading mineral is carbonate. It is localized in the selvages of vein bodies and near clay interlayers (Figure 6F), and it forms nest-like clusters and fills cracks and voids in minerals in earlier associations (Figure 6G). In the fractured areas, carbonate is a cementing material and forms comb structures (Figure 6H). The amount of carbonate in the veins varies from 5 to 15%. Fahlores are represented by tetrahedrite (Figure 7I). They form anhedral grains in quartz veins in association with galena, sphalerite, and chalcopyrite. Carbonate minerals contain acicular crystals and microaggregates of boulangerite and jamesonite. At the Badran and V'yun deposits, bournonite is observed in intergrowths with galena (Figure 7J).

Berthierite–stibnite-type mineralization can be identified at the Malo–Taryn and Badran deposits [21,94]. Minerals occur in the form of individual aggregates and thin veinlets (up to 2–3 cm) in the gold–quartz veins. The superposition of late berthierite–stibnite mineralization on earlier gold–quartz/gold–sulfide–quartz mineralization leads to the development of corrosion (Figure 7K). The intensively cataclazed quartz experienced volumetric dissolution, and small (1–2 mm) euhedral crystals of regenerative quartz formed (Figure 7L) [20].

Figure 6. Photographs showing representative mineral assemblages and textures of the vein minerals in the studied OGDs, central YKMB: (**A**) Milky-white quartz with a massive texture; thread-like interlayers of siltstone composition separate the bands of subhedral quartz from the anhedral quartz; (**B**) banded texture of quartz vein; (**C**) Alternation of banded and massive quartz; (**D**) breccia-like texture of quartz vein, milky-white quartz forms overgrowth textures around the rock fragments; (**E**) spherulitic and cryptogranular quartz cement vein quartz (Qz1); (**F**) Fe-dolomite forms a rhythmically-banded texture in a quartz vein; (**G**) veinlets and nests of intensely oxidized dolomite–ankerite; (**H**) quartz with siltstone xenoliths overgrown with combed dolomite–ankerite with Mg-siderite inclusions; (**I**) in milky-white quartz, a nest of arsenopyrite with dendrite-like gold in cracks. Abbreviations: Gn—galena, Sp—sphalerite, Apy—arsenopyrite, Dol—dolomite, Ank—ankerite, Sd—siderite.

Ag-stibnite-type mineralization is represented by spherulitic, cryptogranular, or collomorphic–nodular quartz with a slight dissemination of fine-grained pyrite with elevated concentrations of Au and Ag and individual crystals of arsenopyrite with an Sb admixture of up to 2–3%. Tetrahedrite, argentite, and acanthite are typomorphic minerals [21,94]. This epithermal mineralization is the latest in the YKMB [99]. It occurs at the Malo–Taryn, Badran, and Khangalas deposits.

Figure 7. Photomicrographs showing the representative mineral assemblages and microtextures of the ore minerals of each vein-type ((**A**)—backscattered electron images, (**B–L**)—reflected light) of the studied OGDs, central YKMB: (**A**) pyrite and arsenopyrite intergrowths in a quartz vein; (**B**) cataclase structures in arsenopyrite grains; cracks are healed with quartz in the Au-polysulfide-quartz paragenesis; (**C**) in cataclazed arsenopyrite aggregate, native gold developed along cracks; (**D**) sphalerite–galena–chalcopyrite aggregate with an allotriomorphic blastic structure. Chalcopyrite emulsions in sphalerite aggregates with a solid solution decomposition structure; (**E**) microtexture from corrosion and substitution of pyrite (Py) crystal caused by minerals of the Au-polysulfide-quartz paragenesis; (**G**) native elongated lumpy gold in interstitial quartz grains; (**H**) lumpy-branched native gold in microcracks; (**I**) tetrahedrite aggregates intergrown with arsenopyrite; (**J**) intergrowths of burnonite with minerals from the Au-polysulfide-quartz paragenesis; (**K**) microtextures from berthierite corrosion due to pyrite crystals; (**L**) euhedral crystals made of regenerative quartz in stibnite aggregates. Abbreviations: Gn—galena, Sp—sphalerite, Ccp—chalcopyrite, Apy—arsenopyrite, Ttr—tetrahedrite, Bnn—burnonite, Brt—berthierite, Sbn—stibnite.

3. Materials and Analytical Methods

3.1. Sample Preparation

Samples for mineralogical–geochemical and isotope–geochemical studies were collected from natural outcrops and the walls and dumps of surfaces and underground workings; 62 thin sections and 36 polished mounts were prepared for mineralogical and microtextural studies. For mineralogical, microtextural, and geochemical studies of disseminated sulfide mineralization, thick polished sections (74 in total) and epoxy-mounted grains (100 sulfide grains in 10 mounts) were prepared.

The microtextural and paragenetic features of the sulfides were studied using a Karl Zeiss Axio M1 optical microscope (Carl Zeiss AG, Jena, Germany) at the Diamond and Precious Metal Geology Institute, Siberian Branch, Russian Academy of Sciences, Yakutsk, Russia. The qualitative chemical and mineral compositions of the samples were studied with the use of a JEOL JSM-6480LV scanning electron microscope (SEM analysis, JEOL Ltd., Tokyo, Japan) equipped with an Energy 350 Oxford energy dispersive spectrometer (20 kV, 1 nA, beam diameter 1 μm) at the Diamond and Precious Metal Geology Institute, Siberian Branch, Russian Academy of Sciences (Yakutsk, Russia).

3.2. Electron Probe Micro-Analyzer (EPMA)

The major element compositions of pyrite and arsenopyrite were determined with standard X-ray spectral analysis on a Camebax-Micro microanalyzer (Cameca, Courbevoie, France) at the Diamond and Precious Metal Geology Institute, Siberian Branch, Russian Academy of Sciences (Yakutsk, Russia). The analytical conditions were as follows: accelerating voltage of 20 kV; beam current of 25 nA; measurement time of 10 s; K series for Fe, Co, Ni, Cu, and S; M series for Au and Pb; L series for As and Sb.; and wavelength-dispersive spectrometer (WDS) with LiF, PET, and TAP crystals. The standards used were FeS_2 for Fe and S, FeAsS for As, Fe-Ni-Co alloy for Co, Ni, Au-Ag alloy for Au and Ag, $CuSbS_2$ for Sb, and PbS for Pb. The detection limit was 0.01%. In each grain, 3 measurements were made: core, midway between core and rim, and rim. In total, 422 analyses were used, of which 293 were used for the first time and 129 were taken from earlier studies [11,12].

3.3. Determination of Gold and Silver Content

The Au and Ag contents were determined using powdery monomineral samples via atomic absorption spectrometry (AAS) with electrothermal atomization on a spectrometer MGA-1000 (LUMEX, St. Petersburg, Russia) at the Diamond and Precious Metal Geology Institute, Siberian Branch, Russian Academy of Sciences (Yakutsk, Russia). The detection limit for gold is 0.02 ppm. In total, 71 analyses were used, of which 34 were used for the first time and 37 taken were from earlier studies [11,12].

3.4. S isotope Analysis

The sulfur isotope composition was analyzed at the Laboratory of Stable Isotopes, Center for Collective Use, Far East Geological Institute, Far East Branch, Russian Academy of Sciences (Vladivostok, Russia). For sulfur isotope analysis of sulfides from the Malo–Taryn, Khangalas, V'yun, and Shumnyi deposits, we used pure sulfide fractions (54 samples) selected by hand and ground into powder. In total, 75 analyses were used, of which 30 were used for the first time and 45 were from earlier studies [11,12]. The analysis was performed using a Flash EA-1112 elemental analyzer (Thermo Scientific, Dreieich, Germany) in the S configuration according to the standard protocol for converting sulfur from sulfide to SO_2. The sample preparation for mass spectrometric sulfur isotope analysis from the Badran deposit (20 samples) was carried out with the local laser method using an NWR Femto femtosecond laser ablation complex [100,101]. The $^{34}S/^{32}S$ isotope ratios were measured on a MAT-253 mass spectrometer (Thermo Fisher Scientific, Germany) in continuous He flux mode. The measurements were performed against a standard laboratory gas, SO_2, calibrated according to international standards IAEA-S-1, IAEA-S-2, IAEA-S-3, and NBS-

127. The results of the $\delta^{34}S$ measurements are provided in reference to the international VCDT standard. Determination accuracy: $\delta^{34}S \pm 0.2‰$ (1σ).

4. Results

4.1. Pyrite and Arsenopyrite Types and Textures

Our [12,26,78] and other works [10,102], beyond any known orebodies and in alteration of the OGDs in the central YKMB, have identified several generations of pyrite (Py1, Py2, Py3) and arsenopyrite (Apy1); a summary of the common textures and pyrite–arsenopyrite classifications for the central YKMB is shown in Table 3. Pyrite is the main and most common ore mineral in sedimentary strata, with diagenetic (Py1), metamorphic (Py2), and metasomatic (Py3) pyrites identified.

Table 3. Summary of common textures and pyrite–arsenopyrite classification OGDs Central YKMB.

Pyrite	Timing	Host Rock	Structure/Texture	Co-Genetic Minerals	Evidence for Timing
Py1	Syn-sedimentary/ diagenetic	Sandstones, siltstones	Framboids, nodular aggregates	Detrital quartz	Along the layering
Py2	Metamorphic	Sandstones, siltstones	Euhedral–subhedral; zoning structure; corroded structure	Sericite, carbonate	Fault zones
Py3	Metasomatic	Sandstones, siltstones, andesite, dacite	Euhedral–subhedral; fine-grained inclusions of galena, sphalerite, chalco-pyrite; zoning structure; corroded structure	Arsenopyrite, sericite, carbonate	Proximal alterations
Apy1	Metasomatic	Sandstones, siltstones, andesite, dacite	Euhedral–subhedral; fine-grained inclusions of galena, sphalerite, chalco-pyrite; zoning structure; corroded structure	Pyrite, sericite, carbonate	Proximal alterations

4.1.1. Syn-Sedimentary/Diagenetic Pyrite (Py1)

The earliest form of pyrite is observed in the Upper Paleozoic clastic sedimentary rocks (Figure 8). Pyrite1 is represented by dust-like and fine-grained spherical (framboidal) particles (Figure 8A) and nodular aggregates ranging in size from 10 to 100 microns (Figure 8B). They are composed of pyrite microcrystals with a zonal structure in a carboniferous–silicon matrix. Nodular aggregates often have a porous texture, which suggests the formation of minerals as a result of rapid crystallization [33]. In pyrite1, marcasite can be observed as anhedral-to-colloform grains. The location of Py1 follows the primary sedimentary texture of rocks (Figure 8C).

4.1.2. Metamorphic Pyrite2

Pyrite2 can be most widely observed in the zones of regional faults (Adycha–Taryn, Charky–Indigirka, Chai–Yureya). It is represented by crystals with subhedral and euhedral shapes, ranging in size from 2–3 microns to 2–4 mm, which form dissemination (Figure 9A), intergrowths, nests (Figure 9B), and thin veinlets (Figure 9C). In individual crystals and aggregates, a xenomorphic porous central part can be observed, formed during diagenesis, surrounded by an idiomorphic shell of metamorphogenic Py2 (Figure 9).

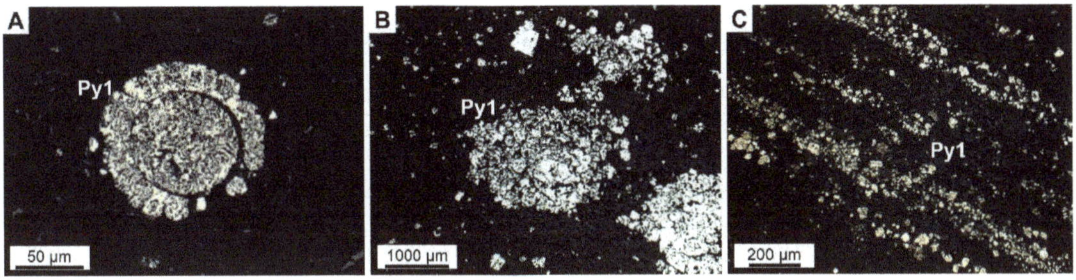

Figure 8. Microtexture of diagenetic Py1 (reflected light) OGDs, central YKMB: (**A**) framboids and zonal structure; (**B**) nodular aggregate pyrite and marcasite as anhedral-to-colloform grains; (**C**) layered arrangement of Py1 crystals in siltstone.

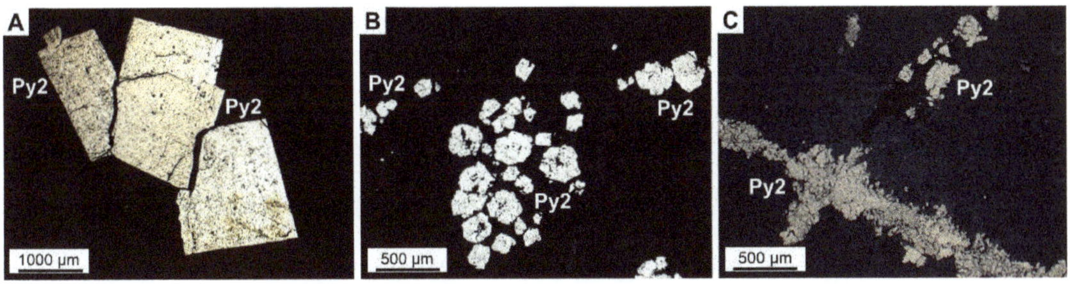

Figure 9. Microtextures of metamorphic pyrite2 (reflected light) OGDs, central YKMB: (**A**) intergrowths of cubic pyrite; (**B**) accumulation of crystals and pyrite intergrowths. The central part of the individual grains is corroded; (**C**) micro-veinlets of pyrite along cracks.

4.1.3. Hydrothermal Pyrite3 and Arsenopyrite1 from Proximal Alteration

Sulfides (Py3 and Apy1) are localized in the proximal sericite-carbonate–quartz alteration and make a major contribution to the economic value of the veinlet-disseminated mineralization containing "invisible" gold. A proximal alteration is characterized by various types of transformation in rocks. Sericitization, carbonation, and silicification developed in the clastic rocks (sandstones, siltstones). Carbonation, chloritization, and argillization manifested in the dykes of basic, intermediate, and felsic composition. The hydrothermal alteration of both clastic and intrusive rocks involves pyrite–arsenopyrite sulfidation. Py3 is cubic and pentagondodecahedral in shape and often has a zonal structure (Figure 10). The second typomorphic mineral of the association is arsenopyrite (Apy1) with a short-prismatic and pseudopyramidal form (Figure 10A). The size of Apy1 crystals ranges from fractions up to 1–1.5 mm or, less often, up to 2–3 mm. Microinclusions in the minerals of the Au-polysulfide-quartz and sulfosalt–carbonate associations can be noted in Py3 and Apy1 crystals (Figure 10B). These minerals are confined to pores, voids, and microcracks, as well as the growth zones of pyrite and arsenopyrite crystals. A rod-like quartz–carbonate rim is often formed at the edges of sulfide crystals (Figure 10C). The content of disseminated sulfide mineralization in proximal alteration can be up to 6–8%. Usually, pyrite prevails at a distance from quartz veins/veinlets, and arsenopyrite prevails near them.

4.2. Chemical Composition of Pyrite and Arsenopyrite

The EPMA analyses revealed distinct trace-element abundances in the various generations of pyrite and arsenopyrite (for the full dataset, see Supplementary Tables S1 and S2).

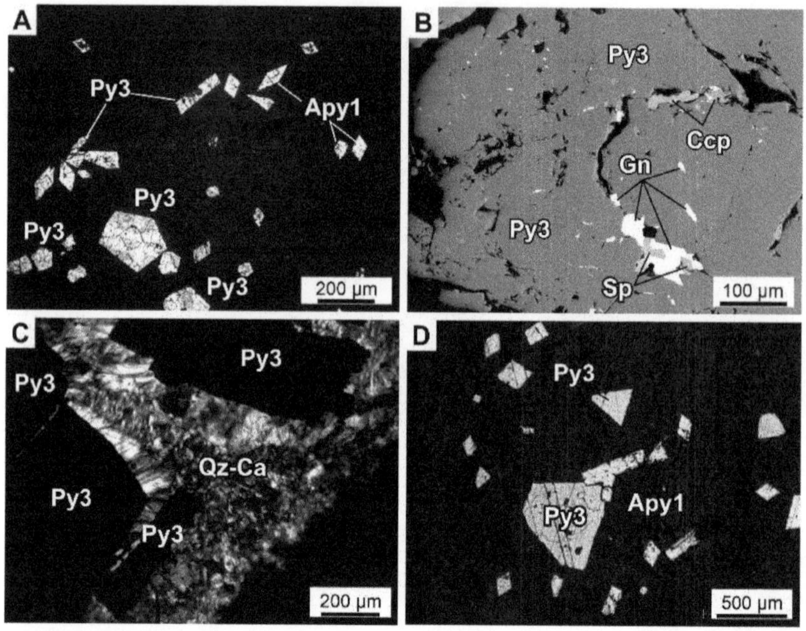

Figure 10. Microtextures of alteration pyrite3 and arsenopyrite1 from proximal alteration rock sediment-hosted OGDs (**A–C**) and from proximal alteration rock intrusion-hosted OGDs (**D**) ((**A**,**C**,**D**) reflected light and (**B**) backscattered electron images), central YKMB. (**A**) Dissemination of Py3 and Apy1: pentagondodecahedral shape and zoning are characteristic of individual large pyrite crystals; (**B**) microinclusions of galena (gn), chalcopyrite (Ccp), and sphalerite (Sp) in pores, voids, and microcracks of pyrite crystals; (**C**) rod-like rims of quartz–carbonate composition; (**D**) dissemination of the alteration Py3 and Apy1 in a dyke.

4.2.1. Chemical Composition of Pyrite in Sediment-Hosted Orogenic Gold Deposits

Most of the analyzed pyrites of the sediment-hosted orogenic gold deposits have a non-stoichiometric composition (in 63% of analyses, $Fe/(S + As) \neq 0.5$) and the trace elements As, Co, Ni, Cu, and Sb, as well as, less often, Pb (Supplementary Table S1). Pyrite is characterized by a deficit of sulfur (in 75–95% of analyses, $S/Fe < 2.0$ at $Cs = 51.16–53.82$ wt%). Concentrations of trace elements vary within wide limits; the values of the coefficients of variation indicate the heterogeneity of the sample ($V\sigma$ from 38% to 138%).

Arsenian Py is an indicator. In the metamorphic Py2, $C_{As} \leq 0.32$ wt%. In alteration Py3, the main trace element is As (from 0.31 to 3.16 wt%) (Figure 11A). For the majority (70%) of the analyzed grains, $C_{As} < 1.5$ wt%. The zonal distribution of As in pyrite crystals is typical. Core (C_{As} to 3.08 wt%), intermediate ($C_{As} < 2.0$ wt%), and rim (C_{As} to 2.20 wt%) zones were found.

Cobalt, Ni, Cu, Sb, and Pb are typomorphic trace elements in Py3 (Supplementary Table S1). Other elements are present in quantities below the detection limit of EPMA analyses. The total trace-element content varies from 0.01 to 0.55 wt%; the entire spectrum of elements can be found in 20% of crystals. Py3 Σ (Co, Ni, Cu, Sb, Pb) < 0.15 wt% prevails (Figure 11B). There is a decrease in the amount of trace-element content in Py3 compared with Py1 and Py2 (Figure 11B). The pyrites of the Badran deposits are characterized by an increase in the concentration of Sb, and at the Malo–Taryn deposit, the Sb is below the detection limit (Figure 12E). In Py3, the common trace element is Co (from 0.01 to 0.22 wt%; in 90% of analyses, $C_{Co} < 0.1$ wt%). In zonal crystals, the maximum concentration of Co is characteristic of core Py3. Ni and Cu are present in 50% of the analyzed grains in quantities exceeding the detection limit. Ni (up to 0.46 wt%) can be observed in the rim Py3. There is no zoning in the distribution of Cu. The Py3 of the Khangalas deposit is characterized by Pb (to 0.11 wt%) in the core zone. The general formula of pyrite in the sediment-hosted orogenic gold deposit alteration (Py3) pyrite is $Fe_{0.98–1.08}(Ni_{0.0–0.01}Co_{0.0–0.01})S_{1.95–2.00}As_{0.01–0.05}$.

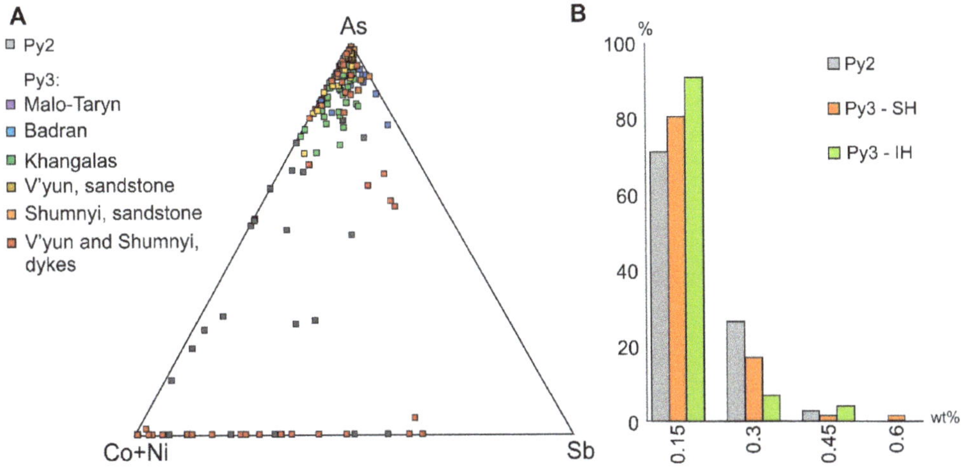

Figure 11. Distribution of trace elements in pyrites of different generations in the studied OGDs, central YKMB: (**A**) ternary diagram As–Co+Ni–Sb for Py2 and Py3; (**B**) content of Σ (Co, Ni, Cu, Sb, Pb).

4.2.2. Chemical Composition of Pyrite in Intrusion-Hosted Orogenic Gold Deposits

Pyrite in the intrusion-hosted orogenic gold deposits is analyzed both in the host clastic rocks and in dykes of intermediate composition. Most of the pyrite analyzed had a nonstoichiometric composition (in 57% of analyses, $Fe/S + As \neq 0.5$) and they showed a deficit of sulfur (in 75–95% of analyses, $S/Fe < 2.0$ at $C_S = 51.01–54.08$ wt%) (Supplementary Table S1). The concentrations of trace elements vary within a large range; the values of the coefficient of variation indicate the heterogeneity of the samples ($V\sigma$ from 42 to 227%). The distributions of trace-element concentrations in Py3 from clastic rocks and from dykes are comparable, but a number of features can be observed.

In the alteration Py3 of the clastic rocks, As is the main trace element (from 0.30 to 3.16 wt%) (Figure 12A). Concentrations of $C_{As} < 0.5$ wt% are only in 14% of grains, for 60% $C_{As} > 1.0$ wt%. High As contents are recorded in individual Py3 grains from dykes ($C_{As} = 0.57–2.02$ wt%); 34% of Py3 have $C_{As} = 0.01–1.00$ wt%, and 44% have non-arsenian varieties (Supplementary Table S1).

Cobalt, Ni, Cu, and Sb are typomorphic trace elements in Py3 from clastic rocks (Figure 12). The remaining trace elements are present in quantities below the detection limit of the EPMA analysis. The total content of these elements varies from 0.01 to 0.42 wt%, but the whole spectrum of elements is fixed in single crystals. Py3 grains of Σ (Co, Ni, Cu, Sb) < 0.15 wt% prevail (Figure 11B). Co, Ni, and Cu are typomorphic trace elements in Py3 from dykes. Antimony is determined in Py3 from some dykes; it is characterized by increased concentrations (to 0.10 wt%,) (Figure 12E). The total content of these elements varies widely from 0.003 to 3.97 wt%, but the whole spectrum of elements is fixed in single crystals. Elevated concentrations of Ni and Co can be observed in individual crystals ($C_{Ni} = 3.52$ wt%; $C_{Cu} =$ to 2.31 wt%) (Supplementary Table S1). The remaining trace elements are present in quantities below the detection limit of EPMA analyses.

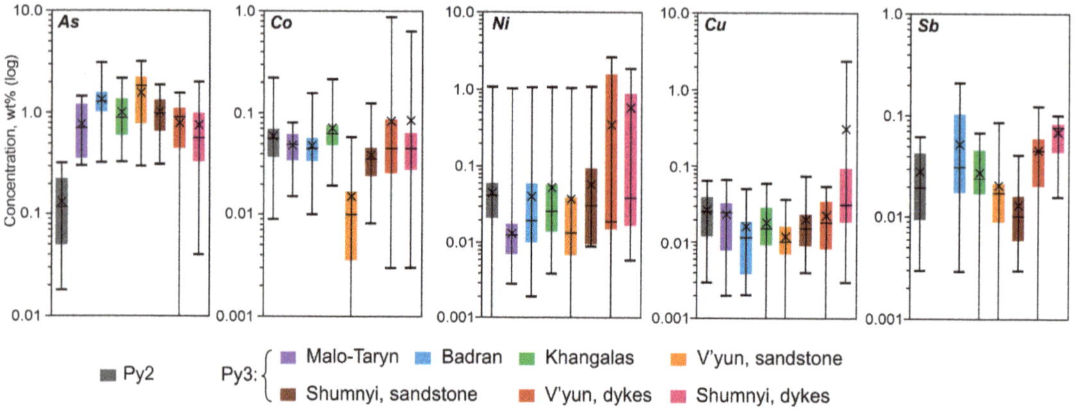

Figure 12. Boxplots showing the compositional difference in Py1-2 and Py3 acquired using an EPMA analysis of the studied OGDs, central YKMB. The lower border of the line shows the minimum value; the upper shows the maximum value. Horizontal lines in the boxes denote the median, the X marks in the boxes denote the mean, and the bottom and top of each box denote the first and third quartiles, respectively. Concentrations are on a logarithmic scale.

4.2.3. Chemical Composition of Arsenopyrite in Sediment-Hosted Orogenic Gold Deposits

Most of the analyzed Apy1 has a nonstoichiometric composition (in 72% of analyses, $Fe/(S + As) \neq 0.5$) and is rich in sulfur (As/S from 0.77 to 0.99). In some cases, the ratio is $As/S > 1.0$. For Apy1, the trace elements Co, Ni, Cu, and Sb are typomorphic (Supplementary Table S2, Figure 13). Other elements are present in quantities below the detection limit of EPMA analyses.

The concentrations of typomorphic elements vary significantly, and in most of the analyzed grains, Σ (Co, Ni, Cu, Sb) is no more than 0.15 wt% (Figures 13A and 14). The ratios of trace elements are individual for different deposits (Figure 13B). Thus, at the Malo–Taryn deposit, Sb accounts for 60% of the trace elements, and at the Khangalas deposit, 45% is Co. Antimony is found in all the analyzed grains; its content is not stable ($V\sigma$ up to 124%), but this is the main trace element in Apy1.

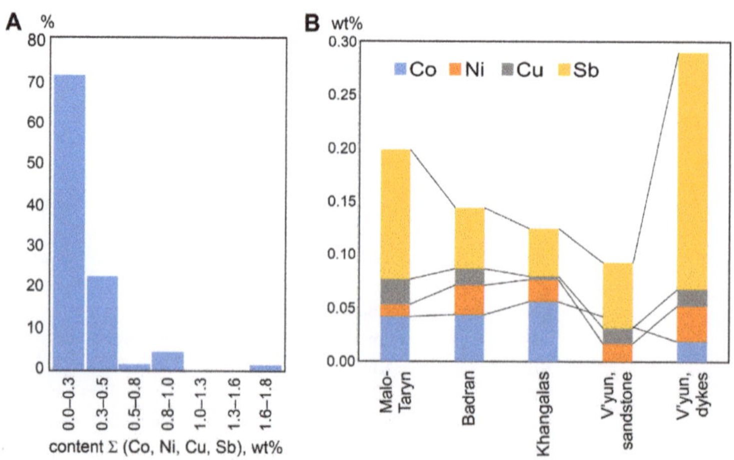

Figure 13. Distribution of trace elements in arsenopyrite1 of the studied OGDs, central YKMB: (**A**) content of Σ (Co, Ni, Cu, Sb); (**B**) the ratio of individual trace elements in the studied deposits.

4.2.4. Chemical Composition of Arsenopyrite in the Intrusion-Hosted Orogenic Gold Deposits

Arsenopyrite1 in intrusion-hosted orogenic gold deposits is rare. Most of the analyzed arsenopyrites from the dykes of the V'yun deposit have a nonstoichiometric composition (in 67% of analyses, Fe/(S + As) $\neq$ 0.5) and are rich in sulfur (As/S from 0.75 to 0.94). For Apy1 intrusion-hosted orogenic gold deposits, the typomorphic element is Sb (Supplementary Table S2), averaging 85% of the total volume of the trace elements. In some samples, the Sb content reaches 1.03–1.8 wt%. Ni and Cu were determined in 65% of the analyzed grains; Co in the Apy1 of the dykes is present within C_{Co} = 0.01–0.07 wt% and in clastic rocks in quantities below the detection limit of EPMA analyses.

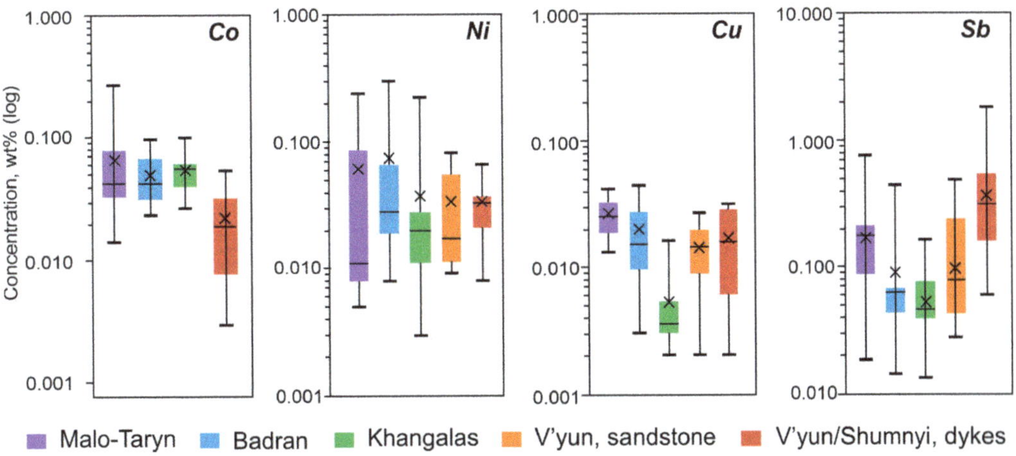

Figure 14. Boxplots showing the compositional difference in the arsenopyrite1 of the studied OGDs, central YKMB, in this study acquired by EPMA. The lower border of the line shows the minimum value; the upper one shows the maximum value. Horizontal lines in the boxes denote the median, the X marks in the boxes denote the mean, and the bottom and top of each box denote the first and third quartiles, respectively. Concentrations are on a logarithmic scale.

4.3. Gold and Silver Content of Sulfides from Proximal Alteration according to AAS Data

Data on the gold and silver content in pyrites and arsenopyrites from the proximal alteration are shown in Table 4 and illustrated in Figure 15.

Table 4. Results of atomic absorption atomic spectrometry (ppm) of sulfides from the proximal alteration of the studied OGDs, central YKMB.

Rock	Mineral	Au, ppm	Ag, ppm	Au/Ag	N	References
Malo–Taryn deposit						
Alteration rock after sandstones	Pyrite	$\frac{0.4-10.1}{5.1/5.4}$ *	$\frac{0.9-8.1}{5.4/6.3}$	$\frac{1.2-5.2}{2.2/1.2}$	6	This study, and [21]
	Arsenopyrite	$\frac{5.0-28.1}{17.1/17.8}$	$\frac{0.1-10.2}{4.2/3.3}$	$\frac{0.9-191.6}{55.7/15.1}$	4	
Badran deposit						
Alteration rock after sandstones and siltstones	Pyrite	$\frac{13.7-155.5}{57.5/41.1}$	$\frac{1.2-16.9}{6.0/3.6}$	$\frac{1.2-83.4}{17.0/9.9}$	12	[11]
	Arsenopyrite	$\frac{34.8-168.5}{66.9/54.6}$	$\frac{1.3-12.6}{4.9/2.9}$	$\frac{9.5-42.6}{21.8/17.5}$	9	

Table 4. *Cont.*

Rock	Mineral	Au, ppm	Ag, ppm	Au/Ag	N	References
		Khangalas deposit				
Alteration rock after sandstones and siltstones	Pyrite	0.8–39.3 11.2/7.4	1.1–17.4 6.2/6.1	0.5–10.9 2.6/1.8	13	This study, and [12]
	Arsenopyrite	12.3–23.8 17.5/16.4	0.4–11.8 6.5/7.2	1.4–28.6 11.1/3.3	3	
		V'yun deposit				
Alteration rock after dykes	Pyrite	0.3–25.8 5.0/0.7	0.9–2.0 1.3/1.1	0.3–16.0 3.1/0.6	6	This study
Alteration rock after sandstones and siltstones	Pyrite	0.3–159.5 35.6/14.4	0.4–23.7 5.6/3.0	0.2–30.5 6.0/2.3	8	
Alteration rock after sandstones and siltstones	Arsenopyrite	28.9, 58.4	2.2, 2.3	12.8, 26.8	2	
		Shumnyi deposit				
Alteration rock after dykes	Pyrite	9.8–53.9 28.8/25.8	0.6–2.8 2.1/2.6	4.0–40.9 18.7/14.9	4	This study
Alteration rock after sandstones	Pyrite	2.3–38.4 13.2/6.0	0.3–2.0 1.1/1.0	3.9–23.1 12.7/12.0	4	

* Minimum–Maximum / Mean/Median.

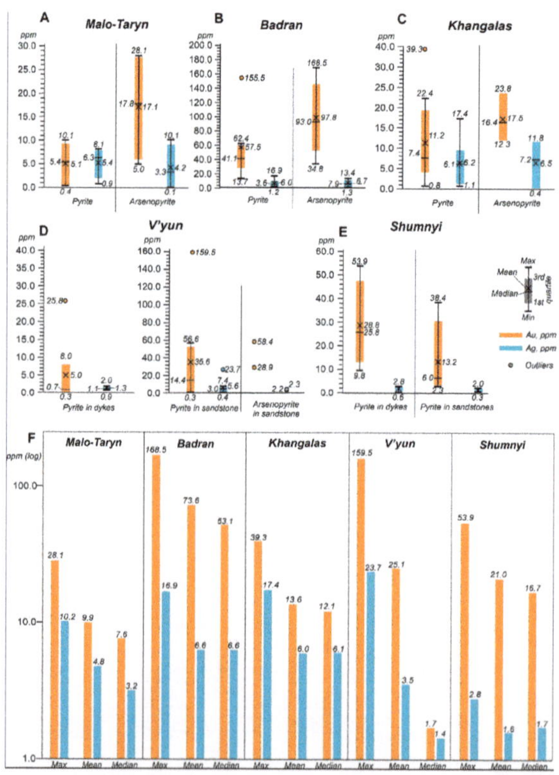

Figure 15. The gold and silver content in sulfides from the alteration of the studied OGDs, central YKMB: (**A**) Malo–Taryn; (**B**) Badran; (**C**) Khangalas; (**D**) V'yun; (**E**) Shumnyi; (**F**) comparative diagrams between the maximum, mean, and median values of the gold and silver content in sulfides from the alteration.

4.4. Sulfur Isotopic Composition of Sulfides

The $\delta^{34}S$ isotopic composition of pyrite and arsenopyrite from the Malo–Taryn, Badran, and Khangalas proximal alterations in the sediment-hosted deposits and the V'yun and Shumnyi proximal alterations in the intrusion-hosted deposit of the central YKMB is shown in Table 5.

Table 5. Sulfur isotope data of pyrite and arsenopyrite from proximal alteration rock sediment-hosted deposits and intrusion-hosted deposits of the studied OGDs, central YKMB.

№	Sample	Deposit	Mineral	Rock	$d^{34}S_{VCDT}$,‰	Reference
1	MT-66-16	Malo–Taryn	Arsenopyrite	Alteration rock after sandstones	−1.4	This study
2	Y-2012/1				−1.6	
3	M-1-16		Pyrite	Alteration rock after siltstones	−2.3	
4	MT-76-16			Alteration rock after sandstones	−5.5	
5	Y-2012/2				1.4	
6	B-24/2-19	Badran *	Arsenopyrite	Alteration rock after siltstones	−0.3/−0.2	[11]
7	B-16-19				0.0/−0.3	
8	B-35/2-19				−0.7/−0.8	
9	B-10/2-19			Alteration rock after sandstones	−0.3/−0.3	
10	B-54/2-19				−0.5/−0.2	
11	B-14/3-19				−0.4/0.1	
12	B-56/2-19				−1.1/−0.6	
13	B-40-19				−0.4/−0.3	
14	B-24/1-19		Pyrite	Alteration rock after siltstones	0.4/1.6	
15	B-44-19				0.3/0.3	
16	B-17-19				1.0/1.2	
17	B-26-19				0.8/1.1	
18	B-35/1-19				−0.7/1.4	
19	B-10/1-19			Alteration rock after sandstones	0.1/1.8	
20	B-54/1-19				0.5/1.8	
21	B-14/1-19				0.6/1.9	
22	B-51-19				1.0/0.7	
23	B-52-19				1.1/1.0	
24	B-41-19				−0.2/−0.3	
25	B-56/1-19				1.5/1.5	
26	B-33-19				−0.5/1.0	

Table 5. *Cont.*

№	Sample	Deposit	Mineral	Rock	$d^{34}S_{VCDT}$,‰	Reference
27	KG-9-19	Khangalas	Arsenopyrite	Alteration rock after sandstones	−1.4	[12]
28	K-4-17				−1.2	
29	KG-26-19				−1.1	This study
30	KG-29-19				−2.1	
31	KG-8-19		Pyrite	Alteration rock after siltstones	−0.8	
32	KG-20-19			Alteration rock after sandstone and siltstones	−1.0	
33	KG-32-19			Alteration rock after sandstones	−1.3	[12]
34	K-9-17/1				−0.6	
35	K-9-17/2				−1.5	This study
36	K-4-17				−1.5	
37	K-14-17				−1.9	
38	V-14-18	V'yun	Pyrite	Alteration rock after dykes	−1.9	This study
39	V-15-18				3.1	
40	V-20-18				−6.4	
41	V-22-18				3.1	
42	V-43-18				−4.6	
43	V-140-18				−4.7	
44	VF-24-18		Arsenopyrite	Alteration rock after sandstones	4.4	
45	VF-27-18		Pyrite	Alteration rock after sandstone and siltstones	3.7	
46	VF-27-18				4.4	
47	VZ-158-18			Alteration rock after sandstones	5.6	
48	V-162-18				2.3	
49	S-42-18/1	Shumnyi	Pyrite	Alteration rock after dykes	2.1	This study
50	S-42-18/2				2.5	
51	SU-22-18				2.4	
52	S-113-18/1				4.8	
53	S-113-18/2				5.1	
54	S-17-18			Alteration rock after sandstones	4.3	
55	S-112-18				5.0	

* The values of the S isotopic composition of pyrite and arsenopyrite for the Badran deposit were determined using local measurement methods at the periphery (denominator) and center (numerator).

4.4.1. Sediment-Hosted Orogenic Gold Deposits

In three analyses of Py grains from the alteration of the Malo–Taryn deposit, $\delta^{34}S$ has a range from −5.5 to +1.4‰; two analyses of Apy grains provided a restricted $\delta^{34}S$ range—−1.6 and −1.4‰. The $\delta^{34}S$ values of Py and Apy from the alteration sampled at various depths (from 587 to 916 m) of the Badran deposit have a narrow range of values, from −1.1 to +1.9‰. The highest variations of $\delta^{34}S$ are determined in 26 analyses of Py (from −0.7 to +1.9‰, mean +0.8‰, median +1.0‰), and 16 analyses yielded a $\delta^{34}S$ Apy range of −1.1 to +0.1‰ (mean −0.4‰, median −0.3‰). The sulfur $\delta^{34}S$ isotopic composition of sulfides from the alteration of the Khangalas deposit is characterized by a narrow range of negative $\delta^{34}S$ values, from −2.1 to −0.6‰, according to seven analyses of Py (from −1.9

to −0.6‰, mean −1.2‰, median −1.3‰) and four analyses of Apy (from −2.1 to −1.1‰, mean −1.5‰, median −1.3‰).

4.4.2. Intrusion-Hosted Orogenic Gold Deposits

At the V'yun deposit, there is a heterogeneous $\delta^{34}S$ in the sulfides. The highest variations in six analyses of $\delta^{34}S$ are determined in Py from dykes (from −6.4 to +3.1‰, mean −1.9‰, median −3.3‰). The highest $\delta^{34}S$ values are determined in Py (from +2.3 to +5.6‰, mean +4.0‰, median +4.1‰) and Apy (+4.4‰) from clastic rocks. Seven analyses of $\delta^{34}S$ in Py from the alteration of the Shumnyi deposit yielded positive values from +2.1 to +5.1‰. There are no evident isotopic variations between $\delta^{34}S$ in Py from sandstones (from +4.3 to +5.0‰) or from dykes (from +2.1 to +5.1‰, mean +3.4‰, median +2.5‰).

5. Discussion

5.1. Composition of Pyrite3 and Arsenopyrite1

5.1.1. Incorporation of Metals and Metalloids in Pyrite3

Microelements in pyrite can be in the form of isomorphic trace elements and in the form of micro- and nanoinclusions. Concentrations of trace elements and their correlations with the major elements indicate the form and conditions of their accumulation in pyrite, as well as the evolution of the ore formation conditions.

Arsenic is the most informative in the composition of pyrite. A number of researchers, analyzing changes in the shape and concentration of As in pyrite from gold deposits of various types, suggest using As to explain the evolution of the hydrothermal system [103]. The chemical zoning of As in pyrites may be the result of physicochemical changes in the composition of the ore-forming fluid [104–108] and the effects of magmatic fluids (magmatic vapor plumes) [105].

It was mentioned in Section 4.2 that the Py3 from sediment-hosted (Malo–Taryn, Badran, Khangalas) and intrusion-hosted (V'yun, Shumnyi) gold deposits are characterized by nonstoichiometric compositions. The S/Fe ratio varies in individual deposits within sufficiently large ranges (V'yun, S/Fe = 1.87–2.04; Badran, S/Fe = 1.88–2.09). According to the obtained results from the EPMA analysis, Py3 differs from the calculated values (Fe = 46.55 wt% and S = 53.45 wt%) (Figure 16) and indicates the presence of vacant positions in the structure of pyrite, which are occupied by trace elements. The As-Fe-S ternary diagram demonstrates the possible variants of the occurrence of microelements in the composition of pyrite (Figure 16) [104]. Mainly, As isomorphically replaces S (As^{1-}) (Figure 16), but it can replace Fe in the crystal lattice Fe (As^{2+} and As^{3+}) [109,110] or may occur in the form of nanoinclusions (As^0) [104]. The isomorphic substitution of Fe is the main mechanism for the inclusion of Co, Ni, Cu, Sb, and Pb (Me^{2+}) in the composition of pyrite [33].

The negative trend between concentrations of elements also confirms the isomorphic substitutions. This is most clearly manifested for the S-As pair (Figure 17A): elevated As contents are characteristic of pyrites with a deficit of sulfur; for the entire sample population, $r_{S\text{-}As} = -0.68$, and it increases in individual objects (Malo–Taryn, $r_{S\text{-}As} = -0.77$; Vyun dykes, $r_{S\text{-}As} = -0.80$). The correlations between Fe, S, and other trace elements are variable and are better expressed in individual objects (Malo–Taryn, $r_{S\text{-}Cu} = -0.54$; Shumnyi, $r_{Fe\text{-}Cu} = -0.65$) (Figure 17). Pyrites from igneous rocks are characterized by a negative correlation for the pairs $Fe^{2+} \rightarrow Co^{2+}$ and $Fe^{2+} \rightarrow Ni^{2+}$ ($r = -0.7$–-0.9) (Figure 17C,D). The similar incorporation of metals and metalloids in pyrite3 from sediment-hosted (Malo–Taryn, Badran, Khangalas) and intrusion-hosted (V'yun, Shumnyi) gold deposits indicates their unified nature.

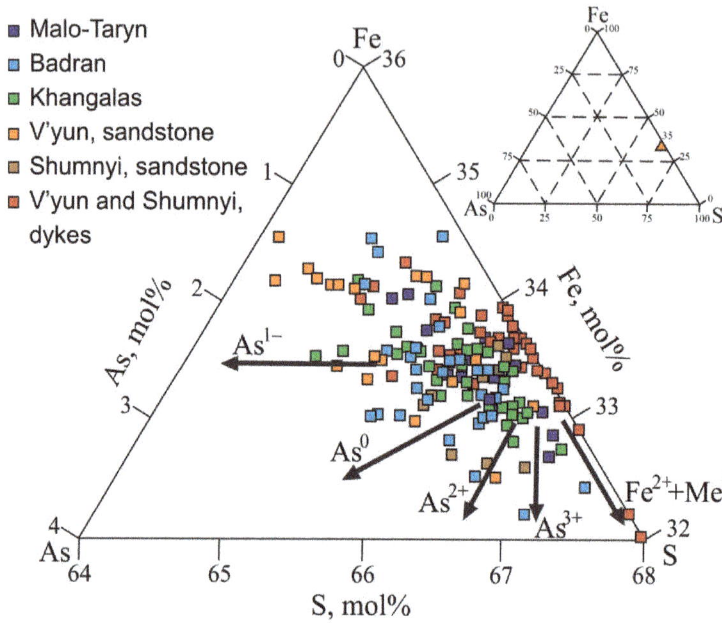

Figure 16. Ternary diagram showing the As-Fe-S composition of pyrite in the studied OGDs, central YKMB. The demarcated fields are after [104].

5.1.2. Incorporation of Metals and Metalloids in Arsenopyrite1

Arsenopyrite is a three-component system and the main isomorphic substitutions occur according to the scheme $S^{1-} \rightarrow As^{1-}$ (r = 0.92–0.99) (Figure 18A) and $Fe^{2+}(Fe^{3+}) \rightarrow As^{2+}(As^{3+})$ (Figure 18B) (r = 0.65–0.89). In the studied Apy1 grains, Fe is characterized by small variations in the content; in 75% of the analyzed grains, C_{Apy} = 33.8–34.8 wt%, which, in the mean, corresponds to stoichiometry. The sulfur content in all grains exceeds the stoichiometric composition (19.7 wt%), which varies between 19–24 wt%, with a mean content of C_S = 21.3 wt%. The As content is less than the stoichiometric content (46.0 wt% with a mean content). Variations in the concentration of S and As are different in the individual deposits (Figure 18C). Arsenopyrite of the Malo–Taryn deposit is closest to the stoichiometric composition (Figure 18 C,D).

A diagram of S/As and (S+As)/Fe [111] clearly reflects both the nonstoichiometry and variability of the compositions, as well as the increased sulfur content of Apy3 in the studied deposits (Figure 18). The deficit of Fe and As is compensated by the trace elements (Co, Ni, Cu, Sb). They can occupy both anionic and cationic vacant positions in the crystal structure, and their distribution in arsenopyrite is less ordered than in pyrite. Slight and moderate negative correlations prevail between the main elements and trace elements (Figure 19). The connection between As and Sb is most clearly manifested (r = −0.45–−0.61) (Figure 19L). For alteration arsenopyrites with nonstoichiometric composition and a predominance (excess) of S, many researchers [39,112,113] specify the increased concentrations of Au. The following trend can also be observed in the deposits studied by us (Figure 18D). Significant variations in the S, As, and Fe and trace-element content in the arsenopyrite also indicate the crystallization of the mineral in an inhomogeneous temperature field [111]. This is reflected in the zonal distribution of both trace elements and the gold content of Apy3.

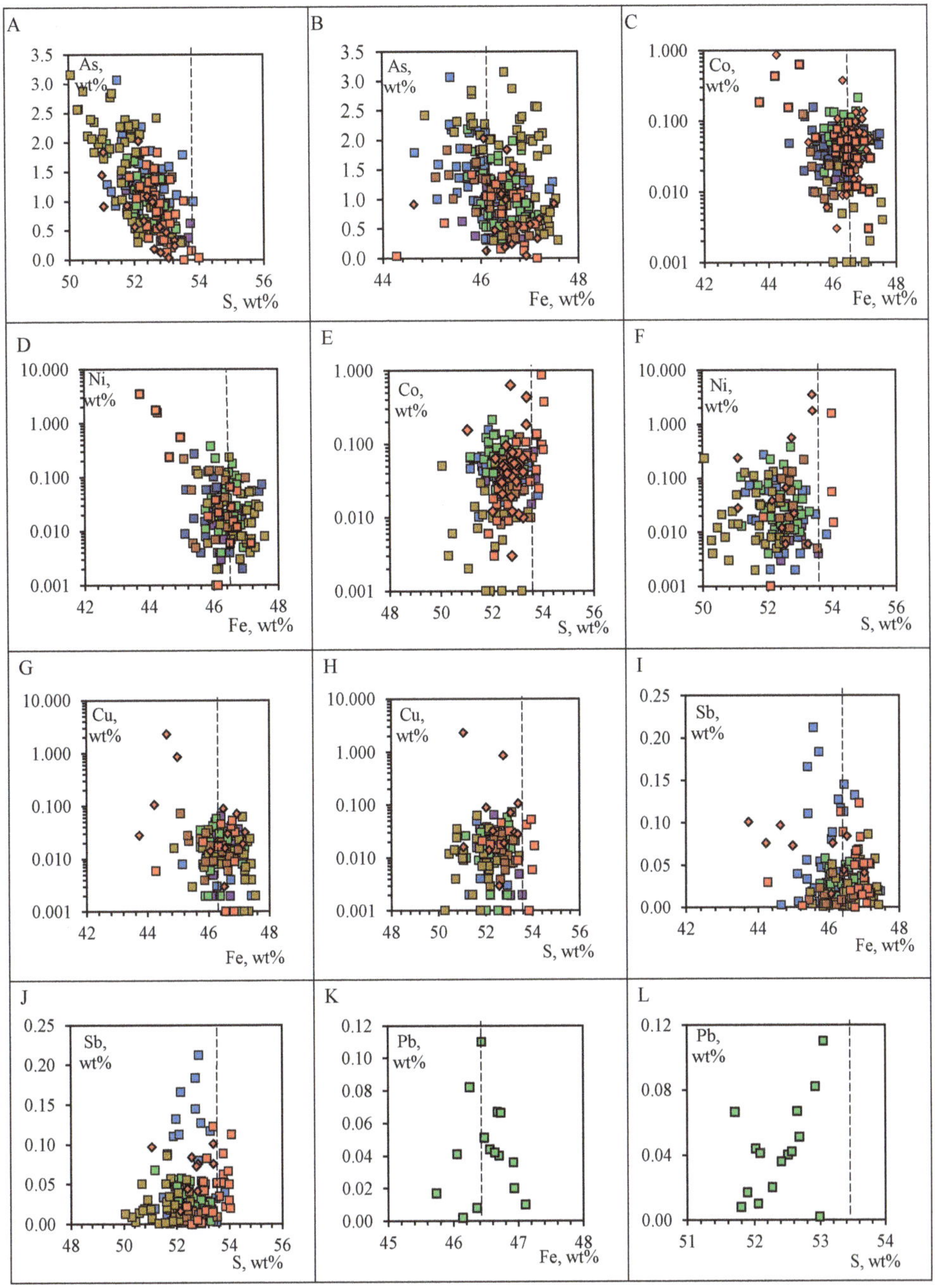

Figure 17. Diagrams of the ratios of concentrations of Fe, S, and trace elements in Py3 in the studied OGDs, central YKMB. (**A**) S vs. As; (**B**) Fe vs. As; (**C**) Fe vs. Co; (**D**) Fe vs. Ni; (**E**) S vs. Co; (**F**) Fe vs. Ni; (**G**) Fe vs. Cu; (**H**) S vs. Cu; (**I**) Fe vs. Co; (**J**) S vs. Cu; (**K**) Fe vs. Pb; (**L**) S vs. Pb. Negative correlation indicates isomorphic substitution. The lines Fe = 46.547 wt% and S = 53.453 wt% correspond to the stoichiometric composition of pyrite. See symbols in Figure 16.

5.1.3. Gold–Cobalt–Nickel Relationships in Pyrite3

The analysis of the Co and Ni content and the ratios of these elements is important for determining the conditions of the formation and genesis of pyrite [29,33,106,114–116]. Py3 is characterized by S/Fe ≠ 2.00 (Figure 20A). In most analyses, pyrites are depleted in sulfur, the deficit of which is compensated by As. Co and Ni in the pyrite structure occur mainly as isomorphic replacements of the types $Fe^{2+} \rightarrow Co^{2+}$ and $Fe^{2+} \rightarrow Ni^{2+}$ [117]. Elevated concentrations of Co and Ni are characteristic of pyrites with Fe content of <46.547 wt%, which corresponds to the stoichiometric composition (Figure 17C,D). The deficit of Fe indicates the presence of cationic vacant positions in the pyrite structure, which are filled with Co and Ni [118]. In grains with high iron content, Co and Ni are part of the pyrite structure and are incorporated there via emplacement isomorphism.

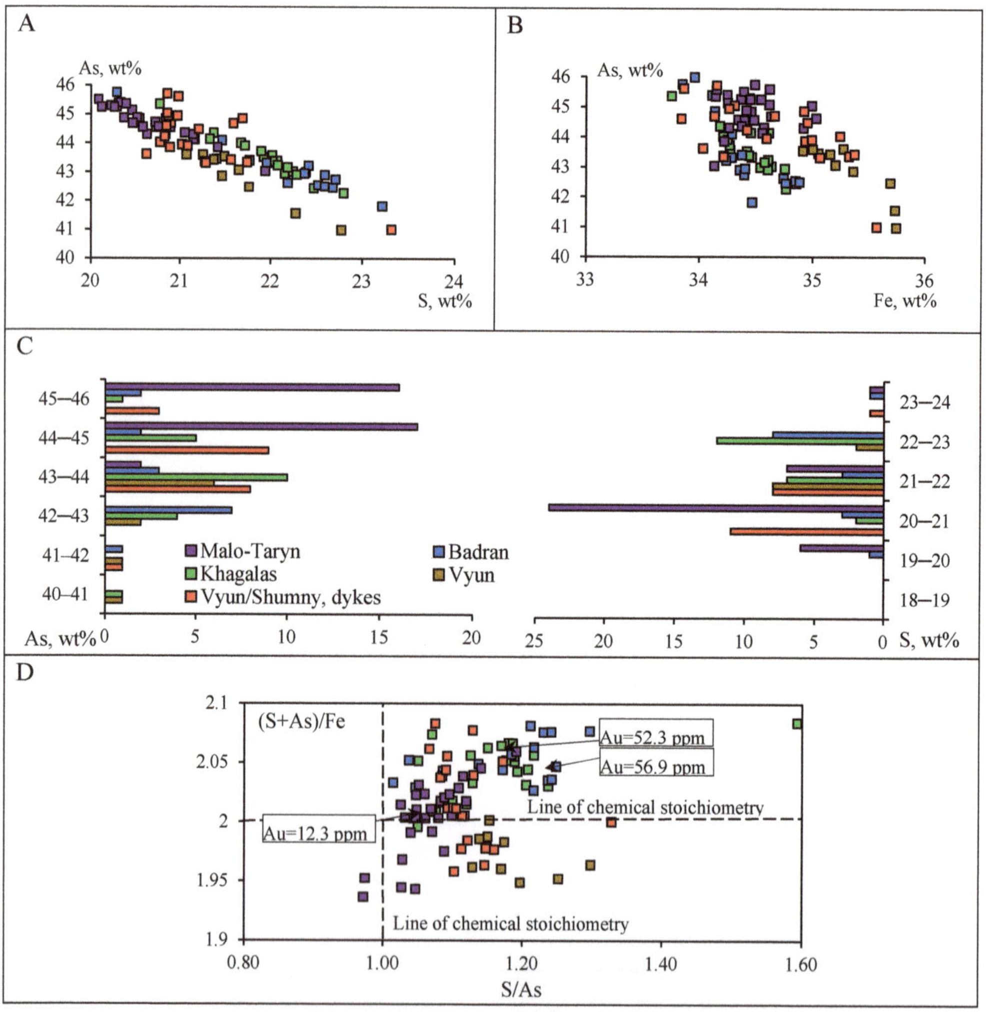

Figure 18. Diagrams of the ratios of concentrations of Fe, S, and As in Apy1 of the studied OGDs, central YKMB. (**A**) As vs. S. (**B**) As vs. Fe. (**C**) Variations in the S and As content in the studied deposits; (**D**) diagrams of the ratios S/As and (S + As)/Fe. The intersection of the lines S/As = 1 and (S + As)/Fe = 2 corresponds to chemical stoichiometry. See symbols in Figure 16.

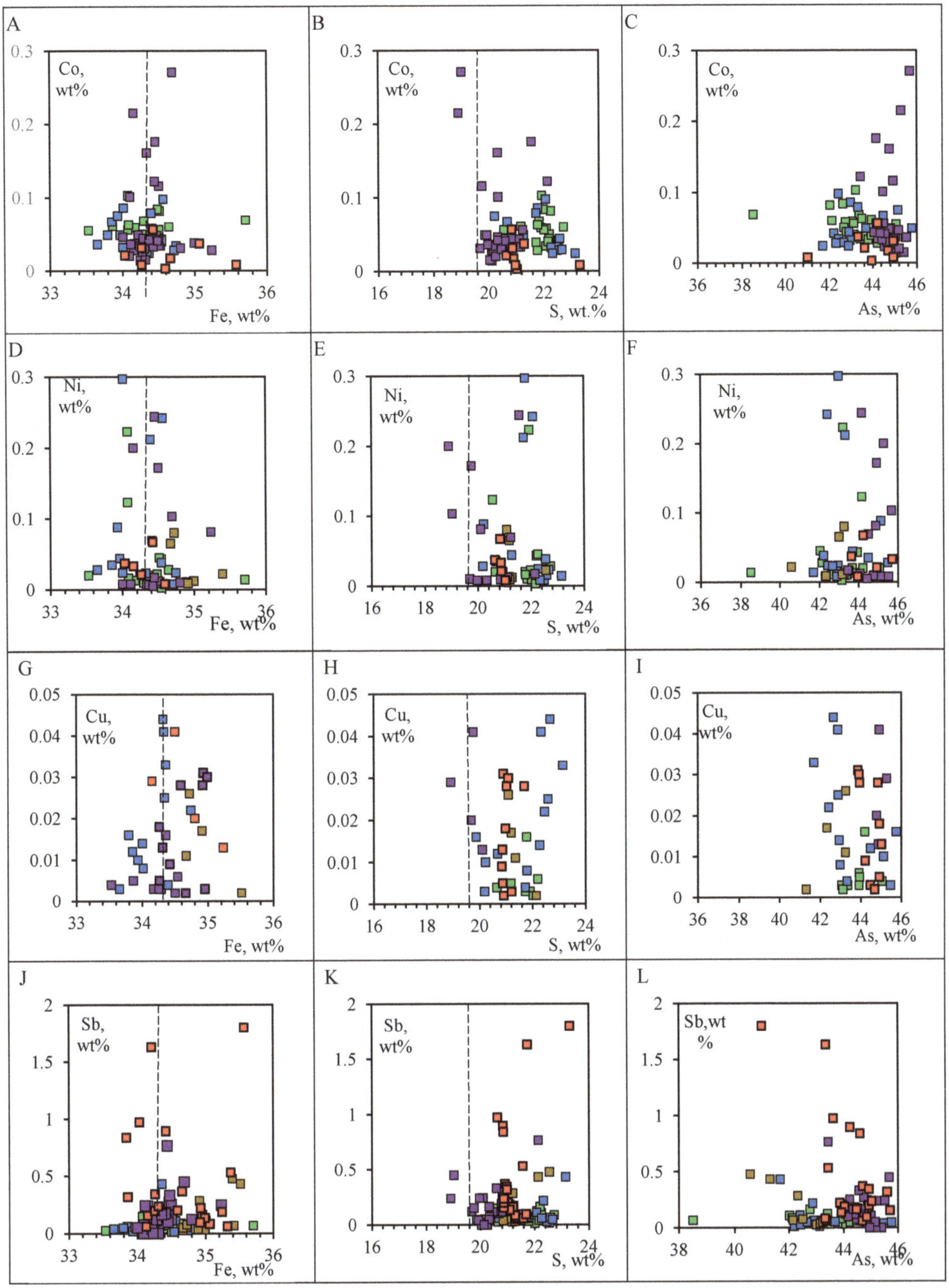

Figure 19. Diagrams of the ratios of concentrations of Fe, S, As, and trace elements in Apy1 of the studied OGDs, central YKMB. (**A**) Fe vs. Co; (**B**) S vs. Co; (**C**) As vs. Co; (**D**) Fe vs. Ni; (**E**) S vs. Ni; (**F**) As vs. Ni; (**G**) Fe vs. Cu; (**H**) S vs. Cu; (**I**) As vs. Cu; (**J**) Fe vs. Sb; (**K**) S vs. Sb; (**L**) As vs. Sb. The lines Fe = 34.30 wt%, S = 19.69 wt%, and As = 46.01 wt% correspond to the stoichiometric composition of pyrite. See symbols in Figure 16.

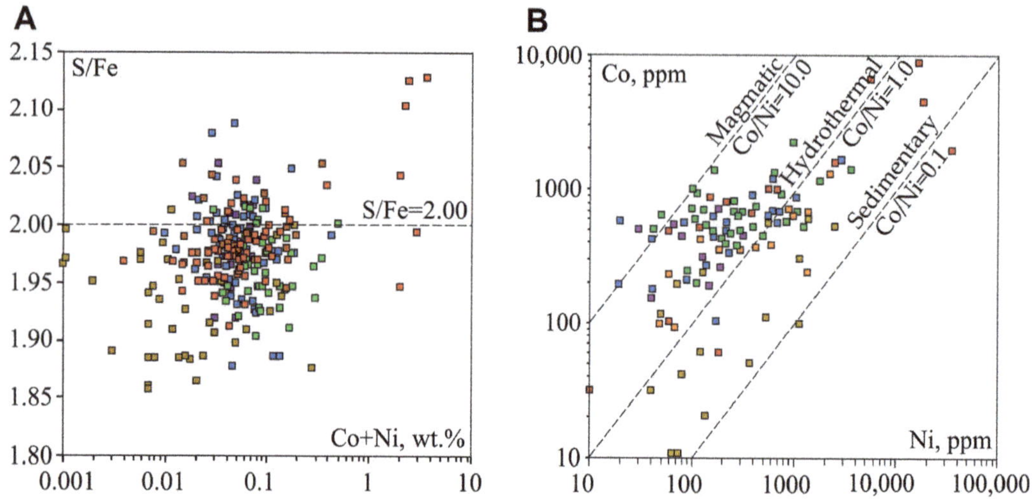

Figure 20. Distribution of the total impurity Co+Ni in Py3 relative to the indicator S/Fe (**A**) of the studied OGDs, central YKMB. (**B**) Diagram of concentration ratios Ni and Co in Py3 of the studied deposits. The demarcated fields are after [115]. See symbols in Figure 16.

Co concentrations vary markedly in the volume of a single grain. Elevated concentrations are characteristic of the central part of the zonal Py3, which can be formed before the formation of mineralization. There is an inverse correlation between Co and Au (r = −0.6) (Figure 21A), which, perhaps, is a temperature effect [119] and indicates the formation of gold-bearing pyrite in low-temperature conditions.

The Co/Ni ratio in Py3 of the studied deposits varies widely (0.1–28.0), but in most analyses, $C_{Co} > C_{Ni}$, and in 90% of analyses, 10.0 > Co/Ni > 0.1, which is typical for hydrothermal pyrite (Figure 20B) [33,102]. High concentrations of Ni in sulfides may indicate [120] the participation of basic and ultrabasic components supplied into hydrothermal fluids and involved in the deposition of sulfides. The Ni/Co ratio can indirectly characterize the gold content of pyrite [119]. Gold is predominantly isomorphic in pyrite with conductivity (Ni/Co > 0.1), and cobalt-rich pyrite (Ni/Co < 0.1) is not gold-bearing (Figure 21B).

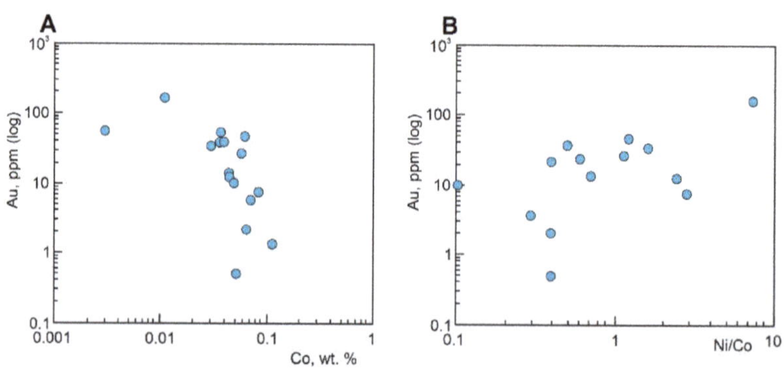

Figure 21. (**A**) Graph showing the relationship between the concentration of Co and Au in Py3 of the studied deposits; (**B**) graph showing the relationship between the concentration of Au and the Ni/Co indicator of the studied deposits.

5.2. Gold Occurrence and Concentration in Sulfides from Proximal Alteration

5.2.1. Gold Concentration

Data on the gold content in pyrite and arsenopyrite from the alteration of the studied sediment-hosted deposits are shown in Table 4 and Figure 15. Minimum mean values of Au were obtained for Py grains from the Malo–Taryn deposit (Au—5.1 ppm); the maximum mean values of Au are determined for arsenopyrite from the Badran deposit (66.9 ppm Au). For the proximal alteration of the V'yun and Shumnyi intrusion-hosted deposits, there are obvious differences between the gold contents in sulfides from clastic rocks compared with dykes. In alteration rocks from dykes in the V'yun deposit, the mean value of Au is 5.0 ppm, whereas, in Py from alterations in clastic rocks, the mean value of Au is 35.6 ppm. In two Apy samples from alterations in sandstones, the following values were obtained: 28.9 and 58.4 ppm Au. The Shumnyi deposit is characterized by the inverse value of variations in the Au content in Py. Thus, in Py from dykes, the mean value of Au is 28.8 ppm, and in Py from sandstones, the mean value of Au is 13.2 ppm.

In general, for all the studied deposits, gold contents in sulfides from proximal alterations were determined from fractions to be 168.5 ppm (Badran deposit, Apy). The highest gold content is found in Py, with 159.5 ppm (V'yun deposit) (Figure 15). Our results are comparable with data on the gold content in sulfides from alterations in many gold deposits in NE Russia, but they are noticeably lower compared with some large deposits with disseminated ores (Natalka—140–482.6 ppm Au, Mayskoe—300–1975 ppm Au [121], Nezhdaninskoe—up to 1400 ppm [28,112], Kyuchus—11.5–440 ppm Au [122]). Nevertheless, these results show the high economic potential of the disseminated mineralization in the studied deposits. For example, for the Khangalas deposit, it was shown that with a minimum gold content in alterations of 0.5 ppm, a length of 1.4 km, a thickness of 50 m, and a depth of 100 m for ore zones, reserves can be increased by 9.1 t Au [12].

5.2.2. Gold in Pyrite3

Gold in sulfides can occur in an isomorphic structurally bound form and in the form of native nano- and microinclusions [31,119,123–125]. The problem of invisible gold has been studied in the most detail in pyrite, which is related to the discovery of a large Carlin deposit in Nevada, where gold is closely associated with arsenian pyrite [126]. However, it is known that pyrite and arsenopyrite with invisible gold occur in deposits of various types (e.g., orogenic, epithermal, intrusion-related, porphyry-Cu, iron-oxide copper-gold, etc.) [32,119]. Over the past 15–20 years, with the advent of new analytical techniques and technologies, a lot of information has been obtained about invisible gold and its form of occurrence in sulfides [31,32,124,127]. It was shown that invisible gold is mainly concentrated in pyrite with a high solid-solution As content—arsenian pyrite [128]. Arsenic can replace either Fe or S in pyrite; at the same time, the highest concentration of As is 8–11% [105,126]. It is assumed that the Au^+ ionic gold replaces Fe by entering distorted octahedral positions, and As replaces S in tetrahedral positions [129], but, nevertheless, the nature of invisible gold occurrence in sulfides is still debated [119]. A negative correlation between Fe and Au in pyrite may indicate the presence of Au in the lattice via the isomorphic substitution of Fe [130]. However, the ionic radius of Au^+ differs from the ionic radius of Fe^{2+}, which makes it impossible to replace Au^+ with Fe^{2+} [131,132]. Chouinard et al. [133] proposed a mechanism of conjugate substitution for the $Au^{3+} + Cu^+ \leftrightarrow 2Fe^{2+}$ type. Gold nanoparticles in arsenian pyrite can be localized at the boundaries of block structures, both as a surface formation and in defects in the crystal lattice [122,129].

It is important to identify [31], for epithermal and Carlin-type deposits, an increase in the solubility of Au in the pyrite structure with an increase in As content and to determine the saturation line of Au on an As vs. Au graph. Based on the data from EMPA, laser ablation inductively coupled plasma mass spectrometry (LA-ICP-MS), secondary-ion mass spectrometry (SIMS), and particle induced X-ray emission with a microfocused beam (micro-PIXE) analyses, Deditius A.P. et al. [32] studied the origin of the inclusion of Au and As and the solubility of gold in pyrite based on Cu-porphyry, Cu-Au, orogenic (OGDs), volcanogenic–massive sulfide (VMS), iron-oxide copper–gold (IOCG), Au Witwatersrand, and coal deposits. They showed that Au^{1+} was the dominant form of Au in arsenian pyrite of the studied deposits, and the authors defined the empirical solid solubility of Au in As-pyrite as $C_{Au} = 0.004 \times C_{As} + 2 \times 10^{-7}$ for a range of temperature between ~150–250 °C (Figure 22). Our recent studies of gold-bearing arsenian pyrite and arsenopyrite from the proximal alteration of the Khangalas deposit have also shown the predominance of structurally related forms of gold Au^{+} in them [12]. These results were confirmed by rather low Au contents in the analyzed Py3. In most samples, Au does not exceed 2.5 ppm [12]. According to [124], the content of the structural form of Au in pyrite does not exceed ~ 5 ppm, and according to [32], it is less than 100 ppm Au. Higher concentrations are mainly related to the presence of nano- and microparticles [134]. The presence of native surface-bound (nano- or micro-) Au^{0} in alteration sulfides is registered in deposits of various types [31,119,122,124].

Table 6 shows the mean arsenic content according to EPMA data, gold, and silver according to AAS data and the Ag/Au ratio in pyrite from the deposits discussed in this study. We used these results to determine the form of gold in sulfides. In the diagrams of the As/Au (Figure 22A) and Ag/Au (Figure 22B) ratios, the mean values of As, Au, and Ag in pyrite from the studied deposits of both types (sediment-hosted and intrusion-hosted) fall into the field of orogenic deposits of different ages and locations, belonging to the Au-As association of the Earth [119].

The Ag/Au ratio in Py3 does not differ significantly in the studied deposits; in the mean, it ranges from 0.07 (1:13.4) to 1.06 (1:0.9) (Table 6). Figure 22B shows that the Ag/Au ratio in the Py3 of the studied deposits is in good conformity with the results [119], which determined that, for deposits of the Au-As association, the disseminated arsenian pyrite has small values for the Ag/Au ratio up to 1, and the sedimentary pyrite has a higher, up to 1000, Ag/Au ratio. Large R. and Maslennikov V. [119] showed that the value of the Ag/Au ratio is a useful criterion to help distinguish disseminated sedimentary pyrite from hydrothermal-alteration-disseminated pyrite in sedimentary rocks. The studied deposits are located in the field of orogenic deposits (Figure 22). The studied orogenic gold in sediment-hosted (Malo–Taryn, Badran, and Khangalas) and intrusion-hosted deposits (V'yun, Shumnyi) have similar ionic structurally bound Au^{+} gold contents in Py3 (Figure 22A).

Table 6. Mean values of As, Au, and Ag in pyrite from the studied OGDs, central YKMB.

Deposit	As, ppm (EPMA)	Au, ppm (AAA)	Ag, ppm (AAA)	Ag/Au
Malo–Taryn	7800/27 *	5.1/6	5.4/6	1.06
Badran	13,340/60	57.5/12	6.0/12	0.1
Khangalas	10,140/47	11.2/13	6.2/13	0.55
V'yun, dyke	15,850/53	5.0/6	1.3/6	0.26
V'yun, sandstone	15,850/53	35.6/8	5.6/8	0.16
Shumnyi, dyke	9970/23	28.8/4	2.1/4	0.07
Shumnyi, sandstone	9970/23	13.2/4	1.1/4	0.08

* Mean values/number of analyses.

5.2.3. Gold in Arsenopyrite1

In arsenopyrite, as in pyrite, gold can be in a native or isomorphic form. The inverse correlation between As and S in Apy (Figure 23) may reflect their conjugate isomorphic substitution in the process of formation [120,135]. Arsenic in arsenopyrite plays an important role as an indicator of the mineralization of Au. The atomic ratio As/S is mainly sensitive to temperature in the sulfur–buffer group, which leads to a higher ratio with an increasing temperature [135–137]. For the arsenopyrite1 of the Vorontsovskoye deposit, rich with sulfur and depleted of gold, it was determined that the ratio is As/S < 1, and for arsenopyrite2, depleted of sulfur and rich with gold, the ratio is As/S >1 [137]. It has been shown [137] that Apy1 crystallizes at a higher temperature and sulfur volatility, whereas for Apy2, these parameters are significantly lower. Our results do not align with these data. The atomic ratio of As/S for the gold-bearing Apy1 of all studied deposits shows that arsenopyrite is rich in sulfur, As/S < 1: Malo–Taryn—from 0.84 to 0.98 (only in two samples: 1.03), mean 0.94; Badran—from 0.77 to 0.99, mean 0.86; Khangalas—from 0.63 to 0.93, mean 0.83; V'yun—from 0.77 to 0.89, mean 0.85; and Shumnyi—from 0.75 to 0.94, mean 0.89 (Table 7).

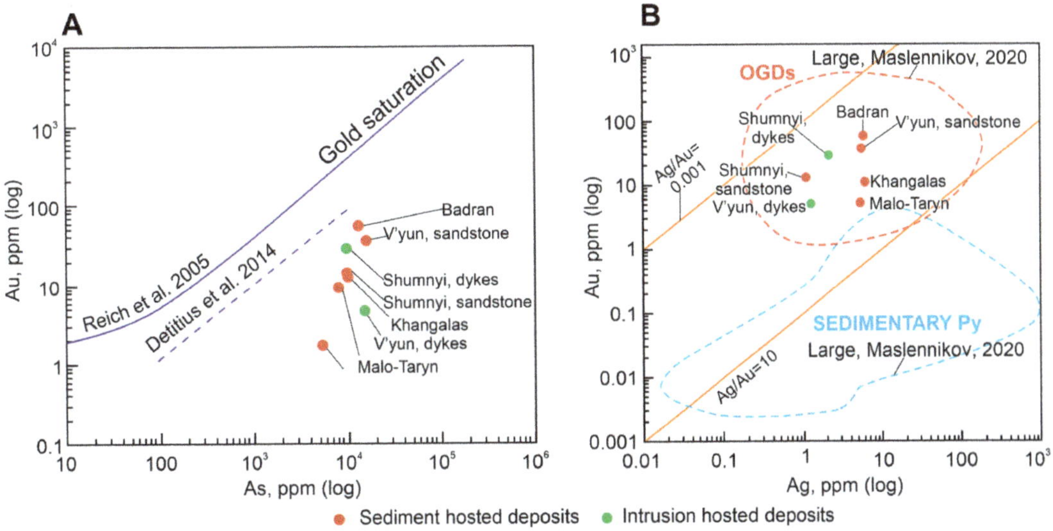

Figure 22. Plot of Py3. (**A**) Au vs. As, the lines of gold saturation after [32,49]; (**B**) Au vs. Ag of the studied OGDs, central YKMB. The demarcated fields are after [119].

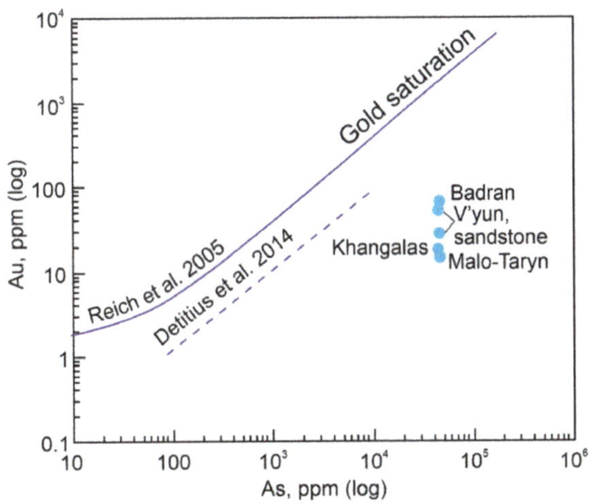

Figure 23. Au-As plot for mean values of Apy1 in the studied OGDs, central YKMB. The lines of gold saturation after [32,49].

The relationship between Au and Fe may show the form of gold occurrence in arsenopyrite [136,137]. In our recent study [12], an inverse dependence of Au and Fe in the Apy1 of the Khangalas deposit was shown, a strong inverse correlation (r = −0.9) with increased contents Au of > 2 ppm was established, and with Au contents of <1 ppm, there is no correlation (r = −0.18).

To determine the form of gold occurrence in arsenopyrite from the alteration of the studied deposits, a correlation diagram with the saturation line Au was used according to [31]. In the diagram, the mean values of As and Au fall into the field of structurally bound Au^+ in Apy1.

Table 7. Mean values of As, Au, and Ag in the arsenopyrite1 of the studied OGDs, central YKMB.

Deposit	As, ppm (EPMA)	Au, ppm (AAA)	Ag, ppm (AAA)	Au/Ag	As/S
Malo–Taryn	44,740/35 *	17.1/4	4.2/4	55.7	0.94
Badran	43,320/16	73.6/9	6.6/9	17.4	0.86
Khangalas	42,920/22	17.5/3	6.5/3	11.1	0.83
V'yun, sandstone	42,470/10	28.9, 58.4	2.2, 2.3	12.8, 26.8	0.85
Shumnyi, dyke	44,230/21	–	–	–	0.89

* Mean values/number of analyses.

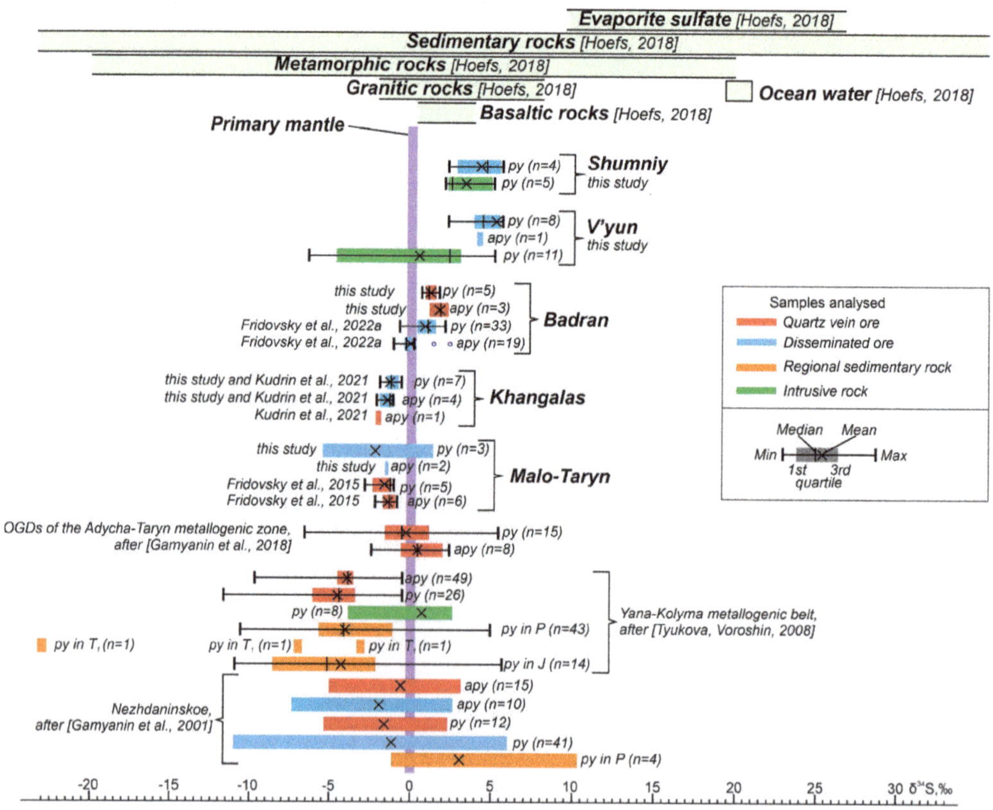

Figure 24. Summary of $\delta^{34}S$ values for orogenic gold deposits in the Yana–Kolyma metallogenic belt and the Nezhdaninskoe deposit, NE Russia. For comparison, materials according to [11,12,21,23,28,39]. Ranges of sulfur isotopic composition in various geological reservoirs according to [146]. In sudfigure: the left border of the line shows the minimum value; the right one shows the maximum value; vertical lines in the boxes denote the median, the X marks in the boxes denote the mean, and the left and right of each box denote the first and third quartiles, respectively.

The qualitative linear scanning of an Apy1 grain from the Khangalas deposit showed that the concentrations of invisible Au, Pb, and Cu tend to decrease from the center to the rim, whereas the Co and Ni contents are noticeably higher in the central zones of the crystal [12]. It has been suggested that the increase in Au, Pb, and Cu is related to a portion of later fluids rich in the Au-polysulfide association.

Thus, the predominant form of invisible Au in the arsenopyrite1 and pyrite3 of the studied deposits is a solid solution of Au^{+} in their crystal lattices. Micro- and nanoparticles of native gold are present in a subordinate amount.

5.3. Sources of Components Based on Stable Isotopes S

There are different ideas about potential sources of OGD fluids and metals, such as in [43] and references therein. Their exhaustive review is given in [138]. The formation of OGDs from ore fluids with very diverse sources is discussed: some deposits have magmatic water components [28,139–141], but some or most of the ore fluid is supplied from metamorphic waters [23,29,54,142,143] and juvenile sources; for example, in [11,43] and references therein. There are data on the source of some metals in the ore-forming system from the host rocks [144]. Based on the example of deposits in the Juneau Gold Belt [145] and, later on, the a comprehensive analysis of the example of Jiaodong Province, China [43], it was shown that sulfur reached ore-forming fluids during the metamorphic transformation of pyrite into pyrrhotite. The sulfur source was supposed to be dispersed syngenetic/diagenetic pyrite in terranes, devolatilized at a depth from the sediment wedge above a subduction zone [138].

Studies of stable isotopes such as $\delta^{34}S$ provide important insights into the fluid source and source(s) of S, which is important for the genetic interpretation of ore deposit formations; see [43,53,54,140,146] and references therein. In general, a large range of $\delta^{34}S$ values in sulfides from −20.0 to +25.0‰ was obtained for the orogenic gold deposits [43]. The variation of $\delta^{34}S$ is related to the involvement of various reservoirs in the formation of ores and variations in physical–chemical parameters during the evolution of ore-forming systems [147]. The values of sulfur sulfide isotopes in metamorphic terranes inside or near transcrustal faults of the Earth's crust according to [2] range between 0 and 10‰, but both higher and lower values have been observed. In addition, the dependence of the isotopic sulfur composition of the Phanerozoic deposits on the age of the host rock should be emphasized [43,51] (Figure 24).

Most of the orogenic gold deposits of the Verkhoyansk–Kolyma orogen have values of $\delta^{34}S$ ranging from −12 to +10‰ (Figure 24) (e.g., [11,15,21,23,39,148]). The mean values of the sulfur isotopes of the arsenopyrite, pyrite, and polysulfide associations from the ore veins of these deposits, as well as from clastic rocks at a distance from the ore-bearing structures, were in a range of −5.0–+6.0‰ (Figure 24). In some deposits, a heavier $\delta^{34}S$ was identified due to the influence of sedimentary diagenetic sulfides from the host rocks [39].

Recently, Gamyanin G.N. et al. [23] have performed a lot of work in studying the conditions of the formation of precious metal mineralization in the Adycha–Taryn metallogenic zone, located in the central sector of the YKMB. In particular, the values of $\delta^{34}S$ in sulfides from the proximal alteration and veins allow the sources of sulfur to be constrained. Thus, for OGD veins, a rather narrow range of values of $\delta^{34}S$, close to 0, can be set: for arsenopyrite, it is from −2.1 to +2.4‰ (mean +0.4‰); for pyrite, it is from −6.6 to +5.4‰ (mean of −0.3‰) (Figure 24) [23]. In our recent work, similar values of $\delta^{34}S$ were obtained for pyrite and arsenopyrite from the alteration of the Badran deposit [11] and arsenopyrite from vein ores and the alteration of the Khangalas deposit [12] (Figure 24). The isotopic composition of sulfide sulfur from the alteration of the Badran deposit studied using the local method showed the value of $\delta^{34}S$ in pyrite to be slightly heavier than in arsenopyrite [11] (Figure 24). Two crystal populations were established in pyrite based on the $\delta^{34}S$ variations. The first population includes isotope-zonal pyrite with sulfur weighing from the center to the rim of the grains (mean in the center of the grains, 0.2‰; mean at the rim of the grains, 1.5‰). The difference in the value of $\delta^{34}S$ is up to 2.1‰. Similar zoning was noted earlier at the Sukhoi

Log deposit, where it is related to the evolution of fluid composition [148]. This can also be caused by variations in the fO_2–pH conditions of ore-forming processes [149]. The second population, taking into account any analysis errors, is isotopically homogeneous pyrite; the values of $\delta^{34}S$ in the center of the crystals differ slightly (0.01–0.33‰) from their rim. It was found that arsenopyrite with values of $\delta^{34}S$ close to zero is more gold-bearing and indicates that sulfur minerals in sediment-hosted orogenic gold deposits were probably derived from the reduction of seawater sulfate [51].

New data on the isotopic composition of sulfide sulfur from the studied sediment-hosted and intrusion-hosted deposits, shown in Table 5 and in Figure 24, have a narrow range of $\delta^{34}S$ values, from −6.4 to +5.6‰ (mean value of about 0‰). These data are consistent with the results of Gamyanin G.N. et al. [23] for OGD vein ores in the Adycha–Taryn metallogenic zone and our recent studies of disseminated sulfide mineralization [11,12]. Such values of $\delta^{34}S$ for the gold deposits in the Yilgarn craton (Australia) are estimated to have a magmatic or mantle source of ore-forming fluid [52]. Similar results were obtained for a number of orogenic gold deposits in Kazakhstan ($\delta^{34}S$ = 0.0 ... −3.3‰), the source of the ore substance of which was determined to be mantle with partial borrowing from crustal sulfur [150,151], and the Nezhdaninskoe OGD, for which the values of $\delta^{34}S$ of pyrite and arsenopyrite from the vein and disseminated ores prevail from −6 up to +0.7‰ (a mean of about +0.6‰) [28].

The plots show that $\delta^{34}S$ in sulfides in the V'yun, Shumnyi, Malo–Taryn, and Badran deposits does not depend on the age of the host rocks and does not correlate with the seawater sulfate curve through the geological time. Near zero mean values of $\delta^{34}S$, together with the mantle signatures for sulfides and native gold grains [11] from the YKMB sediment-hosted orogenic gold deposits, indicate subcrustal and metamorphic sources for the Au-bearing fluid and sulfur. The slightly heavier sulfur isotopic composition of sulfides from the Shumnyi and V'yun intrusion-hosted deposits may indicate a mixture of subcrustal, metamorphic, partially magmatic, and sedimentary sources (Figure 24). A similar isotopic composition of sulfur arsenopyrite and pyrite quartz in the vein ore [23] and disseminated ore may indicate their formation during a single homogeneous hydrothermal event.

6. Conclusions

This investigation of the chemical composition of disseminated sulfide mineralization from the proximal alteration of sediment-hosted (Malo–Taryn, Badran, Khangalas) and intrusion-hosted (V'yun, Shumnyi) orogenic Au deposits from the central part of the Yana–Kolyma metallogenic belt revealed that the gold endowment of these deposits is associated with pyrite3 and arsenopyrite1, and these minerals are an economically important source of gold. The determination of the precise site of invisible gold within Py3 and Apy1 showed the possible prevalence of solid solution Au^{+} in their crystal lattices. Py3 from clastic and igneous rocks has a consistent association of As, Co, Ni, Cu, and Sb, as well as, less often, Pb, and has high conductivity, Co/Ni, and Ag/Au ratio characteristics with respect to hydrothermal pyrites. Both Py3 and Apy1 exhibit a distinct statistical correlation between Au and As and Au and Co contents. In the intrusion-hosted orogenic gold deposits, elevated concentrations of Co and Ni in Py3 were registered, suggesting that the ore fluid reacted with altered volumes of basic dykes. In Apy1, Co, Ni, Cu, and Sb were identified.

A synthesis of the geological, geochemical, and stable isotope data suggests that the invisible gold in the disseminated arsenian pyrite3 and in the arsenopyrite1 is most likely formed from subcrustal and metamorphic hydrothermal systems in the Verkhoyansk–Kolyma orogen.

Supplementary Materials: The following supporting information can be downloaded at: https://www.mdpi.com/article/10.3390/min13030394/s1, Table S1: Compositions (wt%) for pyrite in the proximal alteration rock OGD Central YKMB, analyzed by EPMA.; Table S2: Compositions (wt%) for arsenopyrite in the proximal rock OGD Central YKMB, analyzed by EPMA.

Author Contributions: Conceptualization: V.Y.F.; methodology, V.Y.F., L.I.P. and M.V.K.; validation, V.Y.F., L.I.P. and M.V.K.; writing—original draft preparation, V.Y.F., L.I.P. and M.V.K., writing—review and editing, V.Y.F., L.I.P. and M.V.K.; visualization, M.V.K.; supervision, V.Y.F.; project administration, V.Y.F.; funding acquisition, V.Y.F. and M.V.K. All authors have read and agreed to the published version of the manuscript.

Funding: This work was supported by the Diamond and Precious Metals Geology Institute, Siberian Branch of the Russian Academy of Sciences (DPMGI SB RAS).

Data Availability Statement: Not applicable.

Acknowledgments: The authors are grateful to all engineers for their accurate, timely, and prompt analytical work and, especially, Tarasov Ya.A. for his help in the preparation and selection of the samples. We are sincerely grateful to Shulyak A.P. for translating the text and Dolgopolova A.V. for text editing and valuable comments, which contributed to the improvement of the manuscript.

Conflicts of Interest: The authors declare no conflict of interest.

References

1. Frimmel, H.E. Earth's continental crustal gold endowment. *Earth Planet. Sci. Lett.* **2008**, *267*, 45–55. [CrossRef]
2. Goldfarb, R.J.; Baker, T.; Dube, B.; Groves, D.I.; Hart, C.J.R.; Robert, F.; Gosselin, P. Distribution, character, and genesis of gold deposits in metamorphic terranes. *Econ. Geol.* **2008**, *407–450*. 100th Anniversary Vol. [CrossRef]
3. Mikhailov, B.K.; Struzhkov, S.D.; Aristov, V.V.; Natalenko, M.V.; Tsymbalyuk, N.V.; Tyamisov, N.E.; Uzunkoyan, A.A. Gold potential of the Yana-Kolyma province. *Ores Met.* **2007**, *5*, 4–17. (In Russian)
4. Konstantinov, M.M. *Gold Deposits of Russia*; Konstantinov, M.M., Ed.; Aquarelle: Moscow, Russia, 2010; p. 385. (In Russian)
5. Goldfarb, R.; Taylor, R.; Collins, G.; Goryachev, N.; Orlandini, O. Phanerozoic continental growth and gold metallogeny of Asia. *Gondwana Res.* **2014**, *25*, 48–102. [CrossRef]
6. Goryachev, N.A.; Pirajno, F. Gold deposit and gold metallogeny of Far East Russia. *Ore Geol. Rev.* **2014**, *59*, 123–151. [CrossRef]
7. Goldfarb, R.; Groves, D.I.; Gardoll, S. Orogenic gold and geologic time: A global synthesis. *Ore Geol. Rev.* **2001**, *18*, 1–75. [CrossRef]
8. Goryachev, N.A. Gold deposits in the history of the Earth. *Geol. Ore Depos.* **2019**, *61*, 3–18. [CrossRef]
9. Volkov, A.V.; Sidorov, A.A.; Goncharov, V.I.; Sidorov, V.A. Disseminated gold-sulfide deposits in the Russian Northeast. *Geol. Ore Dep.* **2002**, *44*, 159–174.
10. Goryachev, N.A.; Sotskaya, O.T.; Ignatiev, A.V.; Velivetskaya, T.I.; Goryacheva, E.M.; Semyshev, F.I.; Berdnikov, N.V.; Malinovsky, M.A.; Alshevsky, A.V. About sulfide mineralization of the zones of large faults of the Yana-Kolyma orogenic belt. *Bull. Northeast. Sci. Cent. FEB RAS* **2020**, *1*, 11–29. (In Russian) [CrossRef]
11. Fridovsky, V.Y.; Polufuntikova, L.I.; Kudrin, M.V.; Goryachev, N.A. Sulfur isotope composition and geochemical characteristics of gold-bearing sulfides of the Badran orogenic deposit, Yana-Kolyma metallogenic belt (North-East Asia). *Dokl. Akad. Nauk* **2022**, *502*, 3–9. [CrossRef]
12. Kudrin, M.V.; Fridovsky, V.Y.; Polufuntikova, L.I.; Kryuchkova, L.Y. Disseminated Gold–Sulfide Mineralization in Metasomatites of the Khangalas Deposit, Yana–Kolyma Metallogenic Belt (Northeast Russia): Analysis of the Texture, Geochemistry, and S Isotopic Composition of Pyrite and Arsenopyrite. *Minerals* **2021**, *11*, 403. [CrossRef]
13. Aristov, V.V.; Prokofiev, V.Y.; Imamendinov, B.N.; Kryazhev, S.G.; Alekseev, V.Y.; Sidorov, A.A. Features of ore formation at the Drazhnoe gold-quartz deposit (Eastern Yakutia, Russia). *Dokl. Akad. Nauk* **2015**, *464*, 879–884. [CrossRef]
14. Goryachev, N.A. *Geology of Mesozoic Gold–Quartz Vein Belts of Northeast Asia*; NEISRI FEB RAS: Magadan, Russia, 1998; p. 210. (In Russian)
15. Gamyanin, G.N. *Mineralogical and Genetic Aspects of Gold Mineralization of the Verkhoyansk–Kolyma Mesozoids*; GEOS: Moscow, Russia, 2001; p. 221. (In Russian)
16. Amuzinsky, V.A. *Metallogenic Epochs and Content of Ore Complexes of the Verkhoyansk Folded System*; YSU Publishing House: Yakutsk, Russia, 2005; p. 248. (In Russian)
17. Goryachev, N.A.; Vikentyeva, O.V.; Bortnikov, N.S.; Prokofiev, V.Y.; Alpatov, V.A.; Golub, V.V. The world-class Natalka gold deposit, northeast Russia: REE patterns, fluid inclusions, stable oxygen isotopes, and formation conditions of ore. *Geol. Ore Depos.* **2008**, *50*, 414–444. [CrossRef]
18. Fridovsky, V.Y.; Gamyanin, G.N.; Polufuntikova, L.I. Dora-Pil ore field: Structure, mineralogy and geochemistry of ore formation environment. *Ores Met.* **2012**, *5*, 7–21. (In Russian)
19. Fridovsky, V.Y.; Gamyanin, G.N.; Polufuntikova, L.I. The Sana Au–quartz deposit within the Taryn ore cluster. *Razved. Okhrana Nedr* **2013**, *2*, 3–7. (In Russian)
20. Fridovsky, V.Y.; Gamyanin, G.N.; Polufuntikova, L.I. Gold-quartz and antimony mineralization in the Maltan deposit in northeast Russia. *Russ. J. Geol. Pac. Ocean.* **2014**, *8*, 276–287. [CrossRef]
21. Fridovsky, V.Y.; Gamyanin, G.N.; Polufuntikova, L.I. The structure, mineralogy, and fluid regime of ore formation in the polygenic Malo-Taryn gold field, northeast Russia. *Russ. J. Geol. Pac.* **2015**, *9*, 274–286. [CrossRef]

22. Voroshin, S.V.; Tyukova, E.E.; Newberry, R.J.; Layer, P.W. Orogenic gold and rare metal deposits of the Upper Kolyma region, Northeastern Russia: Relation to igneous rocks, timing, and metal assemblages. *Ore Geol. Rev.* **2014**, *62*, 1–24. [CrossRef]
23. Gamyanin, G.N.; Fridovsky, V.Y.; Vikent'eva, O.V. Noble-metal mineralization of the Adycha–Taryn metallogenic zone: Geochemistry of stable isotopes, fluid regime, and ore formation conditions. *Russ. Geol. Geoph.* **2018**, *59*, 1271–1287. [CrossRef]
24. Aristov, V.V.; Grigorieva, A.V.; Savchuk, Y.S.; Sidorova, N.V.; Sidorov, V.A. Forms of gold and some typomorphic characteristics of native gold of the Pavlik orogenic deposit (Magadan oblast). *Geol. Ore Depos.* **2021**, *63*, 1–33. [CrossRef]
25. Aristov, V.V. Prospects for the discovery of large rich gold-quartz deposits in the Taryn ore-placer cluster. Prospect and protection of mineral resources. *Razved. Okhrana Nedr* **2009**, *6*, 3–11.
26. Fridovsky, V.Y.; Polufuntikova, L.I.; Tarasov, Y.A. Mineralogy, geochemistry and localization of regional pyritization zones–constraints from early mesozoic deposition in the Chay–Yureya fault of the Kular-Nera Terrane, NE Russia. In *Proceeding of the IOP Conference Series: Earth and Environmental Science*; IOP Publishing: Prague, Czech Republic, 2020; Volume 609, p. 012016. [CrossRef]
27. Fridovsky, V.Y.; Goryachev, N.A.; Krymsky, R.S.; Kudrin, M.V.; Belyatsky, B.V.; Sergeev, S.A. The age of gold mineralization in the Yana–Kolyma metallogenic belt, Northeastern Russia: First data of Re–Os isotope geochronology of native gold. *Russ. J. Pac. Geol.* **2021**, *40*, 18–32. [CrossRef]
28. Gamyanin, G.N.; Bortnikov, N.C.; Alpatov, V.V. *The Nezhdaninskoe Gold Ore Deposit Is a Unique Deposit in the North-East of Russia*; GEOS: Moscow, Russia, 2001; p. 230. (In Russian)
29. Large, R.R.; Danyushevsky, L.; Hollit, C.; Maslennikov, V.; Gilbert, S.; Bull, S.; Scott, R.; Emsbo, P.; Thomas, H.; Singh, B.; et al. Gold and trace element zonation in pyrite using a laser imaging technique: Implications for the timing of gold in orogenic and carlin-style sediment-hosted deposits. *Econ. Geol.* **2009**, *104*, 635–668. [CrossRef]
30. Thomas, H.V.; Large, R.R.; Bull, S.W.; Maslennikov, V.; Berry, R.F.; Fraser, R.; Froud, S.; Moye, R. Pyrite and pyrrhotite textures and composition in sediments, laminated quartz veins, and reefs at Bendigo gold mine, Australia: Insights for ore genesis. *Econ. Geol.* **2011**, *106*, 1–31. [CrossRef]
31. Reich, M.; Deditius, A.; Chryssoulis, S.; Li, J.W.; Ma, C.Q.; Parada, M.A.; Barra, F.; Mittermayr, F. Pyrite as a record of hydrothermal fluid evolution in a porphyry copper system: A SIMS/EMPA trace element study. *Geochim. Cosmochim. Acta* **2013**, *104*, 42–62. [CrossRef]
32. Deditius, A.P.; Reich, M.; Kesler, S.E.; Utsunomiya, S.; Chryssoulis, S.L.; Walshe, J.; Ewing, R.C. The coupled geochemistry of Au and As in pyrite from hydrothermal ore deposits. *Geochim. Cosmochim. Acta* **2014**, *140*, 644–670. [CrossRef]
33. Román, N.; Reich, M.; Leisen, M.; Morata, D.; Barra, F.; Deditius, A.P. Geochemical and micro-textural fingerprints of boiling in pyrite. *Geochim. Cosmochim. Acta* **2019**, *246*, 60–85. [CrossRef]
34. Tauson, V.; Lipko, S.; Kravtsova, R.; Smagunov, N.; Belozerova, O.; Voronova, I. Distribution of "invisible" noble metals between pyrite and arsenopyrite exemplified by minerals coexisting in orogenic Au deposits of North-Eastern Russia. *Minerals* **2019**, *9*, 660. [CrossRef]
35. Wu, Y.F.; Fougerouse, D.; Evans, K.; Reddy, S.M.; David, W.S.; Paul, G.; Li, J.W. Gold, arsenic, and copper zoning in pyrite: Record of fluid chemistry and growth kinetics. *Geology* **2019**, *47*, 641–644. [CrossRef]
36. Ishida, M.; Romero, R.; Leisen, M.; Yasukawa, K.; Nakamura, K.; Barra, F.; Reich, M.; Kato, Y. Auriferous pyrite formed by episodic fluid inputs in the Akeshi and Kasuga high-sulfidation deposits, Southern Kyushu, Japan. *Miner. Depos.* **2021**, *57*, 129–145. [CrossRef]
37. Yang, L.; Wang, Q.; Large, R.R.; Mukherjee, I.; Deng, J.; Li, H.; Yu, H.; Wang, X. Fluid source and metal precipitation mechanism of sediment-hosted Chang'an orogenic gold deposit, SW China: Constraints from sulfide texture, trace element, S, Pb, and He-Ar isotopes and calcite C-O isotopes. *Am. Miner.* **2021**, *106*, 410–429. [CrossRef]
38. Ma, Y.; Jiang, S.; Frimmel, H. Deciphering multiple ore-forming processes of the Shuangqishan orogenic gold deposit, Southeast China by in situ analysis of pyrite. *Ore Geol. Rev.* **2022**, *142*, 104730. [CrossRef]
39. Tyukova, E.E.; Voroshin, S.V. Isotopic composition of sulfur in sulfides from ores and host rocks of the Upper Kolyma region (Magadan region). *Russ. J. Pac. Geol.* **2008**, *27*, 29–43. (In Russian) [CrossRef]
40. Sung, Y.H.; Brugger, J.; Ciobanu, C.L.; Pring, A.; Skinner, W.; Nugus, M. Invisible gold in arsenian pyrite and arsenopyrite from a multistage Archaean gold deposit: Sunrise Dam, Eastern Goldfields Province, Western Australia. *Miner. Depos.* **2009**, *44*, 765. [CrossRef]
41. Large, R.; Thomas, H.; Craw, D.; Henne, A.; Henderson, S. Diagenetic pyrite as a source for metals in orogenic gold deposits, Otago Schist, New Zealand. *N. Z. J. Geol. and Geoph.* **2012**, *55*, 137–149. [CrossRef]
42. Cook, N.J.; Ciobanu, C.L.; Meria, D.; Silcock, D.; Wade, B. Arsenopyrite-pyrite association in an orogenic gold ore: Tracing mineralization history from textures and trace elements. *Econ. Geol.* **2013**, *108*, 1273–1283. [CrossRef]
43. Goldfarb, R.; Groves, D. Orogenic gold: Common or evolving fluid and metal sources through time. *Lithos* **2015**, *233*, 2–26. [CrossRef]
44. Finch, E.G.; Tomkins, A.G. Pyrite-pyrrhotite stability in a metamorphic aureole: Implications for orogenic gold genesis. *Econ. Geol.* **2017**, *112*, 661–674. [CrossRef]
45. Hazarika, P.; Mishra, B.; Pruseth, K.L. Trace-element geochemistry of pyrite and arsenopyrite: Ore genetic implications for late Archean orogenic gold deposits in southern India. *Mineral. Mag.* **2017**, *81*, 661–678. [CrossRef]

46. Augustin, J.; Gaboury, D. Multi-stage and multi-sourced fluid and gold in the formation of orogenic gold deposits in the world-class Mana district of Burkina Faso-Revealed by LA-ICP-MS analysis of pyrites and arsenopyrites. *Ore Geol. Rev.* **2018**, *104*, 95–521. [CrossRef]
47. Tarasova, Y.I.; Budyak, A.E.; Goryachev, N.A.; Tauson, V.L.; Skuzovatov, S.Y.; Bryukhanova, N.N.; Parshin, A.V.; Chugaev, A.V.; Abramova, V.D.; Gareev, B.I.; et al. Mineralogical and isotope-geochemical ($\delta^{13}C$, $\delta^{34}S$ and Pb-Pb) characteristics of the Krasniy gold mine (Baikal-Patom Highlands): Constraining ore-forming mechanisms and the model for Sukhoi Log-type deposits. *Ore Geol. Rev.* **2020**, *119*, 103365. [CrossRef]
48. Du, B.; Shen, J.; Santosh, M.; Liu, H.; Liu, J.; Wang, Y.; Xu, K. Textural, compositional and isotopic characteristics of pyrite from the Zaozigou gold deposit in West Qinling, China: Implications for gold metallogeny. *Ore Geol. Rev.* **2021**, *130*, 103917. [CrossRef]
49. Reich, M.; Kesler, S.E.; Utsunomiya, S.; Palenik, C.S.; Chryssoulis, S.L.; Ewing, R.C. Solubility of gold in arsenian pyrite. *Geochim. Cosmochim. Acta* **2005**, *69*, 2781–2796. [CrossRef]
50. Eremin, R.A.; Voroshin, S.V.; Sidorov, V.A.; Shakhtyrov, V.G.; Pristavko, V.A.; Gashtold, V.V. Geology and genesis of the Natalka gold deposit, Northeast Russia. *Inter. Geol. Rev.* **1994**, *36*, 1113–1138. [CrossRef]
51. Goldfarb, R.J.; Miller, L.D.; Leach, D.L.; Snee, L.W. Gold deposits in metamorphic rocks of Alaska. *Econ. Geol.* **1997**, *9*, 151–190. [CrossRef]
52. Xue, Y.; Campbell, I.H.; Ireland, T.R.; Holden, P.; Armstrong, R. No mass-independent sulfur isotope fractionation in auriferous fluids supports a magmatic origin for Archean gold deposits. *Geology* **2013**, *41*, 791–794. [CrossRef]
53. Shanks, W. *Stable Isotope Geochemistry of Mineral Deposits*; Elsevier Ltd.: Amsterdam, The Nederland, 2014. [CrossRef]
54. LaFlamme, C.; Sugiono, D.; Thébaud, N.; Caruso, S.; Fiorentini, M.L.; Selvaraja, V.; Jeon, H.; Voute, F.; Martin, L. Multiple sulfur isotopes monitor fluid evolution of an orogenic gold deposit. *Geoch. Cosmoch. Acta* **2018**, *222*, 436–446. [CrossRef]
55. Seltmann, R.; Goldfarb, R.; Zu, B.; Creaser, R.; Dolgopolova, A.; Shatov, V. Muruntau, Uzbekistan: The World's Largest Epigenetic Gold Deposit. *Soc. Econ. Geol.* **2020**, *23*, 497–521. [CrossRef]
56. Zhen, S.; Wang, Q.; Wang, D.; Carranza, E.J.M.; Liu, J.; Pang, Z.; Cheng, Z.; Xue, J.; Wang, J.; Zha, Z. Genesis of the Zhangquanzhuang gold deposit in the northern margin of North China Craton: Constraints from deposit geology and ore isotope geochemistry. *Ore Geol. Rev.* **2020**, *122*, 103511. [CrossRef]
57. Parfenov, L.M.; Kuzmin, M.I. *Tectonics, Geodynamics and Metallogeny of the Territory of the Republic of Sakha (Yakutia)*; Parfenov, L.M., Kuzmin, M.I., Eds.; Nauka/Interperiodika: Moscow, Russia, 2001; p. 571. (In Russian)
58. Parfenov, L.M. *Continental Margins and Island Arcs of Mesozoides in Northeast Asia*; Nauka: Novosibirsk, Russia, 1984; p. 192. (In Russian)
59. Parfenov, L.M.; Badarch, G.; Berzin, N.A.; Khanchuk, A.I.; Kuzmin, M.I.; Nokleberg, W.J.; Prokopiev, A.V.; Ogasawara, M.; Yan, H. Summary of Northeast Asia geodynamics and tectonics. *Stephan Mueller Spec. Publ. Ser.* **2009**, *4*, 11–33. [CrossRef]
60. Toro, J.; Miller, E.L.; Prokopiev, A.V.; Zhang, X.; Veselovskiy, R. Mesozoic orogens of the Arctic from Novaya Zemlya to Alaska. *J. Geol. Soc.* **2016**, *173*, 989–1006. [CrossRef]
61. Nokleberg, W.J.; Parfenov, L.M.; Norton, I.O.; Khanchuk, A.I.; Stone, D.B.; Scholl, D.W.; Fujita, K. *Phanerozoic Tectonic Evolution of the Circum-North Pacific*; US Geological Survey, Professional Papers: Denver, CO, USA, 2001; p. 123, No. 1626.
62. Fridovsky, V.Y.; Vernikovskaya, A.E.; Yakovleva, K.Y.; Rodionov, N.V.; Travin, A.V.; Matushkin, N.Y.; Kadilnikov, P.I. Geodynamic Formation Conditions and Age of Granitoids from Small Intrusions in the West of the Yana–Kolyma Gold Belt (Northeast Asia). *Russ. Geol. Geophys.* **2022**, *63*, 483–502. [CrossRef]
63. Akinin, V.V.; Prokopiev, A.V.; Toro, J.; Miller, E.L.; Wooden, J.; Goryachev, N.A.; Alshevsky, A.V.; Bakharev, A.G.; Trunilina, V.A. U–PB SHRIMP ages of granitoids from the Main batholith belt (North East Asia). *Dokl. Akad. Nauk* **2009**, *426*, 216–221. [CrossRef]
64. Gertseva, M.V.; Luchitskaya, M.V.; Sysoev, I.V.; Sokolov, S.D. Stages of formation of the main batholith belt in the Northeast of Russia: U–Th–Pb SIMS and Ar–Ar geochronological data. *Dokl. Akad. Nauk* **2021**, *499*, 5–10. [CrossRef]
65. Prokopiev, A.V.; Borisenko, A.S.; Gamyanin, G.N.; Pavlova, G.G.; Fridovsky, V.Y.; Kondrat'eva, L.A.; Anisimova, G.S.; Trunilina, V.A.; Ivanov, A.I.; Travin, A.V.; et al. Age constraints and tectonic settings of metallogenic and magmatic events in the Verkhoyansk–Kolyma folded area. *Russ. Geol. and Geoph.* **2018**, *59*, 1237–1253. [CrossRef]
66. Fridovsky, V.Y.; Yakovleva, K.Y.; Vernikovskaya, A.E.; Vernikovsky, V.A.; Matushkin, N.Y.; Kadilnikov, P.I.; Rodionov, N.V. Geodynamic emplacement setting of late jurassic dikes of the Yana–Kolyma gold belt, NE folded framing of the Siberian craton: Geochemical, Petrologic, and U–Pb Zircon Data. *Minerals* **2020**, *10*, 1000. [CrossRef]
67. Fridovsky, V.Y. Structural control of orogenic gold deposits of the Verkhoyansk-Kolyma folded region, northeast Russia. *Ore Geol. Rev.* **2018**, *103*, 38–55. [CrossRef]
68. Fridovsky, V.Y.; Kudrin, M.V.; Polufuntikova, L.I. Multi-stage deformation of the Khangalas ore cluster (Verkhoyansk–Kolyma folded region, northeast Russia): Ore-controlling reverse thrust faults and post-mineral strike-slip faults. *Minerals* **2018**, *8*, 270. [CrossRef]
69. Fridovsky, V.Y.; Kryazhev, S.G.; Goryachev, N.A. Physicochemical conditions of quartz formation at the Bazov gold-ore deposit (East Yakutia, Russia). *Russ. J. Pac. Geol.* **2019**, *13*, 407–416. [CrossRef]
70. Fridovsky, V.Y. Structures of gold ore fields and deposits of Yana–Kolyma ore belt. In *Metallogeny of Collisional Geodynamic Settings*; Mezhelovsky, N.V., Gusev, G.S., Eds.; GEOS: Moscow, Russia, 2002; Volume 1, pp. 6–241. (In Russian)
71. Fridovsky, V.Y.; Kudrin, M.V.; Polufuntikova, L.I.; Goryachev, N.A. Ore-controlling thrust faults at the Bazov gold-ore deposit (Eastern Yakutia). *Dokl. Akad. Nauk* **2017**, *474*, 617–619. [CrossRef]

72. Groves, D.I.; Goldfarb, R.J.; Gebre-Mariam, M.; Hagemann, S.G.; Robert, F. Orogenic gold deposits: A proposed classification in the context of their crustal distribution and relationship to other gold deposit types. *Ore Geol. Rev.* **1998**, *13*, 7–27. [CrossRef]
73. Savchuk, Y.S.; Volkov, A.V. Large and super-large orogenic gold deposits: Geodynamics, structure, genetic consequences. *Yakutia. Lithosphere.* **2019**, *19*, 813–833. [CrossRef]
74. Anisimova, G.S.; Kondratieva, L.A.; Serkebaeva, E.S.; Ageenko, V.A. Badran gold deposit, Eastern Yakutia. *Ores and metals.* **2008**, *5*, 49–60. (In Russian)
75. Anisimova, G.S.; Protopopov, R.I. Geological structure and composition of ores of the gold-quartz deposit of Vyun, Eastern. *Ores and Met.* **2009**, *5*, 59–69.
76. Fridovsky, V.Y.; Polufuntikova, L.I.; Gamyanin, G.N.; Solovyov, E.E. Orogenic gold deposits with significant resource potential in the central part of the Yano-Kolyma belt. *Explor. Protect. Miner. Res.* **2015**, *11*, 3–9.
77. Polufuntikova, L.I.; Fridovsky, V.Y.; Goryachev, N.A. Geochemical features of ores and host rocks of the orogenic Malo-Tarynskoye gold deposit (Verkhoyano-Kolyma folded region, North-East of Russia). *Russ. J. of Pacific Geol.* **2020**, *39*, 41–55. [CrossRef]
78. Polufuntikova, L.I.; Fridovsky, V.Y.; Tarasov, Y.A.; Kudrin, M.V. Multistages mineralization and transformation of terrigenous rocks in the Vyun ore field, Yana-Kolyma metallogenic belt, Northeast Asia: Insight from the sedimentary, diagenetic and hydrothermal sulfides and geochemistry of ore-hosting rocks. In *Proceeding of the IOP Conference Series: Earth and Environmental Science*; IOP Publishing: Prague, Czech Republic, 2021; Volume 906, p. 012041. [CrossRef]
79. Joint-Stock Company «VEDK» 2020. Available online: https://www.vedk.com/assets/resources (accessed on 16 September 2022).
80. Zapadnaya Gold Mining Ltd. 2011. Available online: https://www.zapadnaya.ru/qa/zapasy_resursy.html (accessed on 16 September 2022).
81. Gold mining 2008–2023. Available online: https://zolotodb.ru/article/12177 (accessed on 16 September 2022).
82. 2022 «Gold and Technology». Available online: https://zolteh.ru/geology/novaya-kontseptsiya-razvitiya-i-osvoeniya-mineralno-syrevoj-bazy-msb-rudnogo-zolota-v-regionah-sibiri-i-dalnego-vostoka/ (accessed on 16 September 2022).
83. Limited Liability Company "Business Media "Far East", 2008–2021. Available online: https://nedradv.ru (accessed on 16 September 2022).
84. Zaitsev, A.I.; Fridovsky, V.Y.; Vernikovskaya, A.E.; Kudrin, M.V.; Yakovleva, K.Y.; Kadilnikov, P.I. Rb-Sr isotopic study of basites of the dike complex of the Taryn ore-magmatic system (Northeast Russia). *Russ. J. Domes. Geol.* **2018**, *5*, 50–61. (In Russian)
85. Zaitsev, A.I.; Fridovsky, V.Y.; Kudrin, M.V. Granitoids of the Ergelyakh intrusion-related gold–bismuth deposit (Kular-Nera Slate Belt, Northeast Russia): Petrology, physicochemical parameters of formation, and ore potential. *Minerals* **2019**, *9*, 297. [CrossRef]
86. Warr, L.N. IMA–CNMNC approved mineral symbols. *Mineral. Mag.* **2021**, *85*(3), 1–30. [CrossRef]
87. Akimov, G.Y. New data on the age of gold-quartz mineralization in the Upper Indigirka region. *Dokl. Akad. Nauk* **2004**, *398*, 80–83. (In Russian)
88. Zaitsev, A.I.; Fridovsky, V.Y.; Kudrin, M.V. Intensive formation parameters and mineragenic potential of the granitoids of the Kurdat and Samyr massifs, Tas-Kysbyt magmatic belt of the Verkhoyansk-Kolyma folded region. *Russ. J. Domes. Geol.* **2017**, *5*, 80–89. (In Russian)
89. Layer, P.W.; Newberry, R.J.; Fujita, K.; Parfenov, L.; Trunilina, V.; Bakharev, A. Tectonic setting of the plutonic belts of Yakutia, northeast Russia, based on 40Ar/39Ar geochronology and trace element geochemistry. *Geology* **2001**, *29*, 167–170. [CrossRef]
90. Fridovsky, V.Y.; Zaitsev, A.I. New data on the timing of gold mineralization in the Upper Indigirka region (Yana-Kolyma belt) from Ar-Ar and Rb-Sr dating. In *Proceeding of "Geological Processes in the Environments of Subduction, Collision, and Sliding of Lithospheric Plates"*; Izdatelstvo Dalnauka: Vladivostok, Russia, 2018; pp. 296–299. (In Russian)
91. Fridovsky, V.Y. Strike slip fault duplexes of the Badran deposit (North-East Yakutia). Izvestiya vysshikh uchebnykh zavedenii. *Geol. Explor.* **1999**, *1*, 60–66. (In Russian)
92. Fridovsky, V.Y. Verkhoyansk metallogenic zone: Metamorphogenic Au quartz deposits. In *Tectonics, Geodynamics and Metallogeny of the Territory of the Republic of Sakha (Yakutia)*; MAIK/Interperiodika: Moscow, Russia, 2001; pp. 353–355.
93. Anisimova, G.S.; Serkebaeva, E.S.; Kondratyeva, L.A. Native gold of the Badran deposit. *Domestic Geol.* **2006**, *5*, 38–47. (In Russian)
94. Obolensky, A.A.; Gushchina, L.V.; Anisimova, G.S.; Serkebaeva, E.S.; Tomilenko, A.A.; Gibsher, N.A. Physical and chemical modeling of mineral formation processes of the Badran gold deposit (Yakutia). *Geol. and Geophys.* **2011**, *52*, 373–392. [CrossRef]
95. Rozhkov, I.S.; Grinberg, G.A.; Gamyanin, G.A.; Kukhtinskiy, Y.G.; Solovyev, V.I. *Late Mesozoic Magmatism and Gold Mineralization of the Upper Indigirsky Region*; Science: Moscow, Russia, 1971; p. 240. (In Russian)
96. Kudrin, M.V.; Zayakina, N.V.; Fridovsky, V.Y.; Galenchikova, L.T. Hydrous ferric sulfate–$Fe(SO_4)(OH)2H_2O$ from the supergene zone of the Khangalas gold deposit, Eastern Yakutia, Russia. *Notes Rus. Miner. Soc.* **2020**, *149*, 126–141. [CrossRef]
97. Kudrin, M.V.; Polufuntikova, L.I.; Fridovsky, V.Y.; Aristov, V.V.; Tarasov, Y.A. Geochemistry and the form of «invisible» gold in pyrite from metasomatites of the Khangalas deposit, North-East of Russia. *Arct. Subarct. Nat. Resources.* **2020**, *25*, 7–14. [CrossRef]
98. Polufuntikova, L.I.; Fridovsky, V.Y.; Akimova, E.D.; Zaitsev, A.I. Hydrothermal-metasomatic changes in the host rocks of the Vyun ore field (Kular-Nera slate belt, North-East Russia). In *Geology and Mineral Resources of the North-East of Russia. Materials of the IX All-Russian Scientific and Practical Conference*; Izd. Dom SVFU: Yakutsk, Russia, 2019; Volume 1, pp. 77–80.
99. Goryachev, N.A.; Gamyanin, G.N.; Prokof'ev, V.Y.; Velivetskaya, T.A.; Ignat'ev, A.V.; Leskova, N.V. Ag-Sb mineralization of the Yana-Kolyma belt, Northeast Russia. *Russ. J. Pac. Geol.* **2011**, *5*, 97–110. [CrossRef]
100. Ignatiev, A.V.; Velivetskaya, T.A.; Budnitskiy, S.Y.; Yakovenko, V.V.; Vysotskiy, S.V.; Levitskii, V.I. Precision analysis of multisulfur isotopes in sulfides by femtosecond laser ablation GC-IRMS at high spatial resolution. *Chem. Geol.* **2018**, *493*, 316–326. [CrossRef]

101. Velivetskaya, T.A.; Ignatiev, A.V.; Yakovenko, V.V.; Vysotskiy, S.V. An improved femtosecond laser-ablation fluorination method for measurements of sulfur isotopic anomalies ($\Delta^{33}S$ and $\Delta^{36}S$) in sulfides with high precision. *Rapid Commun. Mass Spectrom.* **2019**, *33*, 1722–1729. [CrossRef] [PubMed]
102. Sotskaya, O.T.; Semyshev, F.I.; Malinovsky, M.A.; Alshevsky, A.V.; Livach, A.E.; Goryachev, N.A. Pyrite of sulphidization zones of terrigenic complexes of the Yana-Kolyma orogenic belt (Northeast Russia): Generations, typochimism, mineral associations. *Bull. NE Sci. Cent. FEB RAS* **2022**, *1*, 14–30. [CrossRef]
103. Tardani, D.; Reich, M.; Deditius, A.P.; Chryssoulis, S.; Sánchez-Alfaro, P.; Wrage, J.; Roberts, M.P. Copper–arsenic decoupling in an active geothermal system: A link between pyrite and fluid composition. *Geochim. Cosmochim. Acta* **2017**, *204*, 179–204. [CrossRef]
104. Deditius, A.P.; Utsunomiya, S.; Ewing, R.C.; Chryssoulis, S.L.; Venter, D.; Kesler, S.E. Decoupled geochemical behavior of As and Cu in hydrothermal systems. *Geology* **2009**, *37*, 707–710. [CrossRef]
105. Deditius, A.P.; Utsunomiya, S.; Ewing, R.C.; Kesler, S.E. Nanoscale "liquid" inclusions of As-Fe-S in arsenian pyrite. Am. *Mineral* **2009**, *94*, 391–394. [CrossRef]
106. Reich, M.; Simon, A.; Deditius, A.; Barra, F.; Chryssoulis, S.; Lagas, G.; Tardani, D.; Knipping, J.; Bilenker, L.; Sánchez-Alfaro, P.; et al. Trace element signature of pyrite from the Los Colorados Iron Oxide-Apatite (IOA) deposit, Chile: A missing link between Andean IOA and Iron Oxide Copper-Gold systems? *Econ. Geol.* **2016**, *111*, 743–761. [CrossRef]
107. Reich, M.; Lagas, G.; Deditius, A. New advances in trace element geochemistry of ore minerals and accessory phases. *Ore Geol. Rev.* **2017**, *81*, 1215–1217. [CrossRef]
108. Peterson, E.C.; Mavrogenes, J.A. Linking high-grade gold mineralisation to earthquake-induced fault-valve processes in the Porgera gold deposit, Papua New Guinea. *Geology* **2014**, *42*, 383–386. [CrossRef]
109. Qian, G.; Brugger, J.; Testemale, D.; Skinner, W.; Pring, A. Formation of As(II)-pyrite during experimental replacement of magnetite under hydrothermal conditions. *Geochim. Cosmochim. Acta* **2013**, *100*, 1–10. [CrossRef]
110. Deditius, A.P.; Utsunomiya, S.; Renock, D.; Ewing, R.C.; Ramana, C.V.; Becker, U.; Kesler, S.E. A proposed new type of arsenian pyrite: Composition, nanostructure and geological significance. *Geochim. Cosmochim. Acta* **2008**, *72*, 2919–2933. [CrossRef]
111. Sazonov, A.M.; Kirik, S.D.; Silyanov, S.A.; Bayukov, Y.A.; Tishin, P.A. Typomorphism of arsenopyrite of the Blagodatnoye and Olympiada gold deposits (Yenisei Ridge). *Mineralogy* **2016**, *3*, 53–70. (In Russian)
112. Genkin, A.D. Gold-bearing arsenopyrite from gold deposits: The internal structure of grains, composition, growth mechanisms and condition of gold. *Geol. Ore Depos.* **1998**, *40*, 551–557. (In Russian)
113. Cabri, L.J.; Newville, M.; Gordon, R.; Crozier, D.; Sutton, S.; McMahon, G.; De-Tong, J. Chemical speciation of gold in arsenopyrite. *Can. Mineral* **2000**, *38*, 1265–1281. [CrossRef]
114. Bralia, A.; Sabatini, G.; Troja, F. A revaluation of the Co/Ni ratio in pyrite as geochemical tool in ore genesis problems. *Miner. Depos.* **1979**, *14*, 353–374. [CrossRef]
115. Bajwah, Z.U.; Seccombe, P.K.; Offler, R. Trace element distribution, Co:Ni ratios and genesis of the Big Cadia iron-copper deposit, New South Wales, Australia. *Miner. Depos.* **1987**, *22*, 292–303. [CrossRef]
116. Azovskova, O.B.; Utochkina, N.V.; Zubova, T.P. Geochemical features of pyrite and marcasite from weathering crusts and "ancient" loose deposits of the Aktai area (Northern Urals). In *Ezhegodnik-2013, Tr. IGG UrO RAN*; URO RAN: Ufa, Russia, 2014; pp. 238–245, No. 161. (In Russian)
117. Wang, J.; Liu, Z.; Wang, K.; Zeng, X.; Liu, J.; Zhang, F. Typomorphic Characteristics of Pyrites from the Shuangwang Gold Deposit, Shaanxi, China: Index to Deep Ore Exploration. *Minerals* **2019**, *9*, 383. [CrossRef]
118. Onufrienok, V.V. Comparative analysis of the density of vacant positions and the density of impurity atoms in the structure of pyrite. *Adv. Mod. Nat. Sci.* **2013**, *7*, 61–67.
119. Large, R.R.; Maslennikov, V.V. Invisible Gold Paragenesis and Geochemistry in Pyrite from Orogenic and Sediment-Hosted Gold Deposits. *Minerals* **2020**, *10*, 339. [CrossRef]
120. Lee, M.; Shin, D.; Yoo, B.; Im, H.; Pak, S.; Choi, S. LA-ICP-MS trace element analysis of arsenopyrite from the Samgwang gold deposit, South Korea, and its genetic implications. *Ore Geol. Rev.* **2019**, *114*, 103147. [CrossRef]
121. Volkov, A.V.; Genkin, A.D.; Goncharov, V.I. On the forms of gold occurrence in the ores of the Natalka and Mayskoe deposits (North-East of Russia). *Rus. J. of Pac. Geol.* **2006**, *25*, 18–29.
122. Moskvitina, L.V.; Moskvitin, S.G.; Anisimova, G.S. Research of Nanoscale Gold by Methods of Tunneling and Atomic-Powered Microscopy with Chemical and Ion-Plasma Etching in the Kuchus Deposit (Republic Sakha (Yakutia). In *International Science and Technology Conference "Earth Science"-Section One*; IOP Publishing: Prague, Czech Republic, 2019; Volume 272, pp. 1–7, No. 2. [CrossRef]
123. Cook, N.J.; Ciobanu, C.L.; Mao, J.W. Textural control on gold distribution in As-free pyrite from the Dongping, Huangtuliang and Hougou gold deposits, North China craton (Hebei Province, China). *Chem. Geol.* **2009**, *264*, 101–121. [CrossRef]
124. Tauson, V.L.; Kravtsova, R.G.; Smagunov, N.V.; Spiridonov, A.M.; Grebenshchikova, V.I.; Budyak, A.E. Structurally and superficially bound gold in pyrite from deposits of different genetic types. *Russ. Geol. Geophys.* **2014**, *55*, 273–289. [CrossRef]
125. Volkov, A.V.; Sidorov, A.A. Invisible gold. *Bull. RAS* **2017**, *87*, 40–49. [CrossRef]
126. Cline, J.S. Timing of gold and arsenic sulfide mineral deposition at the Getchell Carlin-type gold deposit, north-central Nevada. *Econ. Geol.* **2001**, *96*, 75–89. [CrossRef]
127. Bortnikov, N.S.; Vikentiev, I.V.; Tagirov, B.R.; Stavrova, O.O.; Cabri, L.J.; Mc Mahon, G.; Bogdanov, Y.A. Invisible gold in sufides from seafloor massive sulfide edifices. *Geol. Ore Depos.* **2003**, *45*, 201–212.

128. Cook, N.J.; Chryssouilis, S.L. Concentrations of invisible gold in the common sulfides. *Can. Mineral.* **1990**, *28*, 1–16.
129. Vikentyev, I.V. Invisible and microscopic gold in pyrite: Research methods and new data for pyrite ores of the Urals. *Geol. Ore Depos.* **2015**, *57*, 267–298. [CrossRef]
130. Wang, C.; Shao, Y.; Huang, K.; Zhou, H.; Zhang, J.; Liu, Z.; Liu, Q. Ore-Forming Processes at the Xiajinbao Gold Deposit in Eastern Hebei Province: Constraints from EPMA and LA-ICPMS Analysis. *Minerals* **2018**, *8*, 388. [CrossRef]
131. Voitkevich, G.V.; Miroshnikov, A.E.; Povarenykh, A.S.; Prokhorov, V.G. *A Short Guide to Geochem*, 2nd ed.; Nedra: Moscow, Russia, 1977; p. 184. (In Russian)
132. Belikova, G.I.; Salikhov, D.N.; Berdnikov, P.G. On the question of gold isomorphism in pyrite. *Geol. Collect. GI USC RAS* **2002**, *3*, 190–193. (In Russian)
133. Chouinard, A.; Paquette, J.; Williams-Jones, A.E. Crystallographic controls on trace-element incorporation in auriferous pyrite from the Pascua epithermal high-sulfidation deposit, Chile-Argentina. *Can. Miner.* **2005**, *43*, 951–963. [CrossRef]
134. Vaughan, J.P.; Kyin, A. Refractory gold ores in Archaean greenstones, Western Australia: Mineralogy, gold paragenesis, metallurgical characterization and classification. *Miner. Magaz.* **2004**, *68*, 255–277. [CrossRef]
135. Sharp, Z.D.; Essene, E.J.; Kelly, W.C. A re-examination of the arsenopyrite geothermometer; pressure considerations and applications to natural assemblages. *Can. Mineral.* **1985**, *23*, 517–534.
136. Lentz, D.R. Sphalerite and arsenopyrite at the Brunswick No. 12 massive sulfide deposit, Bathurst camp, New Brunswick: Constraints on P-T evolution. *Can. Mineral.* **2002**, *40*, 19–31. [CrossRef]
137. Kovalchuk, E.V.; Tagirov, B.R.; Vikentyev, I.V.; Chareev, D.A.; Tyukova, E.E.; Nikolsky, M.S.; Bortnikov, N.S. "Invisible" Gold in Synthetic and Natural Arsenopyrite Crystals, Vorontsovka Deposit, Northern Urals. *Geol. Ore Dep.* **2019**, *61*, 447–468. [CrossRef]
138. Groves, D.I.; Santosh, M.; Deng, J.; Wang, Q.; Yang, L.; Zhang, L. A holistic model for the origin of orogenic gold deposits and its implications for exploration. *Miner. Dep.* **2020**, *55*, 275–292. [CrossRef]
139. Bortnikov, N.S.; Gamyanin, G.N.; Vikentieva, O.V.; Prokofiev, V.Y.; Alpatov, V.A.; Bakharev, A.G. Composition and origin of fluids in the hydrothermal system of the Nezhdaninsky gold ore deposit (Sakha-Yakutia, Russia). *Geol. Ore Dep.* **2007**, *49*, 99–145. [CrossRef]
140. Brueckner, S.M.; Kline, A.K.; Bilenker, L.D.; Poole, J.; Whitney, M.S. Mineral Chemistry and Sulfur Isotope Geochemistry from Tonalite-Hosted, Gold-Bearing Quartz Veins at Hog Mountain, Southwestern Appalachians: Implications for Gold Precipitation Mechanism, Sulfur Source, and Genesis. *Econ. Geol.* **2021**, *116*, 357–388. [CrossRef]
141. Sapancı, Ö.; Köprübaş, N.; Çiftçi, E.; Köprübaş, N.; Tokat, G.; Demir, Y. Mineralogy, geochemistry, fuid inclusion, and stable sulfur isotope investigation of the Terziali shear related orogenic gold deposit (Central Anatolia, Turkey): Implications for ore genesis and mineral exploration. *Arab. J. Geosci.* **2022**, *15*, 113. [CrossRef]
142. Large, R.R.; Bull, S.W.; Maslennikov, V.V. A carbonaceous sedimentary source rock model for Carlin-type and orogenic gold deposits. *Econ. Geol.* **2011**, *106*, 331–358. [CrossRef]
143. Li, R.; Chen, H.; Large, R.R.; Zhao, L.; Liu, Y.; Jiao, J.; Xia, X.P.; Yang, Q. Ore-forming fuid source of the orogenic gold deposit: Implications from a combined pyrite texture and geochemistry study. *Chem. Geol.* **2020**, *552*, 119781. [CrossRef]
144. Chernyshev, I.V.; Chugaev, A.V.; Bortnikov, N.S.; Gamyanin, G.N.; Prokopiev, A.V. Pb isotopic composition and metal sources of Au and Ag deposits of the south Verkhoyansk region (Yakutia, Russia) according to high-precision MC-ICP-MS data. *Geol. of Ore Dep.* **2018**, *60*, 398–417. [CrossRef]
145. Goldfarb, R.J.; Leach, D.L.; Rose, S.C.; Landis, G.P. Fluid inclusion geochemistry of gold-bearing quartz veins of the Juneau Gold Belt, southeastern Alaska—Implications for ore genesis. *Econ Geol. Mono.* **1989**, *6*, 363–375.
146. Hoefs, J. *Stable Isotope Geochemistry, Eighth Edition*; Springer International Publishing AG, part of Springer Nature: Cham, Switzerland, 2018. [CrossRef]
147. Ohmoto, H. Systematics of sulfur and carbon isotopes in hydrothermal ore deposits. *Econ. Geol.* **1972**, *67*, 551–578. [CrossRef]
148. Goryachev, N.A.; Budyak, A.E.; Tarasova, Y.I.; Ignat'ev, A.V.; Velivetskaya, T.A. A case history of applying in situ analysis of the sulfur isotopic compositions of sulfides from ores of the largest deposits in the Bodaibo synclinorium (Eastern Siberia). *Dokl. Akad. Nauk* **2019**, *484*, 156–159. [CrossRef]
149. Dubinina, E.O.; Ikonnikova, T.A.; Chugaev, A.V. Heterogeneity of the sulfur isotopic composition of pyrite at the Sukhoi Log deposit and its controlling factors. *Dokl. Akad. Nauk* **2010**, *435*, 1665–1669. [CrossRef]
150. Kovalev, K.R.; Kalinin, Y.A.; Naumov, E.A.; Kolesnikova, M.K.; Korolyuk, V.N. Gold content of arsenopyrite of gold-sulfide deposits of East Kazakhstan. *Geol. Geophys.* **2011**, *52*, 225–242.
151. Kovalev, K.R.; Kuzmina, O.N.; Dyachkov, B.A.; Vladimirov, A.G.; Kalinin, Y.A.; Naumov, E.A.; Kirillov, M.V.; Annikova, I.Y. Gold-sulfide disseminated mineralization of the Zhaima deposit (East Kazakhstan). *Geol. Ore Dep.* **2016**, *58*, 116–133. [CrossRef]

minerals

MDPI

Article

Gold–Sulfide Mineralization in the Manitanyrd Region, Polar Urals, Russia

Tatiana Petrovna Mayorova [1,2,3,*], Sergei Karpovich Kuznetsov [1], Ludmila Ivanovna Efanova [1] and Natalia Vladimirovna Sokerina [1]

1 Yushkin Institute of Geology, Komi Science Center, Ural Branch, Russian Academy of Sciences, 167982 Syktyvkar, Russia
2 Institute of Natural Sciences, P. Sorokin Syktyvkar State University, 167001 Syktyvkar, Russia
3 Institute of Precambrian Geology and Geochronology Russian Academy of Sciences, 199034 Saint-Petersburg, Russia
* Correspondence: mayorova@geo.komisc.ru; Tel.: +7-8812-24-57-65

Abstract: This article describes the characteristics of gold–sulfide–quartz and gold–sulfide (gold–arsenic) ore occurrences in the Manitanyrd region of the Polar Urals. Ore occurrences are confined to NE-trending shear zones and have the common features of a geological structure. The host rocks are metamorphosed volcanic and volcanic–sedimentary rocks. We analyzed the mineral and chemical composition of the ore mineralization in all studied ore occurrences, showing that they belong to the same mineral type—pyrite–arsenopyrite, with a variable ratio of the main minerals. Arsenic pyrite is present in all ore occurrences. Two stages of ore formation were distinguished: early gold–pyrite–arsenopyrite with finely dispersed gold and late gold–galena–chalcopyrite–sphalerite with coarse gold, fahlore, and sulfosalts Pb, Cu, Bi, Sb. Native gold of the first generation, finely dispersed in arsenopyrite and pyrite, had an average to high fineness (800‰–1000‰) with a relatively low dispersion. Native gold of the second generation was larger, and its fineness in ore occurrences varied; in one of them, it varied from 300‰ to 950‰, while in others, it varied from 800‰ to 950‰. The isotope composition of sulfur in sulfides ($\delta^{34}S$) ranged between −0.2‰ and −8.0‰. $\delta^{34}S$ values of sulfides in the range of −0.2‰ to −3.5‰ were similar to meteorite, indicating the participation of a single deep magmatic source of sulfur in the ore formation. According to the study of fluid inclusions, the formation of ore quartz veins occurs in the temperature range of 467–109 °C. The similarity of the geological–structural, mineralogical–geochemical, and isotope–geochemical features of the gold–sulfide–quartz and gold–sulfide occurrences in the area suggest their formation in a single hydrothermal system.

Keywords: native gold; gold–sulfide–quartz (gold–arsenic) and gold–sulfide formations; Polar Urals; Manitanyrd ridge; ore occurrences; ore mineralogy; sulfur isotopes

Citation: Mayorova, T.P.; Kuznetsov, S.K.; Efanova, L.I.; Sokerina, N.V. Gold–Sulfide Mineralization in the Manitanyrd Region, Polar Urals, Russia. *Minerals* **2023**, *13*, 747. https://doi.org/10.3390/min13060747

Academic Editor: Huan Li

Received: 11 April 2023
Revised: 25 May 2023
Accepted: 25 May 2023
Published: 30 May 2023

1. Introduction

Gold deposits are widely developed in the various ore provinces of the world. In the literature, a significant proportion of hydrothermal vein gold deposits are classified as orogenic in metamorphic terranes, and they display large variations in their mineralization age and style [1–5]. A feature of the mineral composition of these deposits is the constant presence of pyrite and arsenopyrite in varying proportions. Arsenopyrite is the most common sulfide mineral in metasedimentary country rocks, while pyrite or pyrrhotite is more typical of metamorphosed igneous rocks [1,2]. According to [6–8], hydrothermal gold deposits are subdivided into gold–quartz, gold–sulfide–quartz, and gold–quartz–sulfide formations. The gold–quartz–sulfide formation is characterized by a high content of sulfides (>15%), with the gold–arsenic subformation standing out. The typical features of the gold–arsenic deposits are (1) a predominance of arsenopyrite or pyrite in ores; (2) a predominant connection between gold and sulfides, with finely dispersed "invisible" gold in ores; (3) a high positive correlation between gold and arsenic; (4) refractory arsenic primary ores [9,10]. In Russia, this type includes bulk mineable gold deposits such as Olimpiada, Mayskoe, Nezhdaninskoe, and Kyuchyus [8,11–16]. Gold–arsenic deposits occur in various rocks, such as carbon-bearing terrigenous rocks and carbonate–terrigenous suites of different ages and formation categories, metamorphosed under conditions of low (not higher than greenschist) facies of regional metamorphism, which are less common in igneous rocks of basalt–andesite–dacite formation [8,17,18]. Typical of such deposits is a paragenetic relationship with minor intrusions of variegated composition and a considerable distance from large granitoid intrusions. Zones of shear, stratification, cleavage, cutting folds, and areas of injection in the crest parts of the folds serve as ore-hosting structures of gold–arsenic–sulfide deposits. Ore bodies, represented by mineralized zones and deposits, have different forms, subject to the elements of lithological and structural control. Gold is distributed rather evenly in ore bodies with relatively moderate mean contents [8].

Gold ore occurrences of the Manitanyrd region belong to the gold–sulfide–quartz and gold–sulfide formations. Their increased content of arsenopyrite allows us to consider them as gold–arsenic–sulfide. A feature of the geological structure of gold manifestations in this area is their localization in volcanogenic and volcanogenic–sedimentary rocks.

The Manitanyrd gold mining area is located on the western slope of the Polar Urals, within the Manitanyrd ridge. Studies of ore mineralization in this region started in the 1960s when the first gold–sulfide–quartz deposit, Verkhniayuskoye-2, along with several closely spaced gold–sulfide occurrences, were discovered. According to exploration data, the reserves of the Verkhniayuskoye-2 deposit amounted to 3.4 tons of gold (up to 4.8 g/t) and 10.9 tons of silver; it was recognized as small, with refractory pyrite–arsenopyrite ores, and unpromising. No further exploration was carried out, although boreholes traced mineralization to a depth of 240 m without signs of wedging out, and with a high gold grade (approximately 14 g/t), i.e., there were prospects for increasing gold reserves at depth, as well as on the flanks of the deposit. As a result of subsequent prospecting and revision works (1988–2009), two more new occurrences of gold and numerous points with signs of gold mineralization were discovered in the area [19]. At present, the Manitanyrd region is considered one of the industrially promising gold-bearing regions in the north of the Urals.

Gold ore occurrences of the Manitanyrd region belong to the gold–sulfide–quartz and gold–sulfide (gold–arsenic) formation types and a gold–pyrite–arsenopyrite mineral type [19,20]. According to previous research [21], ore-grade gold mineralization in this region is polygenetic and polychronous. Early pyrite–arsenopyrite mineralization with finely dispersed gold, accompanied by propylitic alteration, is typical of mesothermal Au–Ag deposits and was formed during the Riphaean age. Late ore-grade mineralization (sphalerite + galena + native gold), confined to beresites, listvenites, and quartz–sericitic metasomatites, is shallow in depth and is likely to be related to the Late Paleozoic collision processes in the Urals. It is assumed that there is a deep-seated source of the substance of gold occurrences in the Manitanyrd region. It is reasonable that the Verkhneniyayuskoye-2 deposit, most thoroughly studied by the data of mineralogical studies, has two prominent stages of ore formation: gold–pyrite–arsenopyrite and gold–chalcopyrite–galena–sphalerite, separated by the period of cataclasis of the primary ores [22].

A basic understanding of the polygenetic and polychronic nature of the ore-grade gold mineralization in the Manitanyrd region is based on geological, structural, textural, mineralogical, and geochemical data and a few determinations of the absolute age of the associated metasomatites.

To date, the gold ore occurrences of the Manitanyrd region have been studied in various degrees of detail. This study reports on new mineralogical and geochemical data—data on the fluid inclusions and sulfur isotopes of the gold–sulfide mineralization of the area in order to decipher the characteristics of ore-forming fluids and to ultimately determine the sources of the substances.

2. Geology of the Area

2.1. Regional Geological Setting

In the tectonic structure of both the Polar Urals and the entire Urals, the Manitanyrd ridge is positioned in the meridionally elongated West Uralian folded zone, involving the shallow- and deep-water sediments of the Paleozoic continent margin [23]. It constitutes the autochthone in the form of the Manitanyrd anticlinorium, a compound anticline fold, northwestwardly overturned with a predominant southwestward dip of the axial planes at a 60–75° angle [24]. The structure is northeasterly oriented. An anticlinal core is built by the Upper Riphaean and Vendian rocks of the Bedamel series, and the Enganepe suite wings by the Vendian–Cambrian deposits of the Manitanyrd series (Figure 1).

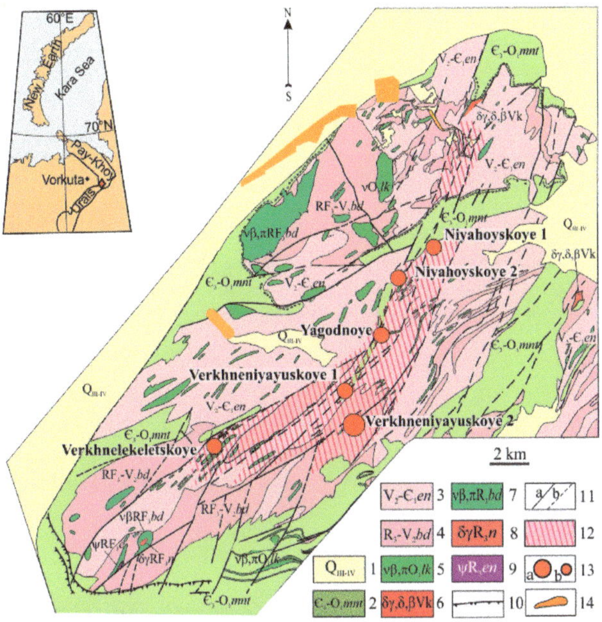

Figure 1. The geological structure and location of gold ore occurrences in the Manitanyrd region (modified from [20]). 1, Quaternary deposits; 2, undivided Manitanyrd series (Upper Cambrian–Lower Ordovician); 3, Enganepe suite (Upper Vendian–Lower Cambrian); 4, undivided Bedamel series (Upper Riphean–Vendian). Intrusions: 5, Lekvozhsky complex: olivine gabbro, dolerites and picrodolerites; 6, Kyzygeiskii complex: diorites, granodiorites, and plagiogranites; 7, Lower Bedamel complex: gabbro and gabbro–dolerites; 8, Niyayu complex: diorites, granodiorites, and plagiogranites; 9, Enganepe complex: serpentinites, apoperidotite, and apopyroxenite; 10, overthrust nappes; 11, faults: a—proved, b—probable; 12, Niyayu ore zone; 13, gold ore deposits (a) and occurrences (b); 14, placer gold occurrences.

The Bedamel series is undivided (R_3–V_2bd). In the deeper section, it is composed of basalts, andesite–basalts, andesites, trachybasalts and their clastolavas and lava breccias, dacites and their tuffs, and tuff–conglomerates and tuff–gritrock. The upper section contains acidic lavas, their clastolavas, and tuffs [24]. The series thickness is 2000–2700 m.

The Enganepe suite (V_2–$Є_1$en) is characterized by rough interbedding of medium- and course-grained sandstones, tuff sandstones, and tuff–conglomerates (coarse-grained arenaceous flysch or molasse). The suite thickness is 1200–1500 m.

The Manitanyrd series (Є–O_1mnt), with angular unconformity, superposes the erosional surface of the Bedamel series and Enganepe suite rocks. It is formed by conglomerates, metaquartzites, and siltstones. The series thickness varies greatly from 500 to 800 m.

The volcanites of the Bedamel series are considered to have been formed during the development of the island arc that, perhaps, had not undergone the final stages of its evolution [24]. The lower age limit of this series is unknown, while the upper limit is identified by the age of the latest geological relationships between the subvolcanic and extrusive rhyolites of the upper part of the series—555–547 Ma [25].

The intrusive rocks in the Manitanyrd region were developed in a minor way (Figure 1). Most numerous are the small bodies and dikes of gabbro and dolerites of the Lower Ordovician Lekvozhsky complex (476 ± 61 Ma [24]). Generally, they are northeasterly elongated, in line with the predominant extension of geological structures. Minor intrusions have been found of serpentized picrites, granodiorites, gabbro, and gabbro–dolerites of the Late Riphean age and quartz diorites of the Vendian age (Figure 1).

The central part of the region is characterized by large, mostly NE-trending faults, which probably occurred in the pre-Ordovician, with subsequent renewal in the Ordovician and at a later time. These faults are confined to the bodies of gabbro and gabbro–dolerites, zones of intense schist-forming and crushing of rocks, and their metasomatic alterations (sericitization, chloritization, and epidotization).

All rocks of the region underwent regional metamorphism of greenschist facies. Commonly developed are hydrothermal quartz, sulfide–quartz, quartz–carbonate, quartz–chlorite, quartz–tremolite, and quartz–epidote veins and veinlets, mostly filling NE-trending cracks, oriented in conformity with the schistosity of the rocks. The veins have a lenticular, slab-shaped form with a thickness that varies from a few centimeters to 20–30 cm. Thus, lode gold mineralization, barren quartz veins, and veinlet-disseminated mineralization in the host rocks are present in the area.

2.2. *Ore Deposit Geology*

All of the examined gold ore occurrences in the Manitanyrd region are located within the 4–5 km-wide northeast-trending Niyayu ore zone, extending for approximately 25 km [19]. Here, Niyahoyskoye-1 and -2, Yagodnoye, Verkhneniyayuskoye-1 and -2, and Verkhnelekeletskoye gold occurrences are positioned from northeast to southwest (Figure 1).

2.2.1. Verkhneniyayuskoye-2 Deposit

The Verkhneniyayuskoye-2 deposit is located in the central part of the Manitanyrd region and the Niyayu ore zone (Figure 1). Within the deposit area, effusive rocks have been found that have the principal composition of the Bedamel series—their tuffs, tuff sandstones, tuff siltstones, and orthoschists [21,26]. Four ore zones stand out in the deposit: Southern, Intermediate, Northern, and Polar, that are northeasterly elongated for 1200 m [19,21,22,27]. The ore zones are 4 to 6 m wide and are traceable downward to a depth of 240 m. The ore zones contain veins and veinlets of sulfide–quartz, quartz, and quartz–chlorite compositions. Quartz veins with sulfide mineralization incorporate lenses of massive sulfide ores with the highest gold content. The thickness of the ore veins is up to 0.2–0.5 m, while that of the veinlets is 4–5 cm. There are distinct boundaries between the veins and host rocks. In selvages, there are traces of slipping, with the intra-ore layers being schistosed and mylonitized. All of the veins are cataclased and often transformed into breccia. The ore bodies are oriented according to the schistosity of the rocks, which have a NE trend (35–50°) and a SE dip at 60–70° [21,26]. In terms of their mineral composition, the ores are classified as arsenopyrite + pyrite, pyrite + sphalerite + arsenopyrite, and sphalerite + arsenopyrite. In the upper part of the deposit, the ores are considerably oxidized.

All other ore occurrences (Niyahoyskoye-1, Niyahoyskoye-2, Yagodnoye, and Verkhneniyayuskoye-1) are confined to the posterior deposits of the Enganepe suite of the

Late Vendian–Early Cambrian; tuff stones, tuff schists, and albite–sericite schists with rare interlayers of basic effusive rocks. Inherent in them is the development of gold-bearing vein zones and veinlet-disseminated sulfide mineralization in the host rocks.

2.2.2. Verkhneniyayuskoye-1 Occurrence

The Verkhneniyayuskoye-1 occurrence is 1.6 km northeast of the Verkhneniyayuskoye-2 deposit. The ore zone is northeasterly elongated as a narrow strip for 1.4 km. The ore mineralization is confined to zones of intense schistose rocks. The thickness of the veinlets varies from several millimeters to 10–15 cm. They occur as an echelon, in accordance with the host rocks, and form zones that are 2–3 m and, rarely, 8–10 m long. Mineralization has been traced by wells to a depth of 32 m. The ore veinlets have a NNE strike (15–20°) and a SE dip at 60–70° [21,26]. The ores in the upper part of the occurrence are also considerably oxidized. In addition to sulfide and quartz–sulfide veinlets, quartz and quartz–carbonate barren veins also occur conformably with the host rocks and are well represented throughout the ore occurrence.

2.2.3. Niyahoyskoye-2 Occurrence

The Niyahoyskoye-2 ore occurrence is located in the north–northeastern part of the Niyayu ore zone (Figure 1). It contains a stripe of mineralized rocks, extending along a NNE strike fault (20–30°) with a SE dip at 50–70°, and is confined to its lower plate near the eastern contact of the dike of dolerites [28]. The ore mineralization is penetrated by up to 122 m-deep wells and has been traced at 160 m with a mean thickness of approximately 14 m. The host tuff stones of the Enganepe suite within the ore zone are strongly cataclased and schistosed, with the hydrothermal alteration involving pyritization, silicification, chloritization, sericitization, and carbonatization. The ore mineralization is represented by quartz–sulfide and sulfide veins and veinlets, oriented according to the schistosity of the rocks. The length of the veins and veinlets reaches 1–3 m, with a thickness that varies from 1 to 30 cm. The ores are pyrite + arsenopyrite. There are also barren quartz, quartz–chlorite, and quartz–carbonate veins and veinlets. A feature of the occurrence is the wide development, along with lode ore veins, of the veinlets and disseminated gold-bearing mineralization.

2.2.4. Niyahoyskoye-1 Occurrence

The Niyahoyskoye-1 ore occurrence is positioned northeast of the Niyahoyskoye-2 occurrence. The ore mineralization is confined to the sites with the most intense deformities and schist formation in the fault zone and is controlled by a system of fine cracks. The host rocks, i.e., deposits of the Enganepe suite and dikes of gabbro–dolerites, are schistosity, and their hydrothermal alterations include silicification, carbonatization, sericitization, and chloritization. The gold-bearing veins are quartz–pyrite–arsenopyrite, which is concordant with the host rocks and cutting them. Barren quartz, quartz–chlorite, and quartz–epidote veins are well represented.

2.2.5. Verkhnelekeletskoye Occurrence

The Verkhnelekeletskoye ore occurrence is located in the SE part of the Niyayu ore zone (Figure 1). The ore occurrence is in the margin part of the dolerite dike of the Lekvozhsky complex and is 4–6 m thick. The ore zone constitutes a stripe of metasomatites of variable thickness (0.5–2 m) parallel to the dike contact, with disseminated sulfide mineralization, a lenticular quartz vein, and numerous veinlets, also containing sulfide (gold + arsenopyrite + pyrite) mineralization. The quartz vein has a maximum thickness of 40 cm and a length of approximately 5 m [29].

2.2.6. Yagodnoye Occurrence

The Yagodnoye ore occurrence is positioned in the northeastern part of the Niyayu ore zone between the Verkhneniyayuskoye-1 and Niyahoyskoye-2 occurrences (Figure 1).

It constitutes a mineralized zone of schistosed, cataclased, and hydrothermally altered tuff stones and tuff siltstones extended along a NE-trending fault. On the surface, the zone thickness is approximately 20 m, increasing to a depth of 30 m. The ore mineralization is mostly represented by disseminated pyrite with a content that is no more than 1%. Single gold-bearing quartz–arsenopyrite veins (2–30 cm thick) have been identified. Quartz, quartz–carbonate, and epidote–quartz veins and veinlets with disseminated sulfides (pyrite, chalcopyrite, pyrrhotite, and, rarely, galena) are represented more extensively. Free native gold has been found in the mineralized zone rocks.

3. Materials and Methods

3.1. Sampling

This research study employed 150 specimens of rock, ore, and vein quartz collected in the course of field works from 1982 to 2009 in the Manitanyrd region, in the gold ore occurrences Verkhneniyayuskoye-1 (10), Verkhneniyayuskoye-2 (50), Niyahoyskoye-1 (10), Niyahoyskoye-2 (30), Yagodnoye (30), and Verkhnelekeletskoye (20). To examine the mineral composition of the ores and the chemical composition of the minerals, 2–4 polished sections, each from the most representative specimens, were used. A total of 60 polished sections were analyzed. Pure free gold separates were handpicked under a binocular microscope after crushing, followed by magnetic and heavy liquid separation. Then, 15 ore specimens from various ore occurrences were selected to determine the isotope composition of sulfur. Pure sulfide separates were handpicked under a binocular microscope after crushing, followed by magnetic and heavy liquid separation. These separates were then ground to an approximate 200 mesh in an agate mortar. To study the fluid inclusions, double-polished plates were made from the quartz specimens. More than 20 crystal plates were analyzed.

3.2. Methods

Most of the analytical studies were carried out in the Institute of Geology, Komi Research Center, Ural Branch of Russian Academy of Sciences (Syktyvkar, Russia), Centre of Isotope Studies of the All-Russian Geological Research Institute (Saint Petersburg, Russia), the Laboratory of Stable Isotopes of the Research Equipment Sharing Centre of the Far East Geological Institute of the Far-Eastern Division of Russian Academy of Sciences (Vladivostok, Russia), and, in part, the Institute of Geology and Mineralogy of the Siberian Branch of Russian Academy of Sciences (Novosibirsk, Russia).

The composition of the native gold and ore minerals in the polished sections and in the individual grains was studied using a JSM-6400 scanning electron microscope (JEOL Ltd., Tokyo, Japan) with a Microspec wave spectrometer, model WDX-400 (City, Microspec Corporation, Fremont, CA, USA), and a Link energy-dispersive spectrometer with the ISIS analysis program. The analytical conditions were as follows: accelerating voltage 20 kV, beam current 30 nA, beam diameter 2 μm, and live spectra acquisition time 30 s. The following X-ray lines were selected: Kα for Fe, Cu, Zn, As, and S; Lα for Ag; Mα for Pb, Bi, Au, Sb, and Hg. Pure metals (Fe, Cu, Zn, Ag, Au, Pb, Bi, and Sb) were used as the standards: FeS_2 for Fe, PbS for S, InAs for As, and CdHgTe for Hg. The detection limits (wt.%) were: 0.1 Fe, 0.15 Cu, 0.25 Ag, Sb, 0.3 As, 0.6 Au, and 0.8 Hg. Part of the chemical analyses of the simple sulfides was performed on a Tescan Vega 3LMN scanning electron microscope (Tescan, Czech Republic) with an X-Max 50 energy dispersive spectrometer (Oxford Instruments, Oxford, UK). The following parameters were used: accelerating voltage 20 kV, beam current 15 nA, beam diameter up to 1 μm, and time to register the spectra of 600,000 counts. The following X-ray lines were selected: Kα for Fe, Cu, Zn, and S; Lα for Ag, Sb, and As; Mα for Pb, Au, Bi, and Hg. Determination error (wt.%): 0.15 Fe, 0.22 As, 0.14 S, 0.29 Cu, 0.69 Pb, 0.51 Bi, 0.2 Sb, 0.5 Au, 0.38 Ag, 0.43 Hg, and 0.2 Zn. Pure metals for Fe, Cu, Zn, Ag, Au, Pb, Bi, and Sb were used as the standards: FeS_2 for Fe and S; PbTe for Pb; InAs for As; HgTe for Hg. The compositions of the fine gold particles (<10 μm) and sulfide microinclusions in pyrite and arsenopyrite were determined with a 190 nm point probe, but the size of the generation region of the X-ray emission in gold and sulfides with an electron beam energy 20 kV was up to 1 μm. Therefore, these analyses cannot be considered quantitative if the minimum size of the object under study is approximately 1 μm. Scanning electron microscopy was used to determine the textural and structural characteristics of the ores and the morphology of the minerals, with the resulting images in back-scattered electrons (BSEs).

The fluid inclusions in the quartz were studied in polished plates with a thickness of 200–300 μm. Homogenization and cryometry of the inclusions were partially carried out in a Linkam THMSG 600 thermal cryo-chamber mounted on an Olympus BX51 microscope (Institute of Geology and Mineralogy, Siberian Branch of the Russian Academy of Sciences, Novosibirsk), with the precision of the measurements being ±1 °C during heating and ±0.2 °C during freezing using a measurement range from −196 to 600 °C. Homogenization of the inclusions was also carried out on a commercial UMTK-3 setup mounted on an Amplival microscope (Carl Zeiss Jena, Germany) with a long-focus objective of 50x/0.50 BD (Institute of Geology, Komi Research Center, Ural Branch, Russian Academy of Sciences, Syktyvkar), a measurement error of ±2 °C, and a range of measurements from room temperature to 600 °C. The salt composition of the fluids in the inclusions was defined by the eutectic temperature of the liquid phase [30]. The salinity of the solutions in the inclusions was measured by the ice melting temperature [31]. The study of fluid inclusions and the interpretation of the obtained data were carried out on the basis of the criteria proposed by Roedder [32].

Isotope analysis of the sulfur in the sulfides was conducted on a Finnigan MAT 253 isotope mass spectrometer (ThermoFinnigan, Bremen, Germany) using the double-puffing system according to Grinenko's methodology [33]. The sulfur isotope ratio is reported as the permil (‰) deviations of the international VCDT standard. The monomineral specimens of sulfides weighing 10 mg were analyzed. The accuracy of $\delta^{34}S$ (2σ) was ±0.20‰.

4. Results

4.1. Mineral Composition of the Ores

The Niyayu ore zone (Figure 1) extends from the northeast to the southwest. In terms of structure, this is a zone of contiguous tectonic disturbances with a north–northeastern strike, characterized by excessive fracturing of the rocks, broadly developed zones of schistose and crushing rocks, and metasomatic alterations (chloritization, epidotization, silicification, and pyritization). All of the major gold occurrences are located within the Niyayu ore zone [19]. Gold–sulfide, gold–sulfide–quartz vein, and vein-disseminated mineralization are located in the pre-Ordovician volcanogenic and volcanogenic sedimentary rocks and are controlled by faults.

The small-scale Verkhneniyayuskoye-2 deposit and gold Niyahoyskoye-2 and Verkhnelekeletskoye occurrences are the most thoroughly studied. Mineralogical–geochemical and isotope–geochemical surveys have been undertaken in the Verkhneniyayuskoye-1, Niyahoyskoye-1, and Yagodnoye ore occurrences.

Verkhneniyayuskoye-2 deposit. Based on mineral composition, the Verkhneniyayuskoye-2 deposit ores were mostly arsenopyrite–pyrite. The textures of the ores were massive and banded (Figure 2a), disseminated, veinlet-disseminated, and taxitic. The content of the sulfides in the ore veins reached 50%. Along with pyrite and arsenopyrite, the ores contained pyrrhotite, pentlandite, sphalerite, galena, chalcopyrite, tetrahedrite and tennantite, cubanite, sternbergite, freibergite, aurostibite, electrum, kustelite, native bismuth, and minor supergene minerals (chalcocite, covellite, goethite, limonite, scorodite, and cerussite) [22].

Pyrite was represented by cubic, pyritohedron-shaped, 1.5–2 mm-sized grains or crystals. Arsenopyrite was seen in up to 0.5 mm- and, in certain cases, 2 mm-sized individuals with columnar, short-columnar, or complex shapes. Arsenic (up to 3.3 wt.%) in pyrite and antimony (up to 3.2 wt.%) in arsenopyrite were sometimes found as impurities. Grains and aggregates of arsenopyrite and pyrite were quite often strongly cataclased, with inclusions of chalcopyrite, galena, sphalerite, and other sulfides, often confined to the cracks (Figure 2). Very typical was the replacement of pyrite with iron hydroxides and arsenopyrite with scorodite. Chalcopyrite, sphalerite, and galena were rather common; however, their content in the ores was lower than that of pyrite and arsenopyrite. Silver (up to 1.5 wt.%), bismuth, and thorium were sometimes found in small amounts in galena. Pyrrhotite, tetrahedrite, and tennantite were rarely observed. The tetrahedrite contained silver, whose content could be rather high (up to 15.4 wt.%), corresponding to the independent mineral phase freibergite.

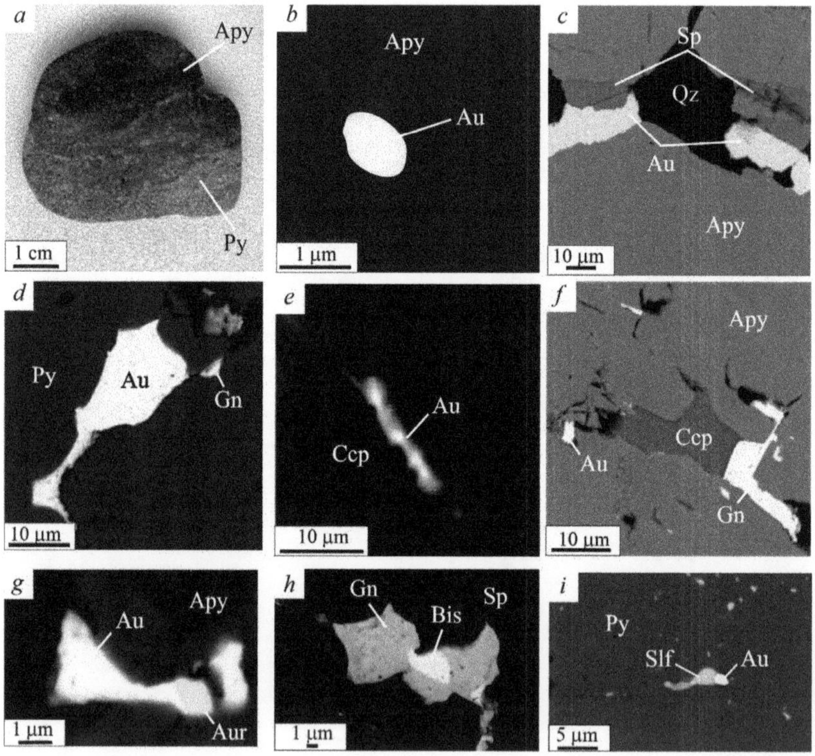

Figure 2. (**a**) Photo of the representative ore hand specimen Verkhneniyayuskoye-2 occurrence; (**b**) submicron inclusions of native gold in arsenopyrite; (**c**) native gold and sphalerite, developed along the cracks in arsenopyrite; (**d**) native gold and galena in pyrite; (**e**) native gold in chalcopyrite; (**f**) chalcopyrite, galena, and native gold, developed along the cracks in arsenopyrite; (**g**) aurostibite intergrown with native gold in arsenopyrite; (**h**) relationship between native bismuth and galena and sphalerite; (**i**) microinclusion of sulfosalt in pyrite. BSE images. Acronyms for minerals: Apy, arsenopyrite; Py, pyrite; Ccp, chalcopyrite; Sp, sphalerite; Gn, galena; Aur, aurostibite; Bis, native bismuth.

Free gold was extracted by crushing the entire samples taken from sulfide-enriched sections of the veins and host rock. This native gold was predominantly small (less than 0.2 mm), though sometimes particles of up to 2–7 mm in diameter were found. The gold particles had a complex shape and were a yellow, greenish-yellow, or reddish-yellow color.

In the polished sections, gold could be seen as the smallest isolated isometric inclusions in arsenopyrite (Figure 2b), pyrite, and, rarely, chalcopyrite (Figure 2e). Such particles of gold were 1–5 µm in size. At the same time, of note were relatively large (up to 10–50 µm) particles with an elongated, irregular shape of late generation, located in the pyrite and arsenopyrite microcracks, often together with chalcopyrite, galena, and sphalerite (Figure 2c,d,f).

From the microprobe analysis results, Ag was the principal impurity in the native gold, with Hg observed. Submicron particles of gold, inherent in pyrite and arsenopyrite, usually have 8–10 wt.%, although sometimes 15–20 wt.%, of Ag (Table 1). The Ag content was generally higher in the native gold and was confined to the microcracks in the arsenopyrite and pyrite grains, being associated with galena, chalcopyrite, and sphalerite (Table 1). Occasionally, grains occurred, comprising more Ag than Au of up to 61 wt.%. Almost without exception, the native gold contained up to 3.6 wt.% of Hg. Cu impurity was observed rarely and in small amounts. The Ag content and other impurities within the gold particles varied to one extent or another. Sometimes, rims were noted with a relatively high content of Ag.

Table 1. Chemical composition of gold from the Verkhneniayuskoye-2 gold deposit (wt.%).

Polished Section No.	Gold Grain No.	Au	Ag	Hg	Total	Fineness, ‰
	Submicron gold inclusions in pyrite and arsenopyrite (gold I)					
2012-6	1	77.17	10.53	0.00	87.70	880
	2	90.55	9.54	0.00	100.09	905
	3	80.42	18.79	0.00	99.21	811
2012-15	1	89.01	7.86	0.00	96.87	919
	2	75.01	20.76	0.00	95.77	783
	3	77.87	8.24	0.00	86.11	904
2012-17	1	86.37	9.48	0.00	95.85	901
	2	86.45	9.77	0.00	96.22	898
	3	86.06	9.65	0.00	95.71	899
2013-1	1	82.03	9.33	0.00	91.36	898
	2	89.45	9.11	0.00	98.56	908
	3	87.23	9.01	0.00	96.24	906
2013-2a	1	94.15	5.97	0.00	100.12	940
	2	90.52	6.04	0.00	96.56	937
2013-4	1	81.33	17.65	0.00	98.98	822
	Gold in association with chalcopyrite, galena, and sphalerite (gold II)					
2012-5	1	30.74	59.42	2.05	92.21	333
	2	29,03	60.95	2.39	92.37	314
2012-8	1	51.83	38.71	0.81	91.35	567
2012-6	1	40.48	48.65	0.00	89.13	454
2012-15	1	58.97	35.85	0.00	94.82	622
	2	86.12	10.14	0.00	96.26	895
	3	84.15	14.3	0.00	98.45	855
2012-17	1	57.22	39.28	0.91	97.41	587
	2	57.15	37.93	1.43	96.51	592
2013-1	1	65.29	6.87	0.00	72.16	905
	2	63.4	35.9	1.72	101.02	628
2013-2a	1	65.78	32.62	1.74	100.16	657
2013-4	1	66.77	27.21	1.21	95.19	701

A specific feature of the ores from this deposit was the presence of bismuth minerals and silver fahlores.

Niyahoyskoye-2 ore occurrence. In terms of mineral composition, the ores were substantially arsenopyrite with a subordinate amount of pyrite. The ores were massive and banded (Figure 3a). Their structures were small- and medium-grained with course-grained plots. The sulfide content reached 70% in the ore veins and veinlets, and 1%–2% in the host rocks. In the near-surface zone, the ores were considerably oxidized.

It has been established that there was chalcopyrite, sphalerite, galena, tetrahedrite, aikinite, native bismuth, and supergene minerals—covellite, scorodite, and oxysulfosalts of lead with bismuth—in the composition of the ores, in association with arsenopyrite and pyrite [29]. Arsenopyrite formed continuous granular masses. Individual grains were elongated columnar or relatively isometric in shape. Most of the arsenopyrite grains were crushed due to cataclasis, with the formation of numerous small cracks across which scorodite had developed (Figure 3c). Pyrite occurred in a subordinate amount, distributed unevenly and forming single disseminations or accumulations of various shapes and sizes in the bulk of the arsenopyrite. Pyrite was represented by crystals of cubic or cuboctahedral shape (Figure 3b) or subhedral grains (Figure 3d,e). It often contained impurities of arsenic (up to 1.46 wt.% As). Pyrite and arsenopyrite often contained microinclusions of sphalerite, galena, tetrahedrite, and gold (Figure 3b,c). Chalcopyrite and tetrahedrite formed relatively large xenomorphic segregations of elongated or close-to-isometric shapes (Figure 3d). Sphalerite and galena occurred rarely, found only as microinclusions in arsenopyrite (or scorodite), pyrite, and quartz.

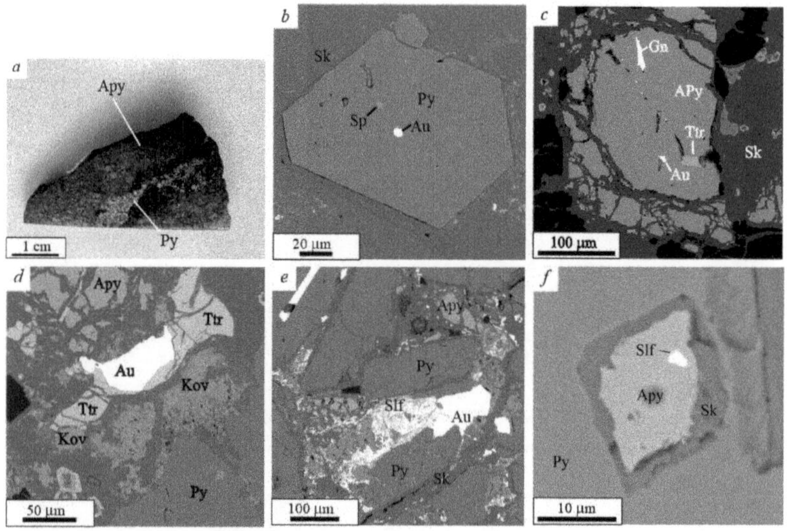

Figure 3. (**a**) Photo of the representative ore hand specimen Niyahoyskoye-2 occurrence. Mineral assemblages: (**b**) crystal of As–pyrite with inclusions of gold and sphalerite; (**c**) galena, gold, and tetrahedrite in arsenopyrite; (**d**) gold in association with tetrahedrite; (**e**) association between gold and pyrite, arsenopyrite, scorodite, and oxysulfosalts Pb and Bi; (**f**) inclusion of sulfosalt in arsenopyrite. BSE images. Acronyms for minerals: Apy, arsenopyrite; Py, pyrite; Ttr, tetrahedrite; Sp, sphalerite; Gn, galena; Sk, scorodite; Kov, covellite; Slf, sulfosalt; Au, native gold.

In close association with sulfides, few, usually 6–200 μm-sized elongated occurrences were rarely found in quartz, close in chemical composition to the sulfosalts of lead and copper with bismuth and antimony [29]. The analyzed occurrences of the sulfosalts had a variable composition with considerable variations in the content of principal elements. Inclusions of sulfosalts (2–10 μm in size) were often found in arsenopyrite and pyrite, forming isometric or elongated, single (Figure 3f), or grouped (cloud-like) occurrences. Due to the small sizes of the inclusions and a strong effect of the matrix of the host mineral (arsenopyrite or pyrite), it was difficult to precisely identify sulfosalts, though they were diagnosed by the set of elements (Pb, Cu, Bi, Sb, and S).

Free gold was found in the intergrowths with arsenopyrite and other sulfides (Figure 3d,e), often observed in the form of microinclusions in pyrite and arsenopyrite (Figure 3b,c). In terms of particle size distribution, the particles mostly qualified as finely dispersed and dust-like (0.001–0.05 mm) or extra fine and fine (0.05–1.00 mm). Medium-grained gold of up to 1.5 mm in size also occurred. The gold particles were plate-like, scaly, lumpy-shaped dendritoids, and also partially faceted crystals were found.

Ag and Hg were the major impurities in the gold. The Ag content varied from 7.11 to 22.61 wt.%. Hg was not always observed; its content amounted to 1.23–8 wt.%. The fineness of the gold varied from 643‰ to 898‰ (Table 2). Medium fineness gold prevailed. Analyzing the surface composition of the gold grains demonstrated that most of them had rims with a lower fineness (700‰–800‰), typical of ore gold. At the same time, some of the gold grains had rims with a high fineness (approximately 990‰), indicative of the effect of supergene processes.

Table 2. Chemical composition of the native gold from the Niyahoyskoye-2 occurrence (wt.%).

Polished Section No.	Gold Grain No.	Au	Ag	Hg	Total	Fineness, ‰
Submicron gold inclusions in pyrite and arsenopyrite (gold I)						
Bez-3-14	1	65.67	22.23	8.00	102.20	643
		67.79	22.61	1.72	98.48	688
	2	64.88	22.16	3.29	97.15	668
Bez-3-17	1	82.43	11.69	1.27	99.97	825
	2	89.74	10.24	0	99.98	898
313701	1	85.10	14.95	0	100.05	847
	2	84.40	15.53	0	99.93	845
Gold in association with arsenopyrite, pyrite, chalcopyrite, and tetrahedrite (gold II)						
19001-3	1	81.64	14.67	0	96.31	841
19001-5	1	75.59	17.37	4.28	97.24	777
	2	74.95	17.84	1.56	94.35	794
19001-6	1	72.99	18.78	2.25	94.02	776
	2	83.55	13.30	0	96.85	863
19001-7	1	84.10	14.60	0	98.70	852
19001-8	1	80.50	18.91	0	99.41	810
	2	77.49	18.15	7.31	102.95	752
	3	76.57	18.45	1.91	96.93	790
19001-9	1	84.62	13.46	0	98.08	863
	2	80.52	14.08	7.62	102.22	788
	3	81.93	14.08	5.76	101.77	805
313503	1	83.24	16.36	0.83	100.43	829
	2	83.83	15.76	0	99.59	842
	3	84.50	15.22	0	99.72	847
		83.63	15.03	1.13	99.79	838
313504	1	82.57	15.79	1.45	99.81	827
	2	81.30	16.72	1.38	99.0	818
	3	83.90	15.22	0	99.12	846

Verkhnelekeletskoye ore occurrence. Based on mineral composition, the ores were essentially pyritic with a subordinate amount of arsenopyrite (Figure 4a). Sphalerite was the principal minor mineral (Figure 4b). Gold was found as submicron inclusions in pyrite, and less frequently in arsenopyrite (Figure 4c). Galena and chalcopyrite occurred only as microinclusions in pyrite and arsenopyrite (Figure 4d). Moreover, tetrahedrite and sulfosalt occurred as microinclusions (Figure 4e,f). Native gold was medium in fineness (816‰–847‰), with silver as the major impurity (Table 3). Thus, a galena + chalcopyrite + sphalerite association was weakly manifested, and, in addition to sphalerite, was represented by micromineralization in pyrite and/or arsenopyrite [29].

Yagodnoye ore occurrence. This occurrence has been poorly studied in the past. Herein, pyrite was shown to be the major ore mineral. Its content in mineralized rocks was not more than 1%–2%. In a few ore veinlets, chalcopyrite, galena, and sphalerite were observed. Free native gold was identified at the ore occurrence, which was extracted from the host rocks and fully disintegrated into a clayish–micaceous mass. The gold grains were 0.1–0.25 mm-sized. The gold was of ore facies, crystallomorphic and complex in shape (Figure 5a,b). The fineness of the gold was 700‰–800‰, with Ag being the major impurity (Table 4). Microinclusions of gold in association with chalcopyrite were found in pyrite

(Figure 5c). Due to the small sizes, it was difficult to identify the gold composition since the analysis results always incorporated the matrix elements (Fe and S). The computed fineness of the gold was 800‰–900‰.

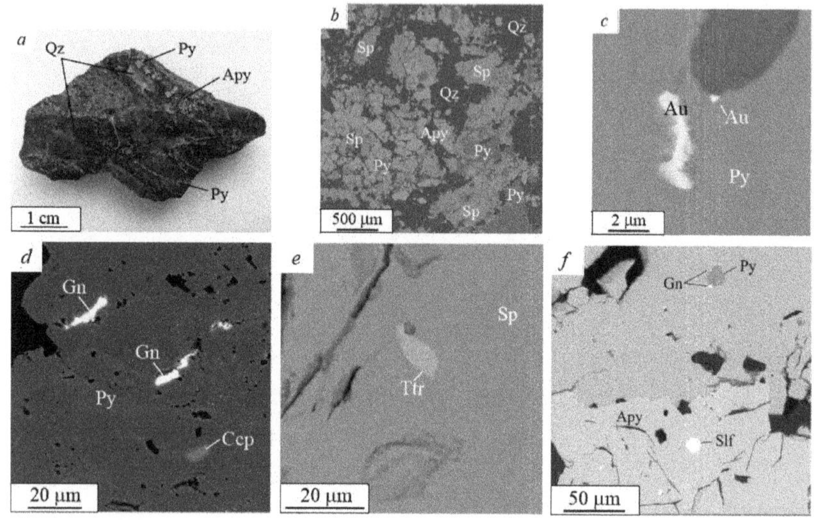

Figure 4. (**a**) Photo of the representative ore hand specimen Verkhnelekeletskoye occurrence. (**b**) Structure of the ore and major mineral assemblages; (**c**) submicron gold inclusions in pyrite; (**d**) microinclusions of galena and chalcopyrite in pyrite; (**e**) inclusion of tetrahedrite in sphalerite; (**f**) submicron inclusions of sulfosalt and galena in arsenopyrite. BSE images. Acronyms for minerals: Qz, quartz; Apy, arsenopyrite; Py, pyrite; Sp, sphalerite; Gn, galena; Ccp, chalcopyrite; Slf, sulfosalt; Au, native gold.

Table 3. Chemical composition of the native gold from the Verkhnelekeletskoye occurrence (wt.%).

Section No.	Grain No.	Au	Ag	Hg	Total	Fineness, ‰
	Submicron gold inclusions in pyrite and arsenopyrite (gold I)					
LK-2	1	81.81	18.21	0	100.03	818
LK-4	1	84.66	15.35	0	100.01	845
	2	81.54	18.34	0	99.88	816

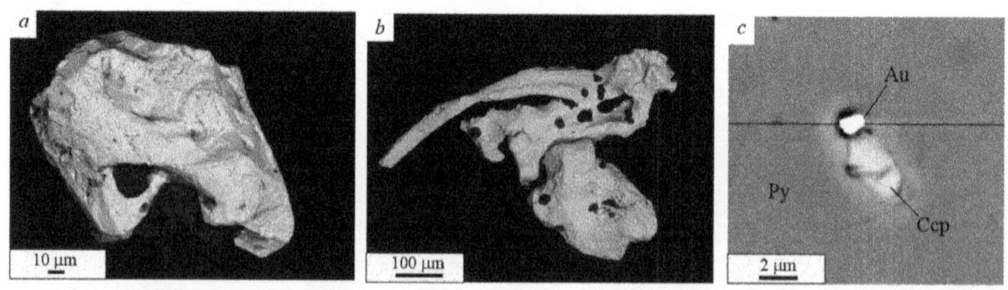

Figure 5. Photos showing the morphology of the free gold from the mineralized zone of the Yagodnoye ore occurrence: (**a**) crystallomorphic and lumpy; (**b**) complex shape; (**c**) microninclusion of gold and chalcopyrite in pyrite. BSE images. Acronyms for minerals: Py, pyrite; Ccp, chalcopyrite; Au, native gold.

Table 4. Chemical composition of the native gold from the Yagodnoye occurrence (wt.%).

Section No.	Grain No.	Au	Ag	Total	Fineness, ‰
	Submicron gold inclusions in pyrite (gold I)				
T-16	1	90.30	8.72	99.02	912
	2	97.26	1.62	98.88	962
	3	98.92	1.16	100.08	988
	4	97.84	1.61	99.45	984
	Native gold (gold II)				
BT-16154	1	90.22	10.61	100.83	895
	2	90.22	10.10	100.32	899
	3	89.36	10.87	100.23	892
	4	92.72	7.76	100.48	923
BT-16155	1	92.27	5.76	98.03	941
	2	93.13	5.35	98.48	946
BT-16157	1	91.68	7.66	99.34	923
	2	88.14	11.00	99.14	889
	3	86.78	11.77	98.55	881
BT-16162	1	94.86	5.25	100.11	948

4.2. Fluid Inclusions

Fluid inclusions in quartz and calcite were studied in the Verkhniayuskoye-2 deposit. As mentioned above, two groups of veins had developed in the deposit: (1) quartz–sulfide and sulfide veins and veinlets with gold mineralization; (2) quartz, quartz–chlorite, quartz–calcite, and other barren veins and veinlets. Their relationship is not entirely clear since both of them were oriented according to the rock schistosity. The vein quartz was milky white, with subhedral quartz grains up to 0.5–1 mm in size. Quartz crystals were rare and filled the voids in the quartz veins. In the massive sulfide ores, quartz filled the interstices between the sulfide grains (Figure 2c) or formed small lenses. Calcite in the quartz veins was present in the form of nests. Specimens of quartz and calcite were taken from the quartz veins with gold–sulfide mineralization (herein called ore veins, ore quartz) and quartz veins with no ore mineralization (herein called barren veins, barren quartz). The study results are given in Table 5.

Table 5. Microthermometric results of the fluid inclusions of the Verkhneniayuskoye-2 gold deposit.

G-Type FI	Th, °C	C-Type FI	Salinity (wt.% NaCl eq.), (*n*)	Te, °C (*n*)	Potential Chemical Systems
			Ore quartz		
P	109–467 (20)	L + V → L ore V	0.5–17 (17)	−23 (10)	NaCl-KCl-H_2O
PS	85–340 (16)	L → L		−35 (1)	NaCl-$MgCl_2$-H_2O
S	109–203 (10)	L → L			
			Barren quartz		
P	116–491 (41)	L + V → L ore V	4.0–16.5 (25)	−23 (16)	NaCl-KCl-H_2O
PS	87–123 (12)	L + V → L		−35 (2)	NaCl-$MgCl_2$-H_2O
S	70–184 (12)	L → L			
			Quartz crystals		
P	135–358 (11)	L + V → L	0.5–2 (3)	~−23 (3)	NaCl-KCl-H_2O
			Calcite		
P	80–287 (12)	L + V → L	15 (1)	~−35 (1)	NaCl-$MgCl_2$-H_2O

Note. Fluid inclusions (FIs): Genetic type (G-type)—P, primary; PS, pseudosecondary; S, secondary. *n*, number of fluid inclusions. Compositional type (C-type): L, liquid aqueous phase; V, vapor phase.

Quartz. Primary, pseudosecondary, and secondary fluid inclusions were analyzed in the quartz. Primary inclusions were isolated or occurred in small groups within individual quartz grains, while pseudosecondary inclusions occurred in a trail inside the quartz grains. Meanwhile, secondary inclusions were located in trails cutting mineral boundaries. The inclusions were usually oval or irregular in shape, sometimes in the form of a negative crystal. Their size reached 80 μm but usually did not exceed 10 μm. The most typical ore and barren quartz comprised one-phase (L) aqueous inclusions of up to 30 μm in size (Figure 6a). Their abundance was over 90%. Biphase inclusions were encountered less frequently (Figure 6b). They consisted of two phases, a liquid aqueous phase and a vapor phase, where the vapor bubble occupied $\leq$ 5% of the volume of the inclusion. In the two phases, liquid–vapor inclusions in which the latter occupies $\geq$ 70% of the volume of the inclusion were rare. The temperature of homogenization of the inclusions in quartz of both types of veins was approximately the same. Most of the inclusions became homogenous up to 394 °C; they usually homogenized into the liquid phase. Higher-temperature inclusions were rare and homogenized into the vapor phase.

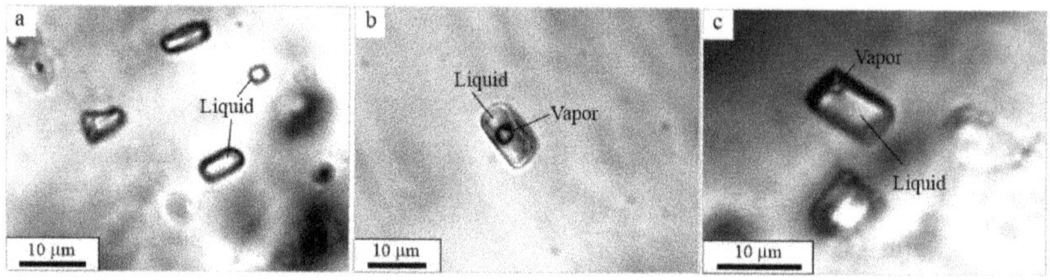

Figure 6. Typical fluid inclusions in quartz: (**a**) One-phase aqueous; (**b**) biphase; (**c**) biphase in calcite.

The crystals were a later generation of quartz, dominated by one-phase inclusions (L) and two-phase (L + V) inclusions, where vapor bubbles occupying 5%–20% of the volume of the inclusion were less common. They displayed subrounded to rectangular or negative crystal shapes. The size of the inclusions reached 80 μm but usually did not exceed 20 μm. Fluid inclusions in the quartz were characterized by low homogenization temperatures (Table 5) in the vapor phase.

Calcite was the latest in the deposit, containing one-phase (L) and biphase (L + V) inclusions, where the vapor bubble occupied approximately 5% of the volume of the inclusion (Figure 6c). The inclusions were often shaped as a negative crystal and were up to 10 μm in size. The homogenization temperature of the fluid inclusions was up to 287 °C, and they were homogenized into the liquid phase.

The fluid inclusions of ore and barren quartz contained aqueous solutions of salts with a eutectic melting point of approximately −23 °C and −35 °C, which corresponds to the salt system NaCl–KCl–H_2O, and NaCl–$MgCl_2$–H_2O in very rare cases. The concentration of salts ranged from 0.5 to 17 wt.% NaCl-equiv. In the fluid inclusions of quartz crystals, the eutectic temperature was −23 °C, which corresponds to the NaCl–H_2O system. The salinity ranged from 0.5 to 2 wt.% NaCl-equiv. In the fluid inclusions of calcite, the eutectic temperature was −35 °C, corresponding to the NaCl–$MgCl_2$–H_2O system. The salinity was approximately 15 wt.% NaCl-equiv. Liquid carbon dioxide was not found in the inclusions.

When examining the bulk composition of the gases in the fluid inclusions of ore and barren quartz, it water vapors and CO_2 were found to be the principal components. CO_2 prevailed among the gases; others were present in dramatically subordinate amounts [26].

4.3. Sulfur Isotopes

The approximate data of the sulfur isotopes of the sulfides are listed in Table 6 and shown in Figure 7. The sulfides in the ores and ore-hosting rocks of the Manitanyrd region occurrences were enriched with light sulfur isotopes and had $\delta^{34}S$ values from −0.2‰ to −8.0‰. The isotope composition of the sulfur in pyrite varied from −0.2‰ to −8.0‰, in arsenopyrite from −0.8‰ to −7.9‰, and in sphalerite from −1.0‰ to −1.9‰.

In the Verkhneniyayuskoye-2 deposit, the isotope composition of the sulfur in the sulfides had a narrow range from −0.2‰ to −2.6‰ with $\delta^{34}S$ values varying from −0.2‰ to −2.6‰ for pyrite, from −0.8‰ to −2.1‰ for arsenopyrite, and from −1.0‰ to −1.9‰ for sphalerite. In the southern zone of the deposit, with mostly gold–pyrite–arsenopyrite-type ores, the sulfur isotopes in the sulfides had the following values of $\delta^{34}S$: from −0.5‰ to −2.6‰ of pyrite, from −0.8‰ to −1.1‰ for arsenopyrite, and from −1.0‰ to −1.9‰ for sphalerite. Pyrite had the maximum range of variations in $\delta^{34}S$ value, including the isotope compositions of arsenopyrite and sphalerite. In the Polar zone of the deposit, with mostly gold–pyritic-type ores, the sulfur isotopes in the sulfides had the following values of $\delta^{34}S$: from −1.6‰ to −1.9‰ for pyrite and −2.1‰ for arsenopyrite. Pyrite had a somewhat lighter sulfur composition (mean −1.7‰) than arsenopyrite (−2.1‰). The pyrite of the wallrock quartz–sericitic metasomatites (samples 14,170 and 14,171) had the lightest sulfur isotope composition (from −0.2‰ to −0.3‰). Disseminated pyrite in the ore-hosting volcanics of the Bedamel series (samples BT-14189 and BT-14193) had $\delta^{34}S$ values from −1.2‰ to −1.5‰ and, thus, in the isotopes, the sulfur was not different from the pyrite of the ore bodies.

Table 6. Sulfuric isotope composition (‰) of the sulfides in the examined ore occurrences in the Manitanyrd region.

Sample	Spot	Mineral	$\delta^{34}S_{VCDT}$, ‰
	Verkhneniyayuskoye-2		
Zr-207-2	Southern zone	Pyrite	−2.5
Zr-207-2A			−2.6
Zr-207-2B			−2.0
Zr-204-2P			−0.5
Zr-204-2R			−1.6
Zr-204-2			−1.1
Zr-204-2		Arsenopyrite	−0.8
Zr-204-2A			−0.8
Zr-204-2S		Sphalerite	−1.8
Zr-204-2F			−1.0
14184	Polar zone	Pyrite	−1.9
14184-2			−1.6
14184-3			−1.6
14170			−0.2
14171			−0.3
BT-14189			−1.5
BT-14193			−1.2
14184-1		Arsenopyrite	−2.1
	Verkhneniyayuskoye-1		
Zr-200-2	Drill Hole 19	Arsenopyrite	−7.2
Zr-200-2A			−7.1
Zr-200-2B			−7.9
Zr-200-2P		Pyrite	−8.0
	Niyahoyskoye-1		
Nkh-1	Zone 1	Pyrite	−8.0
	Niyahoyskoye-2		
313701	Zone 4	Arsenopyrite	−3.9
313701-1			−4.0
313701-2			−4.7
313701-4		Pyrite	−4.1
19001/3	Zone 8	Arsenopyrite	−4.5
19001/3-1			−4.5
	Verkhnelekeletskoye		
LK-1		Pyrite	−1.5
LK-2			−1.2
LK-1a		Arsenopyrite	−2.4
LK-1b			−2.3
LK-1c			−3.5

In the Verkhnelekeletskoye occurrence, gold–pyrite ore-grade mineralization was confined to the quartz vein in the dike of gabbro–dolerites, intruding into the volcanogenic sedimentary rocks of the Enganepe suite. The isotope composition of the sulfur in the sulfides varied slightly from −1.2‰ to −3.5‰ with variations in the δ^{34}S value from −1.2‰ to −1.5‰ for pyrite and from −2.3‰ to −3.5‰ for arsenopyrite, i.e., arsenopyrite had a somewhat lighter sulfur composition.

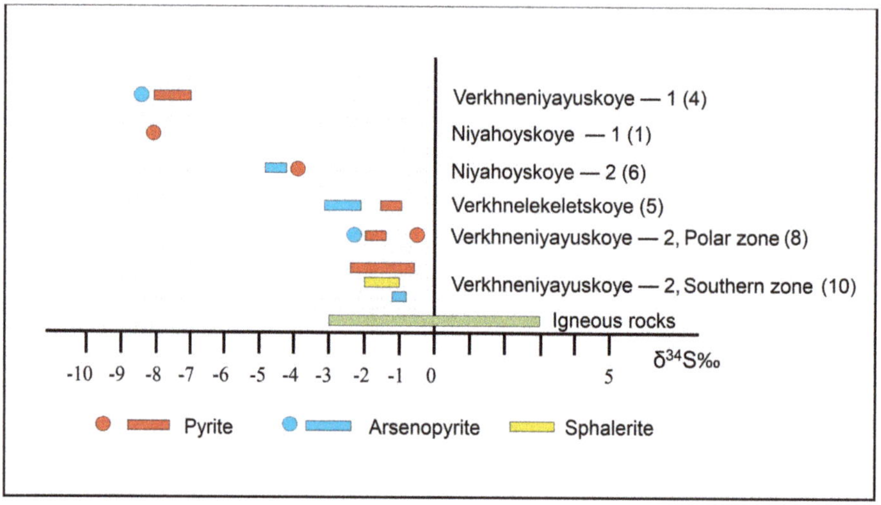

Figure 7. Isotope composition of the sulfur in the sulfides of the gold occurrences of the Manitanyrd region. Circle, δ^{34}S‰ unit values; rectangle, range of variations; bracketed number, the number of analyses.

The Niyahoyskoye-2 ore occurrence, with gold–arsenopyrite-type ores, is also located in the volcanogenic sedimentary rocks of the Enganepe suite. The isotope composition of the sulfur in the sulfides varied from −3.9‰ to −4.7‰, at which there is no difference between arsenopyrite and pyrite in the sulfur isotopes. Compared to the above-described ore occurrences, the sulfur isotope composition in the sulfides from the Niyahoyskoye-2 deposit was somewhat lighter.

In the small-scale Verkhneniyayuskoye-1 and Niyahoyskoye-1 ore occurrences, which constitute mineralized zones of vein-disseminated mineralization in Enganepe suite rocks, the sulfides (pyrite and arsenopyrite) were notable in terms of having a considerably lighter isotope composition of sulfur. The δ^{34}S values varied from −7.1‰ to −8.0‰. Arsenopyrite was enriched with light sulfur isotopes compared to pyrite in all ore occurrences, except for the Verkhneniyayuskoye-2 deposit (Southern zone), where the δ^{34}S values were in the range of the pyrite δ^{34}S variations, corresponding to a light isotope composition.

The isotope composition of the sulfur in the sulfides of the ore occurrences varied within rather narrow ranges: Verkhneniyayuskoye-2 from −0.2‰ to −2.6‰; Niyahoyskoye-2 from −3.9‰ to −4.7‰; Verkhnelekeletskoye from −1.2‰ to −3.5‰. The difference in the δ^{34}S value of the sulfides from different gold occurrences was 2.41‰, 0.8‰, and 2.3‰, respectively.

5. Discussion

All of the studied and other known gold–sulfide–quartz ore occurrences in the Manitanyrd region belong to the same pyrite–arsenopyrite mineral type, with a variable ratio of main minerals.

The conducted studies show that the ore formation was staged. The staging of the ore formation is most clearly manifested in the Verkhneniyayuskoye-2 deposit. Here, pyrite–arsenopyrite mineralization with finely dispersed gold of the early generation was first superimposed on quartz veins and near-vein rocks. Then, there were tectonic shifts that caused cataclasis of the ores. This was followed by the deposition of ore association of the second stage—chalcopyrite, galena, sphalerite, and relatively large gold of the late generation. Fahlore crystallized at the end of this stage. Thus, two stages of ore formation were reasonably distinguished: Gold + pyrite + arsenopyrite and gold + chalcopyrite + galena + sphalerite. In the Niyahoyskoye-2 occurrence, two stages of ore formation were also distinguished: early pyrite + arsenopyrite with finely dispersed gold and late tetrahedrite + chalcopyrite with coarser gold, separated from the early stage by ore cataclasis. Fahlore and sulfosalts crystallized at the end of the late stage. However, galena and sphalerite, the main minerals of the late stage in the Verkhneniyayuskoye-2 deposit, occurred here only as microinclusions in arsenopyrite, pyrite, and, rarely, in quartz.

In the Yagodnoye occurrence, we confidently distinguished an early pyrite stage with finely dispersed gold and, apparently, a late stage with native gold, the paragenetic association of which has not yet been determined. In the Verkhnelekeletskoye occurrence, one stage of pyrite + arsenopyrite (+ sphalerite) with finely dispersed gold was distinguished, comparable to the early stage of ore formation in the Verkhniayuskoye-2 deposit and the Niyahoyskoye-2 occurrence. Minerals of the late stage (Verkhneniyayuskoye-2 deposit)—chalcopyrite + galena + tetrahedrite + lead sulfosalt (presumably boulangerite)—were present only as microinclusions in pyrite and arsenopyrite [29]. A characteristic feature of the ore composition of all studied gold occurrences was the presence of arsenic pyrite (up to 3.3 wt.% As). In the Verkhneniyayuskoye-2 deposit, veins and co-ore quartz crystallized throughout the entire process of hydrothermal mineral formation. The generalized paragenetic sequence of ore minerals in the ore occurrences of the Manitanyrd region is shown in Figure 8.

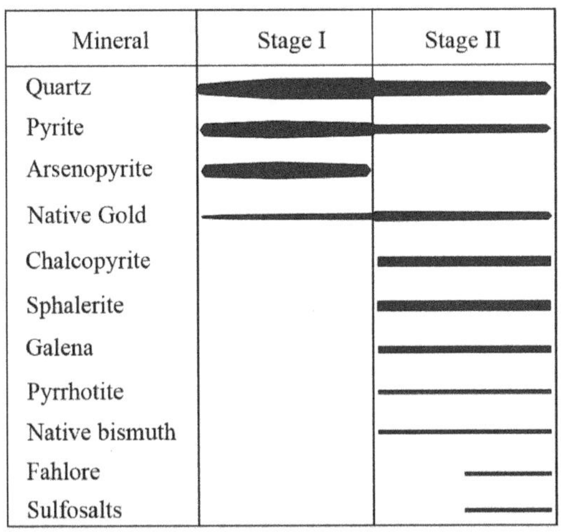

Figure 8. Paragenetic sequence of the ore minerals in the Manitanyrd region gold occurrences. The thickness of the line shows the relative content of minerals.

In all studied ore occurrences of the early-stage minerals, pyrite, and arsenopyrite, finely dispersed gold was present in the form of microinclusions ranging in size from 500 nm to 20 μm, mainly 1–5 μm (gold I). The gold was characterized by medium and high fineness with maxima values in the areas of 800‰–900‰, 900‰–950‰, and 950‰–1000‰ and relatively low dispersion (Figure 9).

Native gold of the late stage had dimensions of more than 0.5 to 1.5–2 mm (gold II). The native gold of the Verkhneniyayuskoye-2 deposit was characterized by a relatively low fineness with a polymodal distribution on the fineness histogram. The scatter of the values of this indicator was in the region of 300‰–950‰ (Figure 9). Native gold from the Niyahoyskoye-2 and Yagodnoye occurrences was characterized by medium and high fineness, unimodal distribution on the fineness histograms, and the distribution of this index in the areas of 800‰–950‰ (Figure 9b,d). Thus, significant differences in the fineness of the gold of the early (gold I) and late (gold II) stages were observed in the Verkhneniyayuskoye-2 deposit, while they were insignificant in the other ore occurrences.

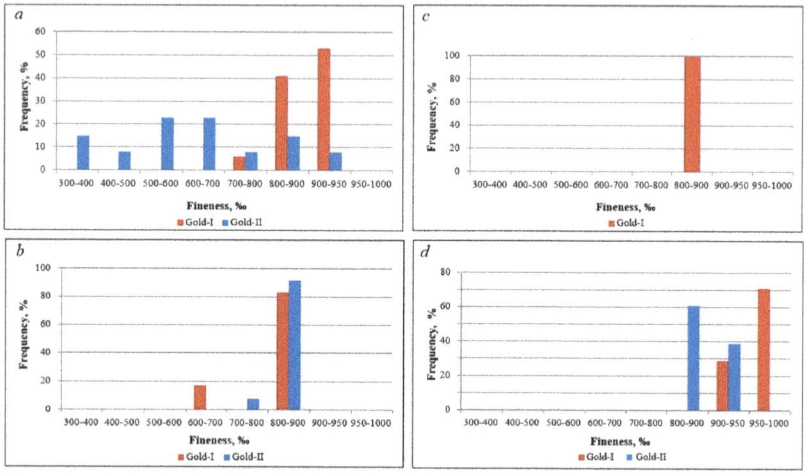

Figure 9. Histograms of gold fineness variations in the occurrences of the Manitanyrd region: (**a**) Verkhneniyayuskoye-2, (**b**) Niyahoyskoye-2, (**c**) Verkhnelekeletskoye, and (**d**) Yagodnoye.

According to the typomorphic features (low fineness, wide scatter of values of this indicator, and polymodal distribution), the native gold of the Verkhneniyayuskoye-2 deposit was similar to the epithermal–volcanogenic deposits, while the Niyahoyskoye-2 and Yagodnoye ore occurrences (medium and high fineness, unimodal distribution with low dispersion) were similar to the gold–arsenic–sulfide ones in the black shale strata of the northeast of Russia [34].

According to the study of fluid inclusions, in the Verkhneniyayuskoye-2 deposit, a wide range of temperatures for homogenization of the primary inclusions (109–467 °C) in the veins of ore quartz was established. The salinity varied over a wide range from 0.5 to 17 wt.% NaCl-equiv., which corresponds to a wide range of inclusion homogenization temperatures. The compositions of the solutions of fluid inclusions correspond to the NaCl–KCl–H_2O and, in rare cases, NaCl–$MgCl_2$–H_2O systems. Numerous studies of fluid inclusions in the quartz of gold deposits of various types have generally shown a narrower temperature range of homogenization of the primary inclusions (<50 °C) and salinity of solutions 5–10 wt.% NaCl-equiv. [35–39].

Our data on fluid inclusions, together with the identified stages of ore formation, likely indicate the duration of the hydrothermal process. Quartz is often cataclased, which could result in overlapping in the same areas of inclusions with different homogenization temperatures. In addition, quartz contains a large number of epigenetic inclusions, indicating repeated healing of cracks. This imposes a limitation on using the data on fluid inclusions in quartz to estimate some parameters of the formation of sulfide mineralization.

The isotope composition of the sulfides in the gold occurrences and host rocks of the Niyayu ore zone was characterized by prevailing light isotopes and negative $\delta^{34}S$ values, varying from −0.2‰ to −8.0‰. Based on the isotope composition of the sulfur in the sulfides, two groups of ore occurrences stood apart. The first group (Verkhneniyayuskoye-2 and Verkhnelekeletskoye) was characterized by the lightest isotope composition of the sulfur in the sulfides ($\delta^{34}S$ = −0.2 ÷ −3.5‰), close to the meteorite standard, narrow range of variations (the difference in the $\delta^{34}S$ value did not exceed 2.4‰), which is usually inherent in systems with magmatic sources of sulfur [40–43]. These ore occurrences belong to the veined gold–sulfide–quartz type.

The second group of ore occurrences (Verkhneniyayuskoye-1 and Niyahoyskoye-1) had the lightest isotope composition of the sulfur in the sulfides of the region ($\delta^{34}S = -7.1 \div -8.0‰$), being well out of sulfur that has a magmatic origin. These ore occurrences belong to the type of mineralized zones with vein-disseminated gold–sulfide ore-grade mineralization and are positioned in the posterior volcanogenic sedimentary rocks. Niyahoyskoye-2 ore occurrence by isotope composition of the sulfur in the sulfides ($\delta^{34}S = -3.9 \div -4.7‰$) occupied an intermediate position; however, the percentage of magmatic sulfur in the formation of sulfides here was rather high. Both types of ore-grade mineralization, namely, the veined gold–sulfide–quartz and vein-disseminated gold–sulfide types of mineralized zones, were combined in this ore occurrence. It should be noted that ore occurrence was also localized in the volcanogenic sedimentary rocks of the Enganepe suite.

The considerably lighter isotope composition of the sulfides ($\delta^{34}S = -7.1 \div -8.0‰$) from the second group of ore occurrences, for which volcanogenic sedimentary rocks were the host rocks, is likely attributed to the mixing of magmatic–hydrothermal fluid with crustal, enriched with sulfur of sedimentary origin [44].

6. Conclusions

Our study results are characteristic of gold–sulfide–quartz and gold–sulfide (gold–arsenic) ore occurrences of the Manitanyrd region of the Polar Urals, which are localized in volcanic and volcanic–sedimentary Upper Precambrian rocks. We revealed that they are confined to the fault zone and have the common features of geological structure. As a result of mineralogical and geochemical studies, the same type of material composition was established for ore mineralization in all studied ore occurrences, belonging to the same mineral type—pyrite + arsenopyrite, with a variable ratio of main minerals. Arsenic pyrite was present in all ore occurrences. Two stages of ore formation were distinguished: early gold + pyrite + arsenopyrite with finely dispersed gold and late gold + galena + chalcopyrite + sphalerite with coarse gold, fahlore, and the sulfosalts Pb, Cu, Bi, and Sb. The sulfur isotope composition of $\delta^{34}S$ sulfides varied from −0.2‰ to −8.0‰. $\delta^{34}S$ values of the sulfides in the range of −0.2‰ to −3.5‰ correspond to the meteorite standard, which indicates the participation of a single deep magmatic source of sulfur in ore formation. According to the study of fluid inclusions, the formation of ore quartz veins occurred in a temperature range from 467 to 109 °C. No substantial differences between ore and barren quartz were found. Their fluid inclusions had similar homogenization temperatures, almost similar salinity of the solutions (equal to 0.5–17 wt.% and 4–16.5 wt.% NaCl-equiv., respectively), and correspond to the same $NaCl–KCl–H_2O$ and $NaCl–MgCl_2–H_2O$ systems. Our data on fluid inclusions, together with the identified stages of ore formation, likely indicate the duration of the hydrothermal process. The similarity of the geological–structural, mineralogical, geochemical, and isotope–geochemical features of the gold–sulfide–quartz and gold–sulfide occurrences of the area suggests their formation in a single hydrothermal system.

Author Contributions: Conceptualization, T.P.M., S.K.K., L.I.E. and N.V.S.; field work, L.I.E.; writing—design of the work, T.P.M. and L.I.E.; acquisition, analysis, and interpretation of data: mineralogy, sulphur isotopes, T.P.M.; mineralogy, S.K.K.; geology, L.I.E.; fluid inclusions, N.V.S.; writing—review and editing, T.P.M., S.K.K., L.I.E. and N.V.S. All authors have read and agreed to the published version of the manuscript.

Funding: The work was carried out according to the state order of the Yushkin Institute of Geology, Komi Science Center, Ural Branch, Russian Academy of Sciences, and with the support of the Russian Science Foundation, project No. 22-77-10088.

Data Availability Statement: Data are available in the references cited.

Acknowledgments: We thank Vasily Filippov, Eugene Tropnikov, and Alexander Shuisky for their assistance with the lab work.

Conflicts of Interest: The authors declare no conflict of interest.

References

1. Groves, D.J.; Goldfarb, R.J.; Gebre-Mariam, M.; Hagemann, S.G.; Robert, F. Orogenic gold deposits: A proposed classification in the context of their crustal distributions and relationship to other gold deposits. *Ore Geol. Rev.* **1998**, *13*, 7–27. [CrossRef]
2. Goldfarb, R.J.; Baker, T.; Dubé, D.; Groves, D.I.; Craig, J.R.; Hart, C.J.; Gosselin, P. *Distribution, Character, and Genesis of Gold Deposits in Metamorphic Terranes*; Society of Economic Geologists, Inc.: Littleton, CO, USA, 2005; pp. 407–450.
3. Nguimatsia, D.F.W.; Bolarinwa, A.T.; Yongue, R.F.; de Dieu Ndikumana, J.; Olajide-Kayode, J.O.; Olisa, O.G.; Abdu-Salam, M.O.; Kamga, M.A.; Djou, E.S. Diversity of Gold Deposits, Geodynamics and Conditions of Formation: A Perspective View. *Open J. Geol.* **2017**, *7*, 1690–1709. [CrossRef]
4. Li, H.; Wang, Q.; Deng, J.; Yang, L.; Dong, C.; Yu, H. Alteration and mineralization styles of the orogenic disseminated Zhenyuan gold deposit, southeastern Tibet: Contrast with Carlin gold deposit. *Geosci. Front.* **2019**, *10*, 1849–1862. [CrossRef]
5. Le Guen, M.; Lescuyer, J.L.; Marcoux, E. Lead-isotope evidence for a Hercynian origin of the Salsigne gold deposit. Southern Massif Central France. *Miner. Depos.* **1992**, *27*, 129–136. [CrossRef]
6. Petrovskaya, N.V. *Native Gold*; Nauka: Moscow, Russia, 1973; p. 348. (In Russian)
7. Petrovskaya, N.V.; Safonov, Y.G.; Sher, S.D. Formations of gold deposits. In *Ore Formations of Endogenous Deposits*; Nauka: Moscow, Russia, 1976; pp. 3–110. (In Russian)
8. Nikolaeva, A.N.; Gavrilov, A.M.; Nekrasova, A.N.; Yablokova, S.V.; Shatilova, L.V. *Free Gold of Ore and Placer Deposits in Russia: Atlas*; Watercolour: Moscow, Russia, 2015; p. 200. (In Russian)
9. Mumin, A.H.; Fleet, M.E.; Chryssoulis, S.L. Gold mineralization in As-rich mesothermal gold ores of the Bogosu–Prestea mining district of the Ashanti Gold Belt, Ghana: Remobilization of "invisible" gold. *Miner. Depos.* **1994**, *29*, 445–460. [CrossRef]
10. Genkin, A.D.; Bortnikov, N.S.; Cabri, L.J.; Wagner, F.E.; Stanley, C.J.; Safonov, Y.G.; McMahon, G.; Friedl, J.; Kerzin, A.L.; Gamyanin, G.N. A multidisciplinary study of invisible gold in arsenopyrite from four mesothermal gold deposits in Siberia, Russian Federation. *Econ. Geol.* **1998**, *93*, 463–487. [CrossRef]
11. Genkin, A.D.; Lopatin, V.A.; Savel'ev, R.A.; Safonov, Y.G.; Sergeev, N.B.; Kerzin, A.L.; Tsepin, A.I.; Amshtuts, K.; Afanas'eva, Z.B.; Wagner, F.; et al. Gold ores of the Olimpiada deposit (Enisei range, Siberia). *Geol. Ore Depos.* **1994**, *36*, 101.
12. Konstantinov, M.M.; Nekrasov, E.M.; Sidorov, A.A.; Struzhkov, S.F. *Gold Mining Giants of Russia and the World*; Scientific World: Moscow, Russia, 2000; p. 272.
13. Bortnikov, N.S.; Gamyanin, G.N.; Alpatov, V.A.; Nosik, L.P.; Naumov, V.B.; Mironova, O.F. Mineralogy, geochemistry, and origin of the Nezhdaninsk gold deposit (Sakha-Yakutia, Russia). *Geol. Ore Depos.* **1998**, *2*, 121–138.
14. Safonov, Y.G. Gold and gold-bearing deposits of the world: Genesis and metallogenic potential. *Geol. Ore Depos.* **2003**, *45*, 305–320.
15. Bortnikov, N.S.; Bryzgalov, I.A.; Krivitskaya, N.N.; Prokof'ev, V.Y.; Vikentieva, O.V. The Maiskoe multimegastage disseminated gold-sulfide deposit (Chukotka, Russia): Mineralogy, fluid inclusions, stable isotopes (O and S), history, and conditions of formation. *Geol. Ore Depos.* **2004**, *6*, 409–440.
16. Sazonov, A.M.; Zvyagina Ye, A.; Silyanov, S.A.; Lobanov, K.V.; Leontyev, S.I.; Kalinin Yu, A.; Savichev, A.A.; Tishin, P.A. Ore genesis of the Olimpiada gold deposit (Yenisei Ridge, Russia). *Geosphere Res.* **2019**, *1*, 17–43. (In Russian) [CrossRef]
17. Golovanov, I.M. (Ed.) *Ore Deposits of Uzbekistan*; FAN: Tashkent, Uzbekistan, 2001; p. 660. (In Russian)
18. Özgür, S.; Nezihi, K.; Emin, Ç.; Necla, K.; Gürkan, T.; Yılmaz, D. Mineralogy, geochemistry, fluid inclusion, and stable sulfur isotope investigation of the Terziali shear-related orogenic gold deposit (Central Anatolia, Turkey): Implications for ore genesis and mineral exploration. *Arab. J. Geosci.* **2022**, *15*, 113. [CrossRef]
19. Efanova, L.I.; Kuznetsov, S.K.; Tarbaev, M.B.; Mayorova, T.P. Gold mineralisation of the Manitanyrd region and prospects for the development of resource potential (Polar Urals). *Russ. Ores Met.* **2020**, *3*, 39–51. (In Russian with English abstract) [CrossRef]
20. Shishkin, M.A.; Astapov, A.P.; Kabatov, N.V.; Kazak, A.P.; Kolesnik, L.S.; Kotelnikov, V.G.; Kudrin, E.N.; Likhotin, V.G.; Mityusheva, V.S.; Molchanova, E.V.; et al. State Geological Map of the Russian Federation, scale 1:1,000,000 (third generation). In *Series Urals. Sheet Q-41–Vorkuta. Explanatory Note Kart*; VSEGEI: St. Petersburg, Russia, 2007.
21. Sazonov, V.N.; Ogorodnikov, V.N.; Koroteev, V.A.; Polenov, Y.A. *Gold Deposits in Urals*; USMU: Yekaterinburg, Russia, 2001; p. 622. (In Russian)
22. Kuznetsov, S.K.; Mayorova, T.P.; Sokerina, N.V.; Filippov, V.N. Gold ore mineralization of the Verkhneniyayuskoye deposit in the Polar Urals. *Russ. Zapiski RMO* **2011**, *4*, 58–71. (In Russian with English abstract)
23. Puchkov, V.N. *Geology of the Urals and Cis-Urals (Actual Problems of Stratigraphy, Tectonics, Geodynamics and Metallogeny)*; Design-PoligraphService: Ufa, Russia, 2010; p. 280. (In Russian)
24. Shishkin, M.A.; Malykh, O.N.; Popov, P.E.; Kolesnik, L.S. State Geological Map of the Russian Federation, scale 1:200,000. In *Series Polar-Urals. Sheet Q-41-V, VI (Vorkuta), Explanatory Note Kart*, 2nd ed.; MF VSEGEI: Moscow, Russia, 2013.
25. Shishkin, M.A.; Malykh, I.M.; Matukov, D.I.; Sergeev, S.A. Rhyolite complexes of the western slope of the Polar Urals. *Geol. Miner. Resour. Eur. Northeast. Russ.* **2004**, *2*, 148–150. (In Russian)
26. Sokerina, N.V.; Zykin, N.N.; Efanova, L.I.; Shanina, S.N.; Simakova, Y.S. Conditions of forming quartz veins of gold ore occurrences in the Manitanyrd region (Nether-Polar Urals). *Lithosphere (Russia)* **2010**, *2*, 100–111. (In Russian)
27. Vakhrushev, S.N.; Makarov, A.B. Mineralogical and geochemical particular features of ores and metasomatites of gold-arsenic occurrences in one of the regions. In *Geochemistry and Mineralogy of Primary and Secondary Aureoles*; FAN: Tashkent, Uzbekistan, 1986; pp. 84–93. (In Russian)

28. Mayorova, T.P.; Efanova, L.I. Niyahoyskoye-2 gold-arsenic occurrence in the Polar Urals (Manitanyrd ridge). *Bull. IG SC RAS* **2019**, *8*, 33–41. (In Russian with English abstract) [CrossRef]
29. Mayorova, T.P.; Efanova, L.I.; Sokerina, N.V. Mineralogical and geochemical features of the ores of the Verkhnelekeletskoye gold-quartzsulfide occurrence in the Polar Urals (Manitanyrd Ridge). *Vestn. Geosci.* **2022**, *6*, 10–25. (In Russian with English abstract) [CrossRef]
30. Borisenko, A.S. Study of salt composition of gaseous-fluid inclusion solutions in the minerals employing cryometry. *Russ. Geol. Geophys.* **1977**, *8*, 16–27.
31. Bodnar, R.J.; Vityk, M.O. Interpretation of microthermometric data for H_2O-NaCl fluid inclusions. In *Fluid Inclusions in Minerals: Methods and Applications*; Pontignano: Siena, Italy, 1994; pp. 117–130.
32. Roedder, E. Fluid inclusions in minerals. *Rev. Miner.* **1984**, *12*, 1–664.
33. Grinenko, V.A. Preparation of sulphur dioxide for isotope analysis. *Russ. J. Inorg. Chem.* **1962**, *7*, 2578–2582.
34. Savva, N.E.; Kravtsova, R.G.; Anisimova, G.S.; Palyanova, G.A. Typomorphism of Native Gold (Geological-Industrial Types of Gold Deposits in the North-East of Russia). *Minerals* **2022**, *12*, 561. [CrossRef]
35. Wilkinson, J.J. Fluid inclusions in hydrothermal ore deposits. *Lithos* **2001**, *55*, 229–272. [CrossRef]
36. Naumov, V.B.; Dorofeeva, V.A.; Mironova, O.F. Physicochemical formation parameters of hydrothermal mineral deposits: Evidence from fluid inclusions. II. Gold, silver, lead, and zink deposits. *Geochem. Int.* **2014**, *52*, 433–455. [CrossRef]
37. Goldfarb, R.J.; Groves, D.I. Orogenic gold: Common or evolving fluid and metal sources through time. *Lithos* **2015**, *233*, 2–26. [CrossRef]
38. Chi, G.; Diamond, L.W.; Lu, H.; Lai, J.; Chu, H. Common Problems and Pitfalls in Fluid Inclusion Study: A Review and Discussion. *Minerals* **2021**, *11*, 7. [CrossRef]
39. Liu, S.; Guo, L.; Ding, J.; Hou, L.; Xu, S.; Shi, M.; Liang, H.; Nie, F.; Cui, X. Evolution of Ore-Forming Fluids and Gold Deposition of the Sanakham Lode Gold Deposit, SW Laos: Constrains from Fluid Inclusions Study. *Minerals* **2022**, *12*, 259. [CrossRef]
40. Faure, G. *Principles of Isotope Geology*; Mir: Moscow, Russia, 1989; p. 590. (In Russian)
41. Ohmoto, H.; Rye, R.O. Isotopes of sulphur and carbon. In *Geochemistry of Hydrothermal Ore Deposits*, 2nd ed.; John Wiley and Sons: New York, MY, USA, 1979; pp. 509–567.
42. Kudrin, M.; Fridovsky, V.; Polufuntikova, L.; Kryuchkova, L. Disseminated Gold-Sulfide Mineralization in Metasomatites of the Khangalas Deposit, Yana–Kolyma Metallogenic Belt (Northeast Russia): Analysis of the Texture, Geochemistry, and S Isotopic Composition of Pyrite and Arsenopyrite. *Minerals* **2021**, *11*, 403. [CrossRef]
43. Fridovsky, V.Y.; Polufuntikova, L.I.; Kudrin, M.V.; Goryachev, N.A. Sulfur isotope composition and geochemical characteristics of gold-bearing sulfides of the Badran orogenic deposit, Yana-Kolyma metallogenic belt (North-East Asia). *Dokl. Akad. Nauk* **2022**, *502*, 3–9. [CrossRef]
44. Fridovsky, V.Y.; Polufuntikova, L.I.; Kudrin, M.V. Origin of Disseminated Gold-Sulfide Mineralization from Proximal Alteration in Orogenic Gold Deposits in the Central Sector of the Yana–Kolyma Metallogenic Belt, NE Russia. *Minerals* **2023**, *13*, 394. [CrossRef]

minerals

MDPI

Article

Ore Mineralogy and Typomorphism of Native Gold of the Spokoininsky Cluster of the Aldan–Stanovoy Gold Province

Larisa A. Kondratieva *, Galina S. Anisimova and Veronika N. Kardashevskaia

Diamond and Precious Metal Geology Institute, SB RAS, 677000 Yakutsk, Russia; gsanisimova1952@mail.ru (G.S.A.); kardashevskaya92@mail.ru (V.N.K.)
* Correspondence: lkon12@yandex.ru; Tel.: +7-4112-33-58-72

Abstract: The ore mineralogy of a new promising target of the Aldan–Stanovoy gold province—the Spokoininsky cluster—is considered. Gold mineralization is represented by a volumetric, nonlinear type, unconventional for the region; it is related to elements of fold structures and reverse fault in the enclosing metamorphic basement rocks. Vein-disseminated sulfide–(pyrite)–quartz ores build up deposit-like bodies in beresites from gneisses and granite gneisses and are associated with Mesozoic igneous rocks of subalkaline formations. Mineralization is characterized by polysulfide (Fe-Cu-Pb); gold–bismuth (Au-Bi) and gold–silver–telluride (Au-Ag-Te) mineral types. Different mineral types have their own typomorphic minerals and typochemistry (fineness and impurities) of native gold. The widespread distribution of telluride mineralization and its great importance in the formation of gold mineralization on the Aldan shield is confirmed. The distribution area of bismuth (including tellurium–bismuth) mineralization in the southern part of the Aldan shield, in the zone of influence of the Stanovoy deep fault, has been identified.

Keywords: gold mineralization; metasomatites; bismuth; tellurides; uytenbogaardtite; bismoclite; cervelleite; native gold; supergene gold; the Spokoininsky cluster; the Aldan-Stanovoy gold province; the Aldan shield

check for updates

Citation: Kondratieva, L.A.; Anisimova, G.S.; Kardashevskaia, V.N. Ore Mineralogy and Typomorphism of Native Gold of the Spokoininsky Cluster of the Aldan–Stanovoy Gold Province. *Minerals* **2023**, *13*, 543. https://doi.org/10.3390/min13040543

Academic Editor: Huan Li

Received: 16 March 2023
Revised: 8 April 2023
Accepted: 10 April 2023
Published: 12 April 2023

1. Introduction

The Spokoininsky ore cluster is a new promising ore target of the Aldan-Stanovoy gold province (ASGP), identified during prospecting in 2020 [1]. The search for ore targets has been undertaken since the 1970s [2] but resulted in only scattered mineralization points with an unclear structural position and mineralogy of ores. The territory of the cluster occupies the upper reaches of the right tributary of the Timpton river—the Ulakhan-Tarakanda River (Bol. Tyrkanda)—and covers the most productive southeastern part of the Tyrkanda gold-rich area, known since the beginning of the last century for rich placers of gold, from which more than 20 tons of gold were extracted. The primary sources of gold feeding the placers have not yet been identified. The largest placer deposit of the region is the placer of the Bol. Tyrkanda River, with gold reserves of more than six tons, which is in development to this day [3]. The placer deposit is localized in the valley of the river, starting near the mouth of the Spokoiny Creek, i.e., the Spokoininsky cluster.

The first mineralogical studies of the Spokoininsky cluster ores revealed a variety of Bi, Te, and Ag minerals, as well as a number of rare minerals, such as bismoclite and uytenbogaardtite. Determination of the connection of gold with certain mineral associations and the comparative analysis of morphological and geochemical properties of native gold of primary ores and supergene gold of eluvial deposits are relevant tasks for understanding the factors of its formation, i.e., the ore genesis.

2. Materials and Methods

Materials used for our research are 72 hand-sampled specimens of primary ores taken from surface mine workings and outcrops, as well as 20 shallow-pit samples from loose detrital clay material of eluvial deposits.

The polished sections made from the samples were optically examined using a Jenavert ore microscope in reflected light. The minerals were analyzed on a Camebax-micro X-ray spectral microanalyzer and a JEOL JSM-6480LV scanning electron microscope with an OXFORD energy spectrometer, and the Back Scattered Electron images were taken at the Diamond and Precious Metal Geology Institute, Siberian Branch, Russian Academy of Sciences (Yakutsk, Russia). The quantitative analysis was carried out using Software INCA Energy (Version 4.17, Oxford Instruments, Abingdon, Oxfordshire, UK) with XPP matrix correction scheme developed by Pouchou and Pichoir. Operating conditions were 20 kV voltage, a beam current of 1.08 nA, a beam diameter of 1 mm, and measurement time of 10 s. Analytical lines: Bi, Au−M; Te, Pb, Ag, Sb, S–L; and Cu, Fe, Zn, S–K. Standards: gold 750–Au, Ag, Bi_2S_3–(bismuthinite)–Bi, HgTe (coloradoite)–Hg, Te, $CuSbS_2$ (chalcostibite)–Cu, Sb, S, ZnS (sphalerite)–Zn, $CuFeS_2$ (chalcopyrite)–Fe, PbS (galena)–Pb, and FeAsS (arsenopyrite)–As. Element detection limits (wt.%) for the X-ray spectral microprobe analyses: Au 0.145, Ag 0.078, As 0.129, Hg 0.137, Cu 0.057, Fe 0.032, Pb 0.076, Bi 0.108, Sb 0.044, and Zn 0.074. Limits of element detection (wt.%) scanning electron microscope equipped with energy spectrometer: Au 1.84, Ag 0.96, Hg 1.6, Cu 1.22, Fe 1.04, Pb 1.78, and Bi 2.7.

3. Geology of the Spokoininsky Cluster

The Spokoininsky cluster is located on the northern slope of the Aldan shield. The territory is characterized by a large depth of the erosion section—the Archean crystalline basement—composed of gneisses and crystalline schists of the Sutamskaya and Kyurikanskaya formations of the Dzheltula series of the Lower Archean, which is exposed throughout the area of the region. The basement rocks are migmatized and contain concordant and transverse bodies of the Archean and Proterozoic granites, ultrabasites, gabbro-diorites, and gabbro-diabases.

The territory is located in the zone of the regional Tyrkanda fault, which has a north–northwestern strike and a thickness of up to 10–15 km. During the Mesozoic tectonic–magmatic events, the Tyrkanda fault zone underwent destruction: the northeastern seams of the extended faults of the Sunnaginskaya system are superimposed on it, which are a reflection of the global North Stanovoy fault located to the south.

The position at the intersection cluster of regional faults determines the intense fault tectonics and magmatism of the Mesozoic activation stage. The Mesozoic igneous rocks are represented by fields of dikes, small stocks, folded rocks of monzonite–syenite, and alkaline–syenite formations. The formation of intrusions involved contact–metasomatic and hydrothermal–metasomatic processes in the form of hornfelsing, chloritization, epidotization, sericitization, and silification of host rocks. The Spokoininsky ore cluster is located within the Mayskaya synclinal structure composed of amphibole and biotite–amphibole gneisses, with interlayers of bipyroxene and diopside gneisses.

The most promising ore fields are Spokoinoe and Mayskoe (Figure 1). The Spokoinoe ore field is located on the left bank of the lower reaches of the Spokoiny creek in the northern part of the cluster, and Mayskoe field covers the watershed of the Spokoiny, Maysky, and Taborny creeks in the south.

The Spokoinoe ore field is confined to the periclinal closure of the Mayskaya synclinal structure, where slight granitization of gneiss is noted. Gneisses are injected by the Early Proterozoic granitoids and intruded by bodies of the Mesozoic igneous rocks (nordmarkites, syenite porphyries, hornblende and biotite porphyries, minettes, spessartites, and vogesites), forming stock-shaped deposit bodies, dikes, and lenticular bodies with sharp swells, confined to the axial parts of complex folds.

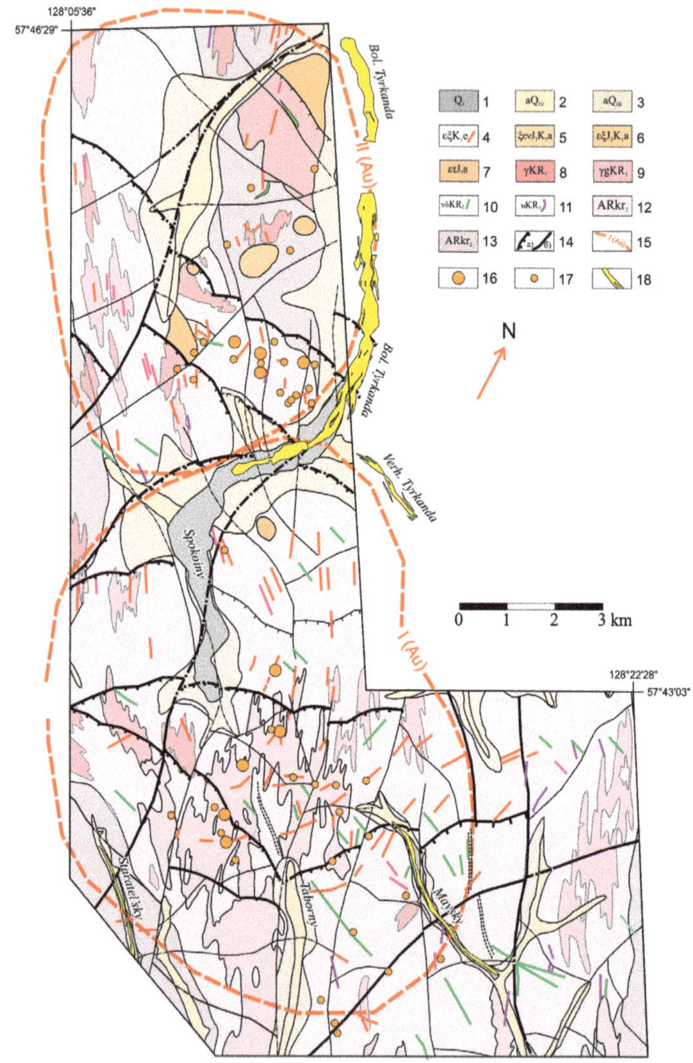

Figure 1. Schematic map of geological structure of the Spokoininsky ore cluster [1] with changes: 1—technogenous formations: rewashed deposits of prospectors polygons; 2—Quaternary: channel and low floodplain alluvium; 3—Upper Quaternary deposits: floodplain, channel, above-floodplain terrace alluvium, and fluvioglacial deposits; 4—dikes of alkaline and alkaline–earth syenites; 5–7—stocks: 5—alkaline–earth syenite, 6—pulaskites, laurvikites, and nordmarkites, 7—alkaline trachytes; 8—biotite, muscovite, and garnet granites, hypersthene–amphibole microcline granites, enderbites, and granites undivided; 9—variously hybridized granitized rocks and granite gneisses; 10—biotite, muscovite, and garnet granites, hypersthene–amphibole microcline granites, enderbites, and granites undivided; 11—variously hybridized granitized rocks and granite gneisses; 12,13—Kurikan Formation: 12—upper subformation (hypersthene and biotite–hypersthene gneisses and crystal schists); 13—middle subformation (biotite–hypersthene and two-pyroxene gneisses and crystal schists); 14—faults: a—reverse (to thrust) faults controlling distribution of highly promising anomalous geochemical fields and b—other faults; 15—ore fields: I (Au)—Mayskoe and II (Au)—Spokoinoe; 16—gold primary occurrences; 17—shallow-pit samples containing free gold; 18—gold placer deposits.

Faults belong to the systems of the Tyrkanda fault: strike-slips, strike-slip reverse faults of the northwestern strike, and the Sunnaginsky fault: normal faults, reverse faults of the northeastern strike. The gentle strippings of the latitudinal and sublatitudinal strikes, characterized as overthrust reverse faults, are less manifested. The latitudinal and northwestern directions within the ore field are well expressed by gold anomalies forming a discrete band, most contrasting at the intersection with the structures of the Sunnaginsky fault.

Gently dipping (10–20°) zones of schistosity of latitudinal and sublatitudinal strike contain gold ore bodies and are transverse, sub-concordant to the general occurrence of gneiss (Figure 2). Morphologically, ore bodies form gently dipping lenticular-ribbon-like deposits in the zones of schistosity, with a wavy surface, alternating with each other echelon-like in the latitudinal direction when wedging out along the strike. Dip angles are gentle, from less than 10° to 30°, and infrequently, 40°. The general azimuth of the dip is to the north. The thickness of individual deposits exposed by ditches ranges from 0.1–0.5 m to 1.6–2.5 m.

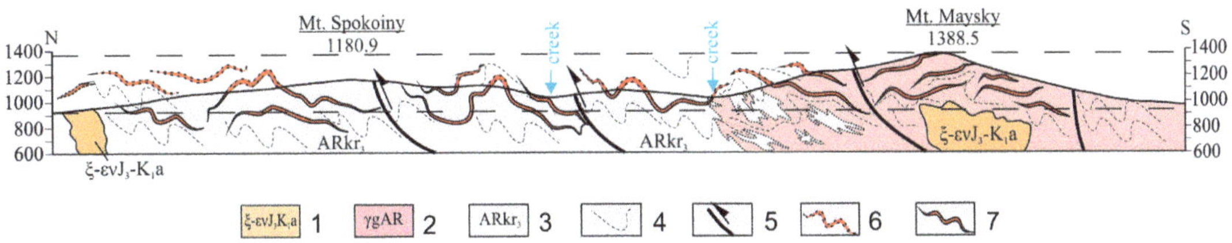

Figure 2. Geological cross section for gold mineralization of the Spokoininsky ore cluster, longitudinal section [1] with changes: 1—stocks of alkaline syenites; 2—granitized gneisses and granite gneisses; 3—Kurikan Formation, upper subformation; 4—folds; 5—reverse (to thrust) faults; 6,7—lodes: a—eroded and b—not eroded.

In contrast to the Spokoinoe ore field, the Mayskoe field is characterized by a significantly higher degree of granitization, especially in the southern, most elevated part of the site and the absence of large outcrops of the Mesozoic igneous rocks. The Mesozoic magmatism occurs in the form of frequent dikes, dike-like bodies of syenite porphyries, and lamprophyres, filling the faults of the northwestern and northeastern strike. In the northern, more eroded part of the ore field, a stock-shaped intrusion of alkaline-earth syenites has been exposed, and due to the presence of high-contrast magnetic anomalies, similar unexposed bodies are assumed in the south.

The correlation of the Mesozoic magmatic formations and gold mineralization is well demonstrated at the Mayskoe field. In the northern part of the site, ore zones are exposed by ditches in contact with the bodies of syenite porphyries and lamprophyries, which are also slightly gold-bearing. In the south are geochemical anomalies of gold trace unexposed intrusions. The main role as an ore-controlling factor also belongs to gentle structures, such as overthrust reverse faults of the latitudinal and sublatitudinal strike. Another important factor is the presence of plicative structures complicating the Mayskaya syncline. Ore mineralization is localized in areas of schistosity, with cleavage confined to various elements of folds, bends of hinges, turns of limbs, and other complicating elements (Figure 2). Ore bodies form gently and steeply dipping lenticular-ribbon-like deposits, bodies of complex shape in the zones of schistosity, with dip angles from 10–20° to 70°. Ore intervals with gold mineralization are 2.0–5.0 m and 7.0–10.0 m. In the identified ore intervals with a gold content, increased Cu contents are recorded (500–5000 g/t), Pb (100–200 g/t), Bi (50–100 g/t), and Ag (60–200 g/t) [1].

In general, a model of gold mineralization of a volumetric, nonlinear type related to elements of plicative structures and reverse fault tectonics in the host-granitized metamorphic rocks of the basement is proposed for the Spokoininsky ore cluster, represented by a combination of deposit-like bodies with vein-disseminated sulfide–(pyrite)–quartz type of mineralization. Mineralization is localized in beresites from gneisses and granite gneisses with vein-disseminated pyrite–quartz gold mineralization and is associated with the Mesozoic igneous rocks of subalkaline formations.

Considering structural features, brecciated, vein–veinlet, and disseminated types of ores are common. The main vein mineral is quartz—white, grayish, rarely honey-yellow, sometimes transparent, opaline silica, fine crystalline, drusoid, massive and brecciate, and chalcedonic in thin veinlets. It forms veinlets, lenses, geodes, and small net-shaped veinlets in beresites and beresitized gneisses, with thicknesses from 0.5 mm to 1.0–15.0 cm.

4. Results

4.1. Mineralogy of Ores

Mineralization of the Spokoininsky gold ore cluster is characterized by the following mineral types: polysulfide (Fe-Cu-As-Pb-Zn), gold–bismuth (Au-Bi), and gold–silver–telluride (Au-Ag-Te) (Table 1).

Table 1. Typification of the ores of the Spokoininsky cluster.

Type of Ore	Minerals
Polysulfide	Pyrite, chalcopyrite, galena, arsenopyrite, sphalerite, native tin, scheelite, and wolframite
Gold–bismuth	Native bismuth, bismuthite, tellurobismuthite, bursaite, matildite, cuprobismutite, smirnite, and bismoclite
Gold–silver–telluride	Krennerite, sylvanite, petzite, hessite, cervelleite, polybasite, native silver, acanthite, and uytenbogaardtite

The polysulfide mineral type is widespread in the area of the ore cluster. The main ore mineral is pyrite. The mineral—to varying degrees oxidized, often large, cubic, and pentagon dodecahedral, and up to 1.0 cm across—grows on quartz druse crystals. Fine-grained pyrite, more often oxidized, forms concentrations, with massive veinlets up to 0.5 cm thick at the base of quartz druses and host beresites. Pyrite is often a matrix for ore microminerals. Chalcopyrite occurs as allotriomorphic particles in quartz and pyrite. All other minerals are micron-sized. Galena forms frequent small particles, represented by thin rounded and irregular shapes. Hypidiomorphic microcrystals (<10 microns) of arsenopyrite are sporadically observed. Sphalerite is rare, characterized by a Fe content of up to 12%.

The ores contain very few grains of native tin, scheelite, and wolframite. Native tin and wolframite form small 10–15 micron interstitial grains in metasomatic quartz and scheelite—oval-rounded and isometric inclusions in pyrite and iron hydroxides replacing pyrite.

Typomorphic minerals of gold–bismuth mineral type are native bismuth, bismuthite (BiS), tellurobismuthite (Bi_2Te_3), bursaite ($Pb_5Bi_4S_{11}$), cuprobismutite ($Cu_8AgBi_{13}S_{24}$), and matildite ($AgBiS_2$) (Figure 3, Table 2). Secondary minerals are represented by smirnite (Bi_2TeO_5) and bismoclite (BiOCl). Bismuth mineralization was identified mainly in ores of the Mayskoe ore field. Tellurobismuthite forms elongated and oval crystals in pyrite, and bursaite forms tabular crystals in quartz. Native bismuth in growth with matildite is enclosed in pyrite. They also revealed matildite with a significant content of Te (7.11 wt.%) as well as an obscure oxygen-containing phase with a spongy surface in which the concentration of Te reached 23.25 wt.%. Cuprobismutite and smirnite were observed in supergene gold: smirnite is represented by a disintegrated mass in the form of a border and cuprobismutite in relic grains. Among the bismuth minerals, matildite and bismoclite were found in the Spokoinoe ore field. Matildite was observed in supergene gold in close growth with galena in the form of a border and had microinclusions in it. Bismoclite is a secondary mineral of bismuth, found in a quartz cavity in the form of a tetragonal crystal with faces truncated at 45°.

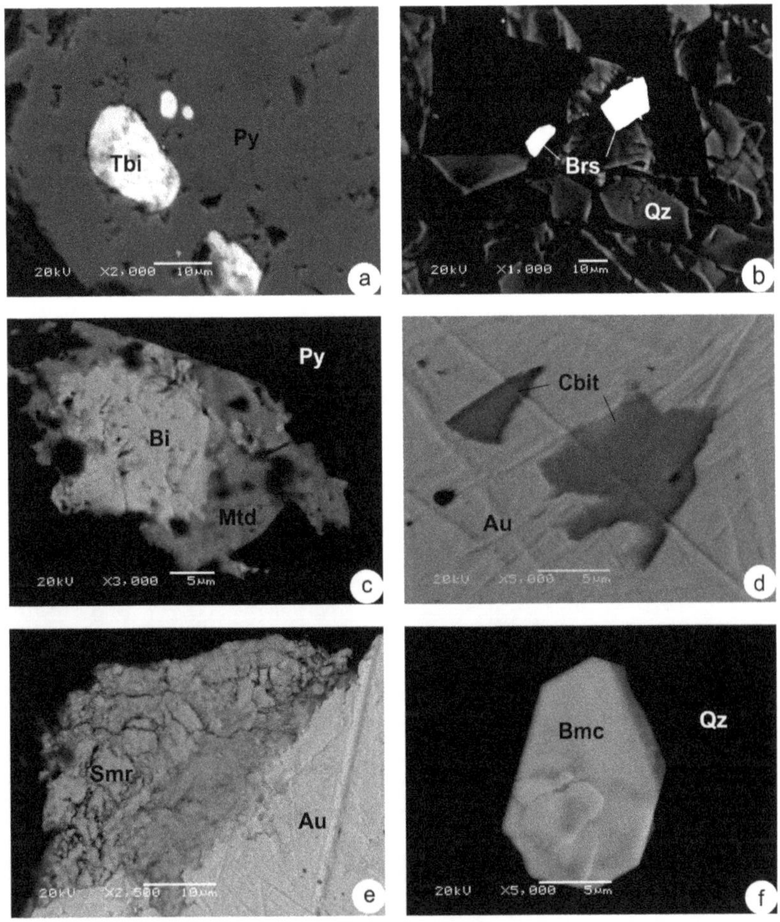

Figure 3. Gold–bismuth mineralization of the Spokoininsky ore cluster: (**a**)—ellipsoidal grains of tellurobismuthite (Tbi) in pyrite (Py), (**b**)—tabular grains of bursaite (Brs) in quartz (Qz), (**c**)—native bismuth (Bi) in growth with matildite (Mtd) in pyrite (Py), (**d**)—relics of cuprobismutite (Cbit) in supergene gold (Au), (**e**)—disintegrated mass of smirnite (Smr) on the edge of supergene gold (Au), (**f**)—bismoclite crystal (Bmc) in quartz (Qz).

Table 2. Chemical composition of bismuth minerals (in wt.%).

Mineral	Bi	Te	Ag	Cu	Cl	Pb	S	Si	Fe	O	Total	apfu
Native Bi	100.33										100.33	$Bi_{1.00}$
Tellurobismuthite	51.43	47.89									99.32	$Bi_{1.96}Te_{3.00}$
Bismuthite	81.50						17.51				99.01	$Bi_{2.13}S_{3.00}$
Bursaite	36.63		3.96			44.92	14.41				99.92	$(Pb_{5.31}Bi_{4.27}Ag_{0.90})S_{11.00}$
Matildite	56.32		27.09				15.33				98.75	$Ag_{1.05}Bi_{1.12}S_{2.00}$
Cuprobismutite	60.36		4.95	12.28			16.34				93.93	$(Cu_{9.10}Bi_{13.54}Ag_{2.16})S_{23.00}$
Bismoclite	71.59				11.87			2.00	2.13	11.55	99.13	$Bi_{1.02}Fe_{0.11}Si_{0.21}Cl_{1.00}$
Smirnite	66.79	19.13							1.86	12.38	100.16	$Bi_{2.06}Te_{0.97}Fe_{0.22}O_{5.00}$

Bismuth minerals are characterized by the presence of a significant impurity of silver (3.96 wt.%) in bursaite and cuprobismutite (4.95 wt.%). Impurities of Fe and/or Si in the composition of the bismoclite and smirnite are due to the influence of the background matrix of quartz and iron hydroxides. There is also a background impurity of iron in smirnite.

Gold–silver–telluride type is represented by krennerite ($AuTe_2$), sylvanite ($(Au,Ag)_2Te_4$), petzite (Ag_3AuTe_2), hessite (Ag_2Te), cervelleite (Ag_4TeS), polybasite, native silver, acanthite (Ag_2S), and uytenbogaardtite (Figure 4, Table 3). The grain size does not exceed 50 microns. In addition, gold bismuth type is more developed in the Mayskoe ore field. Minerals of petzite–hessite paragenesis and polybasite are observed in the form of oval and irregular grains in pyrite. Tellurides of the krennerite group were observed as inclusions in hessite and petzite, gravitating towards the central parts of the grains. There are cases of hessite growths with tellurobismuthite. Acanthite, native silver, and uytenbogaardtite are observed in the ores of the Spokoinoe field, and hessite–cervelleite paragenesis is developed in the form of rounded micron inclusions (no more than 30 microns) in hypergene gold. Along with cervelleite of stoichiometric composition, there is cervelleite with a considerable content of copper up to 5.92%.

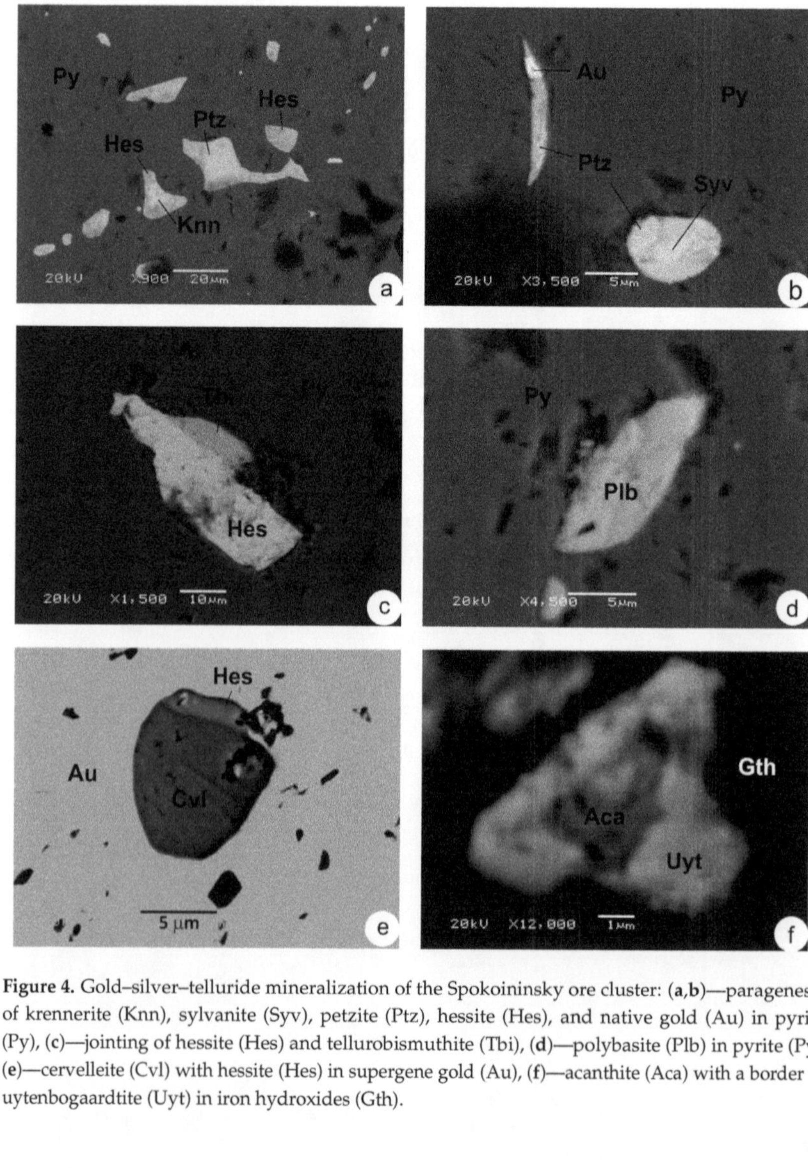

Figure 4. Gold–silver–telluride mineralization of the Spokoininsky ore cluster: (**a**,**b**)—paragenesis of krennerite (Knn), sylvanite (Syv), petzite (Ptz), hessite (Hes), and native gold (Au) in pyrite (Py), (**c**)—jointing of hessite (Hes) and tellurobismuthite (Tbi), (**d**)—polybasite (Plb) in pyrite (Py), (**e**)—cervelleite (Cvl) with hessite (Hes) in supergene gold (Au), (**f**)—acanthite (Aca) with a border of uytenbogaardtite (Uyt) in iron hydroxides (Gth).

Table 3. Chemical composition of Au, Ag, and Te minerals (in wt.%).

Mineral	Ag	Au	Sb	Te	Cu	As	S	Fe	Total	apfu
Native Ag	99.98								99.98	$Au_{1.00}$
Krennerite	3.71	42.60		52.17				2.41	100.89	$Au_{4.23}Ag_{0.67}Fe_{0.84}Te_{8.00}$
Sylvanite	7.08	28.34		63.08					98.50	$Au_{1.16}Ag_{0.53}Te_{4.00}$
Petzite	43.17	21.71		35.60					100.48	$Ag_{2.87}Au_{0.79}Te_{2.00}$
Hessite	62.33			38.40					100.73	$Ag_{1.92}Te_{1.00}$
Cervelleite	68.23			25.12			5.07		98.42	$Ag_{4.00}Te_{1.25}S_{1.00}$
Cu-rich cervelleite	66.14			24.14	5.92		5.38		101.58	$(Ag_{3.65}Cu_{0.56})Te_{1.13}S_{1.00}$
Polybasite	64.57		6.81		8.36	2.26	16.28	2.26	100.53	$[Fe_{0.88}(Ag_{3.97}.Cu_{1.85})_{5.82}(As_{0.65}Sb_{1.21})_{1.86}S_7][Ag_9CuS_4]$
Uytenbogaardtite	56.45	28.53					11.79	4.33	101.10	$Ag_{2.85}Au_{0.79}Fe_{0.43}S_{2.00}$
Acanthite	88.16						8.12	2.91	99.19	$Ag_{3.23}Fe_{0.21}S_{1.00}$

Uytenbogaardtite Ag_3AuS_2 was found in association with spongy and xenomorphic acanthite particles, less often with native gold in oxidized quartz–feldspar metasomatites (Figures 4f and 5). The matrices for minerals are iron hydroxides replacing pyrite. Uytenbogaardtite is characterized by a variable composition (Table 4). It has iron intake from the matrix.

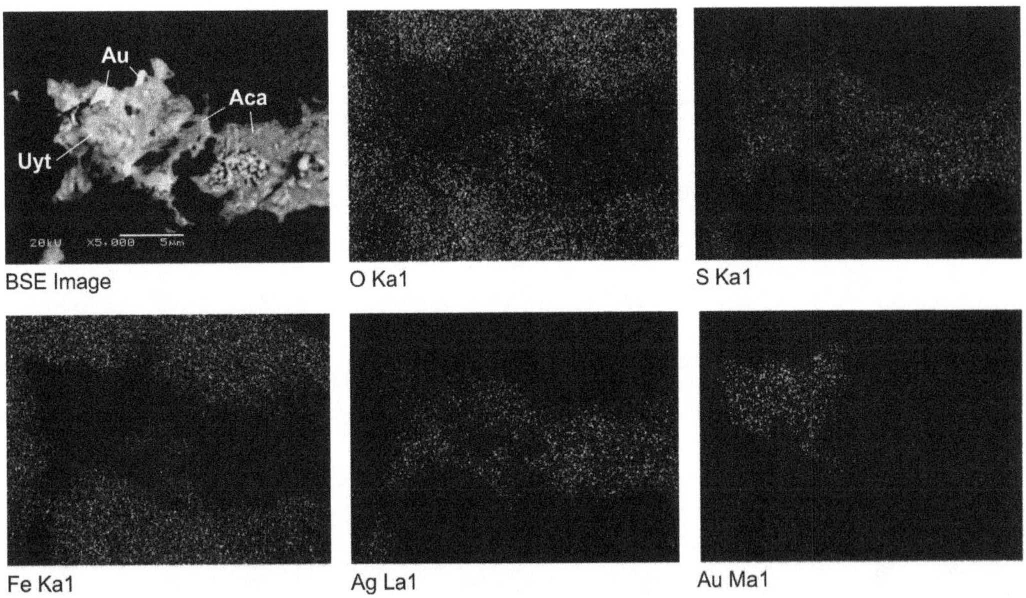

Figure 5. Paragenesis of uytenbogaardtite (Uyt), acanthite (Aca), and native gold (Au) in iron hydroxides: BSE and raster images display the distribution of the elements.

Table 4. Chemical composition of uytenbogaardtite (in wt.%).

No. of Analyses	S	Fe	Ag	Au	Total	apfu
22-5	11.36	6.50	31.66	48.14	97.66	$Ag_{1.66}Au_{1.38}Fe_{0.66}S_{2.00}$
16-5	11.87	4.77	52.14	32.22	101.00	$Ag_{2.61}Au_{0.88}Fe_{0.46}S_{2.00}$
29-1	11.79	4.33	56.45	28.53	101.10	$Ag_{2.85}Au_{0.79}Fe_{0.42}S_{2.00}$
44-3	12.50	3.21	56.15	28.08	99.94	$Ag_{2.67}Au_{0.73}Fe_{0.29}S_{2.00}$
37-2	10.22	3.87	61.92	24.94	100.94	$Ag_{3.60}Au_{0.79}Fe_{0.43}S_{2.00}$
37-1	10.46	4.09	63.71	23.60	101.85	$Ag_{3.62}Au_{0.73}Fe_{0.45}S_{2.00}$

4.2. Typomorphic Features of Gold

Native gold of primary ores and supergene gold of eluvial deposits have been studied.

Granulometry. Native gold in ores is usually finely dispersed and dust-like <5–15 microns, and single grains reach only 0.1 mm.

Supergene gold records the predominance of gold of a very fine fraction (less than <0.1 and 0.1–0.25 mm); medium-sized gold (0.25–0.5 mm) occurs in smaller quantities, and (0.5–1.0 mm) is less common. One small nugget with a size of 5.5 × 3.2 mm was found (Figure 6).

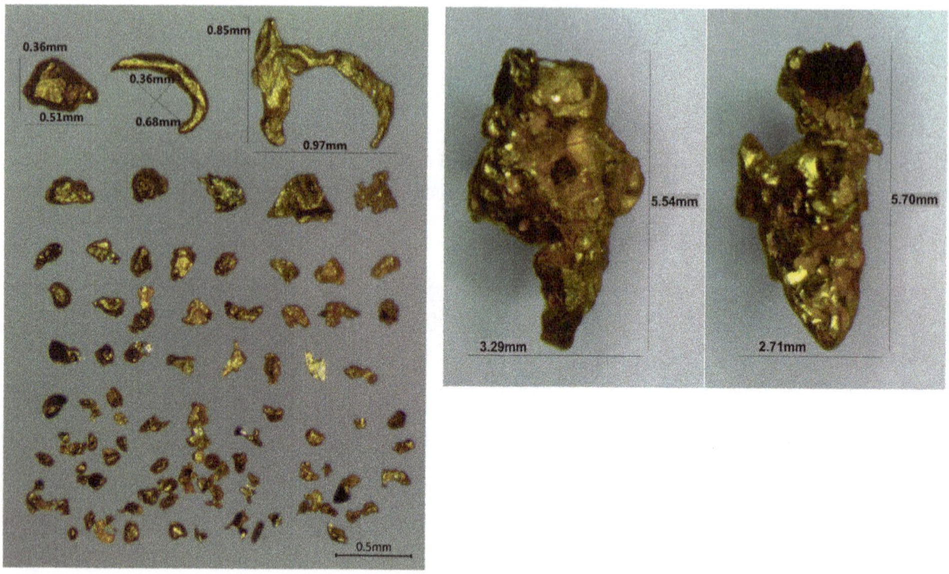

Figure 6. Granulometry of supergene gold.

Morphology. In ores, primary gold has an interstitial and cementing nature. It usually develops in interstices between pyrite grains (Figure 7a) or along cracks in pyrite (Figure 7b), in addition to single pseudoidiomorphic grains of hexagonal appearance (Figure 7c). Microscopically native gold has a porous structure.

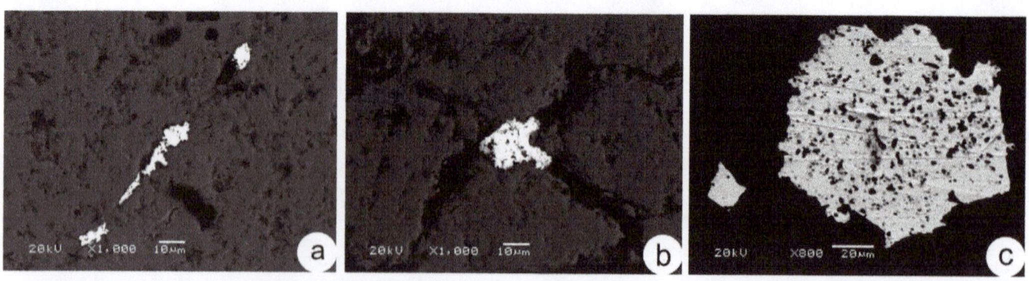

Figure 7. Morphology of native gold of primary ores: (**a**)—interstitial between pyrite grains; (**b**)—cementing along cracks in pyrite; (**c**)—interstitial pseudoidiomorphic grains of hexagonal appearance.

A significant part of supergene gold has a crystalline appearance and is represented by individual crystals or their intergrowths (Figure 8a–e). Crystals form isometric and crystallomorphic grains with shapes close to an octahedron and a combination of a cube and an octahedron, flattened prismatic, elongated needle-like, drusoid, and dendrite-like and coral-like individuals. Striation and growth marks are observed on the crystal faces, and the edges are mostly smoothed. Hemi-idiomorphic gold is also often found—aggregates of grain intergrowths with separate developed cubic and prismatic faces, combined with gold of irregular morphology (Figure 8f). The shape of the grains is of the irregular type, mainly lumpy, scaly, hook-like, lamellar, or wire-shaped (Figure 8h,i). There is a large occurrence of idiomorphic gold in the Mayskoe ore field. In the Spokoinoe ore field, lumpy and scaly gold of irregular morphology was more often observed. Microscopically, gold is usually porous.

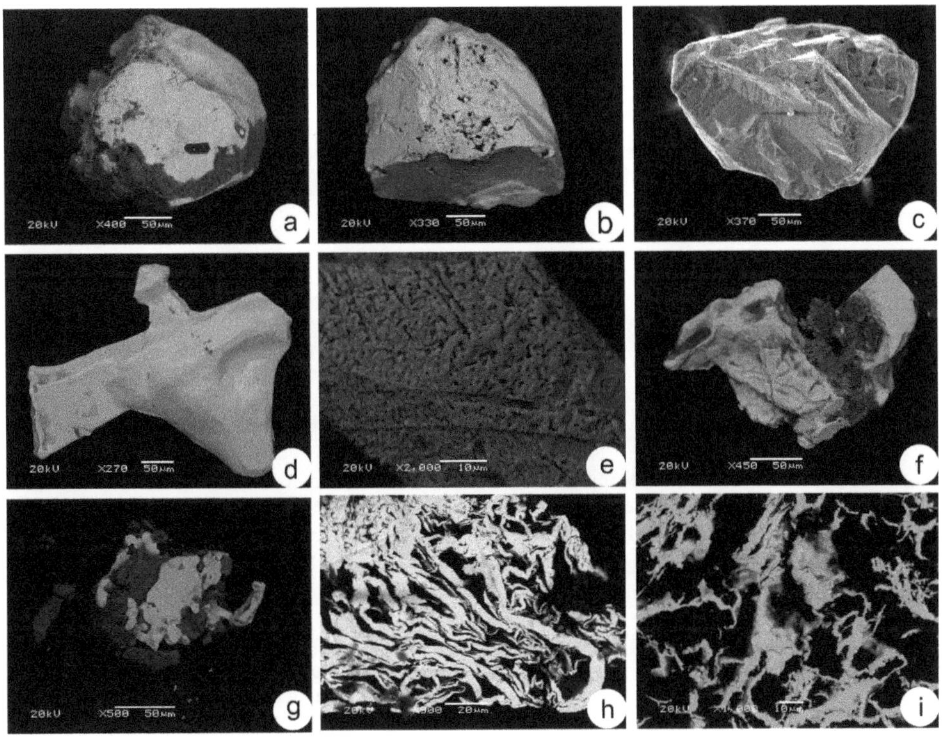

Figure 8. Morphology of supergene gold: (**a**)—isometric crystal of octahedral appearance; (**b**)—isometric crystal with the development of a combination of cube and octahedron shapes; (**c**)—stepped structure of crystallomorphic druzoid gold; (**d**)—clusters of crystals of prismatic and octahedral shapes; (**e**)—prismatic with a microspongy structure; (**f**)—hemi-idiomorphic gold; (**g**)—shell dendritic; (**h**)—lamellar-looped; (**i**)—leafy dendritic.

Often in supergene gold, patterns of the origin of newly formed gold were observed, according to the rhythmiczonal structure of iron hydroxides.

Chemical composition. The main impurity of primary gold according to the microprobe analysis is Ag, and there are single grains with a significant impurity of Cu (6.37–30.04 wt.%). The fineness of native gold has wide variations ranging from 673 to 993‰. In the Spokoinoe field, low-grade gold prevails (average value—753‰); in the Mayskoe field exists medium-grade (average value—810‰) (Figure 9a).

According to the results of the X-ray spectral microanalysis, the content of impurities Zn, As, Hg, Pb, and Sb in supergene gold was below the sensitivity limit (detection). Minor impurities of Fe (up to 0.139 wt.%), Cu (up to 0.231 wt.%), and Bi (up to 0.199 wt.%) were found in single samples. The supergene gold of the Spokoinoe ore field is characterized by low fineness (701‰–800‰) of approximately 60%, and the share of medium-grade gold (801‰–900‰) accounts for 30% (average value—793‰) (Figure 9b). The Mayskoe ore field is dominated by medium-grade gold (800‰–900‰), which accounts for approximately 50% (average value—842‰). Relatively low-grade (700‰–800‰) and high-grade gold (900‰–950‰) amount to 20% and 22%, respectively.

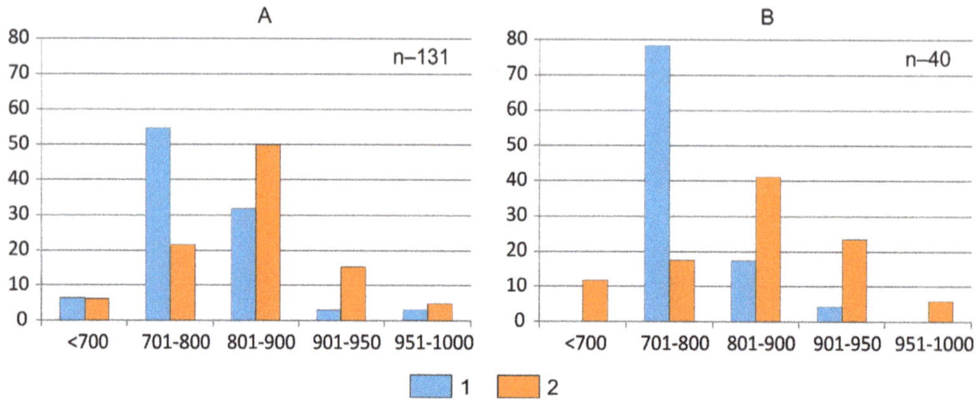

Figure 9. Histograms for the distribution of the fineness of gold: (**A**)—native gold, (**B**)—supergene gold. 1—Spokoinoe ore field, 2—Mayskoe ore field, n—number of analyses.

Compositional heterogeneity. In the same sample, the fineness of primary gold ranges usually from 10 to 130‰, supergene gold usually from 20 to 240‰. Heterogeneous gold, sharply differing in color from bright yellow to a pale whitish color, was identified in shallow-pit samples of the Spokoinoe ore field. Considering morphology, gold is more often thinly lamellar and scaly; in the central part, it has a porous structure, and it is massive on the periphery. The marginal fluctuations of the fineness as a whole are 480‰–999‰ (Figure 10). The overwhelming amount of native gold has a fineness of 750‰–780‰ (average value 778‰). Two types of heterogeneous gold of zonal structure were identified in individual grains. There are gold particles in which the fineness varies from an electrum of 480‰–520‰ on the periphery of the grain to a relatively low-grade 737‰–775‰ in the central parts. Another type of heterogeneous gold is more common—low-grade gold from 668 to 790‰, with a fringe or fragments of a rim of impurity-free native gold. Noteworthy is the complete absence of gold of medium and high fineness in the range 791‰–995‰.

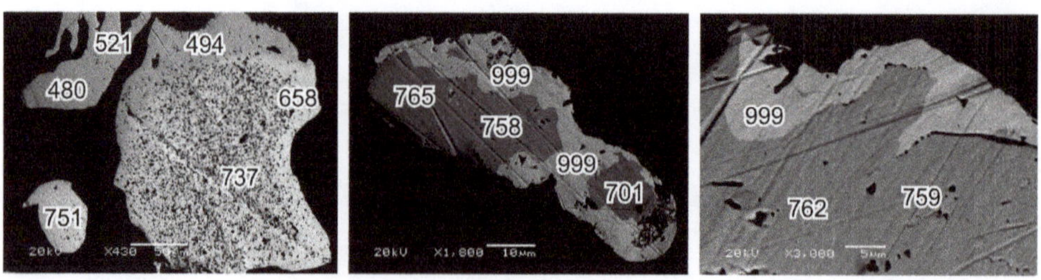

Figure 10. Heterogeneous of supergene gold (the numbers show the degree of fineness).

Relationship with other minerals. Native gold was observed mainly in pyrite and iron hydroxides replacing pyrite and less often in association with hessite, petzite, and acanthite. There is also free gold in quartz and feldspar. In pyrite, native gold is usually located along cracks in the intergranular space or along growth zones. Microporous hessite containing oxygen (Te 28.29–29.90, Ag 54.57–56.75, and O 14.22–16.97) was observed in association with native gold (Figure 11a). Rounded to oval-shaped microinclusions of ore minerals are found in supergene gold: galena, sphalerite, matildite, cuprobismutite, hessite, and cervelleite (Figures 3d, 4e and 11b,c), as well as smirnite from the edge of the grain of gold (Figure 3e). One of the relict grains found in gold is a mixture of hessite and nonequilibrium gold-containing cervelleite-like phase (Te 14.6–27.33, Ag 55.82–59.29, Au 9.97–21.50, and S 3.42–8.62) with microinclusions of cervelleite of stoichiometric composition (Figure 11c). Goethite in the intergrowths with supergene gold often contains an admixture of tellurium up to 2–4 wt.%. In addition, there are cases of location of native gold in supergene minerals similar in composition to plumbojarosite and goethite with rhythmiczonal acanthite threads (Figure 11d).

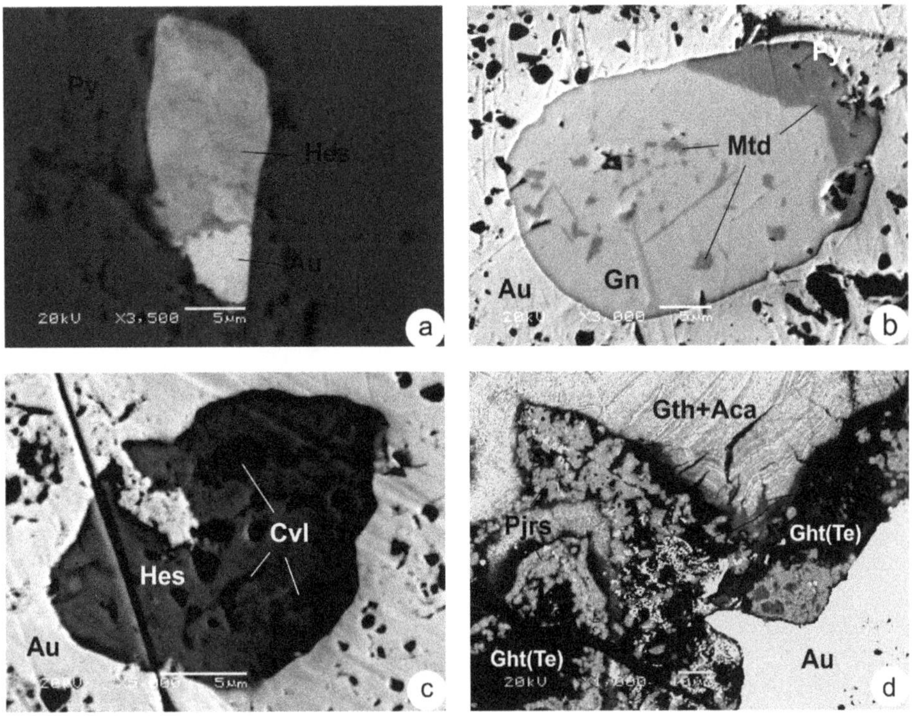

Figure 11. Mineral associations of gold: (**a**)—association of hessite (Hes) and native gold (Au) in pyrite (Py); (**b**,**c**)—relics of minerals in supergene gold: (**b**)—galena (Gn) with phases and border of matildite (Mtd), (**c**)—a mixture of hessite (Hes) with gold-containing cervelleite-like phase and microinclusions of cervelleite (Cvl); (**d**)—the relationship of supergene gold (Au) with plumbojarosite (Pjrs), goethite enriched with tellurium, and goethite with rhythmic-zonal acanthite (Ght + Aca).

5. Discussion

Aldan-Stanovoy gold province is one of the main gold-mining regions in Russia. The gold content of the ASGP is related to hydrothermal–metasomatic processes caused by the Mesozoic tectonic–magmatic activation of the region, involving the intrusion of massifs of subalkaline and alkaline high-potassium igneous rocks of the Jurassic–Cretaceous age [4–11].

The gold mineralization of the ASGP is characterized by a variety of geological and structural positions, hydrothermal–metasomatic formations, and mineral types. Ore-bearing hydrothermal–metasomatic formations are represented by sericite–microcline metasomatites, beresites, gumbaites, jasperoids, and argillizated rocks. The metasomatites of the gumbaite formation are represented by pyrite–carbonate–feldspar metasomatites, which are associated with the main gold and uranium deposits of the Elkon ore cluster. The formation of sericite–microcline metasomatites is associated with the gold–copper–

porphyry (Ryabinovsky) type of mineralization, including veinlet-disseminated sulfide mineralization in alkaline volcano–plutons (Ryabinovoe and Novoe deposits). The jasperoid type of mineralization is represented by deposits formed in three geological structural settings: in contact zones of the Mesozoic alkaline and subalkaline intrusions with carbonate rocks of the Vendian among hydrothermally altered dolomite marbles, magnesian skarns, and syenites (Samolazovsky subtype); in zones of layer-by-layer and intersecting jointing in the Vendian–Lower Cambrian carbonate rocks (Lebedinsky subtype); and at the contact of the Lower Cambrian limestones with the Jurassic sandstones (Kuranakhsky subtype). The gold–argillizite mineralization of the Nimgerkansky type is spatially confined to the intrusions of syenite porphyries of Cretaceous age and is associated with crystal-bearing and amethyst-bearing mineralization. In addition to the Mesozoic mineralization, the Precambrian gold mineralization of the Piniginsky type occurs in the basites of the Medvedev complex [12].

Gold mineralization of the Spokoininsky cluster is represented by a volumetric, nonlinear type, which is unconventional for the region, related to elements of plicative structures and reverse fault tectonics in the enclosing metamorphic basement rocks. Ore zones represent an echeloned system of scalariform deposits grading into gently plunging ore columns in the frontal part of the reverse fault. The connection of gold mineralization with the Mesozoic magmatism is discussed.

For a better understanding of the factors of the formation of gold mineralization, it is necessary to conduct studies of physical–chemical and isotope–geochemical parameters as well as determine the age of mineralization for correlation with magmatic events.

Ore content is related to mineralized zones of crush, cataclase, shear, and schistosity. During the tectonic processing, the porosity of metamorphic and igneous rocks increased, which was a favorable environment for hydrothermal–metasomatic transformations. On the basis of the petrographic study of wallrock changes in igneous and metamorphic rocks and ores of the Spokoininsky cluster, gold-bearing metasomatites can be referred to as low-temperature hydrothermal–metasomatic formations of the beresite and argillized formations. Beresites develop along gneisses and syenites and consist of quartz, sericite, muscovite, hydromica, chlorite, ferruginous carbonates, and fine-grained pyrite. Argillized quartz–feldspar metasomatites are very common in the area of the cluster, often bearing vein–veinlet mineralization of drusoid quartz. The level of gold content is directly dependent on the thickness of wallrock metasomatites and the intensity of silification.

Noble metal mineralization is associated with gold–bismuth (Au-Bi) and gold–silver–telluride (Au-Ag-Te) mineral types.

Gold–bismuth mineralization is locally developed in the deposits of the Aldan shield. Bismuth mineralization has been identified in the ores of deposits close to the Stanovoy plutogenic region in the areas of acidic dikes and small intrusions: Altan-Chaidakh (granodiorites, diorite porphyrites, and dacites) and Bodorono (diorite porphyrites) [13–16]. These deposits, as well as the Spokoininsky cluster, are confined to the zone of the Tyrkanda fault. In addition, bismuth mineralization is developed in the ores of the Lebedinsky cluster located in the center of the magmatogenic structure, where the most intense magmatism occurred in the Central Aldan region. The presence of bismuth mineralization in the ores of the Spokoininsky cluster suggests the influence of acid magma.

The features of the geological structure of the Altan-Chaidakh deposit are determined by the occurrences of the Mesozoic magmatism, very significant in volume and area of distribution, which is related to the Altan-Chaidakh volcanic–tectonic structure. Mineralization is localized in the Lower Jurassic sandstones and the sills of porphyry dacites injecting them. Ore mineralization is represented by complex gold–polymetallic and gold–tellurium–bismuth mineral associations.

Of particular interest are sulfotellurides and tellurides Bi and sulfosalts Pb-Bi, as they are related the groundmass of visible native gold. A wide range of bismuth minerals has been identified, represented by bismuth, tetradymite, bismutoplagionite, bursaite, kosalite, tellurobismuthite, sulfotsumoite, schirmerite, tellurites, and bismuth oxides. Ag, Sb, Cu, and Se are frequent impurities of bismuth minerals. Native gold associated with bismuth minerals has a fineness of 850‰–890‰ [13].

Two producing associations have been identified at the Bodorono deposit: gold–polymetallic and gold–tellurium–bismuth [14–16]. The latter contains lillianite (contains impurities Ag (3.03 wt.%), Sb (1.48 wt.%), and Te (0.52 wt.%)), bismuthite, native bismuth, pilsenite, rare bismuth sulfoselenide, laitakarite (in composition of the mineral, S completely replaced by Se), secondary bismuth mineral smirnite, and tetradymite group minerals—tellurobismuthite, tetradymite, and hedleyite (contains impurity Se (3.72 wt.%)). Native gold with a high fineness of 820‰–940‰ is found in growths with lillianite, bismuthite, and tellurobismuthite.

The predominant spread of both bismuth and telluride mineralization in the area of the Spokoininsky cluster is characterized by the ores of the Mayskoe field. The reason is, apparently, on the one hand, the large granitization of the host complex, the presence of unexposed massifs, as well as a higher hypsometric level of gold mineralization. On the other hand, in the Spokoinoe field, a significant part of the ore bearing the mineralization of Bi and Te underwent denudation, and likely served as a source of placer gold. Secondary bismoclite was observed in the ores, whereas matildite, hessite, and cervelleite were preserved only as relics in supergene gold.

Rare bismuth oxychloride, bismoclite (BiOCl), first described by Mountain (1935), was identified in a sample of eluvial origin, selected on the surface of a pegmatite outcrop at Steinkopf, Namaqualand, Cape Province of South Africa [17]. Subsequently, the bismoclite was identified mainly in the form of alluvial samples near bismuth-containing granite pegmatites or as weathering products of bismuth sulfides in greisen deposits; it was also found in epithermal and volcanic massive sulfide deposits with low and high sulfide content. On the Aldan shield, bismoclite was also found in the Khokhoy karst deposit in the north of the structure, where no other bismuth mineralization was found [18].

The first finding of bismoclite in an ore sample of hydrothermal–magmatic breccia of the San Francisco de los Andes porphyry Bi-Cu-Au deposit, Argentina, is described by [19]. Bismoclite (BiOCl) was found in association with preisingerite ($Bi_3(AsO_4)_2O(OH)$). It is assumed that the bismoclite was formed as a result of weathering of hypogene bismuth-bearing minerals under the influence of meteoric waters containing O_2 and HCl. In addition, the discovery of bismoclite as a mineral phase in the oxidized zone of weathered sediments indicates the existence of hypogene mineralization of Bi at depth.

In our case, bismoclite, along with relic grains of cuprobismutite and matildite, is most likely evidence of the former existence of bismuth mineralization in the ores of the Spokoininsky cluster, which played a significant role in the formation of gold.

Telluride mineralization widely occurs in many gold deposits of the Aldan shield, where it is late, superimposed on early pyrite–quartz mineralization. Au-Bi-Te, Au-Ag-Te, and mixed types of telluride mineralization are found in the distribution of which zoning is identified [20]. In the north of the Aldan shield, the Au-Ag-Te type dominates. In the south, the gold mineralization of the Bodorono and Altan-Chaidakh deposits, bearing various bismuth tellurides, is referred to as the Au-Bi-Te type. The ores of the Spokoininsky cluster are characterized by the presence of mixed Au-Bi-Ag-Te mineralization. Bi-Te minerals include tellurobismuthite and smirnite; and Au-Ag-Te includes krennerite, sylvanite, hessite, petzite, and cervelleite. The existing cases of close growth of tellurobismuthite and hessite are evidence of their paragenetic relationship. While hessite, petzite, and tellurobismuthite are common minerals of the tellurium of the Aldan shield, gold tellurides of the krennerite group are found in a limited number of deposits, and cervelleite is found only in the ores of the Spokoininsky cluster. It was identified as a relict inclusion in supergene gold in association with hessite.

Cervelleite is found in ores of various deposits, including volcanogenic Bambolla mine, Moctezuma, Sonora, (Mexico) (the first finding) [21], Um Samiuki, Egypt [22], deposits of the Southern Urals [23], porphyry San Martin deposit, Argentina [24], Funan Au deposit, China [25], epithermal Mayflower Au-Ag deposit, Montana [26], Eniovche, Bulgaria [27],

Larga, Roşia Montană, Romania [28,29], and skarn Băiţa Bihor and Ocna de Fier, Romania [28,29]. That is, they were identified in ore systems that were genetically related to magmatism.

Cervelleite of the Spokoininsky cluster is characterized by a stable admixture of copper up to 5.92 wt.%. It is not uncommon; a similar cervelleite rich in Cu (up to 6 wt.% Cu) is reported in several works [22,23,30]. On the basis of wide variations in the composition and physical properties of cervelleite-like sulfotellurides, Novoselov et al. (2006) [23] suggest the existence of several new phases which can be distinguished by the Cu content, Te/S ratios, and, presumably, by the crystal structure.

Uytenbogaardtite Ag_3AuS_2 is a rare sulfide of gold and silver. Its formation is possible both in hypogene [31–34] and supergene [35–38] conditions. Uytenbogaardtite of the Spokoininsky cluster has a clearly hypergene nature of formation, as it was identified in association with acanthite in the oxidation zone of quartz–feldspar ore metasomatites. The matrix for minerals is iron hydroxide replacing pyrite, and the chemical composition of uytenbogaardtite is characterized by significant dispersion, which can be explained by the nonequilibrium medium of its formation present in supergene conditions.

The analysis of the obtained data on primary and supergene gold showed the enlargement of gold and the appearance of crystalline forms in eluvial deposits. At the same time, there is a tendency to refine gold, but not to a significant extent. Characteristic edges of pure gold were observed locally. The reason for this was a small degree of oxidation of gold; relics of sulfides were preserved in it.

The difference in the fineness of native gold in the Mayskoe and Spokoinoe ore fields was revealed. Different mineral types are characterized by different typochemistry of native gold. The development of gold–rare-metal mineralization in the Mayskoe ore field explains the higher fineness of native gold. Low-grade gold is characteristic of ores bearing gold–silver–telluride association.

The association of native gold with minerals of tellurium and bismuth minerals indicates their paragenetic relationship. The great importance of bismuth and tellurium mineralization in the formation of gold mineralization has been considered and experimentally proved by many researchers [39–47]. In particular, the model of Au enrichment via the liquid bismuth collector mechanism and Bi/Te control of gold mineralization processes in the study is shown in [39–42], and the substitution of Au-Ag tellurides with native gold in the process of dissolution–reprecipitation is shown in experiments [43–47].

6. Conclusions

The gold mineralization of the Spokoininsky cluster is represented by a volumetric, nonlinear type, which is unconventional for the region, and is related to elements of plicative structures and reverse fault in the enclosing metamorphic basement rocks. Veinlet-disseminated sulfide–(pyrite)–quartz ores form deposit-like bodies in beresites by gneisses and granite gneisses and are related to the Mesozoic igneous rocks of sub-alkaline formations. Noble metal mineralization is characterized by polysulfide (Fe-Cu-Pb), gold–bismuth (Au-Bi) and gold–silver–telluride (Au-Ag-Te) mineral types. Different mineral types have their own typomorphic minerals and typochemistry (fineness and impurities) of native gold.

The widespread distribution of telluride mineralization and its great importance in the formation of gold mineralization on the Aldan shield is confirmed. The distribution area of bismuth (including tellurium–bismuth) mineralization in the southern part of the Aldan shield, in the zone of influence of the Stanovoy deep fault, has been identified.

The conducted mineralogical and geochemical studies show a large commercial prospect of the Mayskoe ore field, less affected by denudation processes, in contrast to the ores of the Spokoinoe ore field, which serve as a source of rich gold-bearing placers.

Author Contributions: Conceptualization, methodology, data curation, and writing—original draft preparation, L.A.K. and G.S.A.; writing—review and editing, L.A.K.; fieldwork, investigation, and visualization, L.A.K. and V.N.K. All authors have read and agreed to the published version of the manuscript.

Funding: This work was supported by the Diamond and Precious Metal Geology Institute, Siberian Branch of the Russian Academy of Sciences (DPMGI SB RAS).

Acknowledgments: We thank N. Khristoforova and S. Karpova for assistance with the SEM analysis (Diamond and Precious Metal Geology Institute, Siberian Branch of the Russian Academy of Sciences). We thank E. Sokolov from JSC "Yakutskgeologiya" for his input in the original work, for the organization of field work, and for his contributions to Figures 1 and 2. We also thank all the editors and anonymous referees.

Conflicts of Interest: The authors declare no conflict of interest.

References

1. Sokolov, E.P.; Babkina, T.G.; Makogonov, I.V.; Linnik, I.A.; Khalgaev, E.U.; Shmatkova, L.E.; Anisimova, G.S.; Kondrat'eva, L.A.; Kardashevskaia, V.N. A new type of gold mineralization in the basement of the Aldan-Stanovoi gold-bearing province. *Ores Met.* **2022**, *2*, 122–140. [CrossRef]
2. Burakov, L.F.; Verevkin, N.I.; Vinichuk, L.Y.; Gomzikov, B.A.; Plotnikov, O.I.; Dik, I.P.; Kalinin, Y.A.; Kolosov, N.P.; Komogortsev, G.P.; Korol'kov, A.P. *Report on Geological Survey Field Works on a Scale of 1:50.000 in the Tyrkandinsky Ore District for 1971–1973*; 1974; Unpublished work.
3. Folomkina, V.A. *Report on the Results of the Detailed Exploration of the a Placer Gold Deposit in the Upper Reaches of the River Bolshaya Tyrkanda for 1991–1996 with Inventory Count*; 1996; Unpublished work.
4. Vetluzhskikh, V.G.; Kim, A.A. Geological and commercial types of gold deposits in Southern Yakutia. *Otechestvennaya Geol.* **1997**, *1*, 16–24.
5. Ugriumov, A.N.; Dvornik, G.P. Altered Formations and Gold Mineralization in the Ore Region of the Mesozoic Tectonic-Magmatic Activation (Aldan Shield). Izvestiya of the Ural State Mining and Geological Academy. *Geol. Geophys. Ser.* **2000**, *10*, 119–128.
6. Vetluzhskikh, V.G.; Kazansky, V.I.; Kochetkov, A.Y.; Yanovskiy, V.M. Central Aldan gold deposits. *Geol. Ore Depos.* **2002**, *44*, 467–499.
7. Kazansky, V.I. The unique Central Aldan gold-uranium ore district (Russia). *Geol. Ore Depos.* **2004**, *46*, 195–211.
8. Kochetkov, A.Y. Mesozoic gold-bearing ore-magmatic systems of the Central Aldan. *Russ. Geol. Geophys.* **2006**, *47*, 850–864.
9. Boitsov, V.E.; Pilipenko, G.N.; Dorozhkina, L.A. Gold and gold-uranium deposits of Central Aldan. In *Proceedings of the Large and Superlarge Deposits of Ore Minerals*; Strategic Types of Ore Raw Materials; IGEM RAS: Moscow, Russia, 2006; Volume 2, pp. 215–240.
10. Maksimov, E.P.; Uyutov, V.I.; Nikitin, V.M. The Central Aldan gold-uranium ore magmatogenic system, Aldan-Stanovoy shield, Russia. *Russ. J. Pac. Geol.* **2010**, *4*, 95–115.
11. Dvornik, G.P. Gold-ore metasomatic formations of the Central Aldan region. *Lithosphere* **2012**, *2*, 90–105.
12. Kravchenko, A.A.; Smelov, A.P.; Berezkin, V.I.; Popov, N.V. *Geology and Genesis of Precambrian Gold-Bearing Metabasites of the Central Part of the Aldano-Stanovoy Shield (on the Example of the P. Pinigin Deposit)*; Ofset: Yakutsk, Russia, 2010; 147p.
13. Anisimova, G.S.; Sokolov, E.P. Altan-Chaidakh—Promising object of the Southern Yakutia. *Otechestvennaya Geol.* **2015**, *5*, 3–10.
14. Anisimova, G.S.; Sokolov, E.P. The Bodorono deposit—New gold ore object of the Southern Yakutia. *Ores Met.* **2014**, *5*, 49–57.
15. Anisimova, G.S.; Sokolov, E.P.; Kardashevskaya, V.N. Gold-rare-metal (Au-Mo-Te-Bi) mineralization of the Upper-Algominsky gold-bearing region (Southern Yakutia). *Otechestvennaya Geol.* **2017**, *5*, 12–22.
16. Kardashevskaya, V.N. *Gold Mineralization of the Algominsky Ore Node of the South Aldan Metallogenic Zone: Mineralogy and Conditions of Ore Formation//ABSTRACT of Thesis for the Degree of PhD in Geological Sciences*; Ofset: Novosibirsk, Russia, 2022; 25p.
17. Mountain, E.D. Two new bismuth minerals from South Africa. *Mineral. Mag.* **1935**, *24*, 59–64. [CrossRef]
18. Kondratieva, L.; Anisimova, G. Minerals of Hg, Tl and As of Khokhoy Deposit (Aldan Shield). In Proceedings of the Geology and Mineral Resources of the North-East of Russia: Materials of the XII All-Russian Scientific and Practical Conference Dedicated to the 65th Anniversary of the Institute of Geology of Diamond and Precious Metals, Siberian Branch of the Russian Academy of Sciences, Yakutsk, Russia, 23–25 March 2022; pp. 184–188.
19. Testa, F.J.; Cooke, D.R.; Zhang, L.; Mas, G.R. Bismoclite (BiOCl) in the San Francisco de los Andes Bi-Cu-Au Deposit, Argentina. First Occurrence of a Bismuth Oxychloride in a Magmatic-Hydrothermal Breccia Pipe and Its Usefulness as an Indicator Phase in Mineral Exploration. *Minerals* **2016**, *6*, 62. [CrossRef]
20. Kondratieva, L.A.; Anisimova, G.S.; Kardashevskaia, V.N. Types of Tellurium Mineralization of Gold Deposits of the Aldan Shield (Southern Yakutia, Russia). *Minerals* **2021**, *11*, 698. [CrossRef]
21. Criddle, A.J.; Chisholm, J.E.; Stanley, C.J. Cervelleite, Ag_4TeS, a new mineral from the Bambolla mine, Mexico, and a description of a photo-chemical reaction involving cervelleite, acanthite and hessite. *Eur. J. Miner.* **1989**, *1*, 371–380. [CrossRef]

22. Helmy, H.M. The Um Samiuki volcanogenic Zn-Cu-Pb-Ag deposit, Eastern Desert, Egypt: A possible new occurrence of cervelleite. *Can. Miner.* **1999**, *37*, 143–158.
23. Novoselov, K.A.; Belogub, E.V.; Zykov, V.V.; Yakovleva, V.A. Silver sulfotellurides from volcanic-hosted massive sulphide deposits in the Southern Urals. *Mineral. Petrol.* **2006**, *87*, 327–349. [CrossRef]
24. Paar, W.; De Brodtkorb, M.K. *Presencia de Cervelleíta y Hessita en la Galena del Yacimiento San Martín, Valcheta, Provincia. de Río Negro. 3° Reunión de Mineralogía y Metalogenia, Instituto de Recursos Minerales;* Universidad Nacional de La Plata, Argentina Publicación 5: Buenos Aires, Argentina, 1996; Volume 5, pp. 173–175.
25. Gu, X.P.; Watanabe, M.; Hoshino, K.; Shibata, Y. New find of silver tellurosulphides from the Funan gold deposit, East Shandong, China. *Eur. J. Miner.* **2003**, *15*, 147–155. [CrossRef]
26. Spry, P.G.; Thieben, S.E. Two new occurrences of benleonardite, a rare silver tellurium sulphosalt, and a possible new occurrence of cervelleite. *Miner. Mag.* **1996**, *60*, 871–876. [CrossRef]
27. Dobrev, S.; Strashimirov, S.; Vassileva, M.; Dragiev, H. *Silver and Silver-Bearing Phases from Chala and Pcheloiad Deposits (Eastern Rhodopes) and Eniovche Deposit (Central Rhodopes);* Annual of the University of Mining and Geology 'St. Ivan Rilski': Sofia, Bulgaria, 2002; Volume 45, pp. 39–44.
28. Cook, N.J.; Ciobanu, C.L. Cervelleite, Ag_4TeS, from three, localities in Romania, substitution of Cu, and the occurrence of the associated phase, Ag_2Cu_2TeS. *N. Jb. Miner. Mh.* **2003**, *7*, 321–336. [CrossRef]
29. Ciobanu, C.L.; Cook, N.J.; Tămaş, C.; Leary, S.; Manske, S.; O'Connor, G.; Minuţ, A. Tellurides-Gold-Base Metal Associations at Roşia Montană: The Role of Hessite as Gold Carrier. In *Au-Ag-Telluride Deposits of the Golden Quadrilateral, South Apuseni Mountains, Romania;* Cook, N., Ciobanu, C.L., Eds.; IAGOD Guidebook Series; Natural History Museum of Landon: London, UK, 2004; Volume 12, pp. 187–202.
30. Voudouris, P.; Spry, P.G.; Sakellaris, G.A.; Mavrogonatos, C. A cervelleite-like mineral and other Ag-Cu-Te-S minerals [Ag_2CuTeS and $(Ag, Cu)_2TeS$] in gold-bearing veins in metamorphic rocks of the Cycladic Blueschist Unit, Kallianou, Evia Island, Greece. *Miner. Petrol.* **2011**, *101*, 169–183. [CrossRef]
31. Nekrasov, I.Y. *Geochemistry, Mineralogy and Genesis of Gold Lodes;* Nauka: Moscow, Russia, 1991; 302p. (In Russian)
32. Warmada, I.W.; Lehmann, B.; Simandjuntak, M. Polymetallic sulfides and sulfosalts of the Pongkorepithermal gold-silver deposit, West Java, Indonesia. *Can. Miner.* **2003**, *41*, 185–200. [CrossRef]
33. Pal'yanova, G.A.; Savva, N.E. Specific genesis of gold and silver sulfides at the Yunoe Deposit (Magadan Region, Russia). *Russ. Geol. Geophys.* **2009**, *50*, 587–602. [CrossRef]
34. Pal'yanova, G.A.; Kokh, K.A.; Seryotkin, Y.V. Formation of gold and silver sulfides in the System Au-Ag-S. *Russ. Geol. Geophys.* **2011**, *52*, 443–449. [CrossRef]
35. Barton, M.D.; Kieft, C.; Burke, E.A.J.; Oen, I.S. Uytenbogaardtite, a new silver-gold sulfide. *Can. Miner.* **1978**, *16*, 651–657.
36. Castor, S.B.; Sjoberg, J.J. Uytenbogaardtite, Ag_3AuS_2, in the Bullford mining district, Nevada. *Can. Miner.* **1993**, *31*, 89–98.
37. Greffié, C.; Bailly, L.; Milési, J.-P. Supergene alteration of primary ore assemblages from low-sulfidation Au-Ag epithermal deposits at Pongkor, Indonesia, and Nazareño, Perú. *Econ. Geol.* **2002**, *97*, 561–571. [CrossRef]
38. Savva, N.E.; Pal'yanova, G.A. Genesis of gold and silver sulfides at Ulakhan deposit (northeastern Russia). *Russ. Geol. Geophys.* **2007**, *48*, 799–810. [CrossRef]
39. Ciobanu, C.L.; Cook, N.J.; Pring, A. Bismuth Tellurides as Gold Scavengers. In *Mineral Deposit Research: Meeting the Global Challenge;* Mao, J.W., Bierlein, F.P., Eds.; Springer: Berlin, Germany, 2005; pp. 1383–1386.
40. Tooth, B.; Brugger, J.; Ciobanu, C.L.; Liu, W. Modelling of gold-scavenging by bismuth melts coexisting with hydrothermal fluids. *Geology* **2008**, *36*, 815–818. [CrossRef]
41. Douglas, N.; Mavrogenes, J.; Hack, A.; England, R. The Liquid Bismuth Collector Model: An Alternative Gold Deposition Mechanism. In Proceedings of the 15th Australian Geological Congress Abstracts, Sydney, NSW, Australia, 3–7 July 2000; Volume 59, p. 135.
42. Feng, H.; Shen, P.; Zhu, R.; Tomkins, A.G.; Brugger, J.; Ma, G.; Li, C.; Wu, Y. Bi/Te control on gold mineralizing processes in the North China Craton: Insights from the Wulong gold deposit. *Miner. Depos.* **2023**, *58*, 263–286. [CrossRef]
43. Zhao, J.; Pring, A. Mineral Transformations in Gold—(Silver) Tellurides in the Presence of Fluids: Nature and Experiment. *Minerals* **2019**, *9*, 167. [CrossRef]
44. Zhai, D.; Liu, J. Gold-telluride-sulfide association in the Sandaowanzi epithermal Au-Ag-Te deposit, NE China: Implications for phase equilibrium and physicochemical conditions. *Miner. Petrol.* **2014**, *108*, 853–871. [CrossRef]
45. Zhao, J.; Brugger, J.; Xia, F.; Ngothai, Y.; Chen, G.; Pring, A. Dissolution-reprecipitation vs. solid-state diffusion: Mechanism of mineral transformations in sylvanite, $(AuAg)_2Te_4$, under hydrothermal conditions. *Am. Miner.* **2013**, *98*, 19–32. [CrossRef]
46. Zhao, J.; Brugger, J.; Grundler, P.V.; Xia, F.; Chen, G.; Pring, A. Mechanism and kinetics of a mineral transformation under hydrothermal conditions: Calaverite to metallic gold. *Am. Miner.* **2009**, *94*, 1541–1555. [CrossRef]
47. Xu, W.; Zhao, J.; Brugger, J.; Chen, G.; Pring, A. Mechanism of mineral transformations in krennerite, Au_3AgTe_8, under hydrothermal conditions. *Am. Miner.* **2013**, *98*, 2086–2095. [CrossRef]

minerals MDPI

Article

The Gold–Palladium Ozernoe Occurrence (Polar Urals, Russia): Mineralogy, Conditions of Formation, Sources of Ore Matter and Fluid

Valery Murzin [1], Galina Palyanova [2,*], Tatiana Mayorova [3] and Tatiana Beliaeva [2]

1 Zavaritsky Institute of Geology and Geochemistry, Ural Branch of Russian Academy of Sciences, Akademika Vonsovskogo Str., 15, 620016 Ekaterinburg, Russia; murzin@igg.uran.ru
2 Sobolev Institute of Geology and Mineralogy, Siberian Branch of Russian Academy of Sciences, Akademika Koptyuga Pr., 3, 630090 Novosibirsk, Russia; zhur0502@igm.nsc.ru
3 Institute of Geology, Federal Research Center, Komi Science Center, Ural Branch, Russian Academy of Sciences, Pervomaiskaya Str., 54, 167982 Syktyvkar, Russia; mayorova@geo.komisc.ru
* Correspondence: palyan@igm.nsc.ru

Abstract: We studied the mineralization and sulfur isotopic composition of sulfides of gold–palladium ores in olivine clinopyroxenites from the Dzelyatyshor massif made up of a continuous layered series of rocks: olivine-free clinopyroxenite–olivine clinopyroxenite–wehrlite. The primary igneous layering of rocks, manifested as different quantitative ratios of clinopyroxene and olivine in them, controls the local trends of variability in the chemistry of mineral-forming medium and the concentrations of ore components, including noble metals, and sulfur in each separate layer during its cooling. The replacement of primary rock-forming minerals by secondary minerals, when the temperature decreases, is a characteristic trend for pyroxenites: (a) olivine → serpentine, secondary magnetite, and (b) clinopyroxene → amphibole, secondary magnetite → chlorite. The deposition of native gold in parageneses with PGM and sulfides at the Ozernoe occurrence took place during the replacement of earlier rock-forming minerals by chlorite. This process completed mineral formation at the deposit and took place at temperatures 150–250 °C and at the high activity of S, Te, Sb, and As of fluid. The variability of mineral formation conditions during chloritization is reflected in the change of native-sulfide forms of Pd by arsenide-antimonide forms and the sulfur isotopic composition of sulfides. The Pd content in native gold increases in the series—Au-Ag solid solution (<1.5 wt.% Pd)—Au-Cu intermetallides (to 6 wt.% Pd)—Cu-Au-Pd solid solutions (16.2–16.9 wt.% Pd). The sulfur isotopic composition of pyrite, chalcopyrite, and bornite varies from −2.1 to −2.9‰. It is assumed that a deep-seated magmatic basic melt was the source of fluid, ore components, and sulfur.

Keywords: clinopyroxenites; palladium gold; platinum group minerals; chlorite geothermometer; sulfur isotopic composition

Citation: Murzin, V.; Palyanova, G.; Mayorova, T.; Beliaeva, T. The Gold–Palladium Ozernoe Occurrence (Polar Urals, Russia): Mineralogy, Conditions of Formation, Sources of Ore Matter and Fluid. *Minerals* **2022**, *12*, 765. https://doi.org/10.3390/min12060765

Academic Editor: Liqiang Yang

Received: 29 April 2022
Accepted: 14 June 2022
Published: 16 June 2022

1. Introduction

The object of this study is the gold–palladium Ozernoe occurrence at the Dzelyatyshor wehrlite–pyroxenite massif in the Polar Urals (Russia). The massif is situated 80 km southwest of the town of Labytnangi in the upper reaches of the Dzelyatyshor stream, left inflow of the Malaya Kharamatalou river. In its geologic, petrographic, and mineralogic-geochemical features, this occurrence is most similar to the so-called "Baronsky" type of gold–palladium mineralization identified among the Uralian deposits [1]. The typical properties of this type of mineralization are: (1) occurrence in olivine pyroxenites of differentiated clinopyroxenite-gabbro massifs with titanomagnetite mineralization; (2) uneven dispersed impregnation of noble metals in rocks at extremely low contents of associated sulfides and (3) absence of visible changes in the appearance or composition of rocks of ore zones, which are outlined only on the basis of sampling results. In the world literature, similar mineralization in the rocks of basic-ultrabasic composition is described as poorly sulfidic with a high content of PGE—«low-S-high platinum-group element deposits» [2–4].

A typical object of "Baronsky" type is the Baron–Klyuevo occurrence at the Volkovsky massif in the Middle Urals (Russia). It occurs in the upper differentiated magmatic sections of this massif in the rocks of dunite-clinopyroxenite-gabbro association. Zones with impregnated mineralization (Pd > Au > Pt, 0.02–0.04 wt.% S) are localized in olivinites, clinopyroxenites, and olivine–anorthite gabbro that smoothly pass into each other in the southern margin of the massif [1,5,6]. Gold–palladium mineralization at the Baronsky occurrence consists of two spatially separated mineral associations: (1) sulfide, with predominant vysotskite PdS, and (2) arsenide–antimonide, with predominant isomertieite $Pd_{11}Sb_2As_2$ [6,7].

The geologic setting of the Ozernoe occurrence at the Dzelyatyshor massif, petrographic and petrochemical characteristics, and mineralogic-geochemical features of rocks and gold–palladium mineralization were reported in previous studies [8–11]. In this study, we conducted a detailed analysis of the mineral composition and mineral parageneses of gold–palladium ores from the Ozernoe occurrence and obtained the first data on the deposition temperature of noble metal minerals, using the chlorite geothermometer, studied the sulfur isotopic composition of sulfides, determined the sources of ore matter and fluid, and revealed the features of the distribution of Pd and Cu in native gold.

2. Brief Description of the Study Area

In the geological aspect, the Dzelyatyshor massif is surrounded by the Voykar–Synya dunite–harzburgite ophiolite massif, which occurs in the polar part of the Main Uralian fault zone (Figure 1). The rocks of the northeastern framing of this massif belong to the Karshor peripheral striped dunite–wehrlite–clinopyroxenite–gabbro complex, whose age is determined as Late Ordovician [12]. In the southeast, the Karshor complex borders granitoids of the Sobsky granodiorite–tonalite complex (D_{1-2}).

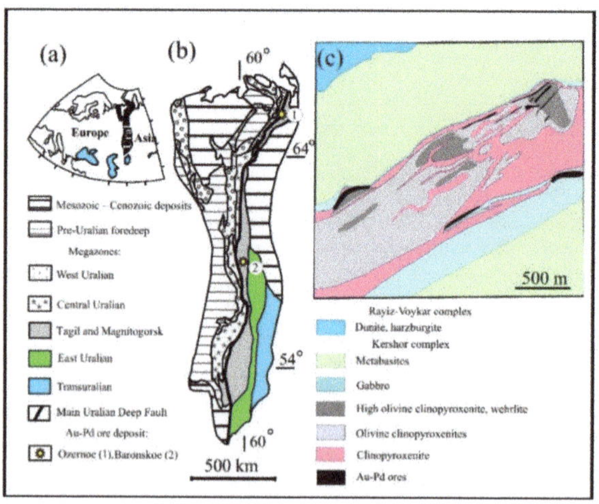

Figure 1. (**a**) Geographic location of the Urals on the Europe–Asia border; (**b**) position of Baronsky type objects in the tectonic structures of the Urals (modified from [13]) (on the right); (**c**) geologic structure and setting of gold–palladium mineralization at the Dzelyatyshor massif.

Clinopyroxenite bodies hosting gold–palladium mineralization of the Ozernoe occurrence were initially attributed to the Karshor complex. In terms of the most modern concepts [10], these clinopyroxenites are identified in the Dzelyatyshor wehrlite-clinopyroxenite massif which is isolated from the Karshor complex. In the petrochemical characteristics, the Dzelyatyshor massif is similar to the wehrlite–clinopyroxenite series of the Platinum-bearing Belt of the Urals and possesses the features of layering displayed in the replacement of olivine by non-olivine rocks from bottom to top. Its contacts with hosting gabbroids of the Karshor complex are tectonic. The Dzelyatyshor massif is composed of a continuous layered series of rocks—non-olivine clinopyroxenite–olivine clinopyroxenite wehrlite [10]. Mineralization belongs to the poor sulfide stratiform type and tends to sheet-like bodies of olivine clinopyroxenites in the upper part of the massif. The bodies consist of 13 ore bodies from 2 to 30 m in thickness [10]. The content of metals in the ores, determined using atomic absorption spectrometry, are: 1.5 wt.% Cu, 2.9 ppm Au, 3.7 ppm Pd, and 1.0 ppm Pt [9,10].

Noble metal mineralization is localized at two levels within the massif. The first lower level occurs in the near-contact part of the massif. The second upper ore level occurs on the horizon of high-olivine pyroxenites with layers of wehrlite. The ore-bearing rocks contain impregnation of magnetite (to 5–7%) and hypogene sulfides, among which chalcopyrite, bornite, pyrrhotite, and pentlandite are predominant. The content of sulfides is typically less than 0.5%, to 1–3% in some parts. Among ores, we identified primary late magmatic minerals, forming segregations of interstitial, and secondary minerals in the form of veinlets and segregations of various shapes, associated with secondary silicates, mainly with serpentine [10].

Noble metal minerals, which amount to 30, are commonly associated with sulfides. The contents of Au and Ag directly correlate with copper content. Geochemical and spatial distribution of the early platinum–metal and later gold–copper mineralization is observed [10].

3. Samples and Research Methods

We studied the samples of olivine-bearing clinopyroxenites with elevated contents of Au and Pd (1–2 ppm), collected from near-surface mine workings (ditches 101, 108, and 126), driven while prospecting for gold by the JSC Yamal Mining Company in 2008. The rocks consist of coarse-grained varieties with olivine content from 1 to 10–15% and with grain sizes of main minerals from 1 to 15 mm. Ore minerals in clinopyroxenites are represented by dispersed impregnations of fine (less than 0.1 mm) and enlarged (to 2–3 mm) magnetite, and sulfides with grain sizes less than 0.5 mm. The content of magnetite in all samples is commonly no more than 2–3%, and that of sulfides is less than 0.5%. In rare cases, the amount of sulfides was 10%, and their grain sizes were more than 2 mm.

The studied samples of clinopyroxenites were separated into three groups according to the content of olivine and sulfides as well as to predominant minerals in sulfide parageneses (Table 1, Figure 2). Sulfides in clinopyroxenites I–III were attributed by us to pyrite–pyrrhotite–chalcopyrite, bornite–chalcopyrite and pyrrhotite–chalcopyrite–pyrite parageneses, respectively. Clinopyroxenites I and III are similar in the specific composition of sulfides, but differ in the content and quantitative ratios of the main sulfide minerals. Clinopyroxenites II and III are characterized by an association of sulfides with magnetite, the crystals of which are enclosed in sulfides or form rims on sulfide segregations (Figure 2c,d).

Table 1. Features of studied clinopyroxenite samples from Ozernoe occurrence.

Samples of Groups	No Sample	Trench Exploration Nor	Olivine Content, Vol.%	Sulfide Content, Vol.%	Sulfide Parageneses
clinopyroxenite I	1545, 1546	K-108, 126	<5	<0.1	Py >> Pyh > Ccp
clinopyroxenite II	101012, 101013	K-101	5–10	0.1–0.5	Bn-Ccp
clinopyroxenite III	211014, 211015	P2110, K-104	>10–15	<10	Pyh > Ccp > Py

Note. Minerals: Py—pyrite, Pyh—pyrrhotite, Ccp—chalcopyrite, Bn—bornite. Abbreviations of minerals are given following the recommendation of IMA [14].

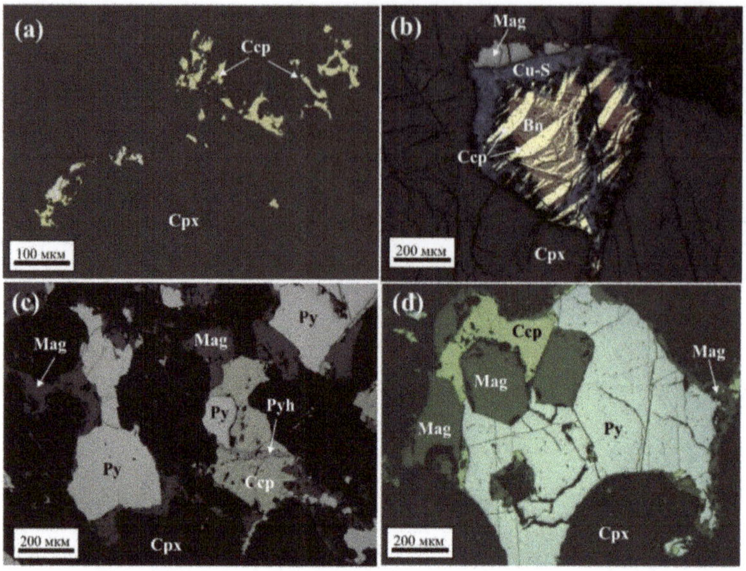

Figure 2. Relationships of main ore minerals in the samples of clinopyroxenites I (**a**), II (**b**), and III (**c**,**d**). Bn—bornite, Ccp—chalcopyrite, Cpx—clinopyroxene, Cu-S—secondary copper sulfide, Mag—magnetite, Py—pyrite, Pyh—pyrrhotite.

Microprobe analyses of minerals of clinopyroxenites I and II were carried out with a CAMECA SX-100 equipped with five WDS spectrometers and a Bruker energy dispersive spectrometer system at the Common Use Center "Geoanalyst" of the Institute of Geology and Geochemistry, Ural Branch of the Russian Academy of Sciences (Ekaterinburg, Russia). Quantitative WDS analyses were performed at 25 kV accelerating voltage and 20 nA sample current, with a beam diameter of about 1 μm.

Analysis of the polished sections of clinopyroxenites III was performed on the scanning electron microscope Tescan Vega 3LMN (Tescan, Czech Republic) with an energy dispersive spectrometer X-Max 50 (Oxford Instruments, Oxford, UK) in the Institute of Geology KomiSC UB of the RAS. Analytical conditions: accelerating voltage was 20 kV, beam current was 15 nA, with the beam diameter up to 1 μm.

Isotopic mass-spectrometric analysis of sulfide sulfur of clinopyroxene II was conducted in the Laboratory of Stable Isotopes in the FEGI FEB RAS. The samples were prepared by laser method using the NWR Femto femtosecond laser ablation [15,16]. The ratio of sulfur isotopes was measured on the MAT-253 mass-spectrometer (Thermo Fisher Scientific, Dreieich, Germany) relative to the laboratory working standard calibrated according to international standards IAEA-S-1, IAEA-S-2 и IAEA-S-3. Results of measurements of $\delta^{34}S$ are reported relative to the international standard VCDT and are given in ppm (‰). The accuracy of analysis was ±0.2‰ (2σ).

The classification of amphiboles is based on the scheme from [17] and that of pyroxenes, on the classical scheme from [18]. The temperature regime of deposition of noble metal minerals was performed using data for chlorite geothermometer, based on the amount of tetrahedral aluminum (Al^{IV}) and molar fraction of iron X_{Fe} by the formula T, °C = $17.5 + 106.2 \times (Al^{IV} - 0.88 \times [X_{Fe} - 0.34])$ [19].

4. Results

4.1. Mineralogy, Mineral Association, and Sequence of Mineral Formation at the Ozernoe Occurrence

4.1.1. Primary (Early) Rock-Forming Minerals of Clinopyroxenites

1. Clinopyroxene

- Primary clinopyroxene with grain sizes from 0.1 to 10 mm in the samples of all groups of clinopyroxenites belongs to the diopside-hedenbergite with a varying ferrosilite mineral (Figure 3a). In clinopyroxenites I and II, it corresponds to diopside with an elevated content of FeO (4.47–7.04 wt.%) and Al_2O_3 (0.79–2.74 wt.%) (Table 2). It contains minor Ti, Mn, Cr, and Na in amounts less than 1 wt.%. Clinopyroxene in clinopyroxenite III is also represented by diopside but with lower contents of FeO (0.66–2.17 wt.%), Al_2O_3 (less than 1.7 wt.%), and other trace elements. This clinopyroxene also contains the thinnest plates of iron oxide, supposedly magnetite, which develop along cleavage cracks (Figure 4a). These plates seem to be the result of the decomposition of the solid solution or oxidation of Fe^{2+}.

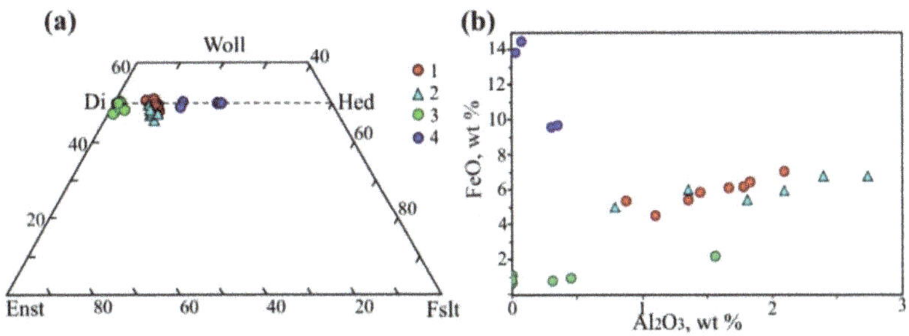

Figure 3. Diagrams of composition of clinopyroxenes (**a**) and contents of FeO-Al_2O_3 (**b**) in primary and secondary clinopyroxenites. Primary clinopyroxene from clinopyroxenites I (1), II (2), and III (3). 4—secondary clinopyroxene from clinopyroxenite I.

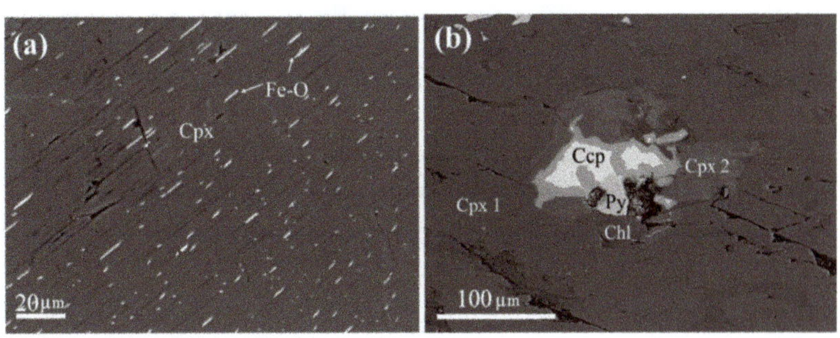

Figure 4. Clinopyroxene in samples of clinopyroxenites III (**a**) and I (**b**). (**a**) Thin plates of Fe-O phase in primary clinopyroxene (Cpx); (**b**) secondary clinopyroxene (Cpx 2) in intergrowths with chlorite (Chl), pyrite (Py), and chalcopyrite (Ccp) in primary clinopyroxene (Cpx 1).

- In spite of considerable variations in the composition of primary clinopyroxene, clinopyroxenites I–III have a direct correlation with the contents of FeO and Al_2O_3 (Figure 3b). An inverse correlation dependence was revealed only for clinopyroxene grains with the highest iron content (Figure 3b, sign 4) (Table 2, No. 9–12). We attributed this clinopyroxene to a later generation. Along with sulfides and chlorite, it fills interstices in the aggregates of early primary pyroxene grains (Figure 4b).

Table 2. Chemical composition of clinopyroxenes in clinopyroxenites I–III (in wt.%).

No	Sample #	SiO_2	TiO_2	Al_2O_3	Cr_2O_3	FeO *	MnO	MgO	CaO	Na_2O	Total
1	1545-2-3	52.68	0.14	1.36	b.d.	5.44	0.25	14.78	25.17	0.14	99.96
2	1545-2-12	52.13	0.16	1.78	b.d.	6.2	0.2	14.44	24.73	0.06	99.70
3	1545-2-14	51.02	0.14	1.67	0.12	6.11	0.21	14.43	23.87	0.09	97.66
4	1545-2-18	52.83	*0.04*	1.1	*0.02*	4.47	0.16	15.36	25.6	0.07	99.65
5	1545-1-35	52.31	0.15	2.1	0.06	7.04	0.11	14.37	23.73	0.06	99.93
6	1545-1-45	51.85	0.1	1.83	b.d.	6.45	0.16	14.11	23.88	0.16	98.54
7	1546-3-8	53.4	0.1	0.87	0.13	5.34	0.16	14.29	25.16	0.11	99.56
8	1546-3-13	53.45	*0.11*	1.45	b.d.	5.84	0.26	14.44	24.46	*0.05*	100.06
9	1546-3-5	52.1	0.01	b.d.	0.01	13.77	1.25	9.21	24.31	*0.03*	100.69
10	1546-3-7	52.08	0.04	0.08	0.04	14.51	0.62	8.78	24.16	*0.04*	100.35
11	1546-3-6	52.91	0.01	0.31	0.09	9.56	0.26	11.92	24.58	0.17	99.81
12	1545-1-41	52.59	*0.09*	0.36	0.03	9.67	0.32	11.98	23.36	1.16	99.56
13	101013-24	51.74	*0.18*	2.74	*0.12*	6.77	0.22	14.8	21.97	0.34	98.88
14	101013-27	53.45	*0.09*	1.36	*0.04*	6.02	0.15	15.41	23.4	0.08	100.00
15	101013-33	54.27	*0.02*	0.79	b.d.	4.98	*0.12*	15.2	24.79	*0.04*	100.21
16	101012-39	53.34	*0.16*	2.1	*0.07*	5.98	*0.11*	15.07	23.52	0.1	100.45
17	101012-48	51.6	0.27	2.4	*0.1*	6.77	0.16	14.2	22.67	0.1	98.27
18	101012-57	52.4	*0.09*	1.81	*0.01*	5.42	0.25	14.72	23.38	0.09	98.17
19	211015-1_3	55.5	b.d.	b.d.	b.d.	1.04	b.d.	18.68	25.57	b.d.	100.79
20	211015-2_6	55.21	b.d.	b.d.	b.d.	0.66	b.d.	18.35	25.86	b.d.	100.08
21	211015-3_3	54.85	b.d.	0.32	b.d.	0.74	b.d.	19.19	24.48	b.d.	99.58
22	211015-4_1	54.31	b.d.	0.46	b.d.	0.93	b.d.	18.61	25.62	b.d.	99.93
23	211015-6_3	54.05	b.d.	1.56	b.d.	2.17	b.d.	17.61	24.47	b.d.	99.86
24	211014-2_2	55.17	b.d.	b.d.	b.d.	1.11	b.d.	18	26.15	b.d.	100.43
25	211014-3_4	55.3	b.d.	b.d.	b.d.	0.85	b.d.	18.23	26.02	b.d.	100.4

Note. Primary clinopyroxene from samples of pyroxenites I (No 1–8), II (No 13–18), and III (No 19–15). No 9–12—secondary clinopyroxene. Here and below in the tables: FeO *—calculated content from measured Fe; definitions with the values of element concentrations below 2θ (standard error of analysis) are highlighted in italics; «b.d.»—below the detection limit.

2. Olivine

Olivine in the rocks under study is characterized by ferruginous fayalite component x(Fa) within 0.13–0.30. The most ferruginous olivine with x(Fa) = 0.25–0.30 is present in the samples of clinopyroxenite I and II, whereas olivine in clinopyroxenite III contains less iron x(Fa) = 0.13–0.15 (Table 3). A constant trace element in olivine is manganese (0.37–0.53 wt.%).

Table 3. Chemical composition of olivine from clinopyroxenites I–III (in wt.%).

	Clinopyroxenite I		II		III		
No	1	2	3	4	5	6	7
Sample #	1545-2-7	1545-2-30	101012-47	101013-26	211015/6	211015/8	211014/4
SiO_2	38.34	37.97	38.06	37.79	40.07	40.25	39.96
TiO_2	b.d.	b.d.	b.d.	*0.03*	b.d.	b.d.	b.d.
Al_2O_3	0.01	b.d.	b.d.	*0.04*	b.d.	b.d.	b.d.
Cr_2O_3	b.d.	b.d.	b.d.	*0.07*	b.d.	b.d.	b.d.
FeO	22.41	23.36	24.82	27.01	13.75	12.34	13.28
MnO	0.43	0.43	0.4	0.37	0.49	0.4	0.53
MgO	38.36	37.8	36.06	34.82	45.43	46.95	46.48
CaO	*0.01*	0.02	b.d.	0.01	0.01	b.d.	b.d.
Na_2O	*0.01*	b.d.	b.d.	b.d.	b.d.	b.d.	b.d.
K_2O	b.d.	b.d.	*0.01*	b.d.	b.d.	b.d.	b.d.
Σ	99.57	99.58	99.35	100.14	99.75	99.94	100.25
x(Fo)	0.75	0.74	0.72	0.70	0.85	0.87	0.87
x(Fa)	0.25	0.26	0.28	0.30	0.15	0.13	0.13

- Magnetite

Xenomorphic segregations of primary magnetite to 2–3 in size fill interstices in the aggregates of clinopyroxene grains. Interstitial magnetite contains chains of small spinel inclusions, oriented in several directions, and rare plates of ilmenite and in some cases is rimmed by secondary silicates—amphibole (magnesio-hornblende) or chlorite. Interstitial magnetite is represented by the Cr-Ti-V-bearing variety. The content of Cr_2O_3 in it attains 2.9 wt.%, TiO_2—2.5 wt.%, V_2O_3—1.1 wt.% (Table 4, # 1–6). Tabular ilmenite in primary magnetite contains Mg (1.66 wt.% MgO), Mn (3.9 wt.% MnO), and V, Cr, Al in amounts less than 0.2 wt.%.

Table 4. Chemical composition of primary interstitial magnetite in clinopyroxenites I and II (in wt.%).

No	# Sample	No. Grain	FeO *	Cr_2O_3	TiO_2	Al_2O_3	MgO	MnO	V_2O_3	Total
	1	1545-2-19	89.8	2.53	0.61	0.52	0.14	0.1	0.94	94.64
I	2	1545-2-26	89.16	2.21	0.77	0.11	0.44	0.16	0.85	93.70
	3	1545-1-36	87.25	1.57	2.46	1.35	0.25	0.15	0.94	93.97
	4	101013-22	87.21	2.89	2.16	1.64	0.16	0.21	0.74	95.01
II	5	101013-23	87.17	2.78	2.44	0.99	*0.06*	0.14	1.05	94.63
	6	101012-49	87.79	2.36	1.74	0.56	0.47	0.21	0.91	94.04

4.1.2. Secondary (Late) Minerals in Clinopyroxenites

Secondary clinopyroxenite minerals are chlorite, serpentine, magnetite, and sulfides, as well as accessory clinopyroxene, titanite, and noble metal minerals. These minerals are present in small quantities (no more than 5–10% in total) and are evenly distributed in the rock mass. Secondary minerals occur in the most permeable zones of granular aggregates of rock-forming minerals. They develop along the cleavage cracks of clinopyroxene, fill microcracks in olivine or clinopyroxene, and are localized in the interstices in olivine–clinopyroxene aggregates.

- Secondary clinopyroxene

Secondary clinopyroxene was observed in clinopyroxene I in single cases. It occurs in association with sulfides (pyrite, chalcopyrite) and chlorite, which fill interstices in the aggregates of primary clinopyroxene grains (Figure 4b). Secondary clinopyroxene in comparison with primary clinopyroxene has a higher iron content (9.6–14.5 wt.% FeO) and a low alumina concentration (less than 0.4 wt.% Al_2O_3) (Figure 3b; Table 2, No 9–12).

- Amphibole

In the samples of clinopyroxentites I and II, amphiboles belong to the group of Ca-amphiboles. They consist of actinolite (No 1–6), magnesio-ferri-hornblende (No 7), magnesio-hornblende (No 8–14), and tremolite (No 15, 16) (Table 5). All amphiboles, except tremolite, contain minor amounts of sodium (from 0.4 to 1.5 wt.% Na_2O) and titanium (to 0.5 wt.% TiO_2) (Table 5). Tremolite contains the least amount of Al_2O_3 и FeO among amphiboles.

Actinolite is developed mainly in the samples of clinopyroxenite I in the form of angular interstitial segregations in the aggregate of clinopyroxene grains (Figure 5a). It is accompanied by small (less than 10 μm) grains of impurity-free secondary magnetite. We have also detected a grain of magnesio-ferri-hornblende virtually completely replaced by chlorite (Figure 6c). Magnesio-hornblende is developed predominantly in clinopyroxenite II in which it, together with chlorite, forms rims on the segregations of primary magnetite and a lattice-like chalcopyrite–bornite aggregate in the interstices of clinopyroxene (Figure 5b). Tremolite is present in all clinopyroxenite samples. It is developed along with the contacts of large clinopyroxene and olivine grains. In the contact zone of these minerals, clinopyroxene is replaced by tremolite, and olivine, by serpentine. Along with tremolite and serpentine, secondary magnetite is formed as symplectic intergrowths with tremolite and larger porous segregations in serpentine (Figure 5c,d). Tremolite and serpentine in clinopyroxene II replace marginal parts of lattice-like bornite–chalcopyrite segregations.

Table 5. Chemical composition of amphiboles in clinopyroxenites I (No 1–5, 7, 14, 16) and II (No 6, 8–13, 15) (in wt.%).

No	Sample #	SiO_2	Al_2O_3	Na_2O	MgO	F	Cl	K_2O	CaO	TiO_2	FeO *	MnO	Cr_2O_3	Total
1	1546-3-9	50.54	6.1	0.96	16.38	0.07	0.03	0.03	13.16	0.17	9.59	0.16	0.13	97.31
2	1546-3-16	51.26	5.01	0.71	16.33	0.09	0.04	0.08	12.43	0.22	10.32	0.16	0.2	96.85
3	1546-3-19	50.85	5.84	1.05	17.52	0.03	0.33	0.01	12.53	0.36	8.01	0.2	0.16	96.91
4	1546-3-20	52.28	5.09	0.63	17.34	0.05	0.03	0.03	12.88	0.28	8.54	0.12	0.2	97.48
5	1546-3-21	53.83	4.23	0.39	18.77	н.о	0.02	0.01	13.00	0.18	6.94	0.14	0.06	97.57
6	101013-31	53.88	4.51	0.59	19.34	н.о	0.03	0.11	13.01	0.13	6.04	0.15	н.о	97.8
7	1546-3-15	47.67	8.12	0.9	15.28	0.15	0.08	0.14	11.48	0.45	12.68	0.24	0.24	97.45
8	101013-32	47.19	10.35	1.4	16.48	0.06	0.03	0.26	12.63	0.48	8.45	0.11	0.25	97.68
9	101012-45	49.05	9.59	1.37	16.82	0.03	0.01	0.01	12.21	0.52	7.11	0.08	0.14	96.95
10	101012-46	48.58	8.68	1.33	16.87	0.08	0.02	н.о	12.24	0.49	8.27	0.1	0.1	96.77
11	101013-25	51.11	7.35	1.08	17.89	0.07	0.02	0.05	12.28	0.28	7.47	0.1	0.05	97.76
12	101012-55	48.67	9.32	1.28	17.11	0.07	0.02	0.04	12.33	0.37	7.46	0.04	0.12	96.83
13	101012-56	48.5	9.82	1.53	17	0.04	0.03	0.03	12.35	0.38	7.63	0.1	0.05	97.45
14	1545-2-4	48.78	7.55	1.01	18.07	н.о	0.01	0.07	12.5	0.35	8.01	0.12	0.06	96.52
15	101012-42	57.72	0.02	b.d.	22.7	0.02	н.о	0.01	13.22	0.07	2.65	0.03	н.о	96.44
16	1545-2-5	56.1	1.14	0.19	22.22	н.о	0.01	0.02	12.96	0.08	3.82	0.17	н.о	96.7

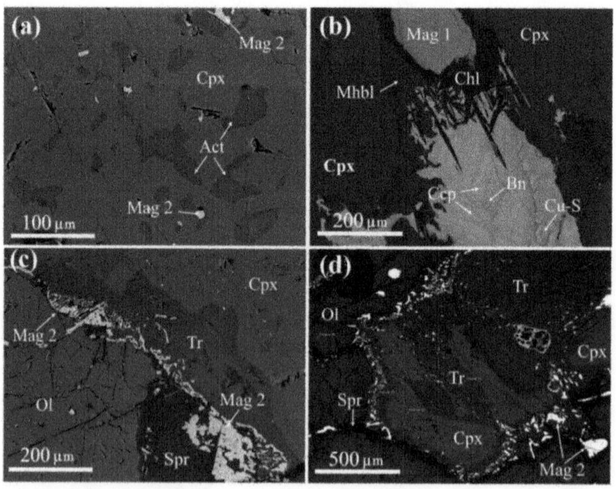

Figure 5. Amphibole in the samples of clinopyroxenite I (**a**,**c**), II (**b**), and III (**d**). (**a**) Interstitial segregations of actinolite (Act) in clinopyroxene (Cpx) and associated grains of secondary magnetite (Mag 2); (**b**) magnesio–hornblende (Mhbl) and chlorite (Chl) in the rim of primary magnetite segregations (Mag 1); aggregate of bornite (Bn), chalcopyrite (Ccp), and secondary copper sulfide (Cu-S) in clinopyroxene matrix (Cpx); (**c**,**d**) development of tremolite (Tr), serpentine (Spr) and secondary magnetite (Mag 2) in the contact zone of clinopyroxene (Cpx) and olivine (Ol).

- Chlorite

In clinopyroxenites I and II, chlorite is developed in the interstices of clinopyroxene (Figure 6a) or along cutting microcracks (Figure 6b). Chlorite replaces primary clinopyroxene, magnesio-ferri-hornblende (Figure 6c), and magnetite (Figure 6d). Tabular chlorite in clinopyroxenite II together with tremolite and serpentine are developed in the marginal parts of segregations of bornite–chalcopyrite lattice-like aggregate (Figure 5b). A typical feature of chlorite in clinopyroxene I is the presence of titanite (Figure 6c,d). In clinopyroxenite III, chlorite is distributed mainly along the cleavage cracks of clinopyroxene.

In the chemical composition, chlorite corresponds to pennine and picnochlorite and is characterized by wide variations in iron content (Figure 7). Chlorite in clinopyroxenite I has the highest iron concentration X_{Fe} = 0.1–0.43 compared to chlorite in clinopyroxenite II and III (X_{Fe} < 0.1) (Table 6). Chlorite with the lowest iron concentration (X_{Fe} < 0.05) was found in clinopyroxenite III. Only in these samples chlorite (pennine) contains minor amounts of potassium (0.5–2 wt.% K_2O).

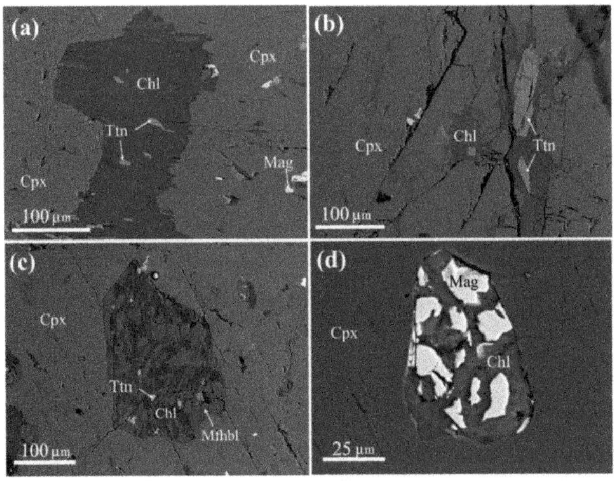

Figure 6. Chlorite in the samples of clinopyroxene I. (**a**) Interstitial chlorite (Chl) in clinopyroxene (Cpx). Chlorite also contains titanite (Ttn) and magnetite (Mag) in clinopyroxene; (**b**) veinlets of chlorite (Chl) with inclusions of titanite (Ttn) in clinopyroxene; (**c**) segregation of magnesio-ferri-hornblende) (Mfhbl) in clinopyroxene. Amphibole is replaced by chlorite with titanite inclusions (Ttn); (**d**) magnetite crystal (Mag) replaced by chlorite (Chl) in clinopyroxene.

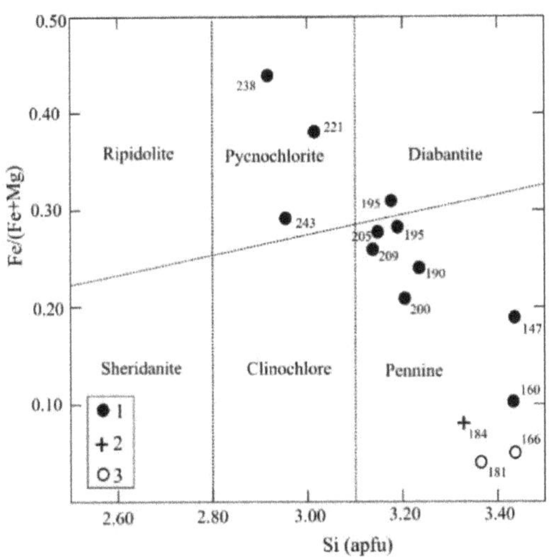

Figure 7. Chlorite composition on the classification diagram from [20]. Analyses of chlorite from clinopyroxenites I (1), II (2), and III (3). Numbers show the formation temperatures calculated with a chlorite geothermometer [19] (Table 6).

Table 6. Chemical composition of chlorite from clinopyroxenites I–III (in wt.%) and calculated temperatures of its formation.

No Grain	1545-2-2	1545-2-13	1545-2-17	1545-2-21	1545-2-22	1545-2-25	1545-1-34	1545-1-39	1545-1-44	1546-3-10	1546-3-17	101012-54	211015/4_2	211014/2_3
						I						II	III	
No	1	2	3	4	5	6	7	8	9	10	11	12	13	14
SiO_2	30.67	31.63	31.59	30.71	35.12	28.65	34.51	27.18	30.96	28.81	30.61	33.82	34.87	35.94
TiO_2	b.d.	b.d.	b.d.	b.d.	b.d.	b.d.	b.d.	b.d.	b.d.	b.d.	b.d.	b.d.	b.d.	0.23
Al_2O_3	13.88	13.61	14.03	15.09	11.75	15.39	10.76	17.53	13.9	17.93	14.21	14.32	12.85	12.18
Cr_2O_3	0.51	0.21	0.68	*0.11*	*0.03*	0.16	0.45	0.28	*0.1*	*0.05*	0.64	0.19	0.14	b.d.
FeO	16.53	13.63	12.13	14.84	6.02	21.29	11.76	23.09	16.35	16.36	17.69	5.05	2.69	3.11
MnO	0.1	0.12	0.16	0.26	0.08	0.17	0.41	0.21	0.21	0.25	0.2	*0.03*	b.d.	b.d.
MgO	24.33	25.69	26.69	24.63	32.25	20	28.24	17.07	24.02	22.38	22.4	31.15	34.04	32.97
CaO	*0.03*	0.11	0.07	0.05	0.06	0.1	0.14	0.08	*0.02*	0.12	0.1	0.06	b.d.	b.d.
Na_2O	b.d.	b.d.	b.d.	b.d.	b.d.	b.d.	b.d.	b.d.	b.d.	b.d.	b.d.	b.d.	b.d.	b.d.
K_2O	b.d.	b.d.	b.d.	b.d.	b.d.	b.d.	b.d.	b.d.	b.d.	b.d.	b.d.	b.d.	0.54	2.10
Σ	86.05	85.00	85.35	85.69	85.31	85.76	86.27	85.44	85.56	85.90	85.85	84.62	85.13	86.53
Crystallochemical coefficients in the chlorite formula (calculation for 20 cations)														
Si	6.29	6.47	6.40	6.27	6.87	6.05	6.91	5.84	6.38	5.92	6.36	6.66	6.72	6.86
Ti	0.00	0.00	0.00	0.00	0.00	0.00	0.00	0.00	0.00	0.00	0.00	0.00	0.00	0.03
Al	3.35	3.28	3.35	3.63	2.71	3.83	2.54	4.44	3.38	4.34	3.48	3.32	2.92	2.74
Cr	0.08	0.03	0.11	0.02	0.00	0.03	0.07	0.05	0.02	0.01	0.11	0.03	0.02	0.00
Fe‴	0.00	0.00	0.07	0.00	0.00	0.13	0.00	0.00	0.00	0.00	0.01	0.00	0.00	0.00
Fe″	2.83	2.33	1.98	2.53	0.98	3.63	1.97	4.14	2.81	2.81	3.05	0.83	0.43	0.50
Mn	0.02	0.02	0.03	0.04	0.01	0.03	0.07	0.04	0.04	0.04	0.04	0.01	0.00	0.00
Mg	7.43	7.83	8.05	7.49	9.40	6.29	8.42	5.47	7.37	6.85	6.93	9.14	9.77	9.37
Ca	0.01	0.02	0.02	0.01	0.01	0.02	0.03	0.02	0.00	0.03	0.02	0.01	0.00	0.00
Na	0.00	0.00	0.00	0.00	0.00	0.00	0.00	0.00	0.00	0.00	0.00	0.00	0.00	0.00
K	0.00	0.00	0.00	0.00	0.00	0.00	0.00	0.00	0.00	0.00	0.00	0.00	0.13	0.51
x(Mg)	0.72	0.77	0.80	0.74	0.90	0.63	0.81	0.57	0.72	0.71	0.69	0.92	0.96	0.95
Al(IV)	1.71	1.53	1.60	1.73	1.13	1.95	1.09	2.16	1.62	2.08	1.64	1.34	1.28	1.14
Al(VI)	1.64	1.76	1.75	1.90	1.58	1.88	1.44	2.28	1.76	2.26	1.84	1.98	1.64	1.59
x(Fe)	0.28	0.23	0.21	0.26	0.10	0.38	0.19	0.43	0.28	0.29	0.31	0.08	0.04	0.05
T, °C	205	190	200	209	160	221	147	238	195	243	195	184	181	166

Note. In calculating crystallochemical formula, Al(IV) supplements Si position to 8 cations, and Al(VI) accounts for the rest part of the total calculated amount of Al.

- Serpentine

In all samples of clinopyroxenites, serpentine actively replaces olivine through the net of microcracks (Figure 8a,b). Among two types of serpentinites are distinguished according to morphology and chemical composition: A (coarse-lamellar, supposedly antigorite) and B (fibrous, apparently lizardite). Serpentine A is developed in clinopyroxenites I and II. Serpentine B is present in clinopyroxenite III. Serpentine A replaces the most ferruginous olivine (22.4–27.9 wt.% FeO) (Table 3, No 1–4) and, in turn, contains the highest iron content (4.6–12.8 wt.% FeO) (Table 7, No 1–5). Serpentine B is less ferruginous (1.1–2.2 wt.% FeO) (Table 7, No 6–10) than serpentine A and replaces olivine with a lower iron content (12.3–13.8 wt.% FeO) (Table 3, No 5–7) X(Fa) = 0.13–0.15 (Table 3, No 6–10). Coarse grains of porous secondary magnetite, tremolite and chalcopyrite are deposited in olivine together with serpentine A (Figure 8b). Serpentine A also corrodes primary magnetite grains and forms symplectic intergrowths with secondary magnetite (Figure 8a). In clinopyroxenite II, serpentine A is developed at the contacts of olivine and clinopyroxene grains. Here, together with tremolite, it is present in the marginal parts of sulfide segregations of bornite-chalcopyrite paragenesis (Figure 8c). In clinopyroxenite III, lamellar (fibrous) aggregates of serpentine B replace pyrite and chalcopyrite grains (Figure 8d).

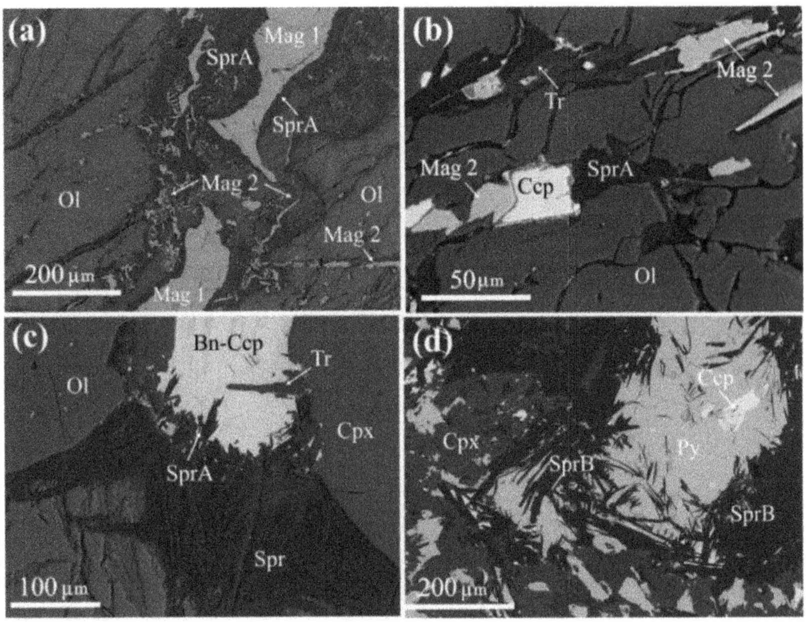

Figure 8. Serpentinite in samples of clinopyroxenites I (**a**,**b**), II (**c**), and III (**d**). (**a**) Relationships of serpentine A (Spr A) with primary (Mag 1) and secondary magnetite (Mag 2) in large olivine grain (Ol); (**b**) microcracks in olivine (Ol) filled with serpentine A (Spr A), tremolite (Tr), chalcopyrite (Ccp), and secondary magnetite (Mag 2); (**c**) localization of serpentine A (Spr A) and sulfides of bornite–chalcopyrite paragenesis (Bn-Ccp) in interstices of olivine (Ol) and clinopyroxene (Cpx) grains. Serpentine A (Spr A) together with tremolite (Tr) corrodes the marginal part of sulfide segregation; (**d**) fibrous aggregates of serpentine B (Srp B) corroding the marginal parts of large segregation of pyrite (Py).

Table 7. Chemical composition of serpentine in clinopyroxenites I–III (in wt.%).

	No	# Grain	MgO	FeO	Al_2O_3	SiO_2	Cr_2O_3	MnO	NiO	Total
I	1	1545-2-55	37.65	4.63	b.d.	44.01	B.d.	B.d.	B.d.	86.29
II	2	101013-29	28.91	12.78	0.54	43.47	*0.01*	0.27	*0.05*	86.17
	3	101013-30	31.43	7.34	*0.01*	50.52	*0.07*	*0.06*	*0.04*	89.84
	4	101012-43	32.86	5.71	*0.03*	49.76	*0.03*	*0.12*	*0.04*	88.83
	5	101012-44	34.38	6.02	*0.02*	45.39	*0*	*0.08*	*0.05*	86.17
III	6	211015-3_1	39.52	2.15	0.49	44.89	B.d.	B.d.	B.d.	87.05
	7	211015-3_2	39.37	1.18	B.d.	45.08	B.d.	B.d.	B.d.	85.63
	8	211015-6_2	40.64	1.12	B.d.	45.33	B.d.	B.d.	B.d.	87.1
	9	211015-8_2	40.3	1.25	B.d.	45.61	B.d.	B.d.	B.d.	87.16
	10	211014-4_2	39.63	1.48	B.d.	44.38	B.d.	B.d.	B.d.	85.5

- Secondary magnetite

Secondary magnetite is present in all studied clinopyroxenites in amounts from 0.5 to 5–10%. In clinopyroxene, it consists of grains less than 0.05 mm in size of isometric or angular shape, which are localized in the cleavage cracks of this mineral, occasionally in the intergrowth with interstitial actinolite (Figure 5a). In olivine, thin veinlets or chains of small secondary magnetite grains fill the central part of cracks filled with serpentine (Figure 8a,b). At the contact of large olivine and pyroxene grains, tiny parts of secondary magnetite "are gathered" in larger porous aggregates in serpentine or form symplectic intergrowths with serpentine or tremolite (Figures 5c,d and 8a). Secondary magnetite in clinopyroxenites II and III overgrows interstitial sulfide segregations or, in clinopyroxene III, its crystals are enclosed in pyrite or chalcopyrite (Figure 2b–d). In contrast to primary magnetite with elevated content of Cr, Ti, and V (Table 4), the contents of trace elements in secondary magnetite are lower and, typically, are below the detection limit or error of analysis (Table 8).

Table 8. Chemical composition of secondary magnetite (in wt.%).

No	# Grain	FeO	Cr_2O_3	TiO_2	Al_2O_3	MgO	MnO	V_2O_3	Total
1	1545-2-9	92.67	b.d.	0.32	*0.02*	0.15	*0.02*	*0.1*	93.28
2	101013-34	92.16	0.44	0.29	*0.07*	0.16	*0.01*	0.28	93.41
3	101012-37	90.14	0.63	0.65	*0.02*	0.1	0.12	0.54	92.20
4	211015_3_9	94.18	b.d.	b.d.	b.d.	b.d.	b.d.	b.d.	94.18
5	211015_5_1	89.89	b.d.	b.d.	b.d.	b.d.	0.37	b.d.	90.26

Note. Samples of clinopyroxenites I (No 1), II (No 3,4) and III (No 5). Analyses: 1,5—in symplectic intergrowths with amphibole; 2—microveinlets in serpentine; 3—inclusion in bornite-chalcopyrite lattice-like aggregate; 4—overgrown on large pyrite segregations.

- Sulfides

In clinopyroxenite I, pyrite, pyrrhotite, and chalcopyrite are in paragenesis with interstitial and veinlet chlorite and are generally typically enclosed in it (Figure 9a–c). Occasionally, sulfide grains are localized in pyroxene but form intergrowths with chlorite (Figure 9d). In clinopyroxenite II, sulfides consisting of decomposed solid solution are localized in the interstices of clinopyroxene grains occasionally together with secondary magnetite (Figure 2b). In the contact zones of olivine and clinopyroxene grains, lattice-like chalcopyrite bornite aggregates in the marginal parts are replaced by tremolite, chlorite and serpentine (Figures 5b and 8c). In clinopyroxenite III, sulfide segregations of interstitial shapes to 2 mm in size are disseminated in the rock mass. Among sulfides, pyrite predominates quantitatively over chalcopyrite, whereas chalcopyrite, over pyrrhotite. Frequently, these sulfides form joint segregations (Figures 2c,d and 9f). Pyrrhotite segregations no more than 0.3 mm in size form intergrowths with pyrite (Figure 9d) or are enclosed in chalcopyrite (Figure 2c). Chalcopyrite was found to contain inclusions of pentlandite and argentopentlandite $Ag(Fe,Ni)_8S_8$ of 10–15 μm in size (Figure 9e, Table 9, No 16, 17).

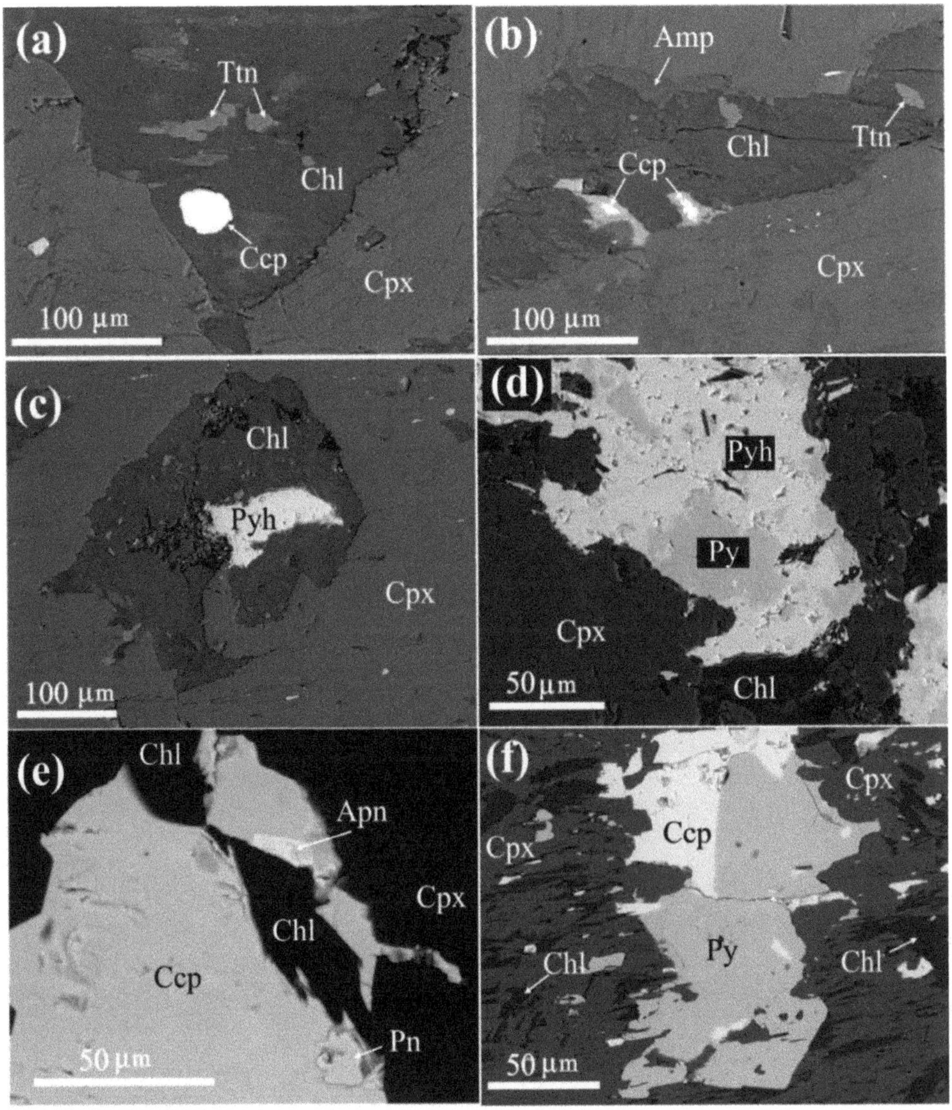

Figure 9. Relationships of sulfide minerals in clinopyroxenites I (**a**–**d**) and III (**e**,**f**). (**a**) Interstitial chlorite (Chl) with inclusions of chalcopyrite (Ccp) and titanite (Ttn) in clinopyroxene (Cpx); (**b**) chlorite veinlets with amphibole rim (Amp) in clinopyroxene. Chlorite contains grains of titanite (Ttn) and partially oxidized chalcopyrite; (**c**) interstitial chlorite with inclusions of pyrrhotite (Pyh) in clinopyroxene; (**d**) intergrowth of pyrite (Py), pyrrhotite, and chlorite in clinopyroxene; (**e**) segregation of chalcopyrite in intergrowth with argentopentlandite (Apn), pentlandite (Pn), and chlorite in clinopyroxene; (**f**) interstitial segregations of pyrite and chalcopyrite in clinopyroxene. Chlorite develops along the cleavage cracks.

Table 9. Chemical composition of sulfides in clinopyroxenites I–III (in wt.%).

No.	# Sample	Cu	Fe	Ni	Co	S	Ag	Total	Mineral
				Pyrite-pyrrhotite-chalcopyrite paragenesis (clinopyroxenite I)					
1	1545-1-33	B.d.	47.61	*0.02*	0.14	53.24	B.d.	101.01	Py
2	1545-1-43	*0.01*	47.09	B.d.	0.09	52.68	B.d.	99.87	Py
3	1546-3-3	0.18	45.43	*0.05*	1.01	51.86	B.d.	98.53	Py
4	1545-2-1	B.d.	62.68	*0.05*	0.13	37.77	B.d.	100.63	Pyh
5	1545-2-16	*0.02*	61.4	B.d.	0.13	38.11	B.d.	99.66	Pyh
6	1546-3-1	34.74	30.34	*0.01*	*0.07*	33.33	B.d.	98.49	Ccp
7	1546-3-2	34.36	30.72	*0.03*	*0.05*	33.31	B.d.	98.47	Ccp
				Bornite–chalcopyrite paragenesis (clinopyroxenite II)					
8	101012-36	35.55	29.73	B.d.	B.d.	33.09	B.d.	98.37	Ccp
9	101012-59	35.71	29.29	*0.02*	*0.06*	33.03	B.d.	98.11	Ccp
10	101012-35	61.64	11.87	B.d.	*0.03*	26.14	B.d.	99.68	Bn
11	101012-40	62.25	11.49	B.d.	*0.04*	24.69	B.d.	98.47	Bn
12	101012-50	61.84	11.45	B.d.	B.d.	25.55	B.d.	98.84	Bn
13	101012-51	70.94	1.04	*0.03*	B.d.	25.66	B.d.	97.67	Spi-Yar
14	101012-52	68.87	0.99	*0.02*	*0.01*	29.72	B.d.	99.61	Spi-Yar
15	101012-60	68.9	1.24	*0.05*	*0.02*	29.84	B.d.	100.05	Spi-Yar
				Pyrrhotite–chalcopyrite–pyrite (clinopyroxenite III)					
16	211014_1_1	B.d.	34.23	19.1	B.d.	32.71	14.72	100.76	Apn
17	211014_1_4	B.d.	31.49	34.39	B.d.	34.07	B.d.	99.95	Pn
18	211014_2_1	B.d.	60.91	B.d.	B.d.	39.34	B.d.	100.25	Pyh
19	211014_4_3	B.d.	61.33	B.d.	B.d.	38.85	B.d.	100.17	Pyh
20	211014_1_5	34.09	31.05	B.d.	B.d.	35.48	B.d.	100.63	Ccp

Note. spionkopite (Spi), yarrowite (Yar).

Chemical composition of sulfides of various parageneses in given in Table 9. The compositions of pyrite FeS_2, chalcopyrite $CuFeS_2$, and bornite Cu_5FeS_4 are similar to stoichiometric composition for these sulfides. Pyrite in clinopyroxene I contains Co (up to 1 wt.%). The composition of pyrrhotite in pyrite–pyrrhotite–chalcopyrite paragenesis (clinopyroxenite I) corresponds to the formula $Fe_{0.96–0.97}S$, whereas pyrrhotite–chalcopyrite–pyrite (clinopyroxenite III), to $Fe_{0.94–0.95}S$. Secondary copper sulfide developed mainly on bornite in clinopyroxenite II (Figure 2b) has an intermediate composition between spionkopite ($Cu_{39}S_{28}$) and yarrowite (Cu_9S_8) (Table 9, No 13–15). Secondary copper sulfide contains Fe (1–1.2 wt.%).

- Noble metal minerals

Gold, silver, palladium, and platinum minerals are represented by particles to 5 μm, rarely to 10–15 μm in size. They occur predominantly in cleavage cracks in clinopyroxene, less frequently in chlorite or in the interstices of clinopyroxene (Figures 10–12). As the sizes of most particles of noble metal minerals are very small, their chemical composition (Table 10, No 3–14) was determined by ruling out the influence of X-ray excitation of matrix mineral and bringing the sum of measured contents of main minerals to 100%.

Samples of clinopyroxenite 1 with pyrite-chalcopyrite paragenesis (sample 1546) were found to contain particles of palladium antimonides of less than 10 μm, which occur in cleavage cracks in pyroxene (Figure 10a) and chlorite veinlets in it (Figure 10b,c). In chemical composition, palladium antimonides correspond to mertieite I ($Pd_{11}(Sb,As)_4$ ($Pd/(Sb + As) = 2.72–2.74$) (Table 10, No 1,2) and stibiopalladinite (Pd_5Sb_2) ($Pd/(Sb + As) = 2.33–2.52$) (Table 10, No 3–5). The largest mertieite I grains contain tiny platinum sulfide inclusions 1–2 μm in size (Table 10, No 6).

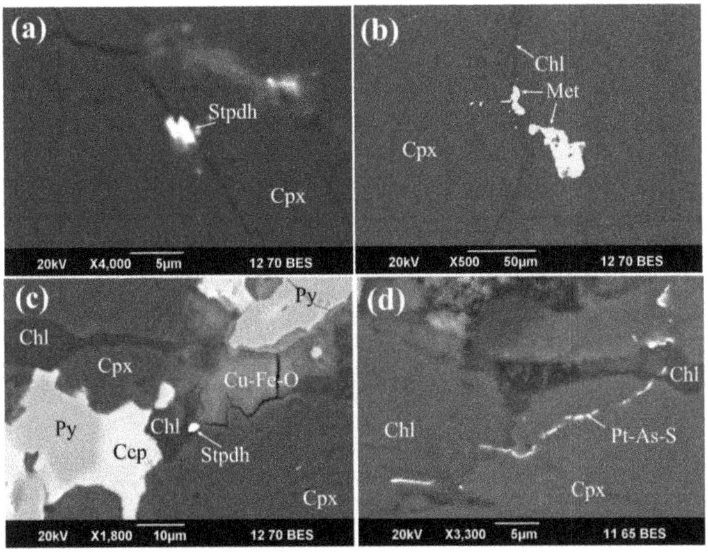

Figure 10. Noble metal minerals in clinopyroxene I. (**a**) Stibiopalladinite (Stpdn) in the cleavage crack in clinopyroxene (Cpx); (**b**) mertieite I (Met) grain in clinopyroxene (Cpx) and chlorite (Chl) veinlet; (**c**) stibiopalladinite (Stpdn) grain enclosed in chlorite. Chlorite is associated with pyrite (Py) and chalcopyrite (Ccp), which are partially replaced by hypergene iron and copper oxides (Fe-Cu-O); (**d**) veinlets of platinum sulfoarsenide (Pt-As-S) cutting clinopyroxene and chlorite.

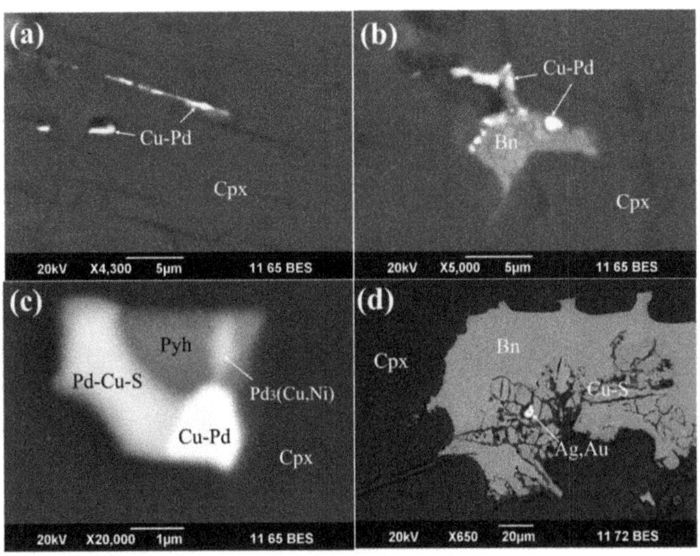

Figure 11. Noble metal minerals in clinopyroxenite II: (**a**) native copper–palladium phases (Cu-Pd) in cleavage cracks in clinopyroxene (Cpx); (**b**) interstitial bornite (Bn) segregation in intergrowth with native copper–palladium phases (Cu-Pd); (**c**) polyphase intergrowth of native Cu-Pd phases (Cu-Pd) (Table 10, No 10), palladium sulfide (Pd-Cu-S) (Table 10, No 12), phases $Pd_3(Cu,Ni)$ (Table 10, No 11) and pyrrhotite (Pyh) in clinopyroxene (Cpx); (**d**) Au,Ag particle (Table 10, No 13) in secondary copper sulfide (Cu-S) developed on bornite (Bn).

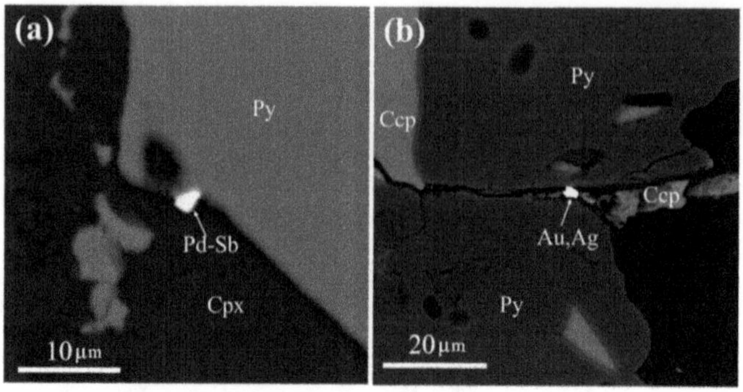

Figure 12. Noble metal minerals in clinopyroxenite III. (**a**) Crystals of palladium antimonide (Pd-Sb) on the surface of pyrite grain (Py); (**b**) Ag-Au particle (Au,Ag) (Table 10, No 14) in chalcopyrite (Ccp) at the contact of two pyrite (Py) grains.

Table 10. Chemical composition of noble metal minerals in clinopyroxenites (in wt.%).

No	# Grain	Pd	Cu	Fe	Ni	Pt	Au	Ag	As	Sb	S	∑	Mineral
1	1546-2-65	71.57	B.d.	*0.06*	B.d.	*0.04*	B.d.	B.d.	2.95	25.07	0.02	99.71	Met-I
2	1546-2-66	71.71	B.d.	*0.1*	B.d.	*0.5*	B.d.	B.d.	3.01	25.24	*0.03*	100.59	Met-I
3	1546-2-107	67.20	0.54	1.16	B.d.	B.d.	B.d.	B.d.	3.86	27.24	B.d.	100	Stpdn
4	1546-2-116	69.67	B.d.	0.72	B.d.	B.d.	B.d.	B.d.	4.3	25.30	B.d.	100	Stpdn
5	1546-2-121	67.7	B.d.	0.85	B.d.	B.d.	B.d.	B.d.	4.22	27.23	B.d.	100	Stpdn
6	1546-2-130	50.73	13.77	21.89	B.d.	B.d.	B.d.	B.d.	1.24	4.01	20.76	100	Pd-Fe-S phase
7	1546-3-16	B.d.	B.d.	B.d.	B.d.	51.65	B.d.	B.d.	45.34	B.d.	3.02	100	Spy
8	1546-3-18	B.d.	B.d.	B.d.	B.d.	51.58	B.d.	B.d.	44.91	B.d.	3.52	100	Spy
9	101013-93	57.38	30.41	2.99	B.d.	3.97	5.25	B.d.	B.d.	B.d.	B.d.	100	Skg
10	101013-99	46.52	31.51	8.24	B.d.	2.13	11.6	B.d.	B.d.	B.d.	B.d.	100	Phase Cu > Pd >> Fe
11	101013-100	83.53	9.98	B.d.	6.49	B.d.	B.d.	B.d.	B.d.	B.d.	B.d.	100	Phase $Pd_3(Cu,Ni)$
12	101013-101	68.05	12.42	5.42	B.d.	B.d.	B.d.	B.d.	B.d.	B.d.	14.11	100	$(Pd,Cu,Fe)_2S$
13	101012-154	B.d.	B.d.	B.d.	B.d.	B.d.	16.6	83.4	B.d.	B.d.	B.d.	100	Ag
14	211015_4-1	B.d.	B.d.	B.d.	B.d.	B.d.	70.36	27.53	B.d.	B.d.	B.d.	97.89	Au

Note. Samples of clinopyroxenites I (No 1–8), II (No 9–13), and III (No 14).

Clinopyroxene I also contains the thinnest microveinlets of sulfoarsenide Pt with composition intermediate between sperrylite ($PtAs_2$) and platarsite (PtAsS) (Table 10, No 7, 8). These microveinlets cut clinopyroxene and are developed along the cleavage cracks of chlorite (Figure 10d).

In clinopyroxenite II with bornite-chalcopyrite sulfide paragenesis, particles of noble metal minerals 3 μm and less in size are also developed in the interstices of clinopyroxene and along its cleavage cracks. They are represented by native Pd-Cu and Pd-Cu-Ni phases (Figure 11,a,b), native gold (kustelite) (Figure 11d), and copper–palladium sulfide. Native and sulfide phases can form polyphase aggregates of grains (Figure 11c). The composition of native Cu-Pd phases is described by the generalized formula $Pd_{0.38–0.48}Cu_{0.43}Fe_{0.04–0.13}Au_{0.02–0.05}Pt_{0.01–0.02}$ (Table 10, No 9, 10), corresponding in stoichiometry to skaergaardite (PdCu). Phase Cu-Pd-Ni (Table 10, No 11) is calculated for the stoichiometry of $Pd_3(Cu,Ni)$ phase which has not been described in literature. The composition of Ag-Au particles ($Ag_{0.76–0.90}Au_{0.1–0.24}$) corresponds to fineness 170–240‰. The composition of copper–palladium sulfide is calculated for the formula of $(Pd,Cu,Fe)_2S$ not described in literature (Table 10, No 12).

In clinopyroxenite III with pyrrhotite–chalcopyrite–pyrite paragenesis, noble metal minerals are rare, their sizes do not exceed 2–3 μm. Among these minerals, we detected Pd-Sb phases, which in the ratios of components are similar to stibiopalladinite and mertieite, and native gold (see Table 10, No 14). Pd-Sb phases are localized on the surface of large pyrite grains in contact with chlorite developed on clinopyroxene (Figure 12a). A particle of native gold (fineness ~720‰) is enclosed in chalcopyrite at the contact of two pyrite grains (Figure 12b).

4.2. Sulfur Isotopic Composition of Sulfides

Sulfur of pyrite, bornite, and chalcopyrite from clinopyroxenite II and III samples of the Ozernoe occurrence has a homogenous isotopic composition $\delta^{34}S = -2.1 \ldots -2.9$ (Table 11). The $\delta^{34}S$ values of bornite and chalcopyrite, which compose lattice-like aggregates and replace copper sulfides (spionkopite–yarrowite), virtually do not differ (Figure 13a,b). Table 11 shows the data on the sulfur isotopic composition of sulfides from the Baronsky occurrence—the typical object of gold–palladium mineralization in the Urals. The $\delta^{34}S = -1.2 \ldots -2.6$‰ values of chalcopyrite from this occurrence (Figure 13e,f) are identical to the sulfur isotopic composition of clinopyroxenite II and III samples from the Ozernoe occurrence.

Table 11. Isotopic composition of sulfur ($\delta^{34}S$) sulfides from the Ozernoe and Baronsky deposits.

Deposit	Mineral (Number of Analyses)	$\delta^{34}S$ (VCDT), ‰
Ozernoe, clinopyroxenite II	Chalcopyrite (3)	−2.3 … −2.6
	Bornite (1)	−2.1
	Bornite, chalcopyrite (4)	−2.1 … −2.6
	Sulfur sulfides (spionkopite–yarrowite) (4)	−0.8 … −2.7
Ozernoe, clinopyroxenite III	Pyrite (5)	−2.1 … −2.5
	Chalcopyrite (5)	−2.5 … −2.9
Baronsky	Chalcopyrite (7)	−1.2 … −2.6

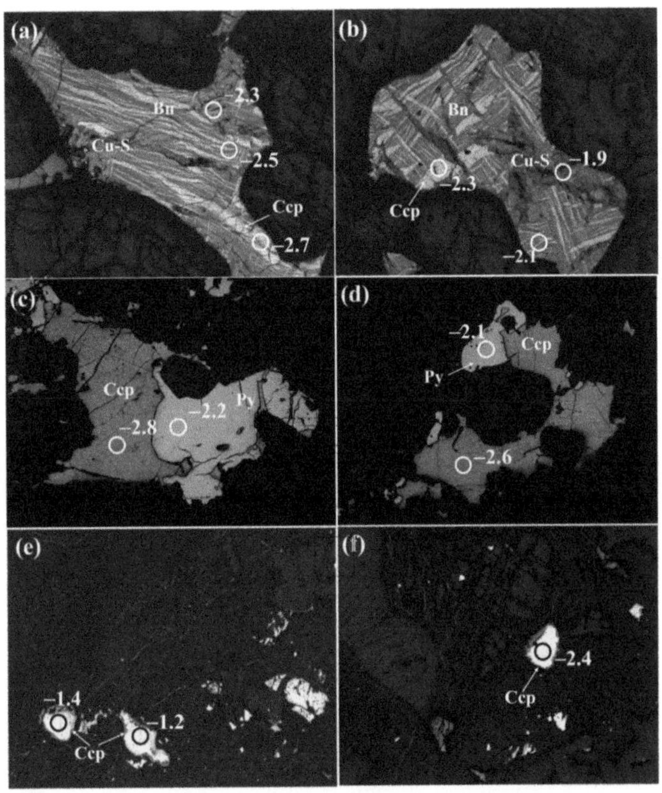

Figure 13. Sulfur isotopic composition of sulfides from Ozernoe (**a**–**d**) and Baronsky (**e**,**f**) deposits. Sites of analysis are shown by circles with a diameter 100 µm, numbers show $\delta^{34}S$ values in ‰. Minerals: Py—pyrite, Ccp—chalcopyrite, Bn—bornite, Cu-S—copper sulfide replacing bornite.

5. Discussion

5.1. Sequence of Mineral Formation at Ozernoe Occurrence

Our data show that the primary minerals of the magmatic stage include the main minerals of studied rocks—clinopyroxene, olivine, and Cr-Ti-V-bearing magnetite (with platelets of ilmenite). Interstitial amphibole (magnesio-ferri-hornblende), which was later replaced by chlorite, could have also been deposited from residual fluid-enriched melt (Figure 6c). Further hydrothermal-metasomatic mineral formation proceeded in the interstitial space and along cleavage cracks of clinopyroxene, contacts of clinopyroxene and olivine and microcracks that cut olivine and clinopyroxene. It started from the deposition of amphibole (actinolite and magnesio-hornblende) and bornite-chalcopyrite solid solutions (iss), which decomposed later. As the temperature decreased and water activity increased, the earlier deposited minerals were replaced by serpentine (on olivine), amphibole (tremolite), secondary magnetite, and chlorite with synchronous titanite. At the contact zone of olivine and clinopyroxene, symplectic aggregates of magnetite with tremolite serpentine were deposited.

The bornite–chalcopyrite solid solution (iss) in clinopyroxene II is the earliest sulfide. This is supported by its further decomposition into a lattice-like aggregate of bornite and chalcopyrite and partial replacement by tremolite (see Figure 8c). The process of mineral formation in clinopyroxenites of all groups led to the formation of chlorite, sulfide parageneses with pyrite, pyrrhotite, and chalcopyrite, as well as platinum group minerals and Au-Ag minerals. Geothermometry of chlorite composition shows that the hydrothermal system by that moment had cooled to 250 °C and continued cooling in parallel with a decrease in the iron content of chlorite (Figure 7).

5.2. Specific Features of the Evolution of Mineral Formation in Clinopyroxenites I–III

A common feature of ore-forming clinopyroxenites I–III is the same specific composition of primary minerals represented by diopside, olivine, and Cr-Ti-V-bearing magnetite. All pyroxenites have the same general tendency of replacement of primary rock-forming minerals by secondary with decreasing temperature: a) olivine $\rightarrow$ serpentine, secondary magnetite, and b) clinopyroxene $\rightarrow$ amphibole, secondary magnetite $\rightarrow$ chlorite.

The determined different ratios of clinopyroxene and olivine contents and different chemical compositions of these minerals indicate variability of the composition of initial melts for clinopyroxenites I–III. This variability reflects the primary magmatic stratification of rocks of the Dzelyatyshor massif, which is, according to [10], manifested as alteration of rocks differing in the composition of clinopyroxene and olivine. A direct dependence of the contents of FeO and Al_2O_3 in primary pyroxene revealed for clinopyroxenites under study (Figure 3b) is typical of the clinopyroxenite–gabbro trend of Uralian massifs and is due to cotectic olivine and clinopyroxene fractionation [21].

Postmagmatic clinopyroxene in clinopyroxenite I has an inverse dependence of FeO and Al_2O_3 contents (Figure 3b). The lowest iron content of clinopyroxene in clinopyroxenite III is most likely related to the postmagmatic "discharge" of part of iron into the oxide phase in the form of thinnest platelets along cleavage cracks (Figure 4a).

The identified higher contents of the main indicator elements (Cr, Ti, V, Mn, Al) in primary magmatic magnetite, compared to secondary hydrothermal magnetite, reflect the ratio of concentrations of these trace elements in the magmatic melt and hydrothermal fluid, which was determined for various ore deposits [22,23].

The revealed differences in the mineral composition of clinopyroxenites I–III and the chemical composition of primary minerals are typical of secondary minerals (Table 12). In particular, clinopyroxenites I–III differ in the set of sulfide minerals and their quantitative ratios (Table 1). In clinopyroxenite I with the lowest content of sulfides (less than 0.1%), chalcopyrite drastically prevails over pyrrhotite and pyrite. The content of sulfides in clinopyroxenite II can reach 0.5%, and intermediate chalcopyrite–bornite solid solution among them prevails over chalcopyrite. In clinopyroxenite III, the content of sulfides in uncommonly high (to 10%). Their main sulfide mineral pyrite predominates over chalcopyrite and pyrrhotite.

Table 12. Comparative characteristics of primary and secondary minerals of various groups of clinopyroxenites.

Minerals	Clinopyroxenite I	Clinopyroxenite II	Clinopyroxenite III
***Primary*:**			
Clinopyroxene 1	Diopside (4.5–7 wt.% FeO)	Diopside (5–6.8 wt.% FeO)	Diopside (0.7–2.2 wt.% FeO)
Olivine	x(Fa) = 0.25–0.26	x(Fa) = 0.28–0.30	x(Fa) = 0.13–0.15
Magnetite 1	Cr-Ti-V-bearing	Cr-Ti-V-bearing	Cr-Ti-V-bearing
***Secondary*:**			
Clinopyroxene 2	Diopside (9.6–14.5 wt.% FeO)		
Amphibole	Actinolite, magnesio-ferri-hornblende, tremolite	Magnesio-hornblende, tremolite	
Chlorite	Pennine, picnochlorite	Pennine	Pennine
Serpentine	Serpentine A	Serpentine A	Serpentine B
Magnetite 2	0.n wt.% TiO_2 и MgO	<1 wt.% Cr_2O_3, MgO, V_2O_3	0.n wt.% MnO
Sulfides	Ccp >> Po > Py	$(Bn\text{-}Ccp)_{ss}$ >Ccp; $\delta^{34}S = -2.1 \ldots -2.7‰$	Py > Ccp >> Po; $\delta^{34}S = -2.1 \ldots -2.9‰$
Pyrrhotite	N_{FeS} = 0.96–0.97		N_{FeS} = 0.94–0.95
Noble metal minerals	Antimonides Pd (mertieite, stibiopalladinite), sulfo-arsenide Pt (sperrylite-platarsite)	Native Pd-Cu and Pd-Cu-Ni phases, native silver (170–240‰).	Mertieite, stibiopalladinite, argentopentlandite, native gold (720 ‰)

Noble metal minerals in clinopyroxenes under study were deposited simultaneously with chlorite. Only platinum sulfoarsenide in clinopyroxene I developed later than chlorite, being deposited in the cleavage cracks of this mineral (see Figure 10d). At the same time, whereas clinopyroxenite II contains native Pd, clinopyroxenites I and III include Pd antimonides. Au-Ag particles of solid solutions detected in clinopyroxenites II and III do not contain palladium (Table 10).

5.3. Association of Noble Metal Minerals at Ozernoe Occurrence

Our study is based on the limited number of samples with gold–palladium mineralization, as a result of which we detected a limited number of noble metal minerals. It was reported in previous works [8,10] that in rocks with a low content of sulfides (no more than 1%), chalcopyrite–bornite segregations similar to those described in clinopyroxenites II are most common. At the same time, some samples contain cubanite, platelets of which can be considered exsolution products of intermediate cubanite–chalcopyrite solid solution. As we did not observe the joint occurrence of bornite–chalcopyrite and cubanite-chalcopyrite segregations, two ore mineral associations were identified—cubanite-pentlandite–pyrrhotite–chalcopyrite and bornite–chalcopyrite [10]. The sets of noble metal minerals in the composition of these mineral associations are virtually the same and consist of native metals, sulfides, antimonides, arsenides, arsenoantimonides, and tellurides (Table 13).

5.4. Features of Native Gold Composition

It was reported in previous research [8,10], which is not in conflict with our data (Table 10), that the composition of native gold in the studied occurrence demonstrates considerable variability. The chemical composition of native gold reported by these authors was recalculated and is presented in Table 14. Some analyses indicate low total contents of elements, which can be due to the small sizes of gold grains and X-ray excitation of matrix minerals. Nevertheless, we present these analyses in the table to demonstrate the ratios of main components and minimum levels of palladium contents in native gold. From the analyses reported in [10] we excluded those with high contents (more than 1 wt.%) of As, Te, and S elements that are not typical of native gold. In Table 14, analyses of gold grains are grouped based on their association with copper sulfides (bornite, chalcopyrite) and arsenides, antimonides, and tellurides of noble metals.

Table 13. Noble metal minerals at Ozernoe occurrence from [8,10].

Classes	Minerals Associations Cubanite-Pentlandite-Pyrrhotite- Chalcopyrite	Bornite-Chalcopyrite
Native metals	Au-Ag solid solution, tetra-auricupride Au-Cu, auricupride Cu_3Au, nielsenite Cu_3Pd, native Pt	Au-Ag solid solution, auricupride, skaergaardite *, phase Cu > Pd >> Fe *, phase $Pd_3(Cu,Ni)$*
Sulfides	Vysotskite PdS, braggite (Pt,Pd)S)	Phase $(Pd,Cu,Fe)_2S$ *
Arsenides	Sperrylite $PtAs_2$, arsenopalladinite $Pd_8As_{2.5}Sb_{0.5}$, atheneite $(Pd,Hg)_3As$, isomertieite $Pd_{11}Sb_2As_2$, majakite PdNiAs	Sperrylite $PtAs_2$, atheneite $(Pd,Hg)_3As$
Antimonides	Stibiopalladinite $(Pd_{5+x}Sb)_{2-x}$	Stibiopalladinite $(Pd_{5+x}Sb)_{2-x}$, mertieite Pd_5Sb_2
Tellurides	Merenskyite $PdTe_2$, moncheite $PtTe_2$	Kotulskite PdTe, hessite Ag_2Te
Tellurobismuthides, bismuthides	*Michenerite* PdBiTe, froodite $PdBi_2$	Froodite $PdBi_2$, sobolevskite PdBi

Note. *—minerals described in this work in the composition of bornite–chalcopyrite paragenesis.

Table 14. Chemical composition of native gold at Ozernoe occurrence from [8,10].

#	Mineral	Au	Ag	Cu	Pd	Fe	Σ	Fineness, ‰
				S (Bornite-chalcopyrite)				
509019	AuCu	74.62	0	23.96	0	0.84	99.42	751
509019	AuCu	74.03	0.94	24.43	0	0.9	100.30	738
509019	Cu_3Au	52.2	0	45.71	0	1.65	99.56	524
509019	AuCu	73.85	0	24.33	0	1.36	99.54	742
509019	AuCu	73.17	0	23.96	0	2.28	99.41	736
509019	AuCu	74.03	0	24.93	0	1.17	100.13	739
509019	AuCu	75.27	0	21.01	0	2.1	98.38	765
509019	AuCu	69.3	6.15	21.2	0	3.23	99.88	694
509019	AuCu	69.61	4.38	22.46	0	3.42	99.87	697
509019	ss(Ag,Au,Cu)	50.59	41.66	2.4	0	1.26	95.91	527
509019	ss(Ag,Au,Cu)	55.87	40.71	1.27	0	1.32	99.17	563
509019	ss(Ag,Au,Cu)	49.42	45.93	1.53	0	1.4	98.28	503
509019	ss(Ag,Au,Cu)	46.48	49.07	3.5	0	1.4	100.45	463
509019	ss(Ag,Au,Cu)	48.09	42.51	3.89	0	0.87	95.36	504
509019	ss(Ag,Au,Cu)	27.34	69.25	3.97	0	0.63	101.19	270
509019	ss(Ag,Au,Cu)	27.28	69.11	3.97	0	0.63	100.99	270
509019	ss(Ag,Au,Cu)	30.49	65.89	4.2	0	0.89	101.47	300
509019	ss(Ag,Au,Cu)	33.26	62.86	5.76	0	0.98	102.86	323
				S-Te-Sb (bornite, chalcopyrite, mertieite, merenskyite, moncheite, native Te)				
1	ss(Au,Cu,Ag,Pd)	58.9	5.5	17.3	2.8		84.50	697
2	ss(Au,Cu,Ag,Pd)	61.2	6.4	18.5	3.2		89.30	685
3	ss(Au,Cu,Ag,Pd)	55.5	2.5	29.3	1.6		88.90	624
4	ss(Cu,Au,Pd)	49.7	0	38.2	3.8		91.70	542
5	ss(Cu,Au,Ag)	44.7	8.3	41.3	0		94.30	474
6	ss(Cu,Au,Pd)	44.3	0	54	2.1		100.40	441
7	ss(Cu,Au,Pd,Ag)	39.7	1.3	53.5	6		100.50	395
8	ss(Cu,Pd,Au,Ag)	7.9	1.4	63.5	16.4		89.20	89
9	ss(Cu,Pd,Au)	5.5	0	64.3	16.2		86.00	64
10	ss(Cu,Pd,Au,Ag)	9.1	2.2	65.7	16.9		93.90	97

Table 14. *Cont.*

#	Mineral	Au	Ag	Cu	Pd	Fe	∑	Fineness, ‰
S-Te-As-Sb (bornite, chalcopyrite stibiopalladinite, kotulskite, braggite, moncheite, atheneite, majakite, mertieite)								
145042	AuCu	66.63	1.54	24.48	4.73	1.58	98.96	673
126010	ss(Au,Ag,Pd)	63.84	17.02	0.00	13.80	1.60	96.26	663
145042	ss(Ag,Au,Cu,Pd)	47.83	44.06	3.84	1.25	1.26	98.24	487
510015	ss(Ag,Au,Pd)	33.49	64.84	0	1.4	0	99.73	336
11	ss(Ag,Au,Cu,Pd)	14.6	80.7	1.2	1	97.50	150	150
12	ss(Ag,Au,Cu)	15.9	83.3	0.9	0	100.10	159	159
13	ss(Ag,Au,Cu)	34.6	62.7	0.8	0	98.10	353	353

Note. ss—solid solution.

In general, native gold at the Ozernoe occurrence is represented by Cu-bearing Au-Ag solid solutions (Cu to 6.5 wt.%) as well as Au-Cu solid solutions and intermetallides (tetra-auricupride and auricupride) with wide variations in fineness (150–750‰). According to the composition of components, it belongs to Au-Cu, Au-Cu-Pd, Au-Ag-Cu, Au-Ag, Au-Ag-Pd, and Au-Ag-Cu-Pd. Gold grains containing no palladium are quantitatively predominant (21 grains of the 35 studied). The absence of palladium in the composition of both Au-Ag, and Au-Cu phases was reported in [10] for sample 509019, in which palladium minerals were not detected and native gold is associated with chalcopyrite and bornite. In the samples, which contained sulfides, tellurides, antimonides, and palladium arsenides, native gold is predominantly palladium-bearing. On the whole, there is a tendency to increase the palladium contents in the series—Au-Ag solid solutions (no more than 1.5 wt.% Pd) and Au-Cu intermetallides—Au-Cu solid solutions (to 6 wt.% Pd) and Cu,Pd,Au solid solutions (16.2–16.9 wt.% Pd) (Figure 14).

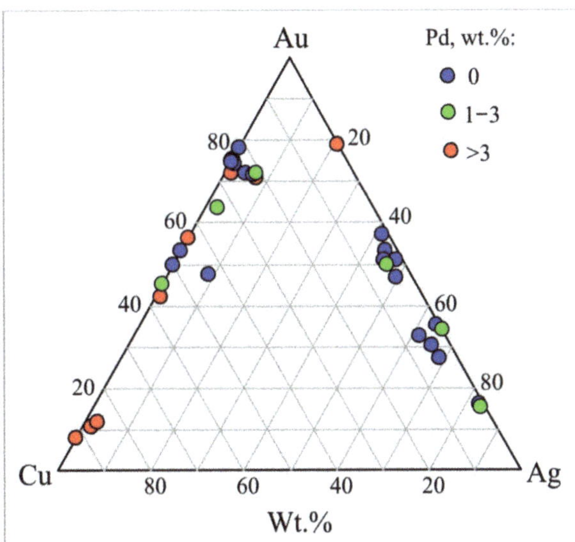

Figure 14. Chemical composition and palladium and copper content of native gold from Ozernoe occurrence on Au-Ag-Cu diagram.

5.5. Reconstruction of Physico-Chemical Parameters of Ore Formation

Following the identified sequence of mineral formation at the Ozernoe occurrence, we determined the stability fields of the main mineral associations and reconstructed the values of sulfur fugacity in the ore-forming system. Log fS_2—T diagram (modified from [24–26]) shows the stability boundaries of iron, copper, gold, silver, and platinum minerals depending on temperature and sulfur fugacity (Figure 15). Early bornite–chalcopyrite solid solution (iss) in clinopyroxenite II is evidence of high temperatures of formation (350–520 °C). As clinopyroxenite II contains magnetite together with bornite–chalcopyrite solid solution (iss), this allows estimation of temperatures and sulfur fugacity in the system. The stability lines of magnetite (Py+Hem)/Mag and solid solution (Bn+Ccp)/iss intersect at 480 °C and $fS_2 = 10^{-3}$, therefore, the stability field of solid solution does not exceed the specified values (Figure 15, field 1). Further in the course of mineral formation, sulfide parageneses with pyrite, pyrrhotite, chalcopyrite, platinum group minerals and Au-Ag minerals were deposited at temperatures 150–250 °C (see Table 6). Chalcopyrite, pyrite, pyrrhotite in clinopyroxenite I together with Pt sulfide in the given range of temperatures are stable at $\log fS_2 = -23 \ldots -11.5$ (Figure 15, field 2). In clinopyroxenite II, native silver (170–240‰) occurs in association with bornite and chalcopyrite, and their stability field is limited by the equilibrium of Uyt/$Ag_{0.75}Au_{0.25}$ and Ccp/(Bn+Pyh) (Figure 15, field 3).

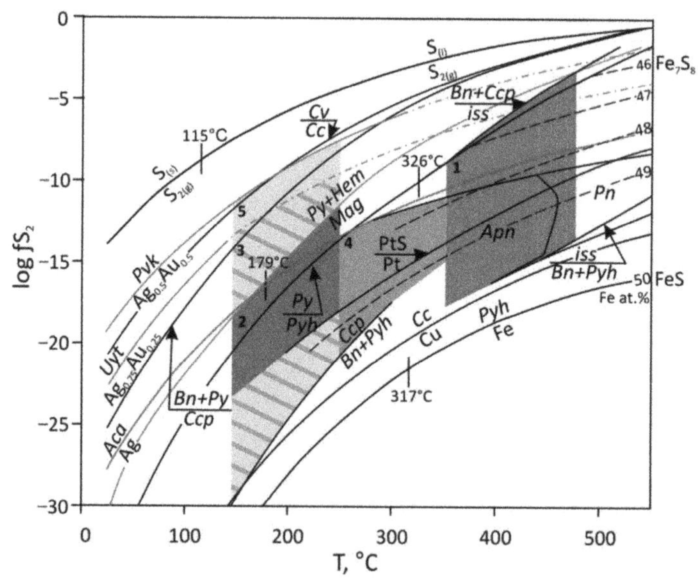

Figure 15. Sulfur fugacity dependence on temperature for Fe-S-O, Cu-Fe-S, Au–Ag–S, and Pt-S systems, and $\log fS_2$ and T estimations for different associations at Ozernoe occurrence. Minerals: Bn—bornite, Ccp—chalcopyrite, iss—bornite-chalcopyrite solid solution, Py—pyrite, Pyh—pyrrhotite, Mag—magnetite, Hem—hematite, Pn—pentlandite, Apn—argentopentlandite, Pvk—petrovskaite; Uyt—uytenbogaardtite; Aca—acanthite; Cc—chalcocite; Cv—covellite.

In clinopyroxenite III, chalcopyrite grains contain pentlandite and argentopentlandite inclusions. According to the diagram (Figure 15), the stability fields of Ni-bearing phases are located in the higher temperature range. Maximum sulfur fugacity values for this association correspond to the stability line of argentopentlandite ($\log fS_2 < -11$), minimum values were estimated from the equilibrium Ccp/(Bn+Pyh) ($\log fS_2 > -21$) (Figure 15, field 4). The association native gold (720 ‰) with chalcopyrite, pyrite and pyrrhotite in clinopyroxenite III is stable at $\log fS_2 = -30 \ldots -7.5$ (Figure 15, field 5).

To estimate the temperatures and values of S_2 fugacity, using a pyrite-pyrrhotite buffer, we analyzed pyrrhotite grains from pyrrhotite-chalcopyrite-pyrite parageneses (clinopyroxenite III). Pyrrhotite composition in this association reaches $N_{FeS} = 0.94$–0.95. The calculations with the help of equations [27,28] showed that during the formation of pyrrhotite of the given composition, the temperature reaches 196 °C and sulfur fugacity log $fS2 = -22,7$, which does not go beyond the bounds of the above-estimated boundary values $fS2$. The results obtained indicate that mineral formation at the Ozernoe occurrence took place at decreasing temperatures and increased sulfur fugacity in the system.

5.6. Model of Formation of Gold–Palladium Mineralization at Ozernoe Occurrence

The earlier proposed model of formation of gold-palladium Ozernoe occurrence suggests that mineralization is polychronous and polygenic, and formed in three stages [10]. The first, magmatogenic, stage is the concentration of PGE in sulfides and olivine-rich rocks. The second, hydrothermally-metasomatic, stage is a transformation of platinoids into sulfide forms and the enrichment of rocks with gold and copper under the influence of Karshor and Soba plutonites. The third, "epithermal", stage was marked by the deposition of arsenide, antimonide, and bismuthide forms of platinoids mainly at low-temperature serpentinization and redistribution of gold in the linear zones of concordant tectonic contacts of the clinopyroxenite massif.

In this manuscript we do not discuss the question of the primary magmatic origin of Fe-Cu sulfide melts enriched with native metals in olivine clinopyroxenites of the Dzelyatyshor massif, accepted for the first stage in the model [10], as we did not detect relics of magmatic sulfides in studied samples. The problem of the influence of As, Bi, and Te on the fractionation of Cu, Fe, Ni, and PGE during crystallization of sulfide magma is also beyond [29,30]). At the same time, the role of low-temperature remobilization of non-ferrous and noble metals as well as S, As, Sb, Te, and Bi from magmatic sulfides by post-cumulus fluids, reported in some publications, is undoubtedly important [3,4,31,32]. The most large-scale enrichment of post-cumulus fluids with copper, palladium, and gold takes place on the cooling of melt of gabbro composition. Fluid discharge leads to the formation of high-sulfide mineralization in olivine gabbro, which belongs to the "Volkovsky" type in the Urals [33]. This type of mineralization is represented by large bornite-chalcopyrite bodies with native gold and tellurides of Pd [34].

The conducted research allows us to suggest a simpler model of deposition of gold-palladium mineralization, in which the only source of ore matter and fluid were cooling portions of basic melt. The absence of another external source of substance and fluid is supported by the homogeneity of the sulfur isotopic composition of sulfides and its closeness to the magmatic zero mark. A slight lightening of the sulfur isotopic composition relative to the magmatic zero mark is most likely related to the depletion of heavy isotope during the separation of fluid from basic melt and kinetics of isotope exchange between fluid and sulfides at a further decrease in temperature.

We think that the primary magmatic layering of rocks manifested in different quantitative ratios of clinopyroxene and olivine in them further controls the local trends in the variability of the chemistry of mineral-forming medium and the concentration of ore components, including noble metals, and sulfur in each layer on its cooling.

Copper and iron sulfides as well as noble metal minerals, including native gold, were deposited among the latest and low-temperature minerals. Mineral forms of metals in each portion (layer) of cooled melt were determined by the variability of the activity of post-magmatic fluid components separated from melt. Most likely, these components that bind native metals into their own minerals, when their activity increases, are S, Sb, As, Te, Se, and Bi. The variations in the activity of these components could be due to the change in redox properties of the fluid and concentrations of metal in it, related to the hydration (serpentinization) of olivine. Absorption of water during serpentinization of olivine provided the residual metal-bearing fluid with reducing properties and stability of native forms of metals.

Within the frames of our model, the sets of sulfide minerals and noble metal minerals attributed to earlier distinguished cubanite–pentlandite–pyrrhotite–chalcopyrite and bornite–chalcopyrite mineral associations [10] could also be a result of the evolution of different compositions of the initial melt.

6. Conclusions

(1) The primary magmatic layering of rocks from the Dzelyatyshor massif, manifested in different quantitative ratios of clinopyroxene and olivine in them, controls the local trends of the variability in the chemistry of the mineral-forming medium and the concentrations of ore components, including noble metals, and sulfur in each separate layer on cooling.

(2) The deposition of native gold in parageneses with PGM and Cu-Fe sulfides at the Ozernoe occurrence proceeded during the replacement of earlier rock-forming minerals by chlorite. This process terminated mineral formation at the deposit and occurred at temperatures 150–250 °C and high activity of S, Te, Sb, and As of the fluid. The calculated variability in the conditions of mineral deposition during chloritization is reflected in the presence of several sulfide parageneses, in the replacement of native-sulfide forms by arsenide–antominide forms, in the considerable variability in the iron content of chlorite (X_{Fe} = 0.04–0.43), and in the sulfur isotopic composition of sulfides.

(3) Palladium content in native gold increases from Au-Ag solid solution (<1.5 wt.% Pd) to Au-Cu intermetallides—Au-Cu solid solution (to 6 wt.% Pd) and Cu,Pd,Au solid solutions (16.2–16.9 wt.% Pd)

(4) The sulfur isotopic composition of pyrite, chalcopyrite, and bornite ($\delta^{34}S$ = −2.1 … −2.9‰) just a little lighter relative to the zero mark of deep-seated magmatic sulfur. The deep-seated magmatic basic melt is assumed to be the source of fluid, ore components, and sulfur.

Author Contributions: Conceptualization, V.M. and G.P.; methodology, V.M.; validation, G.P.; investigation, V.M. and T.M.; writing—original draft preparation, V.M.; writing—review and editing, G.P.; visualization, V.M., T.B. and T.M.; Formal analysis, T.B.; supervision, V.M. All authors have read and agreed to the published version of the manuscript.

Funding: The studies are supported by the Russian Foundation for Basic Research (project No. 20-05-00393a) and are carried out as a part of the IGG UB RAS State assignment (state registration No. 122022600107-1), using the «Geoanalitik» shared research facilities of the IGG UB RAS. The re-equipment and comprehensive development of the «Geoanalitik» shared research facilities of the IGG UB RAS is financially supported by the grant of the Ministry of Science and Higher Education of the Russian Federation (Agreement No. 075-15-2021-680). The studies are carried out within the framework of the state assignment of the Sobolev Institute of Geology and Mineralogy of Siberian Branch of Russian Academy of Sciences and of the FRC Institute of Geology Komi SC UB RAS State assignment (state registration No. 1021062211108-5-1.5.2) financed by the Ministry of Science and Higher Education of the Russian Federation.

Data Availability Statement: The data presented in this study are mainly contained within the article and available in the references listed. To a minor degree, the data presented are not publicly available due to privacy and available on request from the corresponding author.

Acknowledgments: We are very grateful to the reviewers for their comments and suggestions.

Conflicts of Interest: The authors declare no conflict of interest.

References

1. Zoloev, K.K.; Volchenko, Y.A.; Koroteev, V.A.; Malakhov, I.A.; Mardirosyan, A.N.; Khrypov, V.N. *Platinum-Metal Mineralization in the Geological Complexes of the Urals*; Department of Natural Resources for the Ural Region: Yekaterinburg, Russia, 2001; 199p. (In Russian)
2. Maier, W.D.; Barnes, S.-J.; Gartz, V.; Andrews, G. Pt-Pd reefs in magnetitites of the Stella layered intrusion, South Africa: A world of new exploration opportunities for platinum group elements. *Geology* **2003**, *31*, 885–888. [CrossRef]
3. Sluzhenikin, S.F.; Yudovskaya, M.A.; Barnes, S.J.; Abramova, V.D.; Le Vaillant, M.; Petrenko, D.B.; Brovchenko, V.D. Low-sulfide platinum group element ores of the Norilsk-Talnakh camp. *Econ. Geol.* **2020**, *115*, 1267–1303. [CrossRef]
4. Mansur, E.; Barnes, S.-J.; Ferreira Filno, C.F. The effects of post-cumulus alteration on the distribution of chalcophile elements in magmatic sulfide deposits and implications for the formation of low-S-high-PGE zones: The Luanga deposit, Carajas mineral province, Brazil. *Can. Mineral.* **2021**, *59*, 1453–1484. [CrossRef]
5. Zaccarini, F.; Anikina, E.; Pushkarev, E.; Rusin, I.; Garuti, G. Palladium and gold minerals from the Baronskoe-Kluevsky ore deposit (Volkovsky complex, Central Urals, Russia) *Mineral. Petrol.* **2004**, *82*, 137–156. [CrossRef]

6. Anikina, E.V.; Malitch, K.N.; Pushkarev, E.V.; Shmelev, V.R. The Nizhny Tagil and Volkovsky massifs of the Uralian Platinum Belt, and related deposits. In *Field Trip Guidebook. 12th International Platinum Symposium*; IGG UB RAS: Yekaterinburg, Russia, 2014; 48p.
7. Anikina, E.V.; Alekseev, A.V. Mineral-geochemical characteristic of gold-palladium mineralization in the Volkovsky gabbro massif (Platiniferous Urals Belt). *Litosphere* **2010**, *5*, 75–100. (In Russian)
8. Kuznetsov, S.K.; Kotelnikov, V.G.; Onishchenko, S.A.; Filippov, V.N. Copper-gold-palladium mineralization in ultramafic rocks of the Voikaro-Synya massif in the Polar Urals. *Bull. Inst. Geol. Komi Sci. Cent. Ural. Branch Russ. Acad. Sci.* **2004**, *5*, 2–4. (In Russian)
9. Kuznetsov, S.K.; Filippov, V.N.; Onishchenko, S.A.; Kotel'nikov, V.G. Copper-gold-palladium mineralization in ultrabasic rocks of the Polar Urals. *Dokl. Earth Sci.* **2007**, *414*, 501–503. [CrossRef]
10. Pystin, A.M.; Potapov, I.L.; Pystina, Y.I.; Generalov, V.I.; Onishchenko, S.A.; Filippov, V.N.; Shloma, A.A.; Tereshko, V.V. *Low-Sulfide Platinum-Metal Mineralization in the Polar Urals*; Ural Branch of the Russian Academy of Sciences: Yekaterinburg, Russia, 2011; 150p. (In Russian)
11. Pystin, A.M.; Potapov, I.L.; Pystina, Y.I. The low-sulfide gold-platinum ore occurrence at the Polar Urals. *Zap. RMO* **2012**, *141*, 60–63. (In Russian with English abstract)
12. Remizov, D.N.; Petrov, S.Y.; Kos'yanov, A.O.; Nosikov, M.V.; Sergeev, S.A.; Grigoriev, S.I. New age datings of gabbroides of the kershor complex (Polar Urals). *Dokl. Earth Sci.* **2010**, *434*, 1235–1239. [CrossRef]
13. Puchkov, V.N. General features relating to the occurrence of mineral deposits in the Urals: What, where, when and why. *Ore Geol. Rev.* **2017**, *85*, 4–29. [CrossRef]
14. Warr, L.N. IMA–CNMNC approved mineral symbols. *Mineral. Mag.* **2021**, *85*, 291–320. [CrossRef]
15. Ignatiev, A.V.; Velivetskaya, T.A.; Budnitskiy, S.Y.; Yakovenko, V.V.; Vysotskiy, S.V.; Levitskii, V.I. Precision analysis of multisulfur isotopes in sulfides by femtosecond laser ablation GC-IRMS at high spatial resolution. *Chem. Geol.* **2018**, *493*, 316–326. [CrossRef]
16. Velivetskaya, T.A.; Ignatiev, A.V.; Yakovenko, V.V.; Vysotskiy, S.V. An improved femtosecond laser-ablation fluorination method for measurements of sulfur isotopic anomalies (Δ33S and Δ36S) in sulfides with high precision. *Rapid Commun. Mass Spectrom.* **2019**, *33*, 1722–1729. [CrossRef]
17. Leake, B.E.; Woolley, A.R.; Birch, W.D.; Burke, E.A.J.; Ferraris, G.; Grice, J.D.; Hawthorne, F.C.; Kisch, H.J.; Krivovichev, V.G.; Schumacher, J.C.; et al. Nomenclature of amphiboles: Additions and revisions to the International Mineralogical Association's amphibole nomenclature. *Am. Miner.* **2004**, *89*, 883–887.
18. Morimoto, N.C. Nomenclature of pyroxenes. *Am. Miner.* **1988**, *73*, 1123–1133.
19. Cathelineau, M. Cation site occupancy in chlorites and illites as a function of temperature. *Clay Miner.* **1988**, *23*, 471–485. [CrossRef]
20. Hey, M.H. A new review of chlorites. *Miner. Mag.* **1954**, *30*, 277–292. [CrossRef]
21. Fershtater, G.B.; Pushkarev, E.V. Magmatic clinopyroxenites of the Urals and their evolution. *Proc. Acad. Sci. USSR Ser. Geol.* **1987**, *3*, 13–23.
22. Nadoll, P.; Angerer, T.; Mauk, J.L.; French, D.; Walshe, J. The chemistry of hydrothermal magnetite: A review. *Ore Geol. Rev.* **2014**, *61*, 1–32. [CrossRef]
23. Smagunov, N.; Tauson, V.; Lipko, S.; Babkin, D.; Pastushkova, T.; Belozerova, O.; Bryansky, N. Partitioning and Surficial Segregation of Trace Elements in Iron Oxides in Hydrothermal Fluid Systems. *Minerals* **2021**, *11*, 57. [CrossRef]
24. Fleet, M.E. Phase equilibria at high temperatures. *Rev. Mineral. Geochem.* **2006**, *61*, 365–419. [CrossRef]
25. Palyanova, G.A.; Sazonov, A.M.; Zhuravkova, T.V.; Silyanov, S.A. Composition of pyrrhotite as an indicator of gold ore formation conditions at the Sovetskoe deposit (Yenisei Ridge, Russia). *Russ. Geol. Geophys.* **2019**, *60*, 735–751. [CrossRef]
26. Sorokhtina, N.V.; Zaitsev, V.A.; Petrov, S.V.; Kononkova, N.N. Estimation of formation temperature of the noble metal mineralization of the Kovdor alkaline-ultrabasic massif (Kola Peninsula). *Geochem. Int.* **2021**, *59*, 474–490. [CrossRef]
27. Lambert, J.M.; Simkovich, J.R.; Walker, P.L. The kinetics and mechanism of the pyrite-to-pyrrhotite transformation. *Metall. Mater. Trans. B* **1998**, *29B*, 385–396. [CrossRef]
28. Chareev, D.A.; Osadchii, E.G. Pyrrhotite-pyrite equilibria in the Ag–Fe–S system et 245 to 310 °C and standard pressure. *Geochem. Miner. Petrol. Bulgar. Acad. Sci.* **2005**, *43*, 41–46.
29. Sinyakova, E.; Karmanov, N.; Kosyakov, V.; Distler, V. Behavior of Pt, Pd, and Au during crystallization of Cu-rich magmatic sulfide minerals. *Can. Mineral.* **2016**, *54*, 491–509. [CrossRef]
30. Sinyakova, E.F.; Borisenko, A.S.; Kosyakov, V.I. Effect of the presence of As, Bi and Te on the behavior of Pt metals during fractionation crystallization of sulfide magma. *Dokl. Earth Sci.* **2017**, *477*, 1422–1425. [CrossRef]
31. Wood, S.A. The aqueous geochemistry of the platinum group elements with applications to ore deposits. In *Geology, Geochemistry, Mineralogy and Mineral Benefication of Platinum Group Element*; Cabri, L., Ed.; Canadian Institute of Mining, Metallurgy and Petroleum: Ottawa, ON, Canada, 2002; Special Volume 54, pp. 211–249.
32. Holwell, D.; Adeyemi, Z.; Ward, L.A.; Smith, D.J.; Graham, S.D.; Mcdonald, I.; Smith, J.W. Low temperature alteration of magmatic Ni-Cu-PGE sulfides as a source for hydrothermal Ni and PGE ores: A quantitative approach using automated mineralogy. *Ore Geol. Rev.* **2017**, *91*, 718–740. [CrossRef]
33. Poltavetc, Y.A.; Poltavetc, Z.I.; Nechkin, G.S. Volkovsky deposit of titanomagnetite and copper-titanomagnetite ores with accompanying noble-metal mineralization, the Central Urals, Russia). *Geol. Ore Depos.* **2011**, *53*, 126–139. [CrossRef]
34. Murzin, V.V.; Palyanova, G.A.; Anikina, E.V.; Moloshag, V.P. Mineralogy of noble metals (Au, Ag, Pd, Pt) in Volkovskoe Cu-Fe-Ti-V deposit (Middle Urals, Russia). *Lithosphere* **2021**, *21*, 643–659. [CrossRef]

minerals

MDPI

Article

Pd,Hg-Rich Gold and Compounds of the Au-Pd-Hg System at the Itchayvayam Mafic-Ultramafic Complex (Kamchatka, Russia) and Other Localities

Galina Palyanova [1,*], Anton Kutyrev [2], Tatiana Beliaeva [1], Vladimir Shilovskikh [3], Pavel Zhegunov [2], Elena Zhitova [2] and Yurii Seryotkin [1]

[1] Sobolev Institute of Geology and Mineralogy, Siberian Branch, Russian Academy of Sciences, pr. Akademika Koptyuga, 3, 630090 Novosibirsk, Russia; zhur0502@igm.nsc.ru (T.B.); yuvs@igm.nsc.ru (Y.S.)

[2] Institute of Volcanology and Seismology, Far East Branch, Russian Academy of Sciences, Piipa Blvd. 9, 683006 Petropavlovsk-Kamchatsky, Russia; anton.v.kutyrev@gmail.com (A.K.); pavel.zhegunov@bk.ru (P.Z.); zhitova_es@mail.ru (E.Z.)

[3] Centre for Geo-Environmental Research and Modelling (Geomodel), Saint-Petersburg State University, Ul. Ulyanovskaya, 1, 198510 Saint Petersburg, Russia; vvshlvskh@gmail.com

* Correspondence: palyan@igm.nsc.ru

Abstract: The unique minerals of the Au-Pd-Hg system in gold grains from heavy concentrates of the Itchayvayam placers and watercourses draining and ore samples of the Barany outcrop at the Itchayvayam mafic–ultramafic complex (Kamchatka, Russia) were investigated. Gold grains from watercourses draining and heavy concentrates of the Itchayvayam placers contain substitution structures formed by Pd,Hg-rich low-fineness gold ($Au_{0.59-0.52}Pd_{0.24-0.25}Hg_{0.17-0.23}$, 580‰–660‰) and Pd,Hg-poor high-fineness gold ($Au_{0.94-0.90}Pd_{0.02-0.04}Hg_{0.03}$, 910‰–960‰). Potarite (PdHg) without and with impurities (Au < 7.9, Cu < 3.5, Ag < 1.2 wt.%), Ag-poor high-fineness gold ($Au_{0.91}Ag_{0.09}$, 950‰), Ag,Pd,Hg-bearing middle-fineness gold ($Au_{0.75}Ag_{0.08}Pd_{0.09}Hg_{0.08}$—$Au_{0.88}Ag_{0.09}Pd_{0.02}Hg_{0.01}$, 820‰–930‰), and Pd,Hg-rich low-fineness gold with minor contents Ag and Cd ($Au_{0.51-0.55}Pd_{0.25-0.22}Hg_{0.21-0.16}Ag_{0.03-0.06}Cd_{0.01}$, fineness 580‰–630‰) were observed as individual microinclusions in the ore samples of the Barany outcrop. XRD and EBSD study results show that the Pd,Hg-rich low-fineness gold is isotypic to gold and has the same structure type, but different cell dimensions. According to data obtained for the Itchayvayam and some deposits and ore occurrences with Pd,Hg-bearing gold, the stable ternary phases and solid solutions of the following compositions in the Au-Pd-Hg system have been identified: Pd,Hg-poor gold ($Au_{0.94-0.90}Pd_{0.02-0.04}Hg_{0.03}$), Pd,Hg-rich gold ($Au_{0.59-0.52}Pd_{0.24-0.25}Hg_{0.17-0.23}$), Au-potarite ($PdHg_{0.62}Au_{0.38}$—$Pd_{1.04}Hg_{0.96}$—$Au_{0.80}Pd_{0.68}Hg_{0.52}$), and Au,Hg-bearing palladium ($Pd_{0.7}Au_{0.3}Hg_{0.1}$). The genesis of Pd,Hg-rich gold is insufficiently studied. We supposed that the meteoric waters or low-temperature hydrotherms rich in Pd and Hg could lead to the replacement Pd,Hg-poor gold by Pd,Hg-rich gold. High concentrations of Pd in Pd,Hg-bearing gold indicate a genetic relationship with mafic–ultramafic rocks.

Keywords: Pd,Hg-rich low-fineness gold; Au-containing potarite; substitution structure; Itchaivayam mafic–ultramafic complex (Kamchatka, Russia)

Citation: Palyanova, G.; Kutyrev, A.; Beliaeva, T.; Shilovskikh, V.; Zhegunov, P.; Zhitova, E.; Seryotkin, Y. Pd,Hg-Rich Gold and Compounds of the Au-Pd-Hg System at the Itchayvayam Mafic-Ultramafic Complex (Kamchatka, Russia) and Other Localities. *Minerals* **2023**, *13*, 549. https://doi.org/10.3390/min13040549

Academic Editor: Stefano Salvi

Received: 14 March 2023
Revised: 4 April 2023
Accepted: 11 April 2023
Published: 13 April 2023

1. Introduction

Native gold with silver impurity is the most abundant Au mineral in nature [1–9]. Detailed studies of native gold composition show that it contains a wide range of impurities—Cu, Hg, Pd, Pt, Ni, Fe, and other metals [6,10–19]. Native gold containing palladium impurity is called "porpecite" or palladium gold (Au,Pd). The name "porpecite" was not approved by the CNMNC MMA, since the mineral of the composition (Au>>Pd) is a structural analogue and is considered a Pd-rich chemical variety of native gold (www.mindat.org).

Palladium concentrations in native gold usually may reach 20 wt.%, but more often they are 0.9–9 wt.% [15,20–22]. "Mercury" gold (Au,Hg) or gold amalgams contain up to 18 wt.% Hg [23–25]. Experimental and theoretical data on the stability in the Au–Pd–Hg system are known only for mono- and binary phases [26–29] and are absent for ternary ones.

Research on natural gold alloys [30–34] can also provide important information on the composition and structures of ordered phases in complex systems with noble metals. Several minerals are known in the Au-Pd-Hg system (https://www.mindat.org/min-51827.html accessed on 4 April 2022): weishanite (Au,Hg(Ag)), aurihydrargyrumite (Au_6Hg_5), potarite (PdHg), gold, palladium, and mercury. Results of studies of Au-Pd-Hg mineralization from some deposits of noble metals showed the presence of ternary phases: Pd(Hg,Au) and $(Pd,Au)_3Hg_2$ [34]; Pd(Pt)AuHg, Au-rich potarite [10,35]. The list of unnamed minerals in the database https://www.mindat.org/ contains phases of the following composition: $Pd_{0.7}Au_{0.3}Hg_{0.1}$, Pd_3Au_2, Au_3Pd, Au_2Pd; $(Pd,Au)_3Hg_2$ (Table A1).

Our study is based on gold grains and ore samples from the Barany outcrop of the Itchayvayam mafic–ultramafic complex (Kamchatka, Russia) that contain unique minerals of the Au-Pd-Hg system Au-Pd-Hg compounds with low and high contents of Pd and Hg impurities. The preliminary data of native gold composition for the Itchayvayam placers were received by Sidorov et al. (2009, 2019) [36,37]. The Au-Hg-Pd mineral containing 14.5–15.1 wt.% Pd and 19–22 wt.% Hg (this composition is nearly stoichiometric Au_2PdHg or $Au_{0.5}Pd_{0.25}Hg_{0.25}$) and intergrown with cooperate (PtS), vysotskite (PdS), and malanite ($CuPt_2S_4$) was found in the Itchayvayam River. The attempts to obtain a crystal structure for this phase and a relatively close alloy with (Pd + Au)/Hg = 3/2 were unsuccessful [37–39]. Pd- and Hg-enriched gold is rare and poorly characterized [10,13,40,41].

The aim of this work is to determine the composition and crystal structure of Pd,Hg-rich gold and to study the relationships of Au-Pd-Hg phases in separate gold grains and ore samples from the Barany outcrop of the Itchayvayam mafic–ultramafic complex (Kamchatka, Russia), and gold grains from the adjacent placers. Identification and investigation of ternary phases in the Au-Pd-Hg system are of great theoretical and applied importance in geochemistry, metallurgy, chemical technology, and ecology. The unique properties of Pd- and Au-based alloys enable them to be widely used in instrumentation, electrochemical, chemical, jewelry, and other industries [42]. The addition of palladium to gold increases its melting point, modulus of elasticity, strength, hardness, and turns yellow gold white (https://www.azom.com/article.aspx?ArticleID=2372 accessed on 5 August 2022). Au-Pd and Pd-Hg compounds are used as electrocatalysts for oxygen reduction reactions [43]. Despite the wide application of Au-Pd alloys, data on the stable ternary phases in the Au-Pd-Hg system are absent.

Additionally, important tasks of this research are to summarize and analyze data on the composition of the mineral phases of the Au-Pd-Hg ternary system for other objects and the conditions for the formation of Pd,Hg-bearing gold. The concentration levels and ratios of various elements in individual gold grains can provide a geochemical history or "fingerprint" of ore-forming events. These fingerprints can be used to identify sources of gold [17,44–49].

2. Geological Background

The Itchayvayam (or "Itchayvayamskiy") complex is approximately 70 km^2 in area and is the largest zoned mafic–ultramafic complex in the Koryak-Kamchatka Platinum Belt (Russia) (Figure 1). The Belt is located at the Olyutorsky (another name is "Achayvayam-Valaginsky") arc terrain, which is the remnant of the late-Cretaceous Olyutorsky intra-ocean island arc, later collided with the easternmost margin of Paleo-Asia (e.g., [50]). It is located between the upper drainage basins of the Kamenistaya and Itchayvayam Rivers, about 20 km west of Anastasia Bay on the Bering Sea [37]. The host rocks of this massif belong to Vatyna Formation and Achayvayam Formation (Campanian and Maastrichtian stages, respectively). The Vatyna Formation consists of jasper, volcanoclastic rocks, and lavas, the latter of which are primarily basalts with MORB geochemical signatures. According to [51,52], a similar rock assemblage represents the Achayvayam Formation, but jasper is minor, and basalts bear arc geochemical signatures. Hence, the boundary between the formations corresponds to the incipience of the Olyutorsky arc. Isoferroplatinum grains from the placers related to the Itchayvayam complex yielded ^{190}Pt–^{4}He ages from 69.9 $\pm$ 4.1 to 70.6 $\pm$ 4.0 Ma, which corresponds the ages of host sedimentary rocks and the genetically similar Ural-Alaskan complex Galmeonan, that yielded hornblende and biotite ^{40}Ar/^{39}Ar ages ranging from 71.9 to 73.9 Ma [53].

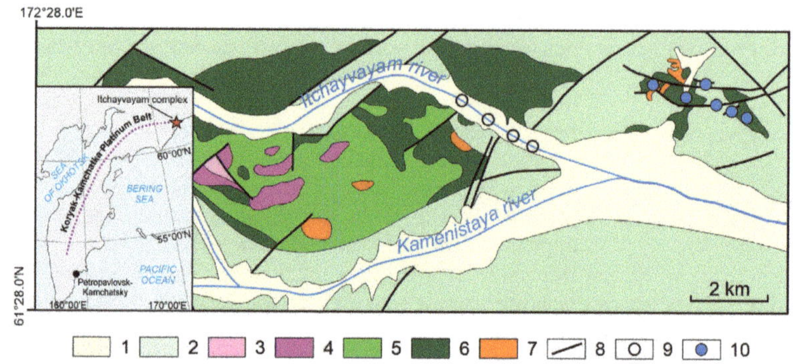

Figure 1. Geological scheme of the Itchayvayam mafic–ultramafic complex and Barany satellite, modified after Razumny et al. (2004) [54]. 1—alluvium, 2—volcanic and sedimentary rocks of Late Cretacous Vatynskaya formation and Achayvayam formation, 3—dunite, 4—wehrlite, 5—pyroxenite, 6—gabbro, 7—monzonite, monzodiorite, monzogabbro, 8—faults, 9—sampling points (from placer, after Sidorov et al. (2019) [37], 10—sampling points (from outcrops, this study).

The Itchayvayam complex is elongated latitudinally and on the surface consists of the main body and two satellite intrusions (Barany and Ailiolan), which according to geophysical data, merge at depth to form a single igneous intrusion [37]. The Itchayvayam complex consists mainly of anorthite gabbro, wehrlite, clinopyroxenite, and dunite. The Barany and Ailiolan satellite intrusions consist of gabbro, which was intruded by later monzonite and monzodiorite bodies, bears stockwork zones with bornite–chalcopyrite mineralization. Native gold and several Pd and Ag tellurides were reported in these zones [54].

Placer PGM occurrences spatially related to the Itchayvayam complex are located on the floodplains and alluvial terraces of the Kamenistaya and Itchaivayam rivers. Three assemblages of PGM, derived from different parts of a complex present: isoferroplatinum assemblage derived from dunite, native platinum assemblage derived from clinopyroxenite and wehrlite, and Au-Pd-Hg assemblage derived from gabbro of either the main body of Itchayvayam complex or its satellite Barany massif [37].

Investigations of the Barany occurrence revealed the following inferred resources: 40,000 tons of Cu, 14.5 tons of Ag, 3 tons of Au, and 0.4 tons of Pd. The prospects occurred in 1995–2000 and in 2016. Later, they were terminated because of the unprofitability of mining the relatively small deposit in conditions of remoteness from existing infrastructure facilities and settlements.

3. Samples and Analytical Methods

Studied samples were collected from alluvial deposits of the Itchayvayam river (61°31′17″ N 172°25′22″ E) and lode sulfide veins in monzogabbro of the Barany satellite of the Itchayvayam mafic–ultramafic complex (61°32′47″ N 172°27′13″ E). The 10 individual gold grains obtained by flushing the watercourses draining the Barany outcrop and 8 gold grains from the heavy concentrates of the placer alluvial deposits of the Itchayvayam river (see sampling points in Figure 1) were studied. Some of the individual gold grains were embedded in epoxy and polished. The polished gold grains examined with a reflected light microscope Olympus BX51 show heterogeneity, different colors, and textures. Chemical analyses of minerals were conducted at the Analytical Center for Multi-elemental and Isotope Research in the Sobolev Institute of Geology and Mineralogy SB RAS in Novosibirsk, using an MIRA 3 LMU electron scanning microscope (Tescan Orsay Holding, Brno, Czech Republic) with the microanalysis system Aztec Energy Xmax-50 (Oxford Instruments Nanoanalysis, Oxford, UK) (analysts Dr. N. Karmanov and M. Khlestov). The composition of native gold and other minerals was studied at the following parameters: accelerating voltage was 20 kV, spectrum recording time was 60 s (total area of spectra > 10^6 counts). The following X-ray lines were selected: $L\alpha$ for Pd, Ag, and $M\alpha$ for Au and Hg. Pure metals (Pd, Ag, Au) and HgTe were used as standards. The lower limits of the determined content (in wt.%) were 0.25 for Ag, 0.6 for Au, 0.8 for Hg, and 0.5 for Pd. Error in determining the main components with the contents higher than 10 wt.% did not exceed 1 rel.%, and when the content of components ranged 2%–10%, the error was no higher than 6–8 rel.%. Close to the lower limit of detection, the error was 15–20 rel.%. In some cases, the live time of spectrum acquisition increased to 120 s, the lower limits of determined contents and the random error of the analysis decreased about 1.4 times. To reduce the effect of microrelief of samples on the quality of analysis, data on the primary homogeneous gold were obtained in the scanning mode of individual sections from 10 × 10 to 100 × 100 μm^2 in size. The composition of small gold grains (<10 μm) was determined with a 10 nm point probe, but the size of the generation region of X-ray emission in gold with the electron beam energy 20 kV was 1 μm. Therefore, the data of analysis cannot be considered quantitative for a single phase if its minimum size is less than 2 μm.

Four polished sections of ore samples of gabbro-monzonites from the Itchayvayam mafic–ultramafic complex (Kamchatka) were also studied in detail. The massif is made up of the main body and a small satellite Barany outcrop. The latter is composed completely of gabbroids intruded by dikes and sill-like bodies of gabbros, diorites, monzodiorites, and quartz monzonites. Sites of hydrothermal-metasomatic processing represented by propilitization, actinolization, and pyritization are widespread within the outcrop. The ore mineralization is restricted to the contact of gabbroids and makes up zones of veined-streaky, stockwork bornite–chalcopyrite mineralization. Ore minerals are represented by the association of sulfides (Figure 2) and noble metals; the main minerals are bornite, chalcopyrite, and chalcocite, and less common minerals are ilmenite, native gold, hessite, naumannite, potarite, and temagamite. The ores also contain cinnabar, acanthite, rutile, titanite, barite, and oxides of rare earth elements.

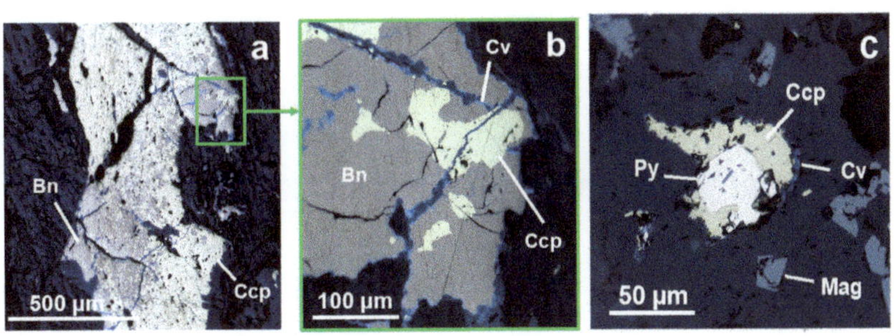

Figure 2. Photos in reflected light of typical ores in altered gabbro-monzonites of the Barany outcrop of Ithchayvam complex: (**a**) bornite-chalcopyrite vein, (**b**) magnified part of photo "a", note covellite altering primary sulfides along the fractures, (**c**) magnetite and sulfides in strongly altered gabbro-monzonite. Mineral abbreviations in Figures and the text: Au—native gold, Bn—bornite, Ccp—chalcopyrite, Cpe—cooperite, Cv—covellite, Ep—epidote, Mag—magnetite, Pmp—pumpellyite, Prh—prehnite, Ptr—potarite, Py—pyrite, Spi—spionkopite, Tem—temagamite, Tnr—tenorite.

Electron probe microanalyses (EPMA) of four polishing sections of ore samples were performed in the Institute of Volcanology and Seismology, Far East Branch of the Russian

Academy of Sciences (analyst V.M. Chubarov), using the TescanVega-3 electron scanning microscope equipped with an energy dispersive spectrometer (EDS X-MAX with detection area 80 mm^2). Some analyses were duplicated using the Camebax 244 microprobe equipped with four wavelength-dispersive spectrometers and energy-dispersive spectrometer X-MAX 50. As standards for noble metal minerals, we used the samples of high-purity 6 metals obtained in the Moscow Institute of 120 Steel and Alloys, which were checked for compliance and composition uniformity. Pure Au, Ag, Ni, Fe, Se, and Sb were used as standards. To determine As, Fe, and S, we used synthetic compounds FeS_2, FeAsS, and InAs. Similarly, Te, Hg, Sb, Bi, Pb, Cd, and Cu were determined using synthetic compounds CdTe, $CuSbS_2$, Bi_2S_3, PbS, HgS, and $CuFeS_2$. Determination of the elements was conducted using the analytical lines of the X-ray spectrum: $K\alpha$ for Fe, Cu, Zn, Ni, Mo, V, Ti, Cr, S, Al, Mg, Ca, Mn, Na, Si, Sc, P, F, and O; $L\alpha$ for Sb, As, Pd, Ag, Se, Te, and Cd; Mα for Au, Hg, and Pb. Analyses were carried out using an accelerating voltage of 20 kV and sample current on the reference standard Ni: 0.7 nA for SEM Vega-3; 20 nA for Camebax 244 electron microprobe. The minimum detection limit connected with the sensitivity of the EPMA analysis is about 0.1 wt.%.

X-ray powder diffraction study of selected samples was performed on a Stoe IPDS-2T diffractometer (Mo$K\alpha$ radiation, graphite monochromator) using Gandolfi mode. Two-dimensional X-ray patterns were radially integrated with the help of the XArea software package. The results were processed in the Stoe WinXPOW 2.21 and Match 3.5.3.109 program packages. Phase analysis was carried out through the PDF-4+ database (PDF-4+—International Centre for Diffraction Data (ICDD), accessed on 12 April 2022).

The gold grains were also investigated using a Rigaku R-Axis Rapid II (St. Petersburg State University, X-ray diffraction Resource Center) diffractometer equipped with a curved image plate detector and a rotating anode X-ray source with Co$K\alpha$ radiation, scan speed 1139°/min, step width 0.02°, 2-theta range 0–140°. The data were integrated using the software package Osc2Tab/SQRay [55]. All X-ray diffraction powder analyses were carried out at room temperature.

EBSD data were acquired in the Geomodel research center of Saint-Petersburg State University using a scanning electron microscope Hitachi S-3400N (Hitachi, Tokyo, Japan) equipped with Oxford NordLys (Oxford Instruments Nanoanalysis, Oxford, UK) Nano electron backscatter diffraction (EBSD) and EDS X-MAX 20 detectors detector under the following conditions: 30 kV accelerating voltage, 2 nA beam current, 40 ms per point dwell time in mapping mode and 200 ms in spot mode, 2 × 2 binning and simultaneous EDX mapping. All the data were handled automatically using Oxford Channel5 software package based on of Au structure (ICSD 44362). Prior to EBSD mapping the samples were polished by Ar plasma using an Oxford Instruments Ionfab300 etcher, an exposition of 10 min, an angle of 45o, an accelerating voltage of 500 V, a current of 200 mA, and a beam diameter of 10 cm (Nanophotonics Resource Center, Scientific Park, St. Petersburg, Russia).

4. Results

4.1. Mineral Au-Hg-Pd Phases of Gabbro-Monzonites of the Barany Outcrop at Itchayvayam Mafic–ultramafic Complex

The individual gold grains obtained by panning the watercourses draining the Barany outcrop and ore samples of gabbro-monzonites contain the minerals of the Au-Pd-Hg system. Gold grains consist of Pd,Hg-bearing high-fineness and low-fineness gold. The ore samples contain potarite and Pd,Hg,Ag,Cd-bearing gold of different fineness. Optical photos (Figures 3a and 4a) clearly show that gold grains are heterogeneous and differ in color with bright yellow, bluish-grey, and greyish-yellow zones. The greyish-yellow zones are represented by substitution textures consisting of tiny bluish-grey inclusions in a yellow matrix. In the SEM photo (BSE mode) (Figures 3b and 4c), these two phases (yellow and bluish-grey) seem to be light- and dark-grey.

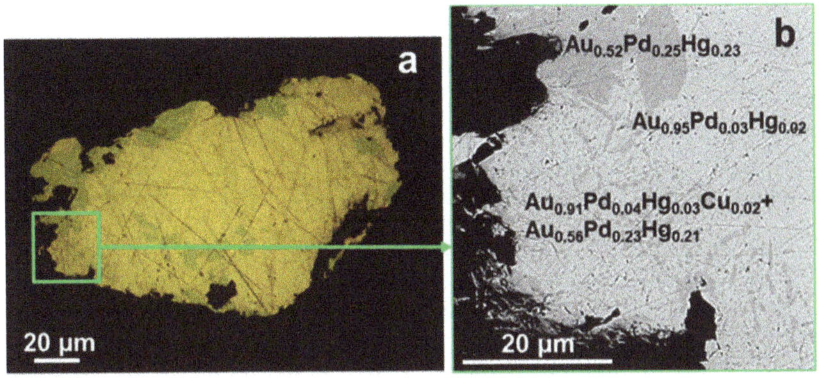

Figure 3. (**a**) Optical photo (in reflected light) of gold grain 1, (**b**) SEM photo (BSE mode) of its left fragment.

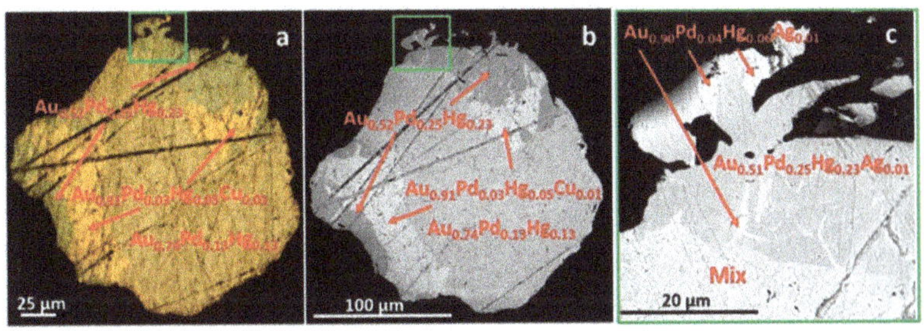

Figure 4. (**a**) Optical photo (in reflected light) of gold grain 2, (**b**) SEM photo (BSE mode) of this grain and (**c**) its left marginal area.

The EPMA data show that the yellow phase consists of high-fineness gold (910‰–964‰) with minor amounts of Hg (2.9–5.9 wt.%), Pd (1.1–3.3 wt.%), and, in places, Cu (0.7 wt.%) or Ag (0.3–1.5 wt.%) that corresponds to $Au_{0.94}Pd_{0.03}Hg_{0.03}$—$Au_{0.89}Pd_{0.06}Hg_{0.05}$, $Au_{0.92}Pd_{0.03}Hg_{0.03}Cu_{0.02}$, or $Au_{0.90}Pd_{0.04}Hg_{0.05}Ag_{0.01}$ (Table 1). Some gold grains contain gold particles without Pd with Ag or Cu impurities (Table 1, grains 1, 3).

Table 1. The composition of heterogeneous gold grains consisting of two phases—Pd,Hg-poor high-fineness and Pd,Hg-rich low-fineness gold (in wt.%, formula, N_{Au}). The individual 5 gold grains obtained by panning the watercourses draining the Barany outcrop.

No. Grain-Spectra	Cu	Pd	Ag	Au	Hg	Σ wt.%	Formula	Nau‰ *
				High-fineness gold (yellow phase)				
1-1	-	1.43	-	95.41	3.46	100.3	$Au_{0.94}Pd_{0.03}Hg_{0.03}$	951
1-2	-	1.27	-	94.55	3.68	99.51	$Au_{0.94}Pd_{0.02}Hg_{0.04}$	950
1-12	0.64	-	-	95.05	2.91	98.59	$Au_{0.95}Hg_{0.03}Cu_{0.02}$	964
1-14	0.7	1.57	-	95.71	3.17	101.16	$Au_{0.92}Pd_{0.03}Hg_{0.03}Cu_{0.02}$	946
1-15	0.66	1.69	-	95.51	3.74	101.6	$Au_{0.92}Pd_{0.03}Hg_{0.03}Cu_{0.02}$	940
2-27	-	2.06	0.56	94.03	5.3	101.95	$Au_{0.90}Pd_{0.04}Hg_{0.05}Ag_{0.01}$	922
2-40	-	2.13	0.34	92.67	5.81	100.94	$Au_{0.90}Pd_{0.04}Hg_{0.05}Ag_{0.01}$	918
2-41	-	1.68	0.36	92.53	5.49	100.07	$Au_{0.91}Pd_{0.03}Hg_{0.05}Ag_{0.01}$	925
2-44	-	2.14	0.34	90.61	5.86	98.96	$Au_{0.90}Pd_{0.04}Hg_{0.05}Ag_{0.01}$	916
3-110	-	-	1.53	91.77	4.44	100.81	$Au_{0.93}Hg_{0.04}Ag_{0.03}$	910
3-111	-	3.07	1.22	93.03	3.45	99.26	$Au_{0.90}Pd_{0.05}Hg_{0.03}Ag_{0.02}$	937
4-334	-	3.25	-	92.56	5.60	101.41	$Au_{0.89}Pd_{0.06}Hg_{0.05}$	913
5-169	-	1.11	-	95.44	2.96	99.52	$Au_{0.95}Pd_{0.02}Hg_{0.03}$	959

Table 1. *Cont.*

No. Grain-Spectra	Cu	Pd	Ag	Au	Hg	Σ wt.%	Formula	Nau‰ *
				Low-fineness gold (bluish-grey phase)				
1-3	-	15.08	-	59.11	26.9	101.1	$Au_{0.52}Pd_{0.25}Hg_{0.23}$	585
1-13	-	15.59	-	59.81	26.08	101.48	$Au_{0.52}Pd_{0.25}Hg_{0.23}$	589
1-16	-	14.82	-	58.7	27.83	101.35	$Au_{0.52}Pd_{0.24}Hg_{0.24}$	579
1-17	-	15.64	0.66	60.1	25.43	101.83	$Au_{0.52}Pd_{0.25}Hg_{0.23}Ag_{0.01}$	590
2-29	-	15.24	-	60.74	25.62	101.6	$Au_{0.53}Pd_{0.25}Hg_{0.22}$	598
2-30	-	15.13	-	58.47	27.02	100.61	$Au_{0.52}Pd_{0.25}Hg_{0.23}$	581
2-39	-	14.73	-	57.22	27.21	99.16	$Au_{0.51}Pd_{0.25}Hg_{0.24}$	577
2-43	-	14.98	0.37	57.34	26.97	99.66	$Au_{0.51}Pd_{0.25}Hg_{0.23}Ag_{0.01}$	575
3-116	-	14.45	-	65.02	19.06	98.53	$Au_{0.59}Pd_{0.24}Hg_{0.17}$	660
4-332	-	14.37	0.35	63.90	21.09	99.70	$Au_{0.57}Pd_{0.24}Hg_{0.18}Ag_{0.01}$	641
5-171	-	15.17	-	61.53	23.65	100.36	$Au_{0.54}Pd_{0.25}Hg_{0.21}$	613
5-172	-	15.90	0.88	61.32	23.43	101.54	$Au_{0.53}Pd_{0.26}Hg_{0.20}Ag_{0.01}$	604

Note: Analyses were conducted at the Analytical Center for Multi-elemental and Isotope Research in the Sobolev Institute of Geology and Mineralogy SB RAS (Novosibirsk) (analysts Dr. N. Karmanov and M. Khlestov). "-" below detection limits. * The fineness of native gold is calculated by equation (1000 × Au/(Au + Me1 + Me2 + and so on), where Au, Ag, Cu, Hg, and Pd as wt.%.

The formula composition $Au_{0.52}Pd_{0.25}Hg_{0.23}$—$Au_{0.59}Pd_{0.24}Hg_{0.17}$ was determined for the bluish-grey phase when calculated for 1 formula unit. The maximum gold concentration in this phase is 65.0 wt.%, the minimum is 57.2 wt.%, the content of Pd is 14.4–15.9 wt.%, and Hg is 19.1–27.8 wt.%, and the rare presence of no more than 0.9 wt.% Ag was also observed (Table 1, grains 1, 2, 4, and 5).

In one of the gold grains, the core contains high-fineness gold, while the peripheral part consists of low-fineness gold with veinlets of high-fineness gold or zones with a substitution structure formed by two Au-Pd-Hg phases (Figure 3a,b). In another gold grain, the core is composed of a substitution structure with an adjacent high-fineness zone, which is replaced by the low-fineness gold in the marginal part (Figure 4a–c).

The ore samples from the Itchayvayam mafic–ultramafic complex (Kamchatka, Russia) were found to contain native gold in the form of individual microinclusions (Figure 5). High-fineness gold (820‰–930‰) with impurities of Ag (4.5–6.1 wt.%), Pd (1.2–5.2 wt.%), and Hg (1.3–8.5 wt.%) ($Au_{0.87}Ag_{0.10}Pd_{0.01}Hg_{0.02}$) is present in epidote (Figure 5a) and in the veinlets with copper sulfate ($Cu_4SO_4(OH)_2(H_2O)_7$) in epidote (Figure 4b and Table 2). More high-fineness gold with minor impurity of Ag (4.6–4.9 wt.%) and without Pd and Hg ($Au_{0.91}Ag_{0.09}$, 950‰) is intergrown with bornite, cooperate, tenorite in epidote, and prehnite (Figure 5c and Table 2).

Table 2. Composition of Ag,Pd,Hg-bearing gold (in wt.%, fineness ‰, formula) and minerals in association with it in polished ore samples of gabbro-monzonites from Barany outcrop of the Itchayvayam mafic–ultramafic complex (East Kamchatka).

No Spectra	Au	Ag	Pd	Hg	Σ wt.%	N_{Au} ‰	Formula	Sample, Area/Minerals in Association/Figure 5
42	91.34	5.06	1.52	2.09	100	913	$Au_{0.87}Ag_{0.09}Pd_{0.03}Hg_{0.02}$	Sample 1, area 10/in epidote/Figure 5a
43	92.58	4.99	1.17	1.25	100	926	$Au_{0.88}Ag_{0.09}Pd_{0.02}Hg_{0.01}$	
44	88.75	4.94	1.22	5.08	100	888	$Au_{0.85}Ag_{0.08}Pd_{0.02}Hg_{0.05}$	
29	90.63	5.95	0.42	2.11	99.11	914	$Au_{0.87}Ag_{0.10}Pd_{0.01}Hg_{0.02}$	Sample 1, area 9/in veinlets with copper sulfate in epidote/Figure 5b
30	90.42	6.12	0.23	1.82	98.59	917	$Au_{0.87}Ag_{0.10}Pd_{0.01}Hg_{0.02}$	
31	80.77	4.51	5.2	8.54	99.01	816	$Au_{0.75}Ag_{0.08}Pd_{0.09}Hg_{0.08}$	
61	92.39	4.79	-	-	97.18	951	$Au_{0.91}Ag_{0.09}$	Sample 1, area 13/with cooperate, tenorite in prehnite/Figure 5c
62	88.62	4.94	-	-	93.56	947	$Au_{0.91}Ag_{0.09}$	
63	85.97	4.61	-	-	90.58	949	$Au_{0.91}Ag_{0.09}$	

Note: Analyses were conducted in the Institute of Volcanology and Seismology, Far East Branch of the Russian Academy of Sciences (analyst V.M. Chubarov). "-" below detection limits.

Table 3. Composition of Ag,Pd,Hg,Cd-bearing low-fineness gold (in wt.%, fineness ‰, formula) with chalcopyrite in epidote in polished ore samples of gabbro-monzonites from Barany outcrop of the Itchayvayam mafic–ultramafic complex (Kamchatka, Russia).

No. Spectra	Pd	Hg	Au	Ag	Cd	Σ wt.%	N_{Au} ‰	Formula
137	15.07	23.3	56.45	1.67	0.62	97.12	581	$Au_{0.51}Pd_{0.25}Hg_{0.21}Ag_{0.03}Cd_{0.01}$
138	13.9	19.43	64.42	3.83	0.96	102.53	628	$Au_{0.55}Pd_{0.22}Hg_{0.16}Ag_{0.06}Cd_{0.01}$
139	14.68	21.85	60.77	1.42	0.6	99.32	612	$Au_{0.54}Pd_{0.24}Hg_{0.19}Ag_{0.02}Cd_{0.01}$

Note: Analyses were conducted in the Institute of Volcanology and Seismology, Far East Branch of the Russian Academy of Sciences (analyst V.M. Chubarov).

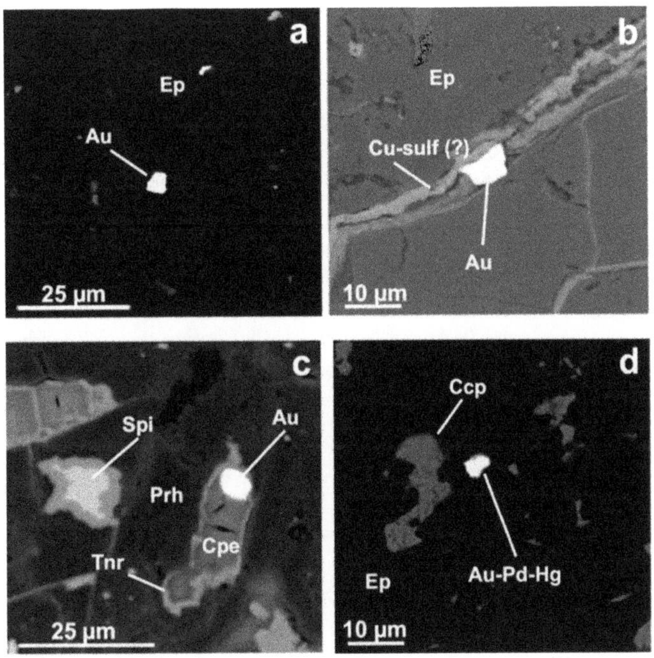

Figure 5. SEM images (BSE mode) of high- and low-fineness gold (see composition in Tables 2 and 3) and associated minerals in polished ore samples: (**a**) high-fineness gold (Au) in epidote (Ep); (**b**) high-fineness gold in Cu-sulphate veinlet in epidote; (**c**) high-fineness gold with cooperate (Cpe), tenorite (Tnr) in prehnite (Prh); (**d**) Pd,Hg-rich low-fineness gold and chalcopyrite in epidote.

The low-fineness gold with high contents of Pd and Hg and low contents of Ag and Cd ($Au_{0.51–0.55}Pd_{0.25–0.22}Hg_{0.21–0.16}Ag_{0.03–0.06}Cd_{0.01}$, fineness 580‰–630‰) (Table 3) and chalcopyrite were found as a microinclusions in epidote (Figure 5d). The ore samples also contain microinclusions of other minerals of palladium (temagamite Pd_3HgTe_3, merenskiite $PdTe_2$, and mertieite II $Pd_8(Sb,As)_3$) and silver (Cd-bearing acanthite Ag_2S; Cd, Se-bearing hessite, Ag_2Te; naumannite Ag_2Se) (Figure 6).

The ore samples contain potarite (PdHg) (Figure 6), frequently with minor amounts of Au, Ag, and Cu. Table 4 shows the composition of potarite and associated minerals. Microinclusions of potarite (4–10 microns) are present in bornite (Figure 6a), occasionally in the intergrowth with pumpellyite (Figure 6b), in the bornite–covellite matrix (Figure 6c) or epidote (Figure 6d). Potarite frequently contains microinclusions of temagamite Pd_3HgTe_3 (Figure 6e,f).

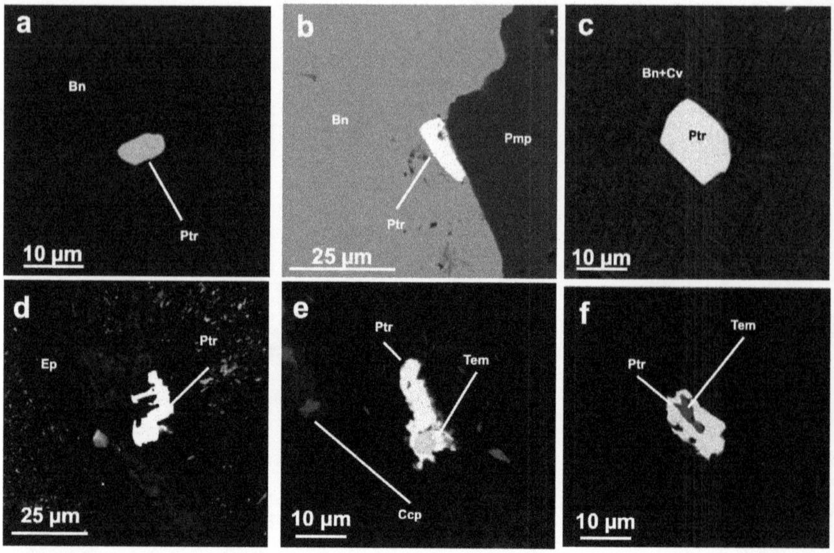

Figure 6. SEM images (BSE mode) of potarite and minerals in association with it in polished ore samples: (**a**) potarite (Ptr) crystal in bornite (Bn); (**b**) potarite inclusion in association with bornite and pumpellyite (Pmp); (**c**) potarite crystal in bornite–covellite matrix (Bn + Cv); (**d**) potarite in epidote (Ep); (**e**,**f**) potarite–temagamite intergrowth.

Table 4. Composition of potarite (in wt.%, formula) and minerals in association with it in polished ore samples from the Barany outcrop of the Itchayvayam mafic–ultramafic complex (Kamchatka, Russia).

No. Spectra	Pd	Hg	Au	Ag	Cu	Σ wt.%	Formula	Sample, Area/Minerals in Association/Figure
71	34.69	65.31	-	-	-	100 *	PdHg	Sample BN64, area 12/inclusion in bornite/Figure 5a
73	35.04	64.96	-	-	-	100 *	PdHg	
91	33.12	54.95	7.92		1.96	97.95	$Pd_{0.94}Hg_{0.84}Au_{0.12}Cu_{0.10}$	Sample BN64, area 6/inclusion in bornite near pumpellyite/Figure 5b
92	33.18	55.89	6.81		2.57	98.46	$Pd_{0.94}Hg_{0.84}Au_{0.10}Cu_{0.12}$	
127	33.15	63.35	-	-	3.53	100.03	$Pd_{0.92}Hg_{0.92}Cu_{0.16}$	Sample 1, area 19/Bornite, covellite (decay structures)/Figure 5c
128	33.29	63.51	-	-	2.62	99.42	$Pd_{0.94}Hg_{0.94}Cu_{0.12}$	
131	33.23	62.82	-	-	3.39	99.45	$Pd_{0.92}Hg_{0.92}Cu_{0.16}$	
144	33.63	62.88	-	0.61	-	97.12	$PdHg_{0.98}Agu_{0.02}$	Sample 1, area 19/in epidote/Figure 5d
146	33.3	62.57	0.51	0.75	-	97.12	$Pd_{0.99}Hg_{0.98}Au_{0.01}Ag_{0.02}$	
148	33.72	55.78	7.16	0.91	-	97.58	$PdHg_{0.86}Au_{0.12}Ag_{0.02}$	
149	33.39	62.58	-	1.24	-	97.2	$Pd_{0.98}Hg_{0.98}Ag_{0.04}$	
2-147	35.3	64.7	-	-	-	100 *	$Pd_{1.02}Hg_{0.98}$	Sample 64, area 9/Intergrowth of potarite and temagamite in silicate/Figure 5e
2-148	34.79	65.21	-	-	-	100 *	PdHg	
2-149	34.9	65.1	-	-	-	100 *	PdHg	
150	34.09	63.09	0.73	-	-	97.91	$PdHg_{0.98}Au_{0.02}$	Sample 1, area 23/potarite with cooperite inclusions in geerite/Figure 5f
155	33.76	62.84	-	-	0.84	97.44	$Pd_{0.98}Hg_{0.98}Cu_{0.04}$	
158	34.21	62.89	0.47	-	0.96	98.54	$Pd_{0.98}Hg_{0.96}Au_{0.02}Cu_{0.04}$	

Note: Analyses were conducted in the Institute of Volcanology and Seismology, Far East Branch of the Russian Academy of Sciences (analyst V.M. Chubarov). * minerals too small for quantitative analysis: normalized to 100% after background subtraction; "-" below detection limits.

The content of Au in potarite is no more than 7.9 wt.% ($Pd_{0.94}Hg_{0.84}Au_{0.12}Cu_{0.10}$). Rarely, potarite contains no more than 1.2 wt.% Ag ($Pd_{0.98}Hg_{0.98}Ag_{0.04}$). The concentrations of Cu in potarite vary from 0.8 to 3.5 wt.%, which corresponds to the formula composition $Pd_{0.98–0.92}Hg_{0.98–0.92}Cu_{0.04–0.16}$. Cu-bearing potarite occurs in intergrowth with copper sulfides (bornite, chalcocite, and covellite). The presence of Cu in potarite may be due to the influence of the copper sulfide matrix (Figure 6a–e).

4.2. Mineral Au-Hg-Pd Phases in the Grains from Heavy Concentrates of the Itchayvayam Placers

The first EPMA results on the grains of heavy concentrates from the placers of the Itchayvayam River showed several Au-Hg-Pd phases (Sidorov et al. 2019) [37]. The most widespread phases are high-fineness gold (910‰–950‰) $Au_{0.94–0.88}Pd_{0.03–0.05}Hg_{0.03–0.04}Ag_{0–0.03}$ ([37], (p. 7), Table 2, no. 5–7 in wt.%: Au—91.8 ÷ 95.6, Hg 3.4–4.4, Pd 1.6–3, Ag 0–1.5) and low-fineness gold $Au_{0.59–0.55}Pd_{0.24–0.25}Hg_{0.17–0.19}Ag_{0–0.01}$ (in wt.%: Au—62.4 ÷ 65, Hg—19.1–22, Pd—14.4–15.1, Ag—0–0.5) ([37], (p. 7), Table 2, no. 1–4). These two phases form intergrowths and are in association with cooperite (PtS) and malanite ($CuPt_2S_4$). Minor amounts of silver are rarely present in high- and low-fineness gold. The low-fineness gold forms thin veins in cooperite and precipitates later than other PGE minerals in the grains.

The gold grain from the heavy concentrates of the Itchayvayam placer contains low-fineness gold replacing high-fineness gold (Figure 7). The composition of low- and high-fineness phases in the grains of placer gold from [36,37] and the composition of these phases from gold grains and ore samples received in this study are similar, which is well seen on the ternary diagram Au(Ag,Cu)-Hg-Pd (Figure 8).

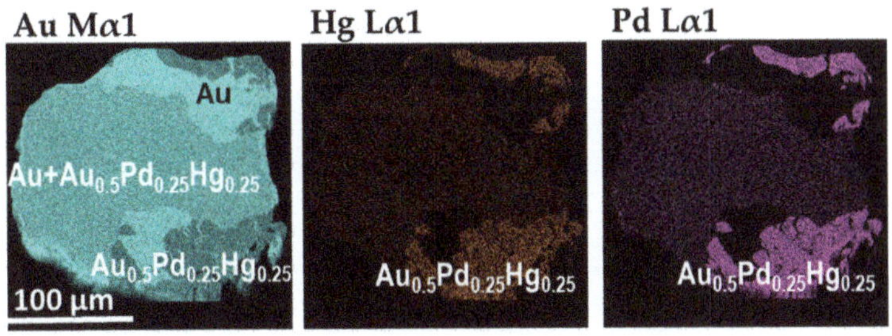

Figure 7. Maps of areal distribution of elements (Au, Hg and Pd) in characteristic rays. The gold grain from the heavy concentrate of the Itchayvayam River placer (Kamchatka, Russia).

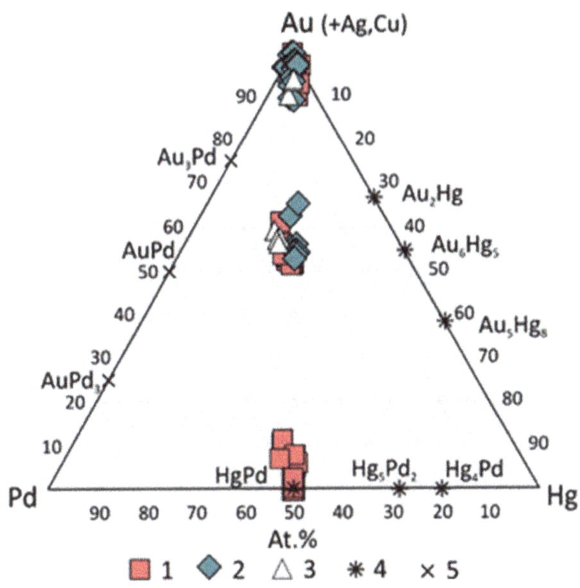

Figure 8. Composition of native gold and Au-Hg-Pd phases (at.%) from Itchayvayam mafic–ultramafic massif (Kamchatka, Russia): 1—this study (Tables 1–4), 2 and 3—from placer of the Itchayvayam River [36,37]. Phases stable in the Au-Hg-Pd system according to experimental and theoretical data: 4—[27,28]; 5—[26].

According to the IMA "50% rule", Pd,Hg-bearing gold should include natural solid solutions of the composition $Au_{1-x-y}Pd_xHg_y$, where x (at. fraction Pd) + y (at. fraction Hg) ≤ 0.5 at. fraction, consequently the Au-Pd-Hg compounds with high contents of Pd and Hg impurities of composition $Au_{0.50}Pd_{0.25}Hg_{0.25}$ can be called native gold—Pd,Hg-rich gold.

4.3. Structure Study Results: XRD and EBSD

Numerous attempts to obtain mineralogical structural data for Pd,Hg-rich low-fineness gold—$Au_{0.50}Pd_{0.25}Hg_{0.25}$ or intermetallic Au_2PdHg were unsuccessful [37–39]. Minerals of a relatively close (though not stable) composition corresponding to Pd-Hg-Au alloy (with (Pd + Au)/Hg = 3/2)) were described for Corrego Bom Sucesso, southern Serra do Espinhaco, Brazil [38,39] as "poorly crystalline" phases.

In the samples studied in the present work, the amount of this phase in the grains was 10%–40% by volume. The diffraction patterns obtained from single-crystal imaging of two grains on a Stoe IPDS-2T diffractometer (Mo*K*α radiation, graphite monochromator) using Gandolfi mode show only lines of gold (Figure 9). The study of 18 grains by the more powerful diffractometer R-Axis Rapid II revealed that all 18 patterns correspond to gold; however, 3 of them also contain additional weak reflections (Figure 10). The interpretation of powder XRD data gave several options: (i) strong reflections correspond to gold, while Au-Pd-Hg alloy is poorly crystalline or X-ray amorphous and does not produce reflections (and weak reflections originate from some other phase); (ii) strong reflections correspond to gold, while weak additional reflections correspond to Au-Pd-Hg alloy or (iii) Au-Pd-Hg alloy is isotypic to gold and their reflections overlap. For the unambiguous assignment of powder XRD data, the EBSD studies of Pd- and Hg-rich areas of Au were undertaken.

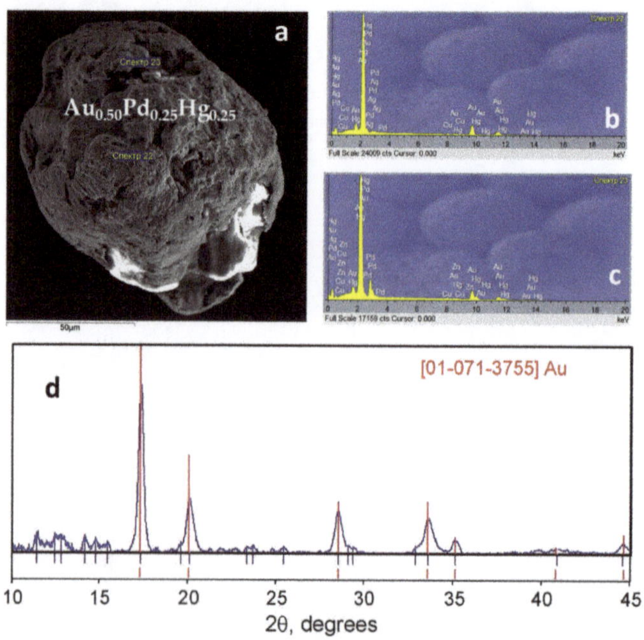

Figure 9. (**a**) SEM image of gold grain 4; (**b**,**c**) EPMA results; (**d**) X-ray diffraction patterns. "спектр" (on Russian) = spectra (on **a**–**c**).

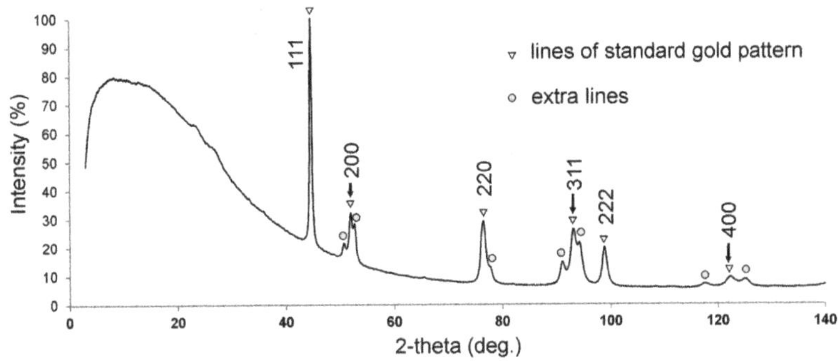

Figure 10. The powder X-ray diffraction pattern with Pd,Hg-poor and Pd,Hg-rich gold. See details in the text.

EBSD study of gold grain 1 from the heavy concentrate of the Itchayvayam River placer shows that it consists of multiple chaotically oriented crystallites with sizes 1–80 μm (Figure 11a). Areas with elevated Pd and Hg contents do not show any decrease in diffraction contrast and crystallinity (Figure 11b) and generate diffraction patterns (Figure 11c,d) very similar to native gold. The orientation map suggests that the neighboring areas with different compositions in many cases have the same crystallographic orientations, which proves their matching or similar structures. However, pattern indexing for Au-Hg-Pd areas has generally slightly higher mean angle deviation (MAD) than for native gold (Figure 11e). The latter can be caused by unit cell distortion leading to slightly different Kikuchi bands positions. Misorientations analysis shows a high degree of mechanical distortions (Figure 11f) within the grain specified by high ductility of gold and multiple deformations during fluvial transport. Pd-Hg-containing crystallites show a lesser degree of distortion compared to nearby native gold, which can be a sign of a higher hardness of the phase. Closer inspection of the solid solution area within the sample corresponding to Figure 2a shows no miscibility among Pd,Hg-poor gold and Pd,Hg-rich gold. Elemental mapping depicts the existence of two pronounced phases forming intergrowths (Figure 11g–i). Diffraction contrast maps reveal the fine subgranular structure and depict the boundaries between Pd,Hg-poor gold and Pd,Hg-rich gold (Figure 11j). Local misorientation maps show that mainly all the disorientations within the solid solution are concentrated on the subgrain boundaries while in native gold they are distributed evenly (Figure 11k). Three grains have three different and independent orientations, while within the grains, all subgrains despite the composition oriented preferably (Figure 11l,m).

EBSD studies of the grain 2 (Figure 4) display granular structure (Figure 11a) where Pd,Hg-poor gold $Au_{0.90}Pd_{0.04}Hg_{0.05}Ag_{0.01}$ is represented by a single crystal and probably represents the primary host protograin subsequently replaced by $Au_{0.52}Pd_{0.25}Hg_{0.23}$ (Figure 11b). The discrepancy between the maps of the distribution of elements and the maps of orientations, namely, the intersection of boundaries of two types, allows us to propose a mechanism for the formation of a phase enriched in palladium and mercury as a result of substitution without loss of structure and not as a result of the decomposition of the protophase with the release of $Au_{0.5}Pd_{0.25}Hg_{0.25}$ in the form of independent grains. The substitution is also supported by the fact that the enriched phases are formed close to the grain surface and can be considered incomplete rims.

The pattern misfit map shows a slightly higher average MAD for $Au_{0.52}Pd_{0.25}Hg_{0.23}$, which matches the data described above (Figure 11e) and almost the same values for $Au_{0.90}Pd_{0.04}Hg_{0.05}Ag_{0.01}$ (Figure 11c). Both Pd-Hg-rich phase is harder than Pd,Hg-poor gold due to higher misorientation accumulation in gold according to the misorientation map (Figure 11d). Since EBSD show that Pd- and Hg-rich areas do not show any decrease in diffraction contrast and crystallinity and generate diffraction patterns very similar to native gold.

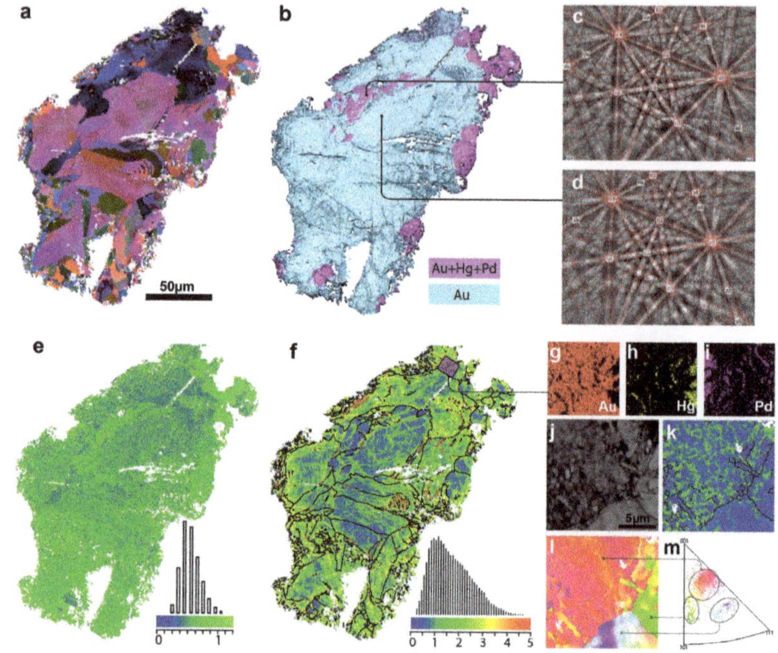

Figure 11. EBSD maps of gold grain 1 (Figure 3): (**a**) orientation map in Euler coloring scheme; (**b**) over-imposed diffraction contrast on EDX map; (**c**,**d**) comparison of diffraction patterns corresponding to different compositions and same orientation; (**e**) EBSD pattern misfit map with mean angle deviation (MAD) chart; (**f**) local misorientation map with misorientation distribution chart; (**g–i**) Inset represents magnified elemental maps; (**j**) diffraction contrast, fine-grained substitution structure; (**k**) local misorientation shows elevated disorientations at subgrain boundaries within grain; (**l**) orientation coloring and (**m**) corresponding orientations distribution inverse pole figure (**m**) with three circles depicting three distinct grains.

Based on that, the appearance of extra reflections in the powder XRD pattern has been interpreted as the cell distortion and the change of the unit cell parameters of Au phase due to admixtures of Pd and Hg. The combination of powder XRD and EBSD data allows to conclude that all reflections at the powder XRD pattern correspond to the structure type of gold, thus these phases are isotypic and some changes in the pattern (appearance of weak split reflections) correspond to the different chemical composition of areas within one grain: Pd,Hg-poor gold and Pd,Hg-rich gold. Despite the fact that our data are interpreted as the presence of only cubic phases with the Au structure, a hypothetical decrease in symmetry in this Au-Pd-Hg system seems possible to us; however, the difference in the unit cell parameters can be small and difficult to detect. The observed increase in pattern fitting error may also be a sign of symmetry lowering.

In other words, the reflection split is due to the appearance of three phases with the same structure type, but different cell dimensions. The calculated unit cell parameters from the positions of split reflections with 200 and 400 indexes are a = 4.03, 4.09, and 4.18 Å.

5. Discussion

5.1. The Composition of Native Gold and Other Phases of the Au-Pd-Hg System

The composition of native gold from the Barany outcrop of the Itchayvayam mafic–ultramafic complex (Kamchatka, Russia) is heterogeneous and unusual. According to EMPA results, it is represented by five varieties:

(I) Pd,Hg-poor gold or high-fineness gold with minor impurities of Pd and Hg ($Au_{0.94–0.90}Pd_{0.02–0.04}Hg_{0.03}$, 910–960) (Table 1);
(II) Pd,Hg-rich gold or low-fineness gold with major contents of Pd and Hg ($Au_{0.59–0.52}Pd_{0.24–0.25}Hg_{0.17–0.23}$, 580‰–660‰, ideal formula $Au_{0.5}Pd_{0.25}Hg_{0.25}$) (Table 1);
(III) usual Ag-poor gold or high-fineness gold with minor contents of Ag ($Au_{0.91}Ag_{0.09}$, ≈950‰);
(IV) Pd,Ag,Hg-poor gold or high-fineness gold with minor or trace impurities of Ag, Pd, and Hg ($Au_{0.75}Ag_{0.08}Pd_{0.09}Hg_{0.08}$—$Au_{0.88}Ag_{0.09}Pd_{0.02}Hg_{0.01}$, 820‰–930‰) (Table 2);
(V) Pd,Hg-rich and Ag,Cd-poor gold or low-fineness gold with major contents of Pd and Hg and low contents of Ag and Cd ($Au_{0.51–0.55}Pd_{0.25–0.22}Hg_{0.21–0.16}Ag_{0.03–0.06}Cd_{0.01}$, fineness 580‰–630‰) (Table 3).

In ore samples, the native gold III (with only Ag impurity) is intergrown with bornite, cooperate, tenorite in epidote, and prehnite, and in the veinlets with copper sulfate ($Cu_4SO_4(OH)_2(H_2O)_7$) in epidote, the native gold IV and V (associated with chalcopyrite) are present in epidote (Figure 5). High- and low-fineness gold (I and II) growing together with each other are found in some gold grains (Figures 2 and 3). The absence of minerals in the intergrowth with Pd,Hg-poor gold I and Pd,Hg-rich gold II makes it difficult to explain their genesis. However, these two phases were previously reported to be in association with cooperate (PtS) and malanite ($CuPt_2S_4$) [37]. The Pd,Hg-rich gold forms thin veins in cooperite and precipitates later than other PGE minerals.

Figure 8 shows the compositions of Au-Pd-Hg phases from the Itchayvayam mafic–ultramafic complex (Kamchatka, Russia) obtained in this study and previously published for the Itchayvayam river placers [36,37]: high-fineness gold, low-fineness gold, and Au-bearing potarite. The silver is absent or does not exceed 0.38 wt.% in minute analytical points of some gold grains. According to these data, the maximum content of Pd in high-fineness gold does not exceed 2.9 wt.%, Hg is less than 6.3 wt.%. Low-fineness gold is characterized by variations in Pd and Hg in the ranges of 9.3–15.1 wt.% and 19.4–27 wt.%. The compositions of Pd,Hg-bearing gold in the samples of gabbro-monzonites from the Itchayvayam mafic–ultramafic massif and Itchayvayam river placers have a similarity that indicates that these rocks are the source of gold in placers.

Potarite from the Itchayvayam river contains impurity of Au up to 7.9 wt.% with Pd and Hg content varying from 33.1 to 35.3 wt.% and 54.9 to 65.3 wt.%, respectively. The intergrowths of Pd,Hg-bearing gold and Au-bearing potarite were not observed in the studied samples from the Itchayvayam. However, dendritic and zoned grains of gold enriched in Pd (up to 10 wt.%), Au-potarite (16 wt.% Pd), or Au-bearing potarite (34 wt.% Pd) from Lower Devonian sediments and minor volcanics in the South Hums district of Devon (southwest England) were reported by Leake et al. (1991) [10]. The results of these authors suggest compositional variations in grains with a complex internal growth structure. One such compositional range is around 34 wt.% Pd, 39–60 wt.% Hg and 0–24 wt.% Au ($PdHg_{0.62}Au_{0.38}$—$Pd_{1.04}Hg_{0.96}$). Other preferred composition, about 57 wt.% Au, 25 wt.% Hg and 16 wt.% Pd ($Au_{0.51}Pd_{0.27}Hg_{0.22}$), occurs in Au-bearing potarite. In addition, a potarite is found as separate sub-grains within an Ag-rich grain overgrowth and as tiny inclusions within inter-sub-granular Ag-enriched films in a Pd–enriched gold. The composition of Pd,Hg-rich low-fineness gold and Au-bearing potarite from the Itchayvayam is close to the composition Au,Pd,Hg-phases from Lower Devonian sediments [10].

The ternary diagram (Figure 12) shows literature data on various samples containing minerals of the Au-Pd-Hg system. Native gold, similar in composition to Pd,Hg-rich low-fineness gold, was also found at several locations: Chudnoe (Russia) [56] (p. 113: among the numerous analytical data, only one point is given; it was an ore sample from the surface), River Dart (England) [20], and Corrego Bom Sucesso (Brazil) [38,39] (Table A2). Pd,Hg-poor gold is more common than Pd,Hg-rich gold. It is known that there are more than ten deposits and ore occurrences with Pd,Hg-poor high-fineness gold (Figure 12). Native gold from Yoko-Dovyren intrusion (Russia) [57,58] and depositions Serra Pelada and Gongo Soco (Brazil) [40,59,60] and alluvial placers Mayt and Big Kuonamka Rivers (Russia) [61] Dziwiszow and Zimnik Creek (Poland) [62,63] are examples of objects with

Pd,Hg-bearing gold with low and high contents of Pd and low contents of Hg (Table A2). The impurity Ag or (and) Cu is often present in such gold.

Both endogenous and exogenous gold has a wide and similar range of Pd content (Table A2, Figure 12). Most of the data in Figure 12 fall into the field of compositions to 10 wt.% Pd (0.16 at. fractions), the highest concentrations of Pd attain 19 wt.% (0.31 at. fraction). The content of Hg in endogenous gold is normally less than 0.5 wt.%, but occasionally reaches 3.5 wt.% (e.g., at the Chudnoe deposit) [56]. It is known the composition of native gold on surfaces at the Chudnoe deposit is 15,19 wt.% Pd, 35,99 wt.% Hg and 46,82 wt.% Au ($Au_{0.43}Pd_{0.25}Hg_{0.32}$) (Borisov, 2005 [56], (p. 113)).

Potarite from placers typically contains higher Au concentrations and wide range of variations in the amount of Pd and Hg (for example, Similkameen River (Canada) [64]; Corrego Bom Sucesso (Brazil) [38,39]. The minor impurity Ag or (and) Cu is often present in Au-potarite. Potarite of stoichiometric composition PdHg (Figure 12a) was detected in the sulfide-bearing mafic and ultramafic rocks of the Yoko-Dovyren Intrusion [57,58]. In potarite from placers and weathering crusts, according to literature data, the amount of Au varies in a wide range (Figure 12b). For example, gold content in potarite from the Bom Sucesso stream alluvium, Minas Gerais, Brazil (Fleet et al. 2002 [34], Table 1. (p. 344)) varies from 2.1 to 18 wt.%, which corresponds to the composition from $Pd_{1.26}Hg_{0.70}Au_{0.04}$ to $Pd_{1.06}Hg_{0.66}Au_{0.28}$. The widest variations in Au concentrations were observed in potarite from Superior Province, Canada [35] from 0 to 39.7 wt.%, which corresponds to composition from PdHg to $Pd_{0.86}Hg_{0.50}Au_{0.64}$ (Chapman et al. 2009 [20], Table 3, (p. 569)). For River Dart, Kester Brook noted that both Pd and Hg may occur individually within gold alloys, both up to a concentration of 12 wt.%, but at higher concentrations they exist as discrete phases. The composition of the Au-Ag-Pd-Hg phase is $Au_{2.10}Pd_{1.35}Hg_{0.43}Ag_{0.12}$ (Chapman et al. 2009 [20], Figure 4f, (p. 571)). Data from [35] show that the content of Hg in potarite (HgPd) varies in a wide range (from 18.4 to 71.2 wt.%). A wide range of variations was revealed for Pd: minimum and maximum calculated values are 21.6 and 78.7 wt.%. Potarite ($Au_{0.80}Pd_{0.68}Hg_{0.52}$) found in the samples picked in the region of Nunavik (Northern Quebec, Canada) contains the highest concentration of Au (46.7 wt.%), 21.6 wt.% Pd and 31.7 wt.% Hg. It is in association with amalgamated gold and cinnabar.

Intergrowths of Pd(Hg,Au) and $(Pd,Au)_3Hg_2$ from Corrego Bom Sucesso are described by Cabral et al. (2009) [39]. Very small potarite grains and a $Pd_{0.7}Au_{0.3}Hg_{0.1}$ grain occur in pentlandite of chalcopyrite veins in chromitites from the Herbeira ultramafic massif in Cabo Ortegal, NW Spain [65].

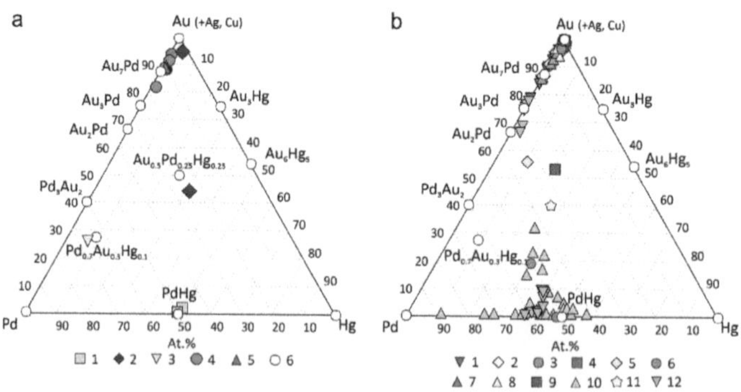

Figure 12. Composition of native gold and Au-Hg-Pd phases (at.%) from worldwide various deposits and ore occurrences: (**a**) endogenous: 1. Yoko-Dovyren intrusion, Russia [57,58], 2. Chudnoe, Russia [56], 3. Herbeira massif, Spain [65], 4. Serra Pelada, Brazil [40], 5. Gongo Soco, Brazil [59,60]. 6. minerals and unnamed minerals; (**b**) exogenous: 1. Mayt River, Russia, 2. Big Kuonamka River, Russia [61], 3. Dziwiszow, Poland [62], 4. Zimnik Creek, Poland [63], 5. River Dart, England [20], 6. The South Hums district of Devon, southwest England [10], 7 Mataganya-Sigiri zone, Guinea [66], 8. Whipsaw Creek, Canada, 9. Similkameen River, Canada [64], 10. Superior Province, Canada [35], 11. Nunavik area, Canada [35], 12. Corrego Bom Sucesso, Brazil [38,39].

Analysis of the above data on the compositions of the mineral phases in the Au-Pd-Hg system for the Itchayvayam and other objects revealed the stable ternary phases and solid solutions of the following compositions: Pd,Hg-poor gold ($Au_{0.94–0.90}Pd_{0.02–0.04}Hg_{0.03}$), Pd,Hg-rich gold ($Au_{0.59–0.52}Pd_{0.24–0.25}Hg_{0.17–0.23}$—these compositions are nearly $Au_{0.5}Pd_{0.25}Hg_{0.25}$), Au-potarite ($PdHg_{0.62}Au_{0.38}$—$Pd_{1.04}Hg_{0.96}$—$Au_{0.80}Pd_{0.68}Hg_{0.52}$), and Au,Hg-bearing palladium ($Pd_{0.7}Au_{0.3}Hg_{0.1}$) (Figure 12). The compositional variations and phase relations of Au-Pd-Hg compounds deserve more study.

5.2. The Structure of Phases in the Au-Pd-Hg System

The X-ray diffraction experiments of Au-Pd-Hg natural phases include (i) Au-bearing potarite grain from drainage sediment at Brownstone (South Hums district of Devon, south-west England) that gave potarite pattern with a slightly larger cell size (Leake et al. 1991 [10] reference to Nancorrow, 1989, personal communications); (ii) Pd(Hg,Au) and $(Pd,Au)_3Hg_2$ compounds did not diffract using the electron-backscattered diffraction (EBSD) and powder X-ray microdiffraction techniques, indicating that they are poorly crystalline [39]. X-ray diffraction also showed that Pd-rich/Ag-rich gold grains belong to gold structure type with decreased/standard unit-cell parameters with a = 4.03–4.06 Å and a = 4.07–4.08 Å, respectively [61].

There are only three natural phases from Au-Pd-Hg system in the Mindat database, and two of those are alloys. However, neither of them has a defined structure and $Pd_{0.7}Au_{0.3}Hg_{0.1}$ is only published as spot analyses with certain compositions, and $(Pd,Au)_3Hg_2$ is reported to give no reasonable EBSD or microX-Ray diffraction (https://www.mindat.org/chesearch.php?inc=Hg%2CAu%2CPd%2C&exc=&class=0&sub=Search+Minerals accessed on 5 August 2022): $Pd_{0.7}Au_{0.3}Hg_{0.1}$; $(Pd,Au)_3Hg_2$. No phases containing all three Hg, Au, and Pd are present in either COD or ICSD catalogues.

When studying the Au grains containing Pd-, Hg-rich fragments by EBSD and powder X-ray diffraction it was found that they belong to the structure type of Au (cubic, space group $Fm3m$, a ~4.07 Å) [67]. This is because the observed pattern (Figure 10) corresponds to Au, while extra reflections are produced as the result of the splitting of 200, 220, 311, and 400 reflections (Figure 10). Our interpretation of the data obtained is that all Au-Pd-Hg phases studied in this work are cubic, $Fm3m$, a = 4.03, 4.09, and 4.18 Å. The splitting of reflections is caused by a change in the unit cell parameters due to the incorporation of Pd (larger than Au) and Hg (smaller than Au) impurities into the structure of Au, i.e., Au/Pd/Hg proportion is responsible for cell metric and therefore reflection splitting. We consider the splitting of reflections (that is, a change in the unit cell parameters) as evidence that the diffraction was obtained precisely for Pd- and Hg-rich Au.

In our work, the most Pd and Hg-enriched fragments reached the composition $Au_{0.52}Pd_{0.25}Hg_{0.23}$ and it is tempting to describe the chemical formula as Au_2PdHg. Thus, when we had powder XRD data but did not yet have EBSD data, we searched for metal compounds of the stoichiometry of Au_2PdHg to find structural analogues. The most interesting (from a structural point of view) is the description of atomic ordering for Cu_2AuPd and Au_2CuPd compounds with the symmetry decrease to tetragonal (or D^1_{4h}-4/mmm) as reported based on electron diffraction data [68,69], although it was argued later [70]. This is important for our work, since the incorporation of Pd and Hg into the Au structure led to fairly wide variations in the unit cell parameters (4.03, 4.09, and 4.18 Å as reported above). If we assume that for certain chemical systems, the ordering of the chemical elements and the stretching and contraction of the cell in one (or two) directions will indeed occur, then this should lead to a decrease in symmetry from cubic to tetragonal (or lower). Despite the fact that our data are interpreted as the presence of only cubic phases with the Au structure, a hypothetical decrease in symmetry in this Au-Pd-Hg system seems possible to us, however, the difference in the unit cell parameters can be small and difficult to detect.

In general, our structural data are consistent with the (limited) literature, which shows that the phases of the Au-Pd-Hg system do not form separate mineral species but are incorporated in the structure as isomorphic impurities in known structure types.

5.3. Deposits with Pd,Hg-Bearing Gold and Genesis of Pd,Hg-Rich Gold

The mechanism of formation of native gold of different chemical compositions is complex and depends on many factors [17,32,44,45]. Native gold can form in different environments—orthomagmatic and hydrothermal systems and exogenous processes [7,32,38,40,60].

Figure 13 shows the locations of the Itchayvayam and some deposits and ore occurrences with Pd,Hg-bearing gold. They are dedicated to different geological and dynamic settings—terranes and orogenic belts and shields. According to age assessments, the processes of formation of Au-Pd mineralization are in Archean, Proterozoic, Paleozoic, and Mesozoic (Table A2).

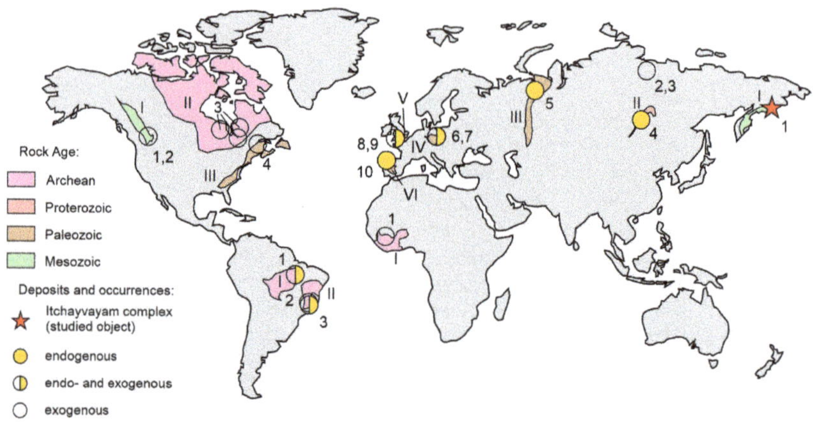

Figure 13. Locations of deposits and ore occurrences with Au-Pd, Hg-bearing gold, and other Au-Pd-Hg phases (Arabic numerals indicate deposits, Roman numerals indicate the main geological structures; the description is shown below). *Eurasia.* 1. Itchayvayam complex [37], 2. Mayt River, 3. Big Kuonamka River [61,71], 4. Yoko-Dovyren intrusion (after [72]), 5. Chudnoe [16] (Russia); 6. Dziwiszow [62], 7. Zimnik Creek [63] (Poland); 8. River Dart [20] (England); 9. Devon, southwest England [10]; 10. Herbeira massif [65] (Spain). Structure: I. Olyutorsky arc terrane (after [73]), II. Baikal-Patom fold belt (after [72]), III. Ural Orogenic belt (after [74]), IV. Bohemian Massif (Variscans orogen) (after [75]), V. British Variscans orogen (after [76]), VI. Central-Iberian zone of the Iberian massif (Variscans orogen) (after [65]). *Africa.* 1. Mataganya-Siguiri zone [77] (Guinea). Structure: I. Kenema-Man shield and Baoule-Mossi domain (West African Craton) (after [77]). *North America.* 1. Whipsaw Creek, 2. Similkameen River [64], 3. Superior Province [35], 4. Nunavik area [35]. Structure: I. Intermontane terranes of the British Columbia (North American Cordilleran Orogenic belt) (after [64]), II. Canadian Shield (after [35]), III. Appalachian orogen (after [35]). *South America.* 1. Serra Pelada [40], 2. Corrego Bom Sucesso [38,39], 3. Gongo Soco [59,60]. Structure (after [78]): I. Central Brazil shield, II. Atlantic shield.

The composition of Pd,Hg-bearing gold, associated minerals at some of these deposits and their brief characteristics are represented in Table A2. Pd,Hg-bearing gold has hydrothermal genesis at many deposits (for example, Fe-oxide-Cu Au-Pd Gongo Soco and Au-Pd-Pt Serra Pelada (Brazil)). The Pd-poor gold assemblage possibly formed later than the Au–Pd–Hg alloys. The ore-forming fluids of the Serra Pelada deposit were characterized by a high redox potential and were reduced by the carbonaceous material of metasiltstone, which hosts most of the ore bodies [79]. Subsequent removal of As, Sb, and Se from PGM led to the formation of Pd–O-bearing compounds, either by low-temperature hydrothermal or meteoric waters. The hematite-gypsum-Pd,Hg-bearing gold from the Fe-oxide-Cu-Au deposit Gongo Soco (Brazil) appears to be of low-temperature hydrothermal origin [60].

The Au–Pd ores with Pd,Hg-poor gold and potarite from infiltration U deposits of the "unconformity" type in the South Devon in southwest England [10,41,80] are characterized by the complete absence of sulfides, the presence of selenides and, sometimes, tellurides and arsenides. Au–Pd mineralization is localized in quartz or carbonate veins with hematite and is formed at low temperatures (<100 °C) from oxidized, slightly acidic chloride or soda brines.

Pd,Hg-bearing gold is found in sulfide and low-sulfide Cu-Ni-PGE ores of layered ultramafic–mafic intrusions with complex metasomatic and hydrothermal transformations as well as in Fe-oxide Cu-Au(U) deposits (IOCG) or Au-Pd deposits associated with a redox barrier in stratified volcanic-sedimentary basins of different ages and placers associated with layered mafic–ultramafic trap complex and metasediment, volcanic rocks.

Pd,Hg-rich low-fineness gold is less common than Pd,Hg-poor high-fineness gold. It has been found in several deposits: the Itchayvayam mafic–ultramafic massifs and placers at Itchayvayam river (Russia) ([36,37], present study), Chudnoe (Russia) [56], Lower Devonian sediments (England) [10] and River Dart (England) [20], and Corrego Bom Sucesso (Brazil) [38,39].

At most deposits minerals of the Au-Pd-Hg system are found in the placers (placer rivers of Anabar, (Russia), Dziwiszow (Poland), Dart (England), Whipsaw Creek and Similkameen (Canada)), and weathering crusts (Corrego Bom Sucesso (Brazil), Matagania-Sigiri Zone (Guinea)) are of exogenous origin (Figure 12b).

Pd-rich gold with minor content of Hg, Cu, and Ag in association with Pt minerals is found in placers of the Mayat and Bolshaya Kuonamka river basins (Anabar massif, northeast of the Siberian Platform) [61]. The Pd-rich gold (up to 18 wt.% Pd and minor content of Hg, Cu, Ag, Pt) recovered from the Córrego Bom Sucesso alluvial deposit lacks evidence of long-distance transport and implies a proximal source area adjacent to the alluvial sediments [38].

The compositions of palladian gold from placers and weathering crusts are of higher fineness (770‰–1000‰). Hg is constantly present in exogenous gold, and its content varies in the range of 1–4 wt.%. According to Spiridonov and Yanakieva (2009) [58], almost all gold amalgams found in placers are technogenic formations. This, however, cannot be the case for the studied Itchayvayam complex and related alluvial deposits because they are located in an area extremely remote from any settlements, and no gold exploration has ever been documented there.

The review of literature data showed wider variations in the composition of potarite from PdHg to $Au_{0.80}Pd_{0.68}Hg_{0.52}$ at placer ore occurrences and deposits of weathering crusts (Figure 12b). The highest content of Au in potarite is 46.7 wt.% [35]. These data indicate the existence of potarite-based solid solutions.

Pd,Hg-bearing low-fineness gold (580‰–660‰) can be a substitution product of Au-Pd-Hg solid solution or crystallize as a separate phase (Figure 5d). Its presence both in the ore samples of rocks and placer ore occurrences of the Itchayvayam suggests its high stability. It occurs in intergrowth with high-fineness gold (Table 3), or as microinclusions in epidote (Table 4).

The discovered substitution structures containing $Au_{0.50}Pd_{0.25}Hg_{0.25}$ in the intergrowth with Pd,Hg-poor high-fineness gold (940‰–964‰) indicate the exogenous genesis (Figures 3, 4 and 7). Based on the mass balance, the formation of a gold-depleted phase can occur with the introduction of Pd and Hg. However, this phase also forms separate grains in epidote (Figure 5d and Table 3), which contain Ag and Cd impurities. The presence of high-fineness gold (with only Ag impurities) (Table 3, No. 61–63) in intergrowth with copper sulfate indicates oxidative low-temperature deposition conditions in the later stages and the participation of meteoric waters. The metastable thiosulphate ions ($S_2O_3^{2-}$) can be produced during the oxidation of sulfide minerals, as an intermediate species in the transformation to dissolved sulphate ions and formation of sulphate minerals [81]. Under circum-neutral pH oxidizing conditions, groundwaters interact with sulfides and enrich in metastable thiosulphate ions that can transport Au and Ag to be precipitated later by either oxidation or reduction [81,82]. The associated Ag is also readily complexed and transported by the thiosulphate ions, so that supergene gold retains a significant Ag content. Silver is more readily dissolved by chloride complexes than gold, and this process can lead to the separation of Au from Ag during supergene processes leaving low-Ag gold [82].

The ore samples of the Itchayvayam mafic–ultramafic complex contain minerals of palladium, mercury, copper, and silver (merenskiite $PdTe_2$, mertieite II $Pd_8(Sb,As)_3$, potarite PdHg, temagamite Pd_3HgTe_3, bornite Cu_5FeS_4, chalcocite Cu_2S, and covellite CuS, Cd-bearing acanthite Ag_2S; Cd, Se-bearing hessite, Ag_2Te; naumannite Ag_2Se). We supposed that the dissolution of minerals of palladium and mercury by meteoric waters or low-temperature hydrotherms could lead to the formation of the Pd,Hg-rich low-fineness gold on high-fineness gold and Au-bearing potarite, while silver and copper partition to sulfides, tellurides, and selenides. The higher contents of Hg in placer gold in this instance is not a result of anthropogenic activity using Hg in placer mining operations though this does occur elsewhere.

6. Conclusions

(1) No miscibility among Pd,Hg-poor gold and Pd,Hg-rich gold, two Au,Pd,Hg phases with different contents Pd and Hg in gold were established. XRD and EBSD study results show that the low-fineness gold with high contents of Pd and Hg ($Au_{0.50}Pd_{0.25}Hg_{0.25}$) is isotypic to gold. This phase has the same structure type, but different cell dimensions.

(2) Similarity of the compositions of among Pd,Hg-poor gold and Pd,Hg-rich gold in the ore samples of gabbro-monzonites from the Itchayvayam mafic–ultramafic massifs and placers at Itchayvayam river indicates that these rocks are the source of placers with Au-Pd-Pt-Ag mineralization.

(3) Based on natural data, stable phases and solid solutions of the following compositions have been identified in the Au-Pd-Hg ternary system: Pd,Hg-poor gold, Pd,Hg-rich gold (this composition is nearly stoichiometric Au_2PdHg or $Au_{0.5}Pd_{0.25}Hg_{0.25}$), Au-potarite (from Pd(Hg,Au) to $(Pd,Au)_3Hg_2$), and Au,Hg-rich palladium. The compositional variations and phase relations of Au-Pd-Hg compounds deserve more study.

(4) Pd,Hg-bearing gold is a good indicator of the possible genetic connection of Au and sulfide and low-sulfide Cu-Ni-PGE ores of layered mafic–ultramafic intrusions with complex metasomatic and hydrothermal transformations as well as in Fe-oxide Cu-Au(U) deposits (IOCG) or Au-Pd deposits associated with a redox barrier in stratified volcanic-sedimentary basins of different ages and placers associated with layered mafic–ultramafic trap complex and metasediment, volcanic rocks. Pd,Hg-rich low-fineness gold is less common than Pd,Hg-poor high-fineness gold. Meteoric waters or low-temperature hydrotherms rich in Pd and Hg could lead to the replacement of high-fineness gold by Pd,Hg-rich low-fineness gold.

Author Contributions: Conceptualization, G.P.; methodology, G.P.; formal analysis, T.B., A.K. and P.Z.; investigation, T.B., V.S. and E.Z.; data curation, T.B., A.K. and P.Z.; writing—original draft preparation, G.P., A.K., E.Z., V.S. and Y.S.; writing—review and editing, G.P., A.K., V.S. and E.Z.; visualization, P.Z., V.S., E.Z. and Y.S.; supervision, G.P.; project administration, G.P. All authors have read and agreed to the published version of the manuscript.

Funding: This research was financially supported by the Russian Foundation for Basic Research (project No. 20-05-00393) and within the framework of the state assignment of Sobolev Institute of Geology and Mineralogy of the Siberian Branch of the Russian Academy of Sciences (Novosibirsk, Russia) (No. 122041400237-8). The work of researchers from the Institute of Volcanology and Seismology has been carried out within the state assignment of Ministry of Science and Higher Education of the Russian Federation theme No. AAAA-A0282-2019-0004. The technical support of XRD Research Center and GEOMODEL Research Center of Saint Petersburg State University (SPbU) through the President of Russian Federation Grant Nsh-1462.2022.1.5 is acknowledged.

Data Availability Statement: Not applicable.

Acknowledgments: Authors are grateful to E.G. Sidorov for samples, N.S. Karmanov, M.V. Khlestov (The Analytical Center for Multi-elemental and Isotope Research in the IGM SB RAS (Novosibirsk, Russia), and V.M. Chubarov (The Institute of Volcanology and Seismology, Far East Branch of the Russian Academy of Sciences) for electron microprobe data.

Conflicts of Interest: The authors declare no conflict of interest.

Appendix A

Table A1. The list of unnamed minerals of system Au-Pd-Hg in the Mindat database.

No.	Formula	Locality	Ref.	
Unnamed	Au_7Pd	Serra Pelada, Carajás, Pará (Brazil)	[40]	[83]
UM2008-11-E: AuPd	Au_3Pd	Serro, Minas Gerais (Brazil)	[38]	https://www.mindat.org/min-47975.html (accessed on 26 October 2022)
Unnamed (Au-Pd Alloy I)	Au_2Pd	Serro, Minas Gerais (Brazil)	[38]	https://www.mindat.org/min-47974.html (accessed on 26 October 2022)
Unnamed (Pd-Au-Hg Alloy)	$(Pd,Au)_3Hg_2$	Córrego Bom Sucesso placers, Serro, Minas Gerais (Brazil)	[39]	https://www.mindat.org/min-47976.html (accessed on 26 October 2022)
UM1999-09-E: AuPd	Pd_3Au_2	Timan Range (Russia)	[84,85]	https://www.mindat.org/min-7735.html (accessed on 26 October 2022)
UM1999-08-E: AuHgPd	$Pd_{0.7}Au_{0.3}Hg_{0.1}$	Herbeira ultramafic massif, A Coruña, Galicia (Spain)	[65]	https://www.mindat.org/locentry-1343331.html (accessed on 26 October 2022)

Table A2. The composition of Pd,Hg-bearing gold, associated minerals in endogenous and exogenous deposits, and their brief characteristics.

Deposit Name (Location), Type	NAu ‰/Impurity Concentrations, wt.%	Associated Minerals	Host Rock, Age	References
		Endogenous deposits		
Serra Pelada, Brazil, Fe-oxide-Cu-Au-Pd(Pt)	920–930/Pd 6.9–7.8, Hg < 0.02, Ag 0.2, Cu 0.07–0.13 (Au_7Pd)	Pd, Pd-Au-Pt-As phase, Svi, Pds, Pd_3As, Pd-oxide, Gth, $Au_{990–996}$	Metasedimentary and volcanic-sedimentary sequences (Archean) with granitic intrusion (Paleoproterozoic)	[40,79]
	910–960 Pd 3.2–9.8, Hg 0.3–1.5, Ag < 0.3, Cu 0.6–0.7	Ah, Pd–Hg–Se, Pd–Bi–Se phase		
	980–990/Pd 1.6–2.4, Hg < 0.02, Ag 0.3–0.4, Cu 0.5–0.6	Au–Pd–Hg, Ism, Mn–Ba oxide		
	990–996/Pd 0.04–0.1, Ag 0.2–0.3, Cu < 0.07	Au_7Pd, Gth (supergene)		

Table A2. *Cont.*

Deposit Name (Location), Type	NAu ‰/Impurity Concentrations, wt.%	Associated Minerals	Host Rock, Age	References
		Endogenous deposits		
Gongo Soco, Brazil, Fe-oxide-Cu Au-Pd	880–890/Pd 5.9–6.1, Hg 0.9, Ag 1.96–2.01, Cu 0.1–2.74	Hem,Gp,Ism,Met, Au-Ptr, (Pd,Cu)O, (Fe,Pd)OOH	Metasediments; Paleoproterozoic	[59,60]
		exogenous deposits (alluvial placers)		
Corrego Bom Sucesso, Brazil, Au–Pd(Pt) Supergene	910–975/Pd 2.4–8.7, Hg < 0.4, Ag 0.1–0.2, Cu 0.1–0.5, Pt < 0.3	Au_3Pd, Au-Ptr	Alluvial placers associated with metasediments rocks; Paleoproterozoic	[38,39]
	810–860/Pd 13.6–15.6, Hg 0.39–0.44, Ag 0.1–0.2, Pt < 0.28 (Au_3Pd)	$Au_{910–975}$		
	Au_2Pd, 770–790/Pd 17.85–18.82, Hg 2.91–3.43, Ag 0.09–0.16, Pt 0.34–0.68			
Anabar river basin, Russia: r.Mayat (a), r.Bolshaya Kuonamka (b), Au-Pt-diamond	a 860–960/Pd 0.8–12.8, Hg 0.01–0.3, Ag 0.9–2.8, Cu < 0.4	Tpdn, Ktu	Alluvial placers associated with sediments, volcanic rocks (Precambrian-Mesozoic) with intrusion rocks (Paleozoic-Mesozoic)	[61]
	b 990–960/Pd 0.73–7.5, Hg 0.04–1.7, Ag 0.6–2.2, Cu 0.1–1.5	Pt		
Dziwiszow, Poland, Au-Pd(Pt)	Centre 960–990/Pd < 0.5, Hg < 2.5, Ag 1.1–3, Cu 0.02–0.1, Pt < 0.04	Hem, Gth, Kln	Alluvial placers associated with granitic rocks; Paleozoic	[62]
	Core 975–997/Pd < 1.4, Hg < 1.9, Ag 0.2–1.8, Cu 0.04–0.1, Pt < 0.03			
Zimnik Creek, Poland, Au-Pd-Hg	core 980/Pd 0.01–0.95, Hg 0.36–0.96, Ag 0.15–1.93	Hem, Mag, Ap, Kln, Ms, Qz	Alluvial placers associated with sediments (conglomerates), volcanic rocks; Paleozoic	[63]
River Dart, England, Au-Pd-Hg	630/Pd 22.7, Hg 13.6, Ag 2.2	$Au_{970–1000}$, Ausb, Ptr, Skg, Stpdn, Carb	Alluvial placers associated with metasediments, volcanics, granodiorite; Paleozoic	[20]
Whipsaw Creek, Canada, Cu-Au(-Ag-PGE)	910–940/Pd 2.46–2.60, Hg 0.20–2.28, Ag 0.72–2.96, Cu 1.07–3.12	Stpdn, Cc, Spy	Alluvial placers associated with alkalic porphyry; Mesozoic	[64]
Similkameen River, Canada, Cu-Au(-Ag-PGE)	970/Pd < 2.1, Hg 0.1–0.7, Cu 0.06–1.8, Ag 0.7	Tem, Bn, Cc, Ccp, Cin, Pd-telluride, Pd-arsenoantimonide	Alluvial placers associated with alkalic porphyry; Mesozoic	[64]
	590/Pd 15.2, Hg 22.9, Ag 3.0, Cu 0.03			

Table A2. *Cont.*

Deposit Name (Location), Type	NAu ‰/Impurity Concentrations, wt.%	Associated Minerals	Host Rock, Age	References
		exogenous deposits (alluvial placers)		
Matagania-Siguiri Zone, Guinea, Au-Pd(-PGE)	920–970/Pd 2.3–8.2, Hg 0.07–0.11, Pt 0–0.09, Ag < 1.25	Spy, Bg	Placers associated with layered mafic–ultramafic trap complex (Mesozoic) and gneiss, supracrustal rocks, greenstone sequences in granite–greenstone terrain (Paleoproterozoic)	[66]

Mineral abbreviations from the list of IMA–CNMNC approved mineral symbols [86]: native gold (Au), native palladium (Pd), native platinum (Pt), muscovite (Ms), fuchsite (Cr-Ms), kaolinite (Kln), quartz (Qz), apatite (Ap), carbonates (Carb), gypsum (Gp), chalcopyrite (Ccp), bornite (Bn), chalcocite (Cc), pentlandite (Pn), cinnabar (Cin), hematite (Hem), magnetite (Mag), goethite (Gth), tetra-auricupride (Taur), auricupride (Auc), potarite (Ptr), skaergaardite (Skg (PdCu)), aurostibite (Ausb ($AuSb_2$)), atheneite (Ah ($(Pd,Hg)_3As$)), mertieite (Met ($Pd_8Sb_{2.5}As_{0.5}$)), isomertieite (Ism ($Pd_{11}Sb_2As_2$)), stibiopalladinite (Stpdn (Pd_3Sb)), kotulskite (Ktu (PdTe)), telluropalladinite (Tpdn (Pd_9Te_4)), temagamite (Tem (Pd_3HgTe_3)), palladseite (Pds ($Pd_{17}Se_{15}$)), sperrylite (Spy ($PtAs_2$)), braggite (Bg), sudovikovite (Svi ($PtSe_2$)).

References

1. Petrovskaya, N.V. *Native Gold*; Nauka: Moscow, Russia, 1973; p. 348. (In Russian)
2. Boyle, R.W. The geochemistry of gold and its deposits. *Geol. Surv. Can. Bull.* **1979**, *280*, 584.
3. Morrison, G.W.; Rose, W.J.; Jaireth, S. Geological and geochemical controls on the silver content (fineness) of gold in gold-silver deposits. *Ore Geol. Rev.* **1991**, *6*, 333–364. [CrossRef]
4. Gammons, C.H.; Williams-Jones, A.E. Hydrothermal geochemistry of electrum; thermodynamic constraints. *Econ. Geol.* **1995**, *90*, 420–432. [CrossRef]
5. Pal'yanova, G.A. Physicochemical modeling of the coupled behavior of gold and silver in hydrothermal processes: Gold fineness, Au/Ag ratios and their possible implications. *Chem. Geol.* **2008**, *255*, 399–413. [CrossRef]
6. Spiridonov, E.; Yanakieva, D. Modern mineralogy of gold: Overview and new data. *ArchéoSciences* **2009**, *33*, 67–73. [CrossRef]
7. Dill, H.G. The "chessboard" classification scheme of mineral deposits: Mineralogy and geology from aluminum to zirconium. *Earth Sci. Rev.* **2010**, *100*, 1–420. [CrossRef]
8. Liang, Y.; Hoshino, K. Thermodynamic calculations of Au_xAg_{1-x} fluid equilibria and their applications for ore-forming conditions. *Appl. Geochem.* **2015**, *52*, 109–117. [CrossRef]
9. Savva, N.E.; Kravtsova, R.G.; Anisimova, G.S.; Palyanova, G.A. Typomorphism of Native Gold (Geological-Industrial Types of Gold Deposits in the North-East of Russia). *Minerals* **2022**, *12*, 561. [CrossRef]
10. Leake, R.C.; Bland, D.J.; Styles, M.T.; Cameron, D.G. Internal structure of Au-Pd-Pt grains from South Devon in relation to low-temperature transport and deposition. *Trans. Inst. Min. Met. Sect. B Appl. Earth Sci.* **1991**, *100*, 159–178.
11. Sluzhenikin, S.F.; Mokhov, A.V. Gold and silver in PGE-Cu-Ni and PGE ores of the Noril'sk deposits, Russia. *Miner. Depos.* **2015**, *50*, 465–492. [CrossRef]
12. Gas'kov, I.V. Major impurity elements in native gold and their association with gold mineralization settings in deposits of Asian fold belts. *Russ. Geol. Geophys.* **2017**, *58*, 1080–1092. [CrossRef]
13. Lalomov, A.V.; Chefranov, R.M.; Naumov, V.A.; Naumova, O.B.; Lebarge, W.; Dilly, R.A. Typomorphic features of placer gold of Vagran cluster (the Northern Urals) and search indicators for primary bedrock gold deposits. *Ore Geol. Rev.* **2017**, *85*, 321–335. [CrossRef]
14. Murzin, V.V.; Chudnenko, K.V.; Palyanova, G.A.; Varlamov, D.A.; Naumov, E.A.; Pirajno, F. Physicochemical model for the genesis of Cu-Ag-Au-Hg solid solutions and intermetallics in the rodingites of the Zolotaya Gora gold deposit (Urals, Russia). *Ore Geol. Rev.* **2018**, *93*, 81–97. [CrossRef]
15. Palyanova, G.A. Gold and Silver Minerals in Sulfide Ore. *Geol. Ore Depos.* **2020**, *62*, 383–406. [CrossRef]
16. Palyanova, G.; Murzin, V.; Borovikov, A.; Karmanov, N.; Kuznetsov, S. Native gold in the Chudnoe Au-Pd-REE deposit (Subpolar Urals, Russia): Composition, minerals in intergrowth and genesis. *Minerals* **2021**, *11*, 451. [CrossRef]
17. Chapman, R.J.; Banks, D.A.; Styles, M.T.; Walshaw, R.D.; Piazolo, S.; Morgan, D.J.; Grimshaw, M.R.; Spence-Jones, C.P.; Matthews, T.J.; Borovinskaya, O. Chemical and physical heterogeneity within native gold: Implications for the design of gold particle studies. *Miner. Depos.* **2021**, *56*, 1563–1588. [CrossRef]
18. Chapman, R.; Torvela, T.; Savastano, L. Insights into Regional Metallogeny from Detailed Compositional Studies of Alluvial Gold: An Example from the Loch Tay Area, Central Scotland. *Minerals* **2023**, *13*, 140. [CrossRef]
19. Nikiforova, Z.S.; Kalinin, Y.A.; Makarov, V.A. Evolution of native gold in exogenous conditions. *Russ. Geol. Geophys.* **2020**, *61*, 1244–1259. [CrossRef]

20. Chapman, R.J.; Leake, R.C.; Bond, D.P.G.; Stedra, V.; Fairgrieve, B. Chemical and Mineralogical Signatures of Gold Formed in Oxidizing Chloride Hydrothermal Systems and Their Significance within Populations of Placer Gold Grains Collected during Reconnaissance. *Econ. Geol.* **2009**, *104*, 563–585. [CrossRef]
21. Varajão, C.A.C.; Colin, F.; Vieillard, P.; Melfi, A.J.; Nahon, D. Early Weathering of Palladium Gold under Lateritic Conditions, Maquiné Mine, Minas Gerais, Brazil. *Appl. Geochem.* **2000**, *15*, 245–263. [CrossRef]
22. Olivo, G.R.; Gauthier, M.; Williams-Jones, A.E.; Levesque, M. The Au-Pd Mineralization at the Conceicao Iron Mine, Itabira District, Southern Sao Francisco Craton, Brazil: An Example of a Jacutinga-Type Deposit. *Econ. Geol.* **2001**, *96*, 61–74. [CrossRef]
23. Vasil'ev, V.I. Mineralogy of mercury. In *Native Metals and Their Solid Solutions Amalgamides, Arsenides, Antimonides, Tellurides and Selenides. Part 1*; Publishing House of SB RAS, Geo Department: Novosibirsk, Russia, 2004; p. 150. (In Russian)
24. Chudnenko, K.; Pal'yanova, G. Thermodynamic properties of Au-Hg binary solid solution. *Thermochim. Acta* **2013**, *566*, 175–180. [CrossRef]
25. Chudnenko, K.V.; Pal'yanova, G.A.; Anisimova, G.S.; Moskvitin, S.G. Ag-Au-Hg solid solutions and physicochemical models of their formation in nature (Kyuchyus deposit as an example). *Appl. Geochem.* **2015**, *55*, 138–151. [CrossRef]
26. Okamoto, H.; Massalski, T.B. The Au-Pd (gold-palladium) system. *Bull. Alloy Phase Diagr.* **1985**, *6*, 229–235. [CrossRef]
27. Okamoto, H.; Massalski, T.B. The Au-Hg (gold-mercury) system. *Bull. Alloy Phase Diagr.* **1989**, *10*, 50–58. [CrossRef]
28. Guminski, C. The Hg-Pd (mercury-palladium) system. *Bull. Alloy Phase Diagr.* **1990**, *11*, 22–26. [CrossRef]
29. Sluiter, M.H.F.; Colinet, C.; Pasturel, A. Ab initio calculation of the phase stability in Au-Pd and Ag-Pt alloys. *Phys. Rev. B* **2006**, *73*, 174204. [CrossRef]
30. Murzin, V.V.; Kudryavtsev, V.; Berzon, R.; Sustavov, S.; Malyugin, A. New data on the instability of natural solid solutions of the system Ag-Au-Cu at <350 °C. *Dokl. Akad. Nauk SSSR* **1983**, *269*, 723–724. (In Russian)
31. Knipe, S.W.; Fleet, M.E. Gold-copper alloy minerals from the Kerr Mine, Ontario. *Can. Miner.* **1997**, *35*, 573–586.
32. Knight, J.; Leitch, C.H. Phase relations in the system Au-Cu-Ag at low temperatures, based on natural Assemblages. *Can. Miner.* **2001**, *39*, 889–905. [CrossRef]
33. Onishchenko, S.A.; Kuznetsov, S.K. Exsolution in the Au-Ag-Cu System in a Gold-Rich Area. *Int. Geochem.* **2022**, *60*, 657–671. [CrossRef]
34. Fleet, M.E.; de Almeida, C.M.; Angeli, N. Botryoidal platinum, palladium and potarite from the Bom Sucesso stream, Minas Gerais, Brazil: Compositional zoning and origin. *Can. Miner.* **2002**, *40*, 341–355. [CrossRef]
35. Makvandi, S.; Pagé, P.; Tremblay, J.; Girard, R. Exploration for platinum-group minerals in Till: A new approach to the recovery, counting, mineral identification and chemical characterization. *Minerals* **2021**, *11*, 264. [CrossRef]
36. Sidorov, E.G. Platinopossibility of Basic-Hyperbasic Complexes of the Koryak-Kamchatka Region. Ph.D. Thesis, Institute of Volcanology and Seismology FEB RAS, Petropavlovsk-Kamchatsky, Russia, 2009. (In Russian).
37. Sidorov, E.G.; Kutyrev, A.V.; Zhitova, E.S.; Chubarov, V.M.; Khanin, D.A. Origin of platinum-group mineral assemblages from placers in rivers Draining from the Ural-Alaskan type Itchayvayamsky ultramafics, Far East Russia. *Can. Miner.* **2019**, *57*, 91–104. [CrossRef]
38. Cabral, A.R.; Tupinambá, M.; Lehmann, B.; Kwitko-Ribeiro, R.; Vymazalová, A. Arborescent Palladiferous Gold and Empirical Au_2Pd and Au_3Pd in Alluvium from Southern Serra Do Espinhaço, Brazil. *Neues Jahrb. Mineral.* **2008**, *184*, 329–336. [CrossRef] [PubMed]
39. Cabral, A.R.; Vymazalová, A.; Lehmann, B.; Tupinambá, M.; Haloda, J.; Laufek, F.; Vlček, V.; Kwitko-Ribeiro, R. Poorly Crystalline Pd-Hg-Au Intermetallic Compounds from Córrego Bom Sucesso, Southern Serra Do Espinhaço, Brazil. *Eur. J. Mineral.* **2009**, *21*, 811–816. [CrossRef]
40. Cabral, A.R.; Lehmann, B.; Kwitko, R.; Costa, C.C. The Serra Pelada Au-Pd-Pt deposit, Carajás mineral province, northern Brazil: Reconnaissance mineralogy and chemistry of very high grade palladian gold mineralization. *Econ. Geol.* **2002**, *97*, 1127–1138. [CrossRef]
41. Shepherd, T.J.; Bouch, J.E.; Gunn, A.G.; McKervey, J.A.; Naden, J.; Scrivener, R.C.; Styles, M.T.; Large, D.E. Permo-Triassic Unconformity-Related Au-Pd Mineralisation, South Devon, UK: New Insights and the European Perspective. *Miner. Depos.* **2005**, *40*, 24–44. [CrossRef]
42. Cai, W.; Gao, T.; Hong, H.; Sun, J. Applications of Gold Nanoparticles in Cancer Nanotechnology. *Nanotechnol. Sci. Appl.* **2008**, *1*, 17–32. [CrossRef]
43. Nugraha, M.; Tsai, M.-C.; Su, W.-N.; Chou, H.-L.; Hwang, B.J. Descriptor Study by Density Functional Theory Analysis for the Direct Synthesis of Hydrogen Peroxide Using Palladium-Gold and Palladium-Mercury Alloy Catalysts. *Mol. Syst. Des. Eng.* **2018**, *3*, 896–907. [CrossRef]
44. Nikolaeva, L.A.; Nekrasova, A.N.; Milyaev, S.A.; Yablokova, S.V.; Gavrilov, A.M. Geochemistry of native gold from deposits of various types. *Geol. Ore Deposit.* **2013**, *55*, 176–184. [CrossRef]
45. Liu, H.; Beaudoin, G. Geochemical signatures in native gold derived from Au-bearing ore deposits. *Ore Geol. Rev.* **2021**, *132*, 104066. [CrossRef]
46. Chapman, R.J.; Moles, N.R.; Bluemel, B.; Walshaw, R.D. Detrital gold as an indicator mineral. In *Recent Advances in Understanding Gold Deposits: From Orogeny to Alluvium*; Torvela, T.M., Chapman, R.J., Lambert-Smith, J., Eds.; Geological Society Publications: London, UK, 2022; Volume 516, pp. 313–336. [CrossRef]

47. Nikiforova, Z. Internal Structures of Placer Gold as an Indicator of Endogenous and Exogenous Processes. *Minerals* **2023**, *13*, 68. [CrossRef]
48. McClenaghan, M.B.; Cabri, L.J. Review of gold and platinum group element (PGE) indicator minerals methods for surficial sediment sampling. *Geochem. Explor. Environ. Anal.* **2011**, *11*, 251–263. [CrossRef]
49. Oberthür, T. The Fate of Platinum-Group Minerals in the Exogenic Environment—From Sulfide Ores via Oxidized Ores into Placers: Case Studies Bushveld Complex, South Africa, and Great Dyke, Zimbabwe. *Minerals* **2018**, *8*, 581. [CrossRef]
50. Konstantinovskaia, E.A. Arc-Continent Collision and Subduction Reversal in the Cenozoic Evolution of the Northwest Pacific: An Example from Kamchatka (NE Russia). *Tectonophysics* **2001**, *333*, 75–94. [CrossRef]
51. Batanova, V.G.; Astrakhantsev, O.V. Island arc mafic–ultramafic plutonic complexes of North Kamchatka. In Proceedings of the 29th International Geological Congress, Kyoto, Japan, 24 August–3 September 1992; Ishiwatari, A., Malpas, J., Ishizuka, H., Eds.; CRC Press: Utrecht, The Netherlands, 1994; pp. 129–143.
52. Batanova, V.G.; Pertsev, A.N.; Kamenetsky, V.S.; Ariskin, A.A.; Mochalov, A.G.; Sobolev, A. V Crustal Evolution of Island-Arc Ultramafic Magma: Galmoenan Pyroxenite-Dunite Plutonic Complex, Koryak Highland (Far East Russia). *J. Petrol.* **2005**, *46*, 1345–1366. [CrossRef]
53. Yakubovich, O.; Kutyrev, A.; Sidorov, E.; Travin, A. 190Pt-4He dating of platinum mineralization in Ural-Alaskan-type complexes in the Kamchatka region: Evidence for remobilization of platinum-group elements. *Miner. Depos.* **2022**, *57*, 743–758. [CrossRef]
54. Razumny, A.V.; Sidorov, E.G.; Sandimirova, E.I. Copper-Gold-Palladuim Mineralization in the Concentric Zoned Massifs of the Koryaksky Upland. *Vestn. KRAUNTS Earth Sci.* **2004**, *1*, 75–80. (In Russian)
55. Britvin, S.N.; Dolivo-Dobrovolsky, D.V.; Krzhizhanovskaya, M.G. Software for processing the X-ray powder diffraction data obtained from the curved image plate detector of Rigaku RAXIS Rapid II diffractometer. *Zap. Ross. Mineral. Obs.* **2017**, *146*, 104–107, (In Russian with English abs.).
56. Borisov, A.V. Geological and Genetic Features of Au-Pd-REE Ore Occurrences in the Maldy-Nyrd Ridge (Subpolar Urals). Ph.D. Thesis, Institute of Geology of Ore Deposits, Petrography, Mineralogy and Geochemistry (IGEM) RAS, Moscow, Russia, 2005. (In Russian).
57. Orsoev, D.A.; Rudashevsky, N.S.; Kretser, Y.L.; Konnikov, E.G. Precious Metal Mineralization in Low-Sulfide Ores of the Ioko-Dovyren Layered Massif, Northern Baikal Region. *Dokl. Earth Sci.* **2003**, *390*, 545–549.
58. Spiridonov, E.M.; Orsoev, D.A.; Ariskin, A.A.; Nikolaev, G.S.; Kislov, E.V.; Korotaeva, N.N.; Yapaskurt, V.O. Hg-and Cd-Bearing Pd, Pt, Au, and Ag Minerals in Sulfide-Bearing Mafic and Ultramafic Rocks of the Yoko-Dovyren Intrusion in the Baikalides of the Northern Baikal Area. *Geochem. Int.* **2019**, *57*, 42–55. [CrossRef]
59. Cabral, A.R.; Lehmann, B.; Kwitko-Ribeiro, R.; Costa, C.H.C. Palladium and Platinum Minerals from the Serra Pelada Au-Pd-Pt Deposit, Carajás Mineral Province, Northern Brazil. *Can. Miner.* **2002**, *40*, 1451–1463. [CrossRef]
60. Cabral, A.R.; Lehmann, B. A Two-Stage Process of Native Palladium Formation at Low Temperatures: Evidence from a Palladian Gold Nugget (Gongo Soco Iron Ore Mine, Minas Gerais, Brazil). *Miner. Mag.* **2003**, *67*, 453–463. [CrossRef]
61. Okrugin, A.; Gerasimov, B. Paragenetic Association of Platinum and Gold Minerals in Placers of the Anabar River in the Northeast of the Siberian Platform. *Minerals* **2023**, *13*, 96. [CrossRef]
62. Wierchowiec, J.; Mikulski, S.Z.; Zieliński, K. Supergene Gold Mineralization from Exploited Placer Deposits at Dziwiszów in the Sudetes (NE Bohemian Massif, SW Poland). *Ore Geol. Rev.* **2021**, *131*, 104049. [CrossRef]
63. Wierchowiec, J.; Zieliński, K. Origin of Placer Gold and Other Heavy Minerals from Fluvial Cenozoic Sediments in Close Proximity to Rote Fäule-Related Au Mineralisation in the North Sudetic Trough, SW Poland. *Geol. Q.* **2017**, *61*, 62–80. [CrossRef]
64. Chapman, R.J.; Mileham, T.J.; Allan, M.M.; Mortensen, J.K. A distinctive Pd-Hg signature in detrital gold derived from alkali Cu-Au porphyry systems. *Ore Geol. Rev.* **2017**, *83*, 84–102. [CrossRef]
65. Moreno, T.; Prichard, H.M.; Lunar, R.; Monterrubio, S.; Fisher, P. Formation of a Secondary Platinum-Group Mineral Assemblage in Chromitites from the Herbeira Ultramafic Massif in Cabo Ortegal, NW Spain. *Eur. J. Mineral.* **1999**, *11*, 363–378. [CrossRef]
66. Bozhko, E.N. To the question about sources of gold-platinoid mineralization of the structural formational zone Matagania-Sigiri (Guinea, West Africa). *Bull. Voronezh Univer.* **2005**, *1*, 193–203. (In Russian)
67. Davey, W.P. Precision Measurements of the Lattice Constants of Twelve Common Metals. *Phys. Rev.* **1925**, *25*, 753. [CrossRef]
68. Nagasawa, A.; Matsuo, Y.; Kakinoki, J. Ordered Alloys of Gold-Palladium System. I. Electron Diffraction Study of Evaporated Au_3Pd Films. *J. Phys. Soc. Jpn.* **1965**, *20*, 1881–1885. [CrossRef]
69. Nagasawa, A. Electron Diffraction Study on Ordered Alloys of the Copper-Gold-Palladium System I. Ordered Alloys of Cu_2AuPd and $CuAu_2Pd$. *J. Phys. Soc. Jpn.* **1966**, *21*, 955–960. [CrossRef]
70. Van Sande, M.; van Tendeloo, G.; van Landuyt, J.; Amelinckx, S. A Study by Transmission Electron Microscopy and Diffraction of the Ternary Alloy System Cu_2AuPd. *Phys. Status Solidi* **1978**, *45*, 553–558. [CrossRef]
71. Okrugin, A.V.; Mazur, A.B.; Zemnukhov, A.L.; Popkov, P.A.; Sleptsov, S.V. The Palladium Gold-PGM Association in the Anabar River Basin, NE Part of Siberian Platform, Russia. *Otechestvennaya Geol.* **2009**, *5*, 3–10. (In Russian)
72. Polyakov, G.V.; Tolstykh, N.D.; Mekhonoshin, A.E.; Izokh, M.Y.; Podlipskii, D.A.; Orsoev, D.A.; Kolotilina, T.B. Ultramafic-Mafic Igneous Complexes of the Precambrian East Siberian Metallogenic Province (Southern Framing of the Siberian Craton): Age, Composition, Origin, and Ore Potential. *Russ. Geol. Geophys.* **2013**, *54*, 1689–1704. [CrossRef]
73. Hourigan, J.K.; Brandon, M.T.; Soloviev, A.V.; Kirmasov, A.B.; Garver, J.I.; Stevenson, J.; Reiners, P.W. Eocene Arc-Continent Collision and Crustal Consolidation in Kamchatka, Russian Far East. *Am. J. Sci.* **2009**, *309*, 333–396. [CrossRef]

74. Plotinskaya, O.Y.; Grabezhev, A.I.; Tessalina, S.; Seltmann, R.; Groznova, E.O.; Abramov, S.S. Porphyry Deposits of the Urals: Geological Framework and Metallogeny. *Ore Geol. Rev.* **2017**, *85*, 153–173. [CrossRef]
75. Awdankiewicz, M.; Kryza, R.; Szczepara, N. Timing of Post-Collisional Volcanism in the Eastern Part of the Variscan Belt: Constraints from SHRIMP Zircon Dating of Permian Rhyolites in the North-Sudetic Basin (SW Poland). *Geol. Mag.* **2014**, *151*, 611–628. [CrossRef]
76. Oplustil, S.; Cleal, C.J. A Comparative Analysis of Some Late Carboniferous Basins of Variscan Europe. *Geol. Mag.* **2007**, *144*, 417–448. [CrossRef]
77. Markwitz, V.; Hein, K.A.A.; Jessell, M.W.; Miller, J. Metallogenic Portfolio of the West Africa Craton. *Ore Geol. Rev.* **2016**, *78*, 558–563. [CrossRef]
78. Heilbron, M.; Cordani, U.G.; Alkmim, F.F. The São Francisco craton and its margins. In *São Francisco Craton 2017, Eastern Brazil*; Springer: Cham, Switzerland, 2017; pp. 3–13.
79. Berni, G.V.; Heinrich, C.A.; Lobato, L.M.; Wall, V. Ore Mineralogy of the Serra Pelada Au-Pd-Pt Deposit, Carajás, Brazil and Implications for Ore-Forming Processes. *Miner. Depos.* **2016**, *51*, 781–795. [CrossRef]
80. Clark, A.M.; Criddle, A.J. Palladium Minerals from Hope's Nose, Torquay, Devon. *Mineral. Mag.* **1982**, *46*, 371–377. [CrossRef]
81. Craw, D.; MacKenzie, D.; Grieve, P. Supergene Gold Mobility in Orogenic Gold Deposits, Otago Schist, New Zealand. *New Zeal. J. Geol. Geophys.* **2015**, *58*, 123–136. [CrossRef]
82. Webster, J.G. The Solubility of Gold and Silver in the System Au-Ag-S-O_2-H_2O at 25 °C and 1 Atm. *Geochim. Cosmochim. Acta* **1986**, *50*, 1837–1845. [CrossRef]
83. Atencio, D. *Type Mineralogy of Brazil*; Clube de Autores: Sao Paulo, Brazil, 2021.
84. Makeev, A.B.; Filippov, V.N. Metallic films on natural diamond crystals (Ichet'yu Deposit, Middle Timan). *Dokl. Earth Sci.* **1999**, *369*, 1161–1165. (In Russian)
85. Smith, D.G.W.; Nickel, E.H. A System of Codification for Unnamed Minerals: Report of the Subcommittee for Unnamed Minerals of the IMA Commission on New Minerals, Nomenclature and Classification. *Can. Mineral.* **2007**, *45*, 983–990. [CrossRef]
86. Warr, L.N. IMA-CNMNC Approved Mineral Symbols. *Mineral. Mag.* **2021**, *85*, 291–320. [CrossRef]

minerals

Review

Palladian Gold: Chemical Composition, Minerals in Association, and Physicochemical Conditions of Formation at Different Types of Gold Deposits

Galina A. Palyanova [1,*], Pavel S. Zhegunov [2], Tatiana V. Beliaeva [1], Valery V. Murzin [3], Andrey A. Borovikov [1] and Nikolay A. Goryachev [4]

[1] Sobolev Institute of Geology and Mineralogy, Siberian Branch of Russian Academy of Sciences, Akademika Koptyuga Pr., 3, 630090 Novosibirsk, Russia; zhur0502@igm.nsc.ru (T.V.B.); borovikov.57@mail.ru (A.A.B.)
[2] Institute of Volcanology and Seismology, Far East Branch of Russian Academy of Sciences, Piipa Blvd. 9, 683006 Petropavlovsk-Kamchatsky, Russia; pavel.zhegunov@bk.ru
[3] Zavaritsky Institute of Geology and Geochemistry, Ural Branch of Russian Academy of Sciences, Akademika Vonsovskogo Street, 15, 620110 Ekaterinburg, Russia; murzin@igg.uran.ru
[4] North-East Interdisciplinary Scientific Research Institute, Far East Branch of the Russian Academy of Sciences, Portovaya Street, 16, 685000 Magadan, Russia; goryachev@neisri.ru
* Correspondence: palyan@igm.nsc.ru

Abstract: This paper reviews and summarizes the available information on the composition of palladian gold with various contents and sets of isomorphic impurities (Ag, Cu, Hg) at 50 deposits and ore occurrences with Au-Pd mineralization. It is revealed that Palladian gold is represented by the systems Au–Pd, Au–Pd–Hg, Au–Pd–Cu, and Au–Pd–Ag–Hg, but more frequently corresponds to Au–Pd–Ag, Au–Pd–Ag–Cu, and Au–Pd–Ag–Cu–Hg. Objects with palladian gold belong to different types of gold deposits and to the deposits at which the main components of ores are PGE, Cr, Cu, Ni, V, and Ti. We propose a classification of the types of deposits with palladian gold: (1) PGE ore deposits related to mafic–ultramafic magmatic complexes (two subtypes—(a) low-sulfide-grade (less than 2%–5% sulfides) Alaskan, and (b) high-sulfide-grade (more than 5% sulfides) Norilsk); (2) orogenic gold deposits (OG); (3) epithermal (porphyry) gold–copper deposits (EPGC); (4) iron oxide copper gold deposits (IOCG); (5) ferruginous quartzite deposits; (6) volcanic exhalation; and (7) gold-PGE placers of five subtypes corresponding to the types of 1–5 primary sources. Physicochemical conditions of the formation of palladian gold at some deposits of type 1 cover two areas—magmatic high-temperature and hydrothermal low-temperature. At the majority of deposits of types 2-4, its formation proceeds with the participation of hydrothermal fluids (300–60 °C) of various salinities (0.2–30 wt.% NaCl eq.). Palladian gold is mainly high-fineness (910‰–990‰), is less frequently medium-fineness, and contains Ag and Cu, but does not contain Hg at the deposits of types 1, 3, and 4. The only exception is the Au-Pd-Hg Itchayvayam ore occurrence (Kamchatka, Russia), for which two varieties of Pd,Hg-bearing native gold (fineness 816‰–960‰ and 580‰–660‰) are determined. Low-fineness palladian gold with the major content of Ag is typical of OG deposits. Medium-fineness palladian gold occurs at ferruginous quartzite deposits and in volcanic exhalations. Hg, Ag, Cu-bearing high-fineness palladian gold is present mainly in placer deposits (type 7). The most common minerals in association with palladian gold are arsenides, stibioarsenides, sulfides, stannides, bismuthides, tellurides, and selenides of Pd and Pt. These are typical of deposit types 1 and 7. The minerals of Au, Ag, and Cu (tetra-auricupride, aurostibite, chalcopyrite, bornite, chalcocite, eucairite, etc.) are in association with palladian gold at OG, EPGC, and IOCG deposits. Hg minerals (cinnabar, tiemannite, coloradoite, potarite) are at some deposits (types 1, 2, 7-1, 7-4). Cu, Fe, and Pd oxides (tenorite, hematite, magnetite, PdO, (Pd,Cu)O) and Fe and Pd hydroxides (goethite, (Fe,Pd)OOH) occur at the deposits of the 3, 4, and 7 groups and indicate the highly oxidizing conditions of ore formation. The most common minerals among host minerals are quartz and muscovite, including fuchsite (Cr-Ms), chlorite, albite, K-feldspar, hornblende, and carbonates (calcite, siderite, etc.). The fineness, content, and set of impurities in palladian gold and minerals in association with it reflect the mineralogy of Au-Pd ores and allow them to be used as indicators for the deposit types.

Citation: Palyanova, G.A.; Zhegunov, P.S.; Beliaeva, T.V.; Murzin, V.V.; Borovikov, A.A.; Goryachev, N.A. Palladian Gold: Chemical Composition, Minerals in Association, and Physicochemical Conditions of Formation at Different Types of Gold Deposits. *Minerals* **2023**, *13*, 1019. https://doi.org/10.3390/min13081019

Academic Editor: Huan Li

Received: 29 June 2023
Revised: 24 July 2023
Accepted: 28 July 2023
Published: 30 July 2023

Keywords: palladian gold; Pd,Cu,Ag,Hg-bearing gold; Au-Pd; Au-Pd-Ag; Au-Pd-Cu; Au-Pd-Hg; Au-Pd-Ag-Cu; Au-Pd-Ag-Hg; Au-Pd-Ag-Cu-Hg systems; fineness; minerals in association; PGE and gold deposits; T,P,X-conditions of ore formation; Au-Pd mineralization types

---

1. Introduction

Pure native gold without impurities (fineness 1000‰) is rare in nature. It contains some elements in the form of isomorphic (isostructural, i.e., included in the crystal structure without changing it) impurities and mineral microinclusions. Depending on their concentrat

ion, impurity-elements may be classified as major (>10 wt.%), minor (1–10 wt.%), and trace (<1 wt.%). Native gold with only silver impurity (argentian gold or Ag-bearing gold) is most widely spread at gold deposits [1–8]. Some review papers report data on the varieties of native gold with copper (cuprian gold), mercury (gold amalgam or mercury gold), and palladium (palladian gold) impurities [1,2,9–12]. On the basis of the occurrence and concentration, Petrovskaya (1973) [1] referred Pd to rare poorly studied impurities, Hg to third-rate, Cu to secondary, and Ag to major elements. A great number of new data on the content of impurities in native gold appeared over the past five decades. All these four elements, depending on their content, can be referred to as major, minor, and trace impurities.

Native gold often contains microinclusions of ore minerals, silicates, oxides, carbonates, etc. Special attention during recent years has been paid to the study of the impurities and minerals in paragenesis with native gold, since these characteristics are important informative signs of deposits with different types of mineralization [8,13–18]. Effective criteria for forecasting and searching for gold deposits can be provided by a study of the mineralogical–geochemical composition of native gold.

Native gold containing palladium was called «porpecite» after the place where it was first discovered in 1798 in the Porpez region of Brazil in the deposits of itabirites (layered quartzites with magnetite and hematite). The name "porpecite" was not approved by the CNMNC MMA, since the Pd-bearing gold or palladian gold (Au,Pd) is a structural analog and is considered as a Pd-rich chemical variety of native gold (www.mindat.org, accessed on 15 January 2022). Palladian gold was found to contain Ag [19,20], Cu [21,22], and Hg [23–25], whereas Fe, Ni, Pt, Rh, and other impurities are rare [21,26,27].

The purpose of this work is to analyze data on the composition and minerals in the association of palladian gold with impurities (Ag, Cu, and Hg) from different deposits, and to summarize the T,P,X-conditions for the formation of Au-Pd mineralization and to classify the deposits. Analysis of literature data and the results of our research show that the great number of existing deposits with palladian gold require classification. Earlier, reviews were made on the composition of Au-Pd mineralization at endogenous [28] and placer deposits with native gold containing Pd and Ag [20]. Since that time, much data on new deposits with palladian gold have been published ([11], and cited references). New data on the compositions of palladian gold and its minerals in the association have been recently obtained by the authors of this paper for such deposits as Chudnoe [29], Volkovskoe [30], Ozernoe [31] (Urals, Russia), and Bleida Far West (Morocco) [32], and for the placers and ore occurrences of the mafic–ultramafic Itchayvayam massif (Kamchatka, Russia) [33], which differ from other deposits. In this paper, we present a review of the composition of palladian gold of the systems Au–Pd, Au–Pd–Ag, Au–Pd–Hg, and Au–Pd–Cu and the more complex systems Au–Pd–Ag–Cu, Au–Pd–Ag–Hg, Au–Pd–Ag–Cu–Hg, etc.

The study of the composition of native gold is of great scientific and practical importance [14,34–38]. Native gold is a significant indicator mineral for Au-bearing deposits [12,14,39]. The composition of native gold is often used as a predictive tool in the study of the genesis of ore deposits and factors that determine the speciation of mineralization [12,40,41]. Chapman et al. (2021) [12], after analyzing the chemical composition of native gold, concluded that the concentrations of Cu, Hg, and Pd can be important indicators of different types of deposits. The level of concentrations and the ratios of various elements in individual gold particles can be responsible for the geochemical history and "prints" of ore-forming processes. Data on the variations in the concentrations of Pd and other elements (Ag, Cu, Hg) in native gold and specific features of the minerals in association and their occurrence in different types of rocks and ore-forming systems can be used for solving geological tasks that arise when promising gold-bearing areas are under study.

2. Geological Setting and Deposits with Palladian Gold

Palladium gold is present in the deposits of various geneses—magmatic, hydrothermal, and sedimentary-exogenous. The latter are the result of the transformation and differentiation of mineral matter of endogenous origin (magmatic or hydrothermal). Palladian gold often occurs in the placers, the sources for many of which have not been determined.

The deposits and ore occurrences with palladian gold are located in Canada, Brazil, England, Scotland, Poland, Spain, USA, Morocco, Guinea, and other countries [11,20,28]. The mindat.org database contains data on findings of palladian gold in Belgium, Germany, and Venezuela. More than twenty deposits and ore occurrences with palladian gold are known in Russia [11]. The geographical spread and geological–structural location of deposits with palladian gold are shown in Figure 1. They occur in different geological and dynamic settings—terranes, orogenic belts, and shields. According to dating assessments, the processes of formation of Au-Pd mineralization took place in the Archean, Proterozoic, Paleozoic, Mesozoic, and Cenozoic.

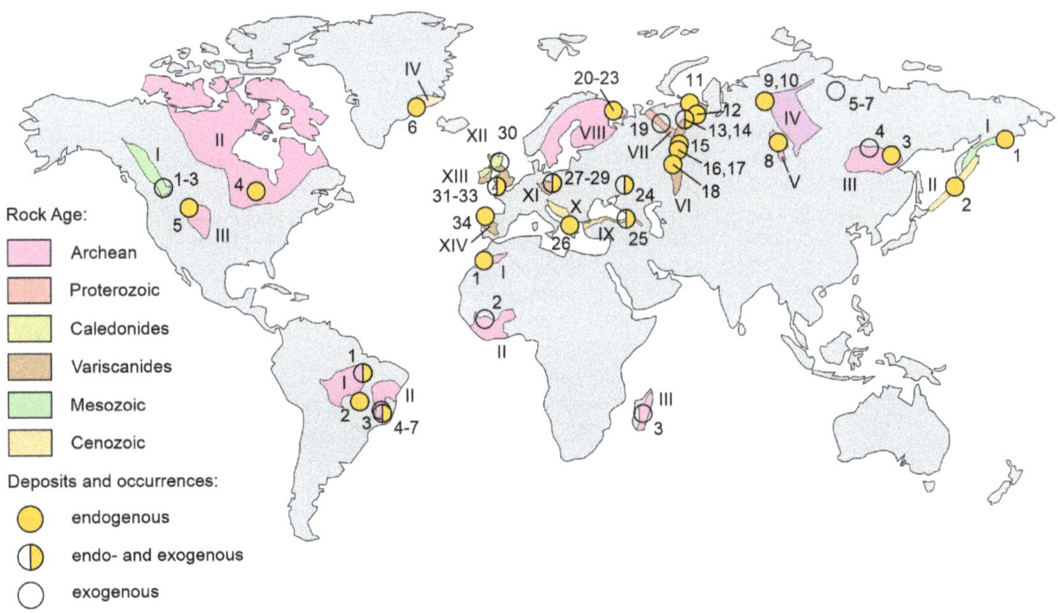

Figure 1. Location of deposits and ore occurrences with palladian gold, and other Au-Pd-Ag-Cu-Hg phases (Arabic numerals indicate deposits, Roman numerals indicate the age of main ore-hosted geological structures; the description is shown below). *Eurasia*: Russia 1–24—1. Itchayvayam complex [33,42]; 2. Ebeko volcano [43]; 3. Konder massif [44,45]; 4. Inagli Pt–Au placer deposit [46]; 5–7. placers of

northern Yakutia [47,48]: 5. Mayat river, 6. Bol'shaya Kuonamka river, 7. Kuoyka river; 8. Uderei [49]; 9,10. Norilsk deposits: 9. Norilsk-1, 10. Talnakh [27,50]; 11. Krutoe, Pay-Khoy Range [51]; 12–19. Ural deposits: 12. Ozernoe [31], 13. Chudnoe and placer Alkes-Voss [29,52,53], 14. Nesterovskoe [53], 15. Serebryansky Kamen [54], 16. Baronskoe [55], 17. Volkovskoe [30], 18. Khamitovskoe [56], 19. Ichet'yu, Timan [57]; 20–23—Fedorovo-Pansky intrusive complex deposits [58]: 20. V. Chuarva, 21. C. Kamennik, 22. Y. Peshempakhk, 23. Fedorova Tundra; 24. Lebedinskoye [59]; 25. Chorokh river, Artvin district, Turkey [60]; 26. Korydallos, Greece [61,62]; 27–29. deposits of the Sudetes Mountains: 27. Zechstein [63], 28. Zimnik Creek [64], 29. Dziwiszów [65]; 30. Lammermuir Hills, Scotland [20]; 31–33. deposits of the south of England: 31. Hope's Nose [66], 32. Dart river [20], 33. Brownstone [20]; 34. Herbeira massif, Spain [67]. Structures: I—Olyutorsky arc terrane, II—Kuril-Kamchatka volcanic belt, III—Aldan shield, IV—Siberian trap province, V—Yenisei ridge, VI—Ural accretionary fold belt (Variscan orogenic belt), VII—Timan folding system, VIII—Baltic Shield, IX—Eastern Pontides-Lesser Caucasus fold-thrust belt, X—Ophiolite belt of Greece, XI—Hercynides of the Sudetes (Variscan orogenic belt), XII—Caledonides of Scotland, XIII—Variscan orogenic belt of Britain, XIV—Central Iberian zone of the Iberian massif (Variscan orogenic belt). *Africa:* 1. Bleida Far West, Morocco [32,68]; 2. Mataganya-Siguiri zone, Guinea [69]; 3. Ambositra District, Madagascar [70]. Structures: I—Anti-Atlas fold belt, II—Kenema-Man shield and Baoule-Mossi domain (West African Craton), III—Precambrian shield of Madagascar. *North America:* 1–3. Deposits of Canada: 1. Friday Creek [71]; 2. Whipsaw Creek [71]; 3. Similkameen river [71]; 4. Marathon [21,72]; 5. Skaergaard massif, Greenland, Denmark [73]; 6. Stillwater, USA [74]. Structures: I—Intermontane terranes of British Columbia (North American Cordilleran orogenic belt), II—Wyoming province, III—Canadian shield, IV—Paleogene platobasalts of eastern Greenland. *South America:* Deposits of Brazil—1. Serra Pelada [25]; 2. Cedrolina Chromitite (Goiás State) [75]; 3. Corrego Bom [76]; 4–7. Itabira deposits: 4. Caue iron mine [26,77], 5. Gongo Soko [25,78], 6. Maquine [28], 7. Conceichao mine [79]. Structures: I—Central Brazilian shield, II—São Francisco craton.

3. Methods

Impurity elements in native gold are determined using chemical, spectral, electron probe microanalyses (EPMA), inductively coupled plasma–mass spectrometry (ICP-MS), and laser ablation–inductively coupled plasma–mass spectrometry (LA-ICP-MS). The chemical and spectral methods as well as ICP-MS require a considerable amount of native gold and its purification from other minerals. Systematic studies of the typomorphism of native gold from different territories of Russia and the Commonwealth of Independent States, with the application of semi-quantitative ICP-MS analysis carried out in Russia [14,80], showed the presence of 70 chemical elements in it. The drawback of the chemical, spectral, and ICP-MS methods is the inability to separate isomorphic impurities and mineral microinclusions in native gold.

EPMA is used for quantitative analysis of the elemental composition of native gold at a micrometer scale. The detection sensitivity of many elements in the local electron raster microprobe analysis is limited by concentrations no less than 0.1–0.01 wt.%. A more complex composition of native gold can be detected by LA-ICP-MS, the sensitivity of which is considerably higher than that of EPMA. With the appearance of the LA-ICP-MS technique [81] for determining the composition of microelements in solid samples, data on micro- and trace elements in native gold were obtained [35,82–90]. The laser ablation–inductively coupled plasma–mass spectrometer has high sensitivity and low detection levels required for measuring microelements in amounts up to ppb (1 per billion). One of the advantages of this technique is the possibility of cleaning the sample surface. LA-ICP-MS is widely used by foreign scientists to study the composition of gold and has not been used for studying gold in Russia yet. According to Chapman et al. (2023) [91], limits of detection (LODs) for Pd and Cu were typically around 900 ppm and 200 ppm, respectively, whilst the avoidance of spectral interference between the HgM_{α} and AuM_{β} X-rays necessitated using the HgM_{β} X-ray line and the associated higher detection limit of 3000 ppm.

In the majority of studies, the chemical composition of native gold was investigated using EPMA, and less frequently, LA-ICP-MS. The concentration of elements Ag, Cu, Pd, and Hg in native gold is commonly measured using EPMA. The spectrum accumulation time in electron probe microanalysis is 20 or more—60–80 s. Macro-impurities in native gold are determined using SEM/EDS, which allows one to determine Ag, Cu, Pd, and Hg with detection limits of 100–1000 ppm.

We analyzed the samples of ores and individual gold grains with Pd-bearing gold from the Chudnoe, Volkovskoe, and Ozernoe deposits (Urals, Russia) and from placers and ore occurrences of the mafic–ultramafic Itchayvayam massif (Kamchatka, Russia) and Au-Pd Bleida Far West deposit (Morocco). Chemical analyses of native gold and other minerals were conducted at the Analytical Center for Multi-elemental and Isotope Research in the IGM SB RAS (Novosibirsk, Russia) by electron probe microanalysis (EPMA) using a MIRA 3 LMU scanning electron microscope (Tescan Orsay Holding, Brno, Czech Republic) equipped with an X-ray energy-dispersive spectrometer (EDS) AZtec Energy XMax-50 (Oxford Instruments Nanoanalysis, Oxford, UK) (analysts Dr. N. Karmanov, M. Khlestov). The composition of native gold was studied at the following parameters: accelerating voltage was 20 kV and live spectrum acquisition time was 60 s (total area of spectra ~10^6 counts). The following X-rays were selected: K series for Fe, Cu, and As and L series for Pd, Ag, Sb, Au, and Hg. We used pure metals (Fe, Cu, Pd, Ag, Au) and InAs for As and HgTe for Hg as the standards. The detection limits (in wt.%) were 0.1 Fe, 0.15 Cu, 0.25 Pd, Ag, Sb, 0.3 As, 0.6 Au, and 0.8 Hg. The error in determining the main components with contents higher than 10 wt.% did not exceed 1 relative (rel.) %, and when the content of components ranged from 2 to 10 wt.%, the error was no higher than 6–8 rel. %. Close to the limit of detection, the error was 15–20 rel.%. In some cases, the spectrum acquisition time increased to 120 s, and the lower limits of determined contents and the random error of the analysis decreased about 1.4 times. This review is primarily based on the EPMA results. A significant amount of EPMA data was taken from the published literature.

4. Results

4.1. Composition, Fineness, and Mineral Associations of Palladian Gold

Palladian gold rarely consists of the binary Au-Pd system; it most frequently contains the ternary Au-Pd-Ag, Au-Pd-Cu, and Au-Pd-Hg, and even more complex systems Au-Pd-Ag-Cu, Au-Pd-Ag-Hg, and Au-Pd-Ag-Cu-Hg. Several varieties of palladian gold were found at many deposits. Along with palladian gold, Ag-bearing gold and other varieties of native gold are present at these deposits. Below, we present data on the composition of palladian gold in the binary (Au-Pd), ternary (Au-Pd-Ag, Au-Pd-Cu, Au-Pd-Hg), quaternary (Au-Pd-Ag-Hg, Au–Pd–Ag–Cu), and five-element (Au–Pd–Ag–Cu–Hg) systems, as well as minerals in association and examples of deposits.

According to the "50 mole % rule" for the mineral nomenclature recommended by the Commission on New Minerals and Mineral Names (CNMMN) of the International Mineralogical Association (IMA) [92], the formula composition of palladian gold can be represented as $Au_{1-x}Pd_x$, where $x < 0.5$ atomic part, when calculating the formula for 1 atomic unit. Au-Pd solid solutions with Ag, Cu, and Hg impurities can be present as $Au_{1-x-y-z-d}Pd_xAg_yCu_zHg_d$, where $0 < x$ (Pd atomic part in native gold) < 0.5; y, z, d in total $\leq 0.5 - x$ atomic parts; and $1 - x - y - z - d > x, y, z, d$. The fineness of native gold is calculated by the equation (1000*Au/(Au + Pd + Me_1 + Me_2 + and so on), where Au, Pd, and other metals (Ag, Cu, Hg, Pt, Ni, Fe) are expressed as wt.%).

Terms used to characterize the amounts of gold in Au-Ag solid solutions—very high-fineness 999‰–951‰, high-fineness 950‰–900‰, medium-fineness 899‰–800‰, relatively low-fineness 799‰–700‰, and low-fineness gold 699‰–400‰; kustelite 399‰–100‰; silver <100‰—were proposed by Petrovskaya (1973) [1]. We used the same terms to characterize palladian gold. The compositions of solid solutions depend on the amounts of impurities of palladium, copper, mercury, and other metals, and the fineness intervals corresponding to native gold can change, which is determined by the atomic weight and the amount of the impurity element. According to the "50 mole % rule", the fineness of palladian gold can be more than 649‰ ($Au_{0.5}Pd_{0.5}$), cuprian gold > 756‰ ($Au_{0.5}Cu_{0.5}$), argentian gold > 630‰ ($Au_{0.5}Ag_{0.5}$), and gold amalgam > 495‰ ($Au_{0.5}Hg_{0.5}$).

Mineral abbreviations used in the tables are taken from [93] albite (Ab), allanite (Aln), altaite (Alt, PbTe), anilite (Ani, Cu_7S_4), anyuiite (Any, $AuPb_2$), apatite (Ap, $Ca_5(PO_4)_3(F,Cl,OH)$), arsenopyrite (Apy, FeAsS), atheneite (Ah, $(Pd,Hg)_3As$), atokite (Ato, $(Pd,Pt)_3Sn$), auricupride (Auc, Cu_3Au), aurostibite (Ausb, $AuSb_2$), biotite (Bt), bohdanowiczite (Boh, AgBiSe), bornite (Bn, Cu_5FeS_4), braggite (Bg, (Pt,Pd,Ni)S), burangaite (Brg, $(Na,Ca)(Fe^{2+},Mg)Al_5(PO_4)_4(OH,O)_6 \cdot 2H_2O$), calcite (Cal, $CaCO_3$), carbonates (Carb), cerussite (Cer, $PbCO_3$), chalcocite (Cc, Cu_2S), chalcopyrite (Ccp, $CuFeS_2$), chlorargyrite (Cag, AgCl), chlorites (Chl), clinopyroxenes (Cpx), chrisstanleyite (Csl, $Ag_2Pd_3Se_4$), cinnabar (Cin, HgS), clausthalite (Cth, PbSe), coloradoite (Clr, HgTe), cooperite (Cpe, (Pt,Pd,Ni)S), empressite (Eps, AgTe), epidote (Ep), eucairite (Eca, AgCuSe), feldspar (Fsp), fischessérite (Fis, Ag_3AuSe_2), fuchsite (Cr-Ms), galena (Gn, PbS), goethite (Gth), gold (Au), gypsum (Gp), hematite (Hem, Fe_2O_3), hessite (Hes, Ag_2Te), isoferroplatinum (Ifpt, Pt_3Fe), isomertieite (Ism, $Pd_{11}Sb_2As_2$); kaolinite (Kln), keithconnite (Kei, $Pd_{3-x}Te$(x = 0.14 to 0.43)), K-feldspar (Kfs), kotulskite (Ktu, PdTe), magnetite (Mag, Fe_3O_4), merenskyite (Mrk, $(Pd,Pt)(Te,Bi)_2$), mertieite-I (Met-I, $Pd_{11}(Sb,As)_4$), mertieite-II (Met-II, $Pd_8(Sb,As)_3$), monazite (Mnz, $REE(PO_4)$), moncheite (Mon, $(Pt,Pd)(Te,Bi)_2$), muscovite (Ms), naumannite (Nau, Ag_2Se), oosterboschite (Oos, $(Pd,Cu)_7Se_3$), padmaite (Pdm, PdBiSe), palladium (Pd), palladoarsenide (Pda, Pd_2As), palladseite (Pds, $Pd_{17}Se_{15}$), paolovite (Plv, Pd_2Sn), pentlandite (Pn, $(Fe,Ni)_9S_8$), perite (Pe, $PbBiClO_2$), platinum (Pt), plumbum (Pb), potarite (Ptr, PdHg), pyracmonite (Pyr, $(NH_4)_3Fe(SO_4)_3$), pyrite (Py, FeS_2), pyrrhotite (Pyh, $Fe_{1-x}S$ (x = 0 to 0.17)), quartz (Qz), silver (Ag), skaergaardite (Skg, PdCu), sobolevskite (Sov, Pd(Bi,Te)), sperrylite (Spy, $PtAs_2$), stannopalladinite (Spdn, $(Pd,Cu)_3Sn_2$), stibiopalladinite (Stpdn, Pd_3Sb), sudburyite (Sdb, PdSb), sudovikovite (Svi, $PtSe_2$), sylvanite (Syv, $AgAuTe_4$), telluropalladinite (Tpdn, Pd_9Te_4), temagamite (Tem, Pd_3HgTe_3), tenorite (Tnr, CuO), tetra-auricupride (Taur, AuCu), tetraferroplatinum (Tfpt, PtFe), tiemannite (Tmn, HgSe), tourmaline (Tur), umangite (Um, Cu_3Se_2), vasilseverginite (Vas, $Cu_9O_4(AsO_4)_2(SO_4)_2$), vysokýite (Vys, $U^{4+}[AsO_2(OH)_2]_4 \cdot 4H_2O$), and vysotskite (Vsk, (Pd,Ni)S).

4.1.1. Palladian Gold and Au-Pd System

Palladian gold of composition $Au_{1-x}Pd_x$, where x ≤ 0.5 is atomic part Pd (<35.1 wt.%) without impurities, was discovered at several deposits and ore occurrences. As an example, Table 1 shows data for four objects: two endogenous—Krutoe ore occurrences in gabbro-dolerites of Pay-Khoy Ridge (Yugor peninsula, Arkhangelsk region, Russia) [51] and Uderei Au-Sb deposit (Yenisei Ridge, Russia) [49], and two placer ore occurrences—Lammermuir Hills (Scotland) [20] and the Chorokh river (Artvin district, Turkey) [60].

Table 1. The composition of palladian gold (in absence of other impurities), its fineness (N_{Au}), and minerals in association at endogenous and exogenous deposits.

Name of Deposit (Location)	N_{Au} ‰/Impurity wt.%	Minerals in Association	Ref.
	Endogenous deposits		
Krutoe ore occurrence (Yugor peninsula, Russia)	949–963/Pd 5.1, 3.7	Qz, Pyh, Ccp, Clr, Alt, Eps, Au-Ag (850–920), Au-Cu-Ag, Au, Taur, Sdb, Spy, Tsp, Hes	[51]
Uderei Au-Sb deposit (Yenisei Ridge, Russia)	970–920/Pd 3.3–7.9	Apy, Py, Au-Ag (760–980), Au, Ausb	[49]
Ebeko volcano (Kuriles, Russia)	787/Pd 21.3 (Au_2Pd)	S	[43]
	Exogenous deposits		
Lammermuir Hills (Scotland)	942/Pd 5.8	Carb, Apy, Gn, Pe, Au_3Cu, Taur, Ausb, Boh, Nau, Um, Tmn, Mrk	[20]
Chorokh river (Artvin district, Turkey)	649–724/Pd 26.5–38.5	Pd, Pd_4S	[60]

Palladian high-fineness gold (Pd content 3.7–5.1 wt.%, fineness 949‰–963‰) with inclusions of argentian gold (Ag content 7.47–15.3 wt.%, fineness 850‰–920‰, sizes from 2 to 100 μm) is intergrown with chalcopyrite in quartz veins and hornblende of gabbro-dolerites in the axial zone of the sill of the Pay-Khoy Ridge (Krutoe ore occurrence, Russia) [51]. Shaibekov et al. (2020) [51] reported that the early (magmatic) association of Au-Pd-Te mineralization is represented by the decomposition products of the ternary Ag—Au—Cu system (tetra-auricupride AuCu and Au-Ag-Cu phases containing Cu from 14.2 to 64.7 wt.% and Ag to 10.2 wt.%), antimonides and stibiotellurides of Pd, and arsenides of Pt (sudburyite, testibiopalladite, and sperrylite), whereas the late (hydrothermal) association, by the minerals of binary Au-Pd and Au-Ag systems, is represented by tellurides of Hg, Pb, and Ag.

Palladian gold from the Uderei deposit contains from 3.3 to 7.9 wt.% Pd and is high-fineness (920‰–970‰) [49]. It occurs in the intergrowth with argentian gold (Ag content from 2.1 to 24.2 wt.%, 760‰–980‰) or pure metal (1000‰). Both varieties of native gold occur in association with arsenopyrite, pyrite, and aurostibite. Gold–antimony ores from this deposit were found to contain grains of zinkenite, chalcostibite, ullmannite, galena, sphalerite, and cinnabar [49].

Native gold from the placers of the Chorokh river (Artvin district, Turkey) contains high concentrations of Pd (26.5–38.5 wt.%, $Au_{0.58}Pd_{0.42}$–$Au_{0.45}Pd_{0.55}$) and, hence, is lower-fineness (649‰–724‰) [60]. In palladian gold, microinclusions of native palladium ($Pd_{0.95}Au_{0.05}$) were detected. The composition of Au-Pd solid solutions with Pd concentrations of more than 35.1 wt.% ($Au_{0.50}Pd_{0.50}$–$Pd_{0.55}Au_{0.45}$) should be attributed to native palladium.

At the ore occurrences of Lammermuir Hills (Scotland) [20], palladian gold contains up to 5.8 wt.% Pd ($Au_{0.9}Pd_{0.1}$, 942‰) and occurs in association with Au-Cu intermetallic compounds—tetra-auricupride and Au_3Cu without Pd impurities, as well as with aurostibite, bohdanowiczite, naumannite, umangite, tiemannite, merenskyite, pyrite, arsenopyrite, galena, and carbonates. Placer gold is heterogeneous and has the characteristics of endogenous and acquired exogenous genesis.

The set of minerals in association with palladian gold from the Uderei deposit (Russia) and placer ore occurrences of Lammermuir Hills (Scotland) are partially similar in that both objects contain aurostibite, arsenopyrite, galena, and argentian gold. Tetra-auricupride is present in paragenesis with palladian gold at the Krutoe ore occurrence (Russia) and the Lammermuir Hills deposit (Scotland), which suggests not only Au-Pd but also Cu-Pd mineralization.

It is worth noting the high concentrations of Pd (21.3 wt.%, Au_2Pd) in gold particles from gray fumaroles and in andesite lavas (Q) of Ebeko volcano (Russia) [43] (Table 1). Au_7Pd (928‰), Au_3Pd (847‰), and Au_2Pd (790‰) phases and Au-bearing palladium are present in the Au-Pd mineralization of the Serra Pelada deposit, Brazil [25]. The Au_3Pd phase as small inclusions was identified in chromite from the Cedrolina Chromitite (Goiás State, Brazil) [75]. Pure gold and Au_3Cu_3Ni alloy were also found at the boundary between chromite and talc or in the silicate matrix of the chromitite ores.

The reported results show that Pd concentrations in gold in the absence of other impurities vary in a wide range from minor (3.3 wt.%, 970‰) up to $Au_{0.50}Pd_{0.50}$ (35.1 wt.%, 649‰). The compositions of this gold are arranged on the Au-Pd side of the ternary diagram of Au-Pd-Ag (Figure 2). This diagram also shows the compositions of palladian gold with and without Ag, which are widespread at other deposits (there are 24 of them; see capture in Figure 2). Summarized literature data on the compositions of palladian gold with the presence of Ag at these deposits are presented in Section 4.1.2.

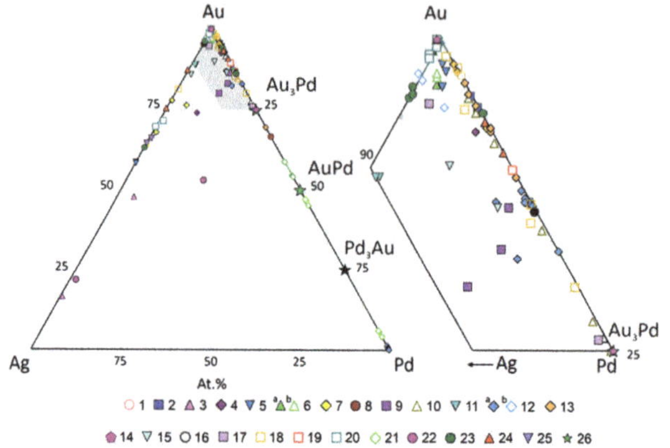

Figure 2. Ag-bearing palladian gold and other phases of Au-Ag-Pd system (in at.%) at the various deposits and ore occurrences: 1. Itchayvayam [33]; Norilsk deposits—2. Norilsk-1, 3. Talnakh, [22]; Ural deposits—4. Baronskoe [55], 5. Chudnoe [53], 6. Nesterovskoe [53]; 7—deposits of the Fedorovo-Pansky intrusive complex [58]; 8—Ebeko Volcano, Kuriles [43]; 9—Stillwater, Montana, USA [74]; 10—Corrego Bom Sucesso, Brazil [76]; 11—Gongo Soco iron mine, Brazil [25]; 12—Serra Pelada, Brazil [25]; 13—Caue iron mine, Brazil [77]; 14—Cedrolina Chromitite (Goiás State, Brazil) [75]; 15—Bleïda Far West, Morocco [68]; 16—Ambositra District, Madagascar [70]; 17—Hope's Nose, Devon, England [66]; 18—Brownstone, Devon, England [20]; 19—Lammermuir Hills, Scotland [20]; 20—placer gold of Zimnik Creek, Poland [64]; 21—Chorokh River, Artvin District, Turkey [60]; 22—Ozernoe [31]; 23—Uderei Au-Sb [49]; 24—Pay-Khoy [51]; 25—KMA (Lebedinskoye) [59]; 26—theoretical compositions of minerals corresponding to their ideal formulas. Shaded markers are endogenous, empty markers are exogenous deposits.

4.1.2. Pd,Ag-Bearing Gold and Au-Pd-Ag System

Palladian gold with Ag impurity (Ag-bearing palladian gold or Pd, Ag-bearing gold) covers the range of solid solutions from Au to $Au_{1-x-y}Pd_xAg_y$, where $y + x \leq 0.5$ atomic fraction, $x \neq 0$, $y \neq 0$, and $1 - x - y > x, y$. As an example, Table 2 contains data on five deposits and occurrences with Ag-bearing palladian gold.

Table 2. The composition of Ag-bearing palladian gold, its fineness (N_{Au}), and minerals in association at endogenous deposits.

Name of Deposit (Location)	N_{Au} ‰/Impurity wt.%	Minerals in Association	Ref.
Brownstone (England)	970–994 (core)/Pd 0.5–2.4, Ag 0.1–0.5 882–924 (border)/Pd 7.4–11.2, Ag 0.01–0.4	Cth, Tmn, Eca	[20]
Hope's Nose (England)	854–997/Pd 0.2–14.2, Ag 0.2–1.9	Csl, Cth, Fis, Cag, Oos, Cal, Cer	[20,66]
Stillwater (USA)	882–922/Pd 6.6–7.3, Ag 1.1–4.8	Vsk, Bg	[74]
Fedorova Tundra (Russia)	770–870/Ag 10.9–19.7, Pd < 2.76, Fe < 0.19	Plv, Pn	[58]
Lebedinskoye (Russia)	780–790/Ag 21.4–22.6, Pd < 0.15, Pt < 0.4, Ni < 0.05, Fe ≈ 1.2	-	[59]

At the Au-Pd Brownstone deposit (England), native gold is present in a vertical quartz-calcite vein hosted by Devonian slates [20] and is represented by two generations (Ag < 0.5 wt.%): high-fineness gold (970‰–994‰) with low content of Pd < 2.4 wt.% (core) and lower-fineness gold (882‰–924‰) with a higher content of Pd up to 11.2 wt.% (border), in association with clausthalite, tiemannite, and eucairite. The low contents of Ag 1.1–4.8 and 0.2–1.9 and the high contents of Pd ≈ 7 wt.% and up to 14.2 wt.% are typical of native gold at the Au-Pd Stillwater (USA) and Hope's Nose (England) deposits (Table 2), respectively. The trace and low contents of Pd < 0.15 and < 2.76 wt.% and high contents of Ag 21.4–22.6 wt.% and 10.9–19.7 are typical of native gold in ferruginous quartzites, accompanying metasomatites, and in the zones of increased sulfidation at the Au–Pd–Pt Lebedinskoye deposit [59] and in the ores of the Fedorova Tundra Pt-metal deposit (Russia) [58], respectively.

Minerals associated with Ag-bearing palladian gold in small calcite veins at the Cu-Ni-sulfide Hope's Nose (England) are chrisstanleyite, clausthalite, fischesserite, chlorargyrite, oosterboschite, calcite, and cerussite [74]. Vysotskite, braggite, paolovite, and pentlandite are associated with Ag-bearing palladian gold at the Stillwater (USA) and Fedorova Tundra (Russia) deposits, respectively. Minor impurities Pt < 0.4, Ni < 0.05, and Fe 0.2–1.3 wt.% are present in Pd,Ag-bearing gold at the deposits Lebedinskoye [59] and Fedorova Tundra (Russia) [58].

Ag-bearing palladian gold has been found at such deposits as Norilsk-1, Talnakh [22], Baronskoe [55], Chudnoe, Nesterovskoe [53], Ozernoe [31], and Krutoe [51]; deposits of the Fedorovo-Pansky intrusive complex [58] in Russia; Corrego Bom Sucesso, Gongo Soco, Caue, and Serra Pelada in Brazil [25,77]; Bleïda Far West in Morocco [68]; and placers of the Ambositra region, Madagascar [70] and Zimnik Creek, Poland [64]. Native gold from these deposits is heterogenous and is represented by two or more varieties of Ag-bearing palladian gold—with Hg and (or) Cu impurities. Section 4.1.6 is devoted to these deposits.

The Au-Pd-Ag diagram (Figure 2) shows the compositions of Ag-bearing palladian gold, and other minerals and phases of this ternary system found in the ores of some deposits.

Ag-bearing palladian gold of the Au-Pd-Ag system depending on the Pd and Ag contents (Figures 2 and 3) can be divided into three varieties:

(1) Au-Pd solid solution in the absence of impurities (fineness > 649‰, LOD < Pd < 50 at.%). Such gold occurs at the following deposits: Uderei, Norilsk-1, Talnakh, Krutoe, Lebedinskoye, Ebeko volcano (Russia), Corrego Bom Sucesso, Caue iron mines (Brazil), and Serra Pelada, and alluvial placers of Ambositra region (Madagaskar), Chorokh river, Artvin district (Turkey), Lammermuir Hills (Scotland), and Brownstone (England), etc.

(2) Au-Pd-Ag solid solutions with high Pd (10–48 at.%) and low Ag content <10 at.% (fineness > 850‰). Native gold of this composition was found at the deposits Stillwater, Montana (USA), Chudnoe (Russia), Serra Pelada, Corrego Bom Sucesso (Brazil), and Bleida Far West (Morocco).

(3) Au-Ag-Pd solid solutions with a high Ag content (10–48 at.%) and a low Pd content <10 at.% (fineness > 630‰). Ag-rich palladian gold is present at the deposits of Talnakh, Baronskoe, Ozernoe, Fedorovo-Pansky intrusive complex, and Nesterovskoe (Russia); Gongo Soco, Caue (Brazil); Hope's Nose, Brownstone (England); and placer Zimnik Creek (Poland). Native silver with low Pd (fineness < 630‰) is found at some of these deposits (for example, Talnakh and Ozernoe).

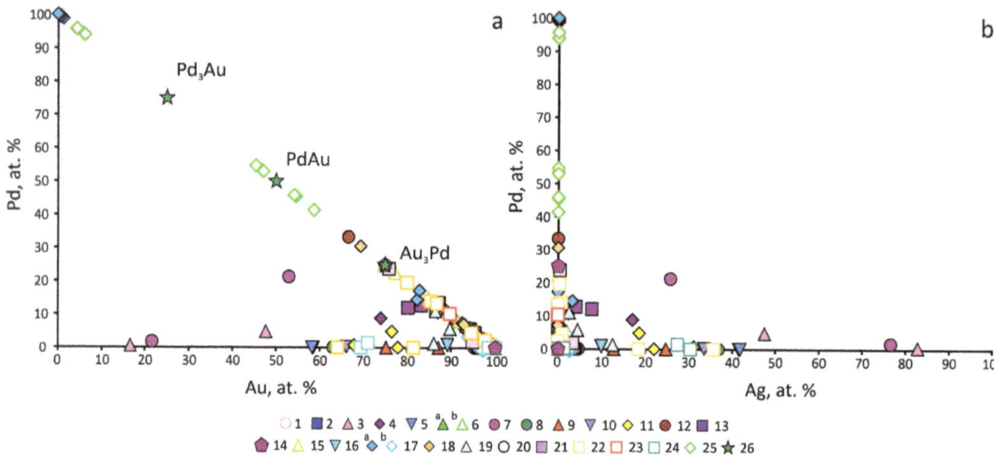

Figure 3. Dependences of the concentrations of Pd on Au (**a**) and Pd on Ag (**b**) in Pd-bearing gold and other Pd-bearing minerals of the Au-Pd-Ag system (in at.%) in various deposits and ore occurrences. Symbols are the same as in Figure 2.

4.1.3. Pd,Cu-Bearing Gold and Au-Pd-Cu System

Palladian gold with Cu impurity (Cu-bearing palladian gold or Pd,Cu-bearing gold) has a composition covering the range of solid solutions from Au to $Au_{1-x-z}Pd_xCu_z$, where $x + z \leq 0.5$ atomic part, $z \neq 0$, $x \neq 0$, and $1 - x - z > x, z$. As an example, Table 3 shows data for two objects with native gold containing Pd and Cu: the Au-Pd-Cu ore occurrence of the Skaergaard massif (Greenland, Denmark) [73] and the placer deposit of the Ambositra region (Madagascar) [70].

Table 3. The composition of Cu-bearing palladian gold, its fineness (N_{Au}), and minerals in association at endogenous and exogenous deposits.

Name of Deposit (Location)	N_{Au} ‰/Impurity wt.%	Minerals in Association	Ref.
Skaergaard massif (Greenland, Denmark)	689–913/Pd 1.3–14.7, Cu 13–19, Pt 0.12–3.6, Fe 0.4–1	Pd-Taur, Pd-Aur, Skg, PdCu, Bn, Cc, Ccp, Ktu, Pda, Vys, Vas	[73]
Ambositra region (Madagascar)	918–978/Pd 1.1–7.9, Cu < 2.8	Taur, Ba-Kfsp, Ca-REE phosphates	[70]

Phases compositionally similar to Pd-Au_3Cu and to Pd-tetra-auricupride in association with skaergaardite, bornite, chalcocite, chalcopyrite, kotulskite, palladoarsenide, vasilseverginite, and vysokýite occur at the Skaergaard layered ultramafic massif (Greenland, Denmark) [73]. In the Au-Cu-Pd-phases, the presence of Pt and Fe was observed (Table 3). Compounds close to stoichiometric $(Au,Pd)_3Cu$ are attributed to the Au_3Cu phase containing on average 24.3 at.% Cu and 15.8 at.% Pd ($n = 76$). A more numerous ($n = 132$) group, referred to as (Au,Cu,Pd) alloys containing 31 at.% Cu and 8 at.% Pd, is closely related to the above group. In the grains with the decomposition structure of the solid solution, the matrix contains 32.8 at.% Cu and 8.3 at.% Pd, and lamellas (AuCu phase) contain 44.1 at.% Cu and 8.5 at.% Pd ($n = 13$) [73].

The Cu content in palladian gold from the placers in the Ambositra region (Madagaskar) is low and varies from 0.6 to 2.8 wt.%, and the Pd content ranges from 1.1 to 7.9 wt.%. Palladium was not detected in tetra-auricupride. In association with tetra-auricupride and Cu-bearing palladian gold, Ba-Kfsp, Ca-REE phosphates were found [70].

Compositions of Pd,Cu-bearing gold and other Au-Cu-Pd phases from different deposits are shown in the diagram of Au-Cu-Pd (Figure 4). Pd,Cu-bearing gold (Pd 3.3–10.3 wt.%,

Cu 6.1–25 wt.%) rarely contains Ag up to 1.1 wt.% (900‰–700‰), Au_3Cu, Pd-tetra-auricupride and Pd-auricupride, Pd-bearing gold without Ag, Cu, and Hg (970‰); and Pd sulfides, arsenides, antimonides, bismuthides, tellurides, stannides, plumbides, and germanides were found in chromitites of dunites from the Konder layered massif [44,45]. Cu,Pd-bearing gold and Pd-tetra-auricupride and Pd-auricupride are typical of such deposits as Norilsk-1, Talnakh, Konder, Chudnoe, Nesterovskoe, Ozernoe (Russia), Serra Pelada, Caue (Brazil), and Marathon (Canada) (Figure 4). At these deposits, such phases as Au–Cu with high Pd and, in places, Pt are widespread. Borisov (2005) [53] found that native gold from the Chudnoe deposit contains inclusions of the Au-Cu phases (Au_3Cu, Au_3Cu_2, and Au_2Cu) with minor Pd.

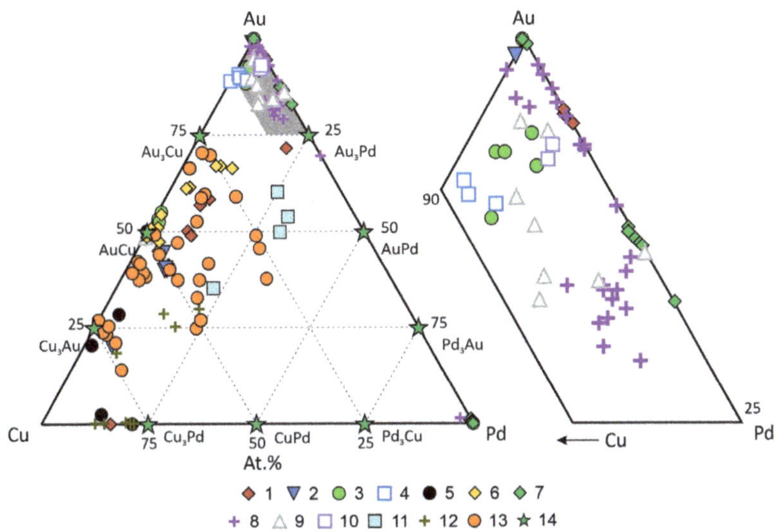

Figure 4. Composition of Pd,Cu-bearing gold and Au-Pd-Cu phases (at.%) from various deposits and ore occurrences. *Russian deposits* (1–6): 1—Norilsk-1 [27]; 2—Talnakh [27]; 3—Chudnoe [53]; 4—Nesterovskoe [53]; 5—Ozernoe [31]; 6—massive Conder [44,45]; *Worldwide deposits* (7–14): 7—Serra Pelada, Brazil [25]; 8—Caue iron mine, Brazil [77]; 9—Ambositra region, Madagascar [70]; 10—Mataganya-Siguiri zone, Guinea [69]; 11—Marathon, Canada [72]; 12—Korydallos Area, Greece [61,62]; 13—Skaergaard intrusion [73]; 14—theoretical composition of minerals in the Cu-Au-Ag system. Shaded markers are endogenous, empty markers are exogenous deposits.

Pd-tetra-auricupride (Au,Pd)Cu and Pd-auricupride $(Au,Pd)Cu_3$ are present in chromitites of the Korydallos Area in the Pindos Ophiolite Complex (NW Greece) [61,62]. The composition of Pd-tetra-auricupride is expressed by the formulae $Cu_{0.53}Au_{0.32}Pd_{0.15}$ and $Cu_{0.56}Au_{0.25}Pd_{0.19}$, and Pd-auricupride is expressed by the formula $Cu_{0.72}Au_{0.18}Pd_{0.10}$.

Cu-bearing palladian gold of the Au-Pd-Cu system depending on the Pd and Cu contents (Figures 4 and 5) can be divided into two varieties:

(1) High- and medium-fineness Cu-bearing palladian gold with low Pd and Cu contents (<10 at.%) (820‰–960‰);

(2) Low-fineness Cu-bearing palladian gold with a high Pd concentration (up to 30 at.%) and 10 at.% < Cu < 50 at.% (580‰–820‰).

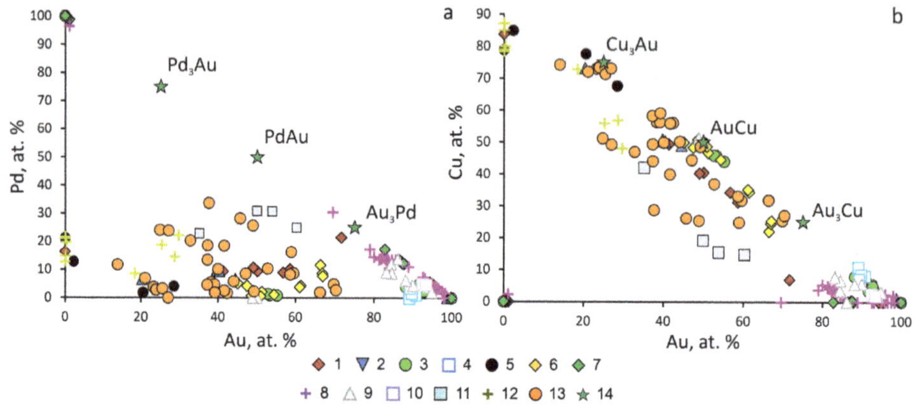

Figure 5. Dependences of the concentrations of Pd (**a**) and Cu (**b**) on Au in Pd,Cu-bearing gold and other minerals of the Au-Pd-Cu system (in at.%) in various deposits and ore occurrence. Symbols are the same as in Figure 4.

The first group includes the Nesterovskoe (Russia), Serra Pelada, and Caue (Brazil) deposits, and placers of the Mataganya-Siguiri zone (Guinea). The second group includes Pd,Cu-bearing gold of the deposits Norilsk-1, Talnakh, Chudnoe, Nesterovskoe, Ozernoe, Konder (Russia), Marathon (Canada), Skaergaard (Denmark), and the placer of Ambositra region (Madagaskar). It is worth noting that the main phases in association with Pd,Cu-bearing gold most commonly are Pd-bearing Au-Cu intermetallic compounds—Pd-tetraauricupride, and Pd-auricupride, as well as Pd sulfides, arsenides, antimonides, bismuthides, and tellurides.

Cu-bearing palladian gold can also contain Ag and (or) Hg. Examples of such deposits are described in Section 4.1.6.

4.1.4. Pd,Hg-Bearing Gold and Au-Pd-Hg System

Palladian gold with Hg impurity (Hg-bearing palladian gold or Pd,Hg-bearing gold) includes palladian gold, covering the range of solid solutions from Au to $Au_{1-x-d}Pd_xHg_d$, where $x + d \leq 0.5$ atomic part, $x \neq 0$, $d \neq 0$, and $1 - x - d > x$, $1 - x - d > d$. Several minerals are known in the Au-Pd-Hg system (https://www.mindat.org/min-51827.html, accessed on 15 January 2022): weishanite (Au,Hg(Ag)), aurihydrargyrumite (Au_6Hg_5), potarite (PdHg), gold, palladium, and mercury. The results of studies of Au-Pd-Hg mineralization from some deposits of noble metals showed the presence of binary and ternary phases: Au_2Pd and Au_3Pd [76]; Pd(Hg,Au) to Pd_3Hg_2 and $(Pd,Au)_3Hg_2$ [94]; Pd(Pt)AuHg, Au-rich potarite [95]. At the Herbeira massif (Spain), the Au-Pd-Hg phase contains Pd 53 wt.% [67] and has formula $Pd_{0.7}Au_{0.3}Hg_{0.1}$, which corresponds to Au,Hg-bearing palladium.

Native gold with Pd and Hg impurities in the absence of Ag is quite rare. Table 4 shows data for two objects with Pd,Hg-bearing gold in the absence and presence of Ag in the Itchaivayam mafic–ultramafic massif (Russia) and Mataganya-Siguiri Zone (Guinea). In the weathering crusts of the Mataganya-Siguiri Zone (Guinea), native gold is represented by two varieties: (1) Hg-bearing high-fineness palladian gold with low contents of Pd~2.7 wt.% and Hg~0.1 wt.%, sometimes Pt up to 0.09 wt.%; (2) Ag,Hg-bearing middle-fineness palladian gold with a high content of Ag~5.5–12.5 and low contents of Pd~1.1–1.8 and Hg up to 0.24 wt.% [69].

Table 4. The composition of Pd,Hg and Pd,Hg,Ag-bearing gold, its fineness (N_{Au}), and minerals in association at endogenous and exogenous deposits.

Name of Deposit (Location)	N_{Au} ‰/Impurity wt.%	Minerals in Association	Ref.
	Endogenous deposits		
Itchayvayam (Russia)	816–960/Ag < 6.1, Pd < 5.2, Hg < 8.5 580–660/Pd 13.9–15.9, Hg 19.1–27.8, Ag < 1.7	Ep, Cpy, Ptr, Cpe, Tnr Ep, Cpy, Cu sulfate, Au	[33,42]
	Exogenous deposits		
Mataganya-Siguiri Zone (Guinea)	960–970/Pd 2.6–2.8, Hg 0.07–0.11, Pt < 0.09 860–920/Pd 1.1–1.8, Hg < 0.24, Ag 5.5–12.5	Spy, Bg -	[69]
Zimnik Creek (Poland)	core 980/Pd 0.01–0.95, Hg 0.36–0.96, Ag 0.15–1.93; border 990–1000/Ag < 0.62, Pd < 0.01	Hem, Mag, Ap, Kln, Ms, Qz	[64]
River Dart (England)	630/Pd 22.7, Hg 13.6, Ag 2.2	Au-Ag > 970, Ausb, Ptr, Skg, Stpdn, Carb	[20]

Palladian gold from the placers that occur in the layered Itchayvayam platinum-bearing mafic–ultramafic massif (northern part of the Koryak–Kamchatka platinum-bearing belt, Russia), according to preliminary data, has varying compositions of Pd,Hg-bearing gold [42]. Au-Pd-Cu ore occurrences of this massif were found to contain high-fineness argentian gold and Au-Hg-Pd phases, which form thin veins in bornite and are in paragenesis with platinum group minerals [24,96].

We studied individual gold grains from the heavy concentrates of the Itchayvayam placers and watercourses draining and ore samples from the Barany outcrop of the Itchayvayam mafic–ultramafic complex (Kamchatka) [33]. Gold grains contain substitution structures formed by Pd,Hg-rich low-fineness gold ($Au_{0.59-0.52}Pd_{0.24-0.25}Hg_{0.17-0.23}$, 580‰–660‰) and high-fineness gold with minor contents of Pd and Hg ($Au_{0.94-0.90}Pd_{0.02-0.04}Hg_{0.03}$, 910‰–960‰) (Figure 6). Potarite (PdHg) without and with impurities Au, Cu, and Ag; high-fineness gold with minor contents of Ag ($Au_{0.91}Ag_{0.09}$, 950‰) or also Pd and Hg ($Au_{0.75}Ag_{0.08}Pd_{0.09}Hg_{0.08}$–$Au_{0.88}Ag_{0.09}Pd_{0.02}Hg_{0.01}$, 820‰–930‰); and Pd,Hg-rich low-fineness gold with minor contents of Ag and Cd ($Au_{0.51-0.55}Pd_{0.25-0.22}Hg_{0.21-0.16}Ag_{0.03-0.06}Cd_{0.01}$, fineness 580‰–630‰) were observed in the ore samples.

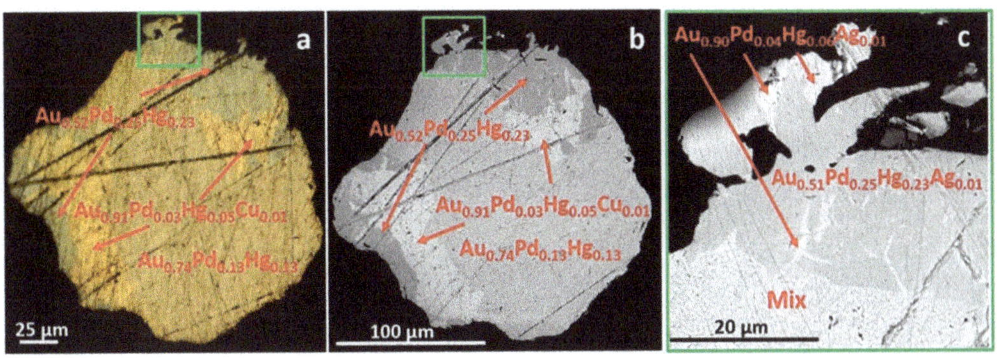

Figure 6. Optical photo (in reflected light) (**a**) and SEM photo (BSE mode) of a gold grain (**b**), as well as SEM photo (BSE mode) of a gold fragment (green box in a,b) (**c**) with low-fineness Pd,Hg-rich gold (580‰–660‰) and high-fineness Pd,Hg-poor gold (910‰–960‰). Ore occurrence of the Itchayvayam mafic–ultramafic complex (Kamchatka, Russia).

Figure 7 shows the composition of Pd,Hg-bearing gold and Au-Pd-Hg phases from the Itchayvayam mafic–ultramafic massif, Chudnoe (Russia) and Herbeira massifs (Spain), and placers of the Mataganya-Siguiri zone (Guinea).

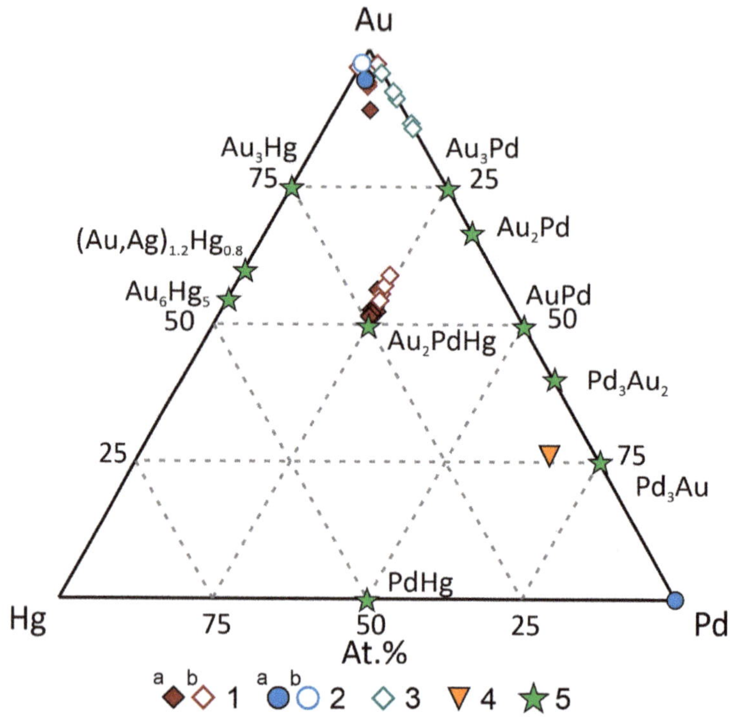

Figure 7. Composition of Pd,Hg-bearing gold and Au-Pd-Hg phases (at.%) from various deposits and ore occurrences. Russian deposits (1, 2): 1—Itchayvayam mafic–ultramafic massif, this study and after Sidorov [42]; 2—Chudnoe [53]; worldwide deposits (3, 4): 3—Mataganya-Siguiri zone, Guinea [69]; 4—Herbeira massif, Spain [67]; 5—theoretical composition of minerals of the Hg-Au-Pd system. Colored markers are endogenous, empty markers are exogenous deposits.

Palladian gold of the Au-Pd-Hg system is represented by two varieties (Figures 7 and 8):

(1) High- and medium-fineness gold with low Hg < 0.02 at.% (<5.2 wt.%) and low Pd contents <0.13 at.% (<8.5 wt.%) (fineness 820‰–960‰);

(2) Low-fineness gold with a high content of Hg 13–27 at.% (13.6–27.8 wt.%) and high Pd ≈25 at.% (13.9–15.9 wt.%) (fineness 580‰–630‰).

Hg-bearing palladian gold of both varieties—low Pd, Hg (1) and high Pd, Hg (2) contents were found at the Itchayvayam deposit. Native gold in the Mataganya-Siguiri Zone (Guinea) placer is an example of ore occurrence with high-fineness Hg-bearing palladian gold (with low Pd and Hg contents) (1).

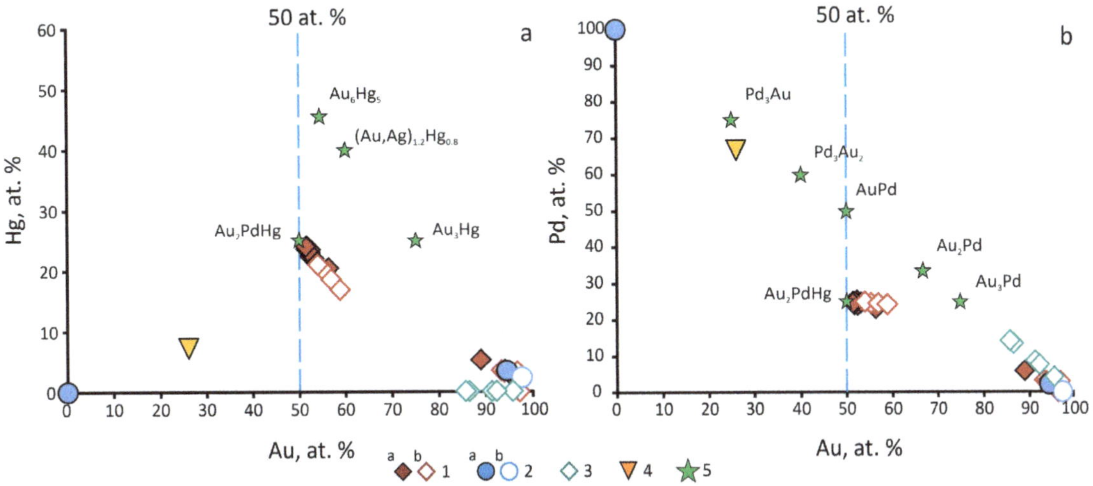

Figure 8. Dependencies of the concentrations of Pd (**a**) and Hg (**b**) on Au in Pd,Hg-bearing gold and other Pd-bearing minerals of the Au-Pd-Ag system (in at.%) in various deposits and ore occurrences. Symbols are the same as in Figure 7.

At some deposits, Pd,Hg-bearing gold has minor contents of Ag. As an example, in Table 4, we provide data on the composition of Pd,Hg,Ag-bearing gold from two alluvial placers related to sediments (conglomerates) and volcanic rocks—Zimnik Creek (Poland) and river Dart (England). The Hg content varies in the range of 0.07 to 13.9 wt.%, and the Ag content is lower at 2.2 wt.%. Native gold with Pd and Hg impurities and higher Ag contents is spread at many other deposits—Serra Pelada (Brazil) [97], Gongo Soco (Brazil) [78,98], Mayat river and Bol'shaya Kuonamka river (Russia) [47], and Dziwiszow (Poland) [65].

The compositions of palladian gold with Hg and Ag impurities cover the range of solid solutions from Au to $Au_{1-x-y-d}Pd_xAg_yHg_d$, where x, y, d in total $\leq$ 0.5 atomic part, $x \neq 0$, $y \neq 0$, $d \neq 0$, and $1 - x - y - d > x, y, d$. Hg,Ag-bearing palladian gold is more widely distributed in nature than Pd,Hg-bearing gold (see Section 4.1.6).

4.1.5. Pd,Cu,Ag-Bearing Gold and Au–Pd–Ag–Cu System

Palladian gold with Cu and Ag impurities (Cu,Ag-bearing palladian gold or Pd,Cu,Ag-bearing gold) covers the range of solid solutions from Au to $Au_{1-x-y}Pd_xAg_yCu_z$, where x, y, z in total $\leq$0.5 atomic part, $x \neq 0$, $y \neq 0$, $z \neq 0$, and $1 - x - y - z > x, y, z$. The group of deposits with Cu,Ag-bearing palladian gold is the most numerous group compared to other groups (Tables 1–4) and includes 18 objects among which there is only one placer deposit—Kuoyka river, basin of the Anabar river (Russia) (Table 5).

Au,Pd,Ag,Cu-solid solutions with Au >> Pd, Ag, Cu (>1 vol%), Au,Ag,Pd-solid solutions, argentian gold (Au > Ag), native silver (Ag >> Au), a Pd-bearing tetra-auricupride (Cu,Ag)(Au,Pd), and phase $(Au,Pd,Ag)_3Cu$ occur in the W Horizon mineralization of the Marathon Cu-PGE deposit, which is one of the highest grade PGE repositories in magmatic ore deposits worldwide [21]. The Marathon deposit is located within the Mesoproterozoic Coldwell Alkaline Complex in the Midcontinent rift of North America. The content of impurities in native gold from this deposit is very high and reaches 22.7 wt.% for Pd, Ag < 37.7 wt.% and Cu < 25.8 wt.%, which corresponds to a fineness of 516‰–835‰ (Table 5).

Table 5. The composition of Pd,Cu,Ag-bearing gold, its fineness (N_{Au}), and minerals in association at endogenous and exogenous deposits.

Deposit Name (Location)	N_{Au} ‰/Impurity wt.%	Minerals in Association	Ref.
	Endogenous deposits		
Caue (Itabira region, Brazil)	800–918/Pd 0.9–19.2, Ag 0.1–1.1, Cu 0.04–4.1	Qz, Fsp, Hem, Gth, Ms, Tur, Ap, Mnz, Carb, Pd-Cu-oxides, Pd	[26,77]
Conceichao mine, - « -	930–940/Pd 0.8–3.2, Ag 0.4–1, Cu 0.7–1, Fe 0.3–6.1 980/Pd 0.9, Ag 0.6	Hem, Mag, Py, Qz, Ms, Met-II, Au-Ag, Kln	[79]
Maquine, - « -	900–948/Ag 3.5–8.2, Pd 1.5–2.75, Cu 0.1–0.8, Pt ≈ 0.03	Hem, Gth, Qz, Spy, Stpdn, Ism	[28]
Marathon (Canada)	516–835/Pd 0.5–22.7, Ag < 37.7, Cu 0.8–25.8	Cpy, Bn, Brg, Ifp, Vas, Vys, Au, Ag, Taur	[21]
Ozernoe (Ural, Russia)	690–700/Cu, Ag > Pd < 3.8	Chl, Cpx, Met-I, Stpdn, Py, Cpy, Bn, Mrk, Taur, Auc, Ag-Cu-Au	[31]
Nesterovskoe (- « -)	970/Pd 0.06–0.91, Ag 0.87–0.96, Cu < 2.01	Qz, Ab, Cr-Ms, Aln, Met, Ah, Mon, Au_3Cu	[53]
Volkovskoe (- « -)	1000	Ti-Mag	[30,55]
	810–914/Ag 8.4–16.2, Pd ≤ 0.3, Cu 0.2–0.4	Mrk, Kei, Syv, Hes, Spy	
Baronskoe (- « -)	840–940/Pd 3.5–9, Ag 1–10.6, Cu 0.6–1.9	Brn, Ccp, Py, Cc, Dg, Vsk, Pd tellurides and arsenides	[55]
	916/Pd 4.3, Ag 1.6, Cu 2.5/$Au_{0.90}Pd_{0.06}Cu_{0.03}$	Pds	
Serebryansky Kamen (-«-)	783/Pd 12.1, Cu 7.5, Ag 2.1	Bn, Vsk, Pd tellurides, Au-Ag (760–970)	[54]
Khamitovskoe (- « -)	880–920/Ag 7.6, 8.1, Cu 1.1, Pd 0.5, 0.9	-	[56]
Bleida Far West (Morocco)	912–993/Pd 0.04–6.29, Ag 0.67–8.34, Cu < 3.1	Pdm, Mrk, Met, Oos, Spy, Pds, Ktu, Taur, Hem, Cc, Ani, Mag, Gth, Kfs, Bt, Chl, Cal, Au-Ag (770–870)	[32,68]
Konder Massif (Aldan, Russia)	974/Pd 2.58	Au_3Cu, Pd-Taur, Pd-Aur, Pd sulfides, arsenides, antimonides, bismuthides, tellurides, stannides, plumbides, germanides	[44,45]
	700–900/Pd 3.3–10.3, Cu 6.1–25, Ag < 1.1		
Norilsk-1 (Russia)	570–880/Ag 6.8–42.1, Pd < 6.2, Cu < 1.5, Pt < 3.9	Pd-Taur, Tfpt, Pyr, Pn	[27]
	950–975/Pd 2.22–2.89, Ag < 3.59, Cu < 0.15, Pt < 1.6	-	
Talnakh (- « -)	160–210/Ag 70.1–82.9, Pd < 3, Cu < 1.5 >980/Pd < 1.25, Cu < 0.54, Ag < 0.38	Pt-Ato, Spdn, Pd-Taur, Pb-Pol, Zvy, Ccp	[27]
Deposit Name (Location)	**N_{Au} ‰/Impurity wt.%**	**Minerals in Association**	**Ref.**
Yu. Peshempakhk (1), V. Chuarvy (2), S. Kamennik (3) (- « -)	(1) 960/Pd 1.90, Ag 0.46, Cu 1.25	-	[58]
	(2) 830–970/Pd 1.44–3.48, Ag 0.78–1.69, Cu 0.23–0.25, Fe 0.2–0.36, Ni < 0.03	-	
	(3) 833–920/Ag 1.5–16.2, Pd < 6.5, Cu < 0.6, Fe < 0.2	Ktu, Pda, Met, Spy	
	Exogenous deposits		
Kuoyka river (Anabar river basin, Russia)	(20) 960/Pd 3.43, Ag 0.39, Cu 0.07	-	[47,48]

The Cauê iron mine, in the ltabira District of Brazil, contains a unique precious-metal-bearing mineral assemblage that includes palladian gold, pure gold, native palladium, palladseite, arsenopalladinite, and Pd-Cu oxide minerals. Palladian gold grains contain up to 19.2 wt.% Pd, while Cu is typically below 4.1 wt.% and rarely reaches 8.3 wt.%, and Ag does not exceed 1.1 wt.% (fineness 808‰–993‰) [77] (Table 5). Palladian gold from the Conceichao deposit in the same region of Brazil has lower concentrations of Pd (<3.2 wt.%) and minor contents of Ag, Cu (fineness 930‰–940‰) or Ag (fineness 980‰) (Table 5) in association with hematite, magnetite, pyrite, quartz, muscovite, mertieite-II, argentian gold (940‰–950‰), and kaolinite. Palladian gold from the Maquine deposit (Brazil) is characterized by lower contents of Pd, Cu, and also trace Pt, whereas the concentrations of Ag amount to 3.5–8.2 wt.% (fineness 900‰–948‰), and minerals in the association are hematite, goethite, quartz, sperrylite, stannopalladinite, and isomertieite.

Palladian gold from some Uralian deposits also has low contents Pd (in wt.%): Volkovskoe < 0.3, Khamitovskoe 0.5–0.9, Nesterovskoe < 2.8, Ozernoe < 3.8. Higher contents of Pd to 9 wt.% were detected in palladian gold from the Baronskoe, and to 12.1 wt.%—from the Serebryansky Kamen deposits (Table 5).

Noble metal mineralization at the Volkovskoe Cu-Fe-Ti-V deposit is concentrated in meso- and melanocratic taxite gabbro containing a dissemination of V-bearing titanomagnetite, chalcopyrite, and bornite. Palladian gold has low concentrations of Cu~0.3 wt.% and high concentrations of Ag from 8.4 to 16.2 wt.% (fineness 800‰–914‰) (Table 5) [30]. Among the minerals of noble metals in copper-sulfide (chalcopyrite, bornite, digenite) ores, there are also tellurides—merenskyite, hessite, and sylvanite, and platinum arsenide—sperrylite [55,99–103].

At the Baronskoe deposit, native gold is composed of Au-Ag, Au-Ag-Pd, Au-Pd-Cu, and Au-Pd-Ag-Cu solid solutions [55,104–106]. Au-Ag-Pd solid solutions (3.5–9 wt.% Pd, to 10 wt.% Ag) form inclusions in chalcopyrite, vysotskite, bornite, Pd tellurides, and arsenides. The Au-Pd-Cu solid solution ($Au_{0.90}Pd_{0.06}Cu_{0.03}$) was found in the intergrowth with palladseite in olivine-magnetite gabbro. Ag,Cu-bearing palladian gold contains Pd 3.4–5.7 wt.%, Ag up to 10.6 wt.%, and Cu 0.6–1.9 wt.% (fineness 840‰–916‰) (Table 5) [55]. Au-Ag solid solutions occur in secondary silicates (serpentine, amphibole, chlorite) and do not contain Pd.

The Ozernoe deposit is known for Au-Ag-Cu solid solutions (to 6.5 wt.% Cu, 7–69 wt.% Ag) and Au-Cu intermetallic compounds (tetra-auricupride, auricupride), in which palladium up to 16.9 wt.% occurs sporadically [31,107,108]. Palladian gold with Cu and Ag impurities has a low fineness (690‰–700‰) (Table 5), and the majority of compositions of Au-Pd-Ag-Cu phases are similar to tetra-auricupride, auricupride, native copper, or silver. Olivine clinopyroxenites from this deposit contain impregnations of sulfides in an amount of no less than 0.5% (in separate zones to 1–3). Palladian gold is present in the cubanite–pentlandite–pyrrhotite–chalcopyrite (1) and bornite–chalcopyrite (2) mineral associations. The first association was found to contain minerals of Pt (native platinum, ferruginous platinum, braggite, sperrylite, and moncheite) and Pd (palladian copper, vysotskite, arsenopalladinite, mayakite, atheneite, isomertieite, stibiopalladinite, michenerite, merenskite, froodite, and palladian silver). In the second association, minerals of Pt (sperrylite), Pd (atheneite, mertieite, stibiopallandinite, kotulskite, tarkianite, sobolevskite, and froodite) and Ag (native silver and hessite) were found. According to Murzin et al. (2022) [31], the Pd content increases from <1.5 wt.% Pd in Au-Ag solid solution to 6 wt.% Pd in Au-Cu intermetallic compounds and to 16.2–16.9 wt.% Pd in Cu-Au-Ag-Pd solid solutions.

Native gold in copper-sulfide ores of the Serebryansky Kamen from gabbro of the Serebryansky massif is heterogeneous and corresponds to Au-Ag (fineness 760‰–970‰) and Au-Cu-Pd-Ag solid solutions with high Pd and lower contents of Cu and Ag (fineness 780‰) (Table 5) [54]. It forms inclusions in bornite and chalcopyrite and frequently occurs in association with Pd sulfides, arsenides, and tellurides.

The ore samples from the Central and Intermediate zones of the Au-Cu epithermal (porphyry) Bleida Far West deposit (Anti-Atlas, Morocco) contain high-fineness palladian gold with low concentrations of Pd and Ag in the absence of Cu, or with minor Cu 0.3–0.8 wt.% (912‰–993‰) [32,68]. Au-Pd-mineralization at this deposit contains no sulfides, and palladian gold occurs in association with mertieite, isomertieite, keitconnite, palladseite, merenskyite, kotulskite, sperrylite, hematite, calcite, quartz, barite, and chlorite (Table 5).

Various alloys of the Au–Ag, Au-Ag-Cu, and Au-Ag-Cu-Pd systems have been found in Cu-Ni–PGE ores from Norilsk and Talnakh deposits (Russia) [27,109]. The Au-Cu-Ag-Pd (Pt) phase is characterized by wide variations in Cu content from LOI up to 42 wt.% and Ag from LOI up to 37 wt.%, Pd from LOI to 14 wt.%, and occasionally Pt < 5 wt.%, which corresponds to Cu,Ag,Pd-bearing gold, $(Au,Pd)_3Cu$, Pd-tetra-auricupride, Pd-auricupride, and Cu,Au,Pd-silver. They are in association with mertieite/isomertieite, keitconnite, palladseite, kotulskite, merenskyite, isoferroplatinum, vysotskite, hematite, calcite, quartz, barite, epidote, chlorite, etc. The fineness of Cu,Ag,Pd-bearing gold changes from 570 to 880 and from 950 to 975‰ for Norilsk-1 and is higher (980‰) for Talnakh (Table 5).

Au–Cu alloys with significant amounts of Pt, Pd, and Ag impurities were reported from platinum placers associated with the Konder alkaline–ultramafic complex, where abundant Au–Cu alloys intergrow with PGM (particularly Pt–Fe alloy) and show a range of compositions close to Au_3Cu and AuCu along with a single Cu_3Au grain [44,45,110]. The Pt and Pd concentrations in these alloys reach 11.8 and 10.3 wt.%.

The contents of Pd in Cu,Ag-bearing palladian gold at Pt-metal deposits of the Fedorovo-Pansky intrusive complex (Russia) change from 0.03 to 6.5 wt.%, Cu < 1.2 wt.%, Ag up to 16.2 wt.%, and the fineness increases from 830 to 970‰ [58]. In places, the presence of impurities of Fe and Ni was observed (Table 5). Native gold from the Kuoyka river (basin of the Anabar river, Russia) contains 3.43 wt.% Pd, whereas Ag and Cu concentrations are low at 0.4 wt.% (fineness 960‰) [47,48].

Figure 9 shows that Ag,Cu-bearing palladian gold of the Au-Pd-Ag-Cu system depending on the Pd, Ag, and Cu contents can be divided into several varieties:

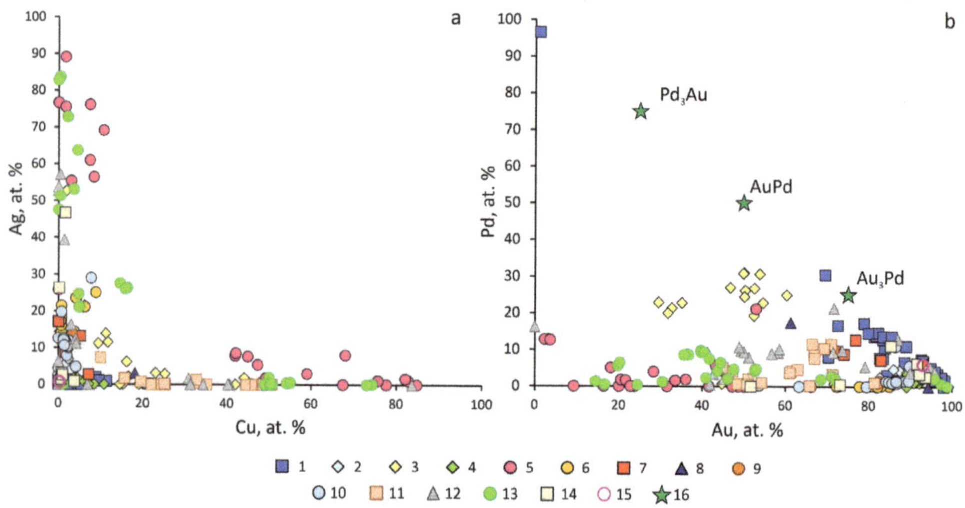

Figure 9. Dependences of the concentrations of Ag on Cu (**a**) and Pd on Au (**b**) in Pd,Ag-Cu-bearing gold and other Pd-bearing minerals of the Au–Pd–Ag–Cu system (in at.%) in various deposits and ore occurrences (Table 5) (filled marker—endogenous, empty—exogenous): 1—Caue iron mine (Itabira, Minas Gerais, Brazil) [26,77]; 2—Maquine (Brazil) [28]; 3—Marathon (Canada) [21]; 4—Nesterovskoe (Polar Urals, Russia) [53]; 5—Ozernoe (Subpolar Ural, Russia) [31]; 6—Volkovskoe (Middle Urals, Russia) [30,55]; 7—Baronskoe (Middle Urals, Russia) [55]; 8—Serebryansky Kamen (Northern Urals, Russia) [54]; 9—Khamitovskoe (Yu.Ural, Russia) [56]; 10—Bleida Far West (Morocco) [32,68]; 11—Konder massif (Aldan, Russia) [44,45]; 12—Norilsk-1, 13—Talnakh (Russia) [27]; 14—Y. Peshempakhk (1), V. Chuarvy (2), S. Kamennik (3) (Russia) [58]; 15—Kuoyka River (Anabar river basin, Russia) [47,48]; 16—theoretical compositions of Au-Pd intermetallic compounds.

(1) High-fineness gold with low contents of Pd, Ag, and Cu < 10 at.% (900‰–970‰). The palladian gold of this composition was found in Conceichao and Maquine mines (Brazil), Nesterovskoe, Khamitovskoe, Kuoyka River (Russia), and Bleida Far West (Morocco).

(2) Middle- and low-fineness gold with a high content of Ag 10–48 at.% and low contents of Pd and Cu < 10 at.%. Volkovskoe and Norilsk (Russia) are examples of deposits with such native gold.

(3) Middle- and low-fineness with high contents of Pd 10–30 at.%, Cu 10–48 at.%, and Ag < 10 at.%. Marathon (Canada), Konder (Russia), and Caue iron mine (Brazil) are examples of deposits with native gold containing high concentrations of Pd and Cu, and low Ag.

4.1.6. Palladian Gold with Ag, Cu, Hg, and Other Metals

Palladian gold commonly contains impurities of all three elements—Ag, Cu, and Hg (Cu,Ag,Hg-bearing palladian gold or Pd,Cu,Ag,Hg-bearing gold) and has a heterogeneous structure and a variable composition of $Au_{1-x-y-z-d}Pd_xAg_yCu_zHg_d$. The Au-Pd-REE Chudnoe (Russia), Fe-oxide-Cu-Au Gongo Soco, Au–Pd–Pt Serra Pelada (Brazil), and Zechstein (Poland) deposits are examples of deposits with such palladian gold (Table 6). Pd,Cu,Ag,Hg-bearing gold is typical of some alluvial deposits: placers of the Mayat and Bol'shaya Kuonamka river basins (Anabar massif, northeast of the Siberian Platform, Russia), Dziwiszow (Poland), Whipsaw Creek, and Similkameen river (Canada) (Table 6).

Table 6. The composition of Pd,Cu,Ag,Hg-bearing gold, its fineness (N_{Au}), and minerals in association at endogenous and exogenous deposits.

Name of Deposit (Location)	N_{Au} ‰/Impurity wt.%	Minerals in Association	Ref.
	Endogenous deposits		
Chudnoe (Ural, Russia)	829–886/Ag 7.7–13.4, Pd 0.5–1.1, Cu 1.1–7.6	Qz, Ab, Kfs, Cr-Ms, Ah, Taur, Au_{950}	[29,53,111–113]
	839–866/Ag 11.9–12.5, Pd 0.5–0.8, Cu 1.7–2.5, Hg 0.6–1	Qz, Ab, Kfs, Cr-Ms, U-Met, Taur, Au_{930}	
	932–986/Ag < 5.64, Pd < 2.78, Cu < 2.78	$Au_{840-860}$	
Serra Pelada (Brazil)	920–930/Pd 6.9–7.8, Ag < 0.2, Cu < 0.1, Hg < 0.02, Fe < 0.05	Pd, Pd-Au-Pt-As phase, Svi, Pds, Pd_3As, Pd-oxide, Gth, $Au_{990-996}$	[97]
	880–940/Pd 3.2–9.8, Ag < 0.3, Cu ≈ 0.7, Hg 0.3–1.5, Fe < 0.05	Ah, Pd–Hg–Se and Pd–Bi–Se phases	
	965–975/Pd 1.6–2.4, Ag 0.4, Cu ≈ 0.6, Hg < 0.02, Fe < 0.1	Ism, Mn–Ba oxide	
	993–997/Ag ≈ 0.3, Pd 0.04–0.1, Cu < 0.07, Fe 0.05–0.3	Au_7Pd, Gth	
Gongo Soco (Brazil)	880–890/Pd 5.9–6.1, Hg 0.9, Ag ≈ 2, Cu 0.1–2.7	Hem, Gp, Ism, Met, Au-Ptr, (Pd,Cu)O, (Fe,Pd)OOH	[25,78]
Zechstein (Poland)	576–795/Ag 12.5–34, Pd 3–6.9, Cu 0.1–0.4, Hg 0.7–6.0, Pt 0.46–1.6, Ir < 0.8	Cat, kerogen, Sov, Pb, Cth, Any, Pd	[63]
	Exogenous deposits		
r.Mayat (1), r. Bol'shaya Kuonamka (2) (Anabar river basin, Russia)	(1) 860–960/Pd 0.8–12.8, Hg < 0.3, Ag 0.9–2.8, Cu < 0.4	Tpdn, Ktu	[47,114]
	(2) 910–970/Pd 0.7–7.5, Hg < 1.7, Ag 0.6–2.2, Cu 0.1–1.5	Pt	
Dziwiszow (Poland)	centre 960–990/Pd < 0.5, Hg < 2.5, Ag 1.1–3, Cu < 0.1, Pt < 0.04	Hem, Gth, Kln	[65]
	core 975–997/Pd < 1.4, Hg < 1.9, Ag 0.2–1.8, Cu < 0.1, Pt < 0.03		
Whipsaw Creek (Canada)	910–940/Pd ≈ 2.5, Hg 0.20–2.28, Ag 0.7–3, Cu 1.1–3.1	Stpdn, Cc, Spy	[71]
Similkameen river (- « -)	970/Pd < 2.1, Hg 0.1–0.7, Cu 0.06–1.8, Ag 0.7	Tem, Bn, Cc, Ccp, Cin, Pd-telluride, Pd-arsenoantimonide	[71]
Friday Creek (- « -)	930/Ag 3.97, Pd 1.94, Cu 0.55, Hg 0.32	Bn, Cc	[71]

Au-Pd-REE ores from the Chudnoe deposit contain palladian gold with various sets of impurity elements: (1) all 4 elements—Ag, Cu, Hg, and Pd; (2) 3 elements—Ag, Cu, Pd or Cu, Hg, and Pd [29]; (3) 2 elements—Hg (3.4–3.5 wt.%) and Pd (1.2–1.6 wt.%) [53]. The presence of Au-Ag (fineness 720‰–730‰) and Au-Ag-Cu-solid solutions (Ag 11.4–14.8 wt.%, Cu 1.4–4.6 wt.%, fineness 836‰–869‰) was also detected in these ores. The content of Ag in palladian gold varies in a wide range, but is no more than 13.4 wt.%, Cu < 7.6 wt.%, Hg < 3.5 wt.% [53], Pd < 2.8 wt.%, and fineness 829‰–986‰ (Table 6). High concentrations of Hg (36 wt.%) and Pd (15.2 wt.%), and low Cu in the absence of Ag were found in the ore samples collected from the surface in weathered material at the Chudnoe deposit [53]—in the composition, this is the Au-Hg phase with Pd and Cu impurities ($Au_{0.41}Hg_{0.31}Pd_{0.25}Cu_{0.03}$). The specific features of this deposit are an unusual association of Cu,Ag,Pd,Hg-bearing gold and palladium minerals (mertieite, isomertieite, atheneite, stillwaterite, and stibiopalladinite) with Cr-bearing muscovite (fuchsite) and REE mineral (allanite) in the absence of sulfides [29,112,113], and the low content of Hg in Cu,Ag,Pd-bearing gold is normally less than 1 wt.% (Table 6).

The Chudnoe deposit is the recharge source for the placer of the Al'kes-Vozh creek and, partly, for other placers downstream of the Balaban-Yu river. The presence of native gold with impurities of Pd (to 2 wt.%), Cu (to 5 wt.%), and Hg (to 0.6 wt.%) in the Al'kes-Vozh placer, as well as the presence of Pd minerals (mertieite, atheneite, stibiopalladinite) in gold grains, served as the basis for identifying a placer-forming type of gold mineralization in the Urals prior to the discovery of its endogenous source [52]. The occurrence of palladian gold in the Al'kes-Vozh placer is 81% and it gradually decreases to 13% in the Balban-Yu placers located below [115].

Ores from the Au–Pd–Pt Serra Pelada deposit (Brazil) contain up to 1 cm coarse-grained gold aggregates that occur in powdery, earthy weathered material [25]. They exhibit a delicate arborescent fabric and are coated by goethite. Four compositional types of palladian gold are recognized at this deposit: (1) "Au_7Pd", the most abundant Au–Pd alloy, hosts Pd arsenides ("guanglinite" and Sb-bearing "guanglinite"), Pd–Pt–Se and Pd–Se phases, sudovikovite, and palladseite; (2) Au–Pd–Hg alloy, characteristically with atheneite and rarely observed Pb-bearing Pd–Hg–Se and Pd–Bi–Se phases; (3) Pd-poor gold—isomertieite—Mn–Ba oxide assemblage; and (4) pure gold with trace Pd (>993‰) in goethite assemblage (Table 6). The Ag content in these types of palladian gold is lower than 0.4 wt.%. The Pd,Ag-poor gold assemblage possibly formed later than the Au–Pd–Hg–Ag alloy. In the ores of the Fe-oxide-Cu-Au Gongo Soco deposit (Brazil), palladian gold with Ag, Cu, and Hg impurities occurs in the form of nuggets, grains, and their aggregates in association with hematite. Gold grains have dark-color coatings composed of Pd–O particles and iron hydroxide. The palladian gold is characterized by a variable Pd content reaching up to 6 wt.% and sporadically observed Hg content up to 1 wt.% [97]. Pd,Hg-bearing gold at these deposits can also contain Cu and Ag impurities to 3 and more wt.% [78]. Palladian gold contains inclusions of platinum group minerals (PGMs): isomertieite, mertieite-II, chrisstanleyite, selenides $Pd_5(Hg,Sb,Ag)_2Se_6$, and $(Pd,Sb,Ag,Hg)_5Se_4$.

Palladian gold (with minor content of Hg, Cu, Ag) in association with Pt minerals is found in the placers of the Mayat and Bol'shaya Kuonamka river basins (Anabar massif, northeast of the Siberian Platform) [116].

Figure 10 shows the dependences of the concentrations of elements in Pd-bearing gold and other Pd-bearing minerals of the Au-Pd-Ag-Cu-Hg system (in at.%) at various deposits and ore occurrences (Table 6). Variations in the concentrations of impurities in native gold range from LOI up to 26 for Pd, <50 Ag, <18 Cu, and <6 Hg (in wt.%). According to Figure 10, the Ag,Cu,Hg-bearing palladian gold depending on the Pd, Ag, and Cu contents can be divided into several varieties:

(1) High-fineness gold with low contents of Pd, Ag, Cu < 10 at.% and Hg < 5.7 at.% (900‰–970‰). The palladian gold of this composition is typical of the deposits Chudnoe (Russia), Serra Pelada and Gongo Soco (Brazil); and alluvial placers Dziwiszow (Poland), Whipsaw Creek, Similkameen River, and Friday Creek (Canada); and Mayat and Bol'shaya Kuonamka river basins (Russia).
(2) Medium- and low-fineness gold with high Ag 10–48 at.% and low contents of Pd, Cu < 10 at.%, and Hg < 5.7 at.%. Chudnoe (Russia) and Zechstein (Poland) are examples of deposits with such Ag,Cu,Hg-bearing palladian gold.
(3) Medium- and low-fineness gold with a high content of Pd 10–28 at.%, and low Cu and Ag < 10 at.% and Hg < 5.7 at.%. Serra Pelada (Brazil), Mayat river basin, and Chudnoe (Russia) are examples of deposits with native gold containing high concentrations of Pd and low Cu, Ag, and Hg.

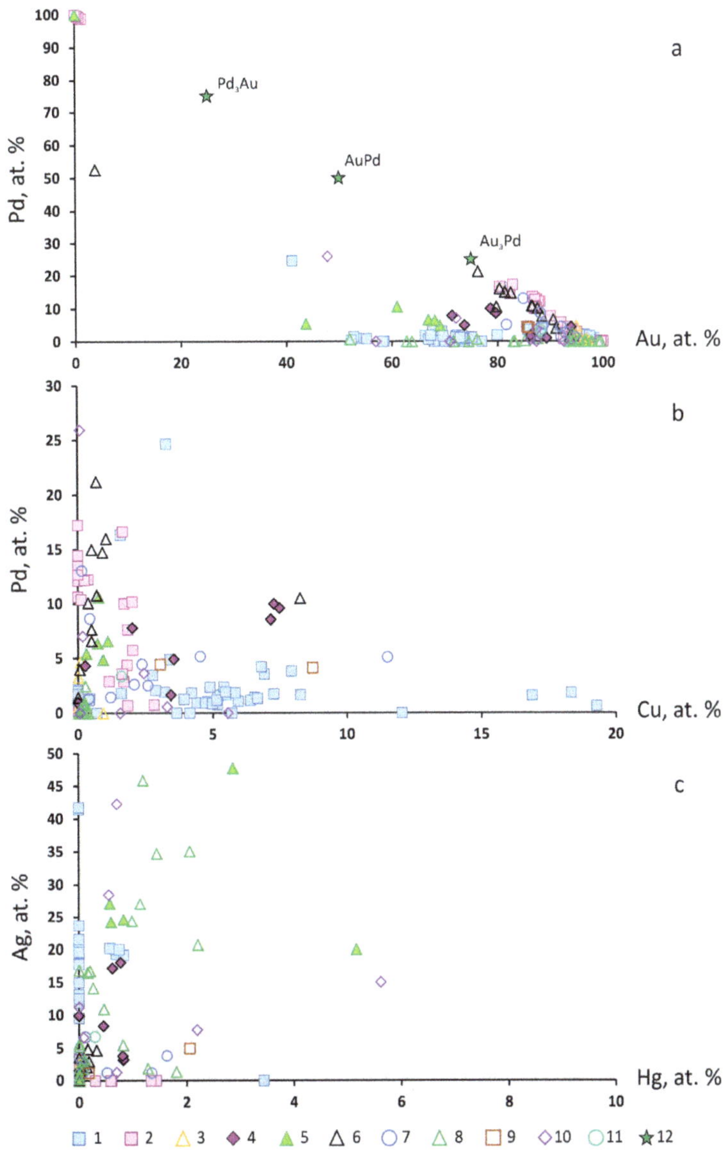

Figure 10. Dependencies of the concentrations of Pd on Au (**a**), Pd on Cu (**b**), and Ag on Hg (**c**) in Pd-bearing gold and other Pd-bearing minerals of the Au–Pd–Ag–Cu–Hg system (in at.%) in various deposits and ore occurrences (Table 6) (filled marker—endogenous, empty—exogenous): 1—Chudnoe (Subpolar Urals, Russia) [29,53,112–114]; 2, 3—Serra Pelada (Brazil) [97]; 4—Gongo Soco (Brazil) [25,78]; 5—Zechstein (Poland) [63]; 6, 7—r.Mayat and r. Bol'shaya Kuonamka, Anabar river basin (Russia) [47,116]; 8—Dziwiszow (Poland) [65]; 9—Whipsaw Creek (Canada) [71]; 10—Similkameen river (Canada) [71]; 11—Friday Creek (Canada) [71]; 12—theoretical compositions of Au-Pd intermetallic compounds.

5. Discussion

5.1. Types of Deposits with Palladian Gold

Native gold occurs in a wide range of gold deposit types and settings [1,2,5–8]. Impurity elements in native gold and minerals in association are indicators of geochemical and mineral types of deposits and regional geochemical conditions [13]. Petrovskaya (1973) [1] reported that the number and concentrations of impurity metals in native gold tend to increase at the deposits with a multistage formation of mineralization or with features of hybridism, and local and uneven changes in the composition of ore bodies with superimposed hydrothermal transformations.

The mechanism of formation of native gold is complex and depends on many factors. A wide range of impurity elements is associated with different geochemical environments in which the mobilization, transportation, and deposition of gold ore mineralization take place. The content of elements in native gold depends on their amounts in hydrothermal solutions, which are controlled by temperature, the presence of ligand elements—Cl, S, and Se and Te to a lesser degree—and also the pH of solutions and oxidation–reduction conditions [20,117]. Metals are largely derived from the mantle or crust by partial melting and fluid-related leaching. Ligands can be provided from the same sources, or from the atmosphere, hydrosphere, and biosphere. Gold and impurity elements can be transported not only by hydrothermal solutions and supercritical fluids, but also by gas mixtures, as well as by sulfide, silicate, and carbonate melts. The spectrum of element impurities and mineral microinclusions often depends on the formational affiliation of the gold deposit, on its connection with any magmatic complex and post-ore processes, and also on the metallogenic features of the gold-bearing provinces.

In this review, based on the literature data and our own studies of the composition of palladian gold from numerous deposits with Au-Pd mineralization, we suggest the following classification of types (Table 7). The studied objects with palladian gold belong not only to gold deposits of different types but to deposits at which the main ore components are PGE, Cr, Cu, Ni, V, and Ti. In the proposed systematization, we do not claim innovations but report them based on materials from various authors for the convenience of presenting results. It includes the main types of PGE-deposits in mafic–ultramafic magmatic complexes [118] within one type with two subtypes—low-sulfide-grade (less than 2%–5% sulfides) Alaskan (1a) and high-sulfide-grade (more than 5% sulfides) Norilsk (1b) (base on Naldrett, 2010 [119]), and gold deposits of 2–4 types: orogenic gold (OG) (type 2a,b), epithermal (porphyry) gold–copper (EPGC) (type 3), and iron oxide–copper–gold (IOCG) (type 4) [119–123] (Table 7). To these types, we added ferruginous quartzites (type 5), which were identified by Dill (2010) [118] within the frames of an independent type of PGE-deposits, and a new non-industrial type of volcanic exhalations (Kamchatka) (type 6), in which gold with Pd impurity was found (Table 1). An important role in the studied objects belongs to gold from placer alluvial deposits and occurrences, which we subdivided into four subtypes (7-1 . . . 7-4) in accordance with the profile of their primary sources and a special subtype of placers associated with horizons of conglomerates and weathering crusts in the cover of large cratons (7-5).

Table 7. Types and subtypes of deposits with palladian gold, its fineness, and impurities.

Type	Subtype (Numbers)	Deposits (Country)	Fineness (Impurities)
1. PGE ore deposits related to mafic–ultramafic magmatic complexes	1a Low-grade sulfide mineralization (includes chromitite and low-grade Cu, Ni)	Skaergaard (Denmark) Stillwater (USA) Konder (Russia) Baronskoe (- " -) Serebryansky Kamen (- " -) Volkovskoe (- " -) Ozernoe (- " -) Khamitovskoe (- " -) 1 V. Chuarvy (- " -), 2 S. Kamennik (- " -), 3 Yu. Peshempakhk(- " -), 4 Fedorova tundra (- " -) Itchayvayam (- " -)	689–913 (Pd,Cu,Ag,Pt,Fe) 882–922 (Pd,Ag) 974 (Pd), 700–900 (Pd,Ag,Cu) 840–940 (Pd,Ag > Cu) 783 (Pd > Cu > Ag) 810–914 (Pd,Ag,Cu) 690–700 (Cu,Ag > Pd) 880–920 (Pd,Ag,Cu) (1) 960 (Pd,Ag,Cu), (2) 830–970 (Pd,Ag,Cu,Fe,Ni), (3) 833–920 (Ag,Pd,Cu,Fe), (4) 770–870 (Ag > Pd > Fe) 816–960, 580–660 (Pd,Hg)
	1b High-grade sulfide mineralization (includes high-grade Cu-Ni)	Norilsk-1 (Russia) Talnakh (- " -) Krutoe (- " -) Marathon (Canada) Cedrolina (Brazil)	570–880, 950–975 (Pd,Ag,Cu,Pt) 980–990, 160–210 (Pd,Ag,Cu) 949–963 (Pd) 663–835 (Cu,Ag > Pd or Cu > Pd) 847 (Au > Pd)
2. Orogenic gold deposits	2a—OGD (Pd,Ag or Pd,Ag,Cu) (5)	Uderei (Russia) Chudnoe (- " -) Nesterovskoe (- " -) Hope's Nose (England) Brownstone (- " -)	920–970 (Pd) 840–990 (Pd,Ag or Pd,Ag,Cu) 955–970 (Pd,Ag,Cu) 854–997 (Pd > >Ag) 970–994, 882–924 (Pd > Ag)
	2b Cu with Organic	Zechstein (Poland)	576–795 (Ag > Pd,Cu,Hg)
3. Epithermal (porphyry) gold–copper deposits	alkaline rocks areas	Bleida Far West (Morocco)	912–993 (Pd,Ag,Cu)
4. Iron Oxide Copper Gold deposits	(5)	Serra Pelada (Brazil) Gongo Soco (- " -) Caue (- " -) Conceichao (- " -) Maquine (- " -)	880–975 (Pd > Ag,Hg,Cu,Fe) 821–967 (Pd,Ag,Cu,Hg) 885–918 (Pd > Ag,Cu) 930–940 (Pd > Ag,Cu) 899–941 (Pd,Ag,Cu)
5. Ferruginous quartzite deposits	(1)	Lebedinskoye (Russia)	780–790 (Ag,Pd,Pt,Ni,Fe)
6. Volcanic exhalation	fumaroles	Ebeko Volcano (- " -)	790 (Pd)
7. Gold-PGE placers	1Pl type 1 areas (2)	Chorokh river (Turkey) river Dart (England)	649–724 (Pd) 630 (Pd > Hg >> Ag)
	2Pl type 2 areas (4)	Ambositra reg. (Madagascar) Lammermuir Hills (Scotland) Zimnik Creek (Poland) Dziwiszow (- " -)	918–978 (Pd > Cu) 942 (Pd) 980–990 (Ag,Pd,Hg) 960–997 (Pd,Ag,Hg,Cu,Pt)
	3Pl type 3 alkaline rocks areas (3)	Friday Creek (Canada) Whipsaw Creek (- " -) Similkameen river (- " -)	930 (Ag > Pd,Hg,Cu) 910–940 (Pd,Ag,Cu,Hg) 970 (Pd,Ag,Cu,Hg)
	4Pl type 4 areas (1)	Corrego Bom Sucesso (Brazil)	770–975 (Pd > Ag,Hg,Cu,Pt)
	5Pl type 5 craton placer areas (4)	Mayat river (Russia) Bol'shaya Kuonamka river (- " -) Kuoyka river (- " -) Mataganya-Siguiri (Guinea)	860–960 (Pd,Ag,Cu,Hg) 910–970 (Pd,Ag,Cu,Hg) 960 (Pd > Ag,Cu) 920–970 (Pd > Hg,Ag,Pt)

According to Figure 1, deposits with palladian gold occur in different geological and dynamic settings—terranes and orogenic belts and shields. Table 8 illustrates that palladian gold is found in the deposits related to different types of magmatism (but mainly with mafic–ultramafic) and in different periods. It is worth noting that the lacuna is more than a billion years old, coinciding with the so-called "dead billion" in the history of the Earth (1.7–0.7 Gya), when the stability conditions existed throughout the planet and a limited number of deposits formed, among which there were virtually no ore-bearing deposits (except IOCG type) [124].

Table 8. Structure, host rock, and mineralization age (Ma) of some endogenous deposits with palladian gold.

Type	Name	Structure (Craton)	Host Rock	Ma	Ref.
1a Low-grade sulfide mineralization (includes chromitite and low-grade Cu, Ni)	Skaergaard, Denmark	East Greenland volcanic province	Mafic	55	[73,125]
	Stillwater, USA	Wyoming Province (N. American)	Mafic–ultramafic	2700–2713	[74,126]
	Itchayvayam	Olyutorsky arc terrane	Mafic–ultramafic	60–74	[42,127]
	Konder	Aldan Shield (Siberian)	Alkaline–ultramafic	130–140	[128,129]
	Baronskoe	Uralian Orogenic belt	Mafic	445–430	[55,125]
	Serebryansky Kamen	- "-	Mafic	415–430	[54,130]
	Volkovskoe	- "-	Mafic	445–430	[30,131]
	Ozernoe	- "-	Mafic–ultramafic	445–455	[31,132]
	Khamitovskoe	- "-	Mafic–ultramafic	587–215	[56,133]
1b High-grade sulfide mineralization (includes high-grade Cu-Ni)	Norilsk-1, Talnakh	Siberian LIP (-"-)	Mafic–ultramafic	250	[27,50]
	Krutoe	Pay-Khoy Fold Belt	Mafic	330–370	[51,134]
	Fedorov-Pana	Baltic Shield (East European)	Mafic–ultramafic	2501–2446	[58,135,136]
	Marathon, Canada	Superior Province (-"-)	Alkaline–mafic–ultramafic	1100	[21]
2 OG	Uderei	Yenisey Fold Belt	Metamorphic	640–710	[49,137]
	Chudnoe, Nesterovskoe	Ural Orogenic belt	Contact metasediments and felsic–mafic volcanites	250–260	[29,53,138]
3 EPGC	Bleida Far West, Morocco	Anti Atlas orogen	Mafic (ophiolite association), felsic, metamorphic	600–550 or 300–330	[32,68]
4 IOCG	Gongo Soco, Caue, Maquine	Atlantic shield (South American)	Metasedimentary, metavolcanic	2600–2400	[25,26,28,77,78]
5 Ferruginous quartzite	Lebedinskoye	(East European)	Metamorphic	2612–2050	[59]

Thus, the above-mentioned facts suggest that Au-Pd mineralization formed in different conditions and from different sources, from mantle riftogenic (oceanic), active margins, and island arcs (subduction) to orogenic accretion-collision and post orogenic (riftogenic).

5.2. Physicochemical Conditions for the Formation of Palladian Gold at Deposits of Different Types

Scarce data are available on the physicochemical conditions of the formation of palladian gold. We summarize the published information on the T,P,X-parameters of the formation of Au-Pd mineralization known for some types of deposits in Table 9.

Table 9. T,P,X-conditions for the formation of Au-Pd mineralization with Pd-bearing gold according to the results of the study of fluid inclusions (FIs) (reference data).

Type of Deposit	Deposit (Country)	T °C; P kbar; wt.% eq. NaCl; FI Composition	Ref.
PGE ore deposits related to mafic–ultramafic magmatic complexes	Au-Pd ores of the Skaergaard massif; Norilsk-1,-2, Volkovskoe, Baronskoe (Russia)	1200–1000; 600–140; -; 23.3–13.6; NaCl	[27,55,106,139]
Orogenic gold deposits	Chudnoe (Russia); Au-Pd deposits in the Permian-Triassic basins (SW England); Zechstein (Poland)	186–60; 0.5–1.15; 0.2–30; Na, Ca, Cl	[29,63,140–142]
Epithermal (porphyry) gold–copper deposits	Bleida Far West (Morocco)	132–80; 0.5–0.04; 1.7–33; Na, Ca, Cl, CO_2	[32,68,143]
Iron oxide copper gold deposits	Serra Pelada, Carajas, Corrego Bom Susesso, Gongo Soco, Conceichao (Brazil)	500–115; 1.2–1.7; 3–30; Na, K, Ca, Mg, Cl, CO_2	[25,78,79,144–147]

Type 1—PGE-deposits in mafic–ultramafic magmatic complexes. In Au-Pd ores at the Skaergaard massif, the formation of Au-Cu-Pd melts and Pd-tetra-auricupride is due to the decomposition of Au-Cu-Pd solid solutions at temperatures of 1200–1000 °C [73]. Au-Pd mineralization of the Volkovskoe and Baronskoe deposits has a hydrothermal-metasomatic genesis, was formed in the range of temperatures of 600–400 °C [55,106,139], and is related to superimposed metamorphic and metasomatic processes (occurrence of amphibole, serpentine, and chlorite) on olivine clinopyroxenites [107,108].

Data on fluid inclusion studies show that the crystallization of palladian gold in Cu-Ni-PGE ores of the Norilsk deposit proceeds with the participation of reduced hydrothermal fluids ($CH_4 \pm C_2H_6$, C_2H_2, and C_3H_8) at temperatures of 270–140 °C and salinities of 23.3–13.6 wt.% eq. NaCl [27].

Type 2—orogenic gold deposits. Au-Pd mineralization at the Chudnoe deposit (Russia) could have been formed by chloride fluids of low and medium salinity at temperatures from 105 to 230 °C and pressures from 0.05 to 1.15 kbar. The salinity of fluids varies from 20.1 to 0.2 wt.% NaCl eq. [29] (Table 9). Au-Pd and Au deposits of the Permian-Triassic basins in SW England (Thorverton) formed in the temperature range from 60 to 155 °C, and the salinity of fluids varied from 2 to 30 wt.% NaCl eq. [140,141]. In the ores from the Au-Pd Zechstein deposit (Poland), an association of palladian gold with chalcosine, digenite, and djurleite was detected, which limits its formation temperature to a range of 93 to 103 °C [63,142,148] (Table 9).

Type 3—epithermal (porphyry) gold–copper deposits. Hematite-quartz veins of the Au-Pd Bleida Far West deposit (Anti-Atlas, Morocco) formed at temperatures from 145 to 325 °C and fluid salinities from 6 to 33 wt.% NaCl eq. The gas phase of fluids included CO_2, CH_4, and N_2 [68,143]. The Au-Pd mineralization at this deposit was formed in the temperature range from 80 to 132 °C with the participation of fluids with high salinity ranging from 18.8 to 29.6 wt.% eq. NaCl. The composition of fluids is dominated by NaCl and $CaCl_2$. Quartz-hematite veins were formed at pressures from 0.76 to 31 kbar; during the formation of quartz veins with Cu-mineralization, the pressure changed from 0.63 to 0.51 kbar. EPMA analysis showed that the mineral-forming fluids contained Cl, Na, Ca, K, Mn, Ba, Sr, Fe, Cr, and S [32].

Type 4—iron oxide copper gold deposits. Hematite-quartz veins from the Au-Pd Gongo Soco mine (Brazil) formed at temperatures of 365–74 °C from fluids with salinities of 10–30 wt.% eq. NaCl. The composition of fluids is dominated by NaCl with $MgCl_2$ impurities [78,146]. Au-Pd mineralization was formed later in the temperature range of 148–229 °C with the participation of medium-salinity fluids of 7.2 to 11.7 wt.% NaCl eq. (Table 9). Iron ores of the Conceichao mine (Brazil) formed at temperatures of 660–600 °C (according to $\Delta\ \delta^{18}O$ hematite/quartz). The deposition of Au-Pd mineralization took place later at lower temperatures (351–115 °C) with the participation of fluids of salinities from 3 to 20 wt.% NaCl eq. [79,147]. Hematite-quartz veins of the Serra Pelada deposit (Brazil) formed in the temperature range from 140 to 185 °C and fluid salinity range from 2.6 to 23.5 wt.% NaCl eq. The fluids contained chlorides Na > K > Ca ≥ Mg. The trapping temperature and salinity of fluid inclusions without and containing gold (50–500 ppb) in hematite are 155–185 °C and 4–6.5 wt.%, and 165–175 °C and 4.5–6 wt.%, respectively [144].

The physicochemical formation conditions of Pd-bearing gold in the ores of deposits of type 1 include two temperature ranges—magmatic high-temperature and hydrothermal low-temperature. In the former case, the formation of Pd-bearing gold takes place as a result of the decomposition of Au-Cu-Pd solid solutions at 1200–1000 °C. In the latter, Au-Pd alloys form with the participation of low-temperature hydrothermal fluids (270–140 °C) of medium salinity (13.6–23.3 wt.% NaCl eq.) (Figure 11A).

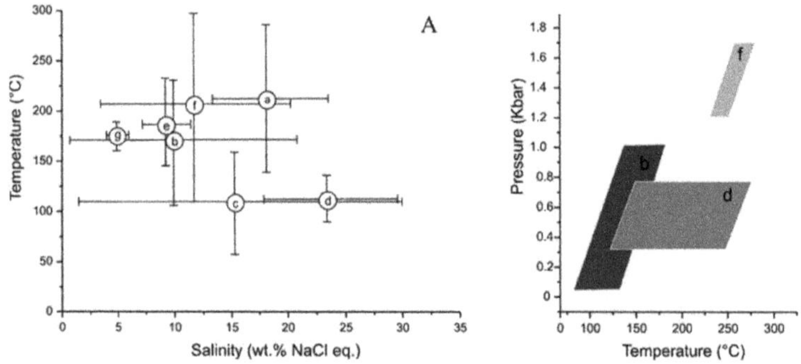

Figure 11. Homogenization temperature and salinity of fluid inclusions in quartz and calcite in association with palladian gold in ores from deposits of different types (**A**): 1c—Norilsk (a); 2—Chudnoe (Russia) (b) and Hope's Nose, Thorverton (England) (c); 3—Bleida Far West (Morocco) (d); 4—Gongo Soko (e), Conceichao (f), and Serra Pelada (Brazil) (g). Probable P-T formation conditions (**B**) for the Au-Pd deposits of Chudnoe (b), Bleida Far West (d), and Serra Pelada (f).

The physicochemical conditions of formation of palladian gold in the ores of deposits of types 2–4 are similar and correspond to the conditions of low-temperature hydrothermal ore formation (Figure 11B). At these deposits, Pd-bearing gold formed in the temperature interval of 300–60 °C, and the salinity of hydrothermal fluids ranges widely from 30 to 0.2 wt.% NaCl eq. (Figure 11A). Estimation of the ore formation pressure at Au-Pd deposits of types 2–4 is based on the results of studies of fluid inclusions in the minerals of early stages that preceded the formation of Pd-bearing gold, and the pressure varies from 1.7 to 0.05 kbar (Figure 11B).

5.3. Impurities, Fineness of Palladian Gold, and Minerals in the Association as Indicators of Deposit Types

The content of Ag, Cu, and Hg in palladian gold and minerals in association with it from deposits of various types has been studied by many researchers. Each of the deposits (Figure 1, Tables 1–7) for which the information was collected and analyzed is unique and has a specific composition of palladian gold and a set of minerals in the association. The heterogeneity of palladian gold forms either during primary crystallization or results from further modification under the effect of chemical and physical factors during subsequent residence in hypogene or surficial environments.

Figure 12a shows data on the palladian gold with various sets of impurities and other co-occurring phases of the Au–Pd–Ag–Cu–Hg system. The Ag impurity was found in palladian gold at 36, Cu—at 32, and Hg—at 20 deposits. The palladian gold is represented by the systems Au–Pd, Au–Pd–Hg, Au–Pd–Cu, and Au–Pd–Ag–Hg, but more frequently corresponds to the Au–Pd–Ag, Au–Pd–Ag–Cu, and Au–Pd–Ag–Cu–Hg. The variations in Ag content in Au–Pd–Ag–Cu solid solutions are considerable and cover the widest range of fineness from 160 to 993. At some deposits, the Pd content exceeds that of Ag and Cu, whereas at others, Cu dominates over Pd and Ag or Ag dominates over Pd and Cu. Low-fineness palladian gold is related to elevated contents of Pd, Ag, Cu, and Hg. At the Marathon deposit (Canada), the fineness of palladian gold is 663–835, and the content of Cu (and Ag) is higher than that of Pd (Table 6). At the Zechstein deposit (Poland), the fineness of palladian gold is low (576‰–795‰) and the content of Ag is higher than that of Pd, Cu, and Hg. The content of Pd is higher than those of Hg and Ag in native gold from the placers of the Dart river (England) (630‰–972‰). Talnakh ores contain palladian silver (160‰–210‰) (Table 5). Wide variations in the concentrations of all four elements and fineness (556‰–931‰) were revealed for palladian gold from placers of the Similkameen river (Canada) (Pd,Ag,Cu,Hg) (Tables 6 and 7). The low fineness of palladian gold is due to the high concentrations of Pd (Chorokh river, Turkey); or Pd and Hg (Itchayvayam (Russia); River Dart (England)); or Pd, Ag, and Cu (Norilsk-1, Talnakh, Ozernoe (Russia); Marathon (Canada); Skaergaard (Denmark)); or Pd, Ag, Cu, and Hg (Zechstein (Poland); Similkameen river (Canada)).

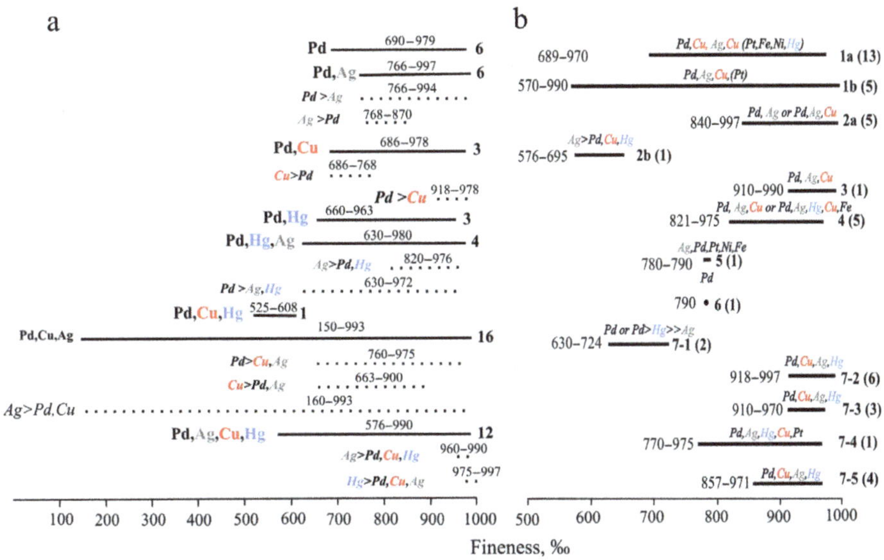

Figure 12. Fineness intervals of palladian gold for deposits with Au-Pd mineralization (**a**) depending on the set and concentrations of impurities (Pd, Ag, Cu, Hg, . . .) in native gold (numerals on the right are the number of deposits) (**b**) for different types and subtypes of deposits.

Figure 12b shows the variations in the fineness and impurity elements of palladian gold from deposits of various types (and subtypes) summarized in Table 7. Palladian gold most commonly occurs in the PGE ore deposits related to mafic–ultramafic magmatic complexes (Table 7, type 1) and contains Ag and Cu impurities; occasionally minor Pt, Fe, and Ni; and does not contain Hg. This is mostly high-fineness gold, is less often medium-, and is rarely low-fineness. The only exception is the Au-Pd-Hg Itchayvayam ore occurrence (Kamchatka, Russia), for which two varieties of Pd,Hg-bearing gold—high- (low content of Pd and Hg) and low-fineness (high content of Pd and Hg) gold—were determined [33,42]. This type is most likely typical of territories with elevated background levels of Hg. It is in the northern part of Kamchatka that a great number of deposits and ore occurrences of the argillisite formation were discovered [149,150]. Ore objects are related to island arc terrigenous and volcanic-siliceous deposits of the Early and Late Cretaceous (terranes of the Olyutorskaya island arc in Figure 1). The Itchayvayam ore occurrence occurs in the deep fault zones and frequently accompanies the massifs of mafic–ultramafic magmatic complexes. The elevated background levels of Hg seem to be because oceanic sediments contain high Hg concentrations compared to clarke. Enrichment of oceanic sediments with Hg could be a result of volcanic activity [150].

For orogenic gold (subtype 2a), epithermal (porphyry) gold–copper (type 3), and iron oxide copper gold (type 4) deposits, palladian gold is mainly high-fineness and contains Ag impurities or Cu too. Low-fineness palladian gold with a major content of Ag is typical of the Zechstein deposit (Poland) (subtype 2b). Palladian gold of medium fineness (780‰–790‰) with Ag, Pd, Pt, Ni, and Fe impurities occurs at the Au–Pd–Pt Lebedinskoye deposit (Russia) (type 5), and was also found in volcanic fumaroles of Ebeko volcano (Russia) (only Pd impurity) (type 6). High-fineness gold (910‰–970‰) with low contents of Pd, Ag, Cu, and Hg is spread over a great part of gold-PGE placers (type 7). Pd,Hg-rich low-fineness gold (630‰–770‰) has been found in some placers of the Chorokh river (Turkey) [60], the Dart river (England) [20], and weathering crusts at the Corrego Bom Sucesso (Brazil) [76].

Tables 1–6 show the minerals that occur in association with palladian gold at the deposits with various types of Au-Pd mineralization. The number of minerals is about 90, apart from unnamed phases. Table 10 contains data on minerals in association with palladian gold from deposits of different types. Minerals (bold) are the most frequent phases. Mineral groups are presented by native metals and intermetallic compounds, oxides, hydroxides, and chalcogenides. Pd, Ag, Cu, and Hg metals that occur in native gold frequently form their minerals and are present in association with palladian gold. In Table 10, these minerals are listed separately by group.

Table 10. Minerals in association with palladian gold from deposits of different types.

Mineral Groups	Minerals	Types of Deposits
Elements and intermetallics	**Pd**, **Pt**, **Cu**, Ag, S, **Taur**, Auc, Skg, Tfpt, Ifpt	Types 1, 2a, 3, 7-2, 6
Pd minerals	Ah, **Met-I/II**, **Ism**, Stpdn, Ktu, Tpdn, Tem, Pds, Ato, Spdn, Plv, **Vsk**, Mrk, Kei, Cpe, Mon, Csl, Oos, Sov, Pdm, **Ptr**	Type 1
Minerals of Au and Ag	Ausb, Any, **Syv**, Fis, **Hes**, Cag, Boh, **Nau**, **Eca**, **Taur**, **Auc**	Types 2, 4, 7-1
Cu minerals	**Ccp**, **Bn**, Cc, Ani, Eca, Um, **Taur**, **Auc**	Types 1, 7-1
Hg minerals	Cin, Tmn, Clr, **Ptr**	Types 1b,d, 7-1, 7-4
Pt minerals	**Spy**, **Bg**, Svi, Pda, Ifpt, Tfpt	Type 1
Oxides, hydroxides	**Hem**, **Mgt**, Tnr, (Pd,Cu)O, PdO, **Gth**, (Fe,Pd)OOH	Types 3, 4, 5, 7-2
Fe, Ni, Pb chalcogenides	**Pn**, **Pyh**, **Gn**, Cth, Alt	Type 1
Host minerals	**Qz**, **Ms** (including fuchsite Cr-Ms), **Chl**, **Ab**, **K-fs**, **Hbl**, Tur, **Carb** (including Cat, Sid and other), S, Kln	Types 1–7

Note: The most common minerals are in bold.

Palladium minerals make up the largest group. Most common among them are arsenides, stibioarsenides, sulfides, stannides, bismuthides, tellurides, selenides, etc. (Table 10). In association with them, Pt minerals—sperrylite, braggite, sudovikovite, palladoarsenide, isoferroplatinum and tetraferroplatinum—are present. Pd and Pt minerals are typical of the deposits of types 1 and 7 (Table 7), which indicates their paragenetic relationship.

Gold and silver minerals frequently occur in association with palladian gold—these are Au-Ag-Cu solid solutions and Ag-Cu intermetallic compounds (tetra-auricupride, auricupride)—and more rare minerals are aurostibite, anyuiite, sylvanite, fischesserite, hessite, chlorargyrite, bohdanowiczite, naumannite, and eucairite. Au-Cu intermetallic compounds were found at more than ten deposits. Tetra-auricupride is spread at the deposits belonging to groups 1 and 7. Aurostibite is in association with palladian gold in the ores of the Uderei deposit (Russia) (type 2a) [49] and placer ore occurrences of Lammermuir Hills (Scotland) and River Dart (England) (type 7a,b) [20].

Copper minerals are represented by sulfides (chalcopyrite, bornite, chalcocite, anilite) and selenides (eucairite, umangite)—these are typical minerals at some deposits of mafic–ultramafic magmatic complexes (type 1). Cu oxides (tenorite, (Pd,Cu)O) were found in exogenous deposits (type 7). The most widespread mercury mineral is potarite, whereas cinnabar, tiemannite, and coloradoite are rare. They were detected at some deposits of types 1, 2b, 7-1, and 7-4.

Fe oxides (hematite, magnetite) and Fe and Pd hydroxides (FeOOH, (Fe,Pd)OOH) are spread at the deposits of 3,4,7 groups and characterize highly oxidizing conditions of ore formation. Chalcogenides of Fe, Ni, and Pb (pentlandite, pyrrhotite, galena, clausthalite, altaite) occasionally occur in association with palladian gold at the deposits of type 1. Arsenopyrite in association with palladian gold was found only at the Uderei (Russia) [49] and Lammermuir Hills (Scotland) deposits [20]. Arsenic often occurs in arsenides and arsenoantimonides of Pd and Pt. Among host minerals containing palladian gold, the most common are quartz, muscovite, including fuchsite (Cr-Ms), chlorite, albite, K-feldspar, kaolinite, hornblende, and carbonates (calcite, siderite).

The above-listed minerals of Au, Pd, Cu, Ag, Hg, Pt, Fe, and Pb are formed together with palladian gold and influence its composition and redistribution of metals between fluids and solid phases. On the whole, minerals in association with palladian gold reflect the mineralogy of ores, which points out their use as indicators of the type of mineralization. PGE at elevated temperatures are typical chalcophiles, and have a high affinity with S, As, Te, Sb, and Bi. After post-magmatic fluid processing, a significant part of the native metals leave the high-temperature solid solutions, and various minerals of PGE, Au, and Ag are formed: native, sulfides, arsenides, tellurides, antimonides, bismuthides, etc.

An important area of research is works focused on studying the mechanism of the mobilization, transfer, and deposition of noble metals. It was proved that during crystallization, hot water fluids are generated, which migrate in the crust [151]. Metals can be transported by brines, altered meteoric water, and metamorphic and magmatogenic fluids. Their deposition from hydrothermal solutions takes place at temperatures from 500 to 100 °C. They can form their own minerals or occur in other minerals as isomorphic impurities. The ability of particular metals to occur in gold depends on their amount in the ore-forming system and the content of other elements with which they can form stable minerals in ore-forming conditions.

The question of the influence of the activity of elements (S, Te, Sb, As, Bi, etc.) that bind palladium into its own minerals on its ability to occur in native gold has not been resolved so far. The solution to this problem is complicated by the fact that hydrothermal fluids that formed Au-Pd mineralization simultaneously contain several of these elements, and native gold is found in parageneses with many minerals of Pd. The Baronskoe in the Middle Urals is an example of deposits at which palladian gold is in paragenesis with Pd sulfides, tellurides, sulfotellurides, arsenides, stibnides, and bismuthides. For this type of deposit, we managed to trace the variations in the contents of Pd in Ag-bearing gold with sulfides (vysotskite), tellurides (kotulskite, telluropalladinite), arsenides Pd and Pt, arsenotellurides (keitconite, As-keitconite), and As-Sb-Te minerals (Te-arsenopalladinite, Te-guanglinite, isomertieite). It was revealed that the occurrence of Pd, as well as Ag, is controlled mainly by the activity of S and Te, and the effect of the activity of As and Sb is negligible (Figure 13).

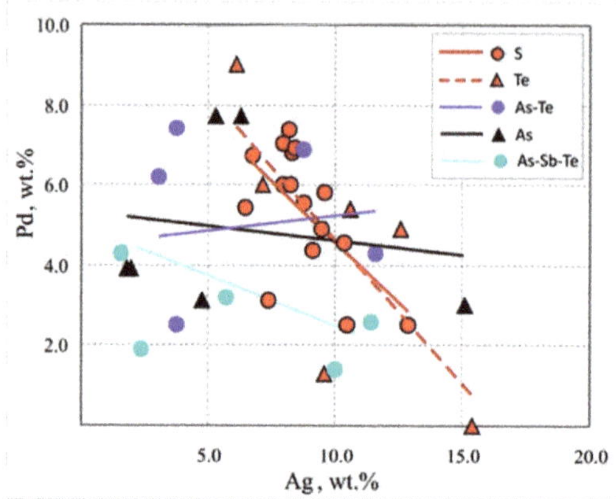

Figure 13. Ratios of Ag and Pd impurities in native gold in parageneses with Pd sulfides (S), tellurides (Te), arsenotellurides (As-Te), arsenides (As), and arseno-stibiotellurides (As-Sb-Te) at the Baronskoe deposit. Ag–Pd linear correlation trends are shown.

What do the Pd impurity and level of its concentrations in native gold indicate? What other isomorphic impurities along with Pd can be present in native gold? Which minerals occur in the intergrowth with palladian gold? At what deposits are Pd-bearing gold spread? What is the source of Pd and other accompanying impurities in native gold? What are the mechanisms and processes of mobilization, transfer, and joint deposition of these noble metals? In what physicochemical conditions and environments does Pd-bearing gold form? We tried to answer these and other questions in this paper. We will make attempts to answer many questions in the future. Impurities are a kind of fingerprint for minerals that make it possible to determine the composition and conditions of ore-forming medium. The composition of native gold is an important typomorphic characteristic of this indicator mineral, bearing fingerprints of formation conditions specific to each deposit. The knowledge of impurity elements in native gold and their content is important for identifying the ore formation to which gold mineralization belongs, for determining the source of placer ore occurrences, developing efficient criteria for forecasting, and searching gold and gold-bearing deposits. The minerals in association with native gold and the sequence of their formation reflecting the specific features of ore formations are the basis for ore-formational analysis [152]. The distribution of these associations determines the contours of deposits. Analysis of data on the deposits with palladian gold showed specific compositional features of the productive mineral associations from various types of deposits located on different continents and in different countries.

Construction of the quantitative genetic models of ore-forming processes at deposits with palladian gold requires thermodynamic data for solid solutions of various compositions. Estimation of the thermodynamic characteristics was performed only for some binary and ternary systems with gold [6,153–156], which makes it impossible to carry out this kind of work at this stage. Model calculations were made for the systems containing Au-Ag [4,6,7], Au-Ag-Hg [155], Au-Ag-Pd [156], Au-Ag-Cu [157], and Au-Ag-Cu-Hg [158]. The composition and set of impurities in palladian gold were found to be more diverse than it was thought to be. Most likely, very soon, it will be possible to confidently determine any region from which its samples were obtained from the composition of impurities in native gold.

6. Conclusions

(1) Depending on the set of impurities, the following varieties of palladian gold are distinguished: Ag-, Cu-, Hg-, Ag,Cu-, Ag,Hg-, Cu,Hg-, and Ag,Cu,Hg-bearing palladian gold. Palladian gold may contain Pt, Fe, Ni, and Cd (in minor quantity).

(2) A classification of types of deposits with palladian gold has been proposed: (1) PGE ore deposits related to mafic–ultramafic magmatic complexes (two subtypes—(a) low-sulfide-grade (less than 2%–5% sulfides) Alaskan and (b) high-sulfide-grade (more than 5% sulfides) Norilsk); (2) orogenic gold deposits; (3) epithermal (porphyry) gold–copper deposits; (4) iron oxide copper gold deposit type; (5) ferruginous quartzite deposits; (6) volcanic exhalation; and (7) gold-PGE placers with five subtypes corresponding to the types of 1–5 primary sources.

(3) Ag,Cu-bearing palladian gold is mainly high-fineness (910‰–990‰), is less frequently medium-fineness, and is rarely low-fineness and does not contain Hg at the deposits of PGE ore deposits related to mafic–ultramafic magmatic complexes, epithermal (porphyry) gold–copper deposits, and iron oxide copper gold deposits (types 1, 3, 4). The only exception is the Au-Pd-Hg Itchayvayam ore occurrence (Kamchatka, Russia) (type 1d) with two varieties of Pd,Hg-bearing gold (high-fineness 816‰–960‰ and low-fineness 580‰–660‰). Low-fineness palladian gold with the major content of Ag is typical for OGD deposit Zechstein (Poland) (type 2), whereas that with the major content of Pd is typical of the placers of the Chorokh river (Artvin district, Turkey) (type 7). Medium-fineness palladian gold with a high content of Pd occurs at deposits and in volcanic exhalations (types 5, 6). Hg,Ag,Cu-bearing high-fineness palladian gold is mainly present in placer deposits (type 7).

(4) The most common minerals in association with palladian gold are Pd and Pt arsenides, stibioarsenides, sulfides, stannides, bismuthides, tellurides, and selenides. They are typical of deposits related to mafic–ultramafic magmatic complexes (types 1, 7). Au, Ag, and Cu minerals (tetra-auricupride, auricupride, aurostibite, chalcopyrite, bornite, sylvanite, hessite, naumannite, eucairite, etc.) occur in association with palladian gold at gold–copper deposits (types 2–4). Cu and Fe oxides (tenorite, hematite, magnetite, (Pd,Cu)O) and Fe and Pd hydroxides (goethite, (Fe,Pd)OOH) are spread at the deposits of 3,4,7 groups and indicate highly oxidizing conditions of ore formation. The most common of Hg minerals is potarite. The main host minerals of palladian gold are quartz, muscovite, including fuchsite (Cr-Ms), chlorite, albite, K-feldspar, kaolinite, hornblende, and carbonates (calcite, siderite).

(5) Palladian gold from many deposits has a heterogeneous composition and occurs in two or more varieties, which suggests unstable deposition conditions, subsequent recrystallization processes, and a long history of formation. Physicochemical conditions of the formation of Pd-bearing gold at some deposits of one type cover two areas—magmatic high-temperature and hydrothermal low-temperature. At the majority of deposits of types 2 and 4, the formation of Pd-bearing gold proceeds with the participation of hydrothermal fluids (300–60 °C) of various salinities (0.2–30 wt.% NaCl eq.).

(6) The fineness, impurity metals, and minerals in association with palladian gold reflect the mineralogy of Au-Pd ores and allow them to be used as indicators for types and subtypes of gold deposits.

7. Future Directions

In the studies of new objects, it is necessary to more thoroughly analyze the chemical composition of palladian gold and minerals in association with it. An important characteristic of palladian gold is the data on the composition of microinclusions obtained by LA-ICP-MS. However, the heterogeneity of its composition and the presence of mineral microinclusions must be taken into account for interpreting the obtained results.

A database of the deposits with palladian gold, as well as argentian, cuprian, and mercury gold must be created. Scarce data are known on the P,T,X-parameters of the formation of native gold of various compositions and sources of both gold and impurity elements (major, minor, and trace). Possible and important directions of further works are a study of different forms of transfer of Au, Pd, Ag, Cu, Hg, and other elements at different P,T,X-parameters of ore-forming conditions, obtaining thermodynamic characteristics of Au-

Pd-Ag-Cu-Hg solid solutions and minerals of these elements and minerals in association with the aim to construct the genesis models of palladian gold and other varieties of native gold.

Author Contributions: Conceptualization, G.A.P.; methodology, G.A.P.; formal analysis, T.V.B., P.S.Z., G.A.P. and V.V.M.; investigation, G.A.P. and V.V.M.; data curation, T.V.B. and P.S.Z.; writing—original draft preparation, G.A.P., V.V.M., A.A.B. and N.A.G.; writing—review and editing, G.A.P., V.V.M. and N.A.G.; visualization, G.A.P., P.S.Z., T.V.B. and V.V.M.; supervision, G.A.P.; project administration, G.A.P. All authors have read and agreed to the published version of the manuscript.

Funding: This research was financially supported by the Russian Foundation for Basic Research (project No. 20-05-00393) and within the framework of the state assignment of Institutes: Sobolev Institute of Geology and Mineralogy of the Siberian Branch of the Russian Academy of Sciences (Novosibirsk, Russia) (No. 122041400237-8); Zavaritsky Institute of Geology and Geochemistry, Ural Branch of Russian Academy of Sciences (No. 123011800011-2); Institute of Volcanology and Seismology of the Far East Branch of Russian Academy of Sciences (No. AAAA-A20-120121090011); and North-East Interdisciplinary Scientific Research Institute, Far East Branch of the Russian Academy of Sciences (No. 121031700301-5).

Data Availability Statement: Not applicable.

Acknowledgments: We are grateful to the anonymous reviewers for polishing the English and the constructive and helpful corrections and comments that improved the manuscript.

Conflicts of Interest: The authors declare no conflict of interest.

References

1. Petrovskaya, N.V. *Native Gold*; Nauka: Moscow, Russia, 1973; 348p. (In Russian)
2. Boyle, R.W. The geochemistry of gold and its deposits. *Geol. Surv. Can. Bull.* **1979**, *280*, 584.
3. Morrison, G.W.; Rose, W.J.; Jaireth, S. Geological and geochemical controls on the silver content (fineness) of gold in gold-silver deposits. *Ore Geol. Rev.* **1991**, *6*, 333–364. [CrossRef]
4. Gammons, C.H.; Williams-Jones, A.E. Hydrothermal geochemistry of electrum; thermodynamic constraints. *Econ. Geol.* **1995**, *90*, 420–432. [CrossRef]
5. Konstantinov, M.M. *Gold Provinces of the World*; Scientific World: Moscow, Russia, 2006; 358p. (In Russian)
6. Pal'yanova, G.A. Physicochemical modeling of the coupled behavior of gold and silver in hydrothermal processes: Gold fineness, Au/Ag ratios and their possible implications. *Chem. Geol.* **2008**, *255*, 399–413. [CrossRef]
7. Liang, Y.; Hoshino, K. Thermodynamic calculations of Au_xAg_{1-x} fluid equilibria and their applications for ore-forming conditions. *Appl. Geochem.* **2015**, *52*, 109–117. [CrossRef]
8. Savva, N.E.; Kravtsova, R.G.; Anisimova, G.S.; Palyanova, G.A. Typomorphism of native gold (Geological-industrial types of gold deposits in the North-East of Russia). *Minerals* **2022**, *12*, 561. [CrossRef]
9. Spiridonov, E.; Yanakieva, D. Modern mineralogy of gold: Overview and new data. *Archéosciences* **2009**, *33*, 67–73. [CrossRef]
10. Gas'kov, I.V. Major impurity elements in native gold and their association with gold mineralization settings in deposits of Asian fold belts. *Russ. Geol. Geophys.* **2017**, *58*, 1080–1092. [CrossRef]
11. Palyanova, G.A. Gold and silver minerals in sulfide ore. *Geol. Ore Depos.* **2020**, *62*, 383–406. [CrossRef]
12. Chapman, R.J.; Banks, D.A.; Styles, M.T.; Walshaw, R.D.; Piazolo, S.; Morgan, D.J.; Grimshaw, M.R.; Spence-Jones, C.P.; Matthews, T.J.; Borovinskaya, O. Chemical and physical heterogeneity within native gold: Implications for the design of gold particle studies. *Miner. Depos.* **2021**, *56*, 1563–1588. [CrossRef]
13. Nikolaeva, L.A.; Gavrilov, A.M.; Nekrasova, A.N.; Yablokova, S.V.; Shatilova, L.V. *Native Gold in Lode and Placer Deposits of Russia. Atlas*; Krivtsov, A.I., Ed.; TsNIGRI: Moscow, Russia, 2003; 184p.
14. Nikolaeva, L.A.; Nekrasova, A.N.; Milyaev, S.A.; Yablokova, S.V.; Gavrilov, A.M. Geochemistry of native gold from deposits of various types. *Geol. Ore Depos.* **2013**, *55*, 176–184. [CrossRef]
15. Chapman, R.J.; Leake, R.C.; Moles, N.R.; Earls, G.; Cooper, C.; Harrington, K.; Berzins, R. The application of microchemical analysis of gold grains to the understanding of complex local and regional gold mineralization: A case study in Ireland and Scotland. *Econ. Geol.* **2000**, *95*, 1753–1773.
16. Chapman, R.J.; Moles, N.R.; Bluemel, B.; Walshaw, R.D. *Detrital Gold as an Indicator Mineral*; Geological Society, London, Special Publications: London, UK, 2022; Volume 516, pp. 313–336.
17. Nikiforova, Z. Internal structures of placer gold as an indicator of endogenous and exogenous processes. *Minerals* **2023**, *13*, 68. [CrossRef]
18. Lalomov, A.; Grigorieva, A.; Kotov, A.; Ivanova, L. Typomorphic features and source of native gold from the Sykhoi Log area placer deposits, Bodaibo gold-bearing district, Siberia, Russia. *Minerals* **2023**, *13*, 707. [CrossRef]

19. Okrugin, A.V. *Platinum-Bearing Placers of the Siberian Platform*; Publishing House YaF SO RAN: Yakutsk, Russia, 2000; 184p.
20. Chapman, R.J.; Leake, R.C.; Bond, D.P.G.; Stedra, V.; Fairgrieve, B. Chemical and mineralogical signatures of gold formed in oxidizing chloride hydrothermal systems and their significance within populations of placer gold grains collected during reconnaissance. *Econ. Geol.* **2009**, *104*, 563–585. [CrossRef]
21. Ames, D.E.; Kjarsgaard, I.M.; McDonald, A.M.; Good, D.J. Insights into the extreme PGE enrichment of the W Horizon, Marathon Cu-Pd deposit, Coldwell Alkaline Complex, Canada: Platinum-group mineralogy, compositions and genetic implications. *Ore Geol. Rev.* **2017**, *90*, 723–747. [CrossRef]
22. Sluzhenikin, S.F. Platinum-copper-nickel and platinum ores of the Norilsk region and their ore mineralization. *Russ. J. Gen. Chem.* **2010**, *81*, 1288–1298. [CrossRef]
23. Krivenko, A.P.; Tolstykh, N.D.; Veselovskiy, N.N.; Mayorova, O.N. Gold-bearing tellurides of PGE and palladium-bearing gold in gabbronorite of Panskiy massif in Kola peninsula. *Dokl. USSR Acad. Sci. Earth Sci. Sect.* **1991**, *319*, 187–191.
24. Tolstykh, N.D.; Sidorov, E.; Laajoki, K.V.O.; Krivenko, A.P.; Podlipskiy, M. The association of platinum-group minerals in placers of the Pustaya river, Kkamchatka, Russia. *Can. Mineral.* **2000**, *38*, 1251–1264. [CrossRef]
25. Cabral, A.R.; Lehmann, B.; Kwitco, R.; Cravo Costa, C.H. Palladium and platinum minerals from the Serra Pelada Au–Pd–Pt deposit, Carajás mineral province, Northern Brazil. *Can. Mineral.* **2002**, *40*, 1451–1463. [CrossRef]
26. Olivo, G.R.; Gauthier, M.; Bardoux, M. Palladian gold from the Caue iron mine, Itabira District, Minas Gerais, Brazil. *Mineral. Mag.* **1994**, *58*, 579–587. [CrossRef]
27. Sluzhenikin, S.F.; Mokhov, A.V. Gold and silver in PGE-Cu-Ni and PGE ores of the Noril'sk deposits, Russia. *Miner. Depos.* **2015**, *50*, 465–492. [CrossRef]
28. Varajão, C.A.C.; Colin, F.; Vieillard, P.; Melfi, A.J.; Nahon, D. Early weathering of palladium gold under lateritic conditions, Maquiné Mine, Minas Gerais, Brazil. *Appl. Geochem.* **2000**, *15*, 245–263. [CrossRef]
29. Palyanova, G.; Murzin, V.; Borovikov, A.; Karmanov, N.; Kuznetsov, S. Native gold in the Chudnoe Au-Pd-REE deposit (Subpolar Urals, Russia): Composition, minerals in intergrowth and genesis. *Minerals* **2021**, *11*, 451. [CrossRef]
30. Murzin, V.V.; Palyanova, G.A.; Anikina, E.V.; Moloshag, V.P. Mineralogy of noble metals (Au, Ag, Pd, Pt) in Volkovskoe Cu-Fe-Ti-V deposit (Middle Urals, Russia). *Lithosphere* **2021**, *21*, 643–659. [CrossRef]
31. Murzin, V.; Palyanova, G.; Mayorova, T.; Beliaeva, T. The gold–palladium Ozernoe occurrence (Polar Urals, Russia): Mineralogy, conditions of formation, sources of ore matter and fluid. *Minerals* **2022**, *12*, 765. [CrossRef]
32. Kalinin, Y.A.; Borovikov, A.A.; Maacha, L.; Zuhair, M.; Palyanova, G.A.; Zhitova, L.M. Au-Pd mineralization and ore-forming fluids of the Bleida Far West deposit (Anti-Atlas, Morocco). *Geol. Ore Depos.* **2022**, *64* (Suppl. 2), S236–S255. [CrossRef]
33. Palyanova, G.; Kutyrev, A.; Beliaeva, T.; Shilovskikh, V.; Zhegunov, P.; Zhitova, E.; Seryotkin, Y. Pd,Hg-rich gold and compounds of the Au-Pd-Hg system at the Itchayvayam mafic-ultramafic complex (Kamchatka, Russia) and other localities. *Minerals* **2023**, *13*, 549. [CrossRef]
34. Townley, B.K.; Hérail, G.; Maksaev, V.; Palacios, C.; Parseval, P.; Sepulveda, T.F.; Orellana, R.; Rivas, P.; Ulloa, C. Gold grain morphology and composition as an exploration tool: Application to gold exploration in covered areas. *Geochem. Explor. Environ. Anal.* **2003**, *3*, 29–38. [CrossRef]
35. Watling, R.J.; Herbert, H.K.; Delev, D.; Abell, I.D. Gold fingerprinting by laser ablation inductively coupled plasma mass spectrometry. *Spectrochim. Acta* **1994**, *49*, 205–219. [CrossRef]
36. Chapman, R.J.; Mortensen, J.K.; Allan, M.M.; Walshaw, R.D.; Bond, J.; MacWilliam, K. A New approach to characterizing deposit type using mineral inclusion assemblages in gold particles. *Econ. Geol.* **2022**, *117*, 361–381. [CrossRef]
37. Petrella, L.; Thébaud, N.; Evans, K.; LaFlamme, C.; Occhipinti, S. The role of competitive fluid-rock interaction processes in the formation of high-grade gold deposits. *Geochim. Cosmochim. Acta* **2021**, *313*, 38–54. [CrossRef]
38. Liu, H.; Beaudoin, G.; Makvandi, S.; Jackson, S.E.; Huang, X. Multivariate statistical analysis of trace element compositions of native gold from orogenic gold deposits: Implication for mineral exploration. *Ore Geol. Rev.* **2021**, *131*, 104061. [CrossRef]
39. Liu, H.; Beaudoin, G. Geochemical signatures in native gold derived from Au-bearing ore deposits. *Ore Geol. Rev.* **2021**, *132*, 104066. [CrossRef]
40. McClenaghan, M.B.; Cabri, L.J. Review of gold and platinum group element (PGE) indicator minerals methods for surficial sediment sampling. *Geochem. Explor. Environ. Anal.* **2011**, *11*, 251–263. [CrossRef]
41. Oberthür, T. The fate of platinum-group minerals in the exogenic environment—From sulfide ores via oxidized ores into placers: Case studies Bushveld Complex, South Africa, and Great Dyke, Zimbabwe. *Minerals* **2018**, *8*, 581. [CrossRef]
42. Sidorov, E.G.; Kutyrev, A.V.; Zhitova, E.S.; Chubarov, V.M.; Khanin, D.A. Origin of platinum-group mineral assemblages from placers in rivers draining from the Ural-Alaskan type Itchayvayamsky ultramafics, far east Russia. *Can. Mineral.* **2019**, *57*, 91–104. [CrossRef]
43. Shevko, E.P.; Bortnikova, S.B.; Abrosimova, N.A.; Kamenetsky, V.S.; Bortnikova, S.P.; Panin, G.L.; Zelenski, M. Trace elements and minerals in fumarolic sulfur: The case of Ebeko volcano, Kuriles. *Geofluids* **2018**, *2018*, 4586363. [CrossRef]
44. Nekrasov, I.Y.; Ivanov, V.V.; Lennikov, A.M.; Sapin, V.I.; Safronov, P.P.; Oktyabr'skii, R.A. Rare natural polycomponent alloys based on gold and copper from the platinum placer in the Konder alkaline-ultrabasic massif, southeastern Aldan shield, Russia. *Geol. Ore Depos.* **2001**, *43*, 406–417.

45. Nekrasov, I.Y.; Lennikov, A.M.; Zalishchak, B.L.; Oktyabrsky, R.A.; Ivanov, V.V.; Sapin, V.I.; Taskaev, V.I. Compositional variations in platinum-group minerals and gold, Konder alkaline-ultrabasic massif, Aldan shield, Russia. *Can. Mineral.* **2005**, *43*, 637–654. [CrossRef]
46. Svetlitskaya, T.V.; Nevolko, P.A.; Kolpakov, V.V.; Tolstykh, N.D. Native gold from the Inagli Pt-Au placer deposit (the Aldan Shield, Russia): Geochemical characteristics and implications for possible bedrock sources. *Miner. Depos.* **2018**, *53*, 323–338. [CrossRef]
47. Okrugin, A.V.; Mazur, A.B.; Zemnuhov, A.L.; Popkov, P.A.; Sleptsov, S.V. The palladium gold-PGM association in the placers of the Anabar River basin, NE part of the Siberian platform, Russia. *Otechestvennaya Geol.* **2009**, *5*, 3–10. (In Russian)
48. Okrugin, A.; Gerasimov, B. Paragenetic association of platinum and gold minerals in placers of the Anabar River in the Northeast of the Siberian Platform. *Minerals* **2023**, *13*, 96. [CrossRef]
49. Rudashevsky, N.S.; Gorbunov, A.A.; Antonov, A.V.; Alikin, O.V.; Rudashevsky, V.N.; Bobrova, O.V. Palladian gold (Au, Pd) in gold and antimony ores of the Uderei deposit, Yenisei ridge. *Proc. Fersman Sci. Sess. GI KSC RAS* **2019**, *16*, 492–496.
50. Spiridonov, E.M.; Serova, A.A.; Korotaeva, N.N.; Zhukov, N.N.; Kulagov, E.A.; Belyakov, S.N.; Kulikova, I.M.; Sereda, E.V.; Tushentsova, I.N. Genetic Pd, Pt, Au, Ag, and Rh mineralogy in Noril'sk sulfide ores. *Geol. Ore Depos.* **2015**, *57*, 402–432. [CrossRef]
51. Shaybekov, R.I.; Sokerina, N.V.; Isaenko, S.I.; Shanina, S.N.; Zykin, N.N. Gold-telluride-palladium mineralization, a new type of mineralization in gabbro-dolerites of the Pai-Khoi ridge (Yugor peninsula, Russia). *Russ. Geol. Geophys.* **2020**, *61*, 268–285. [CrossRef]
52. Malyugin, A.A.; Chervyakovsky, S.G.; Sazonov, V.N. A new placer-forming type of gold mineralization. *Rep. Acad. Sci. USSR* **1986**, *288*, 697–699. (In Russian)
53. Borisov, A.V. Geological and Genetic Features of Au-Pd-REE Ore Occurrences in the Maldy-Nyrd Ridge (Subpolar Urals). Ph.D. Thesis, IGEM RAS, Moscow, Russia, 2005; p. 27. (In Russian)
54. Mikhailov, V.V.; Stepanov, S.Y.; Petrov, S.V.; Shilovskikh, V.V.; Kozlov, A.V.; Palamarchuk, R.S.; Abramova, V.D. New copper–precious metal occurrence in gabbro of the Serebryansky Kamen massif, Ural platinum belt, Northern Urals. *Geol. Ore Depos.* **2021**, *63*, 528–555. [CrossRef]
55. Anikina, E.V.; Alekseev, A.V. Mineral-geochemical characteristic of gold-palladium mineralization in the Volkovsky gabbro massif (Platiniferous Urals Belt). *Litosphere* **2010**, *5*, 75–100. (In Russian)
56. Kovalev, S.G.; Chernikov, A.P.; Burdakov, A.V. First finding of native gold in chromites from rock massifs of Kraka (Southern Urals). *Dokl. Earth Sci.* **2007**, *414*, 526–529. [CrossRef]
57. Makeev, A.B.; Filippov, V.N. Metallic films on natural diamond crystals (Ichet'yu Deposit, Middle Timan). *Dokl. Earth Sci.* **1999**, *369*, 1161–1165. (In Russian)
58. Subbotin, V.V.; Gabov, D.A.; Korchagin, A.U.; Savchenko, E.E. Gold and silver in the composition of PGE ores of the Fedorov-Pana layered intrusive complex. *Her. Kola Sci. Cent. RAS* **2017**, *1*, 53–65. (In Russian with English Abstract)
59. Chernyshov, N.M.; Korobkina, T.P. Peculiarities of distribution and forms of concentration of platinoids and gold in ferruginous quartzites of Lebedyan deposit of KMA. *Bull. Voronezh State Univ. Ser. Geol.* **2005**, *1*, 140–152.
60. Generalov, M.E.; Pautov, L.A. Porpecite of Lieutenant Chernik. *New Data Miner.* **2018**, *52*, 20–24.
61. Kapsiotis, A.; Grammatikopoulos, T.A.; Tsikouras, B.; Hatzipanagiotou, K. Platinum-group mineral characterization in concentrates from high-grade PGE Al-rich chromitites of Korydallos Area in the Pindos Ophiolite Complex (NW Greece). *Resour. Geol.* **2010**, *60*, 178–191. [CrossRef]
62. Kapsiotis, A.N. Origin of mantle peridotites from the Vourinos Ophiolite Complex, Greece, as deduced from Cr-spinel morphological and chemical variations. *J. Geosci.* **2013**, *58*, 217–231. [CrossRef]
63. Kucha, H. Precious metal alloys and organic matter in the Zechstein copper deposits, Poland. *TMPM Tschermaks Mineral. Und Petrogr. Mitteilungen* **1981**, *28*, 1–16. [CrossRef]
64. Wierchowiec, J.; Zieliński, K. Origin of placer gold and other heavy minerals from fluvial Cenozoic sediments in close proximity to rote Fäule-related Au mineralisation in the North Sudetic Trough, SW Poland. *Geol. Q.* **2017**, *61*, 62–80. [CrossRef]
65. Wierchowiec, J.; Mikulski, S.Z.; Zieliński, K. Supergene gold mineralization from exploited placer deposits at Dziwiszów in the Sudetes (NE Bohemian Massif, SW Poland). *Ore Geol. Rev.* **2021**, *131*, 104049. [CrossRef]
66. Paar, W.H.; Roberts, A.C.; Criddle, A.J.; Topa, D. A new mineral, chrisstanleyite, $Ag_2Pd_3Se_4$, from Hope's Nose, Torquay, Devon, England. *Mineral. Mag.* **1998**, *62*, 257–264. [CrossRef]
67. Moreno, T.; Prichard, H.M.; Lunar, R.; Monterrubio, S.; Fisher, P. Formation of a secondary platinum-group mineral assemblage in chromitites from the Herbeira ultramafic massif in Cabo Ortegal, NW Spain. *Eur. J. Mineral.* **1999**, *11*, 363–378. [CrossRef]
68. El Ghorfi, M.; Oberthur, T.; Melcher, F.; Luders, V.; El Boukhari, A.; Maacha, L.; Ziadi, R.; Baoutoul, H. Gold-palladium mineralization at Bleida Far West, Bou Azzer-el Graara inlier, Anti-Atlas, Morocco. *Miner. Depos.* **2006**, *41*, 549–564. [CrossRef]
69. Bozhko, E.N. To the question about sources of gold-platinoid mineralization of the structural formational zone Mataganya-Siguiri (Guinea, West Africa). *Bull. Voronezh Univer.* **2005**, *1*, 193–203. (In Russian)
70. Generalov, M.E.; Pautov, L.A. New occurrence of porpecite and mineralogy features of alluvial gold from the gold-bearing region of Ambositra, Madagascar. *New Data Miner.* **2012**, *47*, 43–46.
71. Chapman, R.J.; Mileham, T.J.; Allan, M.M.; Mortensen, J.K. A distinctive Pd-Hg signature in detrital gold derived from alkali Cu-Au porphyry systems. *Ore Geol. Rev.* **2017**, *83*, 84–102. [CrossRef]

72. Ames, D.E.; Kjarsgaard, I.M.; Good, D.; McDonald, A. Ore mineralogy of Cu-PGE mineralized gabbros, Coldwell Alkaline Complex, Midcontinent rift: Supporting databases, scanning electron microscope and mineral chemistry. *Geol. Surv. Can.* **2016**, 52. Open File 8006. [CrossRef]
73. Rudashevsky, N.S.; Rudashevsky, V.N.; Nielsen, T.F.D. Intermetallic compounds, copper and palladium alloys in Au–Pd ore of the Skaergaard pluton, Greenland. *Geol. Ore Depos.* **2015**, *57*, 674–690. [CrossRef]
74. Cabri, L.J.; Laflamme, J.H.G. Rhodium, platinum, and gold alloys from the Stillwater Complex. *Can. Mineral.* **1974**, *12*, 399–403.
75. Portella, Y.D.M.; Zaccarini, F.; Luvizotto, G.L.; Garuti, G.; Bakker, R.J.; Angeli, N.; Thalhammer, O. The Cedrolina Chromitite, Goiás State, Brazil: A Metamorphic Puzzle. *Minerals* **2016**, *6*, 91. [CrossRef]
76. Cabral, A.R.; Vymazalová, A.; Lehmann, B.; Tupinambá, M.; Haloda, J.; Laufek, F.; Vlček, V.; Kwitko-Ribeiro, R. Poorly crystalline Pd-Hg-Au intermetallic compounds from Córrego Bom Sucesso, Southern Serra Do Espinhaço, Brazil. *Eur. J. Mineral.* **2009**, *21*, 811–816. [CrossRef]
77. Olivo, G.R.; Gauthier, M.; Bardoux, M. Palladium-bearing gold deposit hosted by Proterozoic Lake Superior-type iron-formation at the Caue Iron Mine, Itabira District, Southern Sao Francisco Craton, Brazil: Geologic and structural controls. *Econ. Geol.* **1995**, *90*, 118–134. [CrossRef]
78. Cabral, A.R.; Lehmann, B.; Kwitko-Ribeiro, R.; Jones, R.D.; Rocha Filho, O.G. On the association of Pd-bearing gold, hematite and gypsum in an Ouro Preto nugget. *Can. Mineral.* **2003**, *41*, 473–478. [CrossRef]
79. Olivo, G.R.; Gauthier, M.; Williams-Jones, A.E.; Levesque, M. The Au-Pd mineralization at the Conceicao Iron Mine, Itabira District, Southern Sao Francisco Craton, Brazil: An example of a jacutinga-type deposit. *Econ. Geol.* **2001**, *96*, 61–74. [CrossRef]
80. Nikolaeva, L.A.; Yablokova, S.V. Typomorphic features of native gold and their use in geological exploration. *Ores Met.* **2007**, *6*, 41–57.
81. Gray, A.L. Solid sample introduction by laser ablation inductively coupled plasma mass spectrometry. *Analyst* **1985**, *110*, 551–556. [CrossRef]
82. McCandless, T.E.; Baker, M.E.; Ruiz, J. Trace element analysis of natural gold by laser ablation ICP-MS: A combined external/internal standardisation approach. *Geostand. Geoanalytical Res.* **1997**, *21*, 271–278. [CrossRef]
83. Outridge, P.M.; Doherty, W.; Gregoire, D.C. Determination of trace elemental signatures in placer gold by laser-ablation inductively coupled plasma mass spectrometry as a potential aid for gold exploration. *J. Geochem. Explor.* **1998**, *60*, 229–240. [CrossRef]
84. Penney, G. Fingerprinting Gold Using Laser Ablation Microprobe-Inductively Coupled Plasma-Mass Spectrometry (LAM-ICP-MS): An Exploration Tool. Honors Thesis, Memorial University of Newfoundland, St. John's, NL, Canada, 2001; 65p. *unpublished*.
85. Brown, S.M.; Johnson, C.A.; Watling, R.J.; Premo, W.R. Constraints on the composition of ore fluids and implications for mineralizing events at the Cleo gold deposit, Eastern Goldfields Province, Western Australia. *Aust. J. Earth Sci.* **2003**, *50*, 19–38. [CrossRef]
86. Chapman, R.J.; Mortensen, J.K. Application of microchemical characterization of placer gold grains to exploration for epithermal gold mineralization in regions of poor exposure. *J. Geochem. Explor.* **2006**, *91*, 1–26. [CrossRef]
87. McInnes, M.; Greenough, J.D.; Fryer, B.J.; Wells, R. Trace elements in native gold by solution ICP-MS and their use in mineral exploration: A British Columbia example. *Appl. Geochem.* **2008**, *23*, 1076–1085. [CrossRef]
88. Ehser, A.; Borg, G.; Pernicka, E. Provenance of the gold of the Early Bronze Age Nebra Sky Disk, central Germany: Geochemical characterization of natural gold from Cornwall. *Eur. J. Mineral.* **2011**, *23*, 895–910. [CrossRef]
89. Omang, B.O.; Suh, C.E.; Lehmann, B.; Vishiti, A.; Chombong, N.N.; Fon, A.N.; Egbe, J.A.; Shemang, E.M. Microchemical signature of alluvial gold from two contrasting terrains in Cameroon. *J. Afr. Earth Sci.* **2015**, *112*, 1–4. [CrossRef]
90. Gauert, C.; Schannor, M.; Hecht, L.; Radtke, M.; Reinholz, U. A comparison of in situ analytical methods for trace element measurement in gold samples from various South African gold deposits. *Geostand. Geoanalytical Res.* **2016**, *40*, 267–289. [CrossRef]
91. Chapman, R.; Torvela, T.; Savastano, L. Insights into regional metallogeny from detailed compositional studies of alluvial gold: An example from the Loch Tay Area, Central Scotland. *Minerals* **2023**, *13*, 140. [CrossRef]
92. Nickel, E.H.; Grice, J.D. The IMA commision on new minerals and mineral names: Procedures and guidlines on mineral nomenclature, 1998. *Mineral. Petrol.* **1998**, *64*, 237–263. [CrossRef]
93. Warr, L.N. IMA–CNMNC approved mineral symbols. *Mineral. Mag.* **2021**, *85*, 291–320. [CrossRef]
94. Fleet, M.E.; De Almeida, C.M.; Angeli, N. Botryoidal platinum, palladium and potarite from the Bom Sucesso stream, Minas Gerais, Brazil: Compositional zoning and origin. *Can. Mineral.* **2002**, *40*, 341–355. [CrossRef]
95. Makvandi, S.; Pagé, P.; Tremblay, J.; Girard, R. Exploration for platinum-group minerals in Till: A new approach to the recovery, counting, mineral identification and chemical characterization. *Minerals* **2021**, *11*, 264. [CrossRef]
96. Razumny, A.V.; Sidorov, E.G.; Sandimirova, E.I. Copper-gold-palladuim mineralization in the concentric zoned massifs of the Koryaksky Upland. *Vestn. KRAUNTS Earth Sci.* **2004**, *1*, 75–80. (In Russian)
97. Cabral, A.R.; Lehmann, B.; Kwitko, R.; Costa, C.C. The Serra Pelada Au-Pd-Pt deposit, Carajás mineral province, northern Brazil: Reconnaissance mineralogy and chemistry of very high grade palladian gold mineralization. *Econ. Geol.* **2002**, *97*, 1127–1138. [CrossRef]
98. Cabral, A.R.; Lehmann, B.; Kwitko, R.; Jones, R.D. Palladian gold and palladium arsenide–antimonide minerals from Gongo Soco iron ore mine, Quadrilátero Ferrífero, Minas Gerais, Brazil. *Appl. Earth Sci.* **2002**, *111*, 74–80. [CrossRef]

99. Murzin, V.V.; Moloshag, V.P.; Volchenko, Y.A. Paragenesis of precious metal minerals in copper-iron-vanadium ores of the Volkov type in the Urals. *Dokl. AN SSSR* **1988**, *300*, 1200–1202. (In Russian)
100. Moloshag, V.P.; Korobeinikov, A.F. New Data on Platinoid Mineralization of Copper-Iron-Vanadium Ores. In *Magmatic and Metamorphic Formations of the Urals and Their Metallogeny*; Ural Branch of the Russian Academy of Sciences: Yekaterinburg, Russia, 2000; pp. 90–101. (In Russian)
101. Poltavetc, Y.A.; Sazonov, V.N.; Poltavetc, Z.I.; Nechkin, G.S. Distribution of noble metals in ore mineral assemblages of the Volkovsky gabbroic pluton, Central Urals. *Geochem. Int.* **2006**, *44*, 143–163. [CrossRef]
102. Poltavetc, Y.A.; Poltavetc, Z.I.; Nechkin, G.S. Volkovsky deposit of titanomagnetite and copper- titanomagnetite ores with accompanying noble-metal mineralization, the Central Urals, Russia. *Geol. Ore Depos.* **2011**, *53*, 126–139. [CrossRef]
103. Volchenko, Y.A.; Koroteev, V.A.; Neustroeva, I.I.; Voronina, L.K. New gold-platinum-palladium ore occurrences in the palladium-bearing belt of the Urals. *Tr. IGG UrO RAN* **2007**, *153*, 214–220. (In Russian)
104. Anikina, E.V.; Zaccarini, F.; Knauf, V.V.; Rusin, I.A.; Pushkarev, E.V.; Garouti, J. Palladium and gold minerals in the ores of the Baron ore occurrence, Volkovsky gabbro-diorite massif. *Bull. Ural Dep. Rus. Mineral. Soc.* **2005**, *4*, 5–25. (In Russian with English Abstract)
105. Zoloev, K.K.; Volchenko, Y.A.; Koroteev, V.A.; Malakhov, I.A.; Mardirosyan, A.N.; Khrypov, V.N. *Platinum-Metal Mineralization in the Geological Complexes of the Urals*; Department of Natural Resources for the Ural Region: Yekaterinburg, Russia, 2001; 199p. (In Russian)
106. Zaaccarini, F.; Anikina, E.; Pushkarev, E.I.; Rusin, I.; Garuti, G. Palladium and gold minerals from the Baronskoe-Kluevsky ore deposit (Volkovsky complex, Central Urals, Russia). *Mineral. Petrol.* **2004**, *82*, 137–156. [CrossRef]
107. Kuznetsov, S.K.; Filippov, V.N.; Onishchenko, S.A.; Kotel'nikov, V.G. Copper-gold-palladium mineralization in ultrabasic rocks of the Polar Urals. *Dokl. Earth Sci.* **2007**, *414*, 501–503. [CrossRef]
108. Pystin, A.M.; Potapov, I.L.; Pystina, Y.I.; Generalov, V.I.; Onishchenko, S.A.; Filippov, V.N.; Shloma, A.A.; Tereshko, V.V. *Low-Sulfide Platinum-Metal Mineralization in the Polar Urals*; Ural Branch of the Russian Academy of Sciences: Yekaterinburg, Russia, 2011; 150p. (In Russian)
109. Spiridonov, E.M.; Kulagov, E.A.; Kulikova, I.M. Pt-Pd tetra-auricupride and associated minerals in ores of the Norilsk-1 deposit. *Geol. Ore Depos.* **2003**, *45*, 232–241.
110. Shcheka, G.G.; Lehmann, B.; Gierth, E.; Gömann, K.; Wallianos, A. Macrocrystals of Pt-Fe alloy from the Kondyor PGE placer deposit, Khabarovskiy kray, Russia: Trace-element content, mineral inclusions and reaction assemblages. *Can. Mineral.* **2004**, *42*, 601–617. [CrossRef]
111. Onishchenko, S.A.; Kuznetsov, S.K.; Tropnikov, E.M. Epigenetic alteration of cupreous gold in the Au-Ag-Cu-Pd exsolution texture. *Dokl. Earth Sci.* **2020**, *492*, 418–421. [CrossRef]
112. Onishchenko, S.A.; Kuznetsov, S.K. Exsolution in the Au-Ag-Cu system in a gold-rich area. *Geochem. Int.* **2022**, *60*, 657–671. [CrossRef]
113. Onishchenko, S.A.; Kuznetsov, S.K. Native gold of the Chudnoe gold–palladium deposit (Subpolar Urals, Russia). *Geol. I Geofiz.* **2023**, *64*, 233–254. [CrossRef]
114. Gerasimov, B.B.; Nikiforova, Z.S.; Pavlov, V.I. Mineralogical and geochemical characteristics of placer gold in the Bol'shaya Kuonamka river. *Educ. Sci. Yakutsk Russ.* **2014**, *3*, 74–78. (In Russian)
115. Murzin, V.V.; Malyugin, A.A. *Typomorphism of Gold in the Hypergenesis Zone (on the Example of the Urals)*; UNC AN USSR: Sverdlovsk, Russia, 1987; 96p. (In Russian)
116. Gerasimov, B.B.; Kravchenko, A.A. Ore occurrences of the Anabar placer area—Potential primary sources of gold. *NEFU Bulletin. Earth Sci.* **2020**, *4*, 17–28.
117. McQueen, K.G. Ore deposit types and their primary expressions. In *Regolith Expression of Australian Ore Systems*; Butt, C.R.M., Robertson, I.D.M., Scott, K.M., Cornelius, M., Eds.; CRC LEME: Bentley, WA, Australia, 2005; pp. 1–14.
118. Dill, H.G. The "chessboard" classification scheme of mineral deposits: Mineralogy and geology from aluminum to zirconium. *Earth Sci. Rev.* **2010**, *100*, 1–420. [CrossRef]
119. Groves, D.I.; Goldfarb, R.J.; Gebre-Mariam, M.; Hagemann, S.G.; Robert, F. Orogenic gold deposits: A proposed classification in the context of their crustal distribution and relationship to other gold deposit types. *Ore Geol. Rev.* **1998**, *13*, 7–27. [CrossRef]
120. Naldrett, A.J. Secular Variation of Magmatic Sulfide Deposits and Their Source Magmas. *Econ. Geol.* **2010**, *105*, 669–688. [CrossRef]
121. Goldfarb, R.J.; Groves, D.I.; Gardoll, I. Orogenic gold and geologic time: A global synthesis. *Ore Geol. Rev.* **2001**, *18*, 1–75. [CrossRef]
122. Sillitoe, R.H. Porphyry copper systems. *Econ. Geol.* **2010**, *105*, 3–41. [CrossRef]
123. Goldfarb, R.J.; Groves, D.I. Orogenic gold: Common or evolving fluid and metal sources through time. *Lithos* **2015**, *233*. [CrossRef]
124. Goryachev, N.A. *Ore Deposits in the Earth's History: A Tectonic and Metallogenic Essay*; Dalnauka, IP Serdyuk O.A.: Vladivostok, Russia, 2021; 208p.
125. Tegner, C.; Lesher, C.E.; Larsen, L.M.; Watt, W.S. Evidence from the rare-earth-element record of mantle melting for cooling of the Tertiary Iceland plume. *Nature* **1998**, *395*, 591–594. [CrossRef]
126. McCallum, I.S. The Stillwater Complex, a review of the geology. In *Stillwater Complex, Geology & Guide. Billings: 9th International Platinum Symposium*; Cooper, R.W.L., Zientek, M.L., Boudreau, A.E., McCallum, I.S., Miller, J.D., Geraghty, E.P., Mooney, D.G., Eds.; Johnson Matthey: West Chester, PA, USA, 2002; pp. A1–A25.

127. Bundtzen, T.K.; Sidorov, E.G.; Layer, P.W.; Chubarov, V.M. Geology, geochemistry, and new isotopic ages of selected PGE-Cr and PGE-Ni-Cu bearing mafic ultramafic complexes in the Farewell and Goodnews Bay terranes, Alaska, and Sredinny terrane, Kamchatka Peninsula, Russia Far East. In *Metallogeny of the Pacific Northwest: Proceedings of the Interim IAGOD Conference*; Khanchuk, A.I., Gonevchuk, G.A., Mitrokhin, A.N., Simanenko, L.F., Cook, N.J., Seltmann, R., Eds.; IAGOD: Vladivostok, Russia, 2004; pp. 83–85.
128. Pushkarev, E.V.; Morozova, A.V.; Khiller, V.V.; Glavatskykh, S.P.; Kamenetsky, V.S.; Rodemann, T. Ontogeny of ore Cr-spinel and composition of inclusions as indicators of the pneumatolytic–hydrothermal origin of PGM-bearing chromitites from Kondyor massif, the Aldan shield. *Geol. Ore Depos.* **2015**, *57*, 352–380. [CrossRef]
129. Efimov, A.A.; Ronkin, Y.L.; Malich, K.N.; Lepikhina, G.A. New Sm-Nd and Rb-Sr (ID-TIMS) isotope data for apatite-phlogopite clinopyroxenites from the dunite core of the Konder massif, Aldan shield, Yakutia. *Dokl. Earth Sci.* **2012**, *445*, 956–961. [CrossRef]
130. Popov, V.S.; Belyatsky, B.V. Sm-Nd age of dunite-clinopyroxenite-tylaite association of the Kytlym massif, the platinum belt of the Urals. *Dokl. Earth Sci.* **2006**, *409*, 795–800. [CrossRef]
131. Fershtater, G.B.; Krasnobaev, A.A.; Borodina, N.S.; Bea, F.; Montero, P. Geodynamic settings and history of the Paleozoic intrusive magmatism of the Central and Southern Urals: Results of zircon dating. *Geotectonics* **2007**, *41*, 465–486. [CrossRef]
132. Remizov, D.N.; Petrov, S.Y.; Kos'yanov, A.O.; Nosikov, M.V.; Sergeev, S.A.; Grigoriev, S.I. New age datings of gabbroides of the kershor complex (Polar Urals). *Dokl. Earth Sci.* **2010**, *434*, 1235–1239. [CrossRef]
133. Krasnobaev, A.A.; Fershtater, G.B.; Busharina, S.V. Character of zircon distribution in dunites of the South Urals (Sakharin and East Khabarnin massifs). *Dokl. Earth Sci.* **2009**, *426*, 685–689. [CrossRef]
134. Shaibekov, R.I. Age of the dolerite body of Mount Sopcha (Central Pai-Khoi). *Bull. Inst. Geol. Komi Sci. Cent. Ural. Branch Russ. Acad. Sci.* **2007**, *3*, 11–13.
135. Subbotin, V.V.; Korchagin, A.U.; Savchenko, E.E. Platinum-metal mineralization of the Fedorovo-Pansky ore cluster: Types of mineralization, mineral composition, features of genesis. *Bull. Kola Sci. Cent. Russ. Acad. Sci.* **2012**, *1*, 54–65.
136. Schissel, D.; Tsvetkov, A.A.; Mitrofanov, F.P.; Korchagin, A.U. Basal platinum-group element mineralization in the Federov Pansky layered mafic intrusion, Kola Peninsula, Russia. *Econ. Geol.* **2002**, *97*, 1657–1677. [CrossRef]
137. Nevolko, P.A.; Borisenko, A.S. Antimony mineralization in the gold-sulfide deposits of the Yenisei Ridge. *Razved. I Okhrana Nedr.* **2009**, *2*, 11–14. (In Russian)
138. Moralev, G.V.; Borisov, A.V.; Surenkov, S.V.; Tarbaev, M.B.; Ponomarchuk, V.A. First ^{39}Ar-^{40}Ar datings on micas from the Chudnoe Au-Pd-REE occurrence, Near-Polar Urals. *Dokl. Earth Sci.* **2005**, *400*, 109–112.
139. Volchenko, Y.A.; Koroteev, V.A.; Chashchukhin, I.S. Genetic types of chromite-platinoid mineralization in alpine-type complexes of mobile systems (on the example of the Urals). *Proc. Inst. Geol. Geochem.* **1998**, *145*, 190–193. (In Russian)
140. Shepherd, T.J.; Bouch, J.E.; Gunn, A.G.; McKervey, J.A.; Naden, J.; Scrivener, R.C.; Styles, M.T.; Large, D.E. Permo–Triassic unconformity-related Au-Pd mineralisation, South Devon, UK: New insights and the European perspective. *Miner. Depos.* **2005**, *40*, 24–44. [CrossRef]
141. Scrivener, R.C.; Cooper, B.V.; George, M.C.; Shepherd, T.J. Gold-bearing carbonate veins in the Middle Devonian limestone of Hope's Nose, Torquay. *Proc. Ussher Soc.* **1982**, *5*, 393.
142. Keith, S.B.; Spieth, V.; Rasmussen, J.C. Zechstein-Kupferschiefer mineralization reconsidered as a product of ultra-deep hydrothermal, mud-brine volcanism. In *Contributions to Mineralization*; IntechOpen: Kirkuk, Iraq, 2017. [CrossRef]
143. Barakat, A.; Marignac, C.; Boiron, M.-C.; Bouabdelli, M. Géomatériaux/Geomaterials (Métallogénie/Ore deposits) Caractérisation des paragenèses et des paléocirculations fluides dans l'indice d'or de Bleïda (Anti-Atlas, Maroc). *Comptes Rendus Geosci.* **2002**, *334*, 35–41. [CrossRef]
144. Berni, G.V.; Heinrich, C.A.; Walle, M.; Wall, V.J. Fluid geochemistry of the Serra Pelada Au-Pd-Pt deposit, Carajas, Brazil: Exceptional metal enrichment caused by deep reaching hydrothermal oxidation. *Ore Geol. Rev.* **2019**, *111*, 102991. [CrossRef]
145. Cabral, A.R.; Lehmann, B.; Tupinambá, M.; Wiedenbeck, M.; Brauns, M. Geology, mineral chemistry and tourmaline B isotopes of the Córrego Bom Sucesso area, southern Serra do Espinhaço, Minas Gerais, Brazil: Implications for Au–Pd–Pt exploration in quartzitic terrain. *J. Geochem. Explor.* **2011**, *110*, 260–277. [CrossRef]
146. Romer, R.L.; Lüders, V.; Banks, D.A.; Schneider, J. U-Pb data of Au-Pd-Pt-bearing quartz-hematite veins, Quadrilátero Ferrífero, Minas Gerais, Brazil. In *Mineral Deposit Research: Meeting the Global Challenge*; Mao, J., Bierlein, F.P., Eds.; Springer: Berlin/Heidelberg, Germany, 2005. [CrossRef]
147. Rosière, C.A.; Rios, F.J. The origin of hematite in high-grade iron ores based on infrared microscopy and fluid inclusion studies: The example of the Conceição mine, Quadrilátero ferrífero, Brazil. *Econ. Geol.* **2004**, *99*, 611–624. [CrossRef]
148. Piestrzyński, A.; Pieczonka, J.; Głuszek, A. Redbed-type gold mineralisation, Kupferschiefer, south-west Poland. *Miner. Depos.* **2002**, *37*, 512–528. [CrossRef]
149. Razumny, A.V.; Sidorenko, V.I.; Sapozhnikova, L.P.; Krotova-Putintseva, A.E.; Lazareva, E.I.; Suprunenko, O.I.; Surikov, S.N.; Sidorov, M.D.; Sidorov, E.G.; Badredinov, Z.G.; et al. *State Geological Map of the Russian Federation*; Scale 1:1,000,000 (third generation); Series Koryaksko-Kurilskaya; Sheet P-59-Plowmen; Explanatory letter; Cartographic factory VSEGEI: St. Petersburg, Russia, 2019; 323p.
150. Babkin, P.V.; Baranov, Y.E.; Vasiliev, V.I.; Demidova, N.G.; Kirikilitsa, S.I.; Kuznetsov, V.A.; Nikiforov, N.A.; Obolensky, A.A.; Ozerova, N.A.; Smirnov, V.I.; et al. *Mercury Metallogeny*; Smirnova, V.I., Kuznetsova, V.A., Fedorchuk, V.P., Eds.; Nedra: Moscow, Russia, 1976; 255p.

151. Fontboté, L.; Kouzmanov, K.; Chiaradia, M.; Pokrovski, G.S. Sulfide minerals in hydrothermal deposits. *Elements* **2017**, *13*, 97–103. [CrossRef]
152. Nikolaeva, L.A.; Gavrilov, A.M.; Nekrasova, A.N.; Yablokova, S.V. Generalized indicative models of native gold at deposits of various ore-formational types. *Ores Met.* **2008**, *3*, 62–68.
153. Pal'yanova, G.A.; Shvarov, Y.V.; Shironosova, G.P.; Laptev, Y.V. Methodological approaches to the assessment of gold fineness during thermodynamic modeling of hydrothermal systems. *Geochem. Int.* **2005**, *43*, 1247–1251.
154. Chudnenko, K.V.; Pal'yanova, G.A. Thermodynamic properties of solid solutions in the Ag-Au-Cu system. *Russ. Geol. Geophys.* **2014**, *55*, 349–360. [CrossRef]
155. Chudnenko, K.V.; Pal'yanova, G.A.; Anisimova, G.S.; Moskvitin, S.G. Ag-Au-Hg solid solutions and physicochemical models of their formation in nature (Kyuchyus deposit as an example). *Appl. Geochem.* **2015**, *55*, 138–151. [CrossRef]
156. Chudnenko, K.V. Thermodynamic properties of components in the Ag–Au–Pd system. *Russ. J. Inorg. Chem.* **2020**, *65*, 94–99. [CrossRef]
157. Chudnenko, K.V.; Palyanova, G.A. Thermodynamic modeling of native formation Cu-Ag-Au-Hg solid solutions. *Appl. Geochem.* **2016**, *66*, 88–100. [CrossRef]
158. Murzin, V.V.; Chudnenko, K.V.; Palyanova, G.A.; Varlamov, D.A.; Naumov, E.A.; Pirajno, F. Physicochemical model for the genesis of Cu-Ag-Au-Hg solid solutions and intermetallics in the rodingites of the Zolotaya Gora gold deposit (Urals, Russia). *Ore Geol. Rev.* **2018**, *93*, 81–97. [CrossRef]

minerals

Article

Gold Provenance in Placers from Pureo Area, Southern Chile Coastal Cordillera, and Their Relationship with Paleozoic Metamorphic Rocks

Pablo Becerra [1,*], Pablo Sanchez-Alfaro [1,2], José Piquer [1], Gaëlle Plissart [1], Belén Garroz [1] and Daniela Kunstmann [1]

[1] Instituto de Ciencias de la Tierra, Universidad Austral de Chile, Valdivia 5110566, Chile
[2] Andean Geothermal Center of Excellence (CEGA), Universidad de Chile, Santiago 8370450, Chile
* Correspondence: pablo.becerra01@alumnos.uach.cl

Abstract: Southern Chile placer gold deposits have been known and exploited since Spanish colonial times. Despite this, precise knowledge about their origin is scarce. This work aims to identify possible primary sources of the gold in the Pureo placers by studying the morphological and chemical characteristics of gold particles according to their spatial distribution. The former was determined by measurements and classification under a binocular microscope, allowing us to acquire a set of parameters related to the amount of transport that had affected the samples. The microchemical characteristics were determined by studying gold particles using optical microscopy, scanning electron microscopy (SEM) and electron microprobe (EMPA), where the native gold composition (in terms of major and minor elements) and the suite of mineral inclusions were obtained. The results regarding morphological characteristics suggest a low amount of transport from a primary source (<15 km). Microchemical data from gold particles indicate two compositional sub-populations, distinguished in both native gold composition (<15 Ag wt% with up to 4 Hg wt% and >15 Ag wt% with Hg bellow 1 wt%) and mineral inclusions (pyrite-galena rich and arsenopyrite rich, respectively), indicating two different primary gold sources. These results suggest a local origin of gold in the Coastal Cordillera, where the possible primary sources are associated with (i) massive sulfide deposits present in Paleozoic–Triassic metamorphic rocks and (ii) hydrothermal deposits associated with more recent Cenozoic intrusive activity. These conclusions have implications for the exploration of new placer deposits and of gold-bearing hypogene deposits (e.g., metamorphosed VMS deposits) in unexplored zones of southern Chile Coastal Cordillera.

Keywords: placers; gold; Chile; gold microchemistry; gold morphometry; gold provenance

Citation: Becerra, P.; Sanchez-Alfaro, P.; Piquer, J.; Plissart, G.; Garroz, B.; Kunstmann, D. Gold Provenance in Placers from Pureo Area, Southern Chile Coastal Cordillera, and Their Relationship with Paleozoic Metamorphic Rocks. *Minerals* **2022**, *12*, 1147. https://doi.org/10.3390/min12091147

Academic Editor: Galina Palyanova

Received: 6 August 2022
Accepted: 7 September 2022
Published: 10 September 2022

1. Introduction

Placers are the earliest known and worked gold deposits [1]. Their formation is related to the physical transport and concentration of gold particles from primary sources; therefore, their physical and chemical features can be used to trace the original source of gold mineralization. Systematic study of placer gold particles in a variety of geological environments concurs in terms of two key characteristics of placer gold particles related to their provenance. First, particle morphology evolution is considered to be directly related to transport distance [2,3]. The size and shape of gold particles are affected during their transport in a fluvial environment, being gradually flattened and rounded as they are transported [2,4,5]. On the other hand, the microchemical characteristics of particles have been related to the type of primary deposit from which the particles were eroded [6,7], where native gold composition and mineral inclusions are considered the principal characteristics related directly to the conditions of gold formation in the hypogene environment, which are commonly studied by using SEM and EMPA techniques [6,7]. Several works have used these relationships to investigate the provenance of gold particles in placer deposits worldwide, in terms of their transport distance [3] and the style of primary gold-bearing deposit that originated the particles [8], constituting an exploration tool for unknown primary gold deposits [6].

In southern Chile, placer gold deposits have been known and exploited since Spanish colonial times [9]. Despite this, the source of the gold remains unclear. Previous work about gold particles in southern Chile placers has proposed that a large part of the gold at a regional scale comes from eroded primary deposits in the Andean Principal Cordillera [10]. However, it has also been suggested that some placer deposits in the Coastal Cordillera have a local origin [10,11] that could be related to metamorphic or magmatic-hydrothermal processes in the Paleozoic metamorphic rocks constituting the Coastal Cordillera. The Pureo gold placer area, located in the Coastal Cordillera at ~39° S, offers an opportunity to study the primary sources of gold in southern Chile because: (1) placers are being actively exploited by small-scale artisanal miners, whose mining tunnels allow access to the deposits; and (2) placers are located relatively close (<20 km) to known primary occurrences of gold in metamorphic rocks [12,13]. In this work, we aim to identify possible primary sources of gold for the placer deposits in the Pureo area by studying the morphological and microchemical characteristics of the gold particles according to their spatial distribution in order to identify their relationships with primary source(s) within the metamorphic rocks that constitute this zone of the Coastal Cordillera.

2. Geological Background

The Pureo area is located in the southern Chile Coastal Cordillera, a mountain range along the Pacific Coast [13], composed mainly by a heterogeneous unit of Paleozoic-Triassic, highly deformed metamorphic rocks. In the study area, this basement is made by a low T–high P unit, grouped as the Bahia Mansa Metamorphic Complex (BMMC, Figure 1) [14,15], which corresponds to the oldest geological unit in this zone [16]. The BMMC consists of pelitic to semi-pelitic and mafic schists and minor occurrences of serpentinized ultramafic rocks [14,17]. The BMMC is interpreted as a paleo-accretionary prism associated with subduction that was active during the Late Paleozoic, for a period of 50–100 Ma [15].

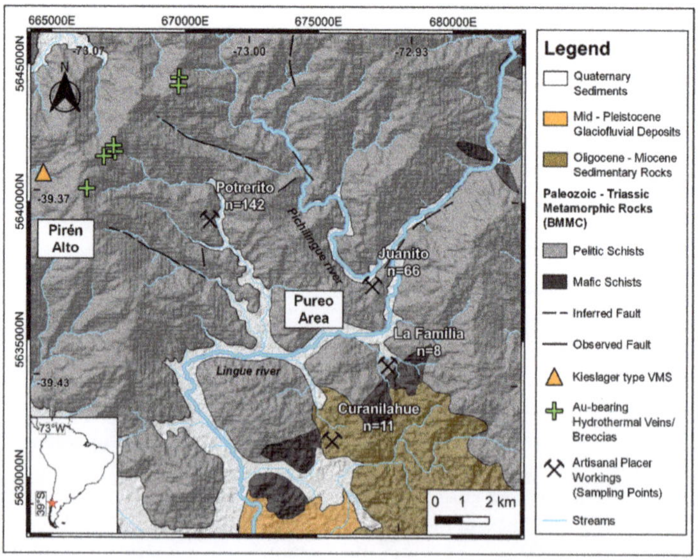

Figure 1. Geological map of the Pureo area. Hypogene mineralizations and active placer artisanal workings with the respective samples are shown. BMMC: Bahía Mansa Metamorphic Complex (modified after [12,13,16,18–20]).

Frequent occurrences of mineralized rocks have been reported in this area, corresponding to metamorphosed deposits associated with BMMC rocks, especially small lenses of Kieslager-type massive sulfide (VMS) deposits (Pirén Alto occurrence, Figure 1, [18,21,22]) and associated exhalites as banded iron formations and "coticules" (spessartine-rich metacherts), which are genetically related to a common sub-marine volcanic-exhalative origin [12,21,23]. In addition, occurrences of gold-bearing hydrothermal veins and breccias associated with probable Cenozoic intrusions have been reported (Pirén Alto prospect, Figure 1; [12,13,24,25]). Both occurrences are located to the northwest of the placer deposits (Figure 1). Detailed information about the deposits that host the placers, such as sedimentary facies, spatial extension, or the specific ages of these auriferous gravel deposits in the Pureo area, remains scarce. Despite this, placer locations near the Pureo area at regional scale are hosted in an early Pleistocene unit of glaciofluvial gravel deposits [26], and consequently, the Pureo placers could be related to a similar unit. Although these deposits are considered fossil placers, we refer to them as placers, considering that the term 'placer' is commonly used for Cenozoic deposits (e.g., [27]). Auriferous gravels in the Pureo area form old subhorizontal terraces over the main channels of the Pichilingue and Lingue rivers [28] which are currently worked by local artisanal miners. These deposits overlay the BMMC rocks (Figure 2) and are almost exclusively formed by clasts of BMMC lithologies [12]. Gold is irregularly distributed in the gravels, but the highest concentrations are restricted to the first 4–5 m above the bedrock [29].

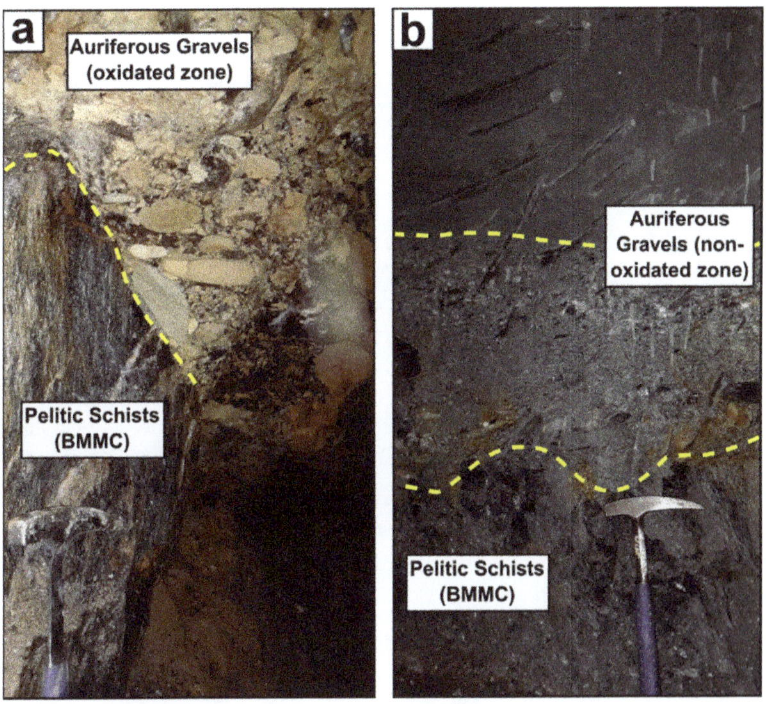

Figure 2. Auriferous gravels overlying pelitic schists of BMMC. Photographs from underground workings, following the boundary between gravels and bedrock (pelitic schists) within the terraced deposits. (**a**) A near-surface oxidated zone of gravels that exhibit reddish colors. (**b**) Farther surface non-oxidated zone of gravels, exhibiting bluish grey colors. In both locations, sediments for hand-panning were taken immediately above the contact with pelitic schists (above the yellow dashed line).

3. Materials and Methods

3.1. Sample Collection

Samples of gold particles were collected manually by hand-panning auriferous gravels from four underground artisanal mines, with the support of local miners (Figure 1). Auriferous gravels were taken in the first 1–2 m above the bedrock. Although these artisanal workings exploit the same gravels, some differences in the oxidation level of the auriferous gravels were observed between them (Figure 2), e.g., in the workings that exploit parts of the deposit close to the surface, the gravels exhibit reddish colors (Curanilahue, Potrerito), while some tunnels that go far away from the surface expose less oxidated zones and exhibit bluish grey colors (Juanito, La Familia). Particles were separated and handpicked under a binocular microscope. A total of 227 particles were recovered. The number of particles obtained for each mine is indicated in Figure 1.

3.2. Morphological Analysis

A morphological analysis of the particles was carried out following the methodology described in [30], which consists of the processing of images of particles taken under an optical microscope using plugins in the *ImageJ* software (version 1.52o) [31]. The output from *ImageJ* consists of the dimensions of each gold particle: length (L), width (W) and thickness (t). Two photographs of perpendicular cross-sections were taken to each group of particles. This information was used to calculate morphological indices which are commonly used in the literature, i.e., the Cailleux Flatness Index [32] and Shilo's Flatness Index [33,34], both of which can quantify the mass redistribution in malleable particles due to progressive hammering during transport in fluvial environments. The CFI and SFI calculations are shown in Equations (1) and (2), respectively:

$$\mathrm{CFI} = (\mathrm{L} + \mathrm{W})/2\mathrm{t}, \tag{1}$$

$$\mathrm{SFI} = [(\mathrm{L} + \mathrm{W})/2\mathrm{t}] - 1, \tag{2}$$

Higher CFI and SFI values reflect greater amounts of transport [2–5,32]. Another calculated index was Hofmann's Shape Entropy [35]. Its calculation is shown in Equation (3):

$$\mathrm{HSE} = -[(\mathrm{pL}*\ln \mathrm{pL}) + (\mathrm{pW}*\ln \mathrm{pW}) + (\mathrm{pt}*\ln \mathrm{pt})]/1.0986, \tag{3}$$

where p is the proportion of each particle axis (L, W, t) and 1.0986 is the maximum possible relative entropy value for a three-component system [35]. This index reflects the particle evolution from equidimensional shapes to flattened disks due to transport and was used by [36] to quantitatively estimate a model for particle shape-transport distance of gold in Klondike placers. Values of HSE lower than 1 reflect progressive flattening and greater transport [30,36,37].

3.3. Microchemical Analysis

A total of 75 gold particles were mounted in epoxy and polished for microchemical analyses. Polished sections were first observed using a reflected light optical microscope to identify target zones for subsequent analysis. SEM-EDS microanalyses were done with a Zeiss EVO-MC10 SEM, with an accelerating voltage of 20 kV and a working distance of 8 mm, at the Universidad Austral de Chile (UACh) in Valdivia, Chile. These microanalyses allowed us to accurately identify mineral inclusions and chemical zoning inside gold particles. Quantitative EMPA analyses were done in previously selected zones of 19 particles with a Cameca SX50 Electron Microprobe at the Institut Français de Recherche pour L'exploitation de la Mer (IFremer) in Brest, France. Both particle cores and rims were analyzed separately, as they represent the original primary native gold composition [6] and supergene altered composition, respectively [2,38,39]. Selected zones were analyzed for Au, Ag, Hg, Cu, Te, Pt, Pd. Complementarily, a gold grain found as an inclusion in a pyrite from a coticule clast from Juanito mine was analyzed as it represents a possible source lithology. EMPA analysis were performed with an accelerating voltage of 20 kV and a beam current of 50 nA with counting times of 30 s for each element at peak and 15 s for background. Estimated detection limits were 0.1 wt% for Au, 0.11 wt% for Ag, 0.02 wt% for Hg, 0.038 wt% for Cu, 0.04 wt% for Te, 0.15 wt% for Pt, and 0.1 wt% for Pd. Punctual analysis, transects and maps were performed in both SEM-EDS and EMPA analysis.

4. Results

4.1. Particle Morphology

Mean gold particle size values for the Pureo area in terms of length (L), width (W) and thickness (T) were 1.14 mm, 0.78 mm, and 0.38 mm respectively. Gold particle sizes and morphological indexes are shown in Figure 3 for each mine.

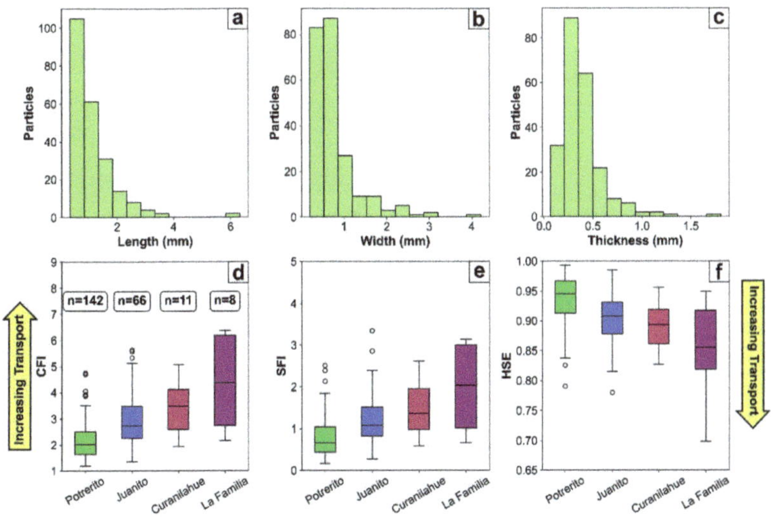

Figure 3. Gold particle size and morphological index values obtained in the Pureo area. (**a–c**) Histograms of particle dimensions (Length (L), Width (W) and Thickness (T)) for all samples. (**d–f**) Boxplots for morphological index values. Boxes represents inter-quartile range, horizontal line in the boxes represents the median, whiskers extend from the box by 1.5x the inter-quartile range and the white dots represents flier values that are outside the whiskers. CFI: Cailleux's flatness index, SFI: Shilo's flatness index. Both CFI and SFI increase with increasing transport; HSE: Hofmann's shape entropy. HSE decreases with increasing transport.

Gold particles from Juanito and Potrerito showed the lowest and the most homogeneous morphological index values. The Potrerito particles showed the lowest CFI and SFI values, varying between 1–3 and 0–2 respectively (Figure 3d,e), with the bulk of particles exhibiting HSE values higher than 0.9 (Figure 3f). Particles from Juanito showed similar values, but slightly higher for CFI and SFI, with values varying between 2–4 and 1–3 respectively (Figure 3d,e), and lower HSE values. The Curanilahue and La Familia particles showed a wider range of morphological index values (Figure 3d,e), probably due to the low number of recovered samples. The highest CFI-SFI and lowest HSE values were found in a particle from La Familia. Another characteristic of the Pureo particles is the common observation of accessory minerals conserved in their surfaces, mainly quartz, Fe-oxides and less commonly, pyrite (Figure 4). Particles showing accessory minerals were documented in all the sampled mines.

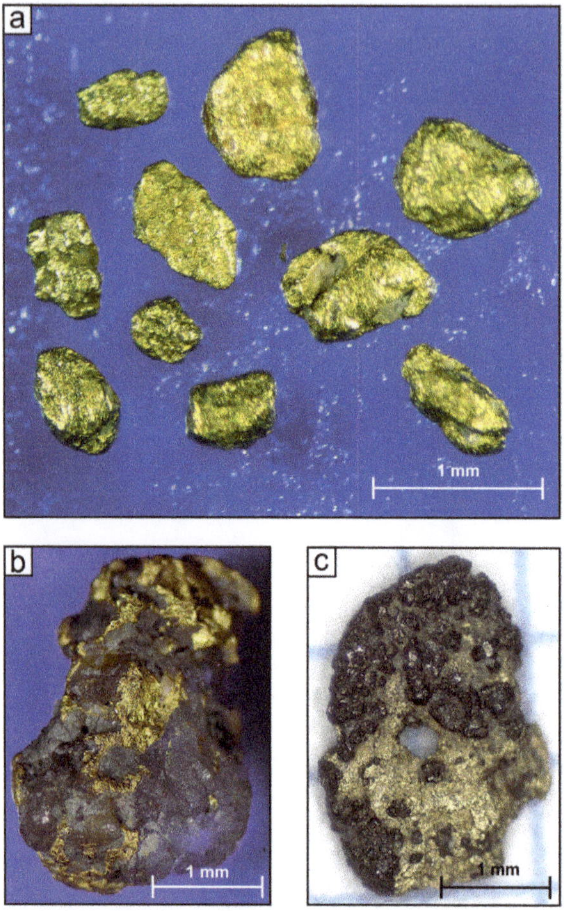

Figure 4. Placer gold particles from the Pureo area. (**a**) Gold particles under binocular microscope. (**b**) Quartz pebble with inclusions of native gold, which has not been released from its quartz matrix, probably from an eroded vein. (**c**) Pyrite coverage over particle surface.

4.2. Microchemical Characteristics

4.2.1. Native Gold Composition and Internal Chemical Zonations

A total of 75 gold particles were analyzed using SEM-EDS in order to determine the semi-quantitative native gold composition. Additionally, 19 of them were analyzed in EMPA to determine the quantitative composition. In general, our microchemical results showed that Pureo native gold particles mainly comprised Au and Ag with minor detectable Hg contents (>0.02 wt%) in the particle cores, recorded in all particles analyzed with EMPA. The contents of all other elements, such as Cu, Te, Pt, and Pd, were below the detection limits.

The semi-quantitative silver compositions of the 75 particle cores, as determined by SEM-EDS analyses, are shown in a cumulative frequency plot (Figure 5a), which shows homogeneous curves in the La Familia and Curanilahue particles. Conversely, the Potrerito and Juanito curves show a slope break at around 15 Ag wt% (850‰ Au) (Figure 5a), and therefore, two compositional sub-populations were distinguished in terms of their Ag contents. Quantitative core compositions, as determined by EMPA punctual analyses, varied between 72.9–94.4 wt% Au, 5.7–28.1 wt% Ag and 0.02–4.02 wt% Hg. These compositions are equivalent to a gold fineness range between 649–955‰ in particle cores. An important observation is that Ag and Hg appear to be slightly inversely correlated (Figure 5b). In other words, the particle cores that showed highest mercury contents were those with the lowest silver contents (<15 Ag wt%).

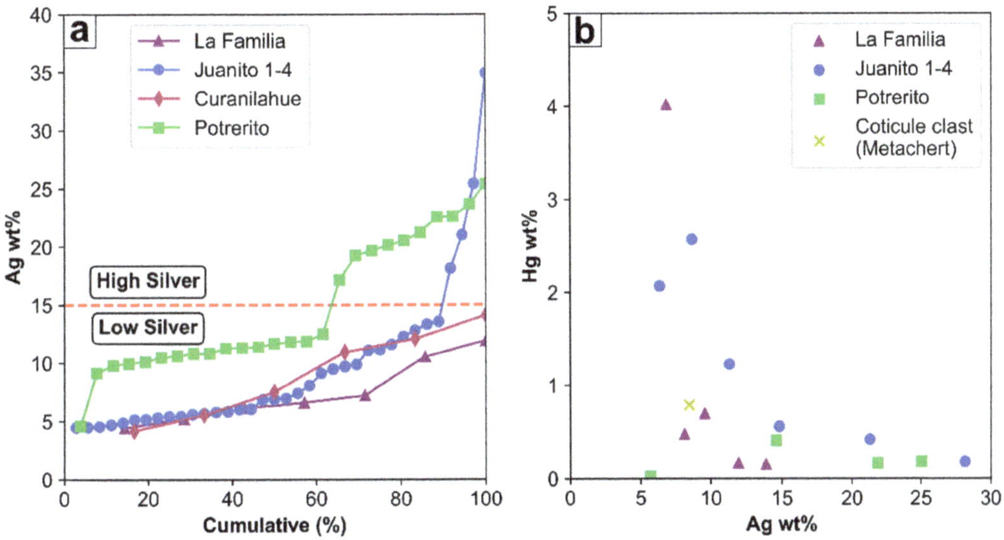

Figure 5. Compositions of native gold particle cores determined with SEM-EDS and EMPA analyses. (**a**) Cumulative frequency plots for semi-quantitative core silver contents obtained with SEM-EDS. (**b**) Variations in quantitative silver and mercury contents obtained with EMPA. The composition of a gold grain found as a pyrite inclusion in a "coticule" (metachert) clast is shown. Particles with lower silver contents show the highest mercury contents.

Chemical zonations associated with different proportions of native gold components were observed. Ag-Hg-depleted rims (Figure 6a) were the most common zonation found, characterized by a depletion of these elements and an enrichment of Au that develops in the rims and fractures of the particles. Rim zonation is limited by marked and abrupt contact between the two compositions without any transitional zones between the cores and the Ag-Hg-depleted rims (Figure 6a). An estimation of the rim proportion was done by measuring two transects over the particle section analyzed in a BSE image, calculating the proportion of the line segment contained in the rim and its length and averaging them (Figure 7). Rim size was observed to be higher in particles recovered from near surface areas of the auriferous gravel deposit, which present marked oxidation (Figure 7). In contrast, in the particles recovered farther from surface with a lower oxidation, the size of the rim is smaller (Figure 7). The oxidation levels of the gravel deposits were observed in artisanal underground workings as an abrupt change in the deposit color (Figure 2). Although gold particle cores are mostly compositionally homogeneous, some particles showed heterogeneities that occur in the form of irregular or 'vein-like' patches with a different composition from that of the surrounding core (Figure 6b). Compositionally heterogeneous cores were observed only in six gold particles from Potrerito that exhibited high silver contents (over 15 Ag wt%).

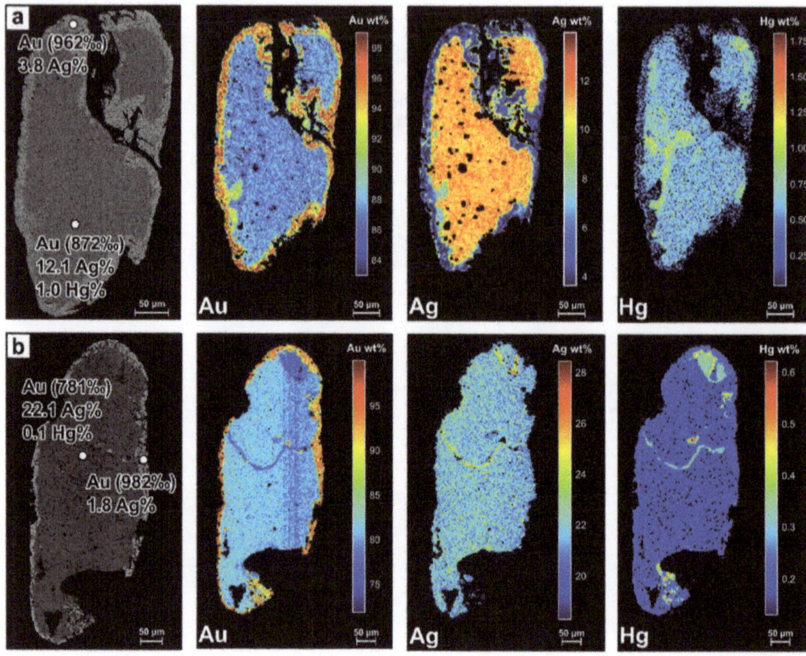

Figure 6. BSE image and EMPA chemical maps for Au-Ag-Hg. Colorscale in chemical maps shows the relative element abundance. Composition and native gold fineness in both core and rim of particle are shown. (**a**) Rim zonation that exhibits Ag-Hg depleted and Au enriched domains toward the edges. There is a sharp contact between the rim and the core. Hg contents seems to be less associated with Au-Ag contents in the particle core (**b**) Particle showing a compositional heterogeneity in the form of a 'vein-like' zone with higher contents of Ag and Hg. In both particles, Hg is absent in particle rims.

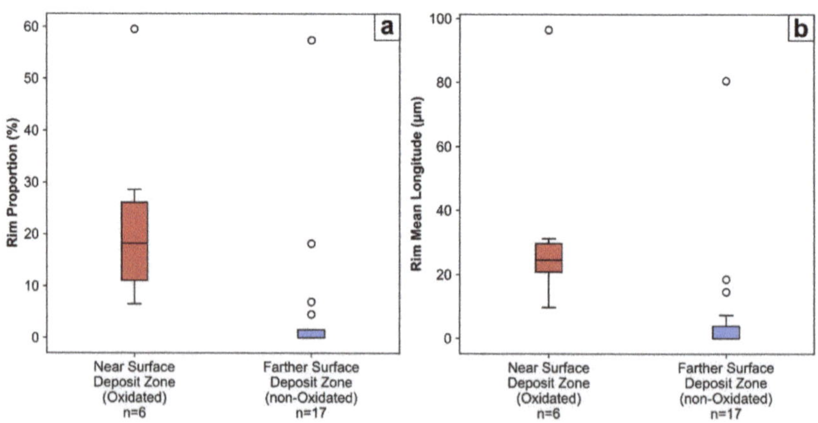

Figure 7. Differences in Ag-Hg depleted rim zonation size observed in different zones of the deposits. (**a**) Proportion of the particle cross-section affected by rim development. (**b**) Mean longitude of the rim. Near surface zones of the deposit (most oxidated), exhibit a higher rim development than the farther surface zones (non-oxidated).

4.2.2. Mineral Inclusions

Mineral inclusions are a very common feature observed in the Pureo gold particles. Fifty-five of the 75 particles studied using optical microscopy and SEM showed mineral inclusions, mainly composed of silicates and ore minerals. Micron-scaled ore mineral inclusions (Figure 8) were recorded in 29 particles from all mines. The ore minerals identified in SEM with EDS were mainly sulfides such as pyrite, galena and, less commonly, arsenopyrite, chalcopyrite, sphalerite, gersdorffite and Sb-bearing sulfosalts as boulangerite and tetrahedrite. Quartz was the most common silicate. A summary of the ore mineral inclusions and their frequencies can be found in Table 1. These minerals are mainly found as hosted inclusions within particle cores, but in some cases, they occur on particle surfaces (Figure 8b).

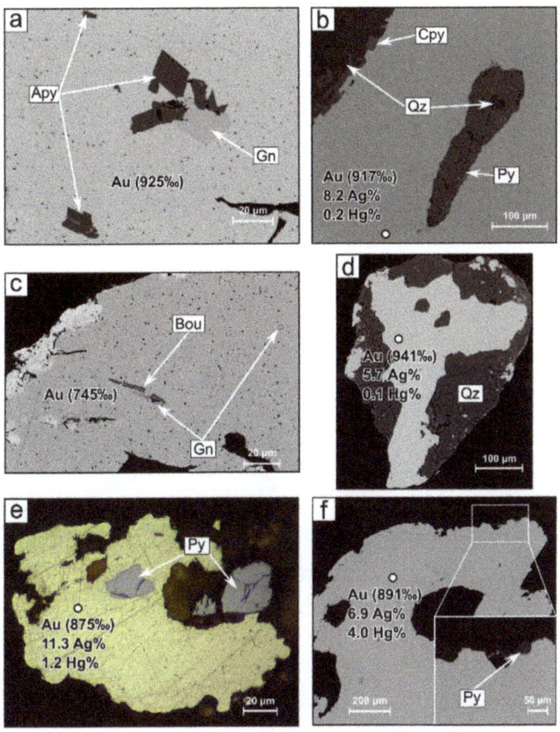

Figure 8. BSE and reflected light microscopy images of mineral inclusions in contact with native gold in Pureo particles. Composition and fineness of the hosting native gold is shown. (**a**) Arsenopyrite (Apy) and galena (Gn) in the particle core. (**b**) Pyrite (Py), chalcopyrite (Cpy) and quartz intergrowths occurring both in core and surface (upper left corner). (**c**) Boulangerite (Bou) and galena (Gn) in particle core. A change in the host native gold composition (rim) is visible as a lighter grey zone in the upper left particle border. (**d**) Quartz present in both core and surface of the particle. (**e**,**f**) Pyrite (Py) inclusions in contact with native gold cores with >1 Hg wt%.

The type of ore mineral found as an inclusion is related to the silver content of its host native gold (Figure 9), indicating that certain ore minerals tend to be more common in a determined composition. This relation confirmed the presence of two populations of native gold, as already evidenced with the slope break in silver cumulative frequency curves (Figure 5a). Pyrite, galena, and chalcopyrite are commonly found in low silver particles (lower than 15 Ag wt%; 850‰ Au), whereas arsenopyrite seems to be more frequent in high silver particles (higher than 15 Ag wt%; 850‰ Au). Also, gersdorffite was only observed in low silver particles, while Sb-bearing sulfosalts as tetrahedrite and boulangerite were only observed in high silver particles.

Table 1. Ore minerals found as inclusions in native gold particles.

Placer Location	Total Particles Analyzed	Particles with Ore Mineral Inclusions	Ore Mineral Inclusions
La Familia	8	6	Py, Gn, Apy, Cpy, Gdf
Juanito	36	11	Py, Gn, Cpy, Gdf
Potrerito	26	11	Py, Gn, Sp, Apy, Cpy, Tth, Blg
Curanilahue	6	1	Py, Apy, Gn

Number of particles analyzed and particles with mineral inclusions are shown. Ore mineral inclusions are listed in order of abundance. Mineral abbreviations: Py–Pyrite, Gn–Galena, Apy–Arsenopyrite, Cpy–Chalcopyrite, Sp–Sphalerite, Gdf–Gersdorffite, Tth–Tetrahedrite, Blg–Boulangerite.

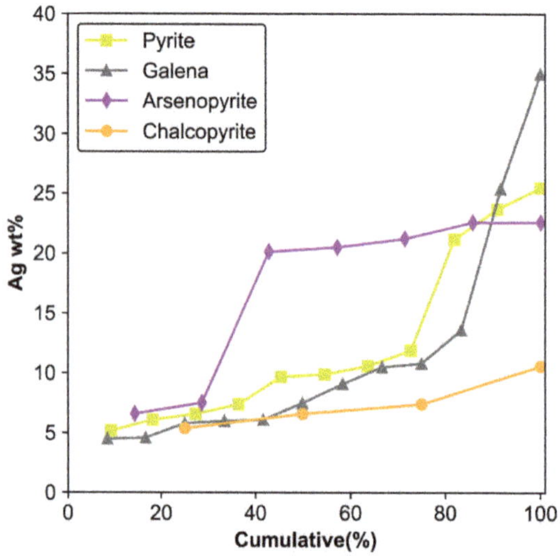

Figure 9. SEM-EDS silver contents in native gold particle cores hosting most commonly observed ore mineral inclusions. Arsenopyrite appears to be more abundant in high-silver–low fineness (>15 Ag wt%) particles, while galena, pyrite and chalcopyrite are more likely to be found in low-silver–high fineness (<15 Ag wt%) particles.

5. Discussion

5.1. Particle Morphology and Transport Distance

The morphological features of placer gold particles are related to their distance of transport [2,3]. One of the main relationships between morphology and transport is particle flatness, which increases with the transport distance, due to progressive particle hammering that occurs during transport [5]. Pureo gold particles show CFI values lower than 6 (Figure 3d), suggesting low transport distances [2,3,5]. CFI values of 1–8.6 would indicate transport distances lower than 1 km [3], a range similar to the one observed in Pureo gold particles. SFI represents the same as CFI, and it has been used by other authors [2,34] to compile flatness and transport distance of placer gold particles in a variety of localities worldwide. These compilations constrain SFI values obtained in Pureo particles to transport distances no further than 15 km. Similarly, HSE values obtained in the particles also point to reduced transport distances, based on the HSE–distance model developed by [30] for Klondike placers. An estimation of transport distance using this model was done with HSE distribution for each mine and weighted according to the number of samples recovered in each mine. Equations and detailed description of the model can be found in [30,36]. The given transport distances for Pureo gold particles are only first-order estimations, since specific sedimentologic and hydrologic variables for Pureo area placer deposits are unknown. This estimation leads to transport distances between 1–8 km (Figure 10), which is consistent with those estimated with CFI and SFI. As shown with this estimation, transport distances between mines are slightly dissimilar (Figure 10), as Potrerito gold particles shows the lowest CFI, SFI and highest HSE (Figure 3d–f) values, suggesting that these particles have the lowest transport distances (Figure 10). The Curanilahue and La Familia mines show the highest transport distances (Figure 10).

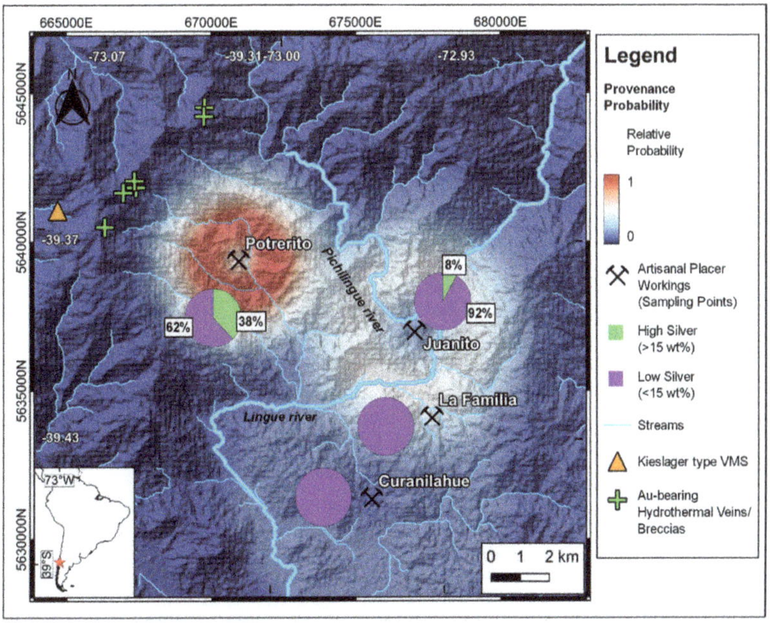

Figure 10. Relative probability (values normalized between 0–1) of provenance according to particle morphology quantified by HSE distributions and distance estimated using [30] shape–distance models for Klondike placers. Probability was weighted with the number of samples recovered in each site and normalized, obtaining a relative probability. Proportions of compositional sub-populations occurring in the placers and hypogene mineralizations are shown.

Low transport distances indicated by morphological indices are consistent with the frequent observation of accessory minerals such as quartz or pyrite in particle surfaces, since these are considered inherited characteristics from their primary sources [40]. According to [3], these features are removed rapidly, i.e., in the first hundreds of meters of transport, due to mechanical effects and the chemical instability of minerals like pyrite in the supergene environment. Thus, our results regarding the morphological features of the Pureo gold particles constrain their transport to distances no further than 15 km, indicating a local origin of the gold in the Coastal Cordillera. Our results are consistent with those presented in previous works [10,11] that reported morphological results compatible with low transport distances and a local origin of gold in BMMC rocks in other placer localities in the Coastal Cordillera, such as Marilán, Bahía Mansa, Hueyusca and Madre de Dios [10,11].

5.2. Microchemistry and Gold Sub-Populations

The co-existence of native gold and mineral inclusions in placer gold particles is considered to be an inherited relict that reflect the formation conditions of primary gold in the source mineralization [6,41]. As previously shown, the silver cumulative curves for particle cores show a marked slope break around 15 Ag wt% (Figure 5a). This break is interpreted as the contribution of two compositional sub-populations with different ranges in their silver contents that could be derived from different mineralization styles as sources of gold or a single hypogene source that shows internal zoning due to fluid evolution or multiple paragenetic stages with generation of different gold types [7]. However, this slope break is also visible in the distribution of ore mineral inclusions, where pyrite, galena and chalcopyrite appear to be mostly present in low silver particles, while arsenopyrite appears to be mostly present in high silver particles (Figure 9). Slope breaks in cumulative plots integrating the abundances of ore mineral inclusions with their host native gold silver contents, have been used to support the presence of compositionally distinct sub-populations in placer gold that comes from multiple sources [42,43]. In addition to this, mercury contents observed in particle cores (Figure 5b) can be interpreted as an additional discriminant between sub-populations and hypogene sources [7,44], whose variation is consistent with slope breaks of cumulative silver content curves. Moreover, the compositionally homogeneous cores exhibited by low silver particles contrast with the occurrence of compositional core heterogeneities only observed in high silver particles. These heterogeneities are interpreted to reflect changes in the hypogene mineralizing environment conditions [45]. Therefore, our results indicate that two populations of gold particles are present in the Pureo area, probably deriving from different types of primary sources. Considering the Pureo area geologic context and the presence of more than one type of gold-bearing mineralization style [12,13], the possibility of different sources originating the compositional sub-populations is possible. However, the possibility of a single zoned source cannot be completely discarded, due to the lack of details about the known gold mineralization styles in this area.

Considering this distinction between gold sub-populations, it is possible to see some differences in their morphologies (Figure 11). The grain morphologies in high silver particles tend to show restricted and lower CFI, SFI and HSE values than the low silver particles. Considering the above discussion about the morphology of gold particles, the difference in morphological index values could indicate a rather restricted and local origin for the high silver particles. The proportions of each compositional sub-population in the locations shown in Figure 10 could implicate that the contributing source for high silver particles is restricted to the N-NW of the study area, explaining the absence of those particles in the Curanilahue and La Familia mines (Figure 10).

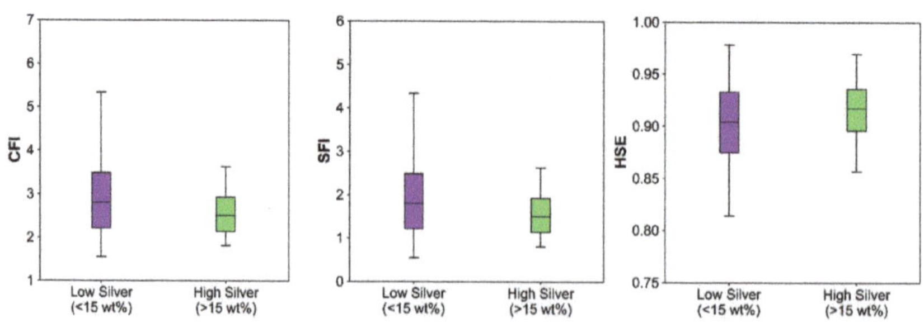

Figure 11. Morphological index variations for the two compositional sub-populations. Low silver population shows a broader range of values and higher maximum values than high silver particles, exhibiting possible differences in their transport. CFI: Cailleux's flatness index, SFI: Shilo's flatness index, HSE: Hofmann's shape entropy.

Other features related with supergene environment processes, such as compositional rim formation, were commonly observed in the Pureo gold particles. Rims are a common feature in placer gold particles in a variety of geological environments; their formation is mostly attributed to the removal of Ag and Hg from the native gold, causing a relative enrichment in Au in the particle rim due to exposure to supergene conditions [2,38,39,46,47]. Rims can also be formed due to the authigenic re-precipitation of gold in a supergene environment, a process that can be important in tropical environments [6,48]. Despite this, the climatic conditions and active tectonics in southern Chile Coastal Cordillera (38° S) do not favor the formation of well-developed weathering fronts as they do in tropical environments [49]. Therefore, this process, although present, may not be so relevant in here. According to [2], the formation of rims in placer gold particles can occur during their transport. Transport is an intermittent process where, for some periods, particles are moving in an active stream while in others, they remain in dormant sediments [2]. Rims are thought to thicken during the latter stages, because as particles move, these rims are removed due to abrasion [2]. In consideration of this, rim formation in the Pureo gold particles was not associated with transport, but rather, with the post-depositional stage of the deposits. Indeed, the differences in the rim development were probably related to their different oxidation levels and the distance to the surface (Figure 7), because oxidation is a key factor in the destabilization of Au-Ag alloys that causes rim formation [39]. In other words, rims in particles extracted from zones of the deposits far away from the surface can be associated with their last exposition to atmospheric conditions before burial, while rims in particles from near surface zones undergo further development and are probably enhanced by the progressive oxidation in these zones.

5.3. Mineralization Styles in the Coastal Cordillera and Possible Sources of Gold

Considering that our morphological data suggest a local origin of gold in the Coastal Cordillera metamorphic rocks, we discuss the possible relationship between placer gold and known primary gold occurrences in these metamorphic rocks.

The nearest possible gold source is the Kieslager VMS mineralization, located at the west of the study area (Figure 1), which corresponds to orebodies composed of pyrite, chalcopyrite, pyrrothite and minor sphalerite, where minor gold has been described [12,22] as well as the associated exhalites. This is supported by the fact that "coticule" clasts in gravel deposits are present in the host placer deposits, and that these coticules are considered parts of a single VMS sub-marine hydrothermal system which was active during the Paleozoic era and is preserved in the BMMC rocks [12,18,21,23]. Moreover, our results show that the microchemical characteristics of a native gold grain found in a coticule clast [20] are similar to those in the low silver population defined above, with detectable contents of Hg in the native gold (Figure 5b). The formation of Au-Ag-Hg alloys in VMS deposits that have undergone regional metamorphism in similar conditions as the BMMC rocks is reported by [50], suggesting that although these alloys can be formed in the submarine hydrothermal stage of the protolith, they can also form during their metamorphism due to metal remobilization and re-precipitation from other minerals which are present in VMS deposits. Similarly, it has been suggested that hydrated, low to medium grade metamorphism and deformation can promote the growth of pre-existent electrum grains in VMS due to metal remobilization from sulfides [51]. Specifically, gold is known to form from sulfides as pyrite [51] by processes enhancing metal remobilization, as syn-tectonic recrystallization and cataclastic fracturing, features reported in pyrites of the Pirén Alto VMS from the BMMC [20,22]. This suggests that gold possibly formed in pyrite and sulfides from the VMS and surrounding rocks (e.g., 'coticules'), due to remobilization and re-precipitation during subduction and accretion metamorphism [20]. The above could point to VMS as a contributor for the low silver compositional population. Additionally, the presence of galena, pyrite and chalcopyrite as mineral inclusions in this compositional population, is compatible with an origin in VMS, as these sulfides are common in these deposits and have been described in the local occurrences of VMS in BMMC [21,22]. Although Pirén Alto is the only orebody of this type described near the Pureo area, these occurrences are common in the metamorphic rocks of Coastal Cordillera, with several similar orebodies recognized to the north of the study area [18,21], therefore it cannot be ruled out that more orebodies of similar characteristics have been eroded away in the area or remain undiscovered, contributing to the gold in Pureo placers.

On the other hand, veins and breccias related to hydrothermal activity in Cenozoic intrusions in the BMMC are considered as another possible contributor [12]. The nearest occurrence of this type of system (Pirén Alto prospect), located around 10 km northwest from the Pureo placers (Figure 1), is characterized by veins and breccias hosted in pelitic and mafic schists of BMMC, with quartz-sericite alteration, where the gold mineralization is associated with pyrite and arsenopyrite [13,25]. This occurrence is probably related to hydrothermal alteration due to felsic sub-volcanic intrusions of unknown age, as observed in drill cores extracted during older explorations of the area [12,24,25]. The association with arsenopyrite in this mineralization style could suggest a relation with the high silver compositional population studied in this work, which is characterized by arsenopyrite abundance as inclusions. However, this relationship is highly uncertain, due to the scarce information of this type of deposits in the Coastal Cordillera. Other possibilities of gold sources from metamorphic rocks of BMMC could be gold-quartz veins found in metamorphic terranes [12]. However, this mineralization style has not been described in BMMC rocks.

6. Conclusions

In this study, we analyzed the morphological and microchemical characteristics of gold particles from the Pureo area. Our results allowed us to conclude a low transport distance of gold particles and to identify two compositional sub-populations that could reflect different types of primary sources for gold. The main conclusions of this study are as follows:

1. The morphological characteristics of Pureo gold particles indicate a limited transport, with distances no further than 15 km, probably in the range of 1–8 km, consistent with a local provenance restricted to the metamorphic rocks of the Coastal Cordillera.
2. Particle microchemical characteristics, such as native gold composition and mineral inclusions, allowed us to identify two compositional sub-populations. These compositional types could be associated with two contrasting primary sources, formed by different types of magmatic-hydrothermal and/or metamorphic processes in the Coastal Cordillera. The possible sources of gold can be correlated with two types of hypogene mineralization styles observed in the study area: metamorphosed hydrothermal deposits of the BMMC (metamorphosed Kieslager massive sulfide deposits and their related exhalites) and subsequent magmatic-hydrothermal processes (Cenozoic intrusive activity and related hydrothermal systems).
3. Both characteristics allowed us to conclude a local origin of the gold, restricted to the erosion of rocks present in the Coastal Cordillera. Because some manifestations of possible primary gold source types remain locally and regionally preserved, it is possible that undiscovered deposits exist in this and other regions of the Coastal Cordillera. The microchemical characteristics defined in this work could provide a basis for further exploration works. Therefore, additional work could focus on morphological and microchemical studies of gold particles from other placer deposits in Coastal Cordillera and placers in active rivers.

Author Contributions: Conceptualization, P.B. and P.S.-A.; methodology, P.B., B.G. and P.S.-A.; investigation, P.B. and B.G.; resources, J.P. and G.P.; writing—original draft preparation, P.B.; writing—review and editing, P.S.-A., G.P. and J.P.; supervision, P.S.-A. and J.P.; funding acquisition, J.P., G.P. and D.K. All authors have read and agreed to the published version of the manuscript.

Funding: This research was funded by FNDR (Fondo Nacional de Desarrollo Regional, Los Ríos, Chile) project "Transferencia Programa de Apoyo a la Pequeña Minería". Additional support was provided by ANID-FONDECYT Regular grant 1201219 to P.S., ANID-FONDECYT grant 11181048 to J.P and Amira Global P1249 project grant to J.P.

Acknowledgments: We would like to thank the invaluable logistical support and deposit knowledge from gold artisanal miners of Pureo. We thank Emilio Henriquez and SEREMI de Minería de Los Ríos for their administrative and logistic support. We also thank Alexandre Corgne and Jessica Langlade for their support with EMPA analyses. We thank the SEM and TEM technical support of the Electron Microscope Unit, a core facility–Zeiss Reference Center of Universidad Austral de Chile. We thank

Mario Pino for his technical support. Also, we would like to thank Jerson Uribe and Harry Barría for his support during fieldwork.

Conflicts of Interest: The authors declare no conflict of interest.

References

1. Boyle, R.W. *Gold: History and Genesis of Deposits*, 1st ed.; Springer: New York, NY, USA, 1987.
2. Knight, J.B.; Morison, S.R.; Mortensen, J.K. The Relationship between Placer Gold Particle Shape, Rimming, and Distance of Fluvial Transport as Exemplified by Gold from the Klondike District, Yukon Territory, Canada. *Econ. Geol.* **1999**, *94*, 635–648. [CrossRef]
3. Townley, B.K.; Hérail, G.; Maksaev, V.; Palacios, C.; de Parseval, P.; Sepulveda, F.; Orellana, R.; Rivas, P.; Ulloa, C. Gold Grain Morphology and Composition as an Exploration Tool: Application to Gold Exploration in Covered Areas. *Geochem. Explor. Environ. Anal.* **2003**, *3*, 29–38. [CrossRef]
4. Hérail, G.; Fornari, M.; Viscarra, G.; Miranda, V. Morphological and Chemical Evolution of Gold Grains during the Formation of a Polygenic Fluviatile Placer: The Mio-Pleistocene Tipuani Placer Example (Andes, Bolivia). *Chron. Rech. Min.* **1990**, *500*, 41–49.
5. Youngson, J.H.; Craw, D. Variation in Placer Style, Gold Morphology, and Gold Particle Behavior down Gravel Bed-Load Rivers; an Example from the Shotover/Arrow-Kawarau-Clutha River System, Otago, New Zealand. *Econ. Geol.* **1999**, *94*, 615–633. [CrossRef]
6. Chapman, R.; Leake, B.; Styles, M. Microchemical Characterization of Alluvial Gold Grains as an Exploration Tool. *Gold Bull.* **2002**, *35*, 53–65. [CrossRef]
7. Chapman, R.J.; Banks, D.A.; Styles, M.T.; Walshaw, R.D.; Piazolo, S.; Morgan, D.J.; Grimshaw, M.R.; Spence-Jones, C.P.; Matthews, T.J.; Borovinskaya, O. Chemical and Physical Heterogeneity within Native Gold: Implications for the Design of Gold Particle Studies. *Miner. Depos.* **2021**, *56*, 1563–1588. [CrossRef]
8. Chapman, R.J.; Leake, R.C.; Bond, D.P.G.; Stedra, V.; Fairgrieve, B. Chemical and Mineralogical Signatures of Gold Formed in Oxidizing Chloride Hydrothermal Systems and Their Significance within Populations of Placer Gold Grains Collected during Reconnaissance. *Econ. Geol.* **2009**, *104*, 563–585. [CrossRef]
9. Rivera, G. *Entre Circas y Cañuelas: Historia de Los Pirquineros Auríferos de Mariquina, Siglos XVI al XXI*, 1st ed.; San José de la Mariquina, Valdivia, Chile, 2017. Available online: https://http://2016.literaturalosrios.cl/catalogo/r1112/entre_circas_y_cauelas_historia_de_los_pirquineros_aurferos_de_mariquina_siglos_xvi_al_xxi (accessed on 6 September 2022).
10. Ordoñez, A. Oro detrítico en la región de Los Lagos (39–42° S): Mecanismos de transporte y proveniencia. *Serv. Natl. Geol. Miner. (Boletin)* **2000**, *56*, 1–40.
11. Zuccone, A. Los placeres auríferos de Madre de Dios y el ambiente metamórfico circundante. Bachelor's Thesis, Universidad de Chile, Santiago, Chile, 1988.
12. Muñoz, V. El Basamento Metamórfico Paleozoico, Serie Occidental en la Hoja Queule, IX y X Región, Chile. Condiciones Presión-Temperatura del Metamorfismo. Master's Thesis, Universidad de Chile, Santiago, Chile, 2007.
13. SERNAGEOMIN. *Estudio Geológico-Económico de la X Región Norte, Chile*; Servicio Nacional de Geología y Minería: Santiago, Chile, 1998; pp. 1–158.
14. Duhart, P.; McDonough, M.; Muñoz, J.; Martin, M.; Villeneuve, M. El Complejo Metamórfico Bahía Mansa en la cordillera de la Costa del centro-sur de Chile (39°30′–42°00′ S): Geocronología K-Ar, 40Ar/39Ar y U-Pb e implicancias en la evolución del margen sur-occidental de Gondwana. *Rev. Geol. Chile* **2001**, *28*, 179–208. [CrossRef]
15. Glodny, J.; Lohrmann, J.; Echtler, H.; Gräfe, K.; Seifert, W.; Collao, S.; Figueroa, O. Internal Dynamics of a Paleoaccretionary Wedge: Insights from Combined Isotope Tectonochronology and Sandbox Modelling of the South-Central Chilean Forearc. *Earth Planet. Sci. Lett.* **2005**, *231*, 23–39. [CrossRef]
16. Quiroz, D.; Duhart, P.; Crignola, P. Geología del área Lanco-Loncoche, regiones de La Araucanía y de Los Ríos, escala 1:100.000, Carta Geológica de Chile, Serie Geología Básica 106. *Serv. Natl. Geol. Miner.* **2007**, *106*, 1–21.
17. Godoy, E.; Kato, T. Late Paleozoic Serpentinites and Mafic Schists from the Coast Range Accretionary Complex, Central Chile: Their Relation to Aeromagnetic Anomalies. *Geol. Rundsch.* **1990**, *79*, 121–130. [CrossRef]
18. Collao, S.; Alfaro, G. Mineralización sulfurada de hierro, cobre y zinc en la Cordillera de la Costa, sur de Chile. *Rev. Geol. Chile* **1982**, *15*, 41–47.
19. Quiroz, D.; Duhart, P. Geología del área Queule-Toltén, regiones de La Araucanía y de Los Ríos, escala 1:100.000, Carta Geológica de Chile, Serie Geología Básica 110. *Serv. Natl. Geol. Miner.* **2008**, *110*, 1–22.
20. Garroz, B. Estudio del Basamento Metamórfico e Identificación de las Principales Fuentes de oro de los Placeres de Mariquina, región de Los Ríos, Chile. Bachelor's Thesis, Universidad Austral de Chile, Valdivia, Chile, 2022.
21. Collao, S.; Alfaro, G.; Hayashi, K. Banded Iron Formation and Massive Sulfide Orebodies, South-Central Chile: Geologic and Isotopic Aspects. In *Stratabound Ore Deposits in the Andes*; Fontboté, L., Amstutz, G.C., Cardozo, M., Cedillo, E., Frutos, J., Eds.; Springer: Berlin, Heidelberg, 1990; pp. 209–219.
22. Schira, W.; Amstutz, G.C.; Fontboté, L. The Pirén Alto Cu-(Zn) Massive Sulfide Occurrence in South-Central Chile—A Kieslager-Type Mineralization in a Paleozoic Ensialic Mature Marginal Basin Setting. In *Stratabound Ore Deposits in the Andes*; Fontboté, L., Amstutz, G.C., Cardozo, M., Cedillo, E., Frutos, J., Eds.; Springer: Berlin, Heidelberg, 1990; pp. 229–251.

23. Willner, A.P.; Pawlig, S.; Massonne, H.-J.; Herve, F. Metamorphic Evolution of Spessartine Quartzites (Coticules) in the High-Pressure, Low-Temperature Complex at Bahia Mansa, Coastal Cordillera of South-Central Chile. *Can. Mineral.* **2001**, *39*, 1547–1569. [CrossRef]
24. Avila, A. *Proyecto Piren, Informe de Avance. Inedit Report*; Minerales Sierra Morena S.A.: Valdivia, Chile, 1990.
25. Troncoso, R.; Cisternas, M.E.; Alfaro, G.; Vukasovic, V. Antecedentes sobre volcanismo terciario, cordillera de la costa, X Región, Chile. In Proceedings of the VII Congreso Geológico Chileno, Concepción, Chile, 17–21 October 1994; Volume 1, pp. 205–209.
26. Duhart, P.; Antinao, J.L.; Clayton, J.; Elgueta, S.; Crignola, P.; McDonough, M. Geología del área Los Lagos-Malalhue, escala 1:100.000, Carta Geológica de Chile, Serie Geología Básica 81. *Serv. Natl. Geol. Miner.* **2003**, *81*, 1–30.
27. Yeend, W.E. *Gold in Placer Deposits*; US Government Printing Office: Washington, DC, USA, 1990; p. 19.
28. Peebles, F. *Estudio Preliminar de las Pertenencias Mineras en los Sectores Madre de Dios y El Roble, Provincia de Valdivia*; Instituto de Investigaciones Geológicas: Santiago, Chile, 1976; pp. 1–8.
29. Agnerian, H. *Report on Madre de Dios Placer Gold Project, Chile. Internal Report*; Global Gold Company: Berlin, Germany, 2007; pp. 1–52.
30. Crawford, E.C.; Mortensen, J.K. An ImageJ Plugin for the Rapid Morphological Characterization of Separated Particles and an Initial Application to Placer Gold Analysis. *Comput. Geosci.* **2009**, *35*, 347–359. [CrossRef]
31. Schindelin, J.; Arganda-Carreras, I.; Frise, E.; Kaynig, V.; Longair, M.; Pietzsch, T.; Preibisch, S.; Rueden, C.; Saalfeld, S.; Schmid, B.; et al. Fiji: An Open-Source Platform for Biological-Image Analysis. *Nat. Methods* **2012**, *9*, 676–682. [CrossRef]
32. Cailleux, A.; Tricart, J. *Initiation à l'étude Des Sables et Des Galets*, 1st ed.; Centre de la Documentation Universitaire: Paris, France, 1959; Volume 1.
33. Shilo, N.A.; Shumilov, Y.V. New Experimental Data on Settling of Gold Particles in Water. *Dokl. Akad. Nauk SSR (Earth Sci. Sect.)* **1970**, *195*, 184–187.
34. dos Santos Alves, K.; Barrios Sánchez, S.; Gómez Barreiro, J.; Merinero Palomares, R.; Compaña Prieto, J.M. Morphological and Compositional Analysis of Alluvial Gold: The Fresnedoso Gold Placer (Spain). *Ore Geol. Rev.* **2020**, *121*, 103489. [CrossRef]
35. Hofmann, H.J. Grain-Shape Indices and Isometric Graphs. *J. Sediment. Res.* **1994**, *A64*, 916–920. [CrossRef]
36. Crawford, E.C. Klondike Placer Gold: New Tools for Examining Morphology, Composition and Crystallinity. Master's Thesis, University of British Columbia, Vancouver, BC, Canada, 2007.
37. Wrighton, T.M. Placer Gold Microchemical Characterization And Shape Analysis Applied As An Exploration Tool In Western Yukon. Master's Thesis, University of British Columbia, Vancouver, BC, Canada, 2013.
38. Groen, J.C.; Craig, J.R.; Rimstidt, J.D. Gold-Rich Rim Formation on Electrum Grains in Placers. *Can. Mineral.* **1990**, *28*, 207–228.
39. Krupp, R.E.; Weiser, T. On the Stability of Gold-Silver Alloys in the Weathering Environment. *Miner. Depos.* **1992**, *27*, 268–275. [CrossRef]
40. Chapman, R.J.; Mortensen, J.K. Application of Microchemical Characterization of Placer Gold Grains to Exploration for Epithermal Gold Mineralization in Regions of Poor Exposure. *J. Geochem. Explor.* **2006**, *91*, 1–26. [CrossRef]
41. Leake, R.C.; Chapman, R.J.; Bland, D.J.; Stone, P.; Cameron, D.G.; Styles, M.T. The Origin of Alluvial Gold in the Leadhills Area of Scotland: Evidence from Interpretation of Internal Chemical Characteristics. *J. Geochem. Explor.* **1998**, *63*, 7–36. [CrossRef]
42. Chapman, R.J.; Leake, R.C.; Moles, N.R. The Use of Microchemical Analysis of Alluvial Gold Grains in Mineral Exploration: Experiences in Britain and Ireland. *J. Geochem. Explor.* **2000**, *71*, 241–268. [CrossRef]
43. Potter, M.; Styles, M.T. Gold Characterisation as a Guide to Bedrock Sources for the Estero Hondo Alluvial Gold Mine, Western Ecuador. *Appl. Earth Sci.* **2003**, *112*, 297–304. [CrossRef]
44. Mackenzie, D.J.; Craw, D. The Mercury and Silver Contents of Gold in Quartz Vein Deposits, Otago Schist, New Zealand. *N. Z. J. Geol. Geophys.* **2005**, *48*, 265–278. [CrossRef]
45. Moles, N.R.; Chapman, R.J. Integration of Detrital Gold Microchemistry, Heavy Mineral Distribution, and Sediment Geochemistry to Clarify Regional Metallogeny in Glaciated Terrains: Application in the Caledonides of Southeast Ireland. *Econ. Geol.* **2019**, *114*, 207–232. [CrossRef]
46. Wierchowiec, J. Morphology and Chemistry of Placer Gold Grains—Indicators of the Origin of the Placers: An Example from the East Sudetic Foreland, Poland. *Acta Geol. Pol.* **2002**, *52*, 563–576.
47. Butt, C.R.M.; Hough, R.M.; Verrall, M. Gold Nuggets: The inside Story. *Ore Energy Resour. Geol.* **2020**, *4–5*, 100009. [CrossRef]
48. Nichol, I.; Hale, M.; Fletcher, W.K. Drainage Geochemistry in Gold Exploration. In *Drainage Geochemistry*; Hale, M., Plant, J.A., Eds.; Handbook of Exploration Geochemistry; Elsevier: Amsterdam, The Netherlands, 1994; Volume 6, pp. 499–557. ISBN 978-0-444-81854-6.
49. Rivera, J.; Reich, M.; Schoenberg, R.; González-Jiménez, J.M.; Barra, F.; Aiglsperger, T.; Proenza, J.A.; Carretier, S. Platinum-Group Element and Gold Enrichment in Soils Monitored by Chromium Stable Isotopes during Weathering of Ultramafic Rocks. *Chem. Geol.* **2018**, *499*, 84–99. [CrossRef]
50. Healy, R.E.; Petruk, W. Petrology of Au-Ag-Hg Alloy and "Invisible" Gold in the Troutlake Massive Sulfide Deposit, Flin Flon, Manitoba. *Can. Mineral.* **1990**, *28*, 189–206.
51. Huston, D.L.; Bottrill, R.S.; Creelman, R.A.; Zaw, K.; Ramsden, T.R.; Rand, S.W.; Gemmell, J.B.; Jablonski, W.; Sie, S.H.; Large, R.R. Geologic and Geochemical Controls on the Mineralogy and Grain Size of Gold-Bearing Phases, Eastern Australian Volcanic-Hosted Massive Sulfide Deposits. *Econ. Geol.* **1992**, *87*, 542–563. [CrossRef]

minerals

Article

Internal Structures of Placer Gold as an Indicator of Endogenous and Exogenous Processes

Zinaida Nikiforova

Diamond and Precious Metal Geology Institute, SB RAS, 677000 Yakutsk, Russia; znikiforova@yandex.ru; Tel.: +7-4112-33-58-72

Abstract: The study of the internal structures of placer gold on the territory of the east of the Siberian platform, overlain by a thick cover of Mz-Kz deposits, where traditional methods of searching for gold deposits are not effective, allowed us to determine, for the first time, the stages of ore formation and conditions of its occurrence. The identified indicators of the internal structures of placer gold (structures of primary recrystallization, secondary recrystallization, thick high-grade shells) indicate that placer gold content is formed mainly due to the supply and repeated redeposition of native gold from ancient gold-bearing deposits of the Precambrian stage of ore formation to younger ones. The discovered coarse-, medium-grained, mono-grained, unclear-zonal, granulation and disintegration structures suggest a supply of gold from nearby ore sources of the Mesozoic stage of ore formation. In the weathering crust, a high-grade shell is formed. In the hydrodynamic environment, the internal structures of gold practically do not change and fully correspond to the internal structures of endogenous gold. In aeolian conditions, the internal structures are transformed. In ancient gold-bearing conglomerates, under the impact of lithostatic pressure, as well as in metamorphogenic conditions, when the PT conditions change, the internal structure changes. Thus, for the first time, on a huge factual material, it is proved that the internal structures contain extensive information both about the endogenous origin of gold (the stages of ore formation—Precambrian and Mesozoic) and about its transformation in various exogenous conditions. The identified indicator of the internal structures of placer gold for certain types of sources contribute to a more correct selection of methods for searching for ore and placer gold deposits in closed territories and assessing their prospects. The use of this method makes it possible to develop criteria for forecasting different sources and types of gold deposits based on internal structures.

Keywords: placer gold; internal structures; stages of ore formation; endogenous and exogenous conditions; coarse-grained structure; high-grade shell; recrystallization; granulation; deformation structures

Citation: Nikiforova, Z. Internal Structures of Placer Gold as an Indicator of Endogenous and Exogenous Processes. *Minerals* **2023**, *13*, 68. https://doi.org/10.3390/min13010068

Academic Editors: Galina Palyanova and Huan Li

Received: 23 November 2022
Revised: 26 December 2022
Accepted: 27 December 2022
Published: 30 December 2022

1. Introduction

The indicators of placer gold (morphology, impurity elements, microinclusions, internal structures, etc.) carry extensive information about the nature of native gold, both in endogenous and exogenous conditions [1–11]. In exogenous conditions, the evolution of placer gold occurs, but it retains the "memory" of its primary sources. In this connection, the idea arose that, based on the results of studying the typomorphism of placer gold, in particular internal structures, in the first approximation, it is possible to assume the type of ore source and the depth of its formation, which can serve as an effective method for determining the genesis of placers and the nature of gold ore sources.

The results of studying the internal structures of placer gold, depending on the conditions of its occurrence, allow us to conclude about the stages of ore formation and exogenous transformations. The internal structures of placer gold were first studied by Petrovskaya and Fastalovich [12]. Then, Nikolaeva [2], Popenko [3], Savva and Preis [4], Nikolaeva et al. [8] and Savva et al. [5,13] studied the internal structures of placer gold. They studied the internal structures of native gold from endogenous sources, as well as zones of hypergenesis and alluvial placers. As a result of the study, it was found that endogenous internal structures of the Mesozoic stage of ore formation are characterized by coarse-grained and medium-grained structures, sometimes structures of granulation, disintegration, zonal and unclear-zonal structures and others. In the oxidation zone, they identified the formation of a high-grade shell on low-grade gold, as well as the presence of high-grade veinlets. According to L.A. Nikolaeva et al. [8], the structures of primary crystallization and structures caused by intra- and post-ore transformations are recognized in ore gold. The authors revealed that the gold of shallow-depth deposits is characterized by zonal structures, with a gold fineness of less than 800‰, and the gold of medium-depth deposits with a fineness of more than 800‰ is characterized by unclear zonal structures. They found that during the crystallization of gold in more stable conditions, simple twins are formed and, in less stable conditions—polysynthetic twins are formed. They refer to ordered structures, the same orientation, polysynthetic often discontinuous twins, spotty heterogeneity, relics of early gold, etc., to the signs of recrystallization. It is proved that the recrystallization structures indicate more intensive multiple recrystallizations.

Analysis of the published foreign literature devoted to the study of the mineralogy of placer gold and its typomorphic features showed that the internal structures of gold have not been studied in depth [14–24]. The published articles present the results of studying the morphology of placer gold, chemical composition and microinclusions in order to determine the type of ore source and its location [14–35]. Their research on the mineralogical and geochemical features of placer gold is mainly aimed at studying the relationship between placers and potential ore sources. This approach allowed them to improve the classical method of exploration based on the interpretation of distribution of placer gold to determine the location and type of source from which the placer was formed. Only in two publications, for example, the articles of Groen et al. [36,37] described the change in the internal structure in placers (gold amalgam and other secondary phases in placers) and also studied the process of formation of a high-grade shell on electrum grains in placers. The authors explained the discovery of a high-grade shell by the dissolution of silver under exogenous conditions. Leake et al. [38] studied the internal structure of Au-Pd-Pt grains with respect to low-temperature transport and deposition. Cabral et al. [39] studied brittle microstructures in gold nuggets. Kerr et al. [40] identified the deformation of placer gold related to transfer into a hydrodynamic medium. Stewart et al. [41] studied low-temperature recrystallization of placer gold in alluvial deposits. Chapman et al. [42] examined crystalloblastic deformations at the grain scale. In general, judging by the published works of foreign researchers, there is scattered information on the internal structures of placer gold and, currently, there are no generalizing works on this problem.

This article presents the results of studying the internal structures of gold, the placer gold content of the south-east of the Siberian platform, in order to identify the genesis of placers and their ore sources. Placer gold content has been known for more than a century and a half, but the types of ore sources and their location have not yet been identified. This is due to the fact that this area is overlain by a thick cover of Mz-Kz sediments, where traditional search methods do not bring positive results. In this regard, it became necessary to develop a new approach to study the mineralogical and geochemical features of placer gold and the mechanisms of its distribution in the studied area in order to determine the genesis of placers and types of ore sources. The mineralogical method developed by us—the study of the mineralogical and geochemical features of placer gold and the mechanisms of its distribution to identify the genesis of placers and their ore sources—was noted as the most important achievement of the Russian Academy of Sciences, Earth Science [43].

2. Materials and Methods

The article is based on the results of a long-term study of internal structures and mechanical dispersion in the east of the Siberian platform (Figure 1). The objects of the study in the south-east of the Siberian platform were placer gold, panned out from alluvial and aeolian deposits of the Lena-Vilyui interfluve (200 sampling points) and from alluvial deposits of the basins of the middle Lena (100 sampling points) and Upper Amga (70 sampling points). For several decades, during field work, tray-type sampling of alluvial and aeolian deposits has been carried out, and placer gold was panned out then extracted by conventional methods and studied under binoculars (shape, size, surface, etc.) and on modern devices in order to identify the chemical composition, microinclusions and internal structures. In total, about 24,000 grains were analyzed, about 2000 grains in the Lena-Vilyui interfluve, in the basin of the rivers of the middle Lena (200 grains), Upper Amga (170 grains).

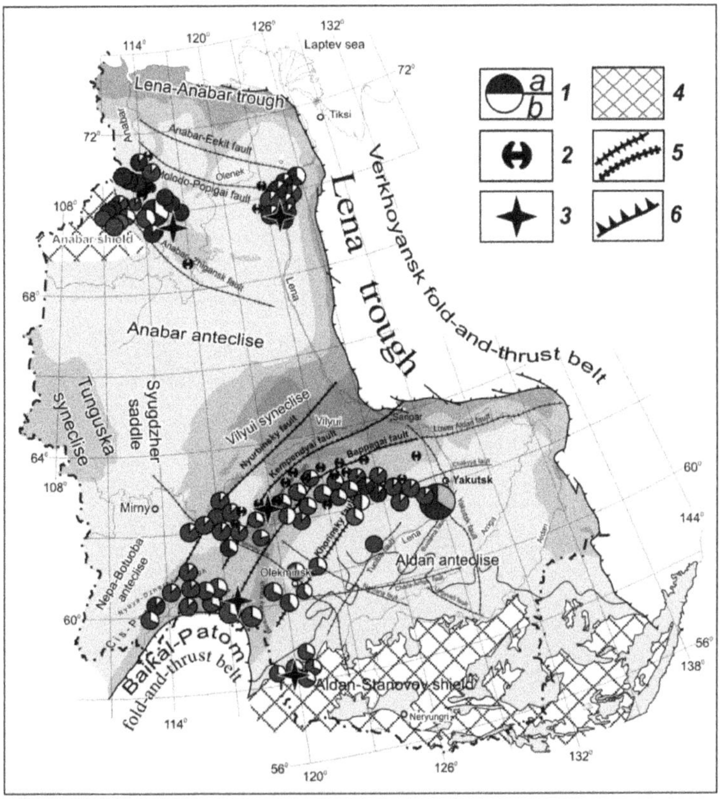

Figure 1. Diagram of the studied area and area of distribution of aeolian and two types of gold in the East Siberian platform: 1—types of gold and their ratio (%): a—I type: size 0.10–0.25 mm, fineness > 900‰, b—II type: size >0.25 mm, fineness 600–800‰; 2—aeolian gold; 3—places of discovery of ore gold; 4—outcrop areas of the rocks of the crystalline basement; 5—faults; 6—front of the Phanerozoic belts.

Typomorphic features (morphology, chemical composition, microinclusions, internal structure) of native gold were studied using well-known mineralogical and geochemical methods. All analytical work was carried out in the laboratory of physical–chemical methods of analysis, DPMGI SB RAS (Yakutsk). The study of the morphology, surface structures and internal structure of the gold particles was carried out using a scanning electron microscope "JEOL JSM-6480LV" (Japanese Electron Optics Laboratory, Tokyo, Japan), stereoscopic microscope "LEICA MZ6" (KaVo, Biberach an der Riss, Germany) and an ore microscope "JENAVERT SL 100" (Carl Zeiss AG, Oberkochen, Germany). The trace element composition of native gold was analyzed on an X-ray microanalyzer, "JXA-50A", "JSM-6480LV" (Japanese Electron Optics Laboratory, Tokyo, Japan). The content of impurity elements was studied via atomic emission spectrography. Microinclusions in native gold were identified using a scanning electron microscope "JEOL JSM-6480LV", with an energy-dispersive spectrometer Energy 350 of Oxford Instruments (London, UK), Software Oxford Instruments INCA the microanalysis Suite Issue v.4.17. Quantitative analysis and processing of the results were carried out using the XPP method in the software INCA Energy.

The internal structures of native gold were studied according to the generally accepted method [12] using a reagent: $HCl + HNO_3 + FeCl_3 \times 6H_2O + CrO_3$ + thiourea + water. The etching of gold was repeated several times. The reagent was applied to the surface of polished gold mounted in epoxy resin. The gold particles were etched from 10 to 30 s, at several approaches. After each etching procedure, the polished section was washed under a strong stream of water, then dried. Polyphase particles were etched differentially: first, the low-grade phases of gold were etched, and then, with a stronger solution, the high-grade phases [4]. Standard reagents were used for etching ($HNO_3 + CrO_3$ – kustelite; $HCl + CrO_3$ + thiourea – electrum; $HCl + CrO_3$ of various concentrations - medium-grade gold; $HCl+4HNO_3$ – high-grade gold). After that, the observed internal structures after special preparation were studied in detail using the NEOPHOT 32 ore microscope and the JEOL JSM-6480LV scanning electron microscope. The interpretation of internal structures was carried out in accordance with the ideas of N.V. Petrovskaya [1], N.E. Savva, V.K. Preis [4] and L.A. Nikolaeva et al. [8].

3. Results

The internal structures of placer gold were studied from materials collected in the southeast of the Siberian Platform in the basins of the rivers of the middle Lena (the mouth of the Vitim, Bol. Patom, Dzherba, Tokko, Torgo rivers), the Upper Amga and the rivers of the Lena-Vilyui interfluve (Tonguo, Chebyda, Kempendyai, Namana, etc.) [9].

Generalization of the results of studying the typomorphic features of placer gold and analysis of the mechanisms of its distribution in the southeastern part of the Siberian platform on the territory of the Lena-Vilyui interfluves, basin of the middle Lena and Upper Amga allowed us, for the first time, to distinguish two types of gold with certain indicators, corresponding to two stages of ore formation—Precambrian and Mesozoic (Table 1, Figure 1) [9].

Table 1. Typomorphic features of two types of placer gold from the east of the Siberian platform.

Types of Gold	Size	Shapes	Surface	Fineness and Impurity Elements	Intergrowths	Microinclusions	Internal Structure
I type	−0.5 mm	Flake, lamellar, lumpy	Shagreen, often with casts of pressing of minerals	900–1000‰; Cu: 0.2–1.2%, Hg: 0.1–0.2%	Pseudo-intergrowths with rounded minerals (ilmenite, zircon, quartz)	Quartz, pyrite, arsenopyrite	Complete recrystallization, lines of plastic deformations, thick high grade-shells (20–30 μm)
II type	0.5–1 mm 1 mm and >	Tabular, lumpy, lumpy-angular of ore habit	Rough-shagreen, pitted-tubercular, porous	800–899‰—30–70%, 700–799‰—10–35%, 600–699‰—up to 15%, Hg: up to 6.2%; As: 0.1%; Pb: 0.005%; Sn: 0.02%; Sb: 0.0008%; Fe: 0.1%; Pt: up to 0.1%	Intergrowths with chalcedony quartz	Quartz, albite, sulfides (pyrite, arsenopyrite, etc.), tellurides (petzite, etc.)	Unchanged mono- or coarse-grained structure with clear grain boundaries, unclear zonal structures, thin high-grade shells (the first microns), sometimes the structures of endogenous transformations are granulation and disintegration

The first type of gold is represented mainly by flake and lamellar forms, with a size of 0.1–0.25 mm, very high fineness (>950‰) with a surprisingly small set of impurity elements and almost complete absence of microinclusions (Table 1). At the same time, it was determined that placer gold of the Precambrian stage of formation is very much changed, has certain internal structures and depends on the duration and conditions of stay in exogenous environment. The internal structure of placer gold is mainly primarily recrystallized and secondarily recrystallized, with deformation structures (Luders lines, etc.) and thick high-grade shells (10–20 μm) (Figure 2).

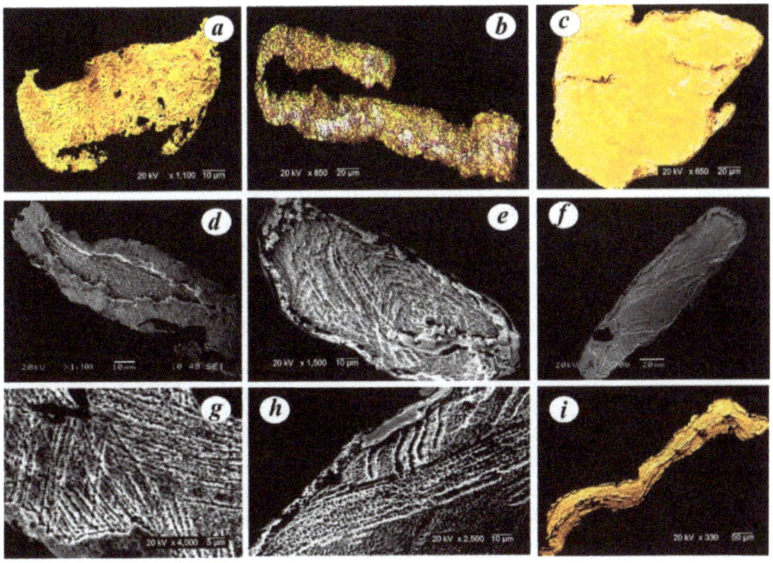

Figure 2. Internal structures of placer flake gold of the Precambrian stage of ore formation: (**a**) primary recrystallization with deformation lines; (**b**) secondary recrystallization; (**c**) complete recrystallization on high-grade gold with fine deformation lines; (**d**) thick high-grade shell (20–30 μm) on low-grade gold; (**e**,**f**) high-grade shells on gold particles with deformation lines; (**g**–**i**) multidirectional deformation lines at different magnifications.

It is known that gold is an inert metal that practically does not change in exogenous conditions; however, we found that when staying in various exogenous conditions for billions of years and repeated redeposition, primary recrystallization and secondary recrystallization of gold occurred with the removal of silver and impurity elements [1,2,4,8]. When studying the internal structure of placer gold from the Quaternary deposits in the east of the Siberian Platform, it was found that almost all flake gold of type I with a size of 0.1–0.25 mm, of high fineness, is characterized by primarily recrystallized and secondarily recrystallized internal structure. Flake gold with such structures of primary recrystallization and secondary recrystallization is found in all watercourses, as in the north-east of the Lena-Anabar interfluve—Bol. Kuonamka, Udzha, Sololi, etc., and in the central and southern part of the east of the Siberian platform of the Lena-Vilyui interfluve (Chebyda, Kempendyai, Tonguo, Namana rivers, etc.) and the basin of the middle Lena with left-bank and right-bank tributaries. The identification of such structures indicates the repeated redeposition of placer gold from the Precambrian sources into younger ones.

At the same time, some regularity in the mineralogy of placer gold of the Precambrian stage of ore formation was revealed. This stage of ore formation is characterized mainly by fine gold with a size of 0.1–0.25 mm, has a very high fineness—more than 900‰—a surprisingly small range of variations in fineness and a low content of impurity elements (Ag, Cu, Hg). This is explained by the fact that with repeated redeposition from ancient gold-bearing deposits of the Precambrian age into younger ones, gold is purified in exogenous conditions, silver and impurity elements are removed with a change in the internal structure. As a result, the internal structure of this gold was significantly transformed; primary recrystallization, secondary recrystallization structures, deformation lines and thick high-grade shells (10–30 μm) were formed in it, indicating a long stay of gold in various exogenous environments—the hypergenesis zone, hydrodynamic, aeolian conditions and others.

When studying the distribution mechanisms of placer gold, it was found that gold with primarily recrystallized and secondarily recrystallized internal structures of deformations (Luders lines, etc.) and thick high-grade shells (10–30 microns) is very widespread (Figure 1).

It is proved that primary recrystallization and secondary recrystallization structures occur on the places with a disturbed crystal structure of small grains (Figure 2a–c). The identified structures are formed mainly during a prolonged stay in exogenous conditions in weathering crusts, in buried placers or during repeated redeposition from ancient levels to younger ones.

It is found that the recrystallization structures with deformation lines are characteristic of the platy gold particles from the basin of the mouth of the Vitim river (Figure 1). The structures of complete recrystallization are characteristic of the flake gold of the Lena-Anabar and Lena-Vilyui interfluves, as well as for the middle Lena basin (Figure 2b,c). Structures of complete recrystallization on high-grade flake gold with thin deformation lines were found in the middle Lena basin, both in left-bank and right-bank tributaries (Figure 2c).

Thick high-grade shells are formed in the hypergenesis zone due to the removal of silver. Further, thick, dense, high-grade shells of 20–40 microns are characteristic of placer gold, which has been repeatedly deposited from ancient levels to younger ones (Figure 2d–f). It has been revealed that such thick high-grade shells are characteristic of type I flake gold, which is very widespread in the east of the Siberian Platform. Thick high-grade shells (20–30 microns) on low-grade flake gold were found both in the northeast of the Siberian Platform in the basin of the Udzha river and in its central part in the Chebyda, Namana, Tonguo rivers and in its southern part in the basin of the middle Lena river (Figure 2d) [9,10]. The identified high-grade shells on gold particles with deformation lines are extremely rare (Figure 2e,f).

Numerous multidirectional deformation lines at various magnifications in placer gold were found only in the basins of the Tokko and Torgo rivers (southwest of the Siberian Platform) (Figure 2g–i). They were found where the river valleys are drained by the Precambrian metamorphogenic formations, granite-gneisses of the Archean and Lower Proterozoic age and where occurrences with a gold content of 0.6–1 g/t in ferruginous quartzites in the Borsalinskaya series of the Archean age have been identified. The formation of such numerous deformation lines probably occurs in metamorphic strata, with known deposits of ferruginous quartzites with a gold content of up to 10 g/t. The presence of such internal structures and high-grade shells on gold particles with numerous deformation lines is an indicator for predicting metamorphogenic gold deposits.

In general, it has been established that the features of the internal structures of placer gold of the flake form of the Precambrian stage of ore formation are an indicator of the multiplicity of redeposition and the duration of stay in various exogenous conditions (Figure 1). The results of studying the distribution mechanisms of type I flake gold with recrystallized structures and thick high-grade shells showed that it is characteristic of all rivers of the east of the Siberian Platform and only at certain objects, along with type I gold, type II gold with sharply different internal structures was found.

The second type of placer gold of the Mesozoic stage of ore formation is characterized by a larger size (>0.25 mm, 1–2 mm), lamellar and lumpy forms, sometimes of gold particles and has mainly a medium, 800–900‰, and low 700–800‰ fineness (less often 500–600‰). In gold of this type, a wide range of impurity elements is noted, indicating its shallow origin (Table 1). Such gold was discovered by us in the south-east of the Siberian Platform at the sources of the Namana, Kempendyai rivers (Vilyui syneclise, Kempendyai dislocations), at the mouth of the Bol. Patom river (Urinsky anticlinorium), as well as in the Upper Amga river basin.

The internal structure of type II placer gold is practically unchanged, characterized mainly by mono-grain and coarse-medium-grained structures; sometimes, very thin high-grade shells are noted (Table 1; Figure 3). In addition, it has been found to contain unclear-zonal and interblock structures, as well as a porous, spongy structure characteristic of gold from near-surface deposits, found at the mouth of the Bolshoi Patom river, indicating the proximity of the primary source [9,10].

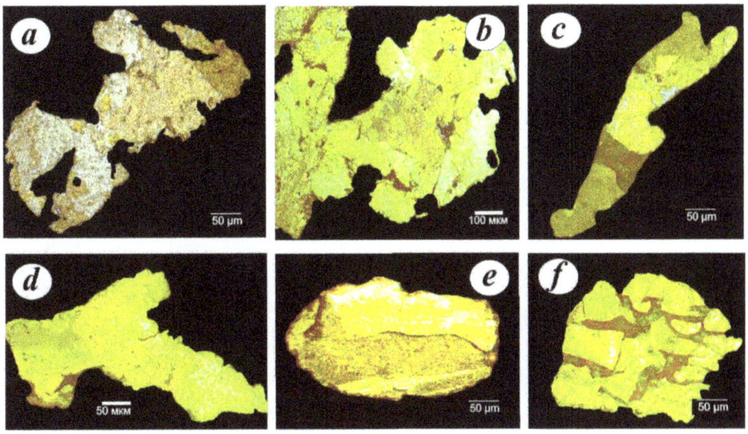

Figure 3. Internal structures of placer gold of the Mesozoic stage of ore formation (Upper Amga river basin): (**a**) coarse-grained structure of high-grade gold; (**b**) inequigranular structure of high-grade gold with manifested granulation; (**c**) the inequigranular structure of high-grade lamellar gold (930–980‰); (**d**) emulsion intergrowth of three phases of gold close to the fineness (light yellow—950‰; yellow—890‰; dark yellow—999‰); (**e**) polygonal intergrowth of two phases of native gold: light yellow—fineness 900‰, dark yellow—999‰; (**f**) polygonal-grained structure of high-grade native gold (940‰) in intergrowth with quartz-hematite-limonite composition.

Type II placer gold has a mainly coarse-grained, inequigranular structure of high-grade gold with manifested granulation, medium-grained internal structure (Figure 3a–c). In the placer gold of the Upper Amga river basin, two- and three-phase gold of various fineness is observed (Figure 3d,e); sometimes, gold forms intergrowths with quartz, potassium feldspar, hematite, limonite and other minerals (Figure 3f).

The placer gold of the Mesozoic stage of ore formation is also characterized by granulation, disintegration structures (Figure 4a–c) and unclear-zonal structures (Figure 4d,e), found in the basin of the middle Lena, along with gold, whose internal structures are the structures of primary recrystallization and secondary recrystallization. Granulation and disintegration structures are extremely rare and have been found at the mouth of the Dzherba and Bol. Patom rivers. The revealed primary structures of granulation and disintegration in placer gold indicate its direct supply from the primary source. Unclear-zonal structures have been identified only in the placer gold of the Tokko and Torgo river basins, which also indicates the proximity of the ore source of the Mesozoic stage of ore formation. In low-grade placer gold, the development of thin high-grade shells is sometimes noted (Figure 4f). The presence of a thin high-grade shell on low-grade placer gold suggests that the placer was formed due to a nearby ore source of the Mesozoic stage of ore formation. The supposed ore sources are formed along the zones of deep faults Bappagaysky (the mouth of the Dzherba river, Bol. Patom), Sensky (Tokko and Torgo rivers), activated in the Mesozoic.

In general, the results of studying the internal structures of type II gold showed that it has a local distribution and is confined to certain geological structures. Placer gold of the Mesozoic stage of ore formation is less changed, as it is finely dispersed and large (up to 1–2 mm or more), it has a low to high gold fineness and a wide range of impurity elements has been found. It is characterized mainly by single grains, coarse- and medium-grained internal structures; it occurs along with type I gold with altered internal structures of primary recrystallization, secondary recrystallization and thick high-grade shells. The results obtained indicate that nearby gold-bearing mineralizations of the Mesozoic stage of ore formation were the ore sources for the formation of such placers.

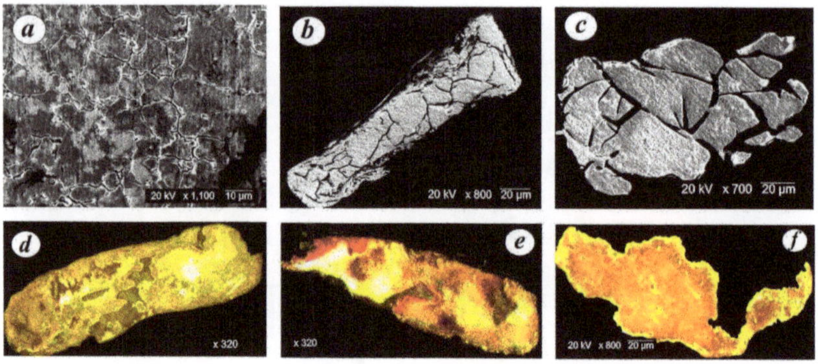

Figure 4. Internal structures of placer gold: (**a**) granulation; (**b**,**c**) disintegration structures (Dzherba, Bol. Patom rivers); (**d**,**e**) unclear-zonal structures (Tokko, Torgo rivers); (**f**) thin high-grade shell on low-grade gold (Torgo river).

Internal structures in exogenous conditions were changed, depending on the conditions where native gold was located [10].

In the hypergenesis zone, morphological forms and material composition of primary gold are changed depending on the mineral type of endogenous mineralization. In the weathering crusts developed along Au-Te and Au-Sb ores, "secondary" gold is formed, which has certain internal structures—porous, kidney-shaped and mustard (Figure 5a,b). Along with the change in the morphology of gold particles, an increase in its fineness is recorded. Simultaneously, with an increase in fineness of, on average, up to 1000 ‰, impurity elements, especially Cu, Pb, Hg, Bi, etc. [26,29,36], are removed in the marginal parts of gold. In this case, a high-grade shell is formed, which has a porous structure and sharply differs from a high-grade dense shell formed in aeolian conditions (Figure 5c).

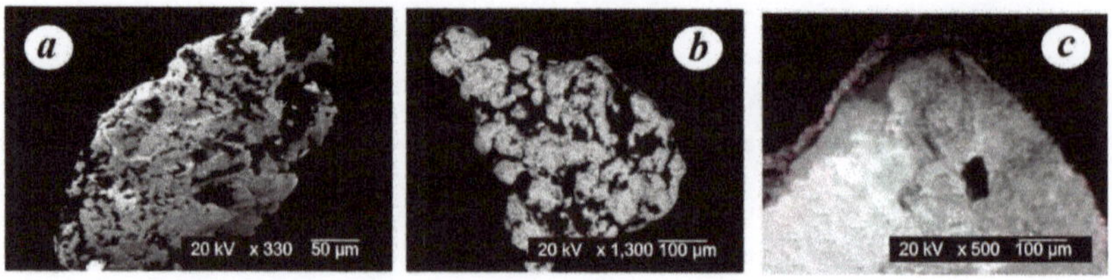

Figure 5. Internal structures of placer gold from the hypergenesis zone: (**a**) porous, mercury removal; (**b**) kidney-shaped, removal of tellurides; (**c**) loose, high-grade shell on mono-grain.

In a hydrodynamic environment, there is practically no change in the internal structure of gold. The internal structure of placer gold depends on the stage of ore formation and on the mineral type of the gold source. Some mechanism of the change in internal structure for placer gold of the Precambrian stage of ore formation in the east of the Siberian Platform has been revealed. This gold is characterized by primary recrystallization, secondary recrystallization structures and thick high-grade shells. This is due to the fact that with repeated redeposition from the Precambrian ancient gold-bearing deposits to younger ones, gold is being purified in exogenous conditions, silver and impurity elements are removed and the internal structure is changed. As a result, the internal structure of this gold is significantly transformed, primary recrystallization, secondary recrystallization structures are formed, as well as thick high-grade shells (10–30 microns) (Figure 2), indicating a long stay of gold in various exogenous environments.

Placer gold of the Mesozoic stage of ore formation is less changed. It is mainly represented by monograins and coarse- and medium-grained internal structures. The identified indicator internal structures of placer gold of the two stages of ore formation practically are not changed in the hydrodynamic environment (river flow, coastal and beach conditions, etc.). Only single simple deformation lines can be found in gold in a river stream [40,42].

In aeolian conditions, there is a change not only in the form, but also in the material composition, and, as a consequence, the internal structure (Figure 6). Aeolian gold has a very wide distribution in the east of the Siberian Platform and was discovered by us in the aeolian deposits of the riverheads of the Lena-Anabar and Lena-Vilyui interfluves (Figure 1).

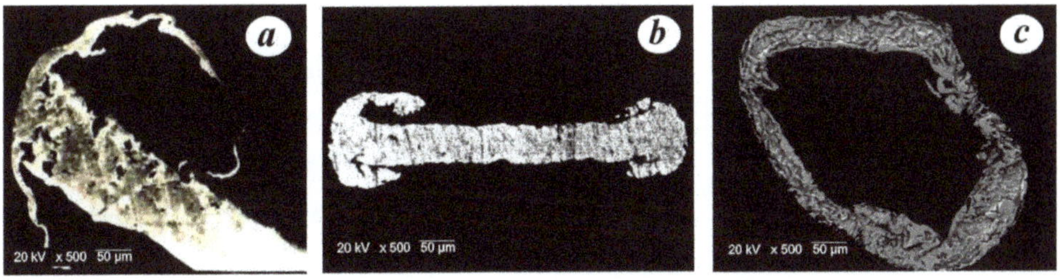

Figure 6. Internal structures of aeolian gold (cross section): (**a**) flake gold particle with a thin ridge along the edge and a recrystallized structure; (**b**) toroidal gold particle with fine-grained recrystallized structure; (**c**) spherical-hollow gold of high fineness 1000‰, the shell of the sphere is completely recrystallized.

Aeolian gold includes toroidal and spherical-hollow gold particles, 0.1–0.25 mm in size [11]. Toroidal gold is usually larger in size and spherical-hollow forms are somewhat smaller. Toroidal gold includes flake gold particles with ridge-like edges, measuring 0.1–0.25 mm. Spherical-hollow gold particles have a partition inside that divides the ball into two chambers. Gold of this shape is characterized by a high degree of grading by size and a rather narrow size range—0.1–0.16 mm. Analysis of the distribution mechanisms of toroidal and spherical-hollow gold according to the literature data revealed that similar forms of gold particles are widespread on all platforms and are found in sediments from the Proterozoic to the Cenozoic, in particular, in the east of the Siberian and European, North American, South American, African and Australian platforms.

It was identified for the first time that the mechanogenic transformation of native gold under aeolian conditions is actively influenced by mechanical and chemical processes and, in this regard, a change was found not only in the shape, but also in the material composition, as well as the internal structure. A clear mechanism has been revealed: an increase in the fineness of gold, a decrease in impurity elements and, as a result, a change in the internal structure.

Toroidal gold is mainly characterized by a fine-grained structure with a partially or completely recrystallized shell. Occasionally, there are gold particles with well-developed fine-grained high-grade shells with preserved relics of an unchanged medium-grained core having a lower fineness. Spherical-hollow forms are usually completely recrystallized with the formation of the fine-grained internal structure of gold.

The results of the study by microprobe analysis of the fineness of individual sections of aeolian gold showed that flakes with thin ridge-like edges with a fineness from 747 to 780‰ have a medium-grained structure. Toroidal shapes with a fineness of 900–970‰ are characterized by a partially recrystallized structure. A completely recrystallized structure was found in toroidal gold with a maximum fineness of 990–1000‰. In spherical-hollow gold of very high fineness, especially in the shell, a complete recrystallization structure is identified. Therefore, for example, in grain (Figure 6c), 17 definitions are 1000‰ [10].

In general, the study of the internal structure of aeolian gold showed that gold with fine-grained structures prevails, mainly of very high fineness, with a minimum content of impurity elements in it (Figure 6a–c). The internal structure of flake gold particles is characterized by the following internal structures—the presence of medium and fine grains, Luders lines (deformations) and well-developed high-grade shells (Figure 6a). Toroidal and spherical-hollow gold particles have mainly recrystallized fine-grained structures (Figure 6b,c).

Thus, during the technogenic process of exposure of sand material to flake gold in aeolian conditions, not only its morphological features are transformed, caused by a complex deformation of the original individual, but also the internal structure. Each stage of complex deformation of flakes in aeolian conditions under the influence of mechanical and chemical processes corresponds to a regular change in its internal structure. The internal structure of flake gold particles is characterized by the presence of medium and fine grains, Luders lines and well-developed high-grade shells. Toroidal and spherical-hollow gold particles have mainly recrystallized fine-grained structures with pronounced high-grade shells.

In ancient gold-bearing conglomerates (buried placers), the internal structures of placer gold (pseudo-ore gold) are transformed. Pseudo-ore gold includes gold of scaly shape in "growth" with quartz, ilmenite, zircon and other minerals of the host deposits, as well as flake gold particles with casts of pressing of minerals, traces of scars, scratches and slickensides on the surface, sometimes with ragged edges or with through holes. The surface of such gold particles is coarse-pitted and finely cellular. The size of the gold particles is 0.1–0.25 mm; it is mainly of high fineness with primarily recrystallized and secondarily recrystallized internal structures. The genesis of such forms of gold is explained by the influence of the lithostatic pressure of the overlying strata on the formed placer, while the minerals of the host deposits are pressed into the gold that has been proven experimentally [10,11].

The study of pseudo-ore gold and its internal structures revealed the following.

Due to the influence of lithostatic pressure of overlying strata on gold in ancient conglomerates, namely, at constant temperatures and pressure, there is a recrystallization of gold with the removal of silver and impurity elements, which leads to the purification of gold, an increase in the fineness of gold particles and the formation of high-grade shells (Figure 7). It is established that during horizontal movements, numerous deformation lines are formed in gold (Figure 7a,b), the predecessors [1,8], explained by the influence of mechanical processes in the hydrodynamic environment. Sometimes, single gold particles with granulation structures are found in gold-bearing conglomerates, which are generated as a result of the substitution of medium-grade gold with the removal of silver (Figure 7c).

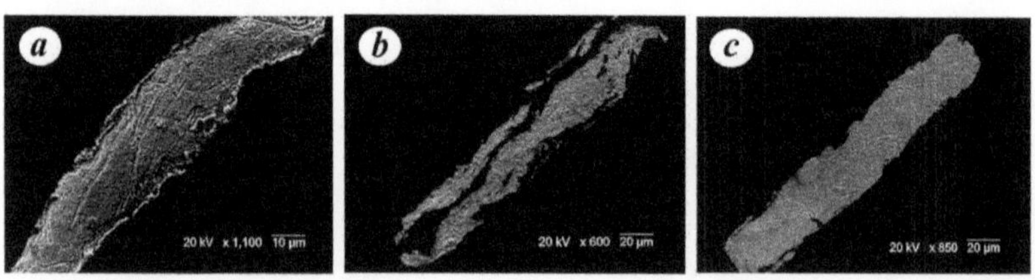

Figure 7. Internal structures of pseudo-ore flake gold (cross section): (**a**,**b**) flake gold particles with numerous lines of deformations; (**c**) granulation structures and numerous deformation lines.

4. Discussion

The study of the internal structures of widespread high-grade flake placer gold in the east of the Siberian Platform with a surprisingly small set of impurity elements, with recrystallization structures, thick high-grade shells and numerous deformation lines allowed us to assume that this gold has been repeatedly redeposited and supplied from ancient Precambrian sources. The sources of type I gold with such internal structures were ores of low-sulfide gold quartz, gold–copper–porphyry, gold–platinum, gold–iron–quartzite mineralization.

It was found that placer gold originating from metamorphic strata of Archean and Early Proterozoic age is characterized by structures of primary recrystallization, secondary recrystallization and numerous deformation lines (Figure 2g–i). Such internal structures are characteristic of metamorphogenic gold, since with the metamorphism of sedimentary strata, under the influence of constant temperature and pressure, gold is transformed [44]. At the same time, there is a process of changing the internal structure of primary gold, involving the removal of silver and impurities, which contributed to the formation of structures of primary recrystallization, secondary recrystallization and with dynamometamorphism (horizontal movements)—numerous plastic deformations. Indeed, according to A.V. Kopeliovich [45], with deep epigenesis—prolonged exposure to constant pressure—a kind of "transit" of impurity elements from the inner parts of minerals to the edges is carried out. In connection with this, the minerals are "self-cleaning" from the impurities contained in them.

The discovery of native gold with structures of primary and secondary recrystallization with numerous deformation lines indicates that the ore sources for placer occurrences of the middle course of the Tokko and Torgo rivers were ferruginous quartzites of the Archean age with a gold content from 0.6 to 1 g/t, giving reason to predict the deposits of gold–ferruginous–quartzite deposits in this area. Gold deposits of ferruginous–quartzite mineralization with a gold content of up to 10 g/t are widespread in southern Yakutia (iron-ore deposits Tarynnakh, Ymalakh, etc.), which confirms this assumption.

Thus, the identification of numerous deformation lines in placer gold indicates that these structures were formed in metamorphic strata and are an indicator for predicting ore deposits of gold–ferrous–quartzite mineralization. We draw attention to the fact that, although metamorphogenic gold has been studied earlier [1,3,4,8], this problem needs to be addressed in additional research.

Thus, the study of the internal structures of placer gold and the mechanisms of its distribution revealed that type I gold with primarily recrystallized and secondarily recrystallized structures has a wide distribution and is found in all watercourses of the east of the Siberian Platform.

Along with transformed primarily recrystallized and secondarily recrystallized structures with numerous deformation lines, indicating the influence of metamorphogenic processes in certain territories, in particular in the southeast of the Siberian Platform in the basin of the Torgo, Tokko rivers, placer gold was found with an unclear-zonal structure, with granulation and disintegration structures indicating the presence of primary sources of the Mesozoic stage of ore formation.

The presence on individual objects of placer gold with coarse-grained, medium-grained and fine-grained structures, multiphase gold sometimes with a thin high-grade shell, with unclear zonal structures, and also with granulation and disintegration structures (Figures 3 and 4) indicates that this gold belongs to the Mesozoic stage of ore formation, the sources of which could be ores of various mineralization (gold–silver, gold–sulfide-quartz, gold–rare metal).

According to L.A. Nikolaeva et al. [8], discovery of multiphase gold (Figure 3d,e) indicates the supply of gold from ore deposits of gold–quartz–sulfide mineralization (Sukhoy Log type), gold–polysulfide–quartz (Darasun type) or gold.

For example, the internal structures found in the gold of the mouth of the Lena-Vilyui interfluve (Kempendyai dislocations, Vilyui paleorift, volcanic activity of andesite–dacite composition), namely, a high-grade shell on low-grade gold with a high silver content up to 25–40% (Figure 8a,b), as well as intergranular high-grade veinlets (Figure 8c) are typical for gold of near-surface deposits. The discovered internal structures indicate the supply of gold from gold–silver sources. The presence of gold with such mineralogical and geochemical features and specific internal structures, namely, high-grade shells on low-grade gold, intergranular high-grade veinlets give reason to assume that gold and silver deposits (such as Cripple Creek) are predicted in this territory.

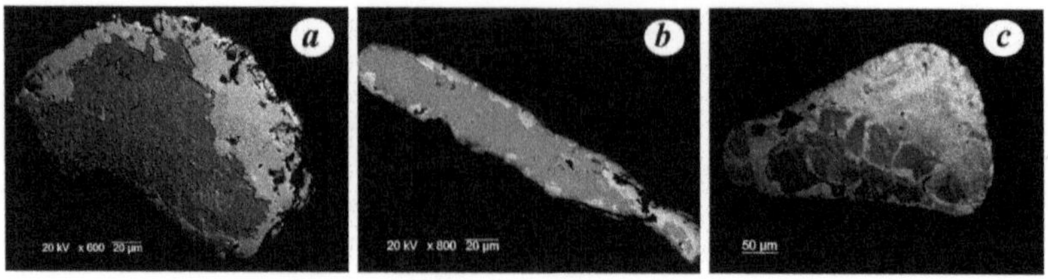

Figure 8. The internal structure of placer gold of the sources of the rivers of the Kempendyai dislocations: (**a**) monocrystalline structure with a dense high-grade shell with a thickness of 20–40 microns; (**b**) very thin fragmentary high-grade shell on low-grade gold; (**c**) intergranular high-grade veinlets.

Discovery of porous spongy placer gold up to 40% (Figure 5a,b) with an increased mercury content (>6%) at the mouth of the Bol. Patom river in the middle Lena basin in the south-east of the Siberian Platform indicates that gold-bearing hydrothermal–metasomatic formations of gold sulfide quartz mineralization were the ore sources of this gold. The following internal structures were found in placer gold: mono- and coarse-grained with thin intermittent high-grade shells, sometimes unclear zoning, spotty heterogeneity, porosity, multiphase nature, granulation and disintegration. The above indicators for internal structures are characteristic of mineralization of the gold sulfide quartz formation. According to the identified complex of mineralogical and geochemical features, placer gold is similar to the gold of the Kuranakh ore-placer cluster. In this regard, the formation of a gold deposit of the Kuranakh type (Carlin) is predicted in this territory.

Thus, it was determined for the first time that native gold has certain internal structures depending on the stage of ore formation and conditions of stay in exogenous environments. Gold of the Precambrian stage of ore formation is characterized by primary recrystallization and secondary recrystallization structures and numerous deformation lines, as well as the presence of a thick high-grade shell. Gold of the Mesozoic stage of ore formation has mainly coarse-medium-grained and inequigranular structures; sometimes, it is marked by multiphase gold, granulation, disintegration and unclear-zonal structures and the development of thin high-grade shells is observed in some of them.

In the hypergenesis zone, the formation of a high-grade shell is observed, gold recrystallization occurs in buried placers or ancient deposits, as well as the formation of deformation lines under the influence of lithostatic pressure of the overlying strata. Pseudo-ore gold with numerous deformation lines is characteristic of all platform areas and occurs in alluvial placers. The discovery of gold particles with such structures can serve as the basis for the search for gold ore metamorphic sources

Recrystallization structures predominate in gold particles in the aeolian medium due to complex deformation. The study of the internal structure of aeolian gold particles and their chemical composition showed that gold with a fine-grained structure prevails, mainly of very high fineness, with a minimum content of impurity elements in it. It is identified that the deformation of gold involves its primary recrystallization with the removal of silver and secondary recrystallization, which contributes to the removal of impurity elements from gold. In aeolian conditions, this process is pronounced and caused by a complex deformation. Each stage of complex deformation of flakes in aeolian conditions under the influence of mechanical and chemical processes corresponds to a regular change in its internal structure due to increased fineness and a decrease in the content of impurity elements. Indeed, due to the intensive transformation of gold during its complex deformation, the thinnest films of gold are stretched, which, overlapping each other, form a shell of spherical shapes. At the same time, the surface of gold increases for active chemical interaction with the components in the medium. Therefore, a long conversion time contributed to the maximum removal of silver and impurity elements and a change in the internal structure.

5. Conclusions

The long-term study of the internal structures of placer gold in the east of the Siberian Platform and the generalization of these results for the first time allowed us to establish the stages of ore formation and the evolution of placer gold under exogenous conditions in hypergenic, hydrodynamic and aeolian environments.

Based on the identification of a complex of indicators of the internal structures of type I placer gold, the sources that made possible the formation of placer gold content (weathering crusts, gold-bearing reservoir rocks, metamorphic strata), were identified for the first time in the east of the Siberian platform. The stages of ore formation (Precambrian and Mesozoic) are also proved.

The indicators of the internal structures of placer gold of the Precambrian stage of ore formation include the presence of structures of primary recrystallization, secondary recrystallization, numerous deformation lines and thick high-grade shells. The indicator signs of the internal structures of placer gold of the Mesozoic stage of ore formation are coarse-medium-grained structures, mono-grains, unclear-zonal, structures of granulation and disintegration.

It is identified that the above-mentioned indicators of the internal structures of placer gold of the Precambrian stage of ore formation are characteristic of almost all placer occurrences of the entire east of the Siberian platform, and indicators of the internal structures of placer gold of the Mesozoic stage of ore formation are characteristic only of certain geological structures.

The revealed internal structures typical for endogenous and exogenous conditions can serve as an effective method for determining the genesis of placers and the nature of gold ore sources. The use of this method makes it possible to develop criteria for forecasting ore sources and types of gold deposits based on internal structures.

Funding: The research was carried out within the framework of the state task of the Diamond and Precious Metals Geology Institute, Siberian Branch, Russian Academy of Sciences, funded by the Ministry of Science and Higher Education of the Russian Federation project № 0381-2019—0004.

Data Availability Statement: Not applicable.

Conflicts of Interest: The author declares that there are no conflict of interest.

References

1. Petrovskaya, N.V. *Native Gold*; Nedra: Moscow, Russia, 1973; 347p. (In Russian)
2. Nikolaeva, L.A. *Genetic Features of Native Gold as a Criterion in the Search and Evaluation of Ores and Placers*; Nedra: Moscow, Russia, 1978; 100p. (In Russian)
3. Popenko, G.S. *Mineralogy of Gold of the Quaternary Placers of Uzbekistan*; FAS: Tashkent, Russia, 1982; 144p. (In Russian)
4. Savva, N.E.; Preis, V.K. *Atlas of Native Gold of the North-East of the USSR*; Nauka: Moscow, Russia, 1990; 293p. (In Russian)
5. Savva, N.E.; Shilyaeva, N.A.; Aleyevskaya, N.L. *Topomineralogy of Constitutional Features of Native Gold of the Lower Amur Placer Area*; SVKNII DVO RAN: Magadan, Russia, 2004; 172p. (In Russian)
6. Samusikov, V.P. Elements-impurities in native gold—Criteria for determining the formational belonging of gold deposits. *RAS Rep.* **2003**, *1*, 99–103. (In Russian)
7. Kalinin, Y.A.; Roslyakov, N.A.; Prudnikov, S.G. *Gold-Bearing Weathering Crusts of Southern Siberia*; "Geo" Academic Publishing House: Novosibirsk, Russia, 2006; 339p. (In Russian)
8. Nikolaeva, L.A.; Gavrilov, A.M.; Nekrasova, A.N.; Yablokova, S.V.; Shatilova, L.V. *Native Gold of Ore and Placer Deposits of Russia*; TsNIGRI: Moscow, Russia, 2015; 200p. (In Russian)
9. Nikiforova, Z.S.; Gerasimov, B.B.; Glushkova, E.G.; Kazenkina, A.G. Gold content of the east of the Siberian platform (placers—primary sources). *Geol. Ore Depos.* **2013**, *55*, 305–315. [CrossRef]
10. Nikiforova, Z.S.; Kalinin, Y.A.; Makarov, V.A. Evolution of native gold in exogenous conditions. *Russ. Geol. Geophys.* **2020**, *61*, 1244–1259. [CrossRef]
11. Nikiforova, Z.S. Criteria for determining the genesis of placers and their different sources based on the morphological features of placer gold. *Minerals* **2021**, *11*, 381. [CrossRef]
12. Petrovskaya, N.V.; Fastalovich, A.I. *Morphological and Structural Features of Native Gold. Materials on the Mineralogy of Gold*; TsNIGRI: Moscow, Russia, 1952; pp. 101–150. (In Russian)

13. Savva, N.E.; Kravtsova, R.G.; Anisimova, G.S.; Palyanova, G.A. Typomorphism of Native Gold (Geological-Industrial Types of Gold Deposits in the North-East of Russia). *Minerals* **2022**, *12*, 561. [CrossRef]
14. Giusti, L.; Smith, D.G.W. An electron microprobe study of some Alberta placer gold. *Tschermaks Mineral. Petrogr. Mitt.* **1984**, *33*, 187–202. [CrossRef]
15. Freyssinet, P.; Zeegers, H.; Tardy, Y. Morphology and geochemistry of gold grains in lateritic profiles of southern Mali. *J. Geochem. Explor.* **1989**, *32*, 17–31. [CrossRef]
16. Knight, J.B.; Mortensen, J.K.; Morison, S.R. Lode and placer gold composition in the Klondike District, Yukon territory, Canada: Implications for the nature and genesis of Klondike placer and lode gold deposits. *Econ. Geol.* **1999**, *94*, 649–664. [CrossRef]
17. Chapman, R.J.; Leake, R.; Moles, N.R.; Earls, G.; Cooper, C.; Harrington, K.; Berzins, R. The application of microchemical analysis of alluvial gold grains to the understanding of complex local and regional gold mineralization: A case study in the Irish and Scottish Caledonides. *Econ. Geol.* **2000**, *95*, 1753–1773. [CrossRef]
18. Chapman, R.J.; Mortensen, J.K. Application of microchemical characterization of placer gold grains to exploration for epithermal gold mineralization in regions of poor exposure. *J. Geochem. Explor.* **2006**, *91*, 1–26. [CrossRef]
19. Dill, H.G. Grain Morphology of Heavy Minerals from Marine and Continental Placer Deposits, with Special Reference to Fe-Ti Oxides. *Sediment. Geol.* **2007**, *198*, 1–27. [CrossRef]
20. Dill, H.G. Geogene and Anthropogenic Controls on the Mineralogy and Geochemistry of Modern Alluvial-(Fluvial) Gold Placer Deposits in Man-made Landscapes in France, Switzerland and Germany. *J. Geochem. Explor.* **2008**, *99*, 29–60. [CrossRef]
21. Dill, H.G. Gems and Placers-A Genetic Relationship Par Excellence. *Minerals* **2018**, *8*, 470.
22. Melchiorre, E.B.; Henderson, J. Topographic gradients and lode gold sourcing recorded by placer gold morphology, geochemistry, and mineral inclusions in the east fork San Gabriel River, California, USA. *Ore Geol. Rev.* **2019**, *109*, 348–357. [CrossRef]
23. Maftei, E.A.; Buzatu, A.; Damian, G.; Buzgar, N.; Dill, H.G.; Apopei, A.I. Micro-Raman—A Tool for the Heavy Mineral Analysis of Gold Placer-Type Deposits (Pianu Valley, Romania). *Minerals* **2020**, *10*, 988. [CrossRef]
24. Chapman, R.J.; Banks, D.A.; Styles, M.T.; Walshaw, R.D.; Piazolo, S.; Morgan, D.J.; Grimshaw, M.R.; Spence-Jones, C.P.; Matthews, T.J.; Borovinskaya, O. Chemical and physical heterogeneity within native gold: Implications for the design of gold particle studies. *Miner. Depos.* **2021**, *56*, 1563–1588. [CrossRef]
25. Solberg, T.N.; Craig, J.R. Chemical variations in gold from the central Appalachians of Virginia. *Microbeam Analysis.* **1981**, 163–166.
26. Mann, A.W. Mobility of gold and silver in lateritic weathering profiles: Some observations from Western Australia. *Econ. Geol.* **1984**, *79*, 38–49. [CrossRef]
27. Falconer, D.M.; Craw, D.; Youngson, B.; Faure, K. Gold and sulfide minerals in tertiary quartz pebble conglomerate gold placers, Southland, New Zealand. *Ore Geol Rev.* **2006**, *28*, 525–545. [CrossRef]
28. Nysten, P. Gold in the volcanogenic mercury- rich sulfide deposit Ldngsele, Skellefte ore district, northern Sweden. *Miner. Depos.* **1986**, *21*, 116–120. [CrossRef]
29. Stoffregen, R. Observations on the behavior of gold during supergene oxidation at Summitville, Colorado, U.S.A., and implications for electrum stability in the weathering environment. *Appl. Geochem.* **1986**, *1*, 549–558. [CrossRef]
30. Craig, J.R.; Callahan, J.E. Paleoplacer gold in the Lilesville sand and gravel deposits, North Carolina. *Ga. Geol. Surv. Bull.* **1989**, *117*, 121–141.
31. Driscoll, A.J.; Hall, D.L.; Craig, J.R. Placer gold: An unrealized potential for characterizing lodes. *Ga. Geol. Surv. Bull.* **1991**, *105*, 95–118.
32. Craw, D.; McLachlan, C.; Negrini, M.; Becker, N. Quantification and prediction of bulk gold fineness at placer gold mines: A New Zealand example. *Minerals* **2017**, *7*, 226. [CrossRef]
33. Dongmo, F.W.N.; Chapman, R.J.; Bolarinwa, A.T.; Yongue, R.F.; Banks, D.A.; Olajide-Kayode, J.O. Microchemical characterization of placer gold grains from the Meyos-Essabikoula area, Ntem complex, southern Cameroon. *J. Afr. Earth Sci.* **2018**, *151*, 189–201.
34. Dunn, S.C.; Bjorn, P.; Rozendaal, A.; Taljaard, R. Secondary gold mineralization in the Amani placer gold deposit, Tanzania. *Ore Geol. Rev.* **2019**, *107*, 87–107.
35. Erasmus, C.S.; Sellschop, J.P.F.; Watterson, J.I.W. New evidence on the composition of mineral grains of native gold. *Nucl. Geophys.* **1987**, *1*, 1–23.
36. Groen, J.C.; Craig, J.R.; Rimstidt, J.D. Gold-amalgam and other secondary phases in placers. *Abstr. Programs—Geol. Soc. Am.* **1986**, *19*, 621p.
37. Groen, J.C.; Craig, J.R.; Rimstidt, J.D. Gold–rich rim formation on electrum grains in placers. *Can. Mineral.* **1990**, *28*, 207–228.
38. Leake, R.C.; Bland, D.J.; Styles, M.T. Internal structure of Au-Pd-Pt grains from south Devon, England, in relation to low-temperature transport and deposition. *Min. Metall.* **1991**, *100*, 159–178.
39. Cabral, A.R.; Suh, C.E.; Kwitko-Ribeiro, R.; Rocha Filho, O.G. Brittle microstructures in gold nuggets: A descriptive study using backscattered electron imaging. *Ore Geol. Rev.* **2008**, *33*, 212–220. [CrossRef]
40. Kerr, G.; Falconer, D.; Reith, F.; Craw, D. Transport-related mylonitic ductile deformation and shape change of alluvial gold, southern New Zealand. *Sediment. Geol.* **2017**, *361*, 52–63. [CrossRef]
41. Stewart, J.; Kerr, G.; Prior, D.; Halfpenny, A.; Pearce, M.; Hough, R.; Craw, D. Low temperature recrystallisation of alluvial gold in paleoplacer deposits. *Ore Geol. Rev.* **2017**, *88*, 43–56. [CrossRef]
42. Chapman, T.; Clarke, G.L.; Piazolo, S.; Robbins, V.A.; Trimby, P.W. Grain-scale dependency of metamorphic reaction on crystal plastic strain. *J. Metamorph. Geol.* **2019**, *3*, 1021–1036. [CrossRef]

43. *Report on the State of Fundamental Sciences in the Russian Federation and on the Most Important Scientific Achievements of Russian Scientists in 2014*; Most Important Achievements; Earth Sciences: Moscow, Russia, 2014; 181p. (In Russian)
44. Moiseenko, V.G. *Metamorphism of Gold Deposits of the Amur Region*; Khabarovsk Publishing House: Blagoveshchensk, Russia, 1965; 127p. (In Russian)
45. Kopeliovich, A.V. *Epigenesis of Ancient Strata of the South-West of the Russian Platform*; Nauka: Moscow, Russia, 1965; 311p. (In Russian)

minerals

MDPI

Article

Typomorphic Features and Source of Native Gold from the Sykhoi Log Area Placer Deposits, Bodaibo Gold-Bearing District, Siberia, Russia

Alexander Lalomov *, Antonina Grigorieva, Alexei Kotov and Lidiya Ivanova

Laboratory of Ore Deposits, Placer Group, Institute of Geology of Ore Deposits Petrography, Mineralogy and Geochemistry of Russian Academy of Science (IGEM RAS), 119017 Moscow, Russia; grig357@mail.ru (A.G.); alekskotov@mail.ru (A.K.); ldsamoshnikova@gmail.com (L.I.)
* Correspondence: lalomov@mail.ru

Abstract: The Bodaibo gold-bearing district in the Lena gold province of Siberia is one of the largest and oldest placer gold-bearing provinces in the world. Approximately 1650 tons of gold has been extracted from the region. Precise studies on the source of these unique placer deposits are lacking and still controversial. Native gold from four different locations was gathered to investigate its morphology, chemical signatures, structure and inclusions. Some data on primary bedrock mineralization were obtained from the published literature. The linear weathering crusts developed along the zones of disjunctive dislocations near the Sukhoi Log gold deposit were researched. If they coincided with zones of low-grade veinlet-disseminated gold–quartz–sulphide mineralization with small gold grain sizes, a supergene replacement of primary mineralization was known to have occurred, accompanied by the formation of gold-rich rims and an increase in the size, content and purity of gold. Such mineralization associated with linear weathering crusts can be a source of local eluvial–proluvial placers, while placers of large valleys are formed due to low-sulphide gold–quartz lodes.

Keywords: placer; native gold; typomorphysm; linear weathered crust; Sukhoi Log deposit

check for updates

Citation: Lalomov, A.; Grigorieva, A.; Kotov, A.; Ivanova, L. Typomorphic Features and Source of Native Gold from the Sykhoi Log Area Placer Deposits, Bodaibo Gold-Bearing District, Siberia, Russia. *Minerals* **2023**, *13*, 707. https://doi.org/10.3390/min13050707

Academic Editor: Huan Li

Received: 24 April 2023
Revised: 13 May 2023
Accepted: 19 May 2023
Published: 22 May 2023

1. Introduction

The Bodaibo gold-bearing district of Lena gold province is located in Siberia, in the Irkutsk region of Russia (Figure 1). The first placer gold of the region was obtained in 1841. It is one of the largest and oldest placer gold-bearing provinces in the world [1,2]. Over the 180 years in which gold prospection and extraction were carried out in the area, approximately 1500 tons of placer gold and more than 150 tons of primary gold (mainly during the last twenty years) have been extracted from the gold deposits of the region. The total potential amount of resources is estimated at about 4000 metric tons, with an annual gold production of over 20 metric tons [3]. In spite of the fact that the main prospects for the gold mining industry in the region are associated with primary ores being developed and prepared for development, placer deposits still provide about half of the gold mined.

Although it has been investigated for over a century, some aspects of gold mineralization are still unclear, in particular concerning the source of this placer gold.

The Bodaibo gold-bearing district is a unique region not only in Russia, but also in the world in terms of placer reserves, a significant part of which has already been mined out [4]. This district also contains large primary bedrock deposits, including the largest deposit in Russia, the Sukhoi Log, whose reserves are associated with scattered vein-disseminated gold–quartz–sulphide mineralization. The main problem is that vein gold–quartz placer-forming deposits with large-size gold have a limited distribution, and large deposits such as the Sukhoi Log are not direct sources of placers because of the fine size of the gold grains. For example, more than 95% of the native gold found in the Sukhoi Log has a grain size of less than 0.1 mm [5]. Native gold of this size is far too small for gravity enrichment and does not accumulate well in the placer deposits. Additionally, the giant gold deposit Sukhoi Log could not be a source of gold placers in the surrounding area because it has a low erosion level; the main ore body is not exposed on the surface [6].

Figure 1. Map of the Bodaibo gold-beating district.

The sulphide-poor quartz–gold-bearing lodes favourable for the formation of placers are of subordinate importance in the territory of the Bodaibo District. We can only assume that the placers were formed due to giant primary, mainly quartz–gold lode deposits, which are now completely eroded [7–10].

This point of view considers only the mechanical processes of placer formation due to the disintegration of bedrock gold deposits. However, a number of researchers believe that under supergene conditions, sulphide-associated gold changes significantly, becomes coarser and transforms into a gravitationally enriched placer-forming class up to the formation of nuggets weighing up to tens of kilograms [11–14]. Such a process can lead to the formation of placer deposits in the absence of significant bedrock gold sources due to poor disseminated gold–sulphide mineralization. One piece of evidence for the growth of gold particles under supergene conditions is the presence of gold-enriched rims on the particles.

The linear weathering crusts (LWCs) developed along the zones of disjunctive dislocations near the Sukhoi Log gold deposit were researched. The deposit consists of yellow-brown clay with weathered shale and quartz detritus. The conducted studies have revealed that even with low concentrations or disseminations of gold in the host rocks, the clay contains gravitationally enriched native gold of economic grade. This gold has evidence of grain growth such as a high-grade gold rim.

The main task of the study was to assess the role of LWCs in the transformation of fine gold of veinlet-disseminated quartz–sulphide mineralization into larger placer-forming classes, as well as the influence of this gold on the processes of placer formation in order to assess the prospects for the exploitation of such placers. Along the way, a conclusion was made about the possible sources of gold for the formation of the unique placers of the Bodaibo gold-bearing district.

2. Geology and Metallogeny of Bodaibo Gold-Bearing District

2.1. Geological Structure, Stratigraphy and Metallogeny of Pre-Cenozoic Stage

The study area is located within a complicated regional structure known as the Bodaibo synclinorium, corresponding geographically to the Patom Highlands. Early, Middle and Late Riphean metasediments of the Bodaibo synclinorium, which correspond to the Bodaibo internal depression (Figure 2), unconformably overlie Archean–Proterozoic metamorphic rocks.

Lower Riphean stratigraphic divisions consist of greenschist facies conglomerates, sandstones, shales and volcanogenic and volcano-terrigenous rocks. Andesite, andesite–basalts and tuffs are intercalated with sandstones and conglomerates and ferruginous quartzites.

The Middle Riphean, which is up to 2500 m thick, consists mainly of terrigenous gravel–sand–siltstone sediments, succeeding calcareous sandstones and limestones. Middle and Late Riphean sedimentary rocks of the Nigry Group are divided into the Buzhuikhta, Ugokhan, Khomolkho and Imnyakh Formations with a total thickness up to 1500 m. There, rocks mainly form the Bodaibo internal depression.

The flysch of the Khomolkho Formation hosts all of the gold mineralization in the Sukhoi Log deposit. The silty and calcareous carbon-bearing shales contain abundant diagenetic sulphides. Rhythmically interbedded sandstones, shales, carbonaceous shales, limestones and sandstones of the Vendian Bodaibo Group up to 2500 m thick terminate the Late Proterozoic sequences.

Terrigenous and carbonaceous layers rich in organic carbon are characteristic of the region. Carbon-bearing sediment deposition reached a maximum in the Middle and Late Riphean. The amount of organic matter increases from carbonate to sandstone to siltstone to pelite, following the normal sequence of a sedimentary succession.

Terrigenous carbonaceous rocks can be found in a continental margin sea basin. Minor amounts of Lower Riphean volcanics are intercalated with coarse-grained clastics and ferriferous quartzites. At the same time, ophiolitic units, including ultramafic rocks and tholeiitic volcanics, formed nearby and developed during the Early Paleozoic Bodaibo synclinorium.

The central part of the synclinorium has been metamorphosed to the biotite subfacies of the greenschist facies. Peripheral rocks have experienced epidote–amphibolite and amphibolite facies regional metamorphism associated with the formation of granite–gneiss domes and palingenetic granitoid plutons. Small-to-middle rank magmatic bodies occur within the Bodaibo synclinorium, but large Paleozoic granite intrusions, such as the Dzhegdakar massif, are exposed on the periphery [15].

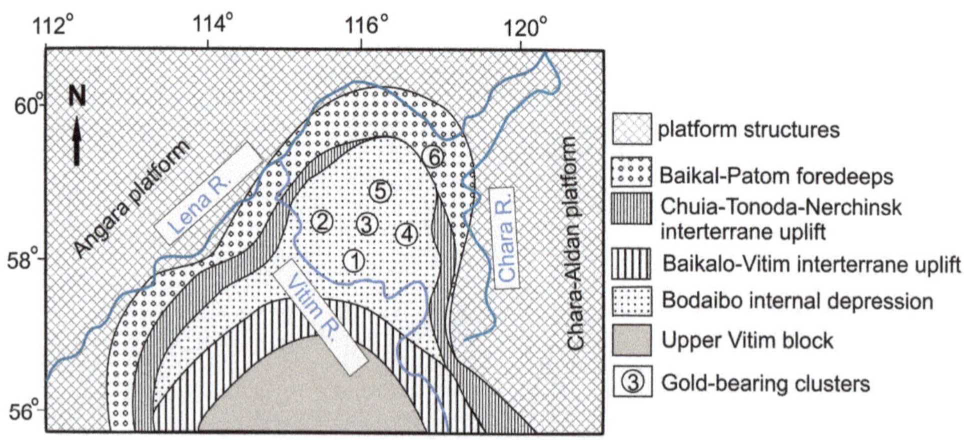

Figure 2. Tectonic scheme of Bodaibo gold-bearing district (modified after [16]). Gold-bearing clusters: 1—Artemovsky, 2—Marakansky, 3—Kropotkinsky (including Sukhoi Log deposit), 4—Svetlinsky, 5—Khomolkho, 6—Dalnetaiginsky.

The two types of primary lode gold mineralization found in the Lena province are the veinlet-disseminated and quartz lode types. The veinlet-disseminated gold–quartz–sulphide mineralization is the most important economic type, which occurs at mine mineral deposits of economic importance, including the Sukhoi Log, Vysochaishee, Verninskoe (Kropotkinsky cluster), Nevskoe (Svetlinsky cluster) and Chertovo Koryto (Khomolkho cluster). This mineralisation is controlled by linear and overturned anticlines, and fault zones in their axial parts and flanks. The largest deposits are hosted in black shales metamorphosed to the sericite–chlorite greenschist facies. Minor gold deposits occur in rocks metamorphosed to the epidote–amphibolite and amphibolite facies, although they are confined to a zone of retrograde metamorphism (Figure 3).

Figure 3. Quartz lode gold mineralization from Verninskoe deposit (Kropotkinsky cluster).

The quartz vein type or gold–sulphide–quartz mineralization is confined to sublatitudinal linear vein zones in the axial parts of anticlines, tectonic faults, shear zones and tension fractures, including those concordant to the bedding plane (Pervenets, Dogaldyn, Krasnoe deposits) and those of cross-cutting settings (Kavkaz). The largest vein clusters are Centralnoe in the Sukhoi Log camp and Dogaldyn in the Artemovskiy camp [6].

2.2. Sukhoi Log Deposit

The geology of the Sukhoi Log is described in much detail elsewhere [3,6,15–19] and the references therein.

The Sukhoi Log is located in the Mama–Bodaibo synclinorium. It is hosted in sedimentary rocks of the Khomolkho Formation and partially in the overlying dolomite-dominated Imnyakh Formation. Igneous rocks in the vicinity of the deposit are confined to the small Konstantinovsky granitoid stock, 6 km south of the deposit [20].

The largest orebody, which has the same name, is about 4 km^2 in area and is composed of veinlet-disseminated gold–quartz–sulphide stockwork-type mineralization, which is confined to the core of the anticline and is surrounded by a wider aureole of barren sulphides. The deposit is hosted mostly in carbonaceous quartz sericite–carbonate shales to sandstones metamorphosed to the sericite and locally sericite–chlorite subfacies of the greenschist facies (Figure 4).

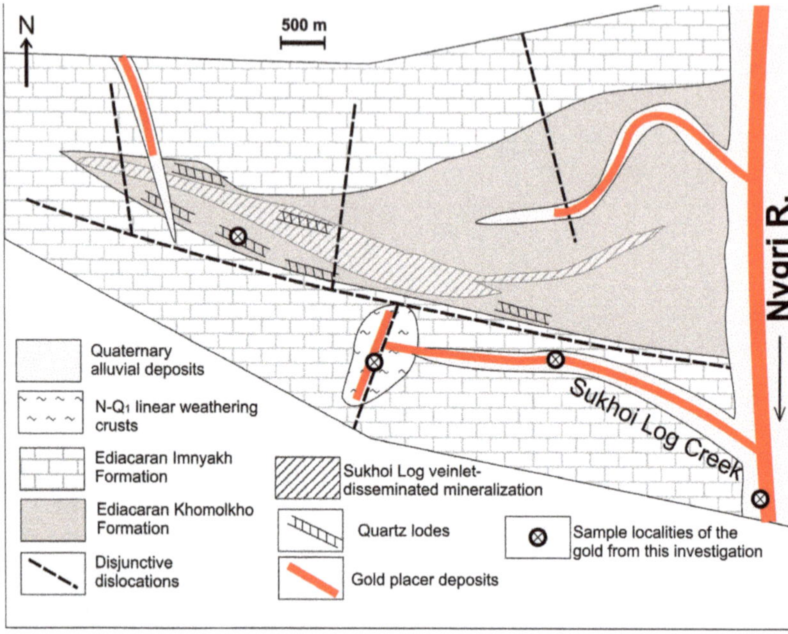

Figure 4. Geological scheme of Sukhoi Log gold deposit with bedrock and placer gold metallogeny (based on [6,16]. Exposure of the veinlet-disseminated gold–quartz–sulphide mineralization on the current surface.

The major gold reserves of the deposit are in quartz–sulphide veinlet-disseminated mineralization (VDM). Low-sulphide gold–quartz veins are of subordinate importance. Outside the economic area the rocks are usually barren or contain only low-grade mineralization.

Gold in veinlet-disseminated ores is found in native rock, mainly in the form of microscopic and thin inclusions in pyrite. The size of gold particles varies from a few microns up to 0.4 mm, while more than 95% of gold is in the order of less than 0.1 mm [16].

The shape of the gold particles is varied and dominated by interstitial secretions, occasionally lumpy, more or less massive interstitial secretions with a complex porous and spongy surface. Plates and scales with uneven edges and a predominantly complex cellular surface are widespread.

The fineness of the studied gold particles varied from 846 to 908. Silver is a permanent component of the alloy, and its content varies from 9 to 14 wt.%. Additionally, trace elements were detected, including mercury (up to 0.25%), iron (up to 0.97%), copper (up to 0.11%) and nickel (up to 0.06%) [5].

2.3. The Structure and Composition of The Sedimentary Cover and Placer Deposits of The Bodaibo District

The area of the Bodaibo District is a deeply dissected lowland. The height of the main part of the watersheds is 850–950 m. The depth of the river valleys is 200–500 m, and the width of their upper contours is 2–4 km. The width of the floodplains is 200–400, occasionally reaching 800 m. The valleys are filled with a sediment sequence of 20–25 to 140–170 m thick. In the buried relief of the valleys, deep thalwegs and terraces are found.

According to Y.P. Kazakevich and M.V. Reverdatto [21], four stages are assigned to the Quaternary history of the development of the present-day relief and placer gold formation (Figure 5).

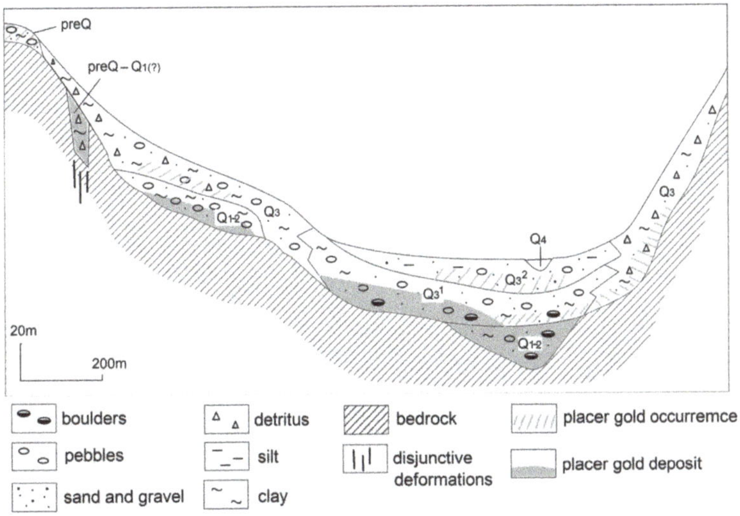

Figure 5. The generalized cross-sections of the deposits of the Nigri Valley (based on [21]. PreQ—uncertain Cenozoic (pre-Quarter) cover gravel; PreQ-$Q_{1(?)}$—supposed Cenozoic–Lower Quaternary linear weathered crusts; Q_1—Lower Quaternary, Q_2—Middle Quaternary, Q_3—Upper Quaternary, Q_4—Holocene alluvial and deluvial deposits.

The first stage was characterized by moderate tectonics, planar denudation and the formation of a peneplain surface covered by alluvial gravel (mostly well-rounded and medium-rounded pebbles of quartz (60%), quartz–feldspar sandstone (30%) and small pebbles of marl and limestone) with a typical red-colour sandy–clay matrix. Fragments of the weathering crust of kaolinite composition can be found. Now, deposits can sometimes be found on the flat watersheds and highlands. The gravel deposits are weakly gold-bearing. The deposits date back to the Pre-Quaternary, presumably the Miocene [21].

Apparently, at the same time, formation of the linear weathering crusts (LWCs) began. This is associated with zones of disjunctive dislocations in the bedrock (Figure 6). At the intersection of these dislocations with gold–sulphide lodes, these linear weathering crusts carry gold mineralization of considerable economic potentiality.

Figure 6. Clay deposit of linear weathering crust.

Y.P. Kazakevich and M.V. Reverdatto (1972) describe this type of placer deposit in the upper part of Sukhoi Log Creek [21], p. 124: "Sukhoi Log Creek, right tributary of the R. Nygri, is a small watercourse, in which two gold-bearing layers are known. One rests on the bedrock of the buried terrace. It is composed of brightly colored rock debris with in yellow and red clay matrix. The formation was distinguished by a very high gold content, present throughout its thickness. The placer was worked out in 1890–1896. The second placer layer lies in the deep thalweg in poorly-sorted alluvial deposits".

The second stage (Early to Middle Pleistocene) began with neotectonic activation and block uplift that resulted in a change in planer denudation to linear deep erosion, which resulted in the cutting of valleys to a depth of 300 to 900 m. This formed a series of erosion terraces covered by pebbles with eroded weathering crusts. The deposits of buried thalwegs of the valleys are also assigned to the second stage. The thalweg deposits contain big blocks and boulders up to 2–3 m in diameter that are evidence of the high hydrodynamic activity of this stage (Figure 7). In the bottoms of the valleys, upon the transition to bedrock, the eluvial horizon is commonly observed to contain angular particles of quartz gruss and clay. The terrasses and thalwegs deposits contain placer gold mineralization of economic importance. Nuggets are often found here.

Figure 7. Cross-section of deposits of the placer of R. Nygri. (**A**) man-made tailing dumps; (**B**) the deposits of the second stage.

In the beginning of the Late Pleistocene, the intensity of the tectonic activity decreased and the change from vertical erosion to lateral erosion resulted in expansion of the valleys, forming the slope sedimentary complex and the main part of the alluvial deposits filling the valleys. Colluvial deposits are represented by detritus of local rocks with a sandy–clay matrix. Alluvium consists of pebbles and boulders mostly of local rocks with a clay–sand matrix. A significant number of the economic placers in the district are associated with alluvial deposits of this stage. In the zones of primary ore mineralization, the slope (colluvial) deposits contain placer gold, which can be a source of alluvial placers but has own economic importance.

During the second part of Late Pleistocene stage, there was a stabilization of the topography and formation of a barren sequence which covers the main gold-bearing strata. It consists of gravel with a sand–silt matrix of alluvial and fluvial–glacial genesis. Holocene sediments form the channels of modern water courses.

Although the total placer gold production from the middle of the 19th century to the present was about 1500 tons, 130 tons of the measured and indicated resources is estimated to be present [22]. The placers continue to be worked by Lena Gold Company and independent prospectors.

3. Sample Collection and Analysis

Native gold from four different locations was obtained and researched (Figure 4). Bulk samples of the sediments each with an approximate weight of 20 kg were taken from (1) the quartz lodes of the Sukhoi Log deposit, (2) the LWC located on the periphery of the Sukhoi Lod gold field, (3) the alluvial deposits of the Sukhoi Log Creek, which erode the LWC and the rocks outside the economic area with low-grade mineralization and (4) the giant Nygri placer downstream of the mouth of the Sukhoi Log Creek. Hand panning was carried out in the field to reduce the weight of the sample and to obtain 20–30 g of heavy mineral concentrates. To obtain the final concentration, the gold grains were separated by heavy liquid (bromoform, 2.89 g/cm^3) in the laboratory of IGEM RAS. All obtained grains were used for research on the grain size of the gold particles.

Twenty gold grains were extracted from every sample for further research with scanning electron microscopy (SEM) and electron microprobe analysis (EMPA) in grains and polished sections. Morphology, grain size analysis and chemical composition of the native gold from veinlet-disseminated gold–quartz–sulphide mineralization was obtained from the geological literature [5,17,23].

Informed speculation on the nature of the source bedrock mineralization and supergene changes is possible through comparison of the microchemical signature of placer grains with the generic characteristics of gold from different styles of mineralization [24]; therefore, the placer gold grains' morphology, inner structure and inclusions were compared with documentation of the available and possible bedrock gold mineralization.

Backscattered electron (BSE) images of the gold were taken for all grains using a scanning electron microscope (SEM), GSM 5610LV, and energy-dispersive spectrometer, INCA-Energy 450 (analyst L. Ivanova).

After SEM investigation, the grains were studied for composition, inclusions and inner structure. The grains were mounted into resin blocks, and the blocks were crushed and polished to provide a cross-section of every grain.

The grains were later analysed with an electron microprobe at the Analytical Laboratory of IGEM RAS (Moscow, Russia) using a JEOL JXA-8200 electron microprobe (Japan) with five wavelength-dispersive spectrometers and an energy-dispersive spectrometer under the following operating conditions (Table 1) by the analyst E. Kovalchuk.

Table 1. Measurement conditions of electron microprobe analysis.

	Ag	Au	Hg	Cu	Fe
Accelerating voltage (kV)	20	20	20	20	20
Sample current (nA)	20	20	20	20	20
Beam diameter (μm)	1	1	1	1	1
X-ray	Lα	Lα	Mβ	Kα	Kα
Crystal analyser	PETH	LIF	PETH	LIF	LIF
Time on peak (s)	20	20	20	20	20
Time on back (s)	10	10	10	10	10
Standard name *	$AgSbS_2$	Au_s	HgS_s	Cu_s	$CuFeS_2$
D.L. 3σ	0.04	0.33	0.11	0.07	0.07

* Approved internal laboratory standards of pure gold without impurities (99.99%) and synthetic materials were used in the analysis.

The grade of the gold in the alloys varied from 71.9 wt.% to 100 wt.%. Silver was present above the detection limit (3 σ) of 0.04 % in all grains. Ag content varied from 0.11% to 28.1%. Cu, Hg and Fe were found above the detection limit (0.07 wt.%, 0.11 wt.% and 0.07 wt.%, respectively) in several grains. Additionally, inclusions and grains with pronounced rim–core zonation were studied in detail with electron microprobe analysis.

The composition of Au grains that had visible core–rim zonation was analysed for every grain in the core and in the rim.

Interpretation of gold compositions was based on a multivariate treatment of the information available. The type of information may vary between sample populations, e.g., minor elements may be detectable or not, and inclusion suites may be more or less representative of the actual population depending on sample size. When available, inclusion suites provide clear evidence of genetic relationships between sample populations and strong indications of the deposit type [25,26]. Ranges of Ag compositions are useful to establish "same or different" criteria and may find application in speculating on zonal relationships between mineralization within the same hydrothermal system, as a consequence of the significant control of the temperature of the depositional environment over the Au-Ag ratio within the alloy. The importance of minor alloying metals (Cu, Hg and Fe) varies according to the degree to which they are present and as well as their concentration [27].

4. Results

Studies of the morphology, chemical composition and mineral inclusions of gold grains have been widely used in gold exploration [28–37].

The native gold from four studied collected samples (quartz lodes, LWC, Sukhoi Log Creek and placer of Nygri Valley) and one from the published geological literature (veinlet-disseminated gold–quartz–sulphide mineralization) differ by morphology, composition, microchemical signature and internal structure, indicating difference in the genesis and supergene changes in the gold [38].

4.1. Gold Particle Size and Morphology

The morphologies of gold grains may indicate the distance they were transported from the source, the chemical compositions and mineral inclusions they contain, may represent the hypogene source and possibly help in distinguishing between primary sources [34,39–41].

The shape and the surface of the gold grains from collected samples are depicted in Figure 8. The size and other typomorphic features are represented on Figure 9 and Table 2.

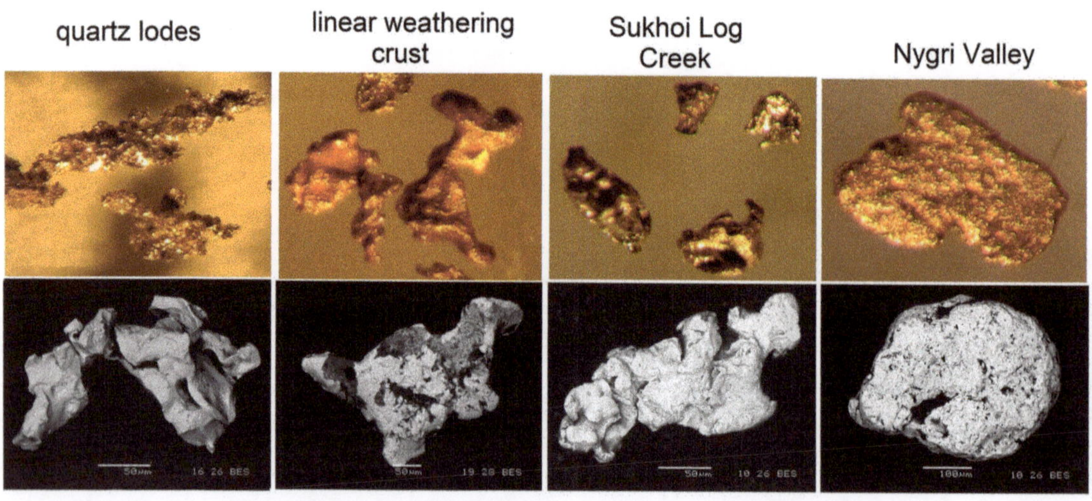

Figure 8. Examples of gold particles from different localities. Optical with binocular (upper line) and SEM (BSE) (lower line) images of examples of native gold from bedrock quartz lodes, weathering crust and placers of Sukhoi Log Creek and Nygri Valley, showing contrasting morphologies indicative of different origins and supergene histories (see text).

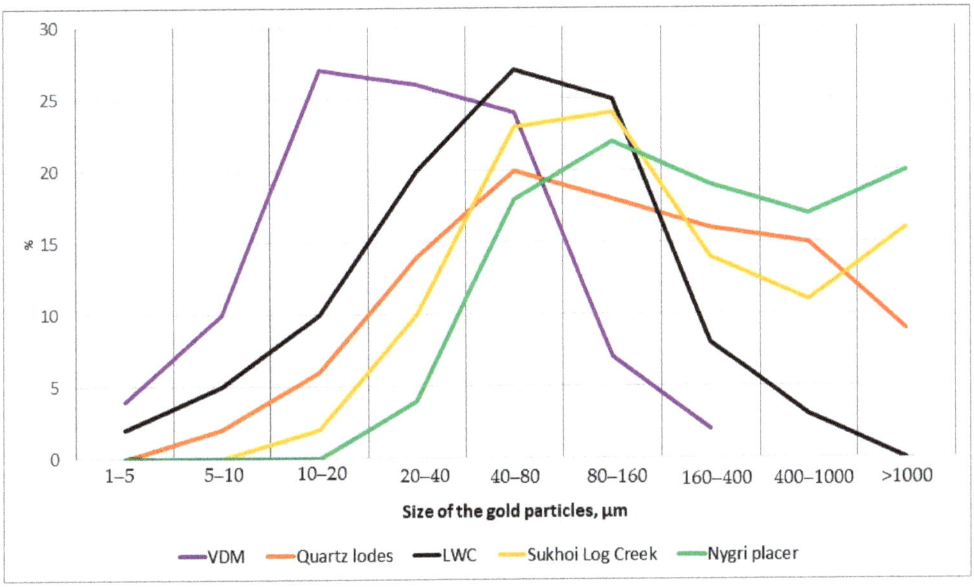

Figure 9. The grain size of the gold grains of studied samples—veinlet-disseminated ore, quartz lodes, linear weathering crusts, Sukhoi Log Creek and Nygri placer.

Table 2. Typomorphic characteristics of native gold from studied types of bedrock and placer mineralization.

		Size of the Grains (mm)		Fineness in the Core		C_h *		Trace Elements Above Detection Limit, wt.%
		min	max	min	max	min	max	
quartz lodes	variation	0.01	>1	868	949	0.97	1.03	Ag (4.5–10.1)—all grains, Cd–1.29 (single), As—0.88 (single)
	average	0.37		910		1.00		
VDM	variation	0.001	0.4	803	895	-	-	Ag (9.5–14.36)—all grains, Cu (0–0.11), Fe (0–0.97), Hg (0–0.25)
	average	0.04		870		1.00(?) **		
LWC	variation	0.02	0.8	771	928	1.00	1.18	Ag (7.2–22.9)—all grains. Cu–0.11 (single), Fe–0.15, 0.22 (two analysis)
	average	0.1		875		1.08		
Sukhoi Log Creek	variation	0.02	>1	719	952	0.97	1.23	Ag (4.8–28.1)—all grains, Hg–0.18 (single) Fe–0.23 (single)
	average	0.48		881		1.05		
Nygri placer	variation	0.04	>1	910	971	0.97	1.03	Ag (2.9–9.0)—all grains
	average	0.6		941		1.00		

* C_h—coefficient of the heterogeneity ratio of the Au content in the core and in the rim of the grains. **—there are no data on the distribution of gold content within the grains, but the grains on polished surfaces (Figure 5 in [5]) display homogeneity of the inner structure.

The gold from low-sulphide–quartz lodes had an irregular shape with angular edges, and the surface was uneven and smoothed, with no traces of supergene influence. Numerous intergrowths with quartz were observed. The grains obtained from quartz lodes of the Sukhoi Log during this study ranged in size from less than 0.01 mm to 1.2 mm (maximum dimension) (Figure 9). According to previous research, the grains from the lodes reach 1–2 cm and more in size [16].

The grains from the LWC had an irregular shape, but smoothed edges and surfaces. The grains were often coated with a film of iron hydroxides and iron oxyhydroxides. Very often, iron hydroxides filled depressions and pits on the surface of grains. The size of the grains of the studied samples ranged from 0.02 mm to 0.8 mm.

The gold grains from the Sukhoi Log Creek were of two different typomorphic types. The first one (up to 60% of the grains) was represented by grains of an irregular shape with smoothed edges and weak traces of roundness. Pits, grooves and traces of dragging were found on the surface of the grains. Iron hydroxide crusts were sometimes observed in the depressions and pits of the grains. The size of the studied grains varied from 0.02 mm to more than 1 mm.

The shape of the second type of grain from the Sukhoi Log Creek varied from irregular to semi-spherical with a crumpled structure. It had no iron hydroxide crusts. The surface had traces of transportation. The size varied from 0.08 mm to more than 1 mm.

The gold grains from the Nygri placer deposit had a spherical to semi-spherical and flattened shape, medium to well-rounded with a smooth pitted surface (Figure 8). The size of the studied grains ranged from 0.04 up to 1 mm, whereas according to the experience of gold miners, 30% of gold grains are more than 2 mm in size, and nuggets weighing hundreds of grams are common [21].

The data on typomorphic features of native gold from veinlet-disseminated gold–quartz–sulphide mineralization were obtained from the published literature [5,6,16,17]. As shown in these studies, native gold in ores is found mainly (85%) in the form of microscopic and thin inclusions in pyrite, and only 15% is located in the host rocks outside the pyrite grains [5]. Native gold is usually observed in the form of individual independent inclusions, and very seldomly it is found intergrown with other common ore minerals, as well as at the boundary with inclusions of vein minerals or in quartz microveinlets [42]. The size of gold grains reaches up to 0.4 mm, and most gold particles have a size of 0.01–0.08 mm. The shape of gold grains is mostly irregular with angular edges and a complex nostril-porous and spongy surface.

The grain size of the gold grains from the studied samples (quartz lodes, LWC, Sukhoi Log Creek and Nygri placer) and those obtained from previous studies on VDM [5,16] are represented in Figure 9 and Table 2.

4.2. Gold Alloy Composition

The four samples of native gold obtained from the studied area were classified by composition, microchemical signature and internal structure.

The fineness of the gold from quartz lodes varied from 868 to 949 (average 910). The alloys contained Ag (4.54–10.1 wt.%, average 7.47 wt.%). In single grains, Cd–1.29 wt.% and As–0.88 wt.% were detected above the detection limit.

Study of Au-Ag alloys from LWCs revealed core–rim zonation of the grains. The average fineness in the core was 875 with variation from 771 to 928. All grains contained Ag (7.20–22.9 wt.% in the core). In a single grain, Cu–0.11 wt.% and Fe (two grains)–0.15 wt.% and 0.22wt.% were detected above the detection limit. The fineness in the rim varied from 976 to 998.

The fineness of the first type (irregular smoothed) of grains from Sukhoi Log Creek was 850–980. All grains contained Ag (4.80–28.1 wt.%). The single grain with maximum Ag content (28.1 wt.%) contained Hg–0.18 wt.%. Additionally, in one grain, Fe was detected above the detection limit (0.23 wt.%). The fineness of the second type (irregular to semi-spherical) was 920–960. All grains contained Ag (4.90–8.12 wt.%). Other trace elements were not detected.

The fineness of the gold alloys from the Nygri placer varied from 910 to 971 (average 941). All grains contained Ag (2.90–9.0 wt.%). Other trace elements were not detected.

The fineness of the gold alloys from VDM was estimated to range from 500 to 980 [16] or 800 to 900 [5], but most of the alloys had a fineness of 840-900.

The cumulative frequency plots of the alloy compositions in the Au grain cores from the studied samples are represented in Figure 10. The cumulative plot for the alloy fineness for quartz lodes and the Nygri placer had a certain conformity. The coefficient of correlation (R) of the fineness for the sample populations was 0.60 (the critical value with a confidence coefficient of 0.95 is 0.504). There is also a strong similarity between the plots of VDM and LWC: the correlation coefficient is 0.66.

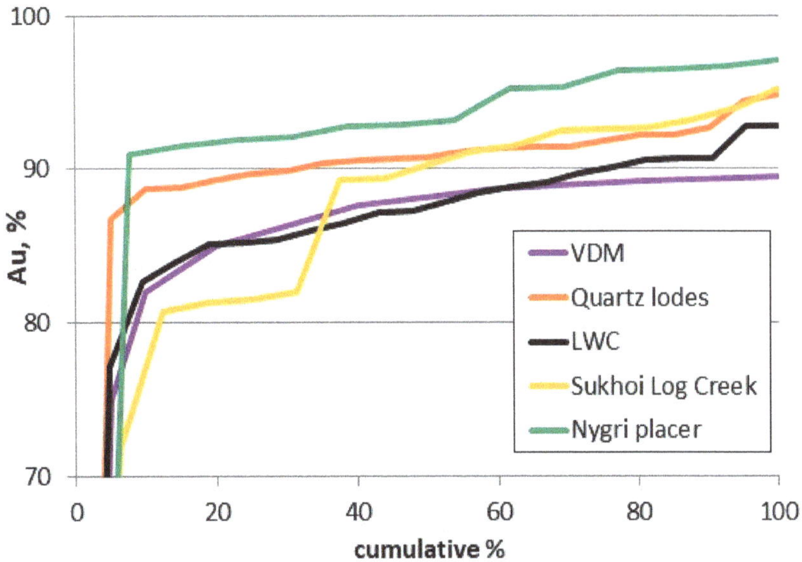

Figure 10. The cumulative frequency plots of the alloy compositions in the Au grain core from quartz lodes, veinlet-disseminated mineralization (VDM), linear weathering crusts (LWC), Sukhoi Log Creek and placer of Nygri Valley.

Gold samples from Sukhoi Log Creek exhibited a fineness profile between these end members. The resulting profile shape with curve kinks suggests two probably overlapping compositional fields based on Au concentrations in the alloy. One of them, located in the field of relatively low-grade gold, corresponded to VDM–LWC alloys, whereas the second, with high-grade gold, was similar in fineness to quartz lode alloys.

The results of research on the inner structure and conclusions about the alloys are shown in Figures 11 and 12 and Table 2. For populations of homogenous gold grains, duplicate sampling generates reproducible data [43]. In order to establish the degree of heterogeneity within both heterogeneous and apparently homogeneous grains, it is necessary to analyse it at several point locations in their sections [39]; therefore, the inner structure of the complex alloys was studied at several points [44].

The analysis of the inner structure of polished grains revealed that there were two main varieties of gold alloys—monotonous and distinct heterogeneous alloys with rim–core zonation and gold-rich inner fracturing zones.

Heterogeneous alloys were obtained mainly from the LWC (70% of the alloys from the sample) and partly from Sukhoi Log Creek (50% of the alloys from the sample) (Figures 11 and 12 C). The concentration of gold in the rim and gold-rich inner fractured zones varied from 98.53 wt.% to 100 wt.% with Ag contents ranging from 0 to 0.85 wt.% (average 0.22 wt.%) and complete absence of Cu, Fe and Hg. The coefficient of heterogeneity (ratio of rim–core concentrations of gold) reached 1.18 for the LWC and 1.23 for Sukhoi Log Creek, with an average of 1.08 and 1.05 for the samples, respectively.

Regarding the alloys from quartz lodes, the Nygri placer, (apparently) and VDM, 40% of the grains from the LWC and 80% of the grains from Sukhoi Log Creek were homogeneous.

	Type of gold	Point	Content, wt. %				
			Au	Ag	Cu	Hg	Fe
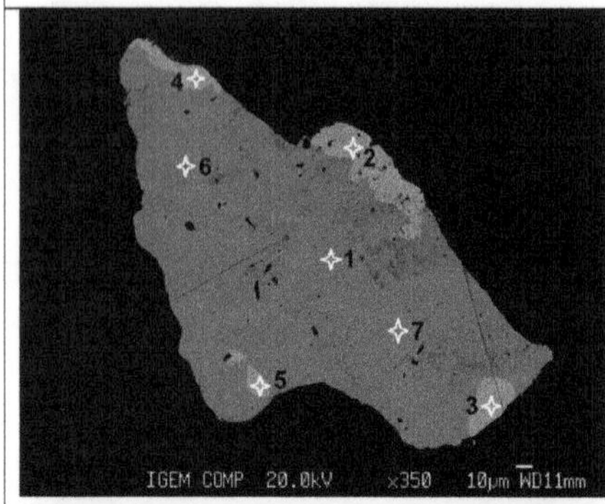	LWC	1	90.06	8.96	0.00	0.00	0.00
		2	99.26	0.65	0.00	0.00	0.00
		3	100.00	0.00	0.00	0.00	0.00
		4	100.00	0.00	0.00	0.00	0.00
		5	98.63	0.14	0.00	0.00	0.00
		6	89.83	8.64	0.00	0.00	0.00
		7	88.87	9.37	0.00	0.00	0.00
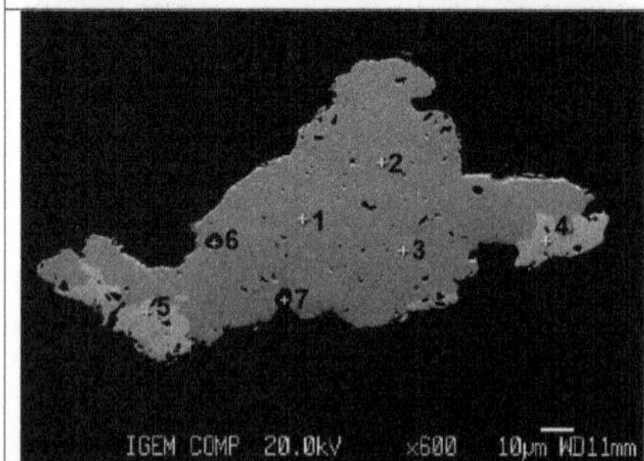	LWC	1	86.19	11.69	0.00	0.00	0.00
		2	86.60	11.93	0.00	0.00	0.00
		3	85.29	11.70	0.00	0.00	0.00
		4	99.69	0.00	0.00	0.00	0.00
		5	99.01	0.12	0.00	0.00	0.00
		6	Iron hydroxide (Fe – 53.4, O – 28.1)				
		7	Iron hydroxide (Fe – 57.7, O – 35.6)				
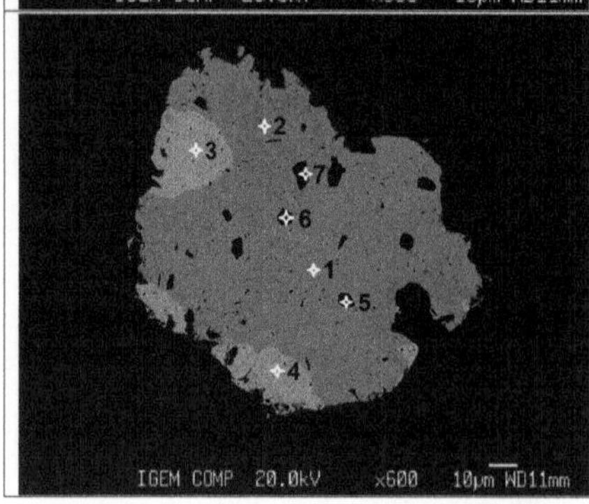	LWC	1	91.47	8.54	0.11	0.00	0.22
		2	90.01	7.65	0.00	0.00	0.15
		3	99.60	0.19	0.00	0.00	0.00
		4	100.00	0.00	0.00	0.00	0.00
		5	Sphalerite (Zn – 56.9, S – 21.2, Fe – 1.69)				
		6	Iron hydroxide (Fe – 57.2, O – 37.6)				
		7	Pyrite (Fe – 46.3, S – 51.6)				

Figure 11. The composition of the placer gold grains from the LWC, with BSE images of the polished sections. Content of detected elements was determined through electron microprobe analysis on different points of the polished surface. The grains had a gold-rich rim and inclusions of sphalerite, pyrite, chlorite and iron hydroxide. Remnants of polishing powder and defects of polishing occasionally resemble inclusions in the SEM images.

	Type of gold	Point	Content, wt. %				
			Au	Ag	Cu	Hg	Fe
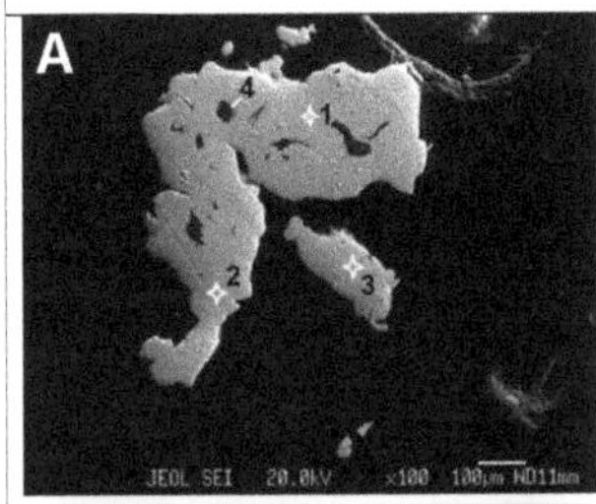	Quartz lode	1	91.48	7.26	0.00	0.00	0.00
		2	90.87	8.73	0.00	0.00	0.00
		3	92.07	7.62	0.00	0.00	0.00
		4	Quartz (Si – 48.30, O – 50.15)				
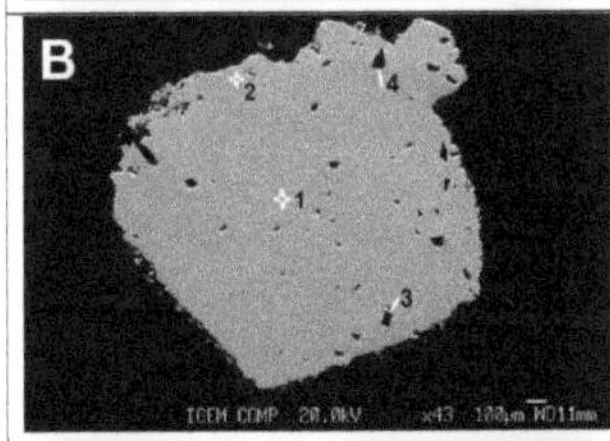	Sukhoi Log Creek	1	71.87	27.38	0.00	0.18	0.00
		2	71.26	28.30	0.00	0.00	0.00
		3	Iron hydroxide (Fe – 52.4, O – 28.8)				
		4	Chlorite				
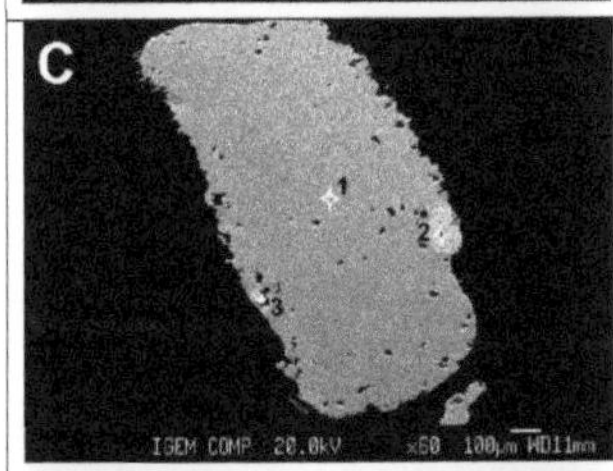	Sukhoi Log Creek	1	89.36	11.10	0.00	0.00	0.00
		2	100.00	0.25	0.00	0.00	0.00
		3	99.65	0.44	0.00	0.00	0.00
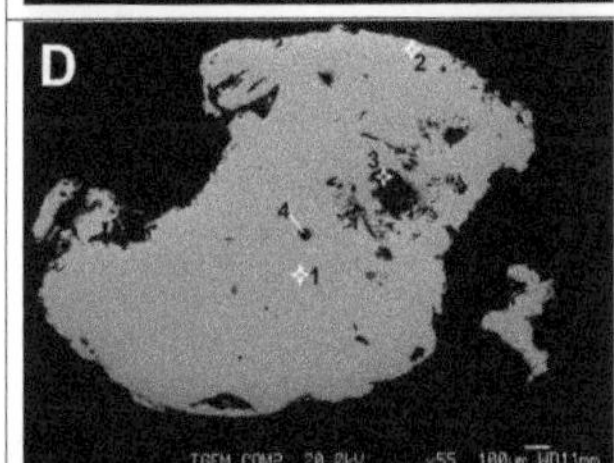	Nygri placer	1	93.15	7.23	0.00	0.00	0.00
		2	92.79	7.39	0.00	0.00	0.00
		3	Iron hydroxide (Fe – 53.4, O – 28.1)				
		4	Quartz (Si – 46.03, O – 50.41)				
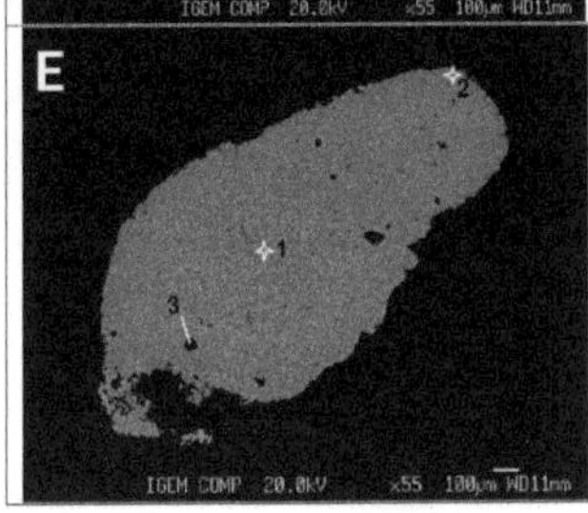	Nygri placer	1	95.87	4.63	0.00	0.00	0.00
		2	95.67	4.67	0.00	0.00	0.00
		3	Pyrite (Fe – 46.32, S – 52.88)				

Figure 12. The composition of the placer gold grains from quartz lode (**A**), Sukhoi Log Creek (**B**,**C**) and the placer of Nygri Valley (**D**,**E**) with BSE images of the polished sections. Content of detected elements was determined through electron microprobe analysis on different points of the polished surface. The grains from Sukhoi Log Creek sometimes had a gold-rich rim. Other grains had a monotonous structure. The grains from the Nygri placer had inclusions of pyrite, quartz and iron hydroxide. Remnants of polishing powder and defects of polishing occasionally resemble inclusions in the SEM images.

4.3. Mineral Inclusion

Most of the sample populations were homogenous, inclusion-free gold particles. The small number of revealed inclusions is not enough to enable statistical processing of the inclusion suite [45], but available data allow preliminary conclusions about the primary mineralization type of the gold.

The most common were iron hydroxide inclusions, which were revealed in the LWC, Sukhoi Log Creek and Nygri placer. The inclusions were located both inside the grains and filling depressions and pits on the surface of grains. The studied sample population of quartz lodes and VDM did not contain iron hydroxide inclusions.

In the studied samples, Au-Ag alloys from quartz lodes contained quartz inclusions only (Si–48.30 wt.%, O–50.15 wt.%). The low-sulphide character of this type of mineralization suggests the presence of sulphide inclusions in these alloys.

Alloys from the VDM were not studied in this research, but based on previous studies of VDM [5], it is assumed that alloys of this type may contain sulphide inclusions and accretions (pyrite, sphalerite and galena).

Alloys from the LWC contained inclusions of iron hydroxide (Fe–53.4–57.7 wt.%, O–28.1–37.6 wt.%), sphalerite (Zn–56.9 wt.%, S–21.2 wt.%, Fe–1.69 wt.%) and pyriteThe analysis of the inner structure of polished grains revealed that there were two main varieties of gold alloys—monotonous and distinct heterogeneous alloys with rim–core zonation and gold-rich inner fracturing zones.

Heterogeneous alloys were obtained mainly from the LWC (70% of the alloys from the sample) and partly from Sukhoi Log Creek (50% of the alloys from the sample) (Figures 11 and 12 C). The concentration of gold in the rim and gold-rich inner fractured zones varied from 98.53 wt.% to 100 wt.% with Ag contents ranging from 0 to 0.85 wt.% (average 0.22 wt.%) and complete absence of Cu, Fe and Hg. The coefficient of heterogeneity (ratio of rim–core concentrations of gold) reached 1.18 for the LWC and 1.23 for Sukhoi Log Creek, with an average of 1.08 and 1.05 for the samples, respectively.

Regarding the alloys from quartz lodes, the Nygri placer, (apparently) and VDM, 40% of the grains from the LWC and 80% of the grains from Sukhoi Log Creek were homogeneous. (Fe–46.3 wt.%–51.6 wt.%) (Figure 11).

Alloys from Sukhoi Log Creek contained undetermined contents of chlorite and iron hydroxide (Fe–52.4 wt.%, O–28.8 wt.%)

The gold grains from the Nygri placer contained inclusions of pyrite (Fe–46.32 wt.%, S–52.88 wt.%), iron hydroxide (Fe–53.4 wt.%, O–28.1 wt.%) and quartz (Si–46.03 wt.%, O–50.41 wt.%) (Figure 12).

4.4. Gold-Rich Rims

A significant number of alloys from the LWC and Sukhoi Log Creek had gold-rich rims. The individual rims of the various grains ranged from 10 to 50 μm in thickness. The boundary between the individual cores and rims was generally sharp, could be visually demonstrated in reflected light and was confirmed by microprobe analyses in polished sections (Figures 11 and 12C).

The relatively high Ag content of the gold cores and the presence of mineral inclusion species unstable in the surface environment indicate that they are relics of a hypogene gold particle source rather than authigenic gold or gold altered by weathering [43]. While the composition of the alloy's homogeneous cores demonstrates source mineralization, the rim–core composition differences expressed by the coefficient of heterogeneity reflect both residual features of a primary grain's structure and supergene transformation of the alloys [46].

Gold grains experience both physical and chemical changes during weathering, transportation and post-depositional processes. During the influence of supergene conditions, grains of placer gold develop an outer rim of nearly pure gold on the more silver-rich electrum core. The rim has very close contact with the core, and appears to be the result of electrochemical processes occurring in weathered crusts, streams and stream sediment conditions. The thickest rims are generally found on grains with a long supergene history [47].

In the studied zonal alloys, the gold content in the rim reached 100.0 wt.% with a silver content of not more than 2.25 wt.% and a Cu and Hg content below the detection limit. Gold-rich rims developed mainly on the projections of grains and sometimes in the gold-rich inner fractured zones.

From the Au content–frequency (n) diagram (Figure 13), it can be seen that the ore and rim assays were completely different and contradict the model of a gradual transformation of the core material into a rim.

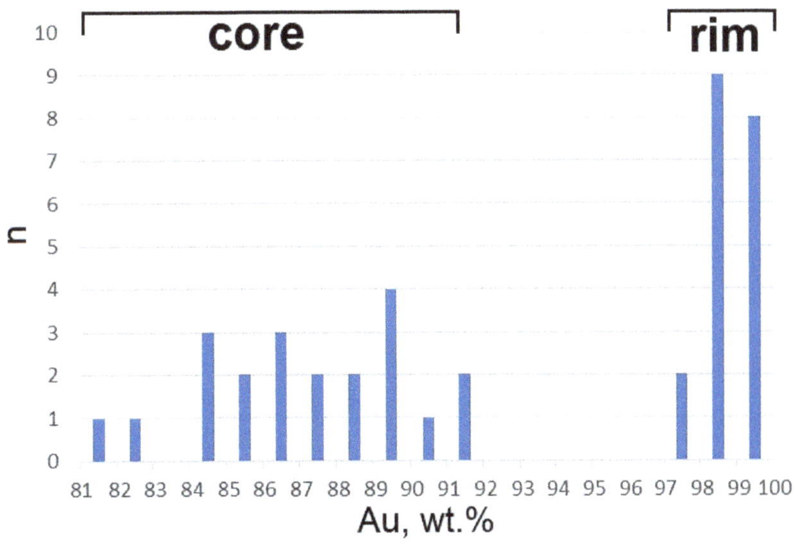

Figure 13. Analysis of the chemical composition of zonal alloys from the LWC and Sukhoi Log Creek for core and rim parts of the gold grains.

4.5. *Main Types of Native Gold*

Additional information for the classification of the placer alloys can be obtained from the plot of Au content (wt.%) vs. coefficient of heterogeneity (C_h). While the fineness of gold in the core is associated primarily with the hypogene factors of its formation, the coefficient of heterogeneity indicates the degree of its supergene transformation [46,48].

The plot illustrates a visual division of the sample population into seven individual fields (Figure 14).

Two primary bedrock types of native gold (low-sulphide gold quartz lodes, field I, and VDM, field II) form fields in their own area of C_h close to 1, which shows an insufficient effect of supergene transformations of these types of gold.

The alloys from the LWC form two separate fields: the first one (field IIIa) is located in the area $C_h \approx 1.0$ and closely corresponds to VDM field II. The second one (with core–rim zonation) is located in the area $C_h > 1.02$.

Alloys from Sukhoi Log Creek also form two separate fields, IVa and IVb, which correspond to LWC fields IIIa and IIIb. The field of the Nygri placer alloys largely coincides with the field of the Au-Ag alloys from low-sulphide quartz lodes.

An interpretation of the plot data will be given in next section, "Discussion".

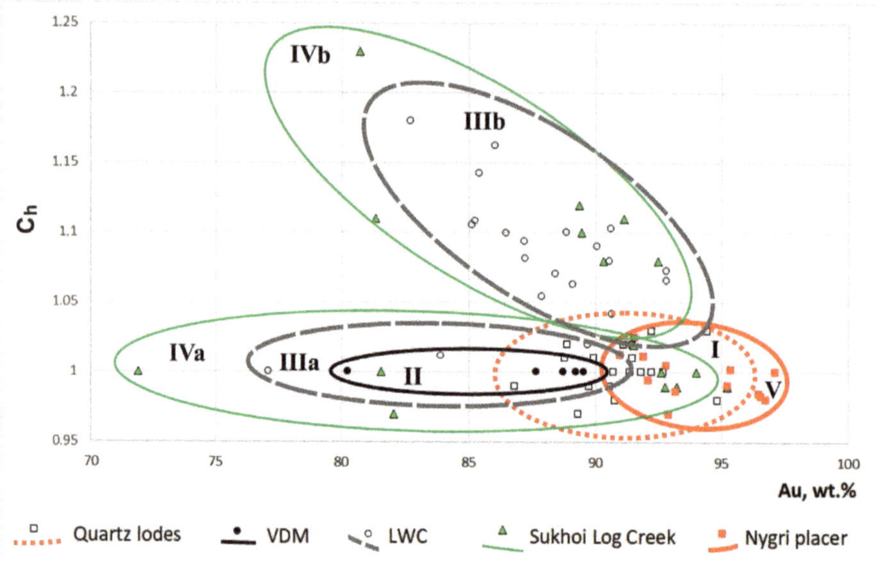

Figure 14. Au content in the core of grains (wt. %) vs. coefficient of heterogeneity (C_h) plot for studied sample population. The diagram distinguishes the types of gold grains both by primary hypogenic characteristics and supergenic features. I—quartz lode field; II—VDM field; IIIa—LWC zonal alloys field; IIIb—LWC homogeneous alloys field; IVa—Sukhoi Log Creek zonal alloys field; IVb—Sukhoi Log Creek homogeneous alloys field; V—Nygri placer field.

5. Discussion

Many gold-bearing regions, even those that have been studied for a long time, have a number of unresolved issues. Bodaibo District is no exception. Three problems are under discussion: (1) the sources of the placer gold; (2) evidence of the transformation of native gold in supergenic conditions and features of this process; (3) the mechanism for the formation of gold-rich rims; (4) the possibility of discovering of new placers in Bodaibo District.

5.1. Sources of the Placer Gold

One of the problems is locating the source of unique alluvial placers, from which approximately 1500 tons of gold has been mined over more than 150 years of operation. Taking into account that only 5 to 20% of gold eroded from bedrock sources accumulates in placers [16], the volume of eroded primary gold ores can be estimated to be from 8 to 32 thousand tons, which is comparable in volume to such giant deposits as Carlin or Witwatersrand [49–51]. At the same time, it should be borne in mind that the most common veinlet-disseminated quartz–sulphide mineralization on the territory of the Bodaibo District contains native gold of a small size, which is poor at forming placer deposits.

Analysis and interpretation of the Au content in the core (wt.%) vs. coefficient of heterogeneity (C_h) plot allows identification of the sources of gold placers in the studied and surrounding area [46].

The close connection between the geochemical signatures of the gold from quartz lodes (Figure 14, field I) and gold alloys from the Nygri placer (Figure 14, field V) indicates that the gold in the placer is largely a result of erosion of this type of primary bedrock mineralization [52,53].

The similar character of the inclusion in the gold grains of quartz lodes and the Nygri placer also indicates a relationship between these types of gold (Figure 12A,D,E). Alloys from the Nygri placer contain inclusions of quartz and pyrite; in our samples of lode alloys, pyrite was not found, but the presence of pyrite was assumed due to the gold–sulphide character of this mineralization. Alloys from the placers contained iron hydroxide inclusions, which apparently are a result of weathering of primary pyrite inclusions. The presence of both iron hydroxide and pyrite inclusions in placer alloys indicates that the weathering process was not deep and prolonged. The lack of Au-rich rims provides no evidence of supergene precipitation, which suggests that the grains were transferred directly from a hypogene source to a surficial and fluvial system [54]. So, the absence of a gold-rich rim on the alloys from the Nygri placer and presence of inclusions of pyrite are evidence of the direct deposition of gold from primary sources into placers, bypassing weathering crusts and intermediate hosts.

The lode mineralization occurring at the Sukhoi Log deposit at the current erosional level is insufficient for the formation of large and rich placers, similar to the placer of the River Nygri. A study on the Verninskoye deposit located in this area, where the gold–quartz lode mineralization is much more pronounced, showed that the vertical range of this mineralization exceeds 300–400 m, being replaced by VDM at greater depths. At the same time, the gold content and the scale of lode mineralization decrease with depth [55].

Drawing an analogy with the Sukhoi Log deposit, we can conclude that the thickness of the eroded ore horizon is several hundred meters, and the gold content in the eroded part is significantly higher than at the level of the current surface. In this scenario, the eroded part of the Sukhoi Log deposit and other ore fields located in this region could provide for the formation of large and unique placers in the Bodaibo District.

Sources of gold for small local placers (such as Sukhoi Log Creek) and eluvial LWC placers will be discussed in the next section.

5.2. Evidence for Transformation of Native Gold in Supergenic Conditions and Features of This Process

More than 90% of VDM gold is in the order of less than 0.1 mm in size and therefore has a very low potential for placer formation. In addition, outside the mineralized area of the Sukhoi Log deposit, the rocks are usually barren or contain only low-grade mineralization. However, within the LWC in this zone, rich placers were formed. They were worked out during the initial stages of development of the Bodaibo District.

Research on native gold from the LWC revealed that 70% of the alloys have core–rim zonation, which is evidence of supergenic transformation (Figure 14, Field IIIb).

The growth of gold grains due to the formation of gold-rich rims leads to an increase in their size. If the main grain size of the VDM ranges within 10–80 μm, then for the LWC, this value is 40–160 μm (Figure 9). This leads to an increase in the placer productivity of the deposits and an increase in the overall fineness of gold.

Some alloys (30%) have no evidence of supergene changes (Figure 14, Field IIIa), which indicates the limited nature of the supergene process and later disintegration of pyrite grains containing Au-Ag alloys. According to their typomorphic characteristics, such grains are similar to VDM gold.

The placer deposit of Sukhoi Lod Creek has, at least, two sources. The first one is the LWC, containing both homogeneous and core–rim zonal alloys. The second source (approximately, 30%) corresponds to gold of quartz lodes exposed on the watershed in the upper course of Sukhoi Log Creek. Single alloys of a relatively low fineness (719) and Hg content of 0.18wt.% (Figure 12B) allow us to make an assumption about the existence of low-temperature hydrothermal mineralization in this zone [43,56–60].

5.3. Origin of Gold-Rich Rims

The gold particles in the bedrock do not have a gold-rich rim, so it is quite reasonably assumed that they were formed in the process of supergene transformations of the alloys [47]. There are two opinions regarding the reason for the formation of this feature of placer gold: selective leaching of Ag and other trace elements from the rim zone [61], and self-electrorefining of placer electrum grains, which probably operates in tandem with dissolution–precipitation (cementation) [47]. Some researchers [12,14] propose that the rims are a result of adhesion to small gold particles as well as the transition and deposition of chemical gold by organic–metallic complexes in supergenic environments.

Observed natural gradients of silver in the rim zone (Figures 11, 12C and 13) illustrate that the profile of silver content contradicts the gradual leaching model. The observed sharp rim–core contact in the gold alloys from the LWC and Sukhoi Log Creek corresponds to the precipitation model. A similar situation was observed in the gold alloys of the Vagran gold-bearing cluster of North Urals [46].

Groen G.C. et al. [47] believe that the development of the rim occurs mainly during gold grain transportation, meaning that the "transportation" is the period in which a surficial environment occurs, whether by actively moving particles or containing them within stream sediments. We believe that taking into consideration that the phase of active movement of the grains in the stream is incomparably shorter than their retention within the sediments, the growth of gold-rich rims in the sediments is obviously more efficient. Moreover, the growth of gold-rich rims is compensated for by the erosion of grains during their transportation in alluvial streams. Additionally, the concentration of gold ions in groundwater is two orders greater than in stream (Figure 9a in [47]). While "a gold-rich rim apparently forms by precipitation of gold from the surrounding solution" [47], p.207, these factors confirm the validity of our opinion that the inner condition in the sediments is more favourable to rim forming than the stream environment itself.

According to [47], p. 224, "The rim generally is thickest on flake-shaped (most transported) grains and thinnest or absent on irregular (least transported) grains". In our case, gold-rich rims were present in grains that did not move (LWC) or had very restricted movement (Sukhoi Log Creek), while most transported grains (Nygri placer) did not have a rim. So, the residence of the grains in an LWC is more favourable to rim formation than their residence in dynamic stream conditions. The stream environment has a significant influence on the shape of the grains, but only a moderate influence on the inner structure of the alloys.

The gold alloys in LWCs do not experience significant transportation, but they do experience long chemical reactions within infiltrating fluid conditions, which results in the growth of a gold-rich rim.

5.4. Perspectives of Discovering of New Placers in Bodaibo District

The study and exploitation of gold placers in the Bodaibo District has been ongoing for more than 150 years. During this period, about 1500 tons of gold was mined, and the measured and indicated resources of the placer gold are estimated to total 130 t. Considering this, the discovery of new placers seems unlikely, and the main prospects for the mining industry are associated with bedrock primary ores.

Scattered gold–sulphide–quartz veins, stockworks and disseminated ores with a low gold content, in the case of preliminary disintegration and enrichment of weak gold-bearing deposits in the weathered crusts, in combination with intermediate hosts, could form a significant source of placer clusters [62,63].

The conducted studies allowed us to identify a new promising type of gold placer in the Bodaibo region, associated with eluvial formations and tectonically controlled by LWCs. Promising objects are associated with disjunctive dislocations located in zones of low-grade disseminated mineralization. In such zones, a supergene transformation of low-grade disseminated sulphide mineralization occurs, accompanied by an increase in the size, content and fineness of gold. There are cases where such placers have been successfully mined [21], but in the presence of large and high-potential alluvial placers, they have rarely attracted the attention of gold miners. Where such features are located on flat watersheds, they have very limited secondary halos and may be missed by standard exploratory surveys. To identify such objects, it is necessary to map disjunctive dislocations (including those without displacement along the fault zone) in zones from the intersection with fields of low-grade disseminated sulphide mineralization. Lithologically, such zones are characterized by the presence of LWCs, represented by multicoloured (brown, yellow, red) clays with weathered rock debris.

Stratigraphic control consists of the localization of this type of mineralization within the distribution of the Khomolkho and Imnyakh formations. For a more accurate and detailed forecast of eluvial placers associated with LWCs, additional studies are needed to map the zones of development of supergenic changes associated with zones of disjunctive dislocations within the most promising formations. It is possible to use digital forecasting methods using the analysis of multifactor spaces and overlay techniques [64].

6. Conclusions

This study of typomorphic features of bedrock gold mineralization and placer gold alloys allowed for the determination of primary sources of gold placers in the Sukhoi Log cluster and surrounding area.

Native gold from the giant Nygri placer as well as other alluvial gold placers in the Bodaibo District contain gold that has significant similarity with the gold from low-sulphide lodes; the absence of a gold-rich rim on the alloys and presence of inclusions of pyrite are evidence of the direct deposition of gold from primary sources into placers, bypassing weathering crusts and intermediate hosts.

The microgeochemical study of gold and mineral inclusions of unconverted alloys from LWCs and composition of the core of zonal alloys indicated a close connection with the gold from the VDM zone.

The appearance of Au-Ag alloys with a gold-rich rim indicates gold particles that have resided for a long time in weathering crusts.

The gold from VDM transformed in LWCs had a significant influence on proximal placer deposits in the small watercourses near the source, but its role was less significant in the main placer valleys.

Data materialized in our research allow us to identify a new promising type of gold placer in the Bodaibo region, associated with eluvial formations and tectonically controlled by LWCs.

Author Contributions: Conceptualization, A.L.; methodology, A.L., A.G. and L.I; software, A.K.; validation, A.G. and L.I.; formal analysis, A.L.; investigation, A.L., A.G., A.K. and L.I.; data curation, A.G. and L.I.; writing—original draft preparation, A.L. and A.K.; writing—review and editing, A.L.; visualization, A.L. and A.K. All authors have read and agreed to the published version of the manuscript.

Funding: This research was funded by Grant 13.1902.21.0018, "Fundamental Problems in the Development of a Mineral Base for High-Technology and Energy-Producing Industries in Russia", agreement 075-15-2020-802.

Data Availability Statement: The data presented in this study are available on request from the corresponding author. The data will not be publicly available until the submission of the PhD thesis of A.K.

Acknowledgments: We are indebted to the management of the Svetly Prospector Company (Lenzoloto association) and the Verninsky Mine (Polyus Gold OJSC) for their allowing us to undertake the sampling that made this work possible. Elena Kovalchuk is also sincerely thanked for assisting with the analytical work. We are also very thankful to the anonymous reviewers who made useful remarks and comments regarding both technical and fundamental issues to improve the paper.

Conflicts of Interest: The authors declare no conflict of interest.

References

1. Smirnov, V.I. *Ore Deposits of the USSR*; Pitman Publishing: London, UK, 1977; Volume II, p. 424.
2. Lalomov, A.V.; Vladimirova, O.V.; Bochneva, A.A. The role of placer deposits in the gold mining industry of Russia. *Gold Tech.* **2022**, *4*, 36–45. (In Russian)

3. Chugaev, A.V.; Budyak, A.E.; Larionova, Y.O.; Chernyshev, I.V.; Travin, A.V.; Tarasova, Y.I.; Gareev, B.I.; Batalin, G.A.; Rassokhina, I.V.; Oleinikova, T.I. 40Ar-39Ar and Rb-Sr age constraints on the formation of Sukhoi-Log-style orogenic gold deposits of the Bodaibo District (Northern Transbaikalia, Russia). *Ore Geol. Rev.* **2022**, *144*, 104855. [CrossRef]
4. Bochneva, A.A.; Lalomov, A.V.; LeBarge, W. Placer mineral deposits of Russian Arctic zone: Genetic prerequisites of formation and prospect of development of mineral resources. *Ore Geol. Rev.* **2021**, *138*, 104349. [CrossRef]
5. Gavrilov, A.M.; Kryazhev, S.G. Mineralogy and geochemistry of ore at the Sukhoi Log deposit. *Razved. Okhrana Nedr.* **2008**, *8*, 3–16. (In Russian)
6. Yudovskaya, M.A.; Distler, V.V.; Prokof'ev, V.Y.; Akinfiev, N.N. Gold mineralisation and orogenic metamorphism in the Lena province of Siberia as assessed from Chertovo Koryto and Sukhoi Log deposits. *Geos. Front.* **2016**, *7*, 453–481. [CrossRef]
7. Goldfarb, R.J.; Groves, D.I.; Gardoll, S. Orogenic gold and geologic time: A global synthesis. *Ore Geol. Rev.* **2001**, *18*, 1–75. [CrossRef]
8. Goryachev, N.; Pirajno, F. Gold deposits and gold metallogeny of Far East Russia. *Ore Geol. Rev.* **2014**, *59*, 123–151. [CrossRef]
9. Safonov, Y.G. Gold and gold-bearing deposits of the world: Genesis and metallogenic potential. *Geol. Ore Depos.* **2003**, *45*, 265–278.
10. Safonov, Y.U.G.; Popov, P.P.; Volkov, A.V.; Zlobina, T.M.; Chaplygin, I.V. Topical problems of gold metallogeny. *Russ. Geol. Geophys.* **2007**, *48*, 977–991. [CrossRef]
11. Dutova, E.M.; Bukaty, M.B.; Nevol'ko, A.I.; Pokrovsky, D.S.; Shvartsev, S.L. Hydrogenic concentration of gold in alluvial placers of the Egor'evskoe area (Salair). *Russ. Geol. Geophys.* **2006**, *47*, 364–376.
12. Roslyakov, N.A.; Kirillov, M.V.; Morozova, N.S.; Zhmodik, S.M.; Kalinin, Y.u.A.; Nesterenko, G.V.; Roslyakova, N.V.; Belyanin, D.K.; Kolpakov, V.V. Unconventional gold placers of the China tectonic depression (Vitim Plateau, East Siberia). *Russ. Geol. Geophys.* **2013**, *54*, 483–494. [CrossRef]
13. Nesterenko, G.V.; Zhmodik, S.M. Alluvial gold placers directly related to primary sources: An important forecast criterion. *Dokl. Earth Sci.* **2014**, *455*, 481–484. [CrossRef]
14. Fernández-Lozano, J.; Blanco-Sánchez, J.A.; García-Talegón, J.; Franco, P. Rubefaction process as prospection criteria within the roman goldmining context of the Eria Gold Mining District (León, NW Spain). *Geo-Temas* **2016**, *16*, 371–374.
15. Distler, V.V.; Yudovskaya, M.A.; Mitrofanov, G.L.; Prokof'ev VYu Lishnevskii, E.M. Geology, composition, and genesis of the Sukhoi Log noble metals deposit, Russia. *Ore Geol. Rev.* **2004**, *24*, 7–44. [CrossRef]
16. Buryak, V.A.; Khmelevskaya, N.M. *Sukhoi Log—One of the Largest World Gold Deposits*; Dalnauka: Vladivostok, Russia, 1997; p. 156. (In Russian)
17. Wood, B.L.; Popov, N.P. The giant Sukhoi Log gold deposit, Siberia. Russ. *Geol. Geophys.* **2006**, *47*, 315–341.
18. Laverov, N.P.; Chernyshev, I.V.; Chugaev, A.V.; Bairova, E.D.; Golzman YuVDistler, V.V.; Yudovskaya, M.A. Formation stages of the large-scale noble metal mineralisation in the Sukhoi Log deposit, East Siberia: Results of isotope-geochronological study. *Dokl, Earth Sci.* **2007**, *415*, 236–241. [CrossRef]
19. Yakubchuk, A.; Stein, H.; Wilde, A. Results of pilot Re–Os dating of sulphides from the Sukhoi Log and Olympiada orogenic gold deposits, Russia. *Ore Geol. Rev.* **2014**, *59*, 21–28. [CrossRef]
20. Lishnevsky, E.N.; Distler, V.V. Deep structure of the earth's crust in the district of the Sukhoi Log gold-platinum deposit (Eastern Siberia, Russia) based on geological and geophysical data. *Geol. Ore Depos.* **2004**, *46*, 76–90.
21. Kazakevich, Y.P.; Reverdatto, M.V. *Lena Gold-Bearing Region: Geology of Cenozoic Sequence and Types of Gold Placers*; Nedra: Moscow, Russia, 1972; p. 152. (In Russian)
22. State report. *On the State and Use of Mineral and Raw Minerals Resources of the Russian Federation in 2019*; VIMS-TSNIGRI: Moscow, Russia, 2020; p. 494. (In Russian)
23. Large, R.R.; Maslennikov, V.V.; Robert, F.; Danyushevsky, L.V.; Chang, Z. Multistage sedimentary and metamorphic origin of pyrite and gold in the Giant Sukhoi Log Deposit, Lena Gold Province, Russia. *Econ. Geol.* **2007**, *102*, 1233–1267. [CrossRef]
24. Chapman, R.J.; Mortensen, J.K. Application of microchemical characterization of placer gold grains to exploration for epithermal gold mineralization in regions of poor exposure. *J. Geochem. Explor.* **2006**, *91*, 1–26. [CrossRef]
25. Dill, H.G.; Klosa, D.; Steyer, G. The Donauplatin: Source rock analysis and origin of a distal fluvial Au-PGE placer in Central Europe. *Mineral. Petrol.* **2009**, *96*, 141–161. [CrossRef]
26. Dill, H.G. Gems and Placers—A Genetic Relationship Par Excellence. *Minerals* **2018**, *8*, 470. [CrossRef]
27. Chapman, R.; Torvela, T.; Savastano, L. Insights into Regional Metallogeny from Detailed Compositional Studies of AlluvialGold: An Example from the Loch TayArea, Central Scotland. *Minerals* **2023**, *13*, 140. [CrossRef]
28. Chapman, R.J.; Leake, R.C.; Moles, N.R.; Earls, G.; Cooper, C.; Harrington, K.; Berzins, R. The application of microchemical analysis of gold grains to the understanding of complex local and regional gold mineralization: A case study in Ireland and Scotland. *Econ. Geol.* **2000**, *95*, 1753–1773.
29. Chapman, R.J.; Mortensen, J.K. Characterization of Gold Mineralization in the Northern Cariboo Gold District, British Columbia, Canada, Through Integration of Compositional Studies of Lode and Detrital Gold with Historical Placer Production: A Template for Evaluation of of orogenic gold districts. *Econ. Geol.* **2016**, *111*, 1321–1345.
30. dos Santos Alves, K.; Barrios S'anchez, S.; G´omez Barreiro, J.; Merinero Palomares, R.; Compaña Prieto, J.M. Morphological and compositional analysis of alluvial gold: The Fresnedoso gold placer (Spain). *Ore Geol. Rev.* **2020**, *121*, 103489. [CrossRef]

31. Fominykh, P.A.; Nevolko, P.A.; Svetlitskaya, T.V.; Kolpakov, V.V. Native gold from the Kamenka-Barabanovsky and Kharuzovka alluvial placers (Northwest Salair Ridge, Western Siberia, Russia): Typomorphic features and possible bedrock sources. *Ore Geol. Rev.* **2020**, *126*, 103781. [CrossRef]
32. Liu, H.; Beaudoin, G.; Makvandi, S.; Jackson, S.E.; Huang, X. Multivariate statistical analysis of trace element compositions of native gold from orogenic gold deposits: Implication for mineral exploration. *Ore Geol. Rev.* **2021**, *131*, 104061. [CrossRef]
33. Liu, H.; Beaudoin, G. Geochemical signatures in native gold derived from Au-bearing ore deposits. *Ore Geol. Rev.* **2021**, *132*, 104066. [CrossRef]
34. Masson, F.-X.; Beaudoin, G.; Laurendeau, D. Quantification of the morphology of gold grains in 3D using X-ray microscopy and SEM photogrammetry. *J. Sediment. Res.* **2020**, *90*, 286–296. [CrossRef]
35. McClenaghan, M.B. Overview of common processing methods for recovery of indicator minerals from sediment and bedrock in mineral exploration. *Geochem. Explor. Environ. Anal.* **2011**, *11*, 265–278. [CrossRef]
36. Svetlitskaya, T.V.; Nevolko, P.A. Au-Pb compounds in nature: A general overview and new evidence from the Inagli Pt–Au placer deposit, the Aldan Shield, Russia. *Ore Geol. Rev.* **2017**, *89*, 719–730. [CrossRef]
37. Gerasimov, B.; Beryozkin, V.; Kravchenko, A. Typomorphic Features of Placer Gold from the Billyakh Tectonic Melange Zone of the Anabar Shield and Its Potential Ore Sources (Northeastern Siberian Platform). *Minerals* **2020**, *10*, 281. [CrossRef]
38. Rasmussen, K.L.; Mortensen, J.K.; Falck, H. Morphological and compositional analysis of placer gold in the South Nahanni River drainage, Northwest Territories. *Yukon Geol. Surv.* **2006**, 237–250.
39. Chapman, R.J.; Mortensen, J.K.; LeBarge, W. Styles of lode gold mineralization contributing to the placers of the Indian River and Black Hills Creek, Yukon Territory, Canada as deduced from microchemical characterization of placer gold grains. *Miner. Deposita* **2011**, *46*, 881–903. [CrossRef]
40. Townley, B.K.; H´erail, G.; Maksaev, V.; Palacios, C.; de Parseval, P.; Sepulveda, F.; Orellana, R.; Rivas, P.; Ulloa, C. Gold grain morphology and composition as an exploration tool: Application to gold exploration in covered areas. *Geochem. Explor. Environ. Anal.* **2003**, *3*, 29–38. [CrossRef]
41. Youngson, J.H.; Craw, D. Variation in Placer Style, Gold Morphology, and Gold Particle Behavior Down Gravel Bed-Load Rivers: An Example from the Shotover/Arrow-Kawarau-Clutha River System, Otago, New Zealand. *Econ. Geol.* **1999**, *94*, 615–634. [CrossRef]
42. Vikentyev, I.V. Invisible and microscopic gold in pyrite: Methods and new data for massive sulphide ores of the Urals. *Geol. Ore Depos.* **2015**, *57*, 237–265. [CrossRef]
43. Chapman, R.J.; Mortensen, J.K.; Crawford, E.C.; LeBarge, W. Microchemical studies of placer and lode gold in the Klondike District, Yukon, Canada: 1. Evidence for a small, gold-rich, orogenic hydrothermal system in the Bonanza and Eldorado creeks. *Econ. Geol.* **2010**, *105*, 1369–1392. [CrossRef]
44. Youngson, J.H.; Wopereis, P.; Kerr, L.S.; Craw, D. Au–Ag–Hg and Au–Ag alloys in Nokomai and Nevis valley placers, northern Southland and Central Otago, New Zealand, and their implications for placer-source relationships. *N. Z. J. Geol. Geophys.* **2002**, *45*, 53–69. [CrossRef]
45. Chapman, R.J.; Banks, D.A.; Styles, M.T.; Walshaw, R.D.; Piazolo, S.; Morgan, D.J.; Grimshaw, M.R.; Spence-Jones, C.P.; Matthews, T.J.; Borovinskaya, O. Chemical and physical heterogeneity within native gold: Implications for the design of gold particle studies. *Miner. Depos.* **2021**, *56*, 1563–1588. [CrossRef]
46. Lalomov, A.V.; Chefranov, R.M.; Naumov, V.A.; Naumova, O.B.; Lebarge, W.; Dilly, R.A. Typomorphic features of placer gold of Vagran cluster (the Northern Urals) and search indicators for primary bedrock gold deposits. *Ore Geol. Rev.* **2017**, *85*, 321–335. [CrossRef]
47. Groen, J.C.; Craig, J.R.; Rimstidt, J.D. Gold-rich rim formation on electrum grains in placers. *Can. Mineral.* **1990**, *28*, 207–228.
48. McTaggart, K.C.; Knight, J. *Geochemistry of Lode and Placer Gold of the Cariboo District, BC*; British Columbia Ministry of Energy, Mines and Petroleum Resources: Victoria, BC, Canada, 1993; 30, p. 25.
49. Arehart, G. Characteristics and origin of sediment-hosted disseminated gold deposits: A review. *Ore Geol. Rev.* **1996**, *11*, 383–403. [CrossRef]
50. Laznicka, P. Geological Studies Related to the Origin and Evolution of the Witwatersrand Basin and Its Mineralization by Members and Collaborators of Econ. Geol. Res. Unit, Univ. Witwatersrand. Special Issue of the South African Journal of Geology, Vol. 93 (1), 1990. Johannesburg, 309 pp. *Ore Geology Reviews* **1992**, *7*, 73–74.
51. Nwaila, G.T.; Durrheim, R.J.; Jolayemi, O.O.; Maselela, H.K.; Jakaitė, L.; Burnett, M.; Zhang, S.E. Significance of granite-greenstone terranes in the formation of Witwatersrand-type gold mineralisation–A case study of the Neoarchaean Black Reef Formation, South Africa. *Ore Geol. Rev.* **2020**, *121*, 103572. [CrossRef]
52. Savva, N.E.; Kravtsova, R.G.; Anisimova, G.S.; Palyanova, G.A. Typomorphism of Native Gold (Geological-Industrial Types of Gold Deposits in the North-East of Russia). *Minerals* **2022**, *12*, 561. [CrossRef]
53. Gerasimov, B. The Use of Typomorphic Features of Placer Gold of the Anabar Region for Determining Its Sources. *Minerals* **2023**, *13*, 480. [CrossRef]
54. Ketchaya, Y.B.; Dong, G.; Santosh, M.; Lemdjou, Y.B. Microchemical signatures of placer gold grains from the Gamba district, northern Cameroon: Implications for possible bedrock sources. *Ore Geol. Rev.* **2022**, *141*, 104640. [CrossRef]

55. Kotov, A.A.; Murashov, K.Y.U. Forecast estimates of the depth of the gold ore bodies of the deposit in the Kropotkin ore cluster (Bodaybo district). In Proceedings of the Scientific and Methodological Foundations of Forecasting, Prospecting, Evaluation of Deposits of Diamonds, Precious and Non-Ferrous Metals, Moscow, Russia, 17–19 April 2019; pp. 108–109. (In Russian).
56. Morrison, G.W.; Rose, W.J.; Jaireth, S. Geological and geochemical controls on the silver content (fineness) of gold in gold-silver deposits. *Ore Geol. Rev.* **1991**, *6*, 333–364. [CrossRef]
57. Chapman, R.J.; Mortensen, J.K.; Crawford, E.C.; LeBarge, W. Microchemical studies of placer and lode gold in the Klondike District, Yukon, Canada: 2. Constrains of the nature and location of regional lode sources. *Econ. Geol.* **2010**, *105*, 1393–1410. [CrossRef]
58. Lalomov, A.V.; Naumov, V.A.; Grigorieva, A.V.; Magazina, L.O. Evolution of the Vagran gold-bearing placer cluster (Northern Urals) and prospects for revealing bedrock mineralization. *Geol. Ore Depos.* **2020**, *5*, 407–418. [CrossRef]
59. Nikiforova, Z. Criteria for determining the genesis of placers and their different sources based on the morphological features of placer gold. *Minerals* **2021**, *11*, 381. [CrossRef]
60. Shelton, K.L.; So, C.-S.; Chang, J.S. Gold-rich mesothermal vein deposits of the Republic of Korea: Geochemical studies of the Jungwon gold area. *Econ. Geol.* **1988**, *83*, 1221–1237. [CrossRef]
61. Petrovsky, D.V.; Silaev, V.I.; Zharkov, V.A.; Petrovsky, V.A. Native gold and companion minerals in the Cenozoic sediments of the Ural Foredeep. *Geol. Ore Depos.* **2012**, *54*, 474–486. [CrossRef]
62. Duk-Rodkin, A.; Barendregt, R.W.; White, J.M.; Singhroy, V. H Geologic evolution of the Yukon River: Implications for placer gold. *Quat. Int.* **2001**, *82*, 5–31. [CrossRef]
63. Lowey, G.W. The origin and evolution of the Klondike goldfields, Yukon, Canada. *Ore Geol. Rev.* **2006**, *28*, 431–450. [CrossRef]
64. Chefranov, R.M.; Lalomov, A.V.; Chefranova, A.V. A Search-Oriented Method of Numerical Forecasting of Rare-Metal Proximal (Close-to-Source) Placers: Evidence from the Lovozero Placer District. *Geol. Ore Depos.* **2023**, *2*, 133–145. [CrossRef]

minerals

MDPI

Article

Native Gold and Unique Gold–Brannerite Nuggets from the Placer of the Kamenny Stream, Ozerninsky Ore Cluster (Western Transbakalia, Russia) and Possible Sources

Sergey M. Zhmodik [1,2,*], Evgeniya V. Airiyants [1], Dmitriy K. Belyanin [1], Bulat B. Damdinov [2], Nikolay S. Karmanov [1], Olga N. Kiseleva [1], Alexander V. Kozlov [3], Alexander A. Mironov [4], Tatyana N. Moroz [1] and Victor A. Ponomarchuk [1]

1 Sobolev Institute of Geology and Mineralogy SB RAS, Novosibirsk 630090, Russia
2 Dobretsov Geological Institute SB RAS, Ulan-Ude 670047, Russia
3 Department of Geology and Exploration of Mineral Deposits, Saint Petersburg Mining University, Saint Petersburg 199106, Russia
4 Sibir Geopoisk, Irkutsk 664017, Russia
* Correspondence: zhmodik@igm.nsc.ru

Abstract: We carried out a comprehensive study of native gold (morphology, composition, intergrowths, and microinclusions) from alluvial deposits of the Kamenny stream (Ozerninsky ore cluster, Western Transbaikalia, Russia). The study showed that there were four types of native gold, which differed significantly in their characteristics and probably had different primary sources from which placers were formed: gold–quartz, oxidized gold–sulfide, gold–silver, and zones of listvenites with copper–gold and gold–brannerite (Elkon-type). Particular attention was paid to the study of unique, both in size and in composition, gold–brannerite nuggets of the Kamenny stream. It was established that the gold in the gold–brannerite nuggets (GBNs) had wide variations in chemical composition and mineral features. According to them, there were five different fineness types of native gold: 750–800‰; 850–880‰; 880–920‰; 930–960‰; and 980–1000‰. The data obtained indicated a multistage, possibly polygenic, and probably polychronous formation of GBN gold–uranium mineralization. The first stage was the formation of early quartz–nasturanium–gold–W–rutile–magnetite association (Middle–Late Paleozoic age). The second was the crystallization of brannerite and the replacement of an earlier pitchblende with brannerite (Late Triassic (T3)–Early Jurassic (J1) age). The third was the formation of the hematite–barite–rutile–gold association as a result of deformation–hydrothermal processes, which was associated with the appearance of zones of alteration in brannerite in contact with native gold with 8–15 wt.% Ag. The fourth was hypergene or the low-temperature hydrothermal alteration of minerals of early stages with the development of iron hydroxides (goethite) with impurities of manganese, tellurium, arsenic, phosphorus, and other elements. The carbon isotopic composition of an organic substance indicates the involvement of a biogenic carbon source. In the OOC area, there were signs that the composition of the GBNs and the quartz–chlorite–K–feldspar-containing rocks corresponded to Elkon-type deposits.

Keywords: gold; brannerite; uranium mineralization; Ozerninsky ore cluster; Western Transbaikalia

check for updates

Citation: Zhmodik, S.M.; Airiyants, E.V.; Belyanin, D.K.; Damdinov, B.B.; Karmanov, N.S.; Kiseleva, O.N.; Kozlov, A.V.; Mironov, A.A.; Moroz, T.N.; Ponomarchuk, V.A. Native Gold and Unique Gold–Brannerite Nuggets from the Placer of the Kamenny Stream, Ozerninsky Ore Cluster (Western Transbakalia, Russia) and Possible Sources. *Minerals* **2023**, *13*, 1149. https://doi.org/10.3390/min13091149

Academic Editor: Huan Li

Received: 7 July 2023
Revised: 18 August 2023
Accepted: 21 August 2023
Published: 30 August 2023

1. Introduction

The causes and timing of the appearance of ore clusters, which often include large deposits of various minerals, are of great interest to researchers. A striking example is the Ozerninsky ore cluster (OOC), with an area of more than 200 km^2 and located in Western Transbaikalia (Russia) as part of the Uda–Vitim structural formation zone, which includes 12 deposits of Pb, Zn, Fe, Cu, baryte, and gold (including one of the largest in Russia, Ozernoe pyrite–polymetallic) and 23 ore occurrences [1,2]. A detailed geological study of the Ozerninsky ore cluster began in 1960 due to iron ore finds. For a short period of time from 1961 to 1964, a series of deposits was discovered in the OOC area: Ozernoe and Ulzutuyskoe polymetallic, Hematitovoe, Magnetitovoe, Gurvunurskoe, Arishinskoe iron ore deposits, Turkulskoe, and Gunduyskoe iron–copper–barite deposits. The geological study of the OOC continued for a long time, but despite this, a consensus on the structure, deep structure, tectonics, magmatism, age, mineralogical–geochemical, and physical–chemical conditions for the formation of deposits and the entire OOC has not yet been developed.

The problem of the gold content of the OOC ores has not been solved to date, despite the fact that many deposits and ore occurrences have encountered ores with elevated gold grades (from 0.2 to the first ppm). In addition, gold grains were found in the alluvial deposits of the Ulzutui, Izvestkovy, and Pravy Surkhebt streams, which have not been sufficiently studied so far. To date, the Nazarovskoe gold deposit (up to 23 ppm Au) has been discovered, as well as a number of gold mineralization points among magnetite, sulfide–magnetite, copper–barite, pyrite–sphalerite, pyrite–quartz ores of most deposits, and the OOC ore occurrences. In 2005, an industrial placer was discovered and exploitation began at the Kamenny stream, with an average content of alluvial gold per layer of –830 mg/m^3. At the beginning of the 21st century, industrial gold placers were discovered virtually throughout the entire area of the OOC, indicating a wider distribution of gold ore than previously thought.

The first studies, using modern analytical methods, revealed a wide variety of typomorphic features of alluvial gold in the OOC area [3–5]. The purpose of this work was a comprehensive study of placer gold [4,5] from the placer of the Kamenny stream (the right tributary of the Levyi Surkhebt stream), located in the central part of the OOC. Studying the Kamenny stream placer was of the greatest interest due to its location in the central part of the OOC, representing the maximum variety of native gold (NG) types and finds of gold–brannerite nuggets [6,7]. Particular attention was paid to the study of gold–brannerite nuggets, unique both in size (up to 200 g) and in mineralogical and geochemical features. We also carried out a comparative analysis of the typomorphic features of native gold, in particular the morphology and composition of both gold itself and the minerals associated with it.

1.1. Brief Description of the Geological Structure of the Ozerninsky Ore Cluster

The studied region is characterized by a highly complex geological structure and the long history of the formation of its structural and material complexes. The region is located in the zone of development of the world's largest Angara–Vitim granitoid batholith. The data on the geological structure of the OOC have been published in a number of papers [8–12], in the last of which many problematic issues were discussed [13–17]. In tectonic terms, this territory is considered as the Paleozoic Uda–Vitim island arc system, which includes the large Eravna volcano tectonic structure. According to the popular view, the structural block containing the OOC is a "remnant" or roof sag (20×10 km^2) of the Angara–Vitim batholith, composed of island arc Lower Paleozoic volcanic, volcanic–sedimentary, and sedimentary rocks of the Oldynda Formation, as well as intrusive and subvolcanic bodies of the Permian age. The formation of the OOC sulfide–iron–oxide mineralization is associated with Lower Cambrian volcanism and hydrothermal–sedimentary ore deposition [11]. But, according to drilling data, no granites have been found within the OOC area at a depth of 1700 m, despite the geophysical forecast. According to the results of gravity studies, the OOC has been described as the "Ozerninskaya group of stratovolcanoes" [18]. The data of magnetotelluric sounding do not confirm the assumptions about the presence of ore-bearing rocks of the Lower Paleozoic age in the roof sag of the Angara–Vitim batholith [19]. The appearance of young datings of intrusive complexes of the Angara–Vitim batholith suggests that the OOC is not sag of the roof of granitoids, but an independent tectonic block injected with younger igneous formations. The main part of the OOC is composed of stratified Paleozoic terrigenous-carbonate-volcanogenic formations intruded by a large number of intrusive (granitoids, syenites) and subvolcanic rocks (dolerites, diorite porphyry, syenite porphyry, etc.) [20]. The age of the material complexes developed in the OOC varies within 530–270 Ma [16,21,22].

A number of researchers substantiate the hydrothermal-metasomatic origin of the ores of the Ozernoe deposit at the late stages of tectonomagmatic activation [23]. At the same time, starting from the Paleozoic, five major stages of magmatism are observed at the OOC [5]. In the Cambrian and Ordovician, island-arc magmatism was intensely manifested; Caledonian tectonic movements ended with powerful granitoid magmatism, which was preceded by mafic intrusions. In the Middle and Late Paleozoic (D-P), two stages of tectonic-magmatic activation occurred due to the initiation and development of large intracontinental rift-like structures, which are associated with felsic volcanism and granitoid magmatism. In the Jurassic and Cretaceous, large outpourings of basaltic lavas occurred, timed to coincide with tectonic depressions. The latest stage of magmatism ended in the Mesozoic after the main stages of granite formation, which may indicate the polychronicity of the volcanotectonic structure and the existence of a magma conduit penetrating the crust and reaching the mantle [24]. According to the latest generalizations, the deposits of the Kurba–Eravninsky ore region, including the OOC, are associated with the formation of the Uda–Vitim island-arc system of the Vendian–Cambrian ensialic type (similar to the Kuril–Kamchatka basalt–andesite–dacite–rhyolitic association with mature island-arc systems) and subsequent processing of its structures in the Middle and Late Paleozoic [20,25].

1.2. Geological Structure of the Kamenny Stream Placer Deposit

Explored placers of gold are located in the central and southeastern parts of the Ozerninsky ore cluster (Figure 1). They are confined to the deposits of river valleys draining the northwestern slope of the Eravna depression. The placers were found in the valley of the Levyi Surkhebt stream, its right tributary the Kamenny stream, the Ulzutui stream, its right tributaries the Hematitovy and Nezametny streams, and the Gunduy-Kholoi stream. As it was said, the Kamenny stream placer is of the greatest interest due to its location in the central part of the OOC, with the maximum variety of native gold types and finds of gold-brannerite nuggets, which are described below. The measured length of the placer is 3.4 km, with an average width of 72 m. The average thickness of the sands is 1.8 m, with an average overburden–1.6 m. The placer is located in the height range of 1130–1070 m and is not contoured in the upper and lower ends along the stream. The gold contents reached 2318 mg/m^3, averaging 724 mg/m^3. The section of loose deposits of the valley of the Kamenny stream contains (top-down):

1. Soil–vegetation layer with blocks, boulders, pebbles, crushed stone, gruss of volcanic rocks and granites (thickness—0–0.4 m);
2. Boulder–block deposits with pebbles, crushed stone, gruss and gravel, cemented by dark brown, brown, reddish brown loam soil (thickness—1.0–2.5 m);
3. Blocks, weakly rounded boulders, crushed stone in brown loam soil (thickness—0–2.5 m);
4. Sand–gravel–pebble deposits in brown and dark brown loam soil, representing a productive layer (thickness—0.4–3.6 m).

The total thickness of loose deposits varies from 2 m in the sides to 4.4 m in the central part of the valley. Productive deposits contain pebble material for 3%–5%, rarely–up to 10%, pebble size–2–3 cm. The content of loam soil in the lower part of the layer reaches 20%–30%. The main part of the layer is made up of uneven-grained sand and slightly rounded gravel. The clastic component of productive deposits, in addition to fragments of host rocks (granitoids and volcanic rocks), contains quartz, magnetite, hematite, limonite, and red jasper. The heavy concentrate fraction is dominated by limonite, hematite, magnetite, and gold; ilmenite, rutile, zircon, leucoxene, and mafic minerals are present in a subordinate amount.

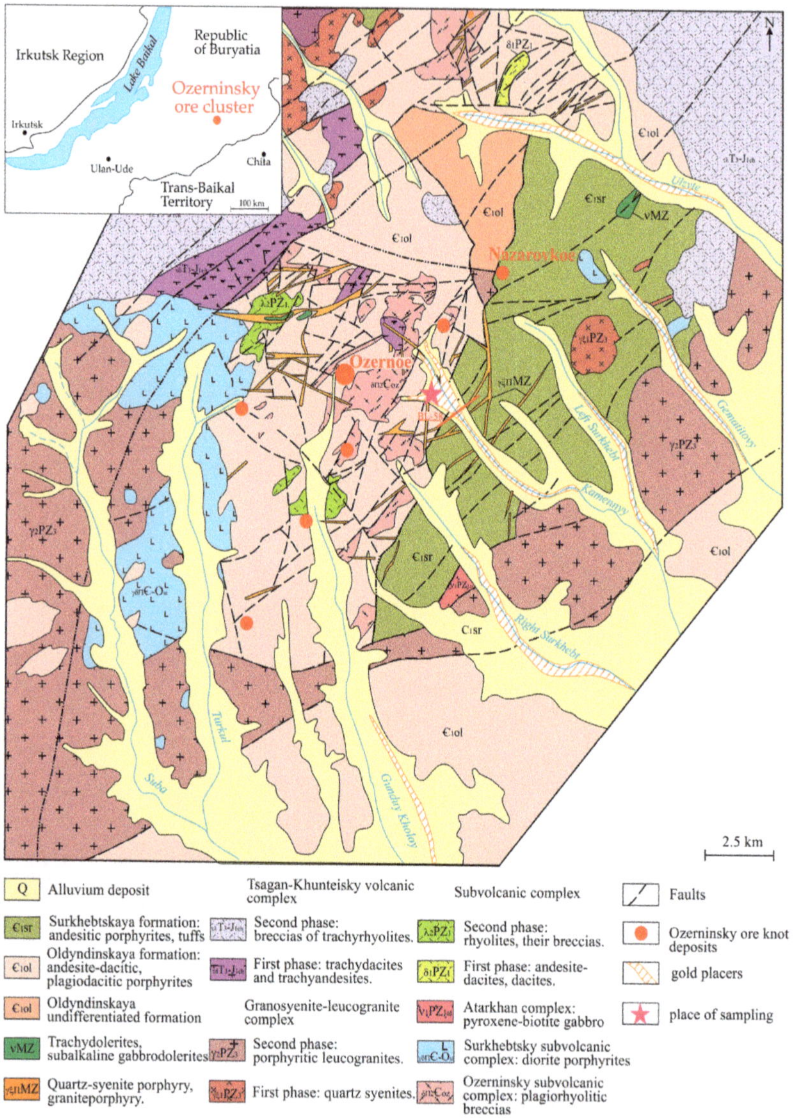

Figure 1. Schematic geological map of the area of the Ozerninsky pyrite-polymetallic deposit. The star marks the location where the nuggets were found. Compiled using materials [1,4].

The productive layer is located in the lowest part of the floodplain, sometimes reaching the sides of the valley. The placer bedrock is composed of rocks of the Surkhebt sequence, granitoids and dark gray volcanic rocks of basic composition, containing gold from traces to 0.6 g/t (average for 25 samples 0.034 g/t). The structure of the shallow placer of the Kamenny stream along the drilling line BL-58 is shown in Figure 2. In the studied section, the productive layer of sand–gravel–pebble deposits has a thickness of 0.4 to 5.6 m (about 1 m on average) and is overlain by "peat" with a thickness of 0.4–1.2 m; there are mostly volcanic rocks of basic composition.

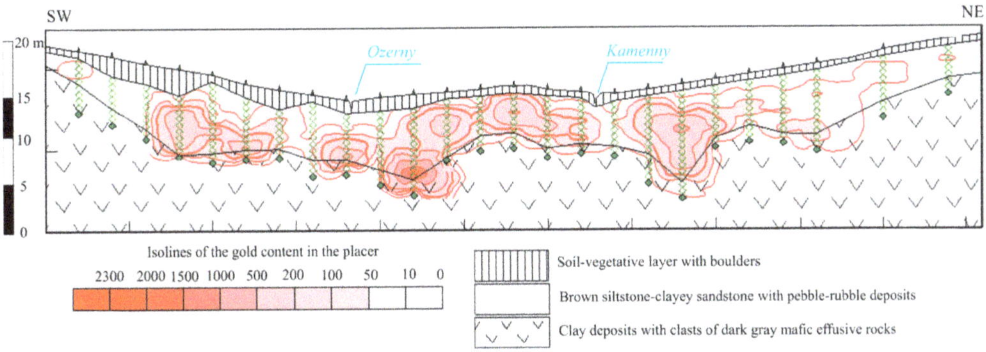

Figure 2. Geological section along the drilling line BL-58 of the Kamenny stream placer with the distribution of gold concentrations in mg/m^3. Compiled using materials [5].

2. Materials and Methods

2.1. Sampling

Native gold and gold–brannerite nuggets were obtained during the mining of the placer of the Kamenny stream, the right tributary of the Levyi Surkhebt stream (OOC) using machinery and equipment for gravity extraction of gold, as well as from drilling wells. Further, the gold grains were selected using binocular microscopes. A comprehensive mineralogical and geochemical study on separate populations of several hundred grains of native gold and seven gold–brannerite nuggets was carried out using various methods.

2.2. Optical and Scanning Electron Microscopy

The chemical composition, morphology and spatial relationships of minerals were studied using scanning electron microscopes (SEM) with energy dispersive spectrometers (EDS): LEO 1430VP (Carl Zeiss, Oberkochen, Germany) with the INCA Energy 350 microanalysis system (Oxford Instruments Nanoanalysis Ltd., High Wycombe, UK) at the Geological Institute, Siberian Branch, Russian Academy of Sciences, Ulan Ude; MIRA 3 LMU (Tescan Orsay Holding, Brno, Czech Republic) with Aztec Energy XMax 50 microanalysis system (Oxford Instruments Nanoanalysis, UK), as well as optical transmission and ore microscopy (AxioScope. A1, Carl Zeiss MicroImaging GmbH, Germany) at the Analytical Center for Multi-Elemental and Isotope Research, Siberian Branch, Russian Academy of Sciences, Novosibirsk. Scanning electron microscope studies were performed at an accelerating voltage of 20 kV and an electron beam current of 0.2–0.3 nA (LEO 1430) and 1.4 nA (MIRA 3). Live spectrum acquisition times were 50 s (LEO 1430) and 20 to 40 s (MIRA 3). In this case, the detection limits of impurities are tenths of a percent and depend, in addition to the acquisition time of the spectra, on the type of the analyzed matrix, on the intensity of the X-ray series used as an analytical signal, and also on the presence of spectral overlaps. The analysis was performed in the mode with the exclusion of the results of analysis below the level of 3 σ background variation, so all the data presented in the work exceed the detection limit. In some cases, to reduce the limits of detection of impurities, the time was increased to 300 s (MIRA 3). Possible spectral overlaps are automatically taken into account by the software of microanalysis systems by means of deconvolution of the spectra of chemical elements when processing energy-dispersive spectra. Both microanalysis systems used the same set of standards—simple synthetic and natural compounds of stoichiometric composition (Al_2O_3, SiO_2, FeS_2, ThO_2, UO_2, $CaMgSi_2O_6$, HgTe, PbTe, BaF_2, phosphates of rare earth elements, etc.), as well as pure metals (Au, Ag, Ti, Fe, Ni, Cu, etc.). In accordance with the recommendation for INCA Energy and AZtec Energy microanalysis systems, the analytical signal was normalized to the intensity of the K-series of Co in the periodically acquired spectra of cobalt metal. Matrix corrections were taken into account by the XPP method implemented in the software of INCA Energy and AZtec Energy microanalysis systems. The analysis was performed in the "All elements analyzed" mode without normalizing the analytical total to 100%. In the final report for oxide minerals, the results were presented as component concentrations with the addition of oxygen concentration, taking into account the valence of cations. It should be noted that the accuracy of the direct determination the oxygen content by the SEM EDS method is high enough to reliably distinguish iron minerals wustite, magnetite, hematite, goethite, siderite by the ratio of Fe and O atomic concentrations (Fe/O $\approx$ 1:1, 3:4, 2:3, 1:2, 1:3). Similarly, SEM EDS oxygen can be used to estimate the degree of oxidation of uranium or to detect the hydration of uranium-bearing minerals. To reduce the influence of the sample microrelief on the analysis accuracy, most analyzes were performed by scanning mineral areas up to 100 μm^2 in size; fine phases with linear sizes less than 3–5 μm were analyzed in the point mode.

2.3. *Raman Spectroscopy*

Micro-Raman spectra were recorded on a Horiba Jobin Yvon LabRAM HR800 spectrometer with a 1024 pixel liquid Nitrogen cooled Charge Coupled Device (LN/CCD) detector using for excitation the wavelengths of 532 nm from a Nd:YAG laser. Raman spectra were collected in a backscattering geometry using an Olympus BX41 microscope. The microscope with an Olympus 50× objective lens of WD = 0.37 mm with 0.75 numerical aperture produces a focal spot diameter of ~2 μm. The power of the laser radiation used was set to about 0.5 mW on the sample to avoid sample heating. The absence of sample heating was tested by comparison with the Raman spectra recorded at 7 and 0.3 mW power of the incident laser beam at the sample. The majority of the spectra were recorded using a neutral density filter, D = 1. The spectrometer was wavenumber calibrated with a silicon standard [26].

2.4. *Measurement of $\delta^{13}C$ Signatures*

The sample was washed with HCl (3 mol/L) to remove carbonates for isotopic analysis. The carbonate-free residues were combusted in a Thermo elemental analyzer (Flash EA 1112) integrated with a Thermo MAT253 isotope ratio mass spectrometer at the Analytical Center for Multi-Elemental and Isotope Research, Siberian Branch, Russian Academy of Sciences, Novosibirsk. The analysis was calibrated using Standard Reference Materials NBS-22 ($\delta^{13}C_{VPDB} = -29.74‰$). The $\delta^{13}C$ results are expressed in the conventional delta (δ) notation as the per mil (‰) deviation from the Vienna Peedee Belemnite (V-PDB). Standard deviation for the measured $\delta^{13}C$ values is better than $\pm 0.1‰$.

3. Results

3.1. *Typomorphic Features of Native Gold*

Morphological characteristics of native gold (NG) from the Kamenny stream placer vary widely [4]. The size of the studied gold particles ranges from 0.04 to 2.5 mm. The granulometric composition of native gold grains (NG) is represented by the classes: +1 mm–34.8 wt%; 1 + 0.5 mm–28.1 wt%; 0.5 + 0.25 mm–16.1 wt%; −0.25 mm–21.0 wt%. The NG shape is lumpy, lamellar, dendritic, club-shaped, irregular shape, fissured-veinlet, and less often intergrowths of crystals (Figure 3). The roundness of the gold is weak, rarely to medium. A distinctive feature of most gold particles is the presence of supergene iron hydroxides on their surface. Among mineral inclusions in NG, only quartz and, rarely, barite are reliably identified macroscopically. Under an optical microscope, several varieties of gold are quite confidently distinguished by color: bright yellow, greenish yellow, grayish greenish white, reddish yellow with reddish rims (Figure 4).

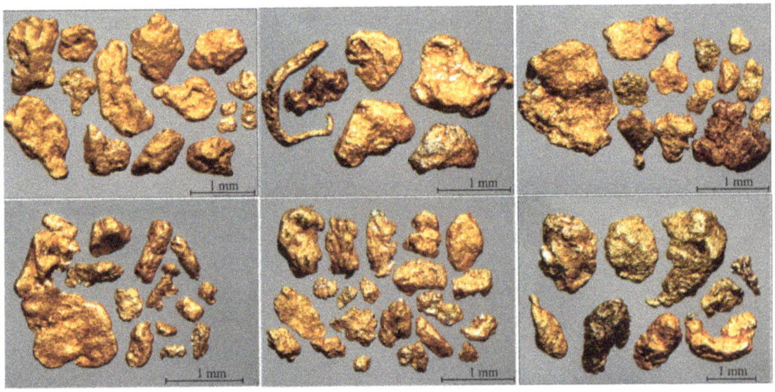

Figure 3. Morphology of native gold grains from the Kamenny stream placer.

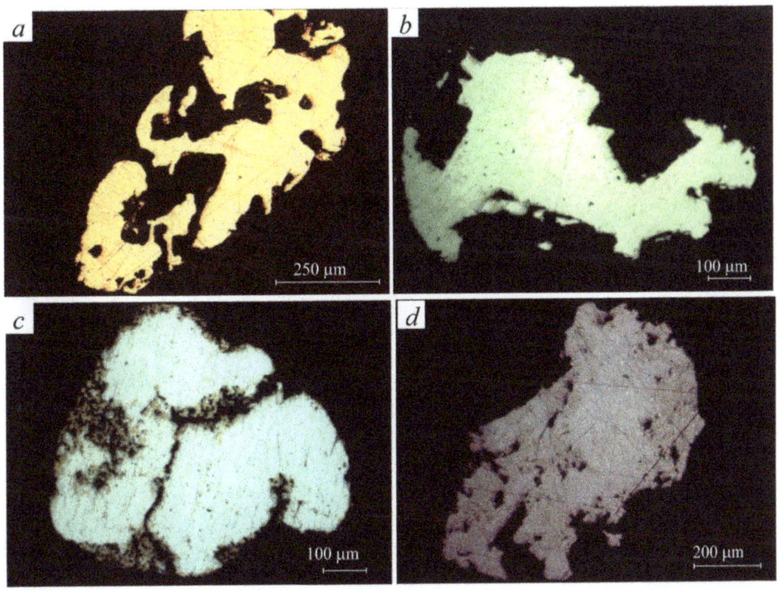

Figure 4. Photos representing gold of different colors (optical microscope): bright yellow (**a**), greenish yellow (**b**), grayish greenish white (electrum) (**c**), reddish yellow (**d**).

The chemical composition of NG is characterized by a limited set of impurity elements, among which only Ag and Cu were found; Te with a content of up to 1.02 wt% was identified in a single grain of NG. Hg is rarely detected from 0.07 to 0.51 wt% Hg. Very significant variations in the concentrations of Ag (0–55.48 wt.%) and Cu (0–22.83 wt.%) have been identified, which is reflected in changes in the color of the gold. The NG placer of the Kamenny stream is characterized by the distribution of very high-grade (960–1000‰) gold in the form of veinlets along the boundaries of individual grains and on the surface of gold particles (Figure 5). The thickness of the veinlets varies from 0.5 to 10, rarely 50 µm. All gold grains were subjected to the process of secondary enrichment under supergene conditions, as evidenced by the cracks and rims of the development of very high-grade gold. Four types of NG are found in the placer of the Kamenny stream. The first type is moderately high-fineness, the second is low-fineness gold, the third is very low-fineness (electrum) and the fourth type is cuprous gold (Figure 6, Table 1). The most common is bright yellow moderately high-fineness gold, with an average fineness of 890‰. It occurs either in the form of small grains of a lamellar, amoeboid, elongated shape. The NG roundness is weak. The main impurities are Ag and Cu, which are present in 50% of the NG, reaching 3.78 wt% (average 0.38 wt%). Low-fineness greenish-yellow gold makes up more than 20% of the placer gold. Its average fineness is 697‰. The size of the NGs ranges from 0.5 to 20 mm, their morphology is predominantly lamellar. The degree of roundness is weak. Cu is present in 50% of the gold grains, reaching 2.06 wt%. The placer also contains (about 10%) very low-grade gold (electrum) (the third type) with an average fineness of 646‰. The size of electrum structures is about 0.5 mm, less often up to 1 mm. The shape of the grains is flattened to isometric, the degree of roundness is medium. This is the most rounded gold of all the studied types of the placer of the Kamenny stream.

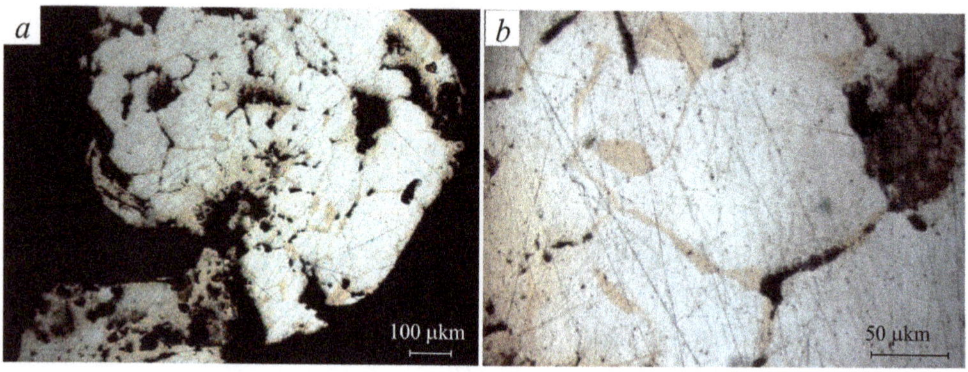

Figure 5. Photos of NG with rims (**a**) and veins of very high-fineness gold (**b**) (optical microscope).

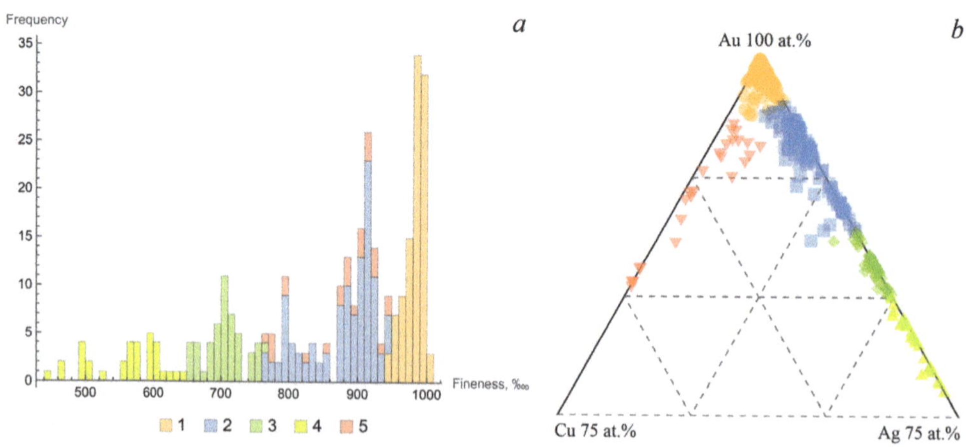

Figure 6. Fineness histogram (‰) (**a**) and ternary chemistry chart (at.%) (**b**) for NG from the placer (1—high-fineness gold from alteration zones (n = 103), 2—moderately fine (n = 113), 3—low-fineness (n = 50), 4—very low-fineness (n = 33), 5—cuprous gold (n = 26)).

Table 1. Typomorphic features of alluvial placer gold.

Type	I	II	III	IV
number of investigated grains	20	15	5	5
fineness (min–max)	765–945	650–764	445–648	769–946
average	890	697	590	919
Cu (min–max)	0–3.78	0–2.06	0–1.14	3.3–22.83
average	0.38	0.19	0.28	4.88
size (mm)	0.1–2.5	0.5–2	0.5–1	0.5–3
color	bright yellow	greenish yellow	grayish greenish white	reddish yellow
morphology	lamellar, amoeba-shaped, elongated	lamellar	flattened, isometric	ellipsoid-flattened, lamellar
roundness	low	weak to medium	medium	weak to medium

There is also copper gold (about 10% of all types) of reddish-yellow and copper-red color. The morphology of such gold is ellipsoid-flattened to lamellar, the size is from 0.5–3 mm. In an optical microscope, copper gold particles have a reddish tint and significant inhomogeneity. The average fineness of the gold varies from 769 to 946‰, with Cu contents from 3.3 to 22.83 wt%.

3.2. Morphologic Features of Gold-Brannerite Nuggets

Finds of gold–brannerite nuggets (GBN) in the placer of the Kamenny stream are of particular interest. The weight of the nuggets ranges from 1–2 to 200 g. Outwardly, these are golden-brown aggregates, and more than 50% of their surface is composed of bright yellow native gold. The rest is represented by a dark brown to black slightly fractured mineral—brannerite (Figure 7). The sections show that the bulk of the nuggets is occupied by brannerite and associated minerals, and the amount of gold does not exceed 15–40 vol.%. In nuggets, gold is located in brannerite in association with hematite (±magnetite), rutile, barite, and less often with muscovite. Sometimes there are quartz, siderite, goethite, nano- and micro-isolations of uraninite, Pb (native or Pb oxide) and tellurides Au, Ag, Bi, Pb (petzite, tellurobismuthite, altaite), and single grains of chalcopyrite. The gamma-spectrometric analysis data testify to the essentially uranium nature of radioactivity, with contents of 14–26 wt% U and 1–6 wt% Th. Based on the study of GBNs, mineralization with different cataclastic textures is observed in the sections. The textures are defined by different morphology and spatial ratios of brannerite, native gold, hematite, rutile and other minerals (Figures 7 and 8). In particular, there are 1—stockworked or veinlet-disseminated texture, which is a network of multidirectional microcracks in brannerite, made by gold; 2—spotted texture represented by(i) gold-hematite aggregate, which includes "relics" of brannerite, and (ii) isometric grains of rutile in gold, slightly elongated with "corroded" edges; 3—"microbrecciaed" texture represented by crushed hematite or rutile cemented with gold (Figure 9).

Figure 7. General view of gold–brannerite nuggets (**a**,**c**,**e**,**g**) and sections of nuggets (**b**,**d**,**f**,**h**–**j**) from the Kamenny stream placer of the OOC, Western Transbaikalia. Appearance of nugget images a, b from [7].

Figure 8. Microphotographs of GBNs in reflected light. Gold in brannerite is represented by disseminated, fractured, veinlet and spotty types of distribution. Gold is located along the cracks and cluster (spots) in brannerite are formed (**a**), gold is located in brannerite along cracks (**b**); a "spot" of gold with cracks diverging from it, filled with gold (**c**); gold-hematite aggregate, which includes "relics" of brannerite (**d**); gold, which includes "relics" of brannerite (**e**); "microbrecciaed" texture represented by crushed hematite and "relics" of brannerite cemented with gold (**f**). Au—native gold; Bnr—brannerite; Brt—barite; Hem—hematite. Images (**b**–**d**) from [7].

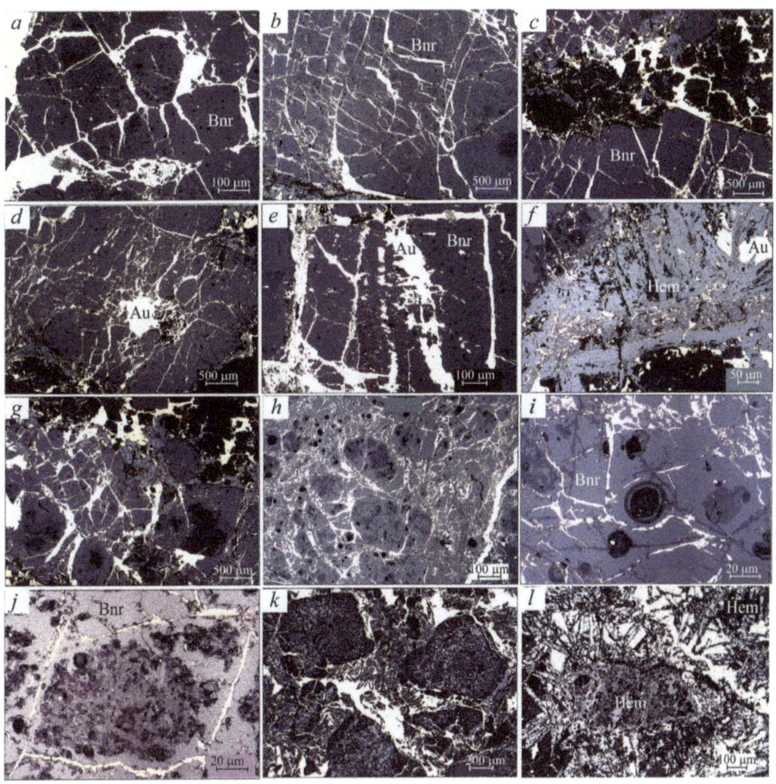

Figure 9. Photos in reflected light ore microscope. Cataclastic GBN textures: NG is located along

cracks breaking brannerite into rectangular blocks (**a–c**); NG is located along multidirectional "bush-like" cracks (**d**); NG is located along cracks of various sizes, chains of gold microparticles in brannerite linearly located in the center (**e**); gold-hematite vein in brannerite with multidirectional systems of microcracks (**f**); combination of block and porphyroclastic structures with separate grains of more preserved brannerite (**g,h**); gold traces brannerite deformation along the microbubble (**i**); a block of partially altered brannerite contoured by cracks with native gold (**j**); porphyroclassic structure composed of grains of partially altered brannerite in gold–hematite cement (**k,l**). Au—native gold; Bnr—brannerite; Hem—hematite.

3.2.1. Gold

Gold in GBNs occurs in the form of spherical micro- and nanoparticles scattered in brannerite, less often as chains of gold microparticles in brannerite linearly located (Figure 9e) or crumpled into isoclinal micro-folds. Most often, gold fills cracks (thickness from a few microns to mm) in brannerite, and both multidirectional "bush-like" crack systems with gold (Figure 9d) and oriented intersecting at an angle close to 90° are common (see Figure 9a–c). Taking into account that brannerite does not have cleavage, this nature of cracks should be related to brittle deformation of brannerite. That is, gold in brannerite is represented to the maximum extent by fractured and veinlet types (Figure 9d,e,g,h). In areas where rutile is replaced by hematite, gold is definitely associated with a fine-grained rutile aggregate, framing it or surrounding individual rutile grains (Figure 10). In muscovite, gold is confined to cleavage planes (Figure 10e). Near relatively large cracks, accompanied by crushing and brecciation, gold acts as a "cementing" mineral, while its dimensions in the polished section reach hundreds of microns. The largest NG structures (up to mm) are observed at the intersections of cracks with gold (Figure 9d–e), rarely unrelated to them.

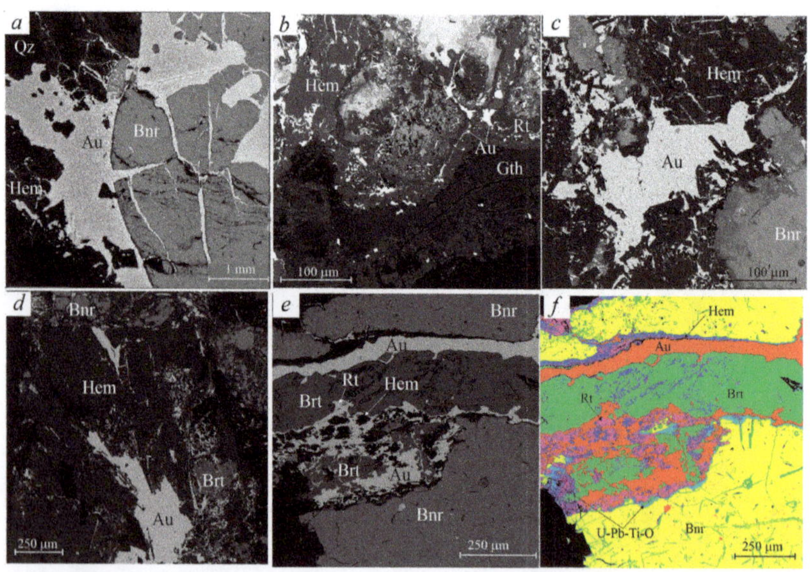

Figure 10. GBN microphotographs according to the results of SEM studies in back-scattered electrons. Deformed brannerite, cracks of which are filled with gold (**a**); rutile is replaced by hematite; gold with a fine-grained rutile aggregate surrounds individual grains of rutile (**b**); gold–hematite assemblage between brannerite grains (**c**); "bush-like" crack systems with gold in hematite (**d**); large rutile–gold–barite veinlet in brannerite (**e**); elemental mapping, mineral distribution is shown in color (**f**). Au—native gold; Bnr—brannerite; Brt—baryte; Hem—hematite; Gth—goethite; Rt—rutile [7].

The composition of NG is determined by Ag impurity and varies from 73 to 100 wt% Au (Figure 11, Table 2). The distribution of Au and Ag in NGs is polymodal, several groups with different modes being definitely distinguished (in ‰): 960–1000 (995); 930–950 (943); 860–925 (885); 820–860 (850) and 750–820 (775). The polymodality of NG fineness is determined by its variation between samples.

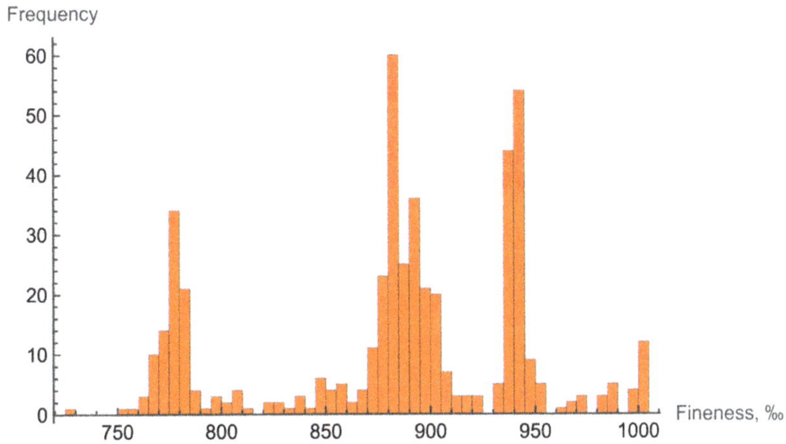

Figure 11. Fineness histogram (‰) for NG in GBNs.

Table 2. Chemical composition of gold in GBNs.

Cu	Ag	Au	Total	Au, ‰
0	5.62	92.38	98	943
0	5.86	93.84	99.7	941
0	21.87	79.6	101.47	784
0	21.8	78	99.8	782
0	22.36	77.42	99.78	776
0	11.89	89.2	101.09	882
0	5.9	93.82	99.72	941
0.25	5.93	94.04	100.22	938
0.3	5.83	93.88	100.01	939
0	9.74	87.49	97.23	900
0	13.09	85.41	98.5	867
0	11.14	86.95	98.09	886
0	1.05	99.21	100.26	990
0	1.74	99.32	101.06	983
0	21.88	76.73	98.61	778
0	22.25	76.44	98.69	775
0	22.11	77.59	99.7	778
0	11.58	86.28	97.86	882
0	13.31	84.14	97.45	863
0	3.23	94.62	97.85	967
0	11.85	87.77	99.62	881
0	4.87	93.64	98.51	951
0	0.38	100.05	100.43	996
0	4.63	94.39	99.02	953

Thus, the composition of NGs corresponds to a very high-fineness, high-fineness, moderate-high-fineness and low-fineness, according to the classification of fineness of native gold according to N.V. Petrovskaya [27]. There is definitely a dependence of the composition of NGs on the host mineral. Moderate fineness gold (86 wt% Au) was found in cracks in brannerite, in association with hematite—87–88 wt% Au and with magnetite—88–89 wt% Au. Gold in association with W-containing rutile, as well as with muscovite, quartz and barite, has a composition of 90–95 wt% Au. Uraninite is found in gold with a composition of 83.34–84.91 wt% Au. Particularly high-grade gold is found in the form of veins, lenses, and isometric micro-sections among NGs with a composition of 85 wt% Au.

3.2.2. Brannerite

Uranium mineralization is represented primarily by brannerite ((U^{4+}, Ca, Th, Y) [$(Ti, Fe)_2O_6$]$\cdot nH_2O$) with submicron inclusions of uraninite. Grains of brannerite are 1–10 mm in size; for some nuggets, brannerite is a matrix containing other minerals. The structure of brannerite is heterogeneous at the macro and micro levels. Brannerite grains are usually cut by a dense network of veinlets filled with native gold and/or brannerite decay products (Figure 12, Table 3). Brannerite is characterized by the presence of rounded cavities 10–20 μm in size. One part of these cavities is free of substances and the other part is filled with either barite (Figure 12f,g) or U-Ti gel (Figure 12c) with a higher content of U and lower Ti and Pb than in brannerite (see Table 3, group "U-(Ti)"). At the micro level, brannerite shows an inhomogeneous block structure, which can be observed as darker and lighter areas when the BSE-image is viewed in detail. In some cases, lighter blocks appear patchy and emulsified. At a high magnification of the BSE-image of brannerite, numerous lighter inclusions with a size of fractions of a micron are visible. It is impossible to unambiguously determine the composition of such small inclusions, but judging by the change in the composition of brannerite, they represent two groups: more enriched in U or more enriched in Pb. Data on the composition of brannerite are given in Table 4. The calculation of the main and impurity components in formula units was carried out for six oxygen atoms and three cations following Colella et al. [28] for a formula without water—AB_2O_6. The water content was estimated from excess oxygen. Distribution of the main and impurity components of brannerite in formula units is shown in Figure 13. For all sections of brannerite analyses, the calculated formula coefficients were deficient in U and Ti, and on average they were 0.74 and 1.72 apfu (atom per formula unit).

There is a slight shift in the amount of cations included in positions A and B in favor of position B: on average, the sum of cations of position A is 0.94 apfu, and 2.06 apfu for position B. Position A along with U (46–54 wt% UO_2) includes constant impurities of Ca (1.6–3.1 wt% CaO) and Pb (1.1–2.6 wt% PbO); in addition, most analyses of brannerite are characterized by an impurity of Th (0–2.6 wt% ThO_2). Position B, along with Ti (32–36 wt% TiO_2), constantly contains admixtures of Fe (3.4–5.8 wt% Fe_2O_3) and Si (0.2–2.9 wt% SiO_2). In the case of long spectrum acquisitions, the appearance of Y in brannerite (max 0.74 wt% Y_2O_3), rarer Mn (max 0.29 wt% MnO) entering A position and V (max 0.60 wt% V_2O_5) belonging to B position is typical. Single analyses of the composition of brannerite show the presence of Ce (max 0.59 wt% Ce_2O_3) and Sr (max 0.61 wt% SrO) belonging to position A, and Al (max 0.61 wt% Al_2O_3) and Zr (max 0.57 wt% ZrO_2) related to position B. In all analyses of brannerite, when calculating the formulas, there is excess oxygen content (0.4–1.6 apfu). If we assume that all excess oxygen is explained by the incorporation of water into the composition of brannerite, then its content can be estimated in the range of 1.8–7.1 wt%. The average uranium valence was estimated as 4.82, which leads to the calculation of the U^{4+} and U^{6+} contents in brannerite as 0.44 and 0.30 apfu, respectively. The formula for the average composition of brannerite according to the SEM-EDX data can be written as $(U^{4+}{}_{0.44}U^{6+}{}_{0.30}Ca_{0.15}Pb_{0.03}Th_{0.02})_{\Sigma 0.94}(Ti_{1.72}Fe_{0.23}Si_{0.10})_{\Sigma 2.06}O_6\cdot 1.15H_2O$.

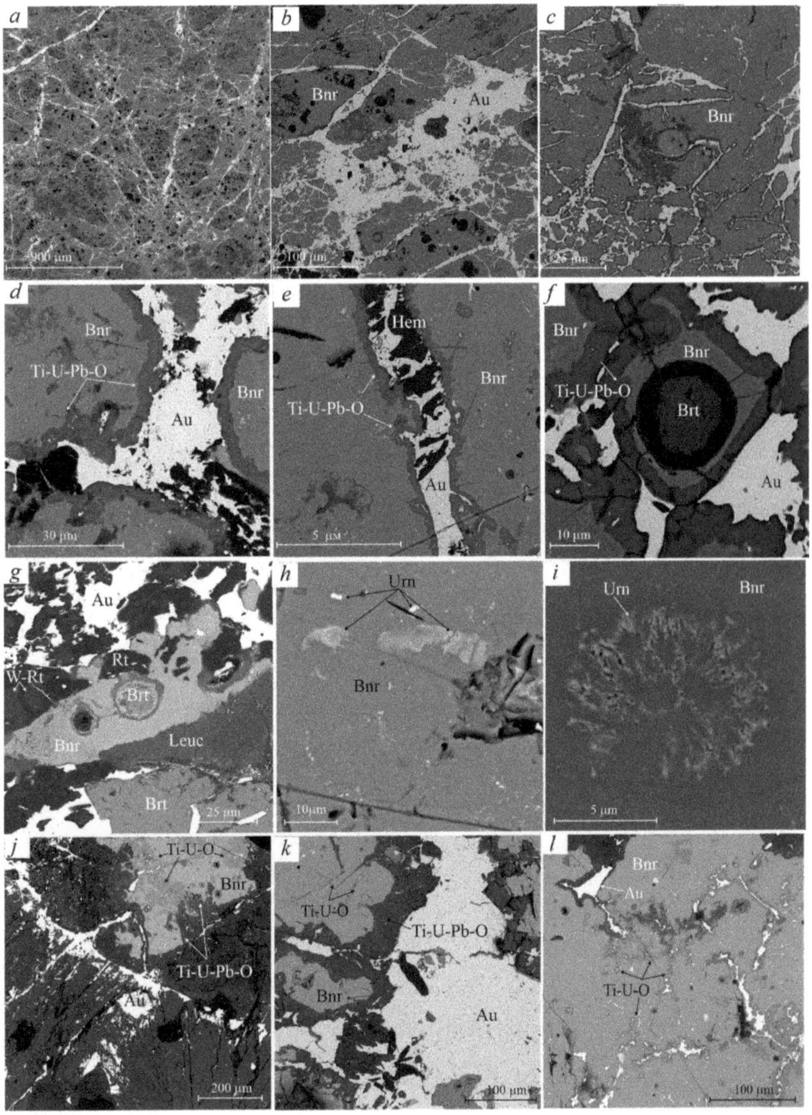

Figure 12. BSE-images of brannerite. Grains of brannerite dissected by a network of veinlets of native gold and numerous areas of altered brannerite (**a**–**c**); (**b**)—a zoomed fragment, numerous cavities are clearly visible, some of which are partially or completely filled with either barite or U-Ti gel and may have a border or zonal structure; (**c**)—a zoomed fragment with a round inclusion of U-Ti gel. Distinct rims of brannerite alteration at the boundary with the gold–hematite assemblage (**d**,**e**); zoned grain of altered brannerite, in the center of which is a barite "core" (**f**); ovoids of barite and W-containing rutile in brannerite grain (**g**); uraninite in brannerite (**h**,**i**); replacement of brannerite with U-Ti gel (**j**–**l**). Au—native gold (NG); Bnr—brannerite; Brt—baryte; Hem—hematite; Leuc—leucoxene; Rt—rutile, W-Rt—tungsten-containing rutile, Urn—uraninite. Images (**g**–**i**) from [7].

Table 3. Elemental composition of brannerite and zones of its alteration according to SEM-EDX data, wt %.

Name	Si	Ti	Al	Fe	Mn	Mg	Ca	Ba	V	Ce	Y	Th	U	W	Pb	P	O	Total
brn-1	0.61	20.54	0.1	3.34			1.55		0.22	0	0.19	1.07	43.96		1.45		28.16	101.19
brn-2	0.74	20.62	0	3.2			1.4		0.22	0	0.29	1.12	43.27		1.42		27.85	100.13
brn-3	0.71	20.6	0.1	3.34			1.5		0.2	0	0.25	1.35	42.68		1.29		27.9	99.92
brn-4	0.43	20.4		2.8	0		1.48		0.2		0.2	0.51	46.14		1.58		27.47	101.21
brn-5	0.85	20		2.98			1.78		0		0.19	0.52	45.14		1.7		27.97	101.13
brn-6	0.84	20.44		3.13			1.72		0		0.27	0.53	44.26	0	1.55		28.35	101.09
brn-7	0.73	20.6	0.13	3.14			1.56		0.21		0	0	44.88		1.54		28.25	101.04
brn-8	0.73	20.68		3.43			1.57		0		0	0.63	43.72		1.39		28.43	100.58
brn-9	0.27	20.72		3			1.09		0.25		0	0.59	46.52		1.48		27.49	101.41
brn-10	0.66	20.12		3.15	0.16		1.74		0		0	0.68	45.27		1.45		27.53	100.76
brn-11	0.7	20.53		3.09			1.81		0		0	0.68	44.41		1.46		28.4	101.08
brn-12	0.56	20.46	0.11	3.08			1.67		0.23		0.19	0.67	44.59		1.48		27.99	101.03
brn-13	0.57	20.06		3.3			1.53		0.18		0.23	0.5	45.36		1.39		28.14	101.26
U-(Ti)-1	0.59	8.25		0.31			1.19	1.99	0.16	0.24	0.09	0.32	55.95		0.2		24.04	93.33
U-(Ti)-2	0.59	8.36		0.39	0.09		1.26	2.06	0.16	0.39	0	0.42	56.23		0.32		23.94	93.66
U-(Ti)-3	0.49	10.7		0.66			1.15	2.37				0.55	49.76		0.62		30.38	96.68
U-(Ti)-4	0.53	9.44		0.64	0		1.14	1.44	0.1	0.21		0.53	51.69		0.31		24.34	89.45
U-(Ti)-5	0.35	10.51	0.06	1.13			1.13	1.81	0.13	0.19		0.3	49.45	1.72	0.34		22.59	88.88
U-(Ti)-6	0.62	8.78		0.34			1.1	2.44	0.1	0.19		0.47	52.24	1.61	0.28	0.18	24.82	90.67
U-(Ti)-7	0.58	8.82	0.13	0.56			1.31	2.2	0			1.61	47.61	1.13	0	0.24	27.85	89.45
U-(Ti)-8	0.29	9.05		0.61			1.49		0		0	0	57.38		0		28.54	94.71
U-Ti-1	0.76	18	0.17	2.89			1.37		0		0	1.39	35.62	1.32	1.25		34.56	97.33
U-Ti-2	0.54	18.03		2.7			0.88		0		0	0.68	40.35		3.25		34.51	100.94
U-Ti-3	0.63	17.3	0.11	2.5			1.5		0		0	0.73	35.94				34.31	93.02
U-Ti-4	0.72	18.27		2.7			0.7		0			0.46	39.4		3.04		34.55	99.84
U-Ti-5	0.75	17.44		2.36			0.82		0			0.67	38.42		1.4		35.76	97.62
U-Ti-6	0.74	21.42		2.72	0.16		1.2		0		0	0.6	35.08		3.55		33.08	98.55
U-Ti-7	0.55	19.14	0.11	4.58	0.62	0.19	0.88	1.78	0.27		0	0.89	30.39	0	1.35		33.7	94.45
U-Ti-8	0.66	21.46		6.21	0.98	0.33	0.96	1.27			0	0.85	25.63		1.69		36.32	96.36
U-Ti-9	0.69	21.43		4.48	0.29	0.27	1.01	1.91	0.28				27.01		1.64		34.29	93.3
U-Ti-Pb-1	1.05	33.28		4.15	0.18		0.58		0			1.04	14.61		4.19		37.61	96.69
U-Ti-Pb-2	0.63	38.96		4.77	0.18		0.51					1.07	11.48		2.47		39.54	99.61
U-Ti-Pb-3	2.94	29.02	0.11	5.17	0.56		0.86	1.49	0.25	0.57		0.84	5.87		9.12	0.44	39.63	96.87
U-Ti-Pb-4	1.35	24.48		7.31	0.99	0.21	0.88					0.79	1.21		4.68	0.27	35.39	77.56
U-Ti-Pb-5	2.62	31.79	0.08	6.78	0.83		1.22		0	0.78		0.7	3.03		6.24	0.39	39.95	94.41
U-Ti-Pb-6	3.66	29.06	0.1	4.35	0.45		1.13	1.9	0.27	0.85		2.76	6.02		7.69	0.59	36.23	95.06
Fe-Ti	0.51	16.28	0.71	34.93	0.24	0	0.13		0.35				2.35	0	0.68	0.34	33.9	90.42
Ti-Fe	0.45	41.1	0.28	11.69	0.2		0.09	0	0.37			1.28	2.76	1.01	0.79	0.28	40.72	101.02
Ti-Fe	1.18	32.53	0.13	11.92	1.41	0.3	1.62		0.2	0.47		2.11	0.82		2.6	0.22	39.62	95.13

Note: brn—original brannerite; U-(Ti)—rounded segregations in brannerite with a higher content of U and a lower content of Ti; U-Ti и U-Ti-Pb–zone 1 and zone 2 of brannerite alteration with a lower content of U and a higher content of O, zone 2 is additionally characterized by a higher content of Ti and Pb; Fe-Ti and Ti-Fe are zones of change with a higher content of Fe or Ti.

Table 4. Statistics of recalculated composition (wt%) and formula of brannerite (apfu), n = 880.

	Min	Median	Max	Mean	S.D.
Wt.%					
UO_2	46.41	49.55	54.09	49.52	1.05
TiO_2	32.02	34.14	36.42	34.19	0.55
CaO	1.56	2.10	3.12	2.14	0.19
SrO	0.00	0.00	0.61	0.00	0.03
MnO	0.00	0.00	0.29	0.00	0.02
MgO	0.00	0.00	0.26	0.00	0.02
PbO	1.12	1.70	2.57	1.73	0.19
ThO_2	0.00	1.06	2.56	1.04	0.52
Y_2O_3	0.00	0.00	0.74	0.10	0.17
Ce_2O_3	0.00	0.00	0.59	0.01	0.06

Table 4. *Cont.*

	Min	Median	Max	Mean	S.D.
Al_2O_3	0.00	0.00	0.61	0.03	0.09
Fe_2O_3*	3.42	4.67	5.79	4.62	0.34
SiO_2	0.18	1.44	2.85	1.44	0.32
V_2O_5	0.00	0.00	0.60	0.06	0.14
ZrO_2	0.00	0.00	0.57	0.00	0.02
H_2O*	1.76	5.24	7.09	5.13	0.73
Apfu (6 O)					
U	0.68	0.74	0.80	0.74	0.02
Ca	0.11	0.15	0.22	0.15	0.01
Sr	0.00	0.00	0.02	0.00	0.00
Mn	0.00	0.00	0.02	0.00	0.00
Mg	0.00	0.00	0.03	0.00	0.00
Pb	0.02	0.03	0.05	0.03	0.00
Th	0.00	0.02	0.04	0.02	0.01
Y	0.00	0.00	0.03	0.00	0.01
Ce	0.00	0.00	0.01	0.00	0.00
Sum(A)	0.87	0.94	1.03	0.94	0.02
(excl. U)	0.16	0.20	0.26	0.20	0.01
Ti	1.64	1.72	1.80	1.72	0.03
Al	0.00	0.00	0.05	0.00	0.01
Fe	0.17	0.24	0.29	0.23	0.02
Si	0.01	0.10	0.19	0.10	0.02
V	0.00	0.00	0.03	0.00	0.01
Zr	0.00	0.00	0.02	0.00	0.00
Sum(B)	1.97	2.06	2.13	2.06	0.02
(excl. Ti)	0.24	0.33	0.44	0.33	0.03
Av. U valency	4.70	4.82	4.95	4.82	0.04
U^{4+}	0.36	0.44	0.51	0.44	0.02
U^{6+}	0.26	0.30	0.35	0.30	0.01
n(H_2O)	0.38	1.17	1.58	1.15	0.17

Note: H_2O recalculated for excess oxygen; S.D.—standard deviation.

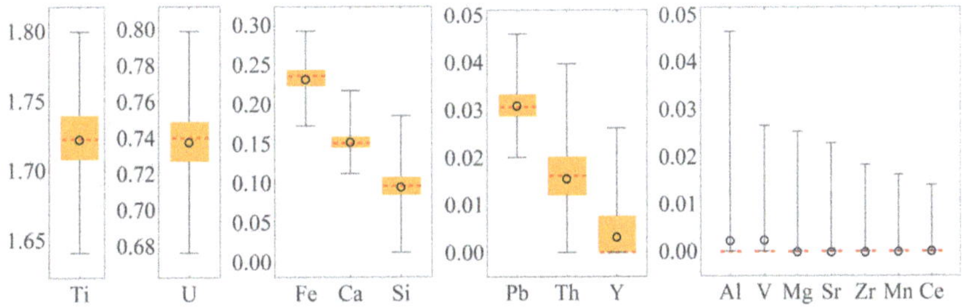

Figure 13. Distribution of brannerite components according to SEM-EDX data, apfu.

In brannerite, there are often areas of alteration, among which there are zones "U-Ti", "U-Ti-Pb", and "Ti-Fe". In most cases, alteration zones with a thickness of 2–3 to 10 μm are recorded in brannerite along the boundary between gold and brannerite (see Figure 12d–f). Depending on the intensity of alteration, there are zones of moderate (zone 1) and strong alteration (zone 2) of brannerite, which are well distinguished both in BSE images and in chemical composition (see Table 3). The areas of moderate alteration (zone 1) are usually located inside brannerite grains along cracks or along the boundary with gold. The elemental composition of this zone is characterized by a lower content of U and a higher content of O compared to the original brannerite (see Table 3, group "U-Ti"). The areas of strong alteration (zone 2), on the one hand, inherit the features of the location and composition characteristic of zone 1, and on the other hand, receive new features. Such areas of strong alteration occur either in veinlets inside grains, usually closer to the periphery, or form rims around grains. The elemental composition of zone 2 is characterized by the lowest content of U and the highest content of O compared with the original brannerite and zone 1. In addition, the content of Ti, Pb, and Fe increases (see Table 3, group "U-Ti-Pb"). Another type or direction of change in the original brannerite substance, possibly reflecting its replacement by oxides or hydroxides of Fe and Ti, is detected both inside brannerite grains and rims (see Table 3, groups "Fe-Ti" and "Ti-Fe"). Thus, in the environment of brannerite, three or more types of alteration sites can be distinguished, differing both in the direction of removal or concentration, and in the list of elements that change the content (U, V, Ca, Ti, Si, Fe, Mn, Ba, Pb, Y, P, O) (Figure 14).

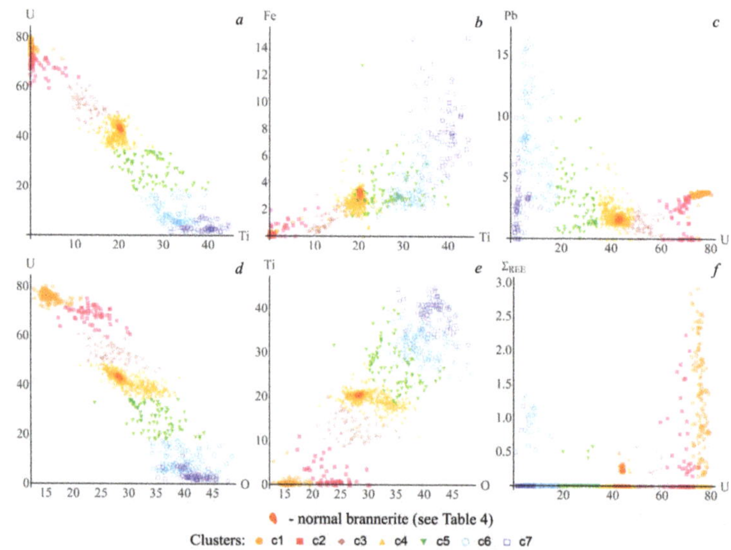

Figure 14. Scatterplots of major chemical elements (**a–f**) in wt.% with the results of K-mean cluster analysis for data on the elemental composition of brannerite, uraninite and areas of alteration. c1 and partially c2—cluster including unaltered and slightly altered uraninite; c3–c7—clusters including altered brannerite: c3 and partially c2—inclusions in brannerite U-Ti gel, see Table 3, group "U-(Ti)"; c4 and partially c5—zone of slight alteration of brannerite, see Table 3, group "U-Ti"; c6 and partially c5—zone of intense alteration of brannerite, see Table 3, group "U-Ti-Pb"; c7—see Table 3, groups "Fe-Ti" and "Ti-Fe". The red marker highlights the brannerite compositions from Table 4.

3.2.3. Uraninite

Uraninite (UO_2) is rare in the studied GBN samples, with the exception of uraninite structures in brannerite in the form of accumulations of nano- and microparticles (from tens and hundreds of nanometers, up to 1–2 μm), as well as dendritic crystals (see Figure 12h,i). Grains of uraninite occur as inclusions in brannerite or in native gold. Uraninite is represented by two varieties (Figure 15). Some uraninite grains correspond in composition to a mixture of UO_2 and UO_3 (Figure 15a). In addition to U and O, Pb, Th, REE, Ca, Ti, Si, and sometimes Fe are constantly present in the composition (Table 5). Judging by the fact that no micro- and nano-inclusions were found in such uraninite, it can be assumed that all the Pb in it is a product of the radioactive decay of uranium. And another part of the uraninite grains, according to the analysis, contains a significant amount of excess oxygen (Figure 15c,d). Such oxidized or hydrated uraninite can be attributed to nasturan (pitchblende, oxidized uraninite) [29]. Morphologically, nasturan is presented in two varieties. The first is individual rounded, isometric or slightly elongated "ovoids" of nasturan (from 2.6 × 3.6 μm to 3.7 × 4.7 μm) with an inhomogeneous internal structure. The second is crystals resembling ordinary uraninite. In some cases, both varieties of uraninite can be observed in the same grain. In this case, oxidized uraninite forms rims around ordinary uraninite (Figure 15b). A notable difference of this hydrated uraninite is the almost complete disappearance of Pb from its composition and the appearance of K in some analyses.

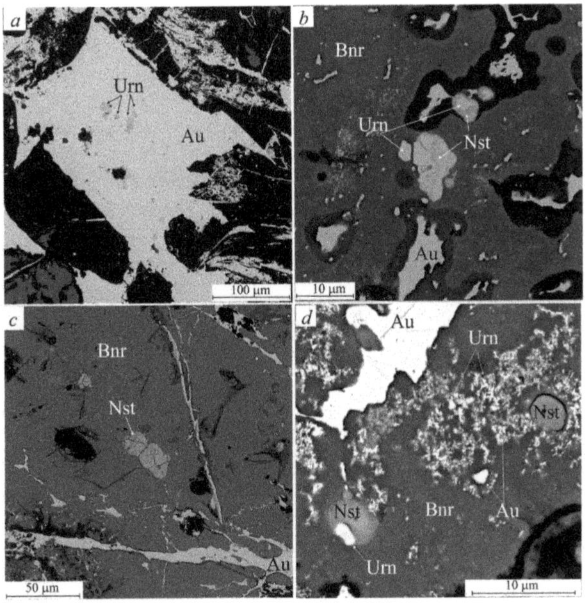

Figure 15. BSE-images of uraninite: inclusions of uraninite inside native gold (**a**); inclusions of "hydrated uraninite" in brannerite (**b**); uraninite with "hydration" rims (**c**); "ovoids" of nasturan in brannerite (**d**).

Table 5. Elemental composition of uraninite by SEM-EDX data, in wt.%.

Name	Si	Ti	Fe	Ca	K	Ce	La	Dy	Y	Th	U	Pb	O	Total
u1-1	0.15	0.34	0	0.35		0.44	0.49	0	0.27	0.78	76.78	3.77	14.18	97.55
u1-2	0.13	0.34	0	0.4		0.5	0.49	0	0.24	1.27	76.41	3.62	14.65	98.05
u1-3	0.13	0.36	0	0.52		0.39	0.44	0	0.31	2.79	74.49	3.67	14.49	97.59
u1-4	0.14	0.34	0.07	0.53		0.4	0.44	0.18	0.35	0.49	76.68	3.68	14.8	98.1
u1-5	0.14	0.36	0.1	0.46		0.35	0.41	0	0.29	0.78	76.82	3.78	14.45	97.94
u1-6	0.13	0.35	0	0.13		0.36	0.41	0	0.3	0.47	77.88	3.76	14.13	97.92
u1-7	0.14	0.32	0	0.42		0.44	0.43	0	0.26	3.74	74.74	3.73	14.09	98.31
u1-8	0.13	0.34	0	0.32		0.44	0.45	0	0.27	3.64	75.58	3.77	14.58	99.52
u1-9	0.14	0.34	0	0.23		0.39	0.41	0	0.27	0.62	78	3.66	14.84	98.9
u1-10	0.12	0.32	0.08	0.2		0.36	0.45	0	0.29	0.57	78.35	3.78	14.55	99.07
u1-11	0.11	0.36	0	0.18		0.31	0.46	0	0.31	0.49	78.33	3.81	14.4	98.76
u1-12	0.16	0.35	0	0.07		0.41	0.49	0	0.25	2.08	77.06	3.69	14.13	98.69
u1-13	0.13	0.36	0.08	0.59		0.35	0.35	0.17	0.53	2.41	74.67	3.67	14.51	97.82
u1-14	0.12	0.35	0.07	0.42		0.43	0.44	0	0.46	2.23	76.04	3.72	14.75	99.03
u1-15	0.19	0.35	0.09	0.05		0.36	0.43	0	0.27	2.05	77.08	3.73	14.13	98.73
u2-1	0.1	0.49	0.1		1.67	0.33				1.55	68.88	0	22.86	95.98
u2-2		0.43	0.08			0.27				1.35	71	0	21.2	94.33
u2-3	0.11	0.48	0.09			0.32				1.37	70.67	0	26.58	99.62
u2-4	0.07	0.41	0.1		1.92	0.29				1.19	72.89	0	22.42	99.29
u2-5	0.07	0.35	0		1.88	0.26				1	70.31	0.13	20.09	94.09

Note: u1—original uraninite; u2—nasturan.

3.2.4. Rutile

Another important GBN mineral is rutile (see Figure 10), containing in its composition iron (from 0–0.5 to 4.95 wt.% FeO_{tot}) and tungsten (max 10.51 wt.% WO_3). Rutile occurs as xenomorphic grains in gold or as elongated gold–rutile intergrowth with signs of their replacement by hematite (see Figures 10b–d and 12g). The association of rutile with gold is observed in various samples, both at the level of large aggregates and in the form of microinclusions of gold less than 1 μm in size in rutile grains. According to the ratio of elements, two types of rutiles are distinguished: without W and W-containing, in the latter, W, together with Fe, replaces Ti (Figure 16). High-fineness gold is associated with W-containing rutile (from 6 to 11 wt.% WO_3). Rutile with lower W contents (from 0 to 3 wt WO_3) is associated with gold, having a composition of 85–85 wt.% Au. High-contrast SEM images (Figure 17) clearly show that rutile with low tungsten content replaces W-containing rutile, primarily in the peripheral parts of grains, along cracks, along cleavage, and also grows on it. Grains with a decomposition structure (Figure 17a), were also found, with lamellae from 70 to 250 nm thick of W-bearing rutile in rutile with low contents of W, confined to the cleavage directions of the mineral according to the "sagenite" type [30]. In isolated cases, V is detected in rutile up to 0.88 wt.% V_2O_3.

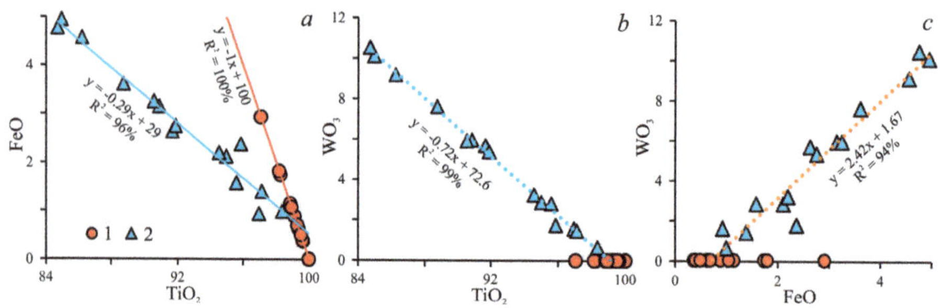

Figure 16. Scatterplots of major oxides of rutile (**a**–**c**) in wt.% showing ratios of TiO_2, FeO and WO_3 in two types of rutile (1—with W; 2—without W) from the gold–brannerite nuggets of the OOC.

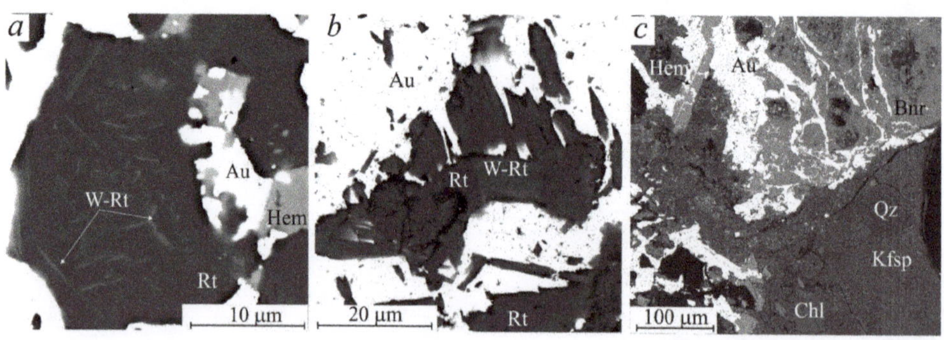

Figure 17. BSE-images of rutile and contact plot of GBN with hosted rock. Grain with a decomposition structure (**a**); rutile with low tungsten content replaces W-containing rutile (**b**); intergrowth brannerite with K-feldspar (Kfsp)-chlorite (Chl)-quartz (Qz) (**c**). Image (**a**) from [7].

3.2.5. Barite

Quite often, GBN contains barite in the form of crystals (see Figures 10c,d and 12f,g), rims along the boundary of brannerite and gold, microveins and clusters in gold, less often as isometric inclusions (up to 10 μm) in brannerite. The appearance of barite in brannerite is accompanied by a concentrically zoned rim (zone) of brannerite alteration and its transition to leucoxene (Figure 12g). Barite does not contain impurities, and Sr (less than 1 wt.% SrO) is detected only in barite inclusions in brannerite. At the same time, the material in the zonal rim is enriched in Pb (up to 10.9–19.8 wt.% PbO), Te (up to 6.3–12.2 wt.% TeO_3), and P (up to 1.2–5.9 wt.% P_2O_5).

3.2.6. Hematite–Magnetite

The mineral composition of GBN contains iron oxides in various forms, which make up to 5%–10% of the volume of individual nuggets. The Fe to O ratio suggests the presence of hematite, magnetite, goethite, and siderite in GBN. Hematite is the most common iron oxide in GBN, judging by the tabular forms of crystals, grouped in parallel or in the form of radially radiant sometimes curved aggregates, color and reflectivity under a microscope (Figure 9). Moreover, hematite is of great importance in the localization of gold, judging by their mutual intergrowths (Figures 9f and 10a,c,f). In many cases, altered brannerite is replaced by hematite (see Figure 12e). According to Raman spectroscopy, hematite is the most abundant in GBN. Raman spectra of the hematite are shown in Figure 18a (curve 1–3). The characteristic bands for identification of hematite are 225, 290–300, 412, 612 and first- and second-order longitudinal optical (LO) bands at ~660 and ~1320, respectively [31–36]. Some Raman spectra exhibit the intensive band of magnetite at 667 cm^{-1} along with an increase in hematite bands of widths (Figure 18, curve 1). Magnetite is characterized by having a main broad band ~668 cm^{-1} and the bands at 540 cm^{-1} and 300 cm^{-1} [32,34]. This bands are present in spectrum, the band at ~300 cm^{-1} as the shoulder (Figure 18, curve 1).

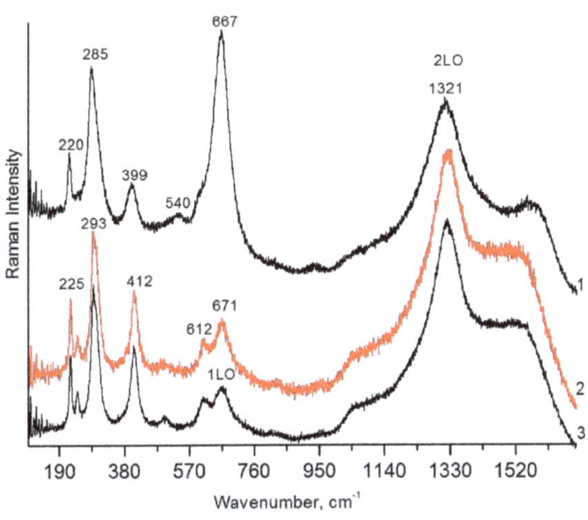

Figure 18. Raman spectra of hematite.

The SEM-EDX data also indicate the presence of magnetite in GBN. Ti magnetites stand out (from 1.56 to 4.94 wt.% Ti), with an insignificant Ti impurity (from 0.11 to 0.72 wt.% Ti) and without it. Approximately one third of the analyzed magnetites contain up to 0.36 wt.% Al. In isolated cases, magnetites contain Co (0.3 wt.%) and W (up to 1.38 wt.%). In hematites, only Ti was found as an impurity from below the detection threshold to 1.61 wt.% Ti. Siderites constantly contain Mn (1.25–6.15 wt.%), Si (0.83–2.25 wt.%), Ca (0.12–0.52 wt.%), P (0.15–0.52 wt.%), as well as Mg and Al in concentrations less than 1 wt.%.

3.2.7. Goethite

Iron hydroxide is represented by goethite and occurs primarily on the surface of GBN and in large (more than 1 mm wide) cracks in weakly altered brannerite (Figures 7a,b,h,i and 10d). The cracks are filled with highly altered aggregates brannerite grains, hematite, rutile, organic matter, goethite, and are cemented with gold. The microtexture inside such cracks can be characterized as porphyroblastic. Brannerite grains have rounded edges and a large number of gas cavities around which leucoxene is formed. Goethite develops after leucoxenized brannerite and rutile. A characteristic feature of iron hydroxides is the constant presence of Mn from 0.5 to 5.2 wt.%, Si from 0.8 to 1.5 wt.%, Ca 0.1 to 0.37 wt.%, P from 0.2 to 0.64 wt.%, Al up to 0.71 wt.%, and Mg up to 0.57 wt.%. In addition, in some locations, Cu up to 2.19 wt.%, Zn up to 0.45 wt.%, and Pb up to 0.75 wt.% are found in the composition of iron hydroxides. The connection of gold with iron hydroxides can be called "relict" inherited from substituted minerals.

3.2.8. K-Feldspar, Quartz and Chlorite

At the edge of one of the GBNs (C6), a contact was found made by ferruginous chlorite (chamosite) with a silicate rock. The rock consists of xenomorphic grains, about 100 microns in size, quartz and K-feldspar and chamosite, which fill the intergranular space (Figure 17c). Gold nanoinclusions are observed among the chlorite. The composition of feldspar fully corresponds to K-feldspar with low contents of Na_2O, FeO_{tot} and BaO (Tables 6 and 7).

Table 6. Chemical composition (wt%) of K-feldspar and quartz from the silicate rock on the contact with GBN.

Sample	Point	SiO_2	Al_2O_3	FeO_{tot}	Na_2O	K_2O	BaO	Total
C6-04	317	65.49	18.71	0.58		16.15	0.56	101.48
C6-04	322	68.01	17.72	0.26	0.19	15.92		102.10
C6-04	325	66.37	17.04	0.26	0.19	15.37		101.23
C6-04	315	99.03	1.21	1.25		0,23		101,72
C6-04	326	100.06	0.70	0.41		0.23		101.40

Note: no data—below detection limit.

Table 7. Chemical composition of chlorites from the silicate rock on the contact with GBN, at the rate of 12%-OH.

S_P	SiO_2	TiO_2	Al_2O_3	FeO_{tot}	MnO	MgO	CaO	K_2O	ZnO	P_2O_5
C6_313	26.02	0.22	17.77	35.61	0.26	2.58	0.2	4.04	0.62	0.69
C6_316	29.48	0.3	19.27	33.54	0.42	1.02	0.7	2	0.31	0.86
C6_318	31.42	0.29	18.31	25.14	0.37	8.11	0.3	2.97	0.68	0.29
C6_323	30.45		17.7	26.47	0.45	10.83	0.22	0.97	0.49	0.32
C6_323	30.9		18.38	24.44	0.44	12.12	0.24	1.11	0.27	

Note: S_P—sample_point.

3.2.9. Organic Matter

In one of the nuggets, there is a crack 0.5 to 1 mm wide, more than 10 mm long, crossing the entire volume of the sample. The crack with "torn" edges is made by a mineral aggregate cemented with native gold and consisting of fragments of leucoxenized brannerite, columnar crystals of hematite, rutile, goethite, and isometric structures (up to 1 mm) of organic matter. Raman spectra confirming the presence of organic matter in leucoxenized brannerite and in normal brannerite (in a small amount) are shown in Figure 19 [37–39]. The broad shoulders around the 1320 cm^{-1} band at ~1140 cm^{-1} and ~1520 cm^{-1} are also assigned to an organic matter in the hematite Raman spectra (Figure 18) [37]. According to isotopic analysis, it was determined that the organic matter has $\delta^{13}C$ −23.6‰.

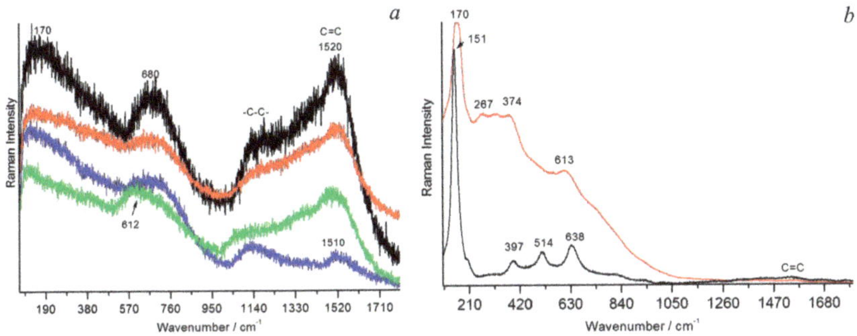

Figure 19. Raman spectra of an organic matter in leucoxenized brannerite (**a**) and normal brannerite (**b**).

4. Discussion

As a result of studies in the alluvial placer of the Kamenny stream of the OOC, four types of primary gold were identified, which differ significantly in their characteristics and, probably, have different primary sources (Table 8). The most common composition of gold corresponds to fineness 676–774‰. This is fully comparable with the fineness of gold (600–880‰), which prevails at the Nazarovsky skarn gold–zinc deposit. [40]. The Nazarovskoe deposit has a powerful oxidation zone, where gold is constantly present. Ores similar to those found at the Nazarovskoe OOC deposit are the primary source of high-grade and relatively low-grade gold in the placer of the Kamenny stream, which is confirmed by some features of the ore (sphalerite inclusions, composition, morphological relics) [5]. Low-depth gold–silver mineralization can be the root sources of low-grade gold, as evidenced by the significant heterogeneity of the composition of gold and the predominance of silver in it. Alloy heterogeneity is present as microfabrics formed either during primary mineralization or by modification of pre-existing alloys by chemical and physical drivers during subsequent residence in either hypogene or surficial environments [41,42]. Veins of chalcedony quartz with finely interspersed pyrite and gold–silver mineralization were established by G.I. Doroshkevich [43] in the zone of tectonic contact of the intrusion of eruptive breccia's of trachyriolites with Lower Cambrian limestones in the northwestern frame of the Ozerninsky ore cluster. The cuprous gold allow us to speak about the possibility of the existence of another primary source, not related to those mentioned above, which are characterized by native gold without copper impurities. The spatial and genetic relationship of cuprous gold with zones of secondary metasomatic alteration of basic and ultrabasic rocks is known [44,45]. Similar formations on the territory of the OOC are still unknown.

Table 8. Possible sources of various types of alluvial placer gold.

Type	I	II	III	IV
fineness	798–945	676–764	445–648	885–946
Possible source	Nazarovskoe gold-zinc skarn deposit		Veins of chalcedony quartz with gold-silver mineralization	Serpentinization zones of basic and ultrabasic rocks

Two exogenous events were established in the OOC area, which influenced the formation of rims of high-fineness gold on NG. One of these events was the formation of an areal weathering crust, which was especially fully manifested in the form of an oxidation zone at the Ozernoye pyrite–polymetallic deposit. The second is the alluvial transport of gold and the formation of a placer. Placer gold is characterized by a wide distribution of high-fineness rims with contrasting boundaries and a rare presence of high-fineness rims with diffusion boundaries, which may indicate a significant duration of gold residence in weathering crusts compared to its presence in alluvium [46]. The association of gold and uranium is well known, rather widespread and characteristic of several types of gold-uranium mineralization [47–51]: iron-oxide-copper-gold-uranium (IOCGU)-Ag deposits and most notably the Olympic Dam in South Australia, associated with hematite breccias, 2.57 to 1.0 billion years old; deposits of the "unconformity" type, Paleo-Meso-Proterozoic age (Alligator Rivers, etc.); Au-U deposits in zones of potassium metasomatism located in areas of tectonic-magmatic activation (TMA) of the Precambrian shields (Elkonian type); and deposits in zones of low-temperature Na-metasomatosis in the TMA structures of folded areas (Shinkolobve and others) (Table 9). In the ores of the listed types of deposits, brannerite is one of the main minerals of uranium, together with uraninite and coffinite. The mineral assemblages of native gold with brannerite are known in such deposits as the Witwatersrand, Blind River and smaller gold–uranium deposits [52–55]. In Russia, a similar gold–brannerite association composes the unique Central Aldan gold–uranium ore region [56]. Here, within the Elkon horst, two types of complex ores have been identified: gold-bearing brannerite mineralization of the Elkon type and brannerite–silver–gold mineralization of the Fedorovsky type [49,50,56]. Gold–brannerite intergrowths are extremely rare and have been found only in ores of the Witwatersrand (South Africa) and Richardson (Canada) in the form of small crystals and grains [54,55]. P. Ramdor (1962) [57] gives a photograph of a polished section (see Figure 288 in p. 332 from [57]) from the Bou-Azzer deposit (Southern Morocco), which shows "cataclastic brannerite is cemented by much gold with a little galena and quartz". The formation of the gold-brannerite association at the Bou Azzer deposit (the polymetallic Bou Azzer Co–As–Fe–Ni($\pm$Ag $\pm$ Au) district) occurred at the early pre-arsenide gold-bearing stage of the hydrothermal system [58]. Also gold–brannerite assemblage is an intermittent feature of gold from various localities [59]. Because of their size, the gold–brannerite nuggets found in the placer of the Kamenny brook are unique in comparison with all previously encountered and described ones [6,7]. GBN is generally characterized by cataclastic structures with a wide variety of microstructures: porphyroclastic, cementitious, reticulate, lattice, zonal, lattice, spotted, radially radiant, and others. The native gold found in GBN has wide variations in composition (from 750 to 1000‰), and at least five types are distinguished: 750–800‰; 850–880‰; 880–920‰; 930–960‰, and 980–1000‰ (Figure 11). NG is found in association with various minerals: uraninite, rutile, brannerite, hematite, barite, carbonaceous matter, goethite. The study of the relationship between the composition of native gold and the associations surrounding it, as well as the identification of the physicochemical conditions of formation, require further research.

Table 9. Types of Au-U mineralization (geological conditions of formation, mineral associations and metasomatic alterations).

Deposit/Ore Mineralization Type	Age, Geochemical Association	Geologic Setting	Secondary Alteration	Mineral Type of the Ores	Deposits, Ore District
Unconformity related deposits: (a)-Fracture controlled; (b)-Clay bounded; (c) endogenic, in zones of structural-stratigraphic unconformities; (d) Proterozoic base overlain by Mesozoic cover	Paleo-, Mezo-, Proterozoic (1700–1600 million years; 1600–900 million years) U, Au, Ni, Cu, Ag, As, Co, Pt, Pb	Paleoproterozoic metamorphic base (pelites, arcoses, metamorphosed in amphibolite facies carbonates); gneisses with graphite, often with weathering crust; overlain by Mesoproterozoic sediments; rocks deformed and brecciated: (a) ores in basement; (b) in overlying deposits	Chloritization, argillitization, seritization, carbonatization	Ore mineralization in breccias, veinlike breccias, disseminated in shales: Uraninite (Urn), Nasturan, rare coffinite (Cof), brannerite (Bnr), organic-uran minerals, gold, pyrite (Py), chalcopyrite (Ccp), graphite (Gr).	Athabasca Basin: McArthur River, Rabbit Lake, Eagle Point, McClean Lake, Dominique-Peter in Canada; Jabiluka, Ranger, Nabarlek, Koongarra (the Alligator River region) in Australia; Kuranah

Table 9. *Cont.*

Deposit/Ore Mineralization Type	Age, Geochemical Association	Geologic Setting	Secondary Alteration	Mineral Type of the Ores	Deposits, Ore District
«Brecciated» Hematite breccia complex deposits (IOCGU): (a) Granite-rich breccias; (b) Hematite-rich breccias	From ~2570 to 1000 million years. U, Au, Cu, Ag, REE, Fe.	They are found in a number of different tectonic settings (rift, subduction zones, basin collapse)	Sericitization, chloritization, hematitization	U-Au-Cu granite-derived matrix; hematite-quartz breccias, hematite breccias; Au, nasturan, bornite (Bn), chalcopyrite (Ccp)	Olympic Dam South Australia; Prominent Hill, Emest Henry, Starra, Osborne in Australia; Candelaria, Salobo, Sossego in South America; Michelin, Sue-Dianne in Canada, Igarapé Bahia Deposit (Brazil)
Deposit/ore mineralization type	Age, geochemical association	Geologic setting	Secondary alteration	Mineral type of the ores	Deposits, ore district
Endogenic in areas of tectonic-magmatic activity of the Precambrian craton Metasomatic U deposits: (a) alkaline metasomatites of sodium series; (b) potassium (elkonites) series; (c) albitization in black shales (shungites)	a–1900–1600 million years: U, Au, Ag, Th, Mo; b–155–123 million years: U, Au; c-Au, PGE, U, V?	a-stockworks and lenses in granites, migmatites and pegmatites; b–tabular-,veinlike-, lenslike-form body in crystalline schists, migmatites, granites; c–albitization zones in shungites	a-albitization, amphibolization, pyroxenitization (aceites); b–kalishpatization (elconites); c-albitization	a–U, Au, Ag, Th-U brannerite-uraninite, coffinite-brannerite; b–Au-U-brannerite, rare Au-U-uraninite	a–Yuzhnoye, Lozovatskoye (Ukraine), Rossing (Namibia), Itatiaia (Brasil), Kokchetavskiy massif; b-Druzhnoye, Kurung, Snezhnoye (Elconskiy ore district); c–Sr.Padma (Onezhskiy trough Baltic craton)
Endogenic in areas of tectonic-magmatic activity of the folded regions. Vein deposits (granite related deposits)	U, Au, Mo, Th, Zr, P_2O_5	Column-shaped, lens- and veinlike deposits in sandstones, carbon shales, diabases, granites and limstones controlled by faulded and crushed zones	Low-temperature Na-metasomatism zones; lightening in black shales	U, P-U, Mo-U; arshinovite-molybdenite-brannerite-nasturan; F-apatite-uraninite	Manybayiskoe, Zaozernoe (Kazakhstan), Shinkolobwe (DR Congo)
Deposits in multiple phase alkaline intrusions Intrusive deposits	Au, U, Cu, Nb, REE, Fe	Disseminated uranothorianite occurs in cupriferous carbonatite complexes.	Sulfidization	REE, U, Th, Au	Palabora (South Africa)
Au-U conglomerates Quartz pebble conglomerate deposits	3070–2200; 1900 million years Au, U, REE	Superimposed troughs in the marginal parts of ancient platforms; oligomictic quartz-pebble conglomerate horizons (termed reef), from 0.5 to >3.5 m thick	Sericitization, chloritization, pyritization	Au-U, Au, U-uraninite, PGM, uranthorite, uranthorianite, monazite, and xenotime are prevailing	Witwatersrand Basin, Jacobina (Brasil), Tarkva (Gana), Blind-River (Canada).

Turani et al. [60] presented data on the composition of brannerite from five hydrothermal and five igneous (pegmatite) ore deposits, and some of these deposits have gold-uranium mineralization (Bou Azzer deposit and Mont Chemin mines). A comparison of these data [60] with the compositions of normal brannerite from GBN OOC is shown in Figure 20. It is noticeable that brannerite from GBN samples are characterized by the lowest values in terms of Ti content and are relatively low in Y and, on the contrary, are among the highest in terms of Pb and Fe. According to the content of U, GBN samples are not distinguished with others, and according to the content of Si, an increased scatter of contents is observed.

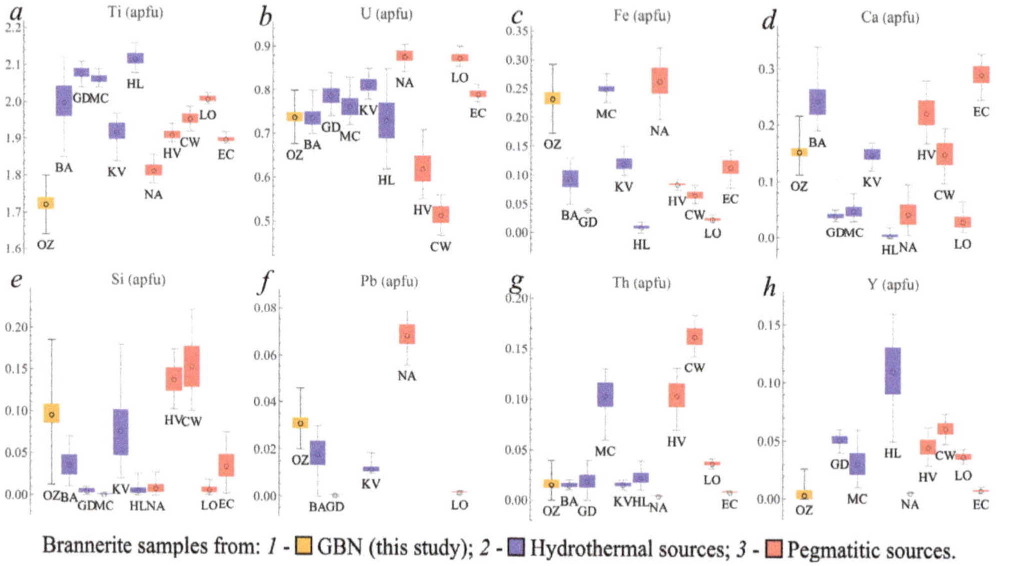

Figure 20. Comparison major (**a**,**b**) and minor (**c**–**h**) element contents for brannerite sample from GBN (1: OZ—Ozerninsky Ore Cluster) with brannerite samples from hydrothermal (2: BA—Bou Azzer, GD—La Gardette, MC—Mont Chemin, KV—Kratka Valley, HL—Himalaya) and pegmatitic (3: NA—Namibia, HV—Hidden Valley, CW—Crocker's Well, LO—Lodrino, EC—El Cabril) sources [60]; distribution of element contents was restored by values of Min, Max, Mean, and S.D. [60].

The data obtained on the composition and relationships of GBN minerals in the OOC area indicate a complex history of their occurrence, which can be represented as at least three stages. The first is the formation of early quartz–nasturan–gold (composition 94–95 wt.% Au)–W-rutile–magnetite assemblage. There are preliminary data on determining the uranium–lead age of uraninite microcrystals by the chemical method [61], indicating its Middle-Late Paleozoic age. The study of impurity elements in rutile enables distinction of rutile in metamorphic-hydrothermal and hydrothermal deposits from rutile in magmatic-hydrothermal deposits and magmatic environments [62]. Rutile from mesothermal and related gold deposits invariably contains W, and in some of the larger and more important deposits, rutile also contains Sb and/or V. Tungsten-bearing detrital rutile grains in the Witwatersrand deposit suggest that there was probably a paleo placer with a mesothermal gold source. Cu-Au deposits tend to contain elevated W and Cu (and sometimes V). The Olympic Dam Au-Cu-U deposit hosts rutile that is enriched in W, Sn and Cu [63]. The second is the crystallization of brannerite and the replacement of an earlier pitchblende by brannerite. The age of brannerite, determined from the data of chemical analysis of the WDX and EDX with a point probe and EDX in a small raster, varies in the range of 200–235 Ma, i.e., corresponds to the Late Triassic (T_3)–Early Jurassic (J_1) time period [7]. The third is the formation of a hematite–barite–rutile–gold association as a result of deformation-hydrothermal processes, which is associated with the appearance of zones of alteration (leucoxenization) in brannerite at contact with native gold with 8–15 wt.% Ag. Fluid separation probably occurred in parallel, which also led to the replacement of brannerite by Fe and Ti hydroxides. The fourth is hypergene or low-temperature hydrothermal alteration of minerals of early stages with the development of iron hydroxides (goethite) with impurities of manganese, tellurium, arsenic, phosphorus and other elements. This stage is associated with the appearance of microveins of very high-fineness gold in lower-fineness gold. Thus, the data obtained indicate a multistage, possibly polygenic, and probably polychronous formation of GBN gold–uranium mineralization.

The relationship of native gold with brannerite shows that it is localized not only along cracks that cross brannerite, but also forms structures of joint precipitation from solutions with it. Moreover, there is no zone of brannerite alteration around isometric micrograins of such gold, in contrast to gold located in cracks, which may indicate different ways of "inflow" of gold into brannerite. The presence of idiomorphic uraninite in native gold makes it possible to consider the possibility of joint transfer of gold and uranium. Intensive dissolution and redeposition of the material of experimental ampoules—gold was found during experimental studies of the Au-UO_2-TiO_2-water fluid system, which were carried out in order to determine the solubility of uraninite in hydrochloric acid solutions at 500 °C, 1000 bar. Analysis and modeling of the discovered phenomenon showed that the redox couple UO_{2+x}/UO_2 oxidizes Au(cr) to Au+(aq), which is then reduced under the action of stronger reducing agents [64,65].

The $\delta^{13}C$ of organic matter from a crack in GBN with an aggregate of leucoxenization brannerite, hematite, and rutile is −23.6‰, which indicates the participation of a biogenic carbon source. It is known that the presence of organic compounds in mineral assemblages and mineral-forming fluids is one of the characteristic features of hydrothermal ore deposits, which indicates the fundamental possibility of participation of organic matter in the transport and deposition of ore matter [66]. Carbonaceous compounds (graphite, anthraxolite, kerite, and others) have been found in many uranium and rare metal deposits, including in association with brannerite and gold. The formation of solid carbonaceous substances occurred from metal-bearing fluids at the end of the main stage of ore formation [67]. The presence of traces of magnetite in hematite from GBN, detected by Raman spectroscopy, i.e., the replacement of magnetite with hematite, may indicate an increase in the redox potential in the system. This may indicate the possibility of high-temperature oxidation of organic matter in the hydrothermal system, which could be accompanied by gold precipitation. The alteration brannerite at the contact with native gold, barite, and numerous "bubbles" of gas-in-fluid inclusions was probably due to an increase in the redox potential.

The results of mineralogical and geochemical studies and the age of GBN from OOC [7] make it possible to carry out a comparative analysis with gold–uranium mineralization of the Mesozoic age of the Central Aldan ore region [48,68] and to predict the development of the described mineralization not only within the OOC, but also the Kurbino–Eravninsky ore district. From the available literature data [69], it is possible to compare normal and altered brannerite in GBN from OOC and uranium–titanium-metagel (UTM) mineralization from Elkon ore district. According to Aleshin et al. [69], in addition to brannerite, the unique reserves of uranium in the Au-U deposits of the Elkon ore district are explained by polyphaser UTM (different groups of UTMs were designated as UTM-1, UTM-2/1, UTM-2/2). Moreover, according to these authors, brannerite, uranium oxides, and silicates make up an insignificant part of the ores. Our study also found that in GBN, along with ordinary brannerite and accessory uraninite, there is a significant proportion of altered brannerite developed along rims or a dense network of veinlets. On the Ti-U diagram, it is noticeable that the zone of weak brannerite alteration "U-Ti" described in this study can be correlated with UTM-1 and the zone of strong brannerite alteration "U-Ti-Pb" can be correlated with UTM-2/1 and UTM-2/2 (Figure 21).

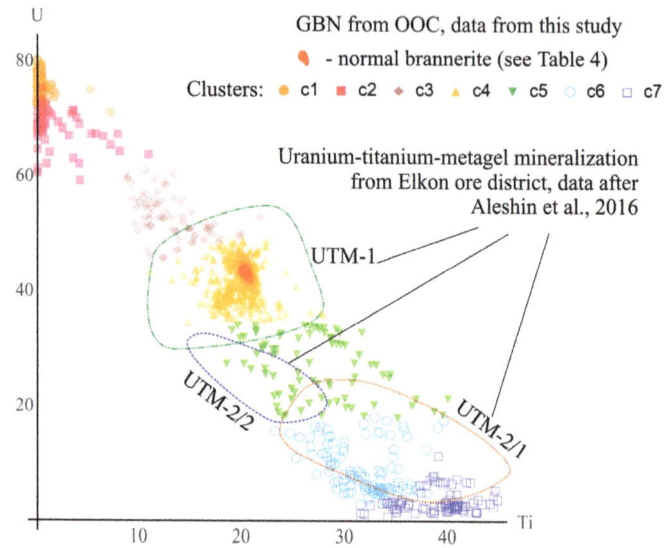

Figure 21. Comparison U-Ti mineralization in GBN from OOC with uranium–titanium-metagel mineralization from Elkon ore district [69]. Cluster denotations see in Figure 14.

In particular, the Yuzhnoe deposit in the Elkon gold–uranium ore region is controlled by a tectonic zone. The zone is formed by one or several dikes of metadiorites alternating with blastomylonites and blastocataclasites surrounded by quartz–K-feldspar metasomatites, chlorite-epidote propylites and beresites. Ores are represented by the gold–brannerite mineral type [47,70]. The GBN brannerite of the Kamenny stream of the OOC is comparable in composition to the brannerite of the Elkon region in composition, which are characterized by low thorium content and an admixture of rare earth elements. In these parameters, brannerite from the OOC and Elkon region differ sharply from brannerite from albitites of Ukraine and pegmatites and conglomerates of the Witwatersrand (South Africa). The results of the study of the GBN of the Kamenny stream of the OOC indicate a wide manifestation of cataclasis processes, with the appearance of a characteristic network and porphyroclastic, cataclastic textures. The discovered fragment of the host quartz–K-feldspar rock with chlorite (chamosite) at the contact with GBN is well correlated with quartz–microcline metasomatites of the Yuzhnoye deposit in the Elkon region.

In addition, another reason for predicting gold–uranium mineralization to a depth is the wide distribution of Mesozoic and younger uranium mineralization, manifested in the OOC area (Eravninsky potential uranium-ore region) and to the north of it (Vitimsky uranium-ore region). The mineralization described in this paper can also be considered as evidence of the existence of "multi-level" gold–uranium mineralization of various ages. The probability of the existence of multilevel uranium (gold–uranium) mineralization in the Vitim–Amalat ore region of Transbaikalia was indicated in the regional forecasting of large-scale mineralization of the "Olympic Dam"-type gold–uranium ores in hematite breccias [71].

An analysis of the composition of GBN rutile indicates the presence of rutiles typical of ore zones (Figure 22) [72,73].

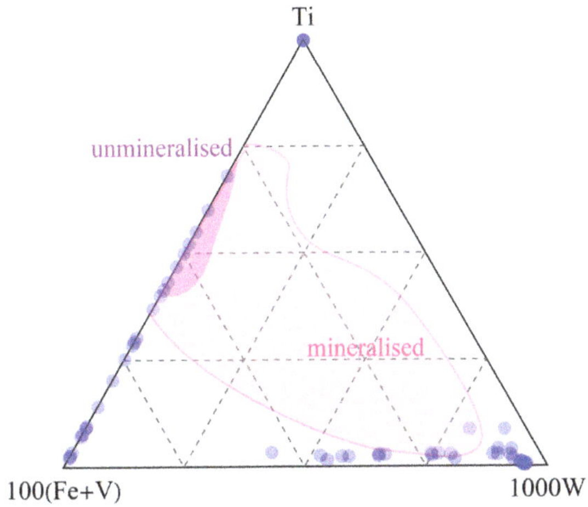

Figure 22. Plot of Ti-100 × (Fe + Cr + V) − 1000 × W for rutile from GBN.

The identified type of uranium ores can be considered as one of the possible sources of the formation of hydrogenic uranium ores in the supergenes zone of the Western Transbaikalia (Khiagda group of hydrogenic deposits). An important conclusion is also the fact that the processes of ore formation within the Ozerninsky ore cluster were diverse and continued for a long time from the Early Paleozoic to the Mesozoic, and the formation of placers continued in the Quaternary.

5. Conclusions

A comprehensive mineralogical and geochemical study of NG and GBN was carried out in the alluvial placer of the Kamenny stream of the OOC (Western Transbaikalia). The obtained data on the morphology, composition and relationships of minerals in NG and GBN testify to the complex history of their formation and various primary sources of gold. Separate areas of the placer, where predominantly unrounded and semi-rounded gold is present, can be considered as eluvial placers.

Based on a set of features, four types of primary NG and GBN were identified, which differ significantly in their characteristics, for which primary sources are predicted. The most common high-fineness and relatively low-fineness gold types are well compared with the gold of the Nazarovsky skarn gold–zinc deposit located in the OOC. For low-fineness gold, shallow gold–silver ores are assumed to be the primary source.

The presence of cuprous and copper-bearing gold in placers suggests the existence of ultramafic complexes with gold-bearing zones of listvenites and rodingites in the OOC territory.

In the OOC area, two exogenous events have been identified that influenced the formation of very high-grade NG: the formation of an areal weathering crust and the alluvial transport of gold and the formation of a placer.

The GBN found in the Kamenny Brook placer is unique in its size, variety of microtextures, and mineralogical features in comparison with all previously encountered descriptions of the co-occurrence of gold and brannerite.

The obtained data on the composition and relationships of GBN minerals in the OOC area indicate a complex history of their occurrence, which can be represented as at least three or four stages: 1—formation of early quartz–nasturan–gold–W-rutile–magnetite(?) assemblage; 2—replacement of earlier pitchblende by brannerite; 3—as a result of deformation-hydrothermal processes, the formation of a hematite–barite–rutile–gold assemblage, which is associated with the appearance of zones of alteration (leucoxenization) in brannerite at contact with gold; 4—hypergene or low-temperature hydrothermal alteration of minerals of early stages with the development of iron hydroxides (goethite) with impurities of manganese, tellurium, arsenic, phosphorus and other elements.

In the OOC area, there are indications that the composition of GBN and host quartz–chlorite–K-feldspar rocks corresponds to the Elkon-type deposits.

Author Contributions: Conceptualization, S.M.Z.; methodology, S.M.Z., N.S.K., T.N.M. and V.A.P.; software, D.K.B.; validation, S.M.Z., E.V.A., N.S.K. and D.K.B.; formal analysis, S.M.Z. and D.K.B.; investigation, S.M.Z., N.S.K., T.N.M., V.A.P., A.V.K. and D.K.B.; resources, A.A.M., A.V.K., S.M.Z. and O.N.K.; data curation, S.M.Z., E.V.A., A.V.K., O.N.K. and D.K.B.; writing—original draft preparation, S.M.Z., E.V.A., N.S.K., O.N.K., A.V.K., T.N.M., V.A.P. and D.K.B.; writing—review and editing, E.V.A., N.S.K., O.N.K., T.N.M., B.B.D., V.A.P. and D.K.B.; visualization, E.V.A., S.M.Z., T.N.M., A.V.K., B.B.D. and D.K.B.; supervision, S.M.Z.; project administration, E.V.A.; funding acquisition, B.B.D., S.M.Z., T.N.M. and V.A.P. All authors have read and agreed to the published version of the manuscript.

Funding: The work was supported by the Russian Science Foundation (grant no. 22-17-00106), thanks to which analytical studies, analysis and interpretation of the obtained data were carried out. Expedition work, research using Raman spectroscopy methods and measurement of $\delta^{13}C$ signatures was carried out with the support of the Ministry of Science and Higher Education of the Russian Federation (government task of the Institute of Geology and Mineralogy of the Siberian Branch of the Russian Academy of Sciences, projects No. 122041400193-7, No. 122041400171-5, 122041400243-9).

Data Availability Statement: The data presented in this study are available in article.

Acknowledgments: The authors would like to thank the Academic Editor and the Reviewers for valuable comments and suggestions that have improved this manuscript.

Conflicts of Interest: The authors declare no conflict of interest.

References and Note

1. Tarasova, R.S.; Babkin, I.N.; Bliznyuk, M.V.; Filko, A.S.; Gorbunov, S.V.; Alkalaev, B.K. *Lake Sulfide Lead-Zinc Deposit. Report for 1964–1969 with Inventory Count*; Buryatia Geological Government: Ulan-Ude, Russia, 1969; 432p. (In Russian)
2. Tarasova, R.S.; Bliznyuk, M.V.; Babkin, I.N. On the formational type and genesis of the Ozernoe lead-zinc sulfide deposit. In *Geology and Genesis of Endogenous Ore Formations of Siberia*; Nauka: Moscow, Russia, 1972; Volume 143, pp. 79–97. (In Russian)
3. Kalinin, Y.A.; Nesterenko, G.V.; Khlybov, V.P.; Zhmodik, S.M.; Kolpakov, V.V.; Popov, V.D. Typomorphism of native gold and its role in the search for non-traditional types of gold ore concentrations within the Eravna depression (Buryatia). In Proceedings of the Noble and rare metals of Siberia and the Far East, Irkutsk, Russia, 24–27 September 2005; pp. 139–142. (In Russian).
4. Nikolaeva, L.A.; Yablokova, S.V.; Shatilova, L.V.; Pozdnyakova, N.N. *Results of the Study of Typomorphic Features of Native Gold in Order to Determine the Primary Source of the Ozerninsky Ore-Placer Node Inform. A Note*; TsNIGRI: Moscow, Russia, 2008; 100p. (In Russian)
5. Kozlov, A.V.; Savichev, A.; Nikitin, D.V.; Pechenkin, M.M.; Alekhina, V.V.; Pokrovskaya, N.V.; Bogatyrev, L.I.; Shelekhov, A.N. *Report on the Results of Prospecting and Appraisal Work for Gold and Other Minerals within the Ozerninsky Ore Cluster and Adjacent Territory in 2008–2010*; SPGGU: St. Petersburg, Russia, 2010; Volume 1, 300p.
6. Mironov, A.G.; Karmanov, N.S.; Mironov, A.A.; Khodyreva, E.V. Gold-brannerite nuggets in placers of the Ozernoe ore cluster. *Russ. Geol. Geophys.* **2008**, *49*, 743–748. [CrossRef]
7. Zhmodik, S.M.; Belyanin, D.K.; Airiyants, E.V.; Karmanov, N.S.; Mironov, A.A.; Damdinov, B.B. Fe-Ti-Au-U-mineralization of the Ozerninsky ore node (Western Transbaikalia, Russia). *Dokl. Earth Sci.* **2022**, *507*, 1050–1056. [CrossRef]
8. Vasiliev, I.L. *Geology of the Eravna Ore Field*; Nauka: Novosibirsk, Russia, 1977; 126p. (In Russia)
9. Distanov, E.G. *Pyrite-Polymetallic Deposits of SIBERIA*; Nauka: Novosibirsk, Russia, 1977; 351p. (In Russian)
10. Distanov, E.G.; Kovalev, K.R. *Textures and Structures of Hydrothermal-Sedimentary Pyrite-Polymetallic Ores of the Ozernoye Deposit*; Nauka: Novosibirsk, Russia, 1975; 172p. (In Russian)
11. Kovalev, K.R.; Buslenko, A.I. *Hydrothermal-Sedimentary Ore Genesis and Polymetamorphism of Ores of the Ozerninsky Ore Node (Western Transbaikalia)*; VO "Nauka"; Siberian Publishing Company: Novosibirsk, Russia, 1992; 214p. (In Russian)
12. Nefediev, M.A. *Volumetric Model and Assessment of the Prospects of the Ozerninsky Ore Node Based on Geophysical Data (Western Transbaikalia)*; BSC SB RAS: Ulan-Ude, Russia, 2009; 184p. (In Russian)
13. Gordienko, I.V.; Bulgatov, A.N.; Minina, O.R.; Klimuk, V.S.; Vetluzhskih, L.I.; Sitnikova, V.S.; Goneger, T.A.; Ruzhentsev, S.V.; Nekrasov, G.E.; Metelkin, D.V.; et al. The late Riphean-Paleozoic of the Uda-Vitim island arc system in the Transbaikalian sector of the Paleoasian ocean. *Russ. Geol. Geophys.* **2010**, *51*, 461–481. [CrossRef]
14. Platov, E.V.; Ignatov, A.M.; Patrakhin, E.G.; Bolshakova, T.V.; Minina, O.R.; Lantseva, V.S. State geological map of the Russian Federation at a scale of 1:200 000. 2nd. Series Barguzino-Vitimskaya. Sheet N-49-XXVIII (Gunda). In *Explanatory Letter*; Cartographic Factory VSEGEI: St. Petersburg, Russia, 2016; 208p. (In Russian)

15. Vikentiev, I.V.; Damdinov, B.B.; Minina, O.R.; Spirina, A.V.; Damdinova, L.B. Classification of polymetallic mining processes and transitional VMS–SEDEX–MV type example of a giant Ozernony Deposit in Transbaikal, Russia. *Geol. Ore Depos.* **2023**, *65*, 201–236. (In Russian) [CrossRef]
16. Minina, O.R.; Gordienko, I.V.; Damdinov, B.B.; Tashlykov, V.S.; Goneger, T.A.; Skripnikov, M.S.; Lantseva, V.S.; Khubanov, V.B.; Kislov, E.V. New data on the age Ozernoe polymetallic deposi (Western Transbaikal). *Lithol. Miner. Resour.* **2023**, *3*, 299–314. (In Russian) [CrossRef]
17. Damdinov, B.B.; Vikentiev, I.V.; Damdinova, L.B.; Minina, O.R.; Zhmodik, S.M.; Sobolev, I.D.; Tyukova, E.E.; Spirina, A.V.; Izvekova, A.D.; Moskvitina, M.L.; et al. Problems of the genesis of deposits of the Ozerninsky polymetallic ore cluster (Western Transbaikalia, Russia). *Natl. Geol.* **2023**, *2*, 73–90. (In Russian) [CrossRef]
18. Tatkov, G.I.; Baderin, A.M. Structure and prospects of the Ozerninsky ore cluster according to the data of medium-scale modeling in a gravitational field. In *Geological and Genetic Models and Local Prediction of Endogenous Mineralization in Transbaikalia: Collection of Scientific Papers GIN BSC SB AS USSR*; OIGGM SB RAS: Novosibirsk, Russia, 1991; pp. 107–113. (In Russian)
19. Kozlov, A.V.; Pechenkin, M.M.; Savichev, A.A.; Bambaev, T.S. New features of deep structure Ozerninskiy ore cluster based on magnetotelluric sounding. *J. Min. Inst.* **2011**, *189*, 260–263. (In Russian)
20. Gordienko, I.V.; Nefed'ev, M.A. The Kurba-Eravna ore district of Western Transbaikalia: Geological and geophysical structure, types of ore deposits, predictive assessment, and mineral-resource potential. *Geol. Ore Depos.* **2015**, *57*, 101–110. [CrossRef]
21. Ruzhentsev, S.V.; Aristov, V.A.; Golionko, B.G.; Nekrasov, G.E.; Minina, O.R.; Larionov, A.N.; Lykhin, D.A. Geodynamics of the Eravna zone (Uda-Vitim fold system of the Transbaikal region): Geological and geochronological data. *Dokl. Earth Sci.* **2010**, *434*, 1168–1171. [CrossRef]
22. Ruzhentsev, S.V.; Minina, O.R.; Nekrasov, G.E.; Aristov, V.A.; Golionko, B.G.; Doronina, B.G.; Lykhin, V.A. The Baikal-Vitim Fold System: Structure and geodynamic evolution. *Geotectonica* **2012**, *46*, 87–110. [CrossRef]
23. Tsarev, D.I.; Firsov, A.P. *The Problem of the Formation of Pyrite Deposits*; Nauka: Moscow, Russia, 1988; 141p.
24. Litvinovsky, B.A.; Postnikov, A.A.; Zanvilevich, A.N.; Zeleny, E.N.; Zoricheva, L.L. New data on magmatism of the Ozerninsky ore cluster (Western Transbaikalia). *Russ. Geol. Geophys.* **1986**, *8*, 56–67. (In Russian)
25. Gordienko, I.V.; Tsygankov, A.A. Magmatism and ore formation in various geodynamic settings of the Sayano-Baikal folded region and adjacent territories. *Prospect. Prot. Miner. Resour.* **2017**, *9*, 36–44. (In Russian)
26. Moroz, T.N.; Edwards, H.G.M.; Ponomarchuk, V.A.; Pyryaev, A.N.; Palchik, N.A.; Goryainov, S.V. Raman spectra of a graphite nontronite association in marbles from Oltrek Island (Lake Baikal, Russia). *J. Raman Spectrosc.* **2020**, *51*, 1885–1893. [CrossRef]
27. Petrovskaya, N.V. *Native Gold (General Characteristics, Typomorphism, Problem of Genesis)*; Nauka: Moscow, Russia, 1973; 348p.
28. Colella, M.; Lumpkin, G.R.; Zhang, Z.; Buck, E.C.; Smith, K.L. Determination of the uranium valence state in the brannerite structure using EELS, XPS, and EDX. *Phys. Chem. Miner.* **2005**, *32*, 52–64. [CrossRef]
29. Dymkov, Y.M. *The Nature of Uranium Resin Ore. Issues of Genetic Mineralogy*; Atomizdat: Moscow, Russia, 1973; 240p. (In Russian)
30. Betekhtin, A.G. *Mineralogy*; State Publishing House of Geological Literature: Moscow, Russia, 1950; 956p. (In Russian)
31. Beattie, I.R.; Gilson, T.R. The single-crystal Raman spectra of nearly opaque materials. Iron(III) oxide and chromium(III) oxide. *J. Chem. Soc. A Inorg. Phys. Theor.* **1970**, 980–986. [CrossRef]
32. El Mendili, Y.; Bardeau, J.-F.; Randrianantoandro, N.; Gourbil, A.; Greneche, J.-M.; Mercier, A.-M.; Grasset, F. New evidences of in situ laser irradiation effects on γ-Fe_2O_3 nanoparticles: A Raman spectroscopic study. *J. Raman Spectrosc.* **2010**, *42*, 239–242. [CrossRef]
33. Hanesch, M. Raman spectroscopy of iron oxides and (oxy)hydroxides at low laser power and possible applications in environmental magnetic studies. *Geophys. J. Intern.* **2009**, *177*, 941–948. [CrossRef]
34. Shebanova, O.N.; Lazor, P. Raman study of magnetite (Fe_3O_4): Laser-induced thermal effects and oxidation. *J. Raman Spectrosc.* **2003**, *34*, 845–852. [CrossRef]
35. Almuslet, N.A.; Ahmed, M.M.; Hassen, S.H. Usage of laser Raman spectroscopy to indentify the unstable compounds of iron oxide. *Int. J. Curr. Adv. Res.* **2017**, *6*, 3470–3473. [CrossRef]
36. Marshall, C.P.; Dufresne, W.J.B.; Rufledt, C.J. Polarized Raman spectra of hematite and assignment of external modes. *J. Raman Spectrosc.* **2020**, *51*, 1522–1529. [CrossRef]
37. Moroz, T.N.; Edwards, H.G.M.; Zhmodik, S.M. Detection of carbonate, phosphate minerals and cyanobacteria in rock from the Tomtor deposit, Russia, by Raman spectroscopy. *Spectrochim. Acta Part A Mol. Biomol. Sppectroscopy* **2021**, *250*, 119372. [CrossRef] [PubMed]
38. Zhang, Y.; Karatchevtseva, I.; Qin, M.; Middleburgh, S.C.; Lumpkin, G.R. Raman spectroscopic study of natural and synthetic brannerite. *J. Nucl. Mater.* **2013**, *437*, 149–153. [CrossRef]
39. Lafuente, R.T.B.; Downs, H.; Yang, N. Stone, The power of databases: The RRUFF project. In *Highlights in Mineralogical Crystallography*; Armbruster, T., Danisi, R.M., Eds.; Walter de Gruyter: Berlin, Germany, 2015; pp. 1–30.
40. Kovalev, K.R.; Baulina, M.V.; Akimtsev, V.A.; Anoshin, G.N. Stratiform zinc-pyrite ores and gold mineralization at the Nazarovskoye deposit. *Russ. Geol. Geophys.* **2003**, *44*, 963–978. (In Russian)
41. Chapman, R.J.; Banks, D.A.; Styles, M.T.; Walshaw, R.D.; Piazolo, S.; Morgan, D.J.; Grim-shaw, M.R.; Spence-Jones, C.P.; Matthews, T.J.; Borovinskaya, O. Chemical and physical heterogeneity within native gold: Implications for the design of gold particle studies. *Miner. Depos.* **2021**, *56*, 1563–1588. [CrossRef]

42. Gammons, C.H.; Williams-Jones, A.E. Hydrothermal geochemistry of electrum; thermodynamic constraints. *Econ. Geol.* **1995**, *90*, 420–432. [CrossRef]
43. Doroshkevich, G.I. Project for conducting prospecting and appraisal work for gold and other minerals within the Ozerninsky ore cluster and adjacent territory, for 2007–2011, Ulan-Ude. 2007.
44. Murzin, V.V.; Kudryavtsev, V.I.; Berzon, R.O.; Sustavov, S.G. Peculiarities of mineralization with cuprous gold associated with rodingitization. *Geol. Ore Depos.* **1987**, *5*, 96–99. (In Russian)
45. Spiridonov, E.M.; Pletnev, P.A. *Copper Gold Deposit Zolota Hora (o "Gold-Rodingite" Formation)*; Scientific World: Moscow, Russia, 2002; 220p. (In Russian)
46. Nesterenko, G.V. *Prediction of Gold Mineralization from Placers (on the Example of Regions of Southern Siberia)*; Publisher "Science"; Siberian Branch: Novosibirsk, Russia, 1991; 191p. (In Russian)
47. Tarkhanov, A.V.; Bugrieva, E.P. The largest uranium deposits of the World. In *Mineral Raw Materials*; VIMS: Moscow, Russia, 2012; Volume 27, pp. 1–118. (In Russian)
48. Mineeva, I.G.; Makarov, A.I.; Oderova, A.V. New methodology for deep prospecting of U and Au-U deposits in paleoriftogenic structures. *Prospect. Prot. Miner. Resour.* **2013**, *7*, 3–11. (In Russian)
49. Boitsov, V.E.; Pilipenko, G.N. Gold and uranium in the Mezozoic hydrothermal deposits of the Central Aldan region (Russia). *Geol. Ore Depos.* **1998**, *40*, 315–328. (In Russian)
50. Terentiev, V.M.; Kazansky, V.I. Elkon uranium ore region on the Aldan shield. *Reg. Geol. Metallog.* **1999**, *8*, 47–58. (In Russian)
51. Corriveau, L. Iron oxide copper-gold (±Ag±Nb±P±REE±U) deposits: A Canadian perspective. In *Mineral Deposits of Canada: A Synthesis of Major Deposit-Types, District Metallogeny, The Evolution of Geological Provinces, and Exploration Methods*; Special Publication; Goodfellow, W.D., Ed.; Geological Association of Canada, Mineral Deposits Division: St. John's, NL, Canada, 2007; Volume 5, pp. 1–23.
52. Pabst, A.; Stinson, M.C. Brannerite with gold from Plumas County, California. *Bull. Geol. Soc. Amer.* **1960**, *71*, 2071–2072.
53. El Goresy, A.; Meixner, H. Brannerite from the iron spar deposits of Olsa near Friesach, Carinthia, Austria. *Neues Jahrb. Für Mineral. Abh.* **1965**, *103*, 94–98. [CrossRef]
54. Steacy, H.R.; Plant, R.; Boyle, R.W. Brannerite associated with native gold at the Richardson Mine, Ontario. *Canad. Miner.* **1974**, *12*, 360–363.
55. England, G.L. The origin of uraninite, bitumen and carbon seams in Witwatersrand gold-uranium-pyrite ore deposits, based on a Permo-Triassic analogue. *Econ. Geol.* **2001**, *96*, 1907–1920. [CrossRef]
56. Kazansky, V.I. The unique Central Aldan gold-uranium ore district (Russia). *Geol. Ore Depos.* **2004**, *46*, 167–181.
57. Ramdor, P. *The Ore Minerals and Their Intergrowth*; Pergamon Press: Oxford, UK, 1969; 1174p. [CrossRef]
58. Bouabdellah, M.; Maacha, L.; Levresse, G.; Saddiqi, O. The Bou Azzer Co–Ni–Fe–As (–Au–Ag) District of Central Anti-Atlas (Morocco): A Long-Lived Late Hercynian to Triassic Magmatic-Hydrothermal to Low-Sulphidation Epithermal System. In *Mineral Deposits of North Africa, Mineral Resource Reviews*; Bouabdellah, M., Slack, J.F., Eds.; Springer International Publishing: Zurich, Switzerland, 2016; pp. 229–247. [CrossRef]
59. Chapman, R.J.; Mortensen, J.K.; Crawford, E.C.; Lebarge, W. Microchemical studies of placer and lode gold in the Klondike District, Yukon, Canada: 1. Evidence for a small, gold-rich, oro-genic hydrothermal system in the Bonanza and Eldorado Creek area. *Econ. Geol.* **2010**, *105*, 1369–1392. [CrossRef]
60. Turuani, M.; Choulet, F.; Eglinger, A.; Goncalves, P.; Machault, J.; Mercadier, J.; Seydoux-Guillaume, A.-M.; Reynaud, S.; Baron, F.; Beafort, D.; et al. Geochemical fingerprints of brannerite (UTi2O6): An integrated study. *Mineral. Mag.* **2020**, *84*, 313–334. [CrossRef]
61. Suzuki, K.; Kato, T. CHIME Dating of Monazite, Xenotime, Zircon and Polycrase: Protocol, Pitfalls and Chemical Criterion of Possibly Discordant Age Data. *Gondwana Res.* **2008**, *14*, 569–586. [CrossRef]
62. Sciuba, M.; Beaudoin, G. Texture and trace element composition of rutile in orogenic gold deposits. *Econ. Geol.* **2021**, *116*, 1865–1892. [CrossRef]
63. Clark, J.R.; Williams-Jones, A.E. *Rutile as a Potential Indicator Mineral for Metamorphosed Metallic Ore Deposits*; Rapport Final de DIVEX, Sous-projet SC2: Montréal, QC, Canada, 2004; p. 17.
64. Ryzhenko, B.N. System Au-UO2-aqueous fluid at 500C, 1 kbar: 2. fluid of the Elkon-type gold-uraniom mineralization. *Geochem. Int.* **2016**, *54*, 996–1004. [CrossRef]
65. Kovalenko, N.I.; Ryzhenko, B.N.; Prisyagina, N.I.; Bychkova, Y.V. Experimental study of uraninite solubility in aqueous HCl solutions at 500 °C and 1 kbar. *Geochemistry* **2011**, *49*, 277–289. (In Russian) [CrossRef]
66. Bannikova, L.A. *Organic Matter in Hydrothermal Ore Formation*; Galimov, E.M., Ed.; Nauka: Moscow, Russia, 1990; 206p.
67. Poceluev, A.A. Carbonaceous substances in hydrothermal uranium and rare metal deposits. *Bull. Tomsk. Polytech. Univ.* **2010**, *316*, 16–23. (In Russian)
68. Boitsov, V.E.; Vercheba, A.A.; Pilipenko, G.N.; Zhdanov, A.V. Metallogenic zoning of the Central Aldan ore region of the Republic of Sakha (Yakutia). *Proc. High. Educ. Establishments. Geol. Explor.* **2010**, *5*, 23–33. (In Russian)
69. Aleshin, A.P.; Kozyr'kov, V.D.; Smirnov, K.M.; Ivanchenko, M.M.; Komarov, V.B.; Griboedova, I.G. Uranium-titanium-metagel mineralization of Au-U deposits of the Elkon ore district (Aldan, Russia) and features of its processing. Proceedings of higher educational establishments. *Geol. Explor.* **2016**, *4*, 50–57. (In Russian) [CrossRef]

70. Terekhov, A.V.; Molchanov, A.V.; Shatova, N.V.; Belova, V.N. Two types of ore-bearing gumbeites of the Elkon gold-uranium cluster (South Yakutia). *Reg. Geol. Metallog.* **2014**, *60*, 71–86. (In Russian)
71. Afanasiev, G.V. Prospects of the "Olympic Dam" type on the territory of Russia. In Proceedings of the III International Symposium "Uranus: Geology, Resources, Production", VIMS, Moscow, Russia, 29–31 May 2013; pp. 17–18.
72. Agangi, A.; Reddy, S.M.; Plavsa, D.; Fougerouse, D.; Clark, C.; Roberts, M.; Johnson, T.E. Antimony in rutile as a pathfinder for orogenic gold deposits. *Ore Geol. Rev.* **2019**, *106*, 1–11. [CrossRef]
73. Maksarov, R.A.; Doroshkevich, A.G.; Prokopiev, I.R.; Redin, Y.O.; Potapov, V.V. V-NB-W-bearing rutile from Karalveem gold deposit as a potential indicator for ore deposit. *Geosph. Res.* **2020**, *3*, 50–59. [CrossRef]

minerals

MDPI

Article

Paragenetic Association of Platinum and Gold Minerals in Placers of the Anabar River in the Northeast of the Siberian Platform

Alexander Okrugin and Boris Gerasimov *

Diamond and Precious Metal Geology Institute, Siberian Branch, Russian Academy of Sciences, 39 Lenin Street, Yakutsk 677000, Russia
* Correspondence: bgerasimov@yandex.ru

Abstract: Areal placers of diamond and precious metals (platinum and gold) of unknown origin are widespread in the Anabar River basin on the northeastern part of the Siberian Platform. This article discusses the typomorphic features of palladium gold (porpezite) and ferroan platinum, which, in addition to fragmented individual grains, sometimes form close growths, which indicates their obvious genetic relationship. This can be used to delimit the primary sources of commercial components of complex placers by their genetic types. The composition of minerals was determined on a Camebax-Micro (Cameca, France) microprobe analyzer, and their microstructural relationships were studied using the scanning microscope JSM-6480LV JEOL. Determination of the structure and parameters of elementary lattices of minerals was carried out on the D8 Discover diffractometer. According to microprobe analysis, the Pd content in porpezite ranges from 0.73% to 12.83%, Ag does not exceed 2.75% and Cu reaches 3–4%. Considering the composition, such a gold–platinum mineral association from placers of the Anabar river is close to precious metals from placers of the Gulinsky massif, as well as Au–PGE ore occurrences related to ultramafic–mafic intrusions of other regions of the world. Complex gold–platinum-metal mineralizations are usually closely related to parent rocks and are often observed in sulfide and chromite ores of layered ultramafic–mafic intrusions with complex metasomatic and hydrothermal transformations. It is shown that in such cases gold and platinum form a magmatogenic paragenesis of minerals that coexist until the separation of hydrothermal solutions from magma, which, as a rule, transports Au and Ag with a small fraction of PGE from the fluid-ore-magmatic system in accordance with the model of the formation of gold–porphyry deposits. Within the Anabar diamond-bearing region, according to modern geophysical data, a significant number of local gravimagnetic anomalies associated with the presence of intrusive massifs of basic and alkaline-ultrabasic rocks in the cover and within the basement have been identified. This allows us to assume that the buried parent rocks of the Anabar Au–Pt placers may be located in both the Precambrian and Phanerozoic strata.

Keywords: platinum group minerals; gold; porpezite; alkaline-ultrabasic rocks; Anabar shield; Siberian Platform

Citation: Okrugin, A.; Gerasimov, B. Paragenetic Association of Platinum and Gold Minerals in Placers of the Anabar River in the Northeast of the Siberian Platform. *Minerals* **2023**, *13*, 96. https://doi.org/10.3390/min13010096

Academic Editor: Maria Economou-Eliopoulos

Received: 15 December 2022
Revised: 4 January 2023
Accepted: 5 January 2023
Published: 7 January 2023

1. Introduction

In the northeast of the Siberian Platform, areal diamond-bearing placers are known [1,2] containing a number of quantities of associated precious metal—gold and platinum [3–8]. Although they were formed as a result of spatial combination (parasteresis) in placers of phases stable under exogenous conditions, it is possible to recognize paragenetic associations of minerals in them. This can be used to delimit the primary sources of commercial components of complex placers by their genetic types. In the last 30 years, we have been conducting regional mineralogical and geochemical typification of platinum group mineral (PGM) associations in placers. On this basis, it is possible to carry out systematic zoning of platinum-bearing placers of the Siberian Platform in order to identify their possible genetic types of primary sources [9,10].

Numerous fragmented placer occurrences of the same mineralogical-geochemical type are sometimes found in vast areas limited by the borders of certain geological structures. This indicates the connection of these placers with autonomous primary sources of the same type, formed during certain tectonic-magmatic stages of activation of platform structures, and therefore they are united into a single platinum-bearing area. If such areas cover regional structural elements, they can even be considered a single province [6].

2. Lena–Anabar Precious-Rare-Metal-Diamond-Bearing Province

In the northeast of the Siberian Platform, the formation of complexes of ultrabasic alkali rocks with carbonatites (UAR), kimberlites, picrites, carbonatites and alkali basites in the Anabar, Olenek and lower Lena River basins took place during several tectonic-magmatic cycles in the Phanerozoic. All these magmatites have a mantle nature and are characterized to varying degrees by diamond, rare-earth and precious-metal specialization. This allows us to identify this ore-bearing area as a single independent Lena–Anabar precious-rare-metal-diamond-bearing metallogenic province (Figure 1), belonging to the polycyclic platform type and requiring a special integrated approach when prospecting and forecasting possible deposits within this vast territory [11]. To discover potential primary sources of diamonds, precious metals and minerals of rare elements in the northeast of the Siberian Platform, having probably different, but paragenetically similar types of primary sources, complex search methods are required, including the study of the typomorphism of indicator minerals [12–18].

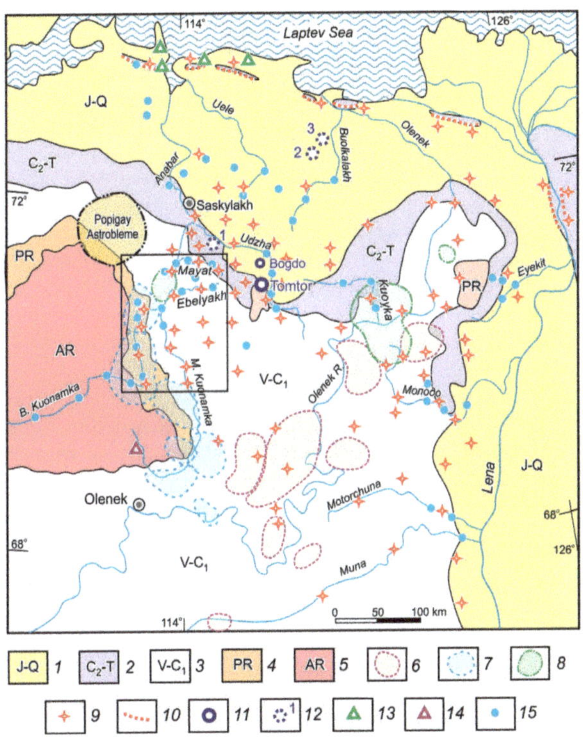

Figure 1. Distribution scheme of diamond, gold, and platinum in placers of the Lena–Anabar province (modified after [11]): 1—Jurassic–Cenozoic sediments; 2—Carboniferous–Triassic sandstones and siltstones;

3—Vendian–Carboniferous dolomites, limestones and sandstones; 4—Proterozoic rocks; 5—Archean metamorphic complexes; 6–8—Fields of kimberlites and carbonatites according to [17]: Middle Paleozoic (6), Early Mesozoic (7) and Late Mesozoic (8); 9—diamonds in placers [17]; 10—diamonds in tuffite of the Carnian stage [18]; 11—Tomtor and Bogdo massifs; 12—massifs assumed according to geophysical data [12]; 13—picrite-basalts, mouth of the Anabar river [13,14]; 14—dike of meimechite-like rocks [15]; 15—gold and platinum in placers [3,16]. The rectangle frame is the location of the research area.

3. Materials and Methods

The purposeful study of precious metals—gold and platinum—by the authors began in the early 1990s as part of the Institute's state financed case studies, as well as during contractual research and production work with the JSCs Nizhne-Lenskoye and Almazy Anabara. Single small (up to 0.5–1 mm) grains of gold and platinum are often found in the alluvial deposits of the Anabar river, but in some places the content of precious metals can reach 100 mg/m^3. During the mining of diamond-bearing placers of the rivers Mayat, Ebelyakh, Khara-Mas, etc. (Figure 1) large grains of gold (up to 5–8 mm) and platinum (up to 3 mm) were found. Here the precious metal content can reach 1–2 g/m^3. In the course of these works, extensive material has been collected on stream sediment samples of watercourses of the Anabar river basin, which are the subject of ongoing specialized research in order to identify potential primary sources of paragenetically related mineral associations. Such typomorphic associations of minerals include palladium gold and platinum group minerals (PGM); their paragenetic link is established not only by their close growths [19–21], but also the repeatability in various known magmatogenic deposits of platinum and gold [22–30]. In this article we present new data on such gold–platinum parageneses from placers of the Anabar river basin.

The composition of minerals was determined using a Camebax-Micro (Cameca, France) microprobe analyzer and their microstructural relationships were studied using the scanning microscope JSM-6480LV JEOL at DPMGI SB RAS; analysts: N.V. Leskova, S.K. Popova, L.M. Popova and N.V. Khristoforova. The surveys were carried out under the following conditions: accelerating voltage 20 kV; probe current 1.09 nA; measurement time 7 s; analytical lines: Au—Mα, Ag—α, other elements—Kα. Standardized minerals, pure metals and their alloys were used as standards X-ray studies on determining the structure and parameters of the elementary lattices of minerals were carried out at the URS-0.3 facility at DPMGI SB RAS (analysts: N.V. Zayakina, T.I. Vasilieva) and also on the D8 Discover diffractometer at MIN SB RAS by L.N. Gorokhova.

4. Results and Discussion

Mineralogical description of the association of platinum and gold in placers of the Anabar river basin, as indicated above, are presented in the works of many researchers who have given a variety of interpretations regarding their supposed primary sources [3–8,19–21,31–34]. Here we will focus in more detail on some typomorphic aspects that prove the paragenetic relationship of gold with PGM.

4.1. Mineralogical Characteristics of PGM and Gold in Placers of the Anabar River

One of the paragenetic associations of minerals is obviously platinum metals and gold with a high Pd content (up to 12.8 wt.%) and Pt (up to 0.4%). Porpezite palladium gold in the northeast of the Siberian Platform was first found in an ultra-heavy concentrate of stream sediment samples in Bolshaya Kuonamka, Anabar, Kuoyka and Molodo rivers [16], then in the diamond-bearing placers of the Mayat and B. Kuonamka rivers [19–21] (Figure 1). The largest amount of porpezite was observed in tail concentrates of the diamond-bearing placer of the Mayat river, where it is usually associated with PGM (Figure 2a) and is represented by separate well-rounded grains of scaly and lamellar habit with a size of 0.1–0.5 mm, and less often up to 1 mm (Figure 2b). Small grains have flattened and lumpy shapes. In contrast to the typical bright yellow native gold (Figure 2c), often found in these placers, porpezite has a dull bronze color and a finely shagreen surface, sometimes giving the metal an earthy appearance.

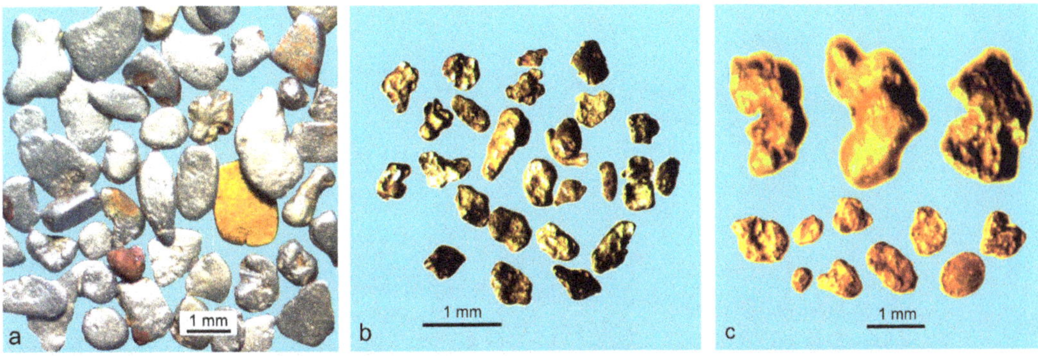

Figure 2. Morphology of platinum and gold mineral grains from the placer of the Mayat river: (**a**) well-rounded flattened grains of platinum (gray) with gold (yellow); (**b**) palladium gold; (**c**) large lumpy (upper row) and small scaly grains (lower row) of gold.

Platinum group minerals from the placer of the Mayat river, as in other placers of the Anabar river basin, are mainly represented by Fe–Pt alloys and minerals of Ru–Ir–Os composition (Table A1), and are found in smaller quantities (95–99% of the total mass of PGM). Fe–Pt alloys are represented by well-rounded silvery-light gray grains with a matte rough surface. The particle sizes mainly range from 0.25 to 0.5 mm, and some of the grains reach 1–3 mm in diameter (Figure 2a). According to microprobe analysis, their Pt content varies from 64.5% to 90.6 wt.%, and Fe from 4.1% to 13.9%. The main impurities of ferroan platinum are Rh up to 12%, Ru—9.7%, Ir—12%, Pd—12%, Os—3.2%, Ni—2.7% and Cu—5.1%.

As X-ray studies have shown, they mainly have a face-centered disordered lattice with unit cell parameters of 3.87–3.88 Å, that is, according to the classification of Fe–Pt alloys [35], these minerals belong to ferroan platinum. Only in one case out of nine definitions, a radiograph of the P-lattice was obtained, i.e., this phase has an ordered structure with a unit cell parameter of 3.86 Å, corresponding to the mineral isoferroplatinum—an intermetallic compound Pt_3Fe.

The triple diagram (Pt + Fe)–(Ir + Os)–(Ru + Rh) with an expanded Pd system shows that in the MPG in the placers of the Anabar river basin (Figure 3A, [14,27,36–41]), high-Ru–Rh platinum of the "Vilyui" type prevails, but a significant amount of platinum is plotted along the "Inagli" Ir-trend. It can be assumed that there are these two types of primary sources or one aggregate complex of rocks combining both types of mineralization, which is most likely. Some points of the compositions fall into the intermediate region occupied by the MPG from chromite ores of the Krasnogorsk massif, belonging to the alpinotype ultramafic island-arc type of the northwestern sector of the Pacific mobile Belt [36].

Minerals of Ru–Ir–Os composition in Anabar placers are observed in the form of small inclusions in ferroan platinum, but they are often found in the form of independent lamellar grains, usually up to 0.5 mm in size, rarely reaching 1–2 mm. Considering the composition according to the nomenclature of Ru–Ir–Os alloys [42], they mainly correspond to osmium, whereas ruthenium and iridium grains are less often found. (Figure 3B). They generally overlap the fields of mineral compositions from the Vilyui and Inagli placers. However, in the placers of the Vilyui River, Ru–Ir–Os alloys are mainly represented by rutheniridosmine, and in the placers of the Inagli massif, osmium and iridium are found in the form of inclusions in isoferroplatinum. Particular osmium grains mainly fall into the field of mineral compositions from placers of the Gulinsky massif, which suggests the possible presence of analogs of the Gulinsky complex of alkaline-ultrabasic rocks in this area.

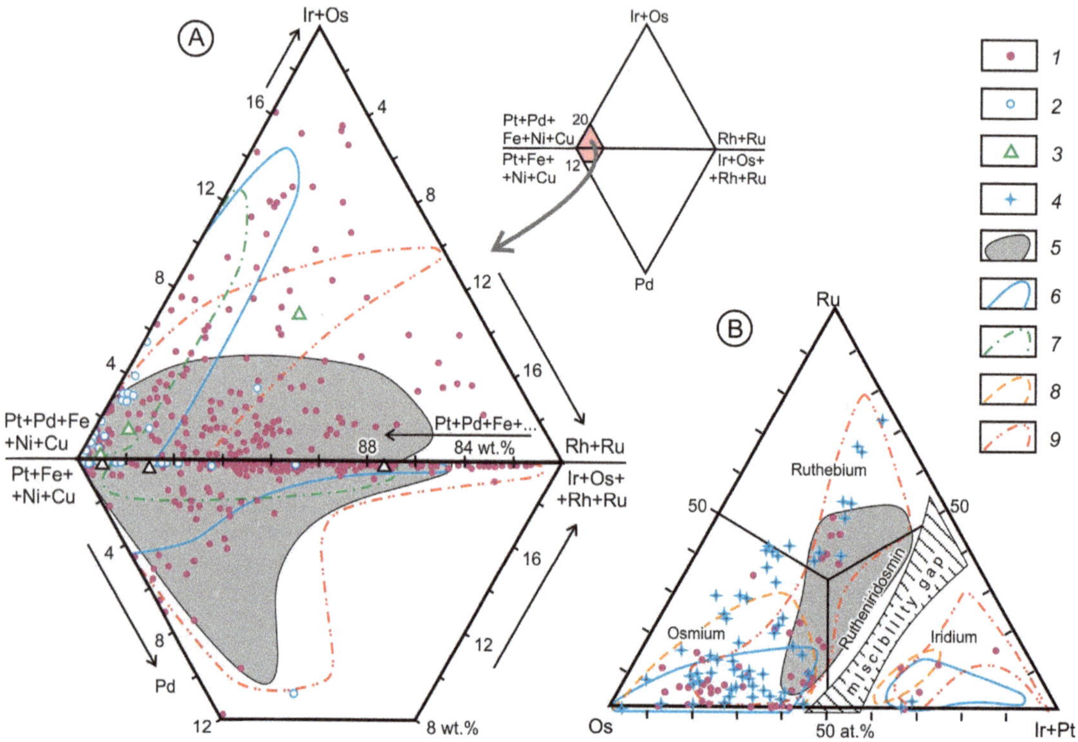

Figure 3. Diagrams of the compositions of Fe–Pt (**A**) and Ru–Ir–Os alloys (**B**). 1—Placers of the Bolshaya, Kuonamka and Anabar rivers [14]; 2—Gulinsky massif [37–40]; 3—placers of the Bor-Uryakh massif [41]; 4—inclusions of Ru–Ir–Os minerals in ferroan platinum from placers of the Anabar river. 5–9—Mineral composition fields from placers of the Vilyui river (5); Inagli massif (6); Ural (7) according to [27], Gulinsky massif (8) according to [37–39] and chromite ores of the Krasnogorsk massif (9) according to [36].

In addition to osmium, ruthenium and iridium, small inclusions are found in ferroan platinum, represented by different sulfides of platinum group elements (PGE): laurite RuS_2, erlichmanite OsS_2, prassoite $Rh_{17}S_{15}$, bowieite Rh_2S_3, vasilite $(Pd, Cu)_{16}S_7$ and others. Arsenides and tellurides PGE are less common, as well as Au-containing (up to 7.4 wt.%) intermetallide PdCu—skaergaardite. Detection of close growth of porpezite with ferroan platinum in the placer of the Bolshaya Kuonamka river [20], along with the presence of similar inclusions of palladium tellurides in the scattered grains of ferroan platinum and porpezite—kotulskite, telluropalladinite and keithconnite (Figure 4), obviously indicates the paragenetic nature of these minerals, which have a common primary source.

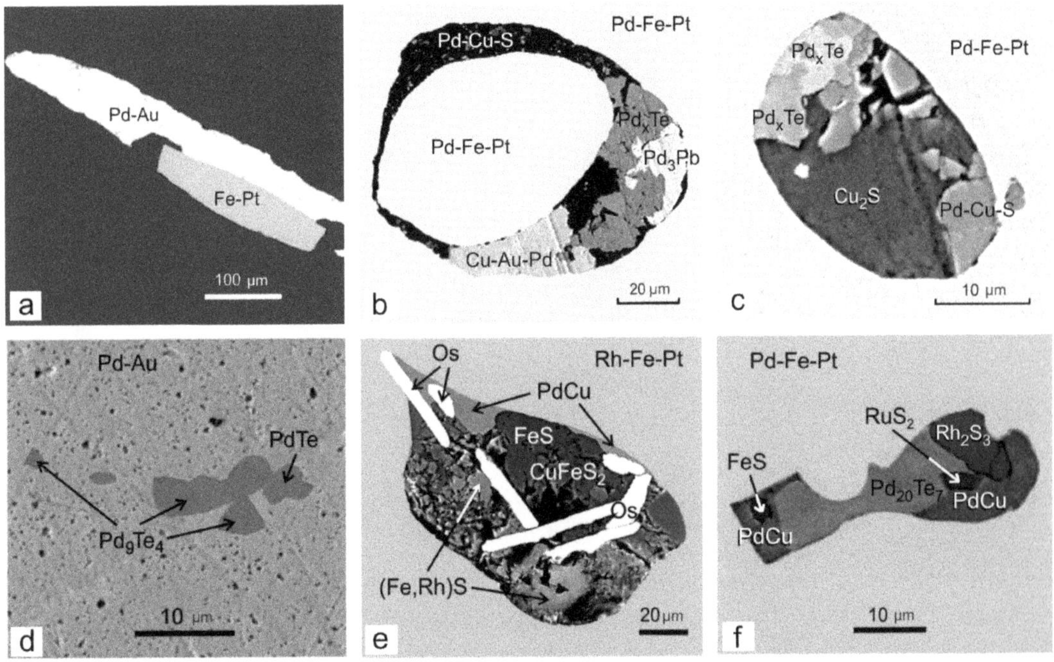

Figure 4. Forms of intergrowths of ferroan platinum and porpezite with inclusions of PGM-containing minerals from the Anabar river placers: (**a**) intergrowth of porpezite (Pd–Au) and ferroan platinum (Fe–Pt), Bolshaya Kuonamka river; (**b**,**c**) inclusions of other minerals in palladium-ferroan platinum (Pd–Fe–Pt) from the placer of the Malaya Kuonamka river; (**d**) inclusions (dark grey) of Pd tellurides in the gray matrix of palladium gold (Pd–Au) of the Mayat river; (**e**,**f**) polymineral inclusions in platinum of the Mayat river. Minerals: Os—osmium; FeS—troilite; $CuFeS_2$—chalcopyrite; Cu_2S—chalcosine; RuS_2—laurite; Rh_2S_3—bowieite; PdCu—skaergaardite; Pd–Cu–S—vasilite; Pd_3Pb—zvyagintsevite; PdTe—kotulskite; $Pd_{20}Te_7$—keithconnite; Pd_9Te_4—telluropalladinite; Cu–Au–Pd alloy composition is given in Table A2—sample MK-11. Backscatter electron (BSE) images.

In porpezite, according to microprobe analysis, the Pd content varies from 0.77% to 12.83%, Ag does not usually exceed 2.75% and the Cu impurity rarely reaches 3% (Table A2). In one grain of ferroan platinum with a high content of Pd (up to 9%) and Cu (up to 4%) from the placer of the Malaya Kuonamka River has a small inclusion (Figure 4b), for which the composition (MK-11 in Table A2) corresponds to the formula ratio $Cu_{0.33}Pd_{0.44}Au_{0.23}$. Such an alloy corresponds to the transition series from Pd–gold to Au-containing skaergaardite—PdCu; this inclusion is often found in ferroan platinum (Figure 4e,f). As X-ray determinations have shown, the parameters of the palladium gold unit cell decrease to a = 4.06–4.03 Å, while silver–gold from the placer of the Mayat river has a standard value for gold a = 4.07–4.08 Å.

Porpezite was previously found by us in the Makylgan and Inagli placers in the Central Aldan region, this mineral on the Aldan shield has also been found in placers and in magnetite–phlogopite–pyroxene rocks of the Kondersky massif [26], where palladium gold has a high Cu content up to the formation of Pd-tetra-auricupride AuCu. In gold–iridium–osmium placers associated with the Gulinsky massif with Ir–Os minerals and gold of different fineness, there is tetra-auricupride and porpezite with a Pd content of up to 8–11 at.% [30]. Comparative compositions of palladium gold from various types of deposits are shown in the diagram (Au + Cu)–Pd–Ag (Figure 5).

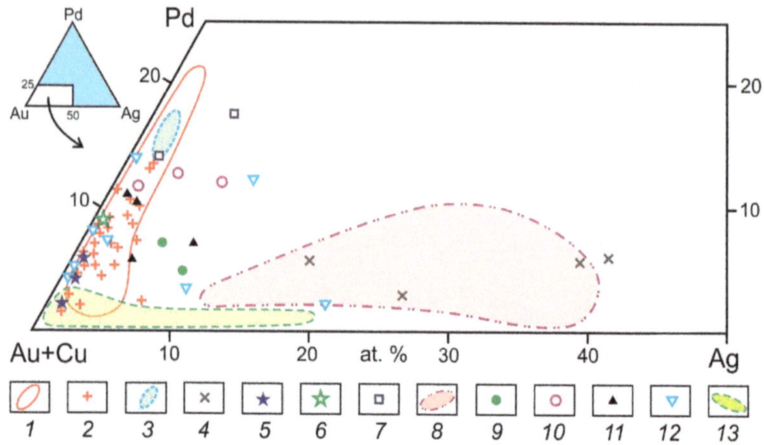

Figure 5. Diagram of palladium gold compositions: 1—Mayat river placer [19]; 2—Bolshaya Kuonamka river (authors data); 3—Guli placer [30]; 4—Norilsk PGE–Cu–Ni deposits, Russia [43]; 5—Inagli placer; 6—Kondersky massif [26]; 7—Itabira placer, Brazil [27]; 8—Merensky Reef, Bushveld, South Africa [23]; 9—Lac des Îles, QC, Canada [24]; 10—Stillwater, OK, USA [22]; 11—Pana massif, Kola Peninsula [25]; 12—Matagania–Sigiri placers, Guinea [29]; 13—Kozhimsky massif, Ural [28].

Porpezite was identified for the first time in 1798 in the Porpez region in Brazil in itabyrite deposits in the form of disseminations in deposits of itabirites—layered quartzites with magnetite and hematite. Here in placers, it is associated with Fe–Pt alloys, tu-lameenite Pt_2FeCu, osmium, arsenopalladinite $Pd_8As_{2.5}Sb_{0.5}$ and atheneite $(Pd, Hg)_3As$. In the primary platinum deposits, porpezite occurs in the form of small disseminations in the ores of layered intrusions of Bushveld, South Africa [23], Lac des Îles, QC, Canada [24], Stillwater, OK, USA [22] and in the Pana massif on the Kola Peninsula [25]. In the streams of Guinea, draining large Mesozoic stratified mafic–ultramafic massifs of the Matagania–Sigiri zone, along with the gold of medium fineness, there is so-called "white gold", with a Pd content of 1.1%–7.71% [29]. Here, in the porpezite, the smallest inclusions of sperrylite $PtAs_2$ and braggite (PdPt)S were identified. In the crushed samples of the Proterozoic metasedimentary rocks within the near-contact parts of the basite–hyperbasite massif, sub-ore gold particles were found with an admixture of Pd up to 1%, Pt 6.6% and Ag 15%, as well as PtFe tetraferroplatinum containing 20.6% Pd. Transition from the early typical magmatic PGE–Cu–Ni mineralization in ultramafic rocks to the late PGE–Cu–Au–Te sulfide mineralization in phlogopite-rich shonkinites was determined in the Mordor dunite–shonkinite–syenitic alkaline complex in Australia [44].

A new Au–Pd type of hydrothermal mineralization was discovered in the Kozhimsky region of the Circumpolar Urals, where visible disseminations (up to 2–8 mm) of Ag-, Cu-, Hg-, and Pd-containing gold in association with PGMs—mertieite, atheneite and an unknown Pd–As phase—were found in fuchsite stringers formed in the foliated Riphean rhyolites [28]. The authors of these studies consider that Au, Pd, Cu, Cr and Ca were redeposited into metasomatically altered rhyolites from underlying basalts and dykes of gabbro-diabases and picrites.

4.2. Possible Primary Sources of MPG and Porpezite in the Anabar River Basin

The complex diamond-bearing placers of the Anabar river are located within the eastern slope of the Anabar shield, where carbonate-terrigenous deposits and volcanogenic formations of the Riphean, Vendian, Cambrian, Permian, Triassic, Jurassic and Cretaceous are developed, intruded by the Late Precambrian, Middle Paleozoic and Mesozoic magmatites, belonging to a variety of mineral types—from kimberlites, alkaline-ultrabasic rocks and carbonatites to subalkaline and tholeiitic basites. Within the Anabar diamond-bearing region, according to modern geophysical data by I.V. Polyansky [19], a significant number of local gravimagnetic anomalies have been revealed, which are associated with the presence of intrusive massifs of basic and alkaline-ultrabasic rocks in the cover and within the basement. In this area, aeromagnetic high-precision survey (AMS-10) revealed 10 probable hidden intrusions (Figure 6), which are located between the metamorphic complex of the Anabar shield and the Riphean rocks of the Udzhinsky uplift [19]. This allows us to assume that the parent rocks of the Anabar Au–Pt placers may be located in both Precambrian and Phanerozoic strata. The buried Precambrian sources in the Phanerozoic could be brought to the surface, within the limits and frames of the Anabar shield and the Udzhinsky uplift, then washed into the area of the Udzhinsky aulacogen, at the same time, stable minerals will be preserved in basal intermediate reservoir rocks and fall from there into modern river valleys, forming halos of dispersions without visible primary sources.

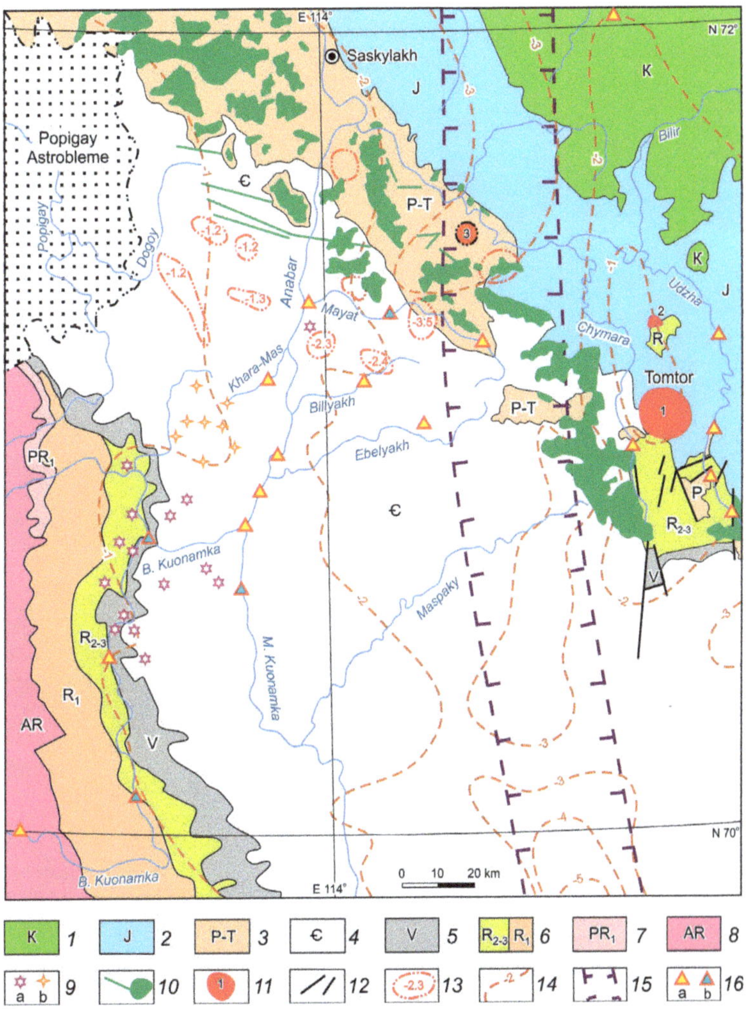

Figure 6. Schematic geological map of the distribution of gold–platinum-bearing placers in the basin of the middle course of the Anabar river (modified after [19]): 1—Cretaceous sands, silts and gravels;

2—Jurassic conglomerates, sandstones, siltstones; 3—Permian and Triassic sandstones, siltstones; 4—Cambrian dolomites, limestones and sandstones; 5—Vendian sandstones, siltstones and conglomerates; 6—middle-upper Riphean ($R_{2\text{-}3}$) and lower Riphean (R_1) conglomerates, sandstones, siltstones, argillite and dolomites; 7—Lower Proterozoic rocks; 8—Archean metamorphic complexes; 9—kimberlite (a) and carbonatite (b) pipes of the Orto–Yargin field; 10—dykes and sills of P–T basites; 11—complexes of alkaline-ultrabasic rocks with carbonatites (1—Tomtor, 2—Bogdo, 3—Chyuempe); 12—faults; 13—anomalies of the potential intrusive massifs of basic, alkaline-ultrabasic rocks in the basement and lower layer of the platform cover (based on geophysical data), figure-estimated depths (-n km) of occurrence; 14—isolines of basement surface (-n km); 15—borders of the Udzhinsky aulacogen; 16—sampling sites of the placer samples of platinum and gold (a) and PGM in association with palladium gold (b). Scheme is based on Geological and Tectonic maps of Yakutia at a scale of 1:1,500,000.

Summarizing the above facts, it can be concluded that the primary sources of the association of palladium gold and ferroan platinum in placers of the Anabar river basin are complex ultramafic–mafic ore-magmatic assemblages, which include alkaline-ultrabasic massifs with carbonatites. They can have a wide time range of development (evolution) from the the Riphean to the Permian–Triassic eruption of alkaline rocks, which covered not only the largest Maimecha–Kotui province, but also the Eastern Anabar region with a very wide range of basitic, alkaline-ultrabasic, carbonatite and kimberlite rocks. The presence of ferroan platinum with polymineral inclusions in the placers of the Anabar river, also including osmium, indicates that the gold–platinum association, along with Pd, contains other PGE, including the highest-temperature Os. In addition, the similarity of the typochemical features of low- and medium-grade gold with a high Cu content and the presence of tetra-auricupride AuCu [16] further brings the paragenetic divergence of the primary sources of the complex gold–platinum placers of the Anabar river with gold–iridium–osmium placers of the Gulinsky massif closer [30].

5. Conclusions

Based on the brief overview of the joint finding of palladium gold, Cu–Au–Pd alloys of different composition with PGM in placers and ores of ultramafic–mafic complexes of the Siberian Platform and other regions of the world, the authors believe that the association of palladium gold and PGM found in placers represents a close paragenetic relationship. Complex gold–platinum-metal mineralizations are usually closely related to parent rocks and are often observed in sulfide and chromite ores of layered ultramafic–mafic intrusions with complex metasomatic and hydrothermal transformations. The concentration of Au and Pd alloys along with sulfides and tellurides of non-ferrous metals and PGE in isolated teardrop inclusions of ferroan platinum (Figure 4) indicates the separation of an excessive number of more easily mobile Au–Pd–Cu–Te–S–As phases from the groundmass of refractory native osmium, iridium and platinum already crystallizing in magmatic conditions. It is obvious that here gold and platinum still form a coexisting magmatogenic paragenesis of minerals; however, at the post-magmatic stage of separation of pneumatic-hydrothermal solutions from magmatic melts due to sharply differing degrees of affinity of PGE and gold to fluid components, these metals are separated. As a result, the groundmass of the PGE–Cu–Ni triad is preserved in the magmatic system itself and its exocontact zone, and with the separating hydrotherms. Au, Ag and a number of non-ferrous metals with Te, As, S and a small fraction of PGE are carried away from the fluid-ore-magmatic system according to the model of formation of gold–porphyry deposits [45]. In typical hydrothermal ores associated with granitoid magmatism, gold is practically sterile relative to PGE. Sometimes small impurities of PGE (up to 0.1%–0.2% Pt and Pd) are detected by spectral analysis in gold from hydrothermal deposits; apparently, they associated with the involvement of deeper basic magmatites in the composition of the parent rocks. Hydrothermal easily mobile Au-containing complexes inherited their distant genetic characteristics of these deep basic magmatites.

Author Contributions: Conceptualization, A.O. and B.G.; methodology, A.O. and B.G.; software, A.O.; validation, A.O. and B.G.; formal analysis, A.O. and B.G.; investigation, A.O.; resources, A.O.; data curation, A.O.; writing—original draft preparation, A.O. and B.G.; writing—review and editing, A.O.; visualization, A.O. and B.G.; supervision, A.O.; project administration, A.O.; funding acquisition, A.O. All authors have read and agreed to the published version of the manuscript.

Funding: The study was conducted within the framework of Russian Science Foundation (regional contest) project No 22-27-20151.

Data Availability Statement: Not applicable.

Acknowledgments: The authors are grateful to A.L. Zemnukhov and R.Y. Zhelonkin from Almazy Anabara JSC for contributing to the collection of the placer material in the field, as well as their colleagues from DPMGI SB RAS who participated in analytical and laboratory work.

Conflicts of Interest: The authors declare no conflict of interest.

Appendix A

Table A1. Representative analyses of platinum group minerals from placers Anabar R., wt.%.

Sample	Pt	Ir	Os	Ru	Rh	Pd	Fe	Ni	Cu	Total
					Fe–Pt alloys					
99/95	87.42	1.25	0.03	0.04	0.03	0.23	9.84	0.13	0.92	99.89
52/95	86.24	3.68	0.13	0.01	0.24	0.00	9.21	0.11	0.76	100.38
49/95	71.68	0.30	0.08	0.05	0.40	10.00	14.66	0.06	1.65	98.88
93/95	83.40	1.54	0.17	0.01	0.59	0.11	11.58	1.13	0.87	99.40
83/95	86.39	0.66	0.18	0.00	0.67	0.05	10.16	0.11	2.49	100.71
87/95	86.35	2.85	1.16	0.32	0.87	0.09	7.27	0.75	0.30	99.96
80/95	78.56	10.05	1.77	0.17	1.21	0.06	6.88	0.26	0.54	99.50
77/95	86.31	0.93	0.25	0.00	1.27	0.20	9.65	0.10	1.01	99.72
67/95	82.20	4.16	0.64	0.39	1.59	0.03	10.57	0.31	0.41	100.30
89/95	70.42	12.53	1.11	1.46	2.46	0.07	9.77	0.53	0.46	98.81
63/95	82.84	0.43	0.11	0.50	2.65	0.07	12.49	0.18	1.02	100.29
59/95	78.75	1.15	0.51	1.92	6.61	0.15	10.43	0.39	0.55	100.46
76/95	73.23	0.97	0.89	4.08	8.23	0.15	11.66	0.17	0.61	99.99
24/16	72.11	0.99	1.49	6.90	6.89	2.73	7.11	0.32	0.40	98.94
25/16	73.63	2.99	1.87	8.38	3.61	1.58	8.15	0.38	0.40	100.98
81/78	75.87	0.00	0.32	2.81	11.12	1.06	7.16	0.26	0.30	98.90
					Ru–Ir–Os alloys					
96/95	0.41	38.28	60.34	0.57	0.28	0.04	0.04	0.02	0.07	100.05
70/95	0.31	25.64	70.63	2.48	0.06	0.15	0.17	0.05	0.06	99.55
32/72	2.23	30.16	63.50	2.56	0.71	0.00	0.26	0.00	0.00	99.42
80/78	2.45	38.20	50.05	7.53	0.36	0.00	0.09	0.03	0.06	98.77
61/95	4.08	38.82	41.99	14.74	0.38	0.06	0.17	0.05	0.03	100.32
56/95	1.86	35.94	39.98	21.89	0.05	0.04	0.26	0.08	0.08	100.18

Table A2. Representative analyses of porpezite, skaergaardite [1] and Cu–Au–Pd alloy [2], wt.%.

Sample	Au	Ag	Cu	Hg	Pd	Total
			Mayat river			
4/158	96.06	2.62	0.00	0.34	0.77	99.79
11/158	93.65	2.75	0.02	0.17	2.20	98.79
7/158	92.59	1.21	0.13	0.03	5.76	99.72
37/154	89.78	0.91	3.00	-	6.37	100.07
29/154	87.91	1.54	0.37	0.06	9.43	99.31
6/158	85.51	1.09	0.25	0.04	12.83	99.72
Ma-1 [1]	7.24	0.10	26.94	-	54.42	98.88 *

Table A2. *Cont.*

Sample	Au	Ag	Cu	Hg	Pd	Total
			Bolshaya Kuonamka river			
97/104	97.10	2.07	0.13	-	0.73	100.03
183/104	95.66	0.65	0.79	0.06	2.50	99.67
225/104	94.86	0.67	0.68	1.40	1.44	99.05
Vp-1–1	94.28	3.94	0.89	0.13	1.45	100.70
Vp-1–2	93.26	0.70	1.54	0.55	2.93	98.98
Vp-1–3	91.69	1.06	4.17	0.07	3.08	100.70
Vp-1–4	93.13	1.68	0.15	0.05	4.96	99.97
Vp-1–5	89.84	1.13	0.05	0.08	7.47	98.58
			Malaya Kuonamka river			
MK-11 [2]	38.50	-	17.37	-	39.37	95.24

Note: dash—content of the element is below the sensitivity of the microprobe analysis method; *—the total also includes 7.43% Pt and 1.64% Fe.

References

1. Rozhkov, I.S.; Mikhalev, G.P.; Prokopchuk, B.I.; Shamshina, E.A. *Diamond-Bearing Placers of the Western Siberian Platform*; Nauka: Moscow, Russia, 1967. (In Russian)
2. Grakhanov, S.A.; Shatalov, V.I.; Shtyrov, V.A.; Kychkin, V.R.; Suleymanov, A.M. *Diamond Placers of Russia*; Academic Publishing House "Geo": Novosibirsk, Russia, 2007. (In Russian)
3. Shpunt, B.R. Platinum minerals in Quaternary deposits of the Anabar-Olenek Uplift. *Geol. Rudn. Mestorozhdenii* **1970**, *2*, 123–126. (In Russian)
4. Shpunt, B.R. Typomorphic features and genesis of placer gold in the north of the Siberian platform. *Geol. Geophys.* **1974**, *9*, 77–78. (In Russian)
5. Okrugin, A.V.; Izbekov, E.D.; Shpunt, B.R.; Leskova, N.V. Platinum metal minerals of anthropogenic deposits of the Vilyui syneclise and Anabar anteclise. In *Typomorphism and Geochemical Features of Minerals of Endogenous Formations of Yakutia*; Publishing House YSC: Yakutsk, Russia, 1985; pp. 40–50. (In Russian)
6. Okrugin, A.V. Placer platinum content of the Lena province. *Otechestvennaya Geol.* **1997**, *9*, 29–32. (In Russian)
7. Okrugin, A.V.; Grakhanov, S.A.; Selivanova, V.V.; Popov, A.A. Gold and platinum in diamond-bearing placers of Western Yakutia. *Sci. Educ. Yakutsk* **2000**, *3*, 19–22. (In Russian)
8. Gerasimov, B.B.; Nikiforova, Z.S. Placer gold content of the Mayat river of the Anabar river basin. *Otechestvennaya Geol.* **2005**, *5*, 38–41. (In Russian)
9. Okrugin, A.V.; Kim, A.A. Topomineralogy of platinoids from placers of the Eastern part of the Siberian platform. In *Rare Native Metals and Intermetallides of the Primary and Placer Deposits of Yakutia*; Publishing House YSC: Yakutsk, Russia, 1992; pp. 77–102. (In Russian)
10. Okrugin, A.V.; Kim, A.A.; Izbekov, E.D.; Shpunt, B.R.; Selivanova, V.V.; Filippov, V.E.; Blinov, A.A.; Mikhailov, V.A.; Zolotonog, A.V.; Surnin, A.A. Platinum-bearing placers of the Siberian Platform and their prospects. In *Platinum of Russia: Problems of Development of the Mineral Resource Base of Platinum Metals in the XXI Century*; Book 2; CJSC Geoinformmark: Moscow, Russia, 1999; Volume 3, pp. 319–333. (In Russian)
11. Okrugin, A.; Tolstov, A.; Baranov, L.; Zemnukhov, A. Lena-Anabar precious-rare-metal-diamond-bearing metallogenic province. In *Geology and Mineral Resources of the North-East of Russia: Materials of the X All-Russian Scientific and Practical Conference with International Participation*; NEFU Publishing House: Yakutsk, Russia, 2020; pp. 272–277. (In Russian)
12. Porshnev, G.I.; Stepanov, L.L. Geology and mineralogy of the Udzhinsky province (north-west of the Yakut ASSR). *Sov. Geol.* **1981**, *12*, 103–106. (In Russian)
13. Milashev, V.A.; Tomanovskaya, Y.I. Occurrences of Alkaline-Ultrabasic Magmatism in the Coastal Part of the Laptev Sea. In *Kimberlite Volcanism and Prospects of the Primary Diamond Content of the Siberian Platform*; Publishing House NIIGA: Leningrad, USSR, 1971; pp. 127–133. (In Russian)
14. Okrugin, A.V.; Zaitsev, A.I.; Borisenko, A.S.; Zemnukhov, A.L.; Ivanov, P.O. Gold-platinum-bearing placers deposits in the river basin of Anabar and their possible relation to alkali-ultrabasic magmatic rocks in the northern Siberian platform. *Otechestvennaya Geol.* **2012**, *5*, 11–20. (In Russian)
15. Muzyka, G.M.; Chumirin, K.G. Manifestation of maimechit analogues on the southern margin of the Anabar massif. In *Geology, Petrography and Mineralogy of Magmatic Formations of the Northeastern Part of the Siberian Platform*; Nauka: Moscow, Russia, 1970; pp. 183–190. (In Russian)
16. Okrugin, A.V. *Platinum-Bearing Placers of the Siberian Platform*; Publishing House YaF SO RAN: Yakutsk, Russia, 2000. (In Russian)

17. Afanasiev, V.P.; Lobanov, S.S.; Pokhilenko, N.P.; Koptil', V.I.; Mityukhin, S.I.; Gerasimchuk, A.; Pomazanskii, B.; Gorev, N. Polygenesis of diamonds in the Siberian platform. *Russ. Geol. Geophys.* **2011**, *52*, 259–274. [CrossRef]
18. Grakhanov, S.A.; Smelov, A.P. Age of predicted primary sources of diamonds in northern Yakutia. *Otechestvennaya Geol.* **2011**, *5*, 56–63. (In Russian)
19. Okrugin, A.V.; Mazur, A.B.; Zemnuhov, A.L.; Popkov, P.A.; Sleptsov, S.V. The palladium gold-PGM association in the placers of the Anabar River basin, NE part of the Siberian platform, Russia. *Otechestvennaya Geol.* **2009**, *5*, 3–10. (In Russian)
20. Gerasimov, B.B.; Nikiforova, Z.S.; Pavlov, V.I. Mineralogical and geochemical characteristics of placer gold in the Bolshaya Kuonamka river. *Educ. Sci. Yakutsk, Russia.* **2014**, *3*, 74–78. (In Russian)
21. Okrugin, A.V.; Gerasimov, B.B. Magmatogenic paragenesis of platinum and palladium gold in placers of the Anbar river on the Siberian platform. In Proceedings of the Geology and Mineral Resources of the North-East of Russia: Materials of the XII All-Russian Scientific and Practical Conference Dedicated to the 65th Anniversary of the Institute of Geology of Diamond and Precious Metals, Siberian Branch of the Russian Academy of Sciences, Yakutsk, Russia, 23–25 March 2022; pp. 224–229. (In Russian).
22. Cabri, L.J.; Laflamme, J.H.G. Rhodium, platinum and gold alloys from the Stillwater Complex. *Can. Mineral.* **1974**, *12*, 399–403.
23. Kingston, G.A.; El-Dosuky, B.T. A Contribution on the platinum-group mineralogy of the Merensky reef at the Rustenburg platinum mine. *Econ. Geol.* **1982**, *77*, 1367–1384. [CrossRef]
24. Cabri, L.J. (Ed.) *Platinum-Group Elements: Mineralogy, Geology, Recovery*; Canadian Institute of Mining and Metallurgy: Montreal, QC, Canada, 1989; Volume 23, p. 267.
25. Krivenko, A.P.; Tolstykh, N.D.; Veselovsky, N.N.; Mayorova, O.N. Gold-bearing tellurides of platinoids and palladium gold in gabbro-norites of the Pana massif, Kola Peninsula. *Rep. AS USSR* **1991**, *319*, 725–729. (In Russian)
26. Nekrasov, I.Y.; Lennikov, A.M.; Oktyabrsky, R.A.; Zalishchak, B.L.; Sapin, V.I. *Petrology and Platinum Potential of Ring Alkaline-Ultrabasic Complexes*; Nauka: Moscow, Russia, 1994. (In Russian)
27. Cabri, L.J.; Harris, D.C.; Weiser, T.W. Mineralogy and distribution of platinum-group mineral placer deposits of the Wold. *Expl. Min. Geol.* **1996**, *5*, 73–167.
28. Tarbaev, M.B.; Kuznetsov, S.K.; Moralev, G.V.; Soboleva, A.A.; Laputina, I.P. New gold-palladium type of mineralization in the Kozhim region of Circumpolar Ural (Russia). *Geol. Ore Depos.* **1996**, *1*, 15–30. (In Russian)
29. Bozhko, E.N. Problem of the hard-rock types of the auriferous-plstiniferous mineralization of the Matagania-Sigiri structural-formational zone (Guinea, West Africa). *Proc. Voronezh State Univ. Geol.* **2005**, *1*, 193–203. (In Russian)
30. Badanina, I.Y.; Malich, K.N.; Goncharov, M.M.; Tuganova, E.V. Precious-metal placers of the Gulinsky massif (north of the Siberian platform): Native gold: Typomorphism of mineral associations and conditions of deposit formation. In *Proceedings of the All-Russian Conference*; IGODPMG RAS: Moscow, Russia, 2010; Volume 1, pp. 56–58. (In Russian)
31. Okrugin, A.V.; Okhlopkov, S.S.; Grakhanov, S.A. Complex placer occurrences of precious metals and gems in the Anabar river basin (northeast of the Siberian Platform). *Otechestvennaya Geol.* **2008**, *5*, 3–13. (In Russian)
32. Gerasimov, B.B.; Nikiforova, Z.S. Assumed mineragenic types of the primary sources of gold of the Anabar region (northeast of the Siberian Platform). *Sci. Educ.* **2017**, *2*, 11–16. (In Russian)
33. Tolstov, A.V. Prospects for the gold content of Anabar anteclise. *Bull. Goskomgeol.* **2002**, *1*, 44–49.
34. Smelov, A.P.; Berezkin, V.I.; Zedgenizov, A.N. New data on the composition and ore content of the Kotuikan zone of tectonic melange. *Otechestvennaya Geol.* **2002**, *4*, 45–49. (In Russian)
35. Cabri, L.J.; Feather, C.Z. Platinum-iron alloys: A nomenclature based on studi of natural end sinthetic alloys. *Can. Mineral.* **1975**, *13*, 117–126.
36. Dmitrenko, G.G. *Platinum-Group Minerals of Alpine-Type Ultramafites*; Publishing House SVKNII: Magadan, Russia, 1994. (In Russian)
37. Balmasova, E.A.; Smol'skaya, L.S.; Lopatin, G.G.; Lazarenkov, V.G.; Malitch, K.N. Native osmium and iridosmine in the Gulinsky massif. *Trans. Russ. Acad. Sci. Earth Sci. Sect.* **1992**, *325*, 154–157.
38. Likhachev, A.P.; Kirichenko, V.T.; Lopatin, G.G.; Kirichenko, A.A.; Deryagina, G.G.; Rudashevskiy, N.S.; Botova, M.L. Distinctive features of platinum occurrences in the massifs of alkali ultrabasic formations. *Zap. Vses. Mineral. Obsh.* **1987**, *1*, 122–125. (In Russian)
39. Sazonov, A.M.; Romanovskii, A.E.; Grinev, O.M.; Lavrent'ev, Y.G.; Maiorova, O.N.; Pospelova, L.N. Precious-metal mineralization of the Gulinskaya intrusion. *Russ. Geol. Geophys.* **1994**, *9*, 51–65.
40. Malitch, K.N.; Rudashevsky, N.S. Bedrock platinum-metal mineralization in chromitite of the Guli massif. *Trans. Russ. Acad. Sciences. Earth Sci. Sect.* **1992**, *327*, 165–169.
41. Malitch, K.N.; Kogarko, L.N. Chemical composition of platinum-group minerals from the Bor-Uryakh massif (Maimecha-Kotui Province, Russia). *Dokl. Earth Sci.* **2011**, *440*, 1455–1459. [CrossRef]
42. Harris, D.C.; Cabri, L.J. Nomenclature of platinum-group-element alloys: Review and revision. *Can. Mineral.* **1991**, *29*, 231–237.
43. Razin, L.V. Minerals—Natural alloys of gold and copper in ores of copper-nickel deposits of the Norilsk type. *Proc. Fersman Mineral. Mus.* **1975**, *24*, 93–106. (In Russian)

44. Holwell, D.A.; Blanks, D.E. Emplacement of magmatic Cu-Au-Te(-Ni-PGE) sulfide blebs in alkaline mafic rocks of the Mordor Complex, Northern Territory, Australia. *Miner. Depos.* **2020**, *56*, 789–803. [CrossRef]
45. Bukhanova, D.S.; Kutyrev, A.V.; Sidorov, E.G.; Chubarov, V.M. The first finding of platinum group minerals in the Malmyzh gold-copper porphyry deposit, Khabarovsk region, Russia. *Zapiski RMO Proc. Russian Miner. Soc.* **2020**, *149*, 54–64. (In Russian)

minerals MDPI

Article

The Use of Typomorphic Features of Placer Gold of the Anabar Region for Determining Its Sources

Boris Gerasimov

Diamond and Precious Metal Geology Institute, Siberian Branch, Russian Academy of Sciences, 39, prosp. Lenina, Yakutsk 677000, Russia; bgerasimov@yandex.ru

Abstract: Typomorphic features of placer gold of the Anabar region were studied as predictive-exploration criteria. The target of the study was to determine the typomorphic features of placer gold related to the intermediate sources (paleo-placers) and the supposed nearby primary ore occurrences. Two varieties of placer gold were identified. The first variety is well-rounded high-fineness lamellar gold with a highly modified internal structure. This native gold is associated with intermediate sources, Neogene–Quaternary watershed pebble beds. The second type includes slightly rounded gold with a wide variation in fineness (494‰–999‰). Its indicator is a block heterophase internal structure. The set of typomorphic features of this variety of placer gold indicates the vicinity of the primary source, what was the prerequisite for constructing prospecting traverses in order to find ore occurrences. As a result of these studies, hydrothermal-metasomatic formations with gold-sulfide mineralization were identified. The main primary substrate for them is fractured near-fault carbonate rocks of the Cambrian and Vendian–Cambrian age. Along with this, hydrothermalites developed on slightly cemented fine-pebble quartz conglomerates of the Middle Permian age were found in the core of exploration wells. Two types of metasomatic rocks are identified: quartz-potassium feldspar and jasperoid. The main ore minerals were galena and pyrite, different ratios by sites were revealed. Gold was identified in the form of small particles in the carbonate and siliceous substrate of hydrothermal-metasomatic formations. The lithological factor was one of the leading favorable factors for the ore formation due to the presence of near-fault highly permeable fractured carbonate and slightly cemented terrigenous rocks. The structural control of the studied ore occurrences is determined by their localization in the Mayat–Logoy and Dogoy–Kuoy faults of the Molodo–Popigay system of discontinuous faults. We assume a two-stage formation of the gold ore occurrences: during the first stage, the ore components in the form of primary hydrothermal-sedimentary ores in the near-gault zones were formed. The second stage was related to the processes of the Mesozoic tectonic-magmatic activation, when the intrusion of basite dikes initiated the mobilization of ore components the gold-sulfide occurrences were formed in the near-fault zone as a result of silicic-potassic metasomatosis of the carbonate and terrigenous rocks.

Keywords: typomorphic features; placer gold; fineness; internal structure; Anabar placer area; hydrothermal-metasomatic formations; K-feldspathization; jasperoids; ore mineralization; fault zone; ore gold; sulfides

Citation: Gerasimov, B. The Use of Typomorphic Features of Placer Gold of the Anabar Region for Determining Its Sources. *Minerals* **2023**, *13*, 480. https://doi.org/10.3390/min13040480

Academic Editor: Nuo Li

Received: 7 February 2023
Revised: 24 March 2023
Accepted: 27 March 2023
Published: 29 March 2023

1. Introduction

Numerous placer occurrences of fine gold are known in the vast territory of the northeastern part of the Siberian platform. The predecessors had no consensus about the age, genetic characteristics, and location of the primary sources of native gold. Most researchers associated the placer gold content of the studied area with the Precambrian quartz and quartz-carbonate veins [1–3]. Along with this, the Mesozoic age of the ore sources has been assumed [4,5]. At the present stage of research, based on the results of studying the typomorphic features of placer gold from ancient intermediate sources and modern alluvium, two types of gold related to primary sources of different ages have been identified [6]. Z.S. Nikiforova and co-authors, based on the study of the mineralogical and geochemical features of native gold, proved two main stages of ore formation for the entire east of the Siberian platform, including the studied territory, Precambrian and Mesozoic [7,8].

However, a few prospecting works have not led to the discovery of gold deposits until recently. This is due to the fact that certain difficulties emerge when predicting and searching for gold ore occurrences in the northeast of the Siberian platform, since the studied area is overlaid by a thick cover of the Cenozoic sediments, where traditional methods of searching for gold deposits are ineffective. Meanwhile, the methodology of studying the mineralogical and geochemical features of placer gold is successfully used to establish the types of gold mineralization and localization of primary sources, which is reflected in many publications [9–32]. In Russian geological science, the main provisions on the typomorphism of native gold were laid down in N.V. Petrovskaya's fundamental monograph "Native Gold, 1973" [9]. The use of a complex of typomorphic features (granulometry, morphology, chemical composition, and internal structure) of placer gold for predicting primary sources is reflected in well-known publications of Russian authors [10–13]. A number of researchers have been studying the typomorphism of placer gold in the northeastern part of the Siberian Platform [2,3,6–8,14]. There are many scientific papers focused on the methodology of studying the microchemical characteristics of placer gold: fineness, trace elements, and mineral inclusions in order to identify the types of gold mineralization, localization of sources, and clarification of regional metallogeny [16–19]. Researchers also pay great attention to the study of changes in native gold in the hypergene environment and paleo-placers [20–26]. In general, studies using the methodology of studying the typomorphism of native gold are carried out in many regions of the world [27–33].

In this regard, it is advisable to study in detail the complex of typomorphic features of placer gold, as a carrier of the most important genetic information necessary for the development of mineralogical criteria for predictive assessment of placer and ore gold content.

The Anabar river basin belongs to the placer area of the same name in the northeastern part of the Siberian platform and is located on the territory of the Lena–Anabar polymineral placer subprovince [34]. The studied area is characterized by a total background placer gold content with a gold content in quaternary sediments from 10 to 20 to 300 mg/m^3. Gold is found in the alluvium of the modern riverbed and terraces, in the Neogene–Quaternary relict pebbles on watersheds, as well as in basal horizons of the Permian age. According to B.R. Shpunt (1974) [2], gold from the Permian conglomerates is represented by small lamellar particles. Gold from the Neogene–Quaternary sediments developed on watersheds is characterized mainly by small fractions of 0.1–0.2 mm, flattened plates with a polished surface [3]. According to the predecessors, the sources of gold in modern placers were ancient paleo-placers of the Neogene–Quaternary and Permian age [3,4]. One of the most important unsolved problems is the identification of the primary sources of fine gold. Until recently, there was no data on the ore occurrences of gold, which served as the primary sources for placer gold. At the same time, the results of our previous studies of the typomorphism of placer gold in the area indicate the presence of a local ore source [6–8].

The publication presents the results of studying the typomorphism of placer gold as search signs for identifying primary gold-bearing sources on the example of two gold-bearing placer occurrences, "Billyah" and "Nebaibyt". Mineralogical features of hydrothermal-metasomatic formations with gold-sulfide mineralization, first discovered in fault zones, were also studied.

The target of the study was to determine the typomorphic features of placer gold related to the intermediate sources and the supposed nearby primary ore occurrences.

2. Materials and Methods

Native gold from alluvial sediments of the Billyakh river, right inflow of the Anabar river (35 samples) with a total weight of 950 mg and the Nebaibyt river, right inflow of the Bolshaya Kuonamka river (20 samples) with a total weight of 728.5 mg, was studied.

Samples were taken from exploration pits made by "Almazy Anabara" JSC in alluvial sediments for the exploration of placer diamonds.

Granulometric sieving of samples was performed using a set of mineralogical sieves, of which cell size corresponds to the classes of the granulometric scale, namely: +2 mm, 1–2 mm, 0.5–1 mm, 0.25–0.5 mm, and −0.25 mm. After that, the sample material was fractionated by density in a heavy liquid (bromoform). Monomineralic fractions of gold were isolated from the resulting ultrafiltration retentate using a binocular microscope.

The prospecting traverses were carried out on artificial sites (artificial channels, landfills of worked placers, and road clearing sites) and natural outcrops of hydrothermally altered dolomites of the Anabar formation of the Middle Cambrian and dolomitic limestones of the Vendian–Cambrian age. Hand specimen and rock samples were taken mainly in crush zones confined to the faults. The weight of the hand specimen samples ranged from 5 to 10 kg. More than 200 hand specimens and samples were picked. A total of 120 polished sections and 80 thin sections were made and analyzed.

The selection and description of core samples were performed at the core storage of the "Mayat" mine of "Almazy Anabara" JSC. The core from the exploration lines of core drilling holes in the watershed of the Mayat and Morgogor rivers were studied too (Figure 1).

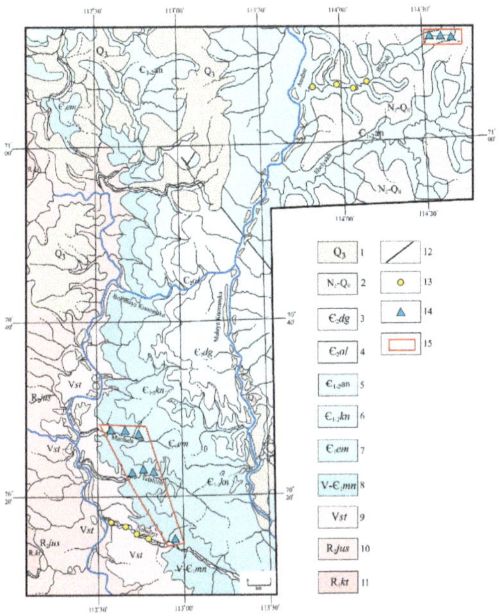

Figure 1. The scheme of the geological structure of the Anabar placer area [35,36]: 1—Quaternary alluvial sediments; 2—Neogene-Quaternary pebbles; 3–7—Cambrian sediments; 3—limestones of the Dzhakhtar formation; 4—clay and dolomitic limestones, marls of the Olenek formation; 5—dolomites of the Anabar formation; 6—clay, bituminous, organogenic-clastic limestones, combustible, siliceous, bituminous shales, siliceous and calcareous claystones, siltstones of the Kuonamskaya formation; 7—clay, dolomitic, algal, silty limestones and marls of the Emyaksin formation; 8—dolomitic limestones of the Manykai formation of the Vendian–Cambrian age; 9—dolomites, clay shales, gritstones and conglomerates of the Starorechenskaya formation of the Vendian; 10—dolomites, sandstones, siltstones, gritstones and conglomerates of the Yusmastakh formation of the Late Riphean; 11—sandy, clay dolomites, sandstones, gritstones, conglomerate-breccias of the Kotuikan formation of the Early Riphean; 12—faults; 13—places of collection of exploratory samples; 14—places of collection of hand specimen samples; 15—areas where hydrothermal-metasomatic formations were found.

Determination of the quantitative composition of native gold, the composition of minerals in polished sections, photographing of samples and mapping of the distribution of elements by area were performed on a scanning electron microscope JEOL JSM-6840LV (Tokyo, Japan) with an energy dispersion spectrometer Energy 350 Oxford Instruments (Abingdon, UK). Quantitative analysis and processing of the results were performed using the XPP method Software INCA Energy version Oxford instruments INCA the microanalysis Suite Issue 4.17. Shooting conditions: accelerating voltage 20 kV, measure current 1.07 nA, spectrum set time during quantitative optimization on cobalt and samples, 7 s. The error of the analysis of the main components is −1%–1.5%. The detection limit for most elements is −0.2%–0.8% and 1% or more for "heavy" elements. Reference samples were used: for Au, Ag-Au850, and Hg-HgTe. Analytical work was performed in the Department of physical–chemical methods of analysis of DPMGI SB RAS (analyst Khristoforova N.V.). Measurements were performed both in the central part and along the periphery of individual gold particles for determining the degree of their chemical transformation, which is expressed in the formation of shells and areas with a fineness of 990‰–1000‰ or in the complete transformation of grains. Analyses of the peripheral parts of native gold grains were not considered in the calculations. The number of determinations included in the calculations is 825.

The gradation of the fineness of gold is provided according to the classification of N.V. Petrovskaya [9].

The internal structure of native gold was studied by etching it in mounted thin sections according to a proven technique [37], using a reagent: $HCl + HNO_3 + FeCl_3 \times 6H_2O + CrO_3$ + thiourea + water (next in text—aqua regia). The reagent was applied to the surface of polished gold mounted in an artificial polished section. The gold particles were etched from 10 to 30 s in several approaches. After each etching procedure, the polished section was washed under a strong stream of water, then dried. After that, the revealed internal structures were studied in detail using a Polar 3 (Observation instruments llc., Saint-Petersburg, Russia) ore microscope and a JEOL scanning electron microscope JSM-6480LV (JEOL Ltd., Tokyo, Japan). The interpretation of the features of internal structures was performed in accordance with the recommendations of N.V. Petrovskaya [9,35], L.A. Nikolaeva [10,12], N.E. Savva, and V.K. Preis [11].

Abbreviations of mineral names are provided according to L.N. Warr [38].

3. Geological and Structural Position of the Region

The geological structure of the studied territory includes the Riphean terrigenous formations, Cambrian carbonate rocks, Permian terrigenous sediments, and Triassic volcanogenic formations overlain by the Neogene and Quaternary poorly consolidated sediments (Figures 1 and 2) [39].

Igneous rocks are represented by intrusive bodies of basic and alkaline-ultrabasic composition of the Triassic age (Figure 2) [39]. The studied placers are located within the Lena–Popigai swell, complicated by structures of the II order, the Ebelyakh uplift and the Billyakh depression, as well as the northeastern framing of the Anabar shield. According to the predecessors, disjunctive faults were of great importance in the development of the area. They form a number of systems of the north-western, north-eastern, latitudinal, and meridional directions, with zones of increased fracturing. It is important to emphasize that the rejuvenation of ancient deep fault systems (Molodo–Popigai, Anabar–Eekit, and Zhigansk) took place in the studied area during the Mesozoic tectonic-magmatic activation, which led to the formation of a whole series of new faults [6,40]. It should be noted that most modern rivers inherited paleovalleys of the Mesozoic watercourses formed on tectonic deformations.

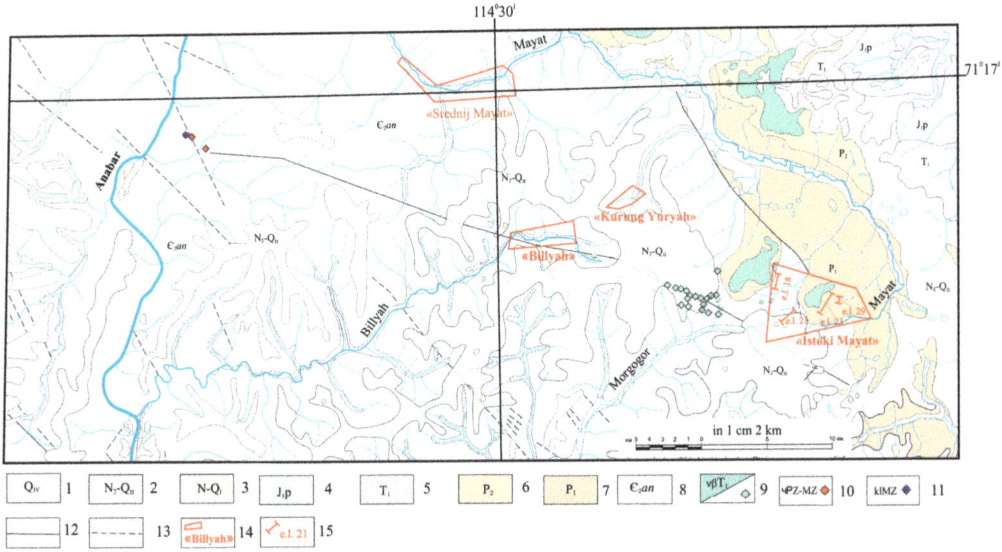

Figure 2. Schematic geological map of the basin of the middle flow of the Anabar river [39]: 1—Quaternary sediments; 2—Upper Pliocene—Middle Quaternary loams, sands, pebbles; 3—Neogene—Lower Quaternary relict pebbles; 4—Jurassic sandstones, siltstones; 5—Triassic basalts and their tuffs; 6—Late Permian sandstones; 7—Early Permian sandstones, conglomerates; 8—dolomites of the Anabar formation of the Cambrian; 9—Mesozoic intrusive formations; 10—kimberlite pipes; 11—dikes of alkaline picrites; 12—determined tectonic deformations; 13—assumed tectonic deformations; 14—areas where hydrothermal-metasomatic formations have been found; 15—exploratory core drilling lines.

4. Results and Discussion of the Study

4.1. Typomorphism of Placer Gold

4.1.1. Native Gold of the Billyakh River Placer

The Billyakh complex gold–platinum–diamond-bearing alluvial placer is located in the middle flow of the watercourse of the same name, the right inflow of the Anabar river (Figure 1). The river valley flows in the dolomites of the Anabar formation of the Middle Cambrian. The productive horizon is represented by sand–pebble material containing silty, clay, and boulder sediments in various ratios. The average thickness of the producing bed in the upper part of the placer is 1.75 m, the average width of the commercial contour is 75.4 m. The slopes of the valley are mostly gentle, swamped, only in the lower reaches, where the riverbed cuts the bedrock. There are steeper areas covered with boulder screes of the dolomites of the Anabar formation. Fragments of alluvial sediments of Middle-Upper Quaternary and Neogene age are observed on the slopes of the valley [39].

The granulometric composition of the studied native gold is as follows: 1–2 mm, 1.5%; 0.5–1 mm, 1.5%; 0.25–0.5 mm, 30%; and −0.25 mm, 67%. The fineness of native placer gold varies very widely (Table 1): 951‰–999‰, 78%; 950‰–900‰, 8%; 800‰–899‰, 8%; 700‰–799‰, 3%; and 400‰–699‰, 3% (Figure 3). The impurity elements Cu and Hg are determined only at the level of the detection limit of the device, 0.2%.

Table 1. The most typical analyses of the chemical composition of native gold of the Billyakh river placer on gradations of gold fineness.

№	Cu	Hg	Au	Ag	Total	The Fineness
1	nd	nd	99.83	0.67	100.5	
2	nd	nd	98.71	0.73	99.40	Very high
3	nd	nd	98.23	0.31	98.50	
4	nd	nd	91.93	8.40	100.30	
5	nd	nd	90.35	9.97	100.30	High
6	nd	nd	92.01	7.81	99.80	
7	nd	nd	87.93	11.59	99.50	
8	nd	nd	88.04	11.17	99.20	Medium
9	nd	nd	85.24	13.95	99.10	
10	nd	nd	76.69	21.57	98.20	
11	nd	nd	71.39	27.18	98.50	Relatively low
12	nd	nd	72.85	25.58	98.40	
13	nd	nd	68.42	29.70	98.10	
14	nd	0.60	68.64	29.26	98.50	Low
15	nd	nd	51.17	48.17	99.50	

Notes: 5 groups; all values in %; nd: not detected.

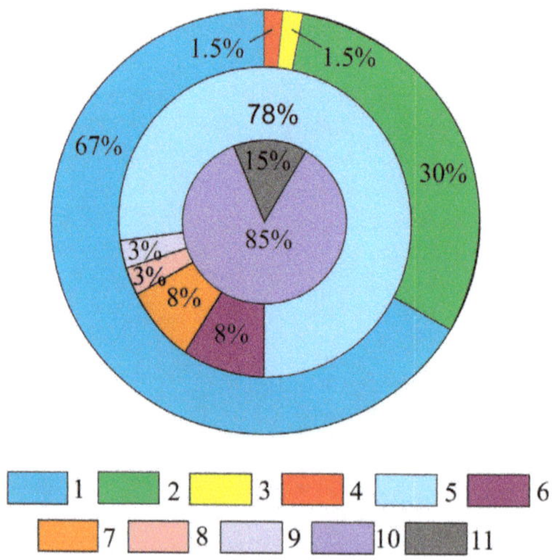

Figure 3. The main typomorphic features of the Billyakh river basin gold: 1–4—granulometric composition: 1—<0.2 mm, 2—0.2–0.5 mm, 3—0.5–1 mm, 4—1–2 mm; 5–9—fineness: 5—951‰–999‰ (very high-fineness), 6—950‰–900‰ (high-fineness), 7—800‰–899‰ (medium-fineness), 8—700‰–799‰ (relatively low-fineness), 9—400‰–699‰ (low-fineness); 10–11—morphology: 10—lamellar and scaly, 11—slightly rounded gold.

Considering the morphological features, the main part of gold is characterized by a well-rounded scaly (Figure 4a) and lamellar shape with a thin, coarse, shagreen surface (Figure 4b). Quite often there are plates with faults and holes (Figure 4b). Lamellar gold has an intensely transformed internal structure, which is expressed by the structures of recrystallization, granulation, and the formation of a thick high-grade shell (Figure 4c–f). In flake gold particles, high-fineness gold almost completely replaces primary lower-fineness gold. In such particles, relict primary gold is preserved only in their central parts (Figure 4g). In general, lamellar and scaly gold is characterized by high fineness (900‰–999‰). The set of the identified typomorphic features of the main part of the studied gold indicates the long-time residence in exogenous conditions and its redeposition from intermediate sources (paleo-placers).

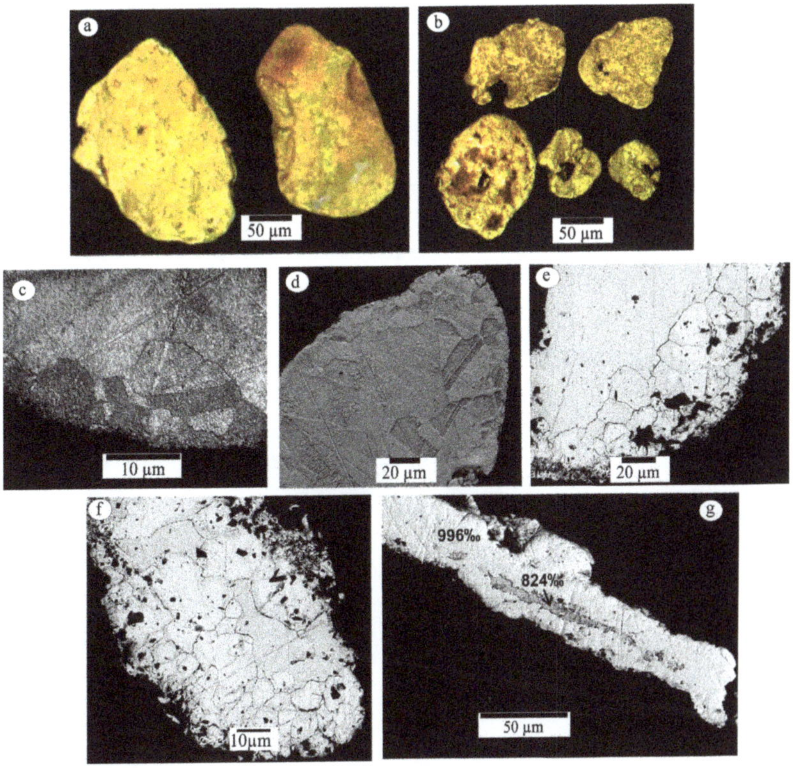

Figure 4. Morphology and internal structure of rounded high-fineness gold from the Billyakh river placer, supplied from intermediate sources: (**a**) scaly gold; (**b**) lamellar gold with casts of pressing of minerals on the surface and with through holes; (**c**,**d**) recrystallization structures; (**e**,**f**) granulation structures, etched with a reagent based on aqua regia; (**g**) relics of medium-fineness gold in the central part of the scale almost completely replaced by high-fineness gold; (**c**–**g**) shot in BSE mode.

Along with this gold, very fine (−0.2 mm) slightly rounded with lamellar and angular, lumpy forms were found in all studied samples, up to 15% (Figure 5a,b). Their surface is rough, spongy, and porous. The fineness varies in very wide intervals, from low (514‰) up to very high (999‰) (Table 1). The analysis of the fineness of gold and its morphological features has shown that exclusively gold particles of ore habit have relatively low (799‰–700‰) and low (699‰–500‰) fineness. The main feature of this gold is its heterogeneous heterophase internal structure, with a block nature. A low-fineness (500‰–600‰) phase was identified in the central part of the gold particles (Figure 5c), and a relatively low-fineness one along the periphery (750‰). An intermittent very thin (the first microns) high-fineness phase, characterized by a lighter shade, is observed along the marginal parts of relatively low-fineness gold (Figure 5d, is shown by yellow arrows). The sharpness and discontinuity of the high-fineness margin indicate a short-time residence of this gold in exogenous conditions. In general, the set of typomorphic features of this gold indicates the presence of nearby primary sources.

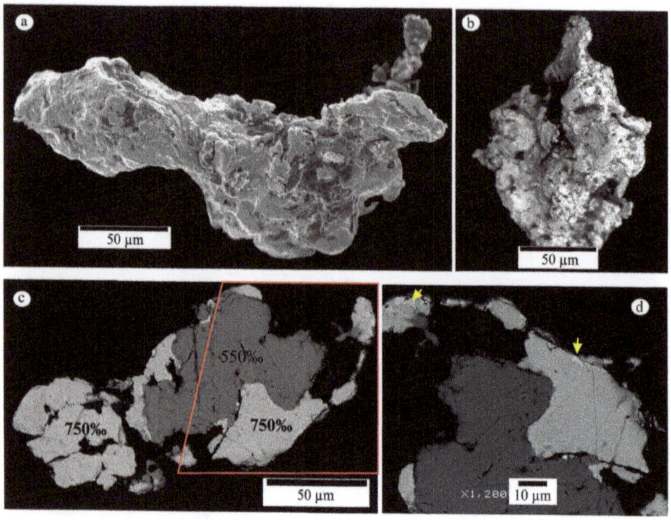

Figure 5. Morphology and features of the internal structure of slightly rounded gold of the Billyakh river placer: (**a**) slightly rounded gold particle; (**b**) angular, lumpy gold with an rough porous surface; (**c**) block nature of the internal structure of multiphase gold (detalized area marked in red); (**d**) detalized area: multiphase gold, the yellow arrows show the high-fineness phase, etched with a reagent based on aqua regia; (**a**,**b**) shot in SEI mode; (**c**,**d**) shot in BSE mode.

4.1.2. Native Gold of the Nebaibyt River Placer

The Nebaibyt river valley drains the northeastern slope of the Anabar massif and flows through dolomitic limestone of the Vendian–Cambrian and terrigenous-carbonate sediments of the Vendian age (Figure 1).

The granulometric composition of the studied gold is as follows: 1–0.5 mm, 5%; 0.25–0.5 mm, 40%; and −0.25 mm, 55%. Its fineness throughout the placer is as follows: 951‰–999‰, 80%; 950‰–900‰, 8%; 800‰–899‰, 4%; 700‰–799‰, 3%; 600‰–699‰, 1%; and <600‰, 4% (Figure 6).

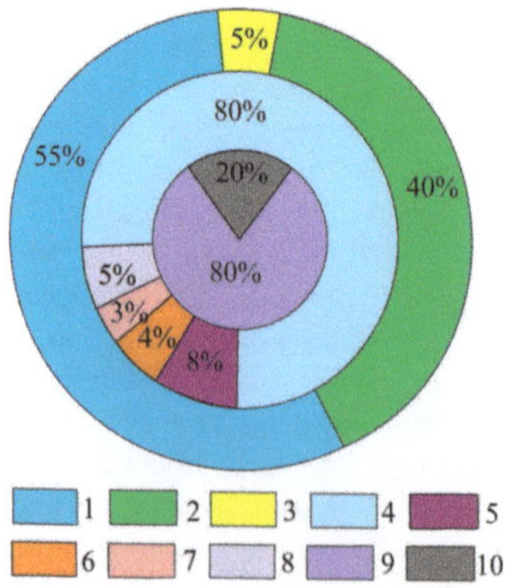

Figure 6. The main typomorphic features of the Nebaibyt river basin gold: 1–3—granulometric composition: 1—<0.2 mm, 2—0.2–0.5 mm, 3—0.5–1 mm; 48—fineness: 4—951‰–999‰ (very high-fineness), 5—950‰–900‰ (high-grade), 6—800‰–899‰ (medium-grade), 7—700‰–799‰ (relatively low-grade), 8—400‰–699‰ (low-grade); 9–10—morphology: 9—lamellar and scaly, 10—slightly rounded gold.

Native gold is mainly represented by a lamellar morphology with a coarse-shagreen surface. Lamellar gold has a high fineness (900‰–999‰) and is characterized by a highly modified internal structure with recrystallization and granulation structures, and by thick high-grade margins and very high-grade intergranular veinlets (Figure 7). This indicates the repeated redeposition of gold from ancient intermediate sources into younger sediments.

At the same time, in all the samples studied, up to 15%, slightly rounded gold was identified, which is represented by isometric crystals, angular-lumpy and lamellar forms, as well as peculiar hook-shaped individuals (Figure 8). As a rule, slightly rounded gold is marked in the −0.25 mm class. Its fineness varies very widely, from low (494‰) to very high (more than 950‰) (Table 2). Its internal structure also has a heterogeneous heterophase nature (Figure 8c), similar to the internal structure of slightly rounded placer gold of the Billyakh river placer (Figure 5c,d). Gold particles have not shown signs of transformation in hydrodynamic conditions, which indicates the vicinity of their primary source.

Table 2. The most typical analyses of the chemical composition of native gold of the Nebaibyt river placer on gradations of gold fineness.

№	Cu	Hg	Au	Ag	Total	The Fineness
1	0.60	nd	99.06	0.34	100.02	Very high
2	nd	nd	97.66	0.84	98.50	
3	nd	nd	99.13	0.87	100.30	
4	nd	nd	92.77	5.89	98.68	High
5	nd	nd	94.94	4.97	99.90	
6	nd	nd	91.22	8.05	99.30	
7	nd	nd	87.79	10.96	98.70	Medium
8	nd	nd	83.65	15.21	98.80	
9	nd	nd	86.22	11.89	98.10	
10	nd	nd	71.23	28.98	100.20	Relatively low
11	nd	nd	77.49	22.48	100.00	
12	nd	nd	74.85	24.58	99.40	
13	nd	nd	66.14	31.87	98.00	Low
14	nd	nd	64.10	36.34	100.40	
15	nd	nd	54.96	46.58	101.50	

Notes: 5 groups; all values in %; nd—not detected.

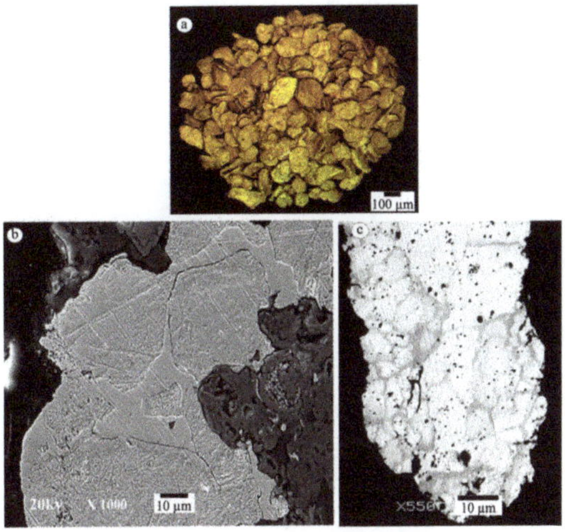

Figure 7. Morphology and internal structure of gold of the Nebaibyt river placer: (**a**) lamellar gold (general view); (**b**) very high-grade (990‰) intergranular veinlets in high-grade gold (950‰); (**c**) granulation structures of high-grade gold, etched with a reagent based on aqua regia; (**b**,**c**) shot in BSE mode.

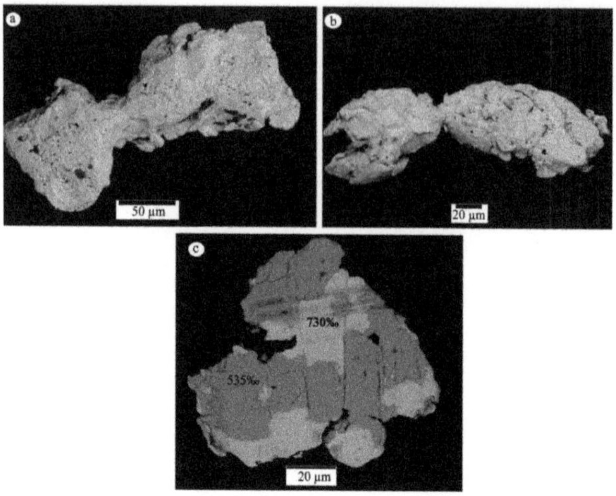

Figure 8. Morphology and internal structure of slightly rounded native gold of the Nebaibyt river placer: (**a**,**b**) slightly rounded gold; (**c**) block nature of the internal structure of multiphase gold, etched with a reagent based on aqua regia; (**a**,**b**) shot in SEI mode; (**c**) shot in BSE mode.

Native gold is mainly represented by a lamellar morphology with a coarse, shagreen surface. Lamellar gold has a high fineness (900‰–999‰) and is characterized by a highly modified internal structure with recrystallization and granulation structures, and by thick high-grade margins and very high-grade intergranular veinlets (Figure 7). This indicates the repeated redeposition of gold from ancient intermediate sources into younger sediments.

No mineral inclusions were found in gold except for small inclusions of quartz.

Since at this stage of research, in gold of the placers of the Billyakh and Neibabyt rivers no ore inclusions were found, we provide data on mineral inclusions in gold of the Mayat river placer, which was studied by us earlier [6,8]. The Mayat river joins the Anabar river just north of the Billyakh river and flows almost parallel to it (Figure 2).

The fineness of gold of the Mayat river varies widely: 951‰–999‰, 35%; 950‰–900‰, 15%; 800‰–899‰, 30%; 700‰–799‰, 10%; and 400‰–699‰, 10%. Inclusions of pyrite, arsenopyrite, galena, quartz, and potassium feldspar were found in slightly rounded gold with a grain size of more than 0.5 mm (Figure 9).

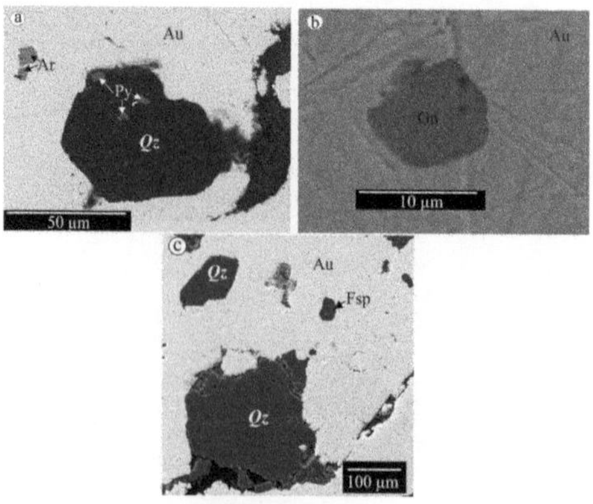

Figure 9. Mineral inclusions in placer gold of the Mayat river (BSE mode): (**a**) quartz (Qz), pyrite (Py) and arsenopyrite (Ar) inclusions in native gold (Au); (**b**) galena inclusion (Gn) in native gold (Au); (**c**) quartz (Qz) and potassium feldspar inclusions (Fsp) in native gold (Au), shot in BSE mode.

As can be seen in Figure 10, three groups of native gold are identified by the nature of the Ag distribution. The first, most common group (about 70%–85% of the whole number of samples) contains up to 15 wt. % silver. The next group with an Ag content of up to 20 wt. % is about from 10% to 20% of the samples. The least represented group (about 5%–7% of the samples) is characterized by a silver content of 20 to 45 wt. %. According to Chapman et al. [28] samples populations that show either bimodal Ag content or a wide range of Ag values, do not necessarily indicate multiple sources. In our case, the totality of the studied typomorphic features allows us to talk about two types of sources of gold in modern placers: (1) an intermediate source (paleo-placers), which can serve the Neogene–Quaternary watershed pebbles (group 1 by Ag content); (2) and nearby ore occurrences (2 and 3 groups by Ag content).

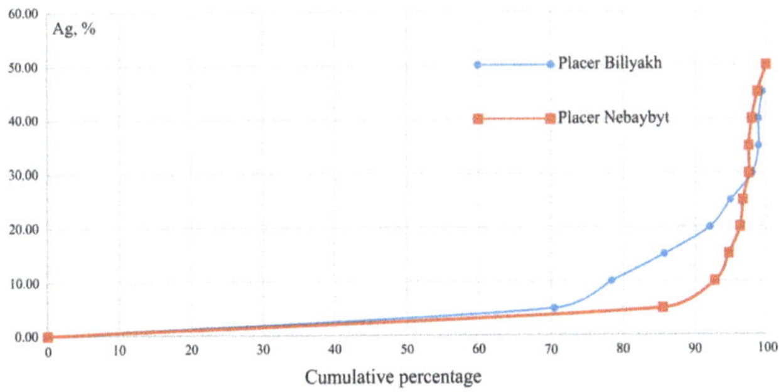

Figure 10. Cumulative Ag content chart. According to the data of 578 (Billyakh placer) and 247 (Nebaibyt placer) determinations of the chemical composition of native gold.

Thus, according to the set of typomorphic features of native gold in the studied watercourses, two varieties of native gold were identified.

The first variety is well-rounded high-grade gold of lamellar and scaly morphology with a naturally treated surface. Its internal structure is intensively modified, represented by recrystallization structures, granulation structures, and thick very high-grade shells (Figures 4c–g and 7b,c). According to L.A. Nikolaeva and co-authors [12], the structure of gold, which was in ancient conglomerates for a long time, reflects profound changes caused by its transformation in conditions of not only ancient crust and placer formation, but also diagenesis and epigenesis of sediments. At the same time, a high-grade shell can reach a thickness of 0.25 mm or more, partially or completely replacing the primary matrix. Mechanical deformations contribute to the processes of recrystallization of the peripheral parts of the gold particles, which is expressed by the formation of a micrograin structure. With a long-time residence at rest, the process of collective recrystallization occurs and leads to an even deeper transformation of the high-grade shell. The most intense changes in the internal structure of gold occur after the burial of sediments and their lithification as a result of temperature exposure, involving intense diffusion of silver. Disintegration of gold grains occurs, which leads to the formation of granulation structures [12]. Stewart et al. [27] explained the presence of relatively thick (>100 microns) gold-rich shells in gold particles from the point of view of recrystallization during physical deformation related to the transfer of gold particles during successive fluvial downcuttings in a tectonically active region. According to their data [27], grains of native gold that were processed through paleo-placers can completely recrystallize. The degree of recrystallization, in their opinion, depends on the amount of deformation of the gold particles. According to Chapman et al. [28], the heterogeneity spatially associated with grain boundaries is formed much later than the time of precipitation of native gold. Thus, significant internal crystallographic transformations in gold particles in the conditions of paleo-placers contribute to the acceleration of their chemical changes. In general, the results of the study of the internal structure of the well-rounded gold of the studied placers indicate its redeposition through intermediate sources. Gold-bearing watershed pebbles the of Neogene–Quaternary age, widely developed in the studied areas, probably served as intermediate sources [5]. According to B.R. Shpunt, the Precambrian gold-bearing quartz veins served as the primary sources of gold for them [2].

The second variety is slightly rounded native gold with a wide variation in fineness: from low (494‰) to very high (999‰). The most specific feature is its block heterophase internal structure (Figure 5c,d; Figure 8c). The heterophase nature of these gold particles can be explained by the specific conditions of formation. Firstly, heterophase nature can be caused by fluctuations in the degree of supersaturation of solutions during their growth in shallow conditions. According to V.P. Samusikov [41], short-term pulsational changes in the degree of supersaturation of solutions occurred in the shallow conditions of ore formation: a sharp increase or decrease, since the thermodynamic conditions were extremely unstable. At the same time, with an increase in the degree of supersaturation of solutions, the concentration of Ag increased, and with a decrease, it decreased [41]. Secondly, sulfide minerals predominate in the composition of the ores of the ore occurrences we discovered, described below, so it can be assumed that sulfide sulfur dominated in ore-bearing solutions. According to I.V. Gaskov [42], as the temperature of solutions decreases and the groundmass of sulfides is deposited, the amount of sulfur and its activity in solutions drops sharply and silver is more concentrated in native gold, which causes a decrease in the fineness of native gold from its early to late generations. According to Chapman [28] and co-authors, the substitution of gold with silver, independent of grain boundary control, is the result of a developing hydrothermal system, whereas Ag-rich intergranular veinlets may form later. According to the same authors, lower ore formation temperatures contribute to a higher Ag content in Au–Ag alloys. Thus, these data indicate a low-temperature shallow environment for the formation of the second variety of gold.

The complex of typomorphic features of this gold (slight rounding, wide variation in fineness, heterophase internal structure unchanged in exogenous conditions) is direct evidence of the vicinity of the primary source formed in shallow conditions. In this regard, it was suggested that in the sources of the Billyakh and Nebaibyt rivers, in the field of development of carbonate rocks in fault zones, gold ore occurrences can be localized, and potential primary sources of very fine gold of the second variety. This became a prerequisite for the execution of prospecting traverses in order to identify ore occurrences. This assumption was confirmed, we identified gold-bearing hydrothermal-metasomatic formations for the first time in the studied area as a result of these traverses. The following sub-chapter provides some of the features of these ore occurrences obtained at this stage of research.

4.2. *Ore Occurrences of Native Gold*

As a result of the prospecting traverses, hydrothermal-metasomatic formations with gold-sulfide mineralization localized in fault zones were identified (Figures 1 and 2). The prospecting traverses were performed on artificial sites (artificial channels, landfills of worked placers, and road clearing sites) and natural outcrops of hydrothermally altered dolomites of the Anabar formation of the Middle Cambrian (Anabar river basin) and dolomitic limestone of the Vendian–Cambrian age (Bolshaya Kuonamka river basin). Hand specimen and rock samples were collected mainly in crush zones confined to the faults. In additional, core samples that were taken by "Almazy Anabara" JSC for the searching of kimberlite bodies in the watershed of the Mayat and Morgogor rivers (Figure 2) were studied too.

4.2.1. Mineralogical Features of Ore-Bearing Hydrothermal-Metasomatic Formations

The initial substrate for hydrothermal-metasomatic formations in the Anabar region was mainly near-fault fractured dolomites of the Anabar formation of the Middle Cambrian and dolomitic limestones of the Vendian–Cambrian ages. In addition, hydrothermalites developed from slightly cemented fine-pebble quartz conglomerates of the Early Permian age were found for the first time in the core of exploration wells (Figure 2). The fragmented nature of sampling does not yet allow us to reliably determine the shape and type of ore body occurrence.

Considering epigenetic mineral associations, quartz-potassium feldspar and siliceous-quartz (jasperoids) hydrothermal-metasomatic formations were identified.

Quartz-potassium feldspar hydrothermal-metasomatic formations have yellowish and rusty-brown tint. Their structure is predominantly banded, net-vein, due to the development of differently oriented veinlets of iron oxides, decomposition products of sulfides (Figure 11a,b). Potassium feldspar in apocarbonate hydrothermalites is represented by very small (1–5 microns) crystals, mainly diamond-shaped, which has a diffuse distribution throughout the groundmass (Figure 11c,d) or it is developed in thin (up to 150 microns) veinlets of quartz-ferruginous-aluminosilicate composition penetrating the dolomite (Figure 11e,f). In the Permian terrigenous rocks, this mineral is characterized by larger sizes (up to 1 mm) and rectangular shapes (Figure 12).

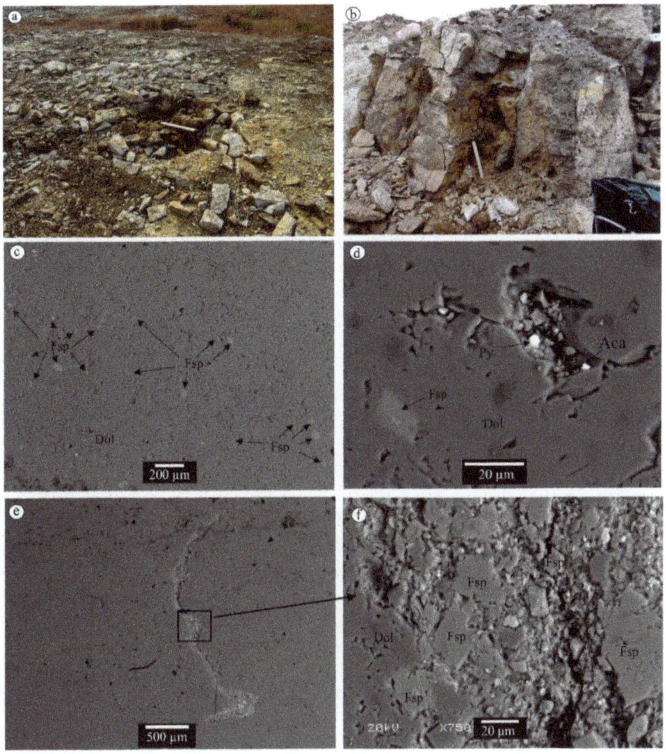

Figure 11. Apocarbonate quartz-potassium feldspar metasomatites: (**a**) the cataclase zone in bedrock of the worked placer, where hydrothermalites are developed (Kurung Yuryakh site); (**b**) weathered hydrothermal-metasomatic formations in the wall of the artificial channel at the sources of the Billyakh river; (**c**,**d**) disseminated potassium feldspar mineralization (BSE mode); (**e**,**f**) potassium feldspar in veinlets of variable quartz-iron-aluminosilicate composition (BSE mode). Fsp—potassium feldspar; Dol—dolomite; Py—pyrite; Aca—acanthite.

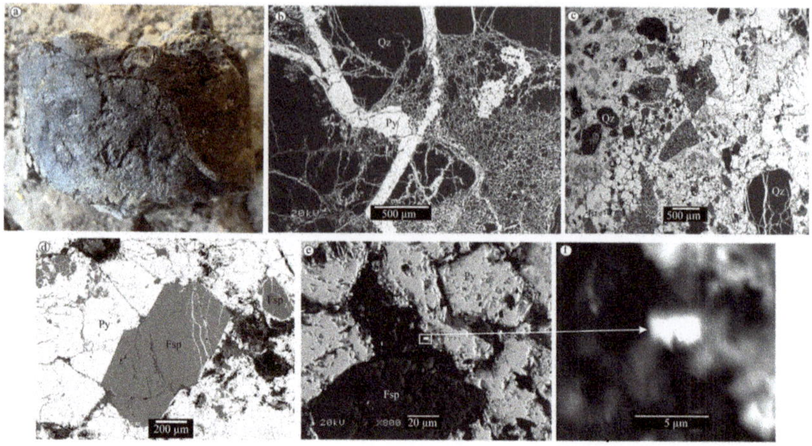

Figure 12. Hydrothermal-metasomatic formations developed by slightly cemented fine-pebble quartz conglomerates of the Early Permian age of the Anabar region: (**a**) core sample of pyritized conglomerate (interval 39.7–40 m of well A-3 of exploration line 21; sample 21-A3-5, microphotography); (**b**,**c**) types of pyrite mineralization (BSE mode): (**b**) net-vein; (**c**) entire; (**d**) pyrite-potassium feldspar mineralization in quartz conglomerates (BSE mode); (**e**,**f**) the smallest gold particle in a mineralized conglomerate (BSE mode). Qz—quartz; Fsp—potassium feldspar; Py—pyrite.

Jasperoids are apocarbonate rocks with a SiO_2 content from 81 to 96.5%. Their characteristic feature is the oolitic structure. In these formations chalcedonic quartz almost completely replaces dolomites (Figure 13). Only very small relics of them are observed in the central parts of the oolites.

Figure 13. Features of jasperoids: (**a**) jasperoid veins developed in the cataclasis zone on the "Billyakh" site; (**b**) polished sections made of jasperoids; (**c**) chalcedonic quartz (thin section); (**d**) veinlets of chalcedonic quartz and siliceous oolites (thin section).

Ore mineralization is represented by sulfides, native gold, and silver. The main sulfide minerals are galena and pyrite, which ratios vary by different sites. At this stage of research, no differences in the composition of ore components in the above-mentioned types of hydrothermal-metasomatic formations were revealed.

Native gold was found in the form of very small (up to 15 microns) isometric grains in microcracks of a carbonate or siliceous substrate (Figure 14a,b). The smallest particle of gold with a size of about 4 microns was found at the border of the growth of pyrite and potassium feldspar in the Permian mineralized conglomerate (Figure 12e,f). Hg and Cu are defined as impurity elements in gold. The maximum content of Hg in gold reaches 2.4%, which is determined at the "Billyakh" site. At the "Istoki Mayat" site, a very fine gold particle with Cu admixture with a content of up to 18% was found in the abundantly pyritized dolomite of the core sample (Figure 14c).

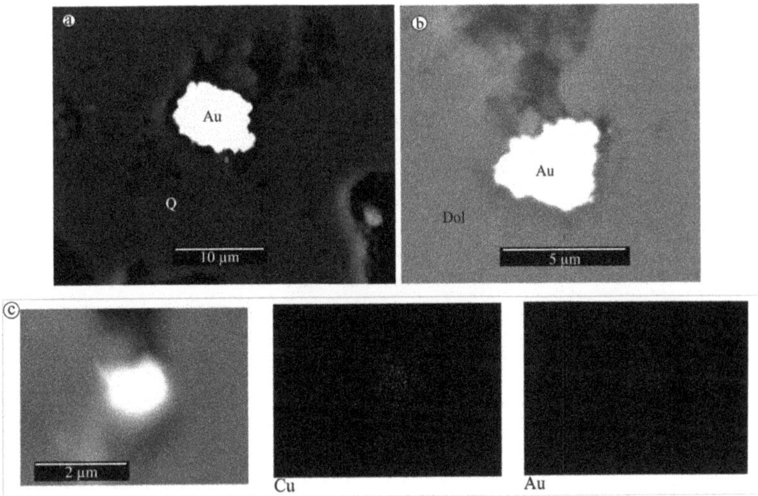

Figure 14. Fine gold (Au) particles in (**a**) jasperoid, (**b**) apocarbonate quartz-potassium feldspar metasomatites (**c**) copper gold (Cu—18%) and impurity elements in gold in the apocarbonate quartz-potassium feldspar hydrothermal-metasomatic formation (shot in X-ray Au and Cu). Q—quartz; Dol—dolomite.

Galena is characterized by very small (the first microns) isometric grains (Figure 15a). Pyrite in modified dolomites occurs in the form of small inclusions (Figure 15a), and in hydrothermal-metasomatic formations formed on the Permian conglomerates, it has a veined and continuous nature of distribution, penetrating into the gaps between quartz pebbles, sometimes completely replacing conglomerate cement and sometimes quartz pebbles (Figure 11b,c). Silver sulfide (acanthite?) is quite widespread in apocarbonate formations and is represented by crystals, and their aggregates are up to 20 microns in size (Figure 15b). Large grains of sphalerite (about 5 mm) were found in the calcite veinlet of quartz-potassium feldspar dolomite (Figure 15c). In peripheral areas and cracks, it is replaced by zincite. Very fine galena and relatively large pyrite were identified as inclusions in sphalerite (Figure 15d). Single smallest (first microns) particles of chalcopyrite, stibnite and arsenopyrite are observed in hydrothermally altered dolomites of all studied sites.

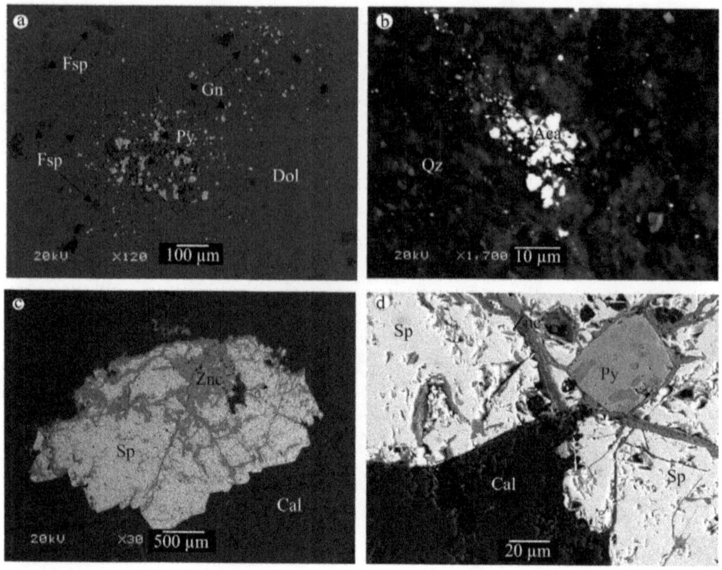

Figure 15. Sulfide minerals of hydrothermal-metasomatic formations (BSE mode): (**a**) inclusions of galena and pyrite in quartz-potassium feldspar dolomite; (**b**) aggregate structure of acanthite in quartz metasomatite; (**c**) large sphalerite developed in calcite veinlet, replaced by zincite along cracks and periphery; (**d**) inclusions of pyrite in sphalerite with thin veinlets of zincite. Qz—quartz; Fsp—potassium feldspar; Dol—dolomite; Py—pyrite; Gn—galena; Sp—sphalerite; Aca—acanthite; Cal—calcite; Znc—zincite.

4.2.2. Lithological and Tectonic Factors of Mineralization

Lithological factors played an important role in the formation of gold mineralization. First, this refers to the degree of permeability of the ore deposition medium, which is determined by the presence of voids and pores in rocks. In our case, the near-fault permeable zones of tectonic fracturing of carbonate rocks and slightly cemented fine-pebble conglomerates were favorable for ore deposition.

The tectonic factor of mineralization control has a great importance. The studied hydrothermal-metasomatic formations are clearly confined to the Dogoy–Mayat and Dogoy–Kuoysky faults of the Molodo–Popigai system of faults (Anabar river basin), as well as the Zhigansky fault zone (Bolshaya Kuonamka river basin). Dikes of the basic rocks of the Mesozoic age are localized within the Dogoy–Kuoysky fault (Figure 2). The latter can be either the actual sources of mineralization, or can serve as a catalyst for the mobilization of dispersed primary hydrothermal-sedimentary ore matter. The first assumption is supported by the findings of pyrite and chalcopyrite in dolerites. According to A.V. Okrugin and co-authors, the presence of sulfide minerals in igneous rocks indicates the saturation of magmatic melts with sulfides and their potential ore content [43]. In addition, the discovery of gold with an admixture of Cu (up to 18%) in apocarbonate hydrothermalites of core samples (Figure 14c) may indicate its connection with basic magmatism [44].

For the section of the middle flow of the Anabar river, two stages of mineralization formation can presumably be distinguished. In the first stage, ore matter in the form of primary hydrothermal ores were formed in the near-fault zone. The second stage is related to the processes of the Mesozoic tectonic-magmatic activation, when dikes of the basic composition intrude along the fault zones, and at the same time, there was a redeposition of ore matter, as well as, possibly, additional supply due to sulfide-saturated magmatic melts. As a result of these processes, gold-sulfide fluid solutions spread along the feathering faults into the discharge zones, forming gold-bearing hydrothermal-metasomatic formations.

5. Conclusions

Thus, our study showed that typomorphic features of native gold from placers can be used as additional criteria in forecasting and searching for primary ore sources of gold. According to the results of the study, the indicators of gold in modern placers are associated with two different sources: intermediate sources (paleo-placers) and nearby ores. These results were a prerequisite for setting prospecting traverses, which resulted in discovery of gold-bearing hydrothermal-metasomatic formations for the first time in the Anabar region.

The identified composition of mineral inclusions in gold of the Mayat river is identical to the mineral paragenesis of gold in considered ore occurrences, this indicates that these ore formations were additional sources of gold for modern placers of the studied area.

Very interesting in our opinion is the identification of specific multiphase internal structures of very fine slightly rounded gold. The fact that they were found at sites remote from each other (more than 120 km), where near-fault apocarbonate gold-bearing hydrothermal-metasomatic formations were formed, allows us to assume that the identified internal structures are indicator signs for native gold of hydrothermal-metasomatic formations of the Anabar region.

Of course, the discovered gold-bearing hydrothermal-metasomatic formations of the Anabar region require further detailed study. Strong turfness, overlapping by the Quaternary cover and, as a result, the fragmented nature of sampling, does not yet allow us to reliably determine the shape and nature of the occurrence of ore bodies. Moreover, the extent of gold-sulfide mineralization was not determined. However, apparently, near-fault hydrothermal-metasomatic formations are quite widespread in the northeast of the Siberian platform. This is evidenced by the discovery of silicified carbonate rocks, similar in composition and structure, with sulfide and precious metal mineralization in the area of the southeastern flank of the Molodo–Popigai fault system. Here, as a result of studying the petrographic features of the pebbles of the alluvial sediments of the Molodo river basin, silicified rocks with an oolitic structure with inclusions of pyrite, silver, and gold were found [45].

Funding: Work is done on state assignment of DPMGI SB RAS.

Data Availability Statement: Not applicable.

Conflicts of Interest: The author declares no conflict of interest.

References

1. Vinogradov, V.A.; Krasilshchikov, A.A.; Gorina, I.G. The sources of gold in the Olenek uplift. *Mater. Geol. Miner. Yakut ASSR. Yakutsk.* **1967**, *15*, 114–119. (In Russian)
2. Shpunt, B.R. Typomorphic features and genesis of placer gold in the north of the Siberian platform. *Geol. Geophys.* **1974**, *9*, 77–88. (In Russian)
3. Yablokova, S.V.; Israelev, L.M. Mineralogy of gold in the sedimentary cover strata of the Olenek uplift of different ages. In *The Works of TsNIGRI*; VTSIOM: Moscow, Russia, 1988; pp. 58–65. (In Russian)
4. Timofeev, V.I.; Nesterov, N.V.; Shpunt, B.R. Gold bearing of Western Yakutia. *Mater. Geol. Miner. Resour. Yakut ASSR. Yakutsk.* **1970**, *17*, 103–110. (In Russian)
5. Shpunt, B.R. Placer occurrences of gold in the Cenozoic sediments of the Lena-Anabar interfluve. In *Placer Gold Content of Central Siberia*; Research Institute of Arctic Geology: Leningrad, Russia, 1973; pp. 31–35. (In Russian)
6. Gerasimov, B.B.; Nikiforova, Z.S. Assumed formational types of primary sources of gold in the Anabar region (northeast of the Siberian platform). *Sci. Educ.* **2017**, *2*, 11–16. (In Russian)
7. Nikiforova, Z.S.; Gerasimov, B.B.; Glushkova, E.G.; Kazenkina, A.G. Gold content of the East of the Siberian platform: Placers—primary sources. *Geol. Ore Depos.* **2013**, *55*, 305–319. [CrossRef]
8. Nikiforova, Z.S.; Gerasimov, B.B.; Glushkova, E.G.; Kazhenkina, A.G. Indicative features of placer gold for the prediction of the formation types of gold deposits (east of the Siberian Platform). *Russ. Geol. Geophys.* **2018**, *59*, 1318–1329. [CrossRef]
9. Petrovskaya, N.V. *Native Gold*; Nauka: Moscow, Russia, 1973; p. 347. (In Russian)
10. Nikolaeva, L.A. *Genetic Features of Native Gold as a Criterion in the Search and Evaluation of Ores and Placers*; Nedra: Moscow, Russia, 1978; p. 100. (In Russian)
11. Savva, N.E.; Preis, K.V. *Atlas of Native Gold of the North-East of the USSR*; Nauka: Moscow, Russia, 1990; p. 292. (In Russian)

12. Nikolaeva, L.A.; Gavrilov, A.M.; Nekrasova, A.N.; Yablokova, S.V.; Shatilova, L.V. *Atlas of Native Gold of Ore and Placer Deposits of Russia*; TsNIGRI: Moscow, Russia, 2003; p. 184. (In Russian)
13. Gerasimov, B.; Beryozkin, V.; Kravchenko, A. Typomorphic Features of Placer Gold from the Billyakh Tectonic Melange Zone of the Anabar Shield and Its Potential Ore Sources (Northeastern Siberian Platform). *Minerals* **2020**, *10*, 281. [CrossRef]
14. Nikiforova, Z. Criteria for determining the genesis of placers and their different sources based on the morphological features of placer gold. *Minerals* **2021**, *11*, 381. [CrossRef]
15. Nikiforova, Z.S.; Tolstov, A.V. Gold-bearing placer assemblages in the east of the siberian platform: Origin and prospects. *Geol. Ore Depos.* **2022**, *64*, 1–25. [CrossRef]
16. Chapman, R.J.; Leake, R.C.; Moles, N.R.; Earls, G.; Cooper, C.; Harrington, K.; Berzins, R. The application of microchemical analysis of gold grains to the understanding of complex local and regional gold mineralization: A case study in Ireland and Scotland. *Econ. Geol.* **2000**, *95*, 1753–1773.
17. Chapman, R.J.; Mortensen, J.K.; LeBarge, W.P. Styles of lode gold mineralization contributing to the placers of the Indian River and Black Hills Creek, Yukon Territory, Canada as deduced from microchemical characterization of placer gold grains. *Miner. Depos.* **2011**, *46*, 881–903. [CrossRef]
18. Chapman, R.J.; Mortensen, J.K. Characterization of gold mineralization in the Northern Cariboo Gold District, British Columbia, Canada, through integration of compositional studies of lode and detrital Gold with historical placer production: A template for evaluation of orogenic gold districts. *Econ. Geol.* **2016**, *111*, 1321–1345.
19. Moles, N.R.; Chapman, R.J. Integration of detrital gold microchemistry, heavy mineral distribution, and sediment geochemistry to clarify regional metallogeny in glaciated terrains: Application in the Caledonides of southeast Ireland. *Econ. Geol.* **2019**, *114*, 207–232. [CrossRef]
20. Chapman, R.J.; Moles, N.R.; Bluemel, B.; Walshaw, R.D. *Detrital Gold as an Indicator Mineral*; Geological Society, London, Special Publications: London, UK, 2022; Volume 516, pp. 313–336.
21. Torvela, T.; Lambert-Smith, J.S.; Chapman, R.J. (Eds.) *Recent Advances in Understanding Gold Deposits: From Orogeny to Alluvium*; Geological Society, London, Special Publications: London, UK, 2022; p. 516. [CrossRef]
22. Groen, J.C.; Craig, J.R.; Rimstidt, J.D. Gold-rich rim formation on electrum grains in placers. *Can. Mineral.* **1990**, *28*, 207–228.
23. Falconer, D.; Craw, D. Supergene gold mobility: A textural and geochemical study from gold placers in southern New Zealand. In: Titley, S.R. (Ed.), Supergene Environments, processes and products. *Econ Geol Special Publ.* **2009**, *14*, 77–93.
24. Craw, D.; Lilly, K. Gold nugget morphology and geochemical environments of nugget formation, southern New Zealand. *Ore Geol. Rev.* **2016**, *79*, 301–315. [CrossRef]
25. Craw, D.; Hesson, M.; Kerr, G. Morphological evolution of gold nuggets in proximal sedimentary environments, southern New Zealand. *Ore Geol. Rev.* **2016**, *80*, 784–799. [CrossRef]
26. Nikiforova, Z.S.; Kalinin, Y.u.A.; Makarov, V.A. Evolution of Native Gold under Exogenous Conditions. *Russ. Geol. Geophys.* **2020**, *61*, 1244–1259. [CrossRef]
27. Stewart, J.; Kerr, G.; Prior, D.; Halfpenny, A.; Pearce, M.; Hough, R.; Craw, D. Low temperature recrystallisation of alluvial gold in paleoplacer deposits. *Ore Geol. Rev.* **2017**, *88*, 43–56. [CrossRef]
28. Chapman, R.J.; Banks, D.A.; Styles, M.T.; Walshaw, R.D.; Piazolo, S.; Morgan, D.J.; Grimshaw, M.R.; Spence-Jones, C.P.; Matthews, T.J.; Borovinskaya, O. Chemical and physical heterogeneity within native gold: Implications for the design of gold particle studies. *Miner. Depos.* **2021**, *56*, 1563–1588. [CrossRef]
29. Lalomov, A.V.; Chefranov, R.M.; Naumov, V.A.; Naumova, O.B.; LeBarge, W.; Dilly, R.A. Typomorphic features of placer gold of Vagran cluster (the Northern Urals) and search indicators for primary bedrock gold deposits. *Ore Geol. Rev.* **2017**, *85*, 321–335. [CrossRef]
30. Alam, M.; Li, S.; Santosh, M.; Yuan, M. Morphology and chemistry of placer gold in the Bagrote and Dainterstreams, northern Pakistan: Implications for provenance and exploration. *Geol. J.* **2019**, *54*, 1672–1687. [CrossRef]
31. Dongmo, F.W.; Chapman, R.J.; Bolarinwa, A.T.; Yongue, R.F.; Banks, D.A.; Olajide-Kayode, J.O. Microchemical characterization of placer gold grains from the Meyos-Essabikoula area, Ntem complex, southern Cameroon. *J. Afr. Earth Sci.* **2019**, *151*, 189–201. [CrossRef]
32. Melchiorre, E.B.; Henderson, J. Topographic gradients and lode gold sourcing recorded by placer gold morphology, geochemistry, and mineral inclusions in the east fork San Gabriel River, California, USA. *Ore Geol. Rev.* **2019**, *109*, 348–357. [CrossRef]
33. Potter, M.; Styles, M.T. Gold characterization as a guide to bedrock sources for the Estero Hondo alluvial gold mine, western Ecuador. *Trans. Inst. Min. Metall. Sect. B Appl. Earth Sci.* **2003**, *112*, 297.
34. Patyk-Kara, N.G. *Mineralogy of Placers: Types of Placer Provinces*; IGEM RAS: Moscow, Russia, 2008. (In Russian)
35. Dukhanin, S.F.; Ehrlich, E.N. *Explanatory Note to the Geological Map of the Scale 1: 200 000 (Sheet R-49-XVII, XVIII—Anabar Series)*; Soyuzgeolfond: Moscow, Russia, 1967; p. 70.
36. Rubenchik, I.B.; Borshcheva, N.A.; Zaretsky, L.M. *Explanatory Note to the Geological Map of the Scale 1: 200 000 (Sheet R-50-VII, VIII)*; Sevmorgeo: Moscow, Russia, 1980; p. 72.
37. Petrovskaya, N.V.; Novgorodova, M.I.; Frolova, K.E. *The Nature of Structures and Substructures of Endogenous Native Gold Particles. Mineralogy of Native Elements*; DVNTS USSR Academy of Sciences: Vladivostok, Russia, 1980; pp. 10–20.
38. Warr, L.N. IMA-CNMNC approved mineral symbols. *Mineral. Mag.* **2021**, *85*, 291–320. [CrossRef]

39. Grakhanov, S.A.; Shatalov, V.I.; Shtyrov, V.A.; Kychkin, V.R.; Suleymanov, A.M. *Diamonds Placers of Russia*; "Geo" Publishing House: Novosibirsk, Russia, 2007; p. 457. (In Russian)
40. Milashev, V.A. *Structures of Kimberlite Fields*; Nedra: Leningrad, Russia, 1979; p. 183.
41. Samusikov, V.P. Mechanisms of the concentration of isomorphic impurity elements in minerals during hydrothermal ore formation. *Russ. Geol. Geophys.* **2010**, *51*, 338–352.
42. Gas'kov, I.V. Major impurity elements in native gold and their association with gold mineralization settings in deposits of Asian folded areas. *Russ. Geol. Geophys.* **2017**, *58*, 1080–1092. [CrossRef]
43. Okrugin, A.V.; Zemnukhov, A.L.; Zhuravlev, A.I. Copper-nickel sulfide ore occurrence in dolerites of the eastern slope of the Anabar shield. *Nat. Resour. Arct. Subarctic.* **2021**, *4*, 16–28. [CrossRef]
44. Savva, N.E.; Shilyaeva, N.A.; Alevskaya, N.L. *Topomineralogy of Constitutional Features of Native Gold of the Lower Amur Placer Area*; SVKNII FEB RAS: Magadan, Russia, 2004; p. 173.
45. Gerasimov, B.B.; Kravchenko, A.A. Ore occurrences of the Anabar placer area—Potential primary sources of gold. *NEFU Bulletin. Earth Sci.* **2020**, *4*, 17–28. [CrossRef]

minerals MDPI

Article

Insights into Regional Metallogeny from Detailed Compositional Studies of Alluvial Gold: An Example from the Loch Tay Area, Central Scotland

Robert Chapman *, Taija Torvela and Lucia Savastano

Ores and Mineralization Group, School of Earth and Environment, The University of Leeds, Leeds LS2 9JT, UK
* Correspondence: r.j.chapman@leeds.ac.uk

Abstract: Compositional features of a total of 1887 gold alluvial particles from six localities to the south of Loch Tay in central Scotland were interpreted to establish different types of source mineralization. Populations of gold particles from each locality were grouped according to alloy and inclusion signatures. Inclusion suites provided the primary discriminant with gold from Group 1 localities showing a narrow range of simple sulphide and sulphoarsenide inclusion species, whereas a wide range of minerals including molybdenite, bornite and various Bi and Te- bearing species were identified in gold from Group 2 localities. Whilst the range of Ag in alloys in all populations was similar, higher incidences of measurable Hg and Cu were detected in Group 1 and Group 2 gold samples respectively. The application of compositional templates of gold from other localities worldwide indicated that Group 1 gold is orogenic and Group 2 gold is a mixture of porphyry and epithermal origin; a result that is sympathetic to the spatial relationships of sample localities with local lithologies. This approach both provides an enhanced level of understanding of regional gold metallogeny where in situ sources remain undiscovered, and permits clearer targeting of deposit types during future exploration.

Keywords: alluvial gold; indicator mineral; compositional studies

Citation: Chapman, R.; Torvela, T.; Savastano, L. Insights into Regional Metallogeny from Detailed Compositional Studies of Alluvial Gold: An Example from the Loch Tay Area, Central Scotland. *Minerals* **2023**, *13*, 140. https://doi.org/10.3390/min13020140

Academic Editor: Galina Palyanova

Received: 12 December 2022
Revised: 10 January 2023
Accepted: 11 January 2023
Published: 18 January 2023

1. Introduction

1.1. Application of Gold Compositional Studies to Understanding Regional Metallogeny

Stream sediment geochemistry is a widely used tool in early stage exploration [1,2], and in many cases field teams are taught to record the presence of visible gold in panned concentrates. Regional scale reconnaissance yields important information on the relative distribution of alluvial gold that can in some cases inform follow up study, e.g., [3], but in areas of complex geology different deposit types may be present in the same drainage system, and therefore the discovery of particulate gold in the pan is not indicative of a specific exploration target. The ability to differentiate between gold particles derived from different deposit types would, consequently, be extremely useful. The compositional characteristics of alluvial gold particles that may be related to conditions of formation are preserved after liberation [4,5] and hence provide a route by which this may be achieved.

The development of gold studies as a tool in mineral exploration has been described previously in terms of composition [4,6], and particle morphology [7,8] and the reader is referred to those texts for full accounts. This study focuses on compositional characteristics of gold particles but clear evidence from alluvial gold morphology indicating of a lack of fluvial transport and hence a proximal influx is also an important consideration. In summary, mineralogical features of gold particles comprise the metallic elements other than gold present in the alloy and the suite of inclusions of other minerals observed in the polished section. These features provide criteria by which to characterize a population of gold particles collected from a specific locality (henceforward referred to as 'a sample population'). Mineralogical characteristics of gold can provide two important types of information in these circumstances. Firstly, characterization of sample populations from different localities permits evaluation of whether or not they are likely to be derived from genetically related sources. Such information is valuable when establishing the extent of mineralization that contributes to an alluvial occurrence. This aspect is particularly important when investigating the source of gold from economic placers because these are likely to be derived from substantial lode occurrences [9]. Secondly, compositional signatures of gold particles may be specific to their hard-rock source, i.e., the deposit type. The development of methodologies to develop gold as an indicator mineral have been reviewed [4] and generic deposit-specific inclusion suite signatures have been reported [10]. In other words, specific mineralogical features of gold within a sample population can point towards a specific deposit type; equally, a range of features within a single population can indicate the presence of several different deposit types within an area. The approach of mineralogical characterization of alluvial gold has been used to advantage in various studies to elucidate the placer-lode gold relationships in orogenic gold districts e.g., Klondike, Yukon [11] and the Cariboo Gold District, British Columbia [9]. Similarly, gold formed in magmatic hydrothermal systems was the focus of studies undertaken in Ecuador [12], British Columbia [13] and Yukon [14]. The same principles were applied to infer the source(s) of gold in economic placers of the Ural Mountains, [15,16] and Salair Ridge [17], Russia. In this paper, we show how alluvial gold characterization can offer valuable insights into the regional metallogeny and gold mineralization styles in the Grampian Terrane of Scotland.

1.2. The Study Area

The Highlands of Scotland (Figure 1A) are world famous for their role in the development of geology as a science, particularly for understanding thrust tectonics and metamorphism, e.g., [18,19], but there is also a long history of small scale metalliferous mining [20]. In Perthshire, historic mining was undertaken for lead and silver at Coire Buidhe, with some native gold recorded as well; and for copper at Tomnadashan (Figure 1B) [21]. Whilst gold from hard-rock sources has historically only been produced as a by-product of mining other metals, some fairly notable alluvial mines operated in the 1500s and the widespread existence of alluvial gold in the drainage of the central Highlands and elsewhere in Scotland has been long recognized [22]. Other accounts, presumably drawn from oral traditions [23,24] of artisanal gold seeking, describe a 'Central Goldfield' comprising auriferous areas in Perthshire, Angus and Aberdeen that define a belt to the north of the Highland Boundary Fault (HBF; Figure 1A). Some large gold nuggets have been recorded, e.g., the Turrich Nugget from Glen Quaich, which forms part of the collections of the Natural History Museum, London. Bedrock sources for the widespread alluvial gold in Scotland remained undiscovered until the latter part of the 20th century, when a concerted exploration effort combined traditional prospecting, geophysics, and stream sediment geochemical approaches. These ultimately led to the discovery of in-situ gold-bearing mineralization at Cononish near Tyndrum (Figure 1A; [25]) which is currently the focus of mining operations by Scotgold Ltd. but remains the only hard-rock gold mine ever to have operated in Scotland. The Cononish deposit has attracted some research interest and many aspects of ore genesis have been clarified [26,27]. In addition to Cononish, bedrock occurrences of gold-bearing quartz veins have been identified at localities near Loch Tay (Calliachar-Urlar and Tombuie Figure 1B; [28–31]). In 2020, Green Glen Minerals Ltd. also discovered a new bedrock occurrence of Au-Ag-Pb-Zn quartz veins at Lead Trial (Figure 1B) [32]. The Loch Tay area where these gold occurrences have been identified is the focus of our study because the mutual genetic relationships of gold localities in the Loch Tay area, their classification in the context of modern deposit models and their relationship to the tectonomagmatic history of central Scotland remain unclear.

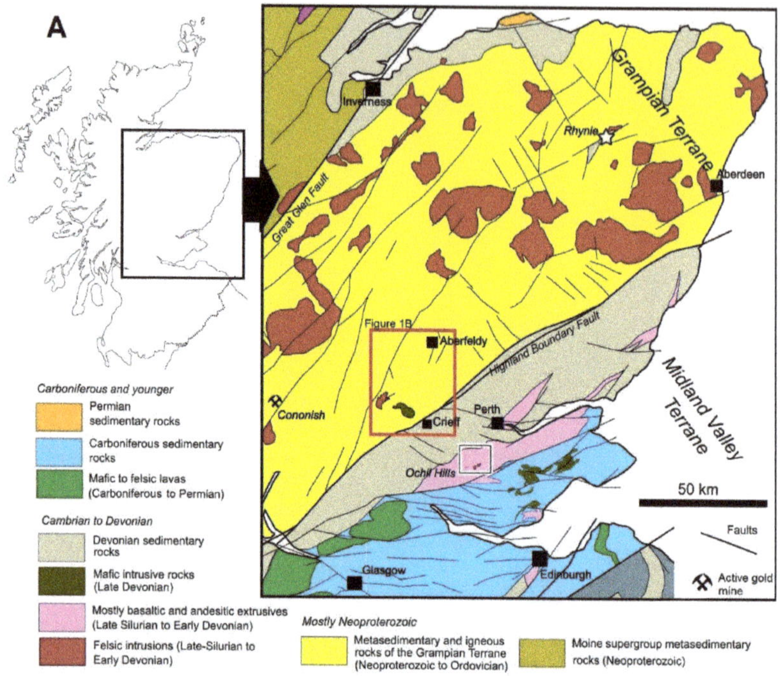

Figure 1. *Cont.*

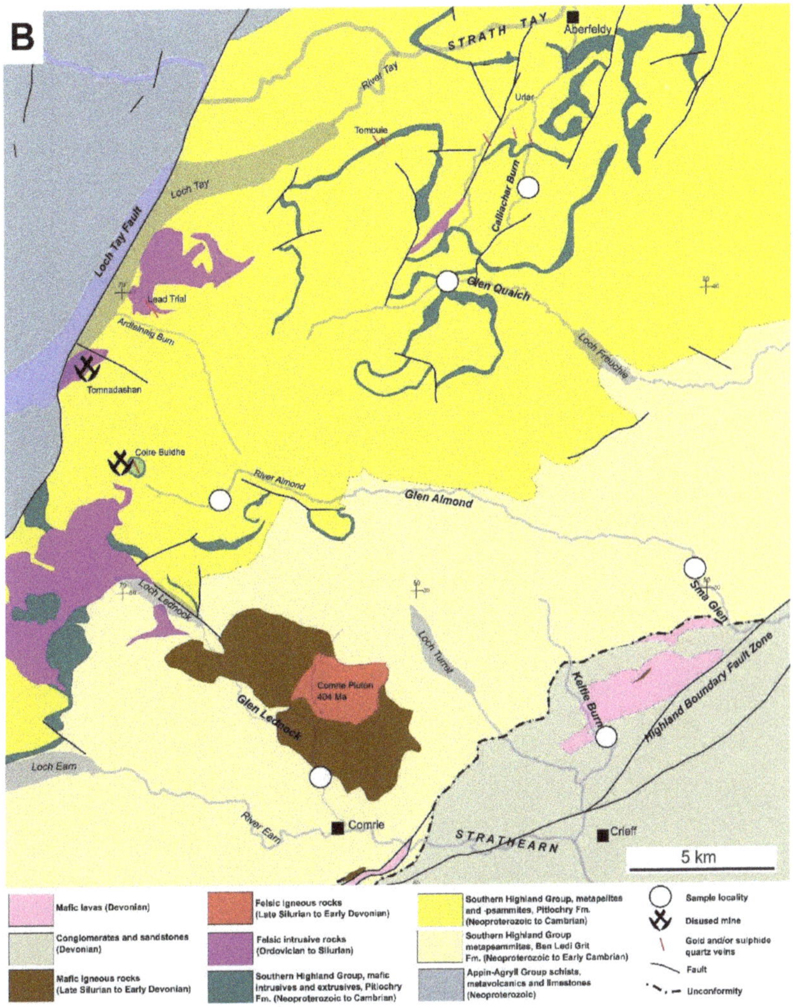

Figure 1. (**A**) Simplified geological map of central Scotland. The location of the study area is indicated with a red rectangle (Figure 1B overleaf); the area of the Ochil Hills gold particle study area of Chapman et al. (2005) [33] is indicated by a white box. The Early Devonian Rhynie gold occurrence is also indicated. (**B**) Geological map of the study area. The sampling localities are indicated, along with the location of the known gold- and sulphide-bearing quartz vein systems.

The study area comprises large tracts of hilly upland moor incised by drainage systems that ultimately discharge in the Firth of Tay at Perth, either via Strathearn in the south, Glen Almond and Glen Quaich in the center, or Strath Tay in the north (Figure 1B). Exposure is generally poor and often confined to the fluvial incisions along the valley floor. The valley sides normally exhibit till deposits and moraines resulting from glaciation, and much of the sediment in the present channels comprises reworked glacial sediment. Consequently, in general, the spatial correlation of alluvial gold with the in-situ source is difficult to establish because of the probable involvement of both glacial and fluvial processes. Glaciers were confined to valleys in the last stages of glaciation [34], and therefore the current fluvial sediment load (and by inference gold particles) comprises material derived from within the current drainage basins. Based on the authors' field work, alluvial gold was found to be present in all of the main drainage systems, i.e., the Tay, Quaich, Almond and Earn, and therefore it is almost certain that in situ gold mineralization is widespread.

A summary of the geological history of central Scotland and the known aspects of the gold mineralization are provided in the subsequent section. However, it is important to highlight that the deposit styles of the gold occurrences at Loch Tay are currently unknown. The geological complexity of the area means several different gold deposit types are possible: igneous rocks provide potential for porphyry and associated epithermal mineralization, whilst the metasedimentary and metaigneous packages are globally common host rocks for orogenic gold deposits. In this contribution we use gold compositional studies to demonstrate that not only do different deposit types occur, but they can be identified in specific drainage systems. Resolving the deposit type(s) present in an area is not a trivial academic exercise as differentiation of the deposit type(s) is important both for guiding exploration approaches and prospect development [35]. As some of our samples were collected upstream from the known vein occurrences, our work also demonstrates that new, still unknown potential targets exist within the study area.

1.3. Geological Background

The study area lies mostly within the Grampian Terrane of Scotland (Figure 1). The Grampian Terrane consists of chiefly Neoproterozoic to Cambrian sedimentary and igneous rocks deposited onto the Laurentian passive margin, e.g., [36]. Most of the study area comprises the Southern Highland Group of the Dalradian Supergroup, consisting predominantly of metapelites and metapsammites with some layers of mafic and felsic igneous rocks (Figure 1B; [37,38]). The Grampian Terrane was pervasively deformed during the Ordovician-Silurian Caledonian orogeny. The first Caledonian collisional event, often referred to as the Grampian phase in Scotland and Ireland, was characterized by a ~NW-SE relative convergence of Laurentia vs. Baltica with crustal thickening through thrust and nappe tectonics [39–41]. This thickening phase was relatively short-lived, culminating at c. 475–465 Ma with granitoid magmatism and regional prograde metamorphism (e.g., [42–44]). The Grampian phase was followed by a period of uplift and unroofing until c. 430 Ma [45–48] and various authors have suggested that cooling through 300 °C occurred between 460 and 430 Ma [46,48].

The second phase of the Caledonian orogeny, the Scandian phase, resulted in the final closure of the Iapetus Ocean and the docking of Avalonia against Laurentia between c. 435 and 410 Ma, although its duration is somewhat debated, with estimates ranging from 10 to 35 Ma (e.g., [40,49,50]). Within central Scotland, the main effects of the Scandian event were transpressional strike-slip movements along major transcurrent orogen-parallel faults and their subsidiary fault systems, resulting from an approximately N-S orientated maximum principal stress [41]. These structures, which include the Loch Tay Fault in Figure 1B, are widely recognized throughout the Grampian Terrane; many of the largest faults are probably long-lived structures but at least some fault activity has been dated at 416–395 Ma [39]. A second phase of significant igneous activity took place around the same time, from c. 425 Ma onwards, interpreted to be mostly related to slab roll-back and subsequent break-off [43,51,52]. This magmatism was dominantly felsic, but most exhibit some degree of bimodalism, particularly the c. 404 Ma Comrie pluton which has a voluminous, early dioritic stage, followed by a slightly younger granitic stage [26,43,53]. It has been proposed that the emplacement of the Late Silurian–Early Devonian granitoids occurred in a transtensional setting, with a WSW-ENE directed regional extension [26]. Certainly there was a transition from transpression to transtension near this time: from c. 410 Ma onwards, widespread deposition of sedimentary rocks and both extrusive and intrusive magmatism, mostly mafic, occurred in association with pull-apart basin formation across central Scotland (Figure 1A) [54–56]. The final convergent event in the region is related to the ~N-S directed Acadian event around 400–390 Ma [40]. The effect of this event was mostly felt in the southern UK and the extent and style of its expression in Scotland is still unclear, but various fault reactivation features, interpreted to be transpressional, of fault zones in the Grampian Terrane show mid-Devonian radiometric dates [56]. The dominant tectonic style from Devonian to Permian was, however, that of extension/transtension.

Various localities with gold-bearing quartz veins have been identified in the Grampian Terrane in Scotland: in addition to Cononish, Calliachar-Urlar, Tombuie, and Lead Trial mentioned earlier, the epithermal hot-spring Rhynie gold occurrence is known NW of Aberdeen (Figure 1A) [57]. Which phase of the Caledonian-Acadian tectonism resulted in the formation of the veining and the gold mineralization in the Grampian Terrane is partly an open question as few of the vein systems have been directly dated, but at least some of the mineralization seems to be Early Devonian. The fault gouge and hydrothermal K-feldspar associated with the gold-bearing veins of the Cononish gold mine was dated at c. 410 Ma, [26] with probable indications of partial setting of the isotope system around 340 Ma (K-Ar and $^{40}Ar/^{39}Ar$ ages). The paragenesis, particularly with respect to the relationship between the dated material and gold, was not described, but the 410 Ma age was interpreted to represent the gold mineralization age, and the later age was suggested to reflect a Carboniferous dextral reactivation of the fault zone (associated with galena precipitation, without gold). The 410 Ma age coincides with the post-Scandian transition from transpression to transtension [58]. The only other age that exists for gold-bearing rocks in Scotland is the 407.1 $\pm$ 2.2 Ma age for the Rhynie cherts, obtained by sampling hydrothermal K-feldspar from two veins that represent feeder conduits and a hydrothermally altered andesite wall rock [58]. A detailed paragenetic interpretation is not provided, but SEM imaging shows a gold particle spatially associated with feldspars and quartz in hydrothermally altered and fractured andesite. The gold mineralization at Rhynie has also been linked with the extensional tectonics and pull-apart volcanosedimentary basin formation [58]. For both Cononish and Rhynie, studies to date suggest dominantly magmatic systems but with complex, evolving fluid sources. Rhynie has been categorized as an epithermal deposit, reporting both meteoric water and magmatic fluid signatures in the hydrothermal alteration minerals associated with the mineralization [57]. At Cononish, the early fluids responsible for the gold mineralization were magmatic but evolved to involve a strong isotopic signature from the surrounding Dalradian metasediments [27].

In the Loch Tay area, little information exists on the known vein systems. Some "deposit"-scale work has been completed on basic vein structures, mineralogy and preliminary parageneses of the veins at Calliachar-Urlar, Tombuie and Coire Buidhe [21,29,31]. No published data exists for the Lead Trial veins, but they are similarly orientated as the Tombuie, Calliachar-Urlar and Coire Buidhe veins, i.e., subvertical with fairly consistent NNW-SSE strikes (Figure 1B). Likewise, very little compositional information describing gold from the study area has been published. Studies of a gold particles observed in ore, liberated from crushed vein samples and collected from adjacent alluvial localities at Calliachar Burn all showed similar Ag and Hg profiles [28,29], with relatively high maxima of 30 wt% and 40 wt% respectively. The inclusion suite was reported in terms of metal element components (Fe: 55%, (Pb + Zn): 20%, Cu: 15% and Co: 10%) and non-metal element components (S: 65%, As: 35%) [28]. For Tombuie, high Ag and Hg values in Au-Ag alloy were reported for seven gold particles in a mineralized in-situ sample [31]. Alluvial gold from Glen Lednock yielded a distinctive inclusion signature containing molybdenite, chalcocite and bornite, not observed in the other localities [59]. The authors interpreted this mineralogical association as indicative of a genetic relationship between the gold and the Comrie pluton (Figure 1B).

Even this limited amount of gold compositional data from Loch Tay, derived from small sample suites, provides an indication of variable gold deposit types and the potential for further gold compositional studies to illuminate regional metallogeny. The results to date demonstrate the genetic relationship between gold in the Calliachar vein system and the adjacent placer, and hint at a genetic relationship with the Tombuie veins on account of the distinctive Ag and Hg contents of the Au alloy. By contrast, gold from Glen Lednock shows clear geochemical markers suggestive of mineralization formed in a magmatic hydrothermal system. We now expand from these previous gold characterization studies to investigate in more detail the range of gold compositional and inclusion data and their wider implications, with samples from different drainages and from upstream of the known vein occurrences at Calliachar-Urlar.

2. Materials and Methods

2.1. Collection of Gold Particles

The alluvial gold particles were collected using the field techniques previously described [59]. Sampling campaigns focused on obtaining sufficient particles to underpin subsequent interpretations of the gold signatures. The sample from Glen Lednock comprises material previously reported [59] augmented by gold particles collected during the present study. Characterization of the inclusion suites have proved to be very effective in establishing the type of gold mineralization [10]; however, the abundance of inclusions varies substantially between sample populations [60], and was unknown at the time of sampling. The average abundance of inclusions in polished sections is 10%, whilst at least 15 inclusion-bearing particles are considered necessary to generate a robust characterization [60]. Consequently field sampling aimed to collect at least 150 gold particles in a single sample population, but the relatively high abundance of gold at some localities facilitated collection of substantially larger numbers.

Details of the sample localities and sample populations are provided in Figure 1B and Table 1.

Table 1. Sample locations and sample characteristics.

Sample	Grid Reference	No. of Particles	Size Range (mm)	
			Min	Max
Calliachar Burn	NN 84011 45090	335	0.1	>10
Glen Quaich	NN 81863 40280	378	0.1	3
Glen Almond	NN 73798 33629	376	0.1	>10
Sma Glen	NN 90561 29437	393	0.1	1
Glen Lednock	NN 76808 23704	194	0.1	2
Keltie Burn	NN 86808 25013	211	0.1	2

2.2. Analytical Approach

Compositional characterization of the sample suites was carried out at the University of Leeds according to the methodology previously described [56]. Particles were mounted according to size on double sided tape, set in resin and polished to optical flatness. All particles within the resin blocks were visually inspected to record the presence of inclusions, using scanning electron microscope (SEM) imaging by a FEI Quanta 650 FEG-ESEM. Inclusion species were identified using an energy dispersive X-ray spectroscopy system (EDS) of the SEM. The Au, Ag, Cu and Hg contents of gold alloys were determined with electron probe microanalyzer (EPMA) JEOL 8230 Superprobe, using 20 kV accelerating voltage, 50 nA beam current, and a combined on- and off-peak count time of 1 min per element. Limits of detection (LOD) for Cu and Pd, were typically around 200 ppm and 900 ppm respectively, whilst avoidance of spectral interference between the HgMα and AuMβ X-rays necessitated using the Hg Mβ x-ray line and the associated higher detection limit of 3000 ppm.

The Ag profile of a sample population was based on a single analysis per sample. In cases where alloy heterogeneity was observed, spatial relationships between the different alloys were used to deduce the earliest alloy stage, and this value was adopted for profiling purposes.

2.3. Data Treatment

Approaches to depicting gold compositional data have been discussed previously [4,10]. Cumulative percentile vs. increasing metal concentration in alloy plots permit direct comparison of sample populations that comprise different numbers of particles and the covariance of metals within Au-Ag alloy may be depicted using binary plots to identify distinct compositional fields. For both plot types, logarithmic scales may be used where appropriate. Data describing minor alloying elements are frequently discussed in terms of the proportions of particles that contain the element to concentrations of >LOD and the proportion in which the element is quantifiable ($3 \times$ LOD = limit of quantitation LOQ).

Graphical depiction of inclusion suites is challenging because of the wide range of mineral species that have been observed. Two approaches have been used to depict inclusion suites, one based on inclusion mineralogy and the other on inclusion mineral chemistry. The use of radar diagrams based on mineral chemistry [14] are particularly useful where numerous inclusion species contain multiple cations or anions, as is commonly observed on gold formed in magmatic–hydrothermal systems [14]. The method involves attributing a 'metal' and 'non-metal' score to each inclusion species observed within a gold particle, with the total metal and non- metal scores each being 1. For example pyrite is scored Fe = 1 and S = 1, and chalcopyrite Fe = 0.5, Cu = 0.5, S = 1. The aggregate metal and non- metal scores for each element are expressed as proportions and plotted on the radar diagrams. This chemical approach to inclusion suite characterization can identify elemental components of the ore fluid even if they are represented in several mineral species. In other cases, specific mineral species may be a useful diagnostic tool and mineralogical data are best represented by spider plots, where tailored arrangement of mineral species on the x-axis highlights differences between inclusion suites in gold from different localities. The vertical axis depicts the proportion of inclusion-bearing particles within the population that contain a specific mineral inclusion. Spider diagrams were used in the current study to aid interpretation of inclusion assemblages where distinctive inclusions were present, and radar diagrams are presented to permit comparisons with previous work that examined the signatures of different deposit types globally.

2.4. Interpretation of Compositional Signatures

Interpretation of gold compositions is based on a multivariate treatment of the information available. The type of information may vary between sample populations, e.g., minor elements may be detectable or not, and inclusion suites may be more or less representative of the actual population depending on sample size. When available, inclusion suites provide clear evidence of genetic relationships between sample populations and strong indications of the deposit type. Ranges of Ag compositions are useful to establish 'same or different' criteria and may find application in speculating on zonal relationships between mineralization within the same hydrothermal system, as a consequence of the significant control of the temperature of the depositional environment over the Au-Ag ratio within the alloy. The importance of minor alloying metals (Cu, Hg and Pd) varies according to the degree to which they are present and their concentration.

3. Results

3.1. Gold Particle Size and Morphology

The size range of gold particles from the various sampling localities is indicated in Table 1 with examples of panned particles shown in Figure 2. The data can identify large differences between populations, but they are also influenced by the choice of the specific sampling locality as sedimentary environment has significant control over dense particle accumulation both in terms of the number and size of gold particles [2]. Nevertheless, with large sample populations such as these, the datasets clearly show that very coarse gold (up to >10 mm) is present in Calliachar Burn and the River Almond, whereas gold in Glen Quaich, Glen Lednock, and Keltie Burn was recorded to a maximum of 2–3 mm. Gold particles from Sma Glen are even smaller, with most being <1 mm.

Gold particles from the River Almond and Calliachar Burn are often equant and many have rough textures (Figure 2A–C). By contrast, the gold collected in Glen Quaich exhibited a more flaky, flat morphology (Figure 2D). A range of particle morphologies was evident in gold from Glen Lednock and Keltie Burn. The populations of small gold particles from

Sma Glen were predominantly equant and rough/hackly with some showing perfectly flat, reflective faces, presumably following detachment from other vein minerals (e.g., Figure 2E). Some very flaky and waterworn gold was also present (e.g., Figure 2F) but visual inspection showed these to comprise less than 20% of the sample population.

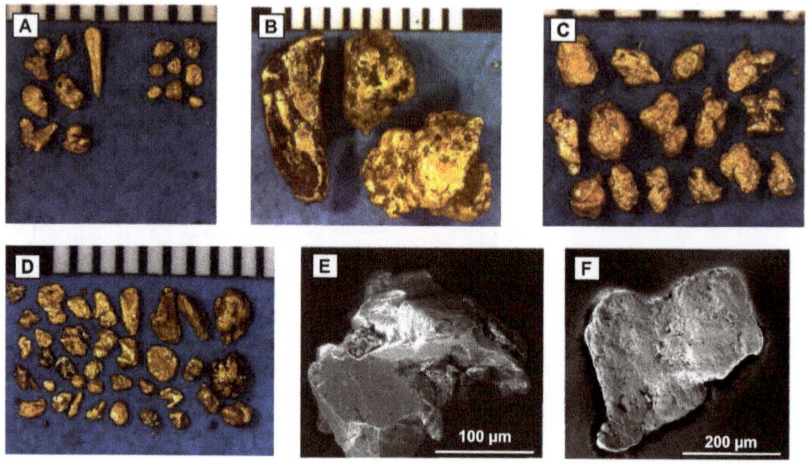

Figure 2. Examples of gold particles from different localities in the study area. (**A**,**B**) Images of alluvial gold particles from Calliachar Burn; (**C**) Glen Almond; and (**D**) Glen Quaich. (**E**,**F**) SEM SE images of gold particles from Sma Glen, showing contrasting morphologies indicative of different transport histories (see text).

3.2. *Gold Alloy Composition*

The Ag profiles of the sample populations from the study are depicted in Figure 3A. All sample populations exhibit a similar range of Ag, typically between 0 and 40 wt%, however the profile shapes differ substantially, with over 75% of the particles in sample populations from Calliachar Burn and the River Almond containing >20 wt% Ag, whereas Ag contents of >20 wt% are observed in less than 30% of the particles from Keltie Burn and Glen Lednock sample populations. The resulting profile shapes of these two groups reflect the dominance of high-Ag alloy (convex curves) and low-Ag alloy (concave curves), respectively. Gold samples from Glen Quaich and Sma Glen exhibit a Ag profile between these end members.

Hg concentrations in Au-Ag alloy are presented in Figure 3B. Sample populations of gold from Calliachar Burn Glen Quaich and Glen Almond show quantifiable Hg in 20, 13 and 8% of particles, respectively. Around 8% of particles from Sma Glen contain detectable Hg, but Hg values in gold from Keltie Burn and Glen Lednock are below LOD.

The cumulative plot for Cu content in the sample alloys is shown in Figure 3C. Copper was detectable in the majority of gold particles from all localities except Sma Glen where only 50% of particles show Cu values over the detection limit of 0.02 wt%. Gold from Keltie Burn and Glen Lednock exhibited the highest proportion of particles containing quantifiable Cu (40% and 25% of the sample, respectively). Only about 5% of the sample populations from Sma Glen, the River Almond, Glen Quaich, and Calliachar Burn contained quantifiable Cu.

The Ag-Hg binary plot for samples with Hg above LOD is presented in Figure 4. The data suggest two, probably overlapping, compositional fields according to Ag and Hg concentrations in the alloy (indicated in Figure 4 with dashed lines). Gold containing between 10 and 15 wt% Ag is far less likely to show high Hg values than gold containing over 15 wt% Ag. Gold from Glen Quaich dominates the low-Ag compositional field whereas all localities are represented in the high Ag-Hg set. The Ag profile for gold from Glen

Quaich (Figure 3A) shows a slight break in the slope at around 15 wt% Ag, corresponding to the point of increased abundance of elevated Hg. Within the high Ag-Hg set there does not appear to be any covariance of Ag and Hg.

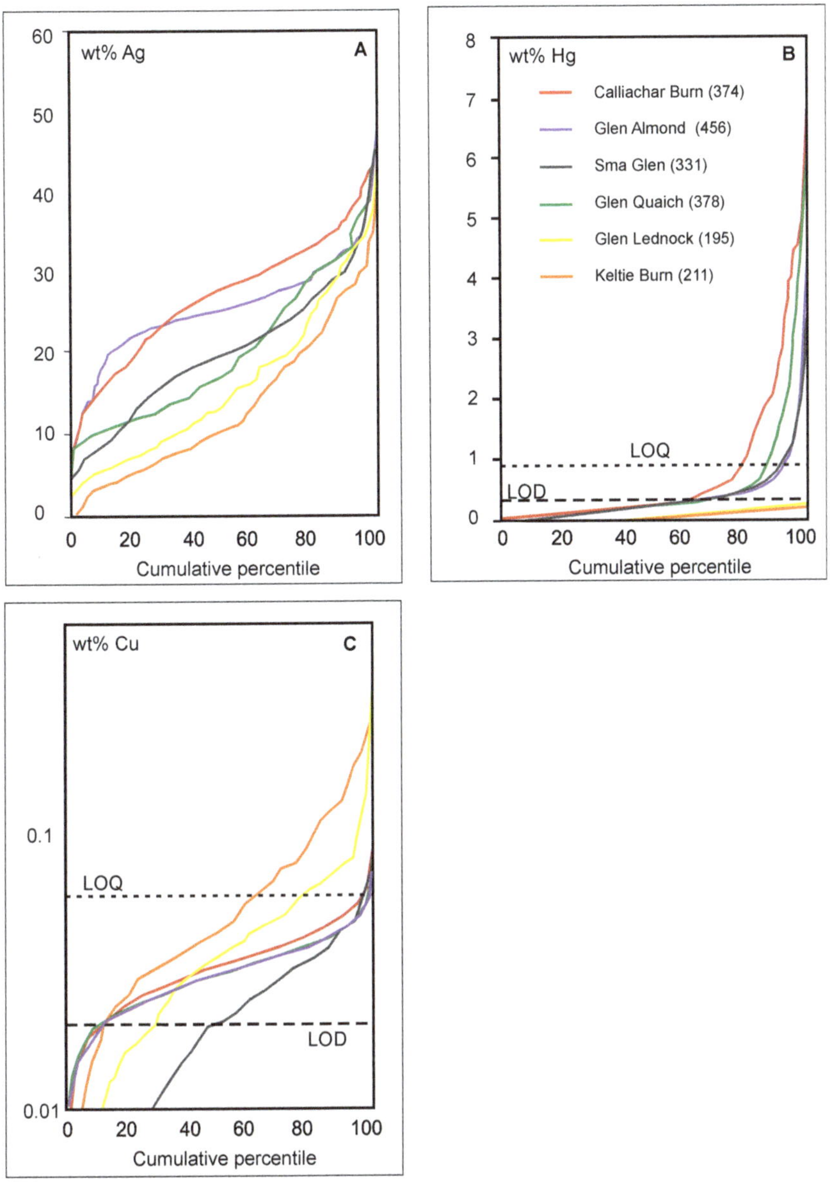

Figure 3. Gold alloy composition. (**A**) Silver content of Au-Ag alloys; (**B**) Mercury content of Au-Ag alloys; (**C**) Cu content of Au-Ag alloys. Note logarithmic scale for y axis of Cu plot. Figures in parentheses refer to number of particles in the sample population. Key in Figure 3B applies to all.

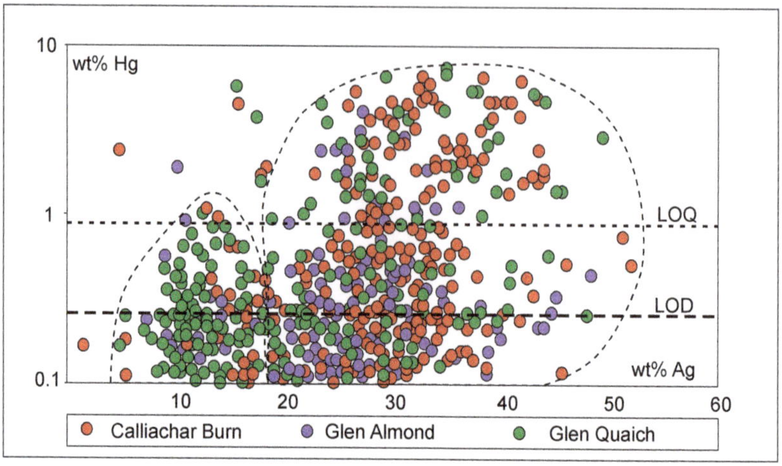

Figure 4. Ag vs. Hg binary plot. Data for gold from Keltie Burn, Sma Glen and Glen Lednock have been omitted as all Hg values are below LOD (Figure 3B). Compositional fields discussed in the text are identified by dashed lines.

Figure 5 shows the covariance of Cu and Ag for the two sample populations where some gold particles contain Cu to >LOQ (Keltie Burn and Glen Lednock). The general inverse relationship between Cu and Ag appears to be a generic feature [60] and overall the two populations appear broadly equivalent.

The sample populations from Glen Lednock and Keltie Burn each contained two gold particles with values of Pd > 4 wt%. Palladium was undetectable in all other cases.

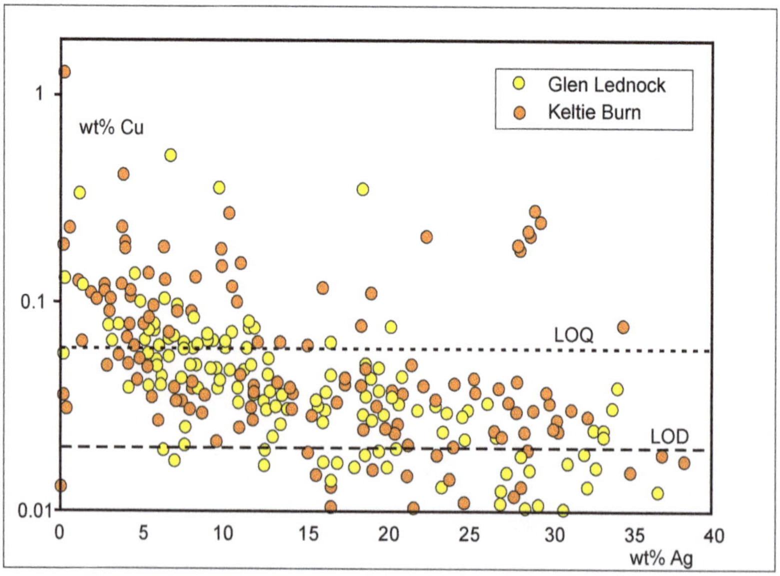

Figure 5. Ag vs. Cu binary plot for gold particles from Glen Lednock and Keltie Burn. Data for gold from other localities has been omitted, as all is <LOQ.

3.3. Inclusion Suites

A representative set of SEM images of the analyzed gold grains is shown in Figure 6. The inclusion data were used to generate both spider diagrams to depict mineralogy (Figure 7) and radar plots to characterize inclusion suites according to mineral chemistry (Figure 8).

Sample populations from Calliachar Burn, the River Almond and Glen Quaich all exhibit inclusion suites of the same relatively limited number of mineral species, with the non- metal signature confined to sulphides and sulphoarsenides (Figures 7A and 8A). We grouped the sample populations showing these inclusion characteristics to 'Group 1'. Pyrite is the dominant sulphide within the inclusion suites in Group 1, with minor contributions from various sulphoarsenides and simple metal sulphides. The main differences between inclusion assemblages within Group 1 derive from the relative abundance of arsenopyrite inclusions in the gold from Glen Quaich and the apparent absence of sphalerite inclusions in the gold from Calliachar Burn. The radar plots depicting the inclusion assemblages (Figure 8A) also depict sulphide and sulphoarsenide minerals with occasional contributions from sulphosalts. In general, gold samples from the River Almond and Glen Quaich contain a greater proportion of sulphides other than pyrite than the gold from Calliachar Burn.

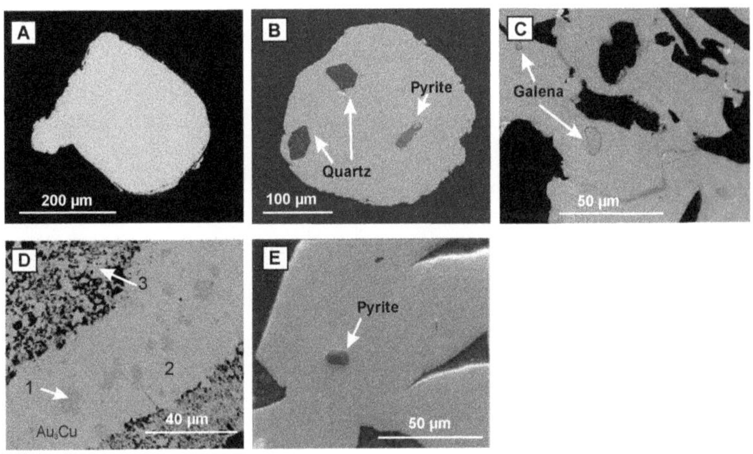

Figure 6. Internal compositional characteristics of gold revealed by SEM imaging. (**A**) An example BSE image of homogenous, inclusion-free gold particle (Glen Quaich); (**B**) Homogenous alloy, containing inclusions of quartz and pyrite (Calliachar Burn); (**C**) Films of Ag-rich Au alloy (dark grey) and Ag-poor Au alloy(very light grey) in a gold particle containing galena inclusions (Glen Quaich); (**D**) Compositional heterogeneity within agold particle containing Pd as an alloy component, (Glen Lednock); 1: Au: 87.4%, Cu: 8.4%, Pd: 4.3%, 2: Au: 91.8%, Ag: 0.2%, Cu: 1.3%, Hg: 0.6%, Pd: 5.6%, 3: Au: 95.1%, Cu: 1.5%, Pd: 3.4% (all wt%); (**E**) example of typical small pyrite inclusion found in gold from localities of Group 1.

The inclusion suites of gold from samples from Sma Glen, Keltie Burn and Glen Lednock differ from those of group 1 in the more extensive range of minerals present and their similar mineralogical and chemical profiles (Figures 7B and 8B). These general similarities permit classification into a single second Group. The inclusion suites in Group 2 samples may be immediately distinguished from those of Group 1 by the strong and consistent Te-signature (Figures 7B and 8B). In addition, the presence of Bi-bearing minerals provides a discriminant, although the speciation varies according to locality. A distinctive bornite-chalcocite-molybdenite inclusion suite contributes to the inclusion suite from Glen Lednock, where the abundance of chalcopyrite is higher than any other locality. Analysis of these inclusion suites by the radar diagrams (Figure 8B) highlights the diversity of metallic element contributions, and the contribution from Bi and Ag in the cases of gold from Sma Glen and Keltie Burn.

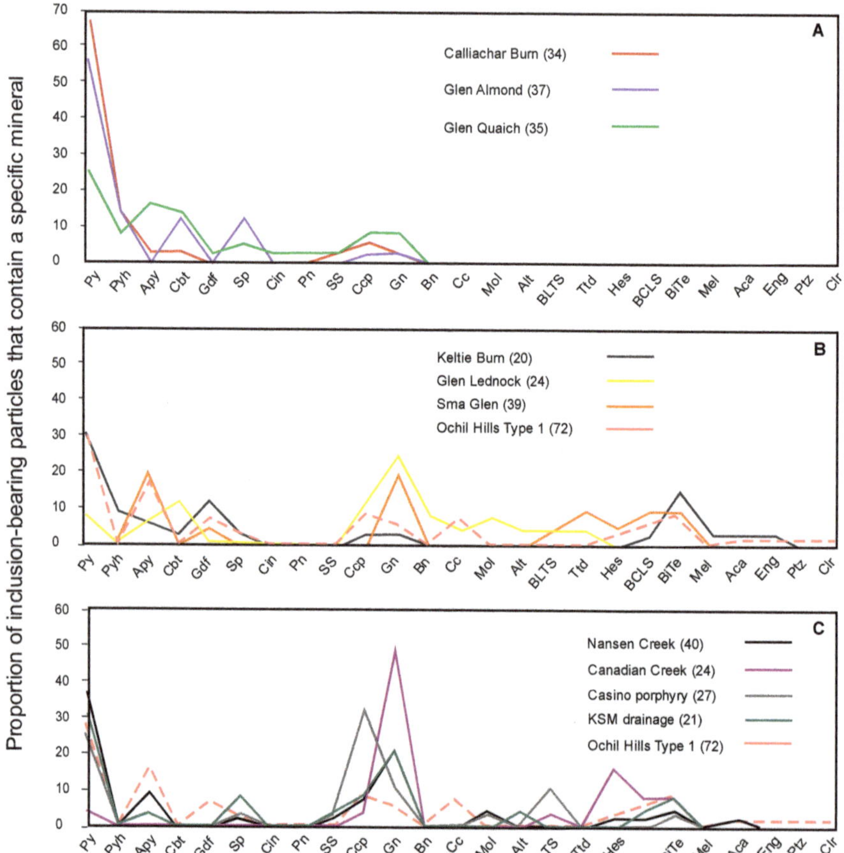

Figure 7. Spider plots showing inclusion suites. (**A**) Sample populations within Group 1 showing similar, restricted inclusion signatures (Calliachar Burn, the River Almond, Glen Quaich); (**B**) Sample populations within Group 2, showing a more variable signature with a wider range of inclusion mineralogy (Sma Glen, Keltie Burn, Glen Lednock) together with the signature of the gold from the Ochil Hills [56]. (**C**) Comparison of inclusion signatures in gold from Canadian porphyry and epithermal localities, [10] with those of the most widespread gold type from Ochil Hills, Scotland [32]. 'KSM' = 'Kerr-Sulphuretes-Mitchell'.

Mineral abbreviations used in Figure 7: Py = pyrite, Pyh = pyrrhotite, Apy = arsenopyrite, Cbt = cobaltite, Gdf = gersdorffite, Sp = sphalerite, Cin = cinnabar, Pn = pentlandite, SS = various sulphosalts, Ccp = chalcopyrite, Gn = galena, Bn = bornite, Cc = chalcocite, Mol = molybdenite, Alt = altaite, BLTS = undifferentiated Bi-Pb-tellurosulphides, Ttd = tetradymite, Hes = hessite, BCLS = undifferentiated Bi-Cu = Pb sulphides, BiTe = undifferentiated Bi tellurides, Me = melonite, Aca = acanthite, Eng = enargite, Ptz = petzite, Clr = coloradoite.

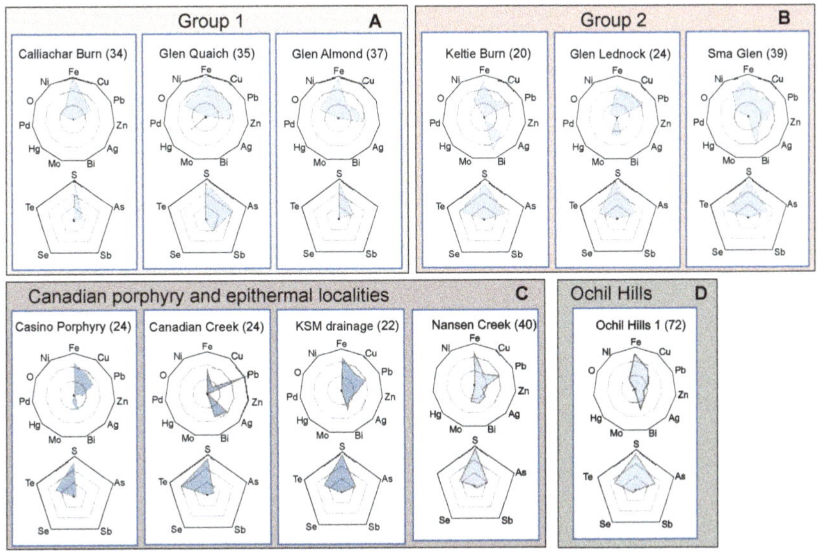

Figure 8. Radar plots showing the elemental components of the various inclusion suites, according to group and locality. Data origins: as per Figure 7: (**A**) Sample localities of Group, (**B**) Sample localities of Group 2, (**C**) Inclusion signatures in gold from Canadian porphyry and epithermal localities, [10], (**D**) Inclusion signature of gold from Ochil Hills, Scotland [32].

4. Discussion

Interpreting the implications of the various gold signatures will be undertaken in two ways: firstly a broad approach that aims to identify any major relationships between signature and local lithologies and geological history, and secondly via a more detailed examination of the individual alloy and inclusion signatures and their relationship to other sources of information. It is worth noting that the large sample populations available to the study provide an excellent platform to characterize inclusion suites, even though the overall inclusion abundance is not high.

4.1. Relationship between Gold Signatures and Local Lithologies

The sample localities that comprise Group 1 (Calliachar Burn, Glen Quaich, Glen Almond) are all located in the Pitlochry Formation of the Southern Highland Group (Figure 1B). For Group 2 sampling localities, Glen Lednock and Sma Glen are within the Ben Ledi Grit Formation, with the Glen Lednock sampling locality being situated very close to the Early Devonian Comrie Pluton. The third Group 2 locality, Keltie Burn, lies south of the unconformity between the deformed rocks of the Southern Highland Group and the overlying Devonian sedimentary and igneous rocks. The sample of gold from Keltie Burn was collected a short distance downstream from the contact between the Devonian sedimentary rocks and mafic igneous rocks (Figure 1B).

There are some clear correlations between the signatures of the various sample populations gold reported above and their geological setting. Sample localities in Group 1 yield broadly comparable inclusion signatures (as described above), but in contrast, signatures of gold from localities in Group 2 show a wide compositional variability. The dominance of Early Devonian igneous rocks in the catchments of the River Lednock and Keltie Burn provides a first order explanation of the differences between signatures of the Group 1 and 2 localities. For Sma Glen, the interpretation is slightly more complicated. The inclusion suite in gold from Sma Glen is clearly more similar to the inclusion suites of the other Group 2 localities than to those in Group 1, although the alloy signature is less distinctive.

Demonstrating a clear association with a potential magmatic source in geological terms is less straightforward for Sma Glen as no igneous rocks have been identified within the Ben Ledi Grit Fm. upstream from the sampling locality. The significance of the sample population from Sma Glen and its likely geological origins is discussed in more detail in a later section. Despite the level of uncertainty regarding the origins of gold from Sma Glen, the broad differences in the compositional signatures in gold in Groups 1 and 2 clearly correlate with geological differences within the study area, permitting confident speculation on the nature of the source mineralization that is yet to be discovered.

4.2. Interpretation of Gold Origins According to Compositional Signatures

4.2.1. Group 1 Localities

The Group 1 localities have been considered together because of the overall similarities in their inclusion suites, alloy compositions and the general geological setting. At Calliachar Burn, previous gold compositional work has focused on the relationship between the vein mineralogy and signature of detrital gold collected downstream of the vein outcrops [6,28,29]. The gold from Calliachar Burn used in the present study was collected approximately 1km upstream of the known vein occurrences but exhibits the high-Ag-and-Hg signature (Figure 3A,B) previously reported of gold from lower Calliachar Burn [28,29].

The setting of Group 1 localities is within metasedimentary sequences with no igneous intrusions located nearby. The closest intrusion is the granite intrusion of unknown age at Loch Lednock c. 5 km west of the Glen Almond sampling site (Figure 1B). The general absence of late-Silurian-Early Devonian igneous rocks near the sampling sites in the northern part of the study area suggests that any gold mineralization here did not originate from magmatic fluids, i.e., they are more likely to be "classical" orogenic gold occurrences. Therefore, a comparison with compositional signatures from other orogenic systems globally may be informative. The Ag profiles of the sample populations from Group 1 localities show a range between 10 and 40 wt% Ag, with mean values of 27.0, 25.5 and 20.2 wt% Ag for gold from Calliachar Burn, Glen Almond and Glen Quaich respectively. Such a large degree of variability within a single sample population from an alluvial occurrence may be ascribed to either variation in the parameters that influence alloy composition within a single mineralizing event [61,62] or the physical mixing of gold populations of a narrower Ag range each relating to different mineralizing episodes. These different episodes may be genetically unrelated but present within the fluvial catchment area or reflect multiple fluid pulses during the evolution of a single mineralizing system. Consideration of the mean Ag contents of populations of multiple samples of gold from Phanerozoic orogenic deposits showed a bimodal distribution with a small high-Ag group yielding values of between 18 and 30 wt% [60]. The Ag profiles of Group 1 gold are most similar to those classified as epizonal orogenic systems [10] both in terms of Ag range and distribution. However textural evidence for epizonal textures in gold from the Calliachar veins has not been reported [29]. Detailed paragenetic studies of the Eas Anie Vein at the Cononish deposit [CSJ] 30 km to the west of the study area showed clear ranges of Ag content within gold of different paragenetic stages, and the Ag range of 10–50 wt% in a sample of alluvial gold from the adjacent Cononish River [59] reflects those contributions. However the Cononish mineralizing system has been linked with magmatic fluids, [27] and appears to be genetically unrelated to the known vein arrays in the study area.

The inclusion suite previously reported in samples of gold from gossan and alluvial localities [59] shows a similar profile to that recorded in the present study, but with a larger contribution from Pb, Zn and Cu. Overall the inclusion suites of pyrite, pyrrhotite, and simple sulphides are most closely aligned with those reported form various Phanerozoic orogenic gold systems worldwide [10]. Gold from the River Almond exhibits a Ag profile very similar to that of Calliachar Burn gold, with high Ag contents dominating the sample population. Around 10% of the gold particles contain Hg to quantifiable concentrations, but Cu is almost entirely below the level of quantification. The Ag profile of gold from

Glen Quaich (Figure 3A) contains a relatively low Ag component when compared with the signatures of gold from Glen Almond and Calliachar Burn. Consideration of the covariance of Ag and Hg reveals a distinct low Ag, low Hg compositional field identified in Figure 4, whilst the remainder of the population conforms to the compositional field of gold from Glen Almond and Calliachar Burn. Inspection of the inclusion suites in the two subsets defined by alloy composition of the Glen Quaich sample did not reveal any clear affiliations, implying that all gold in this sample population is probably genetically related.

The strong similarities in the dataset from Group 1 overall suggests that there is a genetic link between these gold sources and also with the Calliachar veins and adjacent alluvial locality [28,29]. Subtle differences between the individual alloy profiles and inclusion suites are noted, but we suggest that these variations are a consequence of small compositional differences in the source fluids (perhaps generated by contact with a common fluid with different lithologies) or the prevailing physicochemical conditions at the point of precipitation. This hypothesis is consistent with the small degree of variation in the vein mineralogy within the array of veins at Calliachar-Urlar [29]. Furthermore, all known auriferous veins in the northern part of the study area region exhibit a consistent NNW-SSE orientation which further supports a genetic commonality. The relatively simple inclusion suite mineralogy points towards a hydrothermal system typically associated with "orogenic gold" deposits for which both metamorphic and mantle fluid sources have been suggested [63]. This may explain the abundance of Ni and Co-bearing minerals in the inclusion suite of gold from the Group 1 localities. As for Hg, there is a suggestion of a systematic decrease in the Hg content of gold from north to south, i.e., between Calliachar Burn, Glen Quaich and Glen Almond, (Figure 3B), but previous attempts to infer the causality of the Hg contents of gold in relation to wider geological considerations have been unsuccessful [11]. The overall similarity between the signatures of gold from localities in Group 1 and their similarity to the compositional characteristics of in-situ gold from the Calliachar-Urlar area reported by other workers, strongly suggests that exploration in the upper Calliachar, Almond and Quaich catchments should focus on identifying similar vein systems.

4.2.2. Group 2 Localities

This section will first consider the gold from Glen Lednock and Keltie Burn where there is a close spatial relationship between sampling location and igneous lithologies, before moving on to discuss the potential origins of gold in Sma Glen.

In magmatic-hydrothermal systems, the spatial and temporal evolution of hydrothermal systems may be pronounced, encompassing the formation and modification of porphyry mineralization and also the subsequent transition to an epithermal environment [64]. Consequently, the Ag profiles of populations of gold particles precipitated at specific localities within the system also vary [61,62]. Populations of gold particles collected from drainages in the environs of porphyry mineralization are, therefore, likely to include contributions from several of these different populations, with the resulting profile governed by the interplay of drainage, erosion profile of the system and sampling locality. For this reason, Ag profiles do not necessarily constitute a useful primary discriminant for the identification of specific deposit types in the porphyry-epithermal environment, nor are they infallible as a comparator to establish 'same or different' for sample populations collected from different localities. Nevertheless the Ag profiles exhibited by the sample populations from Group 2 localities show Ag ranges which are compatible with those previously reported from porphyry-epithermal systems [60,65,66].

Only around 10% of gold particles from Sma Glen had Hg values above the level of quantitation, and all gold particles from Keltie Burn and Glen Lednock had Hg values below LOD. Consequently a compositional field for Group 2 gold has not been included in Figure 4.

Consideration of the inclusion signature has proved a far more robust approach by which to establish genetic relationships between sample populations from different localities [10], particularly where the ranges of Ag contents of alloy overlap substantially. The inclusion suite of gold from Glen Lednock contains chalcocite, molybdenite and bornite, (Figure 8B) all of which are mineralogically compatible with a porphyry genesis and the location of the sampling site close to the Comrie pluton (Figure 1B). Ore fluids of high Te fugacity are indicated at Glen Lednock and Keltie Burn by the presence of altaite (PbTe) and Pb-Bi telluride inclusions [67]. Chalcopyrite inclusions were not recorded in gold from Keltie Burn, whereas both chalcopyrite and bornite were present in the gold from Glen Lednock (Figure 8B). Interpretation of these features is aided by previous work that has focused on gold from calc-alkalic porphyry systems and related epithermal mineralization: here, we compare our findings with previously reported signatures of gold from magmatic hydrothermal systems in the Canadian Cordillera [10,14] and another within a region of Devonian igneous rocks in the Ochil Hills, Scotland, (Figure 1A) where widespread alluvial gold has been recorded [56].

Inclusion signatures of gold from calc alkaline porphyry systems and related low sulphidation epithermal systems in Yukon [14] and British Columbia [10] were depicted using both spider plots and radar diagrams and these are reproduced in Figures 7C and 8C, respectively. The spider diagrams show the same general form as those describing gold from Glen Lednock and Keltie Burn, in that pyrite, chalcopyrite and galena are the dominant sulphides, with contributions from sulphoarsenides and an array of tellurides, whose speciation varies between localities. The radar diagrams, likewise, exhibit mutually consistent forms owing to minerals containing Bi, Pb, S and Te with varying contributions from Cu, Ag, Mo and Zn.

Four distinct compositional signatures were identified in populations of alluvial gold from the Ochil Hills (Figure 1A) during a regional study of detrital gold [56]. The most common signature was interpreted to be derived from low-sulphidation epithermal mineralization. The inclusion signatures of this gold type (labeled 'Ochil Hills type 1' in Figures 7 and 8) are compared to those of gold from the Canadian localities specified earlier (Figures 7C and 8C). In all cases the same general range of mineralogical characteristics and elemental signature can be observed although the Ochil Hills gold has a slightly less complex array of metals. The elemental signature of gold from Group 2 in our study (Figure 8B) shows many similarities with both the Canadian and the Ochil Hills localities, particularly with regard to S, As and Te. A high Te fugacity in the ore fluid of gold in mineralization in the Ochil Hills is indicated by the presence of coloradoite (HgTe) and petzite (AuAg)Te as inclusion species.

Some useful information is available from the Pd content of the Au-Ag alloy. Gold alloys have been reported from a number of geological settings, often associated with ultrabasic rocks [60,68,69], or the scavenging of various elements including Au and Pd from such rocks [69]. Palladium- bearing Au alloys are also formed in low temperature oxidizing chloride hydrothermal systems [70] and in association with Hg- bearing Au-Ag alloys in late stage hydrothermal activity associated with alkalic porphyry systems [13]. In the present study Pd-bearing Au was identified in sample populations from both Keltie Burn (Figure 6D) and Glen Lednock. Gold alloys of similar composition has previously been reported in detrital gold in the Ochil Hills from drainages also within Devonian Volcanic rocks [56]. Therefore, derivation from veins of this type is the most likely explanation of palladian gold occurrences within our study area.

The Hg content of gold from Keltie Burn and Glen Lednock is well below detection, unlike gold from the Ochil Hills where values to percent levels were sporadically encountered [56]. The proportion of gold particles containing quantifiable Cu in gold from Keltie burn (40%) appears higher than that from Glen Lednock (23%), although the reasons for this difference are unclear.

Overall, the mineralogical signatures of gold from Glen Lednock and Keltie Burn strongly indicate derivation from a magmatic hydrothermal system, with the mineralogy of the inclusion assemblages suggesting clear contributions from porphyry-derived gold at Glen Lednock (bornite-molybdenite-chalcocite) and low-sulphidation epithermal gold in

Keltie Burn (an apparent absence of chalcopyrite and a general similarity to gold from the Ochil Hills 20 km to the south).

The interpretation of the compositional characteristics of gold from Sma Glen is slightly more complicated. Gold from Sma Glen exhibits some features that are very closely aligned with those of other Group 2 gold signatures, both in terms of the elemental and mineralogical profiles of the inclusion suites (Figures 7B and 8B). However, the alloy profiles of Sma Glen gold more closely resemble the gold of Group 1, particularly for Ag, although the profiles do not perfectly conform (Figure 3). It is important to note that the Sma Glen locality is within the Glen Almond catchment, 10 km downstream of the Glen Almond sampling locality (Figure 1B). Consequently, it is probable that some gold present at Sma Glen has been fluvially transported from upstream, and the presence of some topologically water-worn grains supports this (Figure 2F). This likely mixture of gold from the large catchment area may account for some similarity between the alloy signature of Group 1 gold and gold from Sma Glen. However the inclusion signature of the Sma Glen sample is less likely to exhibit characteristics of Group 1 gold because of the obliteration of inclusions with gold particle deformation during prolonged transport [5]. Consequently both the morphology of gold particles at this site (Figure 2E) and the inclusion signature (Figure 8A,B) provide clear evidence of a local influx of gold.

The Te signature of the inclusion suite of gold from Sma Glen is pronounced, and is in part due to the presence of both altaite (PbTe) and melonite (NiTe), both indicative of high-Te fugacity environments [67]. The Bi signature, often in the form of Bi-telluride minerals is also very clear, providing another point of difference with gold from Group 1. As discussed above, it seems reasonable to assume that gold particles derived from porphyry, epithermal and transitional environments may be represented in populations of gold particles collected from the sampling points in Glen Lednock and Keltie Burn and therefore minerals reflecting these environments may be present within the overall inclusion assemblages. The Au-Bi-Te assemblage has been linked to 'deep seated epithermal' mineralization at several localities [56,71,72]. The associated, lower temperature mineralization may exhibit a Au-Ag-Te signature [56,71]. Evidence from inclusion signatures of gold from both Keltie Burn and Glen Lednock indicates that the overall signature is that of a multistage magmatic-hydrothermal system.

With regards to a plausible geological setting for local mineralization, proximity to the northern margin of the Devonian igneous rocks and the unconformity below them, now eroded at the Sma Glen sampling locality (Figure 1B), raises the possibility that the generally poor exposure has prevented the detection of Devonian igneous activity in the vicinity of Sma Glen. It is entirely feasible that some parts of the Devonian magmatic-hydrothermal system permeated farther north into the Ben Ledi Grit Formation.

5. Conclusions

Detailed characterization of alluvial gold particles adds a detailed layer of information over and above that gained from merely recording the presence, abundance and morphology of visible gold during routine stream sediment sampling. Characterization of gold particles also adds value to understanding gained through the study of known bedrock sources, as it can establish whether specific types of mineralization are more widespread than currently recognized. In the area south of Loch Tay, we have identified two distinct gold deposit signatures using analyses of alloy compositions and mineral inclusion suites. The first signature (Group 1) is defined by a large degree of replication in the gold signature from three separate sampling localities in the northern part of the study area. These signatures identify a regional type whose characteristics are compatible with mineralizing fluids generated by deep crust/mantle dehydration ("orogenic gold"). Similarities between these signatures indicate that future exploration in this part of the study area should focus on vein types similar to those already known at Calliachar-Urlar and Tombuie.

The second signature (Group 2), observed in the southern part of the study area in close proximity to Devonian igneous rocks, strongly indicates an association with magmatic hydrothermal systems. No known bedrock sources exist in this area but the comparison of the robust signatures gained from a large sample set with previously determined deposit type templates enabled the identification of the dominant source of gold. In addition to the sampling localities in the vicinity of exposed Devonian magmatic rocks, this approach identified the influx of gold from a presently unknown magmatic-hydrothermal source at Sma Glen in the mid-reaches of the River Almond, clearly distinguishable from alluvial gold present upstream in Glen Almond.

This study provides an example of the application of gold characterization studies during the exploration of areas of complex and poorly outcropping geology. In cases such as this where several deposit types are present, the outcomes of the approach permit more detailed subsequent targeting on a drainage-by-drainage basis, informed by an understanding of the likely deposit type. Further ongoing work will refine the characterization of gold from this study area and more widely, to further develop our understanding of the metallogeny of the Scottish Highlands in the Loch Tay region and in the contiguous auriferous areas along the HBF margins.

Author Contributions: Conceptualization, R.C. and T.T.; methodology, R.C., T.T., L.S.; formal analysis, L.S., R.C.; investigation, R.C., L.S., T.T.; data curation, R.C.; writing—original draft preparation, R.C., T.T.; writing—R.C., T.T., L.S.; visualization, R.C., T.T. All authors have read and agreed to the published version of the manuscript.

Funding: LS is in receipt of a NERC student award.

Data Availability Statement: The data presented in this study are available on request from the corresponding author. The data are not currently publicly available before the submission of the PhD thesis of LS.

Acknowledgments: We are indebted to the Ardtalnaig, Auchnafree, Garrows, Invergeldy, Monzie and Urlar Estates for permission to undertake the sampling that made this work possible. Aiden Lavelle, Kevin Dalton and Gregor Donaghy of Green Glen Minerals are thanked for their assistance in accessing field locations in their license area. Richard Walshaw is also sincerely thanked for assisting with the analytical work.

Conflicts of Interest: The authors declare no conflict of interest.

References

1. McClenaghan, M.B.; Cabri, L.J. Review of gold and platinum group element (PGE) indicator minerals methods for surficial sediment sampling. *Geochem. Explor. Environ. Anal.* **2011**, *11*, 251–263. [CrossRef]
2. Moles, N.R.; Chapman, R.J. Integration of detrital gold microchemistry, heavy mineral distribution, and sediment geochemistry to clarify regional metallogeny in glaciated terrains: Application in the Caledonides of southeast Ireland. *Econ. Geol.* **2019**, *114*, 207–232. [CrossRef]
3. Aherne, S.; Reynolds, N.A.; Burke, D.J. Gold mineralization in the Silurian and Ordovician of South Mayo. In *The Irish Minerals Industry 1980–1990*; Bowden, A.A., Earls, G., O'Connor, P.G., Pyne, J.F., Eds.; Irish Association for Economic Geology: Dublin, Ireland, 1992; pp. 39–49.
4. Chapman, R.J.; Moles, N.R.; Bluemel, B.; Walshaw, R.D. Detrital gold as an indicator mineral. In *Recent Advances in Understanding Gold Deposits: From Orogeny to Alluvium*; Torvela, T.M., Chapman, R.J., Lambert-Smith, J., Eds.; Geological Society Publications: London, UK, 2022; Volume 516, pp. 313–336.
5. Nikiforova, Z.S.; Kalinin, Y.A.; Makarov, V.A. Evolution of native gold in exogenous conditions. *Russ. Geol. Geophys.* **2020**, *61*, 1244–1259. [CrossRef]
6. Chapman, R.J.; Leake, R.C.; Moles, N.R.; Earls, G.; Cooper, C.; Harrington, K.; Berzins, R. The application of microchemical analysis of gold grains to the understanding of complex local and regional gold mineralization: A case study in Ireland and Scotland. *Econ. Geol.* **2000**, *95*, 1753–1773.
7. Nikiforova, Z.S. Criteria for determining the genesis of placers and their different sources based on the morphological features of placer gold. *Minerals* **2021**, *11*, 381. [CrossRef]
8. Masson, F.X.; Beaudoin, G.; Laurendeau, D. Multi-method 2D and 3D reconstruction of gold grain morphology in alluvial deposits: A review and application to the Rivière du Moulin (Québec, Canada). In *Recent Advances in Understanding Gold Deposits: From Orogeny to Alluvium*; Torvela, T.M., Chapman, R.J., Lambert-Smith, J., Eds.; Geological Society Publications: London, UK, 2022; Volume 516, pp. 337–352.

9. Chapman, R.J.; Mortensen, J.K. Characterization of gold mineralization in the northern Cariboo gold district, British Columbia, Canada, through integration of compositional studies of lode and detrital gold with historical placer production: A template for evaluation of orogenic gold districts. *Econ. Geol.* **2016**, *111*, 1321–1345.
10. Chapman, R.J.; Mortensen, J.K.; Allan, M.M.; Walshaw, R.D.; Bond, J.; MacWilliam, K. A New Approach to Characterizing Deposit Type Using Mineral Inclusion Assemblages in Gold Particles. *Econ. Geol.* **2022**, *117*, 361–381. [CrossRef]
11. Chapman, R.J.; Mortensen, J.K.; Crawford, E.C.; Lebarge, W. Microchemical studies of placer and lode gold in the Klondike District, Yukon, Canada: 1. Evidence for a small, gold-rich, orogenic hydrothermal system in the Bonanza and Eldorado Creek area. *Econ. Geol.* **2010**, *105*, 1369–1392. [CrossRef]
12. Potter, M.; Styles, M.T. Gold characterization as a guide to bedrock sources for the Estero Hondo alluvial gold mine, western Ecuador. *Trans Inst. Min. Metall.* **2003**, *112*, 297–304.
13. Chapman, R.J.; Mileham, T.J.; Allan, M.M.; Mortensen, J.K. A distinctive Pd-Hg signature in detrital gold derived from alkalic Cu-Au porphyry systems. *Ore Geol. Rev.* **2017**, *83*, 84–102. [CrossRef]
14. Chapman, R.J.; Allan, M.M.; Mortensen, J.K.; Wrighton, T.M.; Grimshaw, M.R. A new indicator mineral methodology based on a generic Bi-Pb-Te-S mineral inclusion signature in detrital gold from porphyry and low/intermediate sulfidation epithermal environments in Yukon Territory, Canada. *Miner. Deposita* **2018**, *53*, 815–834. [CrossRef]
15. Lalomov, A.V.; Chefranov, R.M.; Naumov, V.A.; Naumova, O.B.; Lebarge, W.; Dilly, R.A. Typomorphic features of placer gold of Vagran cluster (the Northern Urals) and search indicators for primary bedrock gold deposits. *Ore Geol. Rev.* **2017**, *85*, 321–335. [CrossRef]
16. Nevolko, P.A.; Kolpakov, V.V.; Nesterenko, G.G.; Fominykh, P.A. Alluvial placer gold of the Egor'evsk district (northern-Western Salair): Composition characteristics, types and mineral microinclusions. *Russ Geol. Geophys.* **2019**, *60*, 67–85. [CrossRef]
17. Fominykh, P.A.; Nevolko, P.A.; Svetlitskaya, T.V.; Kolpakov, V.V. Native gold from the Kamenka-Barabanovsky and Kharuzovka alluvial placers (Northwest Salair Ridge, Western Siberia, Russia): Typomorphic features and possible bedrock sources. *Ore Geol. Rev.* **2020**, *126*, 103781. [CrossRef]
18. Barrow, G. On an intrusion of muscovite-biotite gneiss in the south-east Highlands of Scotland, and its accompanying metamorphism. *Quart. J. Geol. Soc. Lond.* **1893**, *49*, 330–358. [CrossRef]
19. Peach, B.N.; Horne, J.; Gunn, W.; Clough, C.T.; Hinxman, L.W.; Teall, J.J.H. *The Geological Structure of the North-West Highlands of Scotland*; HMSO: London, UK, 1907.
20. Wilson, G.V.; Flett, J.S. Vol 17: The Lead, Zinc, Copper and Nickel Ores of Scotland. In *Special Reports on the Mineral Resources of Great Britain*; HMSO: London, UK, 1921; Volume 17.
21. Pattrick, R.A.D. Sulphide mineralogy of the Tomnadashan copper deposit and the Corrie Buie lead veins, south Loch Tayside, Scotland. *Min. Mag.* **1984**, *48*, 85–91. [CrossRef]
22. McClaren, M.J. The occurrence of gold in Great Britain and Ireland. *Trans. Inst. Min. Eng.* **1903**, *25*, 435–508.
23. Calvert, J. *The Gold Rocks of Great Britain and Ireland*; Chapman and Hall: London, UK, 1853.
24. Lindsay, W.L. The gold and gold-fields of Scotland. *Trans Edinburgh Geol. Soc.* **1868**, *1*, 105–115. [CrossRef]
25. Parker, R.T.; Clifford, J.A.; Meldrum, A.H. The Cononish gold-silver deposit, Perthshire, Scotland. *Trans. Inst. Min. Metall* **1989**, *98*, 51–54.
26. Treagus, J.E.; Pattrick, R.A.D.; Curtis, S.F. Movement and mineralization in the Tyndrum fault zone, Scotland and its regional significance. *J. Geol. Soc.* **1999**, *156*, 591–604. [CrossRef]
27. Spence-Jones, C.; Jenkin, G.; Boyce, A.; Hill, N.; Sangster, C. Tellurium, magmatic fluids and orogenic gold: An early magmatic fluid pulse at Cononish gold deposit, Scotland. *Ore Geol. Rev.* **2018**, *102*, 894–905. [CrossRef]
28. Leake, R.C.; Bland, D.J.; Cooper, C. Source Characterization of alluvial gold from mineral inclusions and internal compositional variation. *Trans. Instn. Min. Metal.* **1993**, *102*, 65–82.
29. Ixer, R.A.; Pattrick, R.A.; Stanley, C.J. Geology, mineralogy and genesis of gold mineralization at Calliachar-Urlar Burn, Scotland. *Trans. Instn. Min. Metall.* **1997**, *106*, 99–108.
30. Mason, J.; Pattrick, R.A.D.; Gallagher, M.J. Auriferous vein mineralization near Aberfeldy, Scotland. In *Exploration and the Environment*; Institution of Mining and Metallurgy Meeting: Edinburgh, UK, 1991; pp. 50–52.
31. Corkhill, C.; Ixer, R.A.; Mason, J.S.; Irving, D.; Pattrick, R.A. Polymetallic auriferous vein mineralization near Loch Tay, Perthshire, Scotland. *Scot. J. Geol.* **2010**, *46*, 23–30. [CrossRef]
32. Green Glen Minerals. Available online: https://greenglenminerals.com (accessed on 10 November 2022).
33. Chapman, R.J.; Shaw, M.H.; Leake, R.C.; Jackson, B. Gold mineralisation in the central Ochil Hills, Perthshire, UK. *Trans. Instn. Min. Metall.* **2005**, *114*, 53–64. [CrossRef]
34. Thompson, K.S.R. The Last Glaciers in Western Perthshire. Ph.D. Thesis, University of Edinburgh, Edinburgh, UK, 1972.
35. Torvela, T.; Chapman, R.J.; Lambert-Smith, J. The importance of multi-method approaches and developing a characterisation. In *Recent Advances in Understanding Gold Deposits: From Orogeny to Alluvium*; Torvela, T.M., Chapman, R.J., Lambert-Smith, J., Eds.; Geological Society Publications: London, UK, 2022; Volume 516.
36. Stephenson, D.; Mendum, J.; Fettes, D.; Leslie, A. The Dalradian rocks of Scotland: An introduction. *Proc. Geol. Assoc.* **2013**, *124*, 3–82. [CrossRef]
37. Tanner, P.; Thomas, C.; Harris, A.; Gould, D.; Harte, B.; Treagus, J.; Stephenson, D. The Dalradian rocks of the Highland Border region of Scotland. *Proc. Geol. Assoc.* **2013**, *124*, 215–262. [CrossRef]

38. Roberts, J.L.; Treagus, J.E. Polyphase generation of nappe structures in the Dalradian rocks of the southwest Highlands of Scotland. *Scot. J. Geol.* **1977**, *13*, 237–254. [CrossRef]
39. Chew, D.; Strachan, R. The Laurentian Caledonides of Scotland and Ireland. *Geol Soc. Lond.* **2014**, *390*, 45–91. [CrossRef]
40. Corfu, F.; Gasser, D.; Chew, D.M. New Perspectives on the Caledonides of Scandinavia and Related Areas. *Geol. Soc. Lond.* **2014**, *390*, 467–511. [CrossRef]
41. Baxter, F.B.; Ague, J.J.; Depaolo, D.J. Prograde temperature–time evolution in the Barrovian type-locality constrained by Sm/Nd garnet ages from Glen Clova, Scotland. *J. Geol. Soc. Lond.* **2002**, *159*, 71–82. [CrossRef]
42. Oliver, G.J.H.; Wilde, S.A.; Wan, Y. Geochronology and geodynamics of Scottish granitoids from the late Neoproterozoic break-up of Rodinia to Palaeozoic collision. *J. Geol. Soc. Lond.* **2008**, *165*, 661–674. [CrossRef]
43. Mark, D.F.; Rice, C.M.; Hole, M.; Condon, D. Multi-chronometer dating of the Souter Head complex: Rapid exhumation terminates the Grampian Event of the Caledonian Orogeny. *Earth Environ. Sci. Trans. R. Soc. Edinb.* **2020**, *2*, 95–108. [CrossRef]
44. Dempster, T. Uplift patterns and orogenic evolution in the Scottish Dalradian. *J. Geol. Soc. Lond.* **1985**, *142*, 111–128. [CrossRef]
45. Soper, N.J.; Ryan, P.D.; Dewey, J.F. Age of the Grampian Orogeny in Scotland and Ireland. *J. Geol. Soc. Lond.* **1999**, *156*, 1231–1236. [CrossRef]
46. Oliver, G.J.H. Reconstruction of the Grampian episode in Scotland: Its place in the Caledonian Orogeny. *Tectonophys.* **2001**, *332*, 23–49. [CrossRef]
47. Dempster, T.J.; Hudson, N.F.C.; Rogers, G. Metamorphism and cooling of the NE Dalradian. *J. Geol. Soc. Lond.* **1995**, *152*, 383–390. [CrossRef]
48. Dewey, J.F.; Mange, M. Petrography of Ordovician and Silurian sediments in the western Irish Caledonides: Tracers of a short-lived Ordovician continent—Arc collision orogeny and the evolution of the Laurentian Appalachian—Caledonian margin. *Geol Soc. Lond.* **1999**, *164*, 55–107. [CrossRef]
49. Dallmeyer, R.D.; Strachan, R.A.; Rogers, G.; Watt, G.R.; Friend, C.R.L. Dating deformation and cooling in the Caledonian thrust nappes of north Sutherland, Scotland: Insights from 40Ar/39Ar and Rb–Sr chronology. *J. Geol. Soc. Lond.* **2001**, *158*, 501–512. [CrossRef]
50. Jacques, J.M.; Reavy, R.J. Caledonian plutonism and major lineaments in the SW Scottish Highlands. *J. Geol. Soc. London* **1994**, *151*, 955–969. [CrossRef]
51. Neilson, J.C.; Kokelaar, B.P.; Crowley, Q.G. Timing, relations and cause of plutonic and volcanic activity of the Siluro-Devonian post-collision magmatic episode in the Grampian Terrane, Scotland. *J. Geol. Soc. Lond.* **2009**, *166*, 545–561. [CrossRef]
52. Stephenson, D.; Bevins, R.E.; Millward, D.; Highton, A.J.; Parsons, I.; Stone, P.; Wadsworth, W.J. *Caledonian Igneous Rocks of Great Britain*; Joint Nature Conservation Committee 17: Peterborough, UK, 1999.
53. Rippon, J.; Read, W.A.; Park, R.G. The Ochil Fault and the Kincardine basin: Key structures in the tectonic evolution of the Midland Valley of Scotland. *J. Geol. Soc. Lond.* **1996**, *153*, 573–587. [CrossRef]
54. Dewey, J.F.; Strachan, R.A. Changing Silurian–Devonian relative plate motion in the Caledonides; sinistral transpression to sinistral transtension. *J. Geol. Soc. Lond.* **2003**, *160*, 219–229. [CrossRef]
55. Monaghan, A.A.; Pringle, M.S. 40Ar/39Ar geochronology of Carboniferous-Permian volcanism in the Midland Valley, Scotland. *Geol. Soc. Lond.* **2004**, *223*, 219–241. [CrossRef]
56. Mendum, J.R.; Noble, S.R. Mid-Devonian sinistral transpression on the Great Glen Fault: The rise of the Rosemarkie Inlier and the Acadian Event in Scotland. *Geol. Soc. Lond.* **2010**, *335*, 161–187. [CrossRef]
57. Rice, C.M.; Ashcroft, W.A.; Batten, D.J.; Boyce, A.J.; Caulfield, J.B.D.; Fallick, A.E.; Hole, M.J.; Jones, E.; Pearson, M.J.; Rogers, G.; et al. A Devonian auriferous hot spring system, Rhynie, Scotland. *J. Geol. Soc. Lond.* **1995**, *152*, 229–250. [CrossRef]
58. Mark, F.D.; Rice, C.M.; Fallick, A.E.; Trewin, N.H.; Lee, M.R.; Boyce, A.; Lee, J.K.W. 40Ar/39Ar dating of hydrothermal activity, biota and gold mineralization in the Rhynie hot-spring system, Aberdeenshire, Scotland. *Geochim. et Cosmochim. Acta* **2011**, *75*, 555–569. [CrossRef]
59. Leake, R.C.; Chapman, R.J.; Bland, D.J.; Condliffe, E.; Styles, M.T. Microchemical characterization of alluvial gold from Scotland. *Trans. Instn. Min. Metal.* **1997**, *106*, 85–98.
60. Chapman, R.J.; Banks, D.A.; Styles, M.T.; Walshaw, R.D.; Piazolo, S.; Morgan, D.J.; Grimshaw, M.R.; Spence-Jones, C.P.; Matthews, T.J.; Borovinskaya, O. Chemical and physical heterogeneity within native gold: Implications for the design of gold particle studies. *Miner. Deposita* **2021**, *56*, 1563–1588. [CrossRef]
61. Gammons, C.H.; Williams-Jones, A.E. Hydrothermal geochemistry of electrum; thermodynamic constraints. *Econ. Geol.* **1995**, *90*, 420–432. [CrossRef]
62. Palyanova, G.A. Gold and silver minerals in sulfide ore. *Geol. Ore Dep.* **2020**, *62*, 383–406. [CrossRef]
63. Goldfarb, R.J.; Groves, D.I. Orogenic gold: Common or evolving fluid and metal sources through time. *Lithos* **2015**, *233*, 2–26. [CrossRef]
64. Gammons, C.H.; Williams-Jones, A.E. Chemical mobility of gold in the porphyry-epithermal environment. *Econ. Geol.* **1997**, *92*, 45–59. [CrossRef]
65. Morrison, G.W.; Rose, W.J.; Jaireth, S. Geological and geochemical controls on the silver content (fineness) of gold in gold-silver deposits. *Ore Geol. Rev.* **1991**, *6*, 333–364. [CrossRef]
66. Savva, N.E.; Kravtsova, R.G.; Anisimova, G.S.; Palyanova, G.A. Typomorphism of Native Gold (Geological-Industrial Types of Gold Deposits in the North-East of Russia). *Minerals* **2022**, *12*, 561. [CrossRef]

67. Affifi, A.M.; Kelly, W.C.; Essene, E.J. Phase relations among tellurides, sulfides, and oxides; I, Thermochemical data and calculated equilibria. *Econ. Geol.* **1988**, *83*, 377–394. [CrossRef]
68. Palyanova, G.; Murzin, V.; Borovikov, A.; Karmanov, N.; Kuznetsov, S. Native gold in the chudnoe Au-Pd-ree deposit (Subpolar Urals, Russia): Composition, minerals in intergrowth and genesis. *Minerals* **2021**, *11*, 451. [CrossRef]
69. Murzin., V.; Palyanova, G.; Mayorova, T.; Beliaeva, T. The Gold–Palladium Ozernoe Occurrence (Polar Urals, Russia): Mineralogy, Conditions of Formation, Sources of Ore Matter and Fluid. *Minerals* **2022**, *12*, 765. [CrossRef]
70. Chapman, R.J.; Leake, R.C.; Bond, D.P.; Stedra, V.; Fairgrieve, B. Chemical and mineralogical signatures of gold formed in oxidizing chloride hydrothermal systems and their significance within populations of placer gold grains collected during reconnaissance. *Econ. Geol.* **2009**, *104*, 563–585. [CrossRef]
71. Kondratieva, L.A.; Anisimova, G.S.; Kardashevskaia, V.N. Types of Tellurium Mineralization of Gold Deposits of the Aldan Shield (Southern Yakutia, Russia). *Minerals* **2021**, *11*, 698. [CrossRef]
72. James, L.P.; Fuchs, W.A. Exploration of the Exciban gold-copper-tellurium vein system, Camarines Norte, Philippines. *J. Geochem. Expl.* **1990**, *35*, 363–385. [CrossRef]

minerals

MDPI

Article

Compositional Signatures of Gold from Different Deposit Types in British Columbia, Canada

Rob Chapman [1,*], James Kenneth Mortensen [2] and Rory Murphy [1]

[1] Ores and Minerals Group School of Earth and Environment, University of Leeds, Leeds LS2 9JT, UK
[2] MDN Geosciences Ltd., Salt Spring Island, BC V8K 1P5, Canada; mortdunord@gmail.com
* Correspondence: r.j.chapman@leeds.ac.uk

Abstract: A study of both in situ and detrital gold from different deposit types in British Columbia was undertaken to establish deposit-specific compositional characteristics in terms of alloy composition and suites of mineral inclusions. The study is based on 11,840 particles from 160 localities in which nine gold deposit types are represented, although there is a strong bias towards gold of orogenic, low-sulphidation epithermal, and alkalic porphyry origin. In general, Ag values in gold alloys are not a powerful discriminator for deposit type, but minor metals may prove useful where detectable, e.g., Cu in gold from ultramafic associations and Pd and Hg in gold from alkalic porphyry systems. The characterization of inclusion suites is far more illuminating, as they correlate strongly with the mineralogy of auriferous ores from different deposit types. This outcome has confirmed the validity of designing an indicator methodology based on inclusion suites and has permitted the prediction of inclusion suites for gold from other deposit types where data are more scarce. The compositional templates generated in the study were applied to identify the source deposit type(s) of gold from 41 localities (a total of 2916 detrital gold particles) where gold genesis was previously unknown.

Keywords: detrital gold; gold alloy composition; mineral inclusions; indicator minerals; gold deposit types; British Columbia

check for updates

Citation: Chapman, R.; Mortensen, J.K.; Murphy, R. Compositional Signatures of Gold from Different Deposit Types in British Columbia, Canada. *Minerals* **2023**, *13*, 1072. https://doi.org/10.3390/min13081072

Academic Editor: Galina Palyanova

Received: 4 July 2023
Revised: 8 August 2023
Accepted: 10 August 2023
Published: 13 August 2023

1. Introduction

The liberation of mineral particles from host lithologies by weathering generates a mineralogical and geochemical footprint whose extent is governed by transport in the prevailing surficial environment [1,2]. Spatial variations in chemical response or mineral abundance can act as vectors to the in situ source [3], and the approach can be particularly powerful when based on specific mineral-ore deposit style relationships, e.g., kimberlite [4], magmatic Ni–Cu–PGE e.g., [1,5], and gold [6,7]. Particular attention has been given to specific erosional products of copper porphyry mineralization, e.g., magnetite [8,9], apatite [10–12], and tourmaline [13]. These minerals are useful because subtle differences in their chemical composition may be linked to specific settings within a mineralized system, and their physical durability and chemical stability ensure longevity in the surficial environment, such that the geochemical/mineralogical anomaly is not ephemeral.

The presence of detrital gold in surficial sediments is generally accepted as the best evidence for a gold-bearing source, and consideration of the gold morphology and geomorphological process may permit speculation on the likely location of the source [14]. While the characterization of dispersion trains of gold particles in till has successfully been employed as vectors to source [6,15], the composition of gold particles and implications for source type have not found routine application in exploration. In contrast, an increasing number of academic studies have sought to utilize the compositional signatures of placer gold to either illuminate the evolution of economically important placers or speculate on the nature of the source(s). Several placer mining districts in Russia have been the focus of robust studies [16–21], in which distinct signatures of sub-populations of gold have been identified through the study of large numbers of gold particles. Similar approaches have been adopted in remote areas of geological complexity, e.g., South America [22] and Northern Pakistan [23,24]. If gold compositional studies are to find regular application in exploration projects, readily available compositional templates describing the generic features of gold from different deposit types are essential, but these are rarely generated in studies where the focus is a specific placer. In contrast, other studies have sought to identify generic compositional signatures that can subsequently be applied more widely [25–28], and while some clear diagnostic signatures for gold from different deposit types emerged, there are two main knowledge gaps. First, as more data are collected, the potential compositional ranges in gold corresponding to specific deposits are extended; e.g., even large studies of gold particles from different magmatic hydrothermal systems [29] subsequently proved inadequate as compositional templates [30]. Similarly, early attempts to ascribe distinguishing features to gold from a wider range of deposit types [31] were completely revised [32] after a further period in which several relevant studies were published. Second, our understanding of the compositional characteristics and range of gold from some specific deposit types is underrepresented, either as a consequence of a lack of focused studies or because the small particle size of gold commonly associated with some deposit types precludes collection by standard field techniques.

Placer gold is widespread in British Columbia, Canada (BC), as evidenced by the large amount of historical mining activity [33]. However, in many placer gold-producing areas, the in situ source(s) remain undiscovered. Surface exposure is commonly obscured by surficial deposits, and exploration approaches using indicator minerals have found favor [34]. Parallel studies in Yukon, Canada, have demonstrated the potential for placer gold compositions to establish source type and hence contribute to an improved understanding of regional metallogeny [28,35]. The presence of detrital gold, however, is not confined to sites of current or historic placer working, and exploration activities on all scales could collect gold particles and benefit from the interpretation of their compositional signature.

The Cordilleran Orogen that underlies BC is a complex assemblage of terranes that vary considerably in terms of age, composition, and tectonic history. The summary presented here is based on two references that address both tectonic history and metallogeny [36,37]. The region includes "pericratonic" terranes that display a largely continental affinity, some of which (e.g., Yukon–Tanana and Kootenay terranes) are thought to have originally been part of the Northwestern Laurentian margin, as well as continental margin arc terranes (e.g., Stikine and Quesnel terranes), and terranes such as the Cache Creek and Slide Mountain terranes that represent rock units formed in a mainly oceanic environment. These various terranes were assembled into their current configuration through a series of tectonic events that ranged in age from the latest Paleozoic through Early Tertiary time and included both collisions of exotic terranes against the Laurentian margin and each other as well as tectonic shuffling along major, late, dextral (and minor sinistral), crustal-scale strike-slip faults. Individual terranes comprise varying proportions of volcanic and sedimentary rocks and commonly include intrusive rock units that are coeval and comagmatic with the volcanic rocks. In addition, late and post-accretion intrusions are present within most terranes and locally crosscut many of the terrane boundaries. The metamorphic grade that has affected many of the terranes is generally low to moderate.

In addition to the geological complexity of the BC Cordillera, this region also displays a wide range of mineral deposit styles, including many variations on intrusion-related mineralization (porphyry, skarn, epithermal), as well as volcanogenic massive sulphide (VMS) and sedimentary exhalative (SEDEX) deposits and base and precious metal carbonate replacement deposits. The location of the localities mentioned in the text is provided in Figure 1. Gold (and silver) represent the major economic commodities in many of the deposit types in BC, including several subtypes of mainly late-tectonic orogenic gold deposits (e.g., Cariboo, Bralorne, Cassiar, Atlin, and Zeballos camps); epithermal vein deposits (Blackdome, Silbak Premier, Brucejack); and some rare gold-rich VMS deposits (Eskay Creek) [38]. Gold is also an important by-product in a wide variety of other deposit types in BC, including Cu–Au skarns (Hedley), Cu–Au alkalic porphyry deposits (Mt. Milligan, Mt. Polley, Copper Mountain, Galore Creek), and some calc-alkaline porphyry deposits (e.g., Red Chris, Kemess, Highland Valley). Gold is present in at least trace amounts in a very large proportion of mineral deposit types in BC, highlighting its potential to be used as a discriminant between deposit styles.

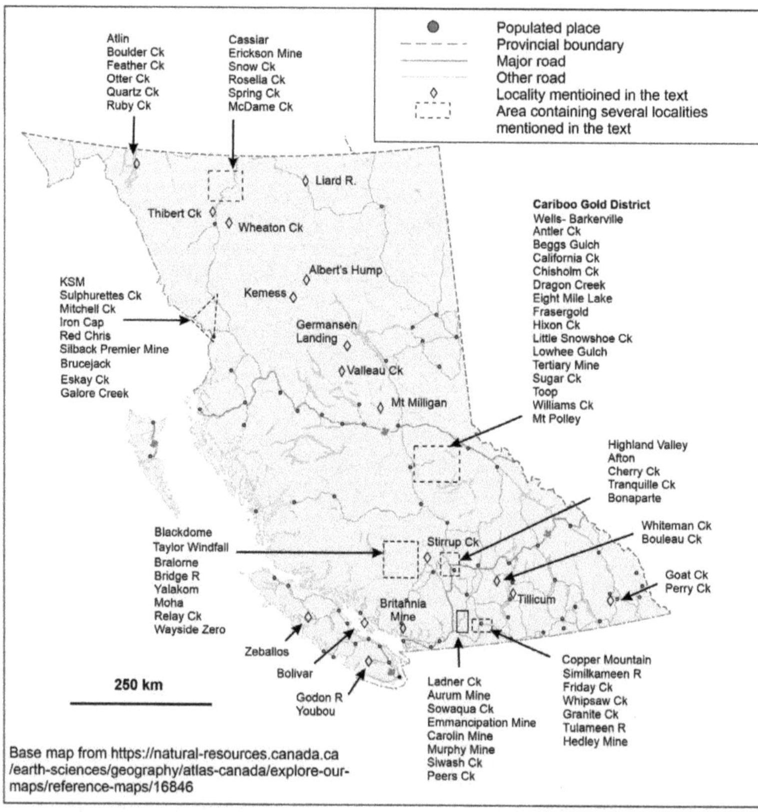

Figure 1. Locations mentioned in the text. Grid references for all sample locations are provided in Appendices A and B.

Gold particles exhibit compositional and microtextural features that are a consequence of their genesis and subsequent residence in their hypogene setting. These features persist post-liberation and erosion and have utility in interpreting the origins of detrital gold particles. This subject has been discussed in detail previously [39,40], and a brief overview is presented here.

Gold is almost always an alloy of Au and Ag, although other minor metals such as Cu and Hg may be detectable by electron microprobe (EMP) analysis. Some gold particles are compositionally homogeneous, but many are heterogeneous due to the presence of microfabrics caused by alloys of different compositions (usually variations in Ag) and/or inclusions of other minerals [32]. The origins of various microfabrics have been classified according to the time of formation with respect to the initial mineralizing event using a dual approach of compositional and crystallographic study [32]. In this way, it has been possible to ensure that the analysis of gold particles generates data pertaining only to the ore-forming stage rather than that resulting from subsequent modifications in residence within either the hypogene or surficial environments.

Differences between the mineralogy of different types of gold mineralization (e.g., low- and high-sulphidation epithermal deposits, calc-alkalic porphyries, and gold from orogenic deposits) are well known, and these are reflected in the suite of mineral inclusions observed in polished sections of gold particles from these different deposit types [32]. Furthermore, the physico-chemical environment of gold precipitation influences the Au–Ag ratio of the resulting alloy [41], together with the concentrations of other minor metals such as Cu, Hg, and Pd [32]. Consequently, the broad controls on ore fluid and mineralization environment (ore deposit type) have a major influence on the gold signature, with further variation arising as a consequence of specific conditions that influence alloy composition. In addition, the temporal and spatial evolution of a hydrothermal mineralizing event can generate a compositional range between gold particles within the overall population, and therefore a sufficient amount must be analyzed to generate a robust compositional signature of gold from a single mineralizing event. The term 'sample population' is used to denote a population of gold particles collected from a specific site. In the overwhelming majority of cases, a sample population exhibits a compositional range, which is effectively a proxy for the stability of the mineralizing environment or an indication of multiple mineralizing episodes.

Gold particle studies may consider sample populations collected either from in situ or placer sources. In situ mineralization may comprise multiple episodes that may or may not have been emplaced under similar conditions. Thus, it is possible that gold from a single in situ locality may exhibit more than one signature [42,43]. Erosion and transport of gold from a single locality generate detrital gold whose composition is faithful to that of the source, but populations of placer gold may contain particles from multiple sources. In order to establish the nature of the contributing gold types, sufficient particles must be available, and these must exhibit sufficient diagnostic criteria to permit interpretation. Despite these challenges, various recent studies have identified groups of compositional characteristics that are exhibited by gold from specific deposit types. Examples include the Pd–Hg inclusions (and alloy) signature of gold from alkalic Cu–Au porphyry systems in BC [26] and the Bi–Te–Pb–S inclusion signature of gold formed in calc-alkalic porphyry systems in Yukon [27]. Gold signatures from mineralized orogenic systems are characterized by a broader array of features within which particular inclusion associations commonly occur, namely a simple base metal signature associated with sulphides $\pm$ (sulpharsenides or tellurides) $\pm$ sulphosalts [28].

Much of the early pioneering work on gold composition was carried out in BC between 1985 and 1993 [44–48]. These studies focused on the relationships between the alloy compositions of gold from known lode sources in Southern and central BC and those of the surrounding placers. The origins of gold in the Fraser River were discussed in terms of potential contributions from the Cariboo Gold District (CGD) and Bridge River area, and a similar approach was applied to the Coquihalla drainage. Two main compositional groups were identified on the basis of Cu and Hg levels in the Au–Ag alloy. The high Cu group was attributed to gold associated with ultrabasic lithologies, whereas the presence of Hg was interpreted as indicative of orogenic gold sources. Within these groups, there were compositional overlaps that could not be resolved through the study of alloy compositions by EMP alone. Nevertheless, examination of microfabrics within the high-Cu population greatly refined the characterization of gold with an ultrabasic association [49].

A study of over 2000 gold particles from placer and lode settings in the CGD [43] augmented the alloy composition data previously reported [47] with both inclusion data for those samples and new material collected for the study. Comparison of mineralogical descriptions of lode occurrences with mineral suites helped refine the classification of gold types in the Wells–Barkerville area, in particular distinguishing between a low-Ag type associated with cosalite inclusions occurring around Wells and a more Ag-rich regional type with an inclusion suite dominated by pyrite and arsenopyrite. The correlation of the Ag contents of these gold types with bulk fineness data from historical placer mining activities permitted the evaluation of the most economically important gold types.

Gold compositional studies in the Northern Cordillera in BC, Yukon, and Alaska have also established generic compositional signatures associated with gold from specific mineralizing environments. Gold from alkalic Cu–Au porphyries in BC yields a Pd–Hg signature [26], while similar work in Yukon showed that gold from calc-alkalic systems shows a Bi–Pb–Te–S signature in the inclusion suite [27]. A perceived disadvantage to this approach was the number of gold particles required to establish the signature, and consequently [50] investigated whether the larger range of detectable elements afforded by the use of laser ablation inductively coupled plasma mass spectrometry (LA-ICP-MS) could generate a consistent signature from fewer gold particles. The aim was to evaluate whether the small number of gold particles generated in stream sediment surveys could find utility in an indicator mineral context. This work developed during the time that the large degree of heterogeneity of trace and ultra-trace elements within gold was becoming clear, and it is now apparent that analysis of only a few particles could produce highly unrepresentative results [40].

The large numbers of gold particles from BC analyzed and inspected prior to the present study revealed internal microfabrics and alloy compositions that were entirely compatible with the detrital model of placer gold, i.e., a model in which eroded gold particles remain largely intact within fluvial sediments. Nevertheless, it is important to note that other workers have reached different conclusions through consideration and interpretation of different information. The apparent discrepancies between both the particle size and bulk fineness of gold in lodes and placers in the Cariboo Gold District have been cited as evidence for gold nugget growth in the supergene environment [51–53]. These assertions resonate with the widely held perception that gold is chemically active in surficial environments to the extent that placer gold may be compositionally distinct from that recovered from the proximal lodes owing to an entirely different genesis. The argument for gold growth in the placer environment has also been advocated more recently in a number of papers, e.g., [54], that propose that the commonly observed micron-scale precipitation of gold onto pre-existing particles through biogenic activity is an ongoing process that results in particle size increases. If gold modification in the surficial environment is widespread and bulk compositions are indeed modified, the potential use of gold as an indicator mineral would be fatally undermined. The subject has recently been discussed at length in a study [39] that considered over 40,000 sections of gold particles from localities worldwide and concluded:

i. Gold particles can increase their mass in specific supergene (not fluvial) settings of circumneutral groundwaters where both Au and Ag are transported as thiosulphate complexes.
ii. Hypogene gold exhibits specific microfabrics and inclusion assemblages, and the identification of these features in placer gold particles confirms a detrital origin.
iii. Gold from the overwhelming majority of placer localities globally exhibits such features, whereas microfabrics consistent with a process of nugget growth have not been recorded in any of the 40,000 placer particles studied.

On the basis of the scale and detail encompassed by this study, we assert that the internal compositions of placer gold particles are faithful to those within the lode source and therefore represent a platform on which to develop a robust indicator methodology.

In this study, we have demonstrated the close correlation between mineral inclusion suites and the mineral assemblages associated with gold in different ore deposit types. In tandem with substantial new alloy and inclusion data describing gold from many localities, we have developed compositional templates for gold from orogenic, low-sulphidation epithermal, and alkalic porphyry settings. There have been substantial advances in characterizing gold from other magmatic hydrothermal and orthomagmatic environments, facilitated both by the data set at our disposal and petrographic studies of auriferous mineralization. The Synthesis of these data sets has generated deposit-specific compositional templates against which 'unknowns' may be compared. In this way, it has been possible to identify the type(s) of source mineralization for some gold localities where this was previously unknown.

2. Materials and Methods

The major objective of producing a database of gold compositions depends on access to sufficient populations of gold particles representing both geological and geographical spread within the province. This project has taken advantage of gold collections from both the University of British Columbia (UBC) and the University of Leeds (UoL), and the geographical spread of gold sample populations examined in the study is shown in Figure 1.

The database describing gold from localities where the deposit type is known comprises 11,520 particles from 133 localities. The UBC collections comprise placer and hypogene gold collected over several years in the 1980s and 1990s. Polished sections of both placer gold populations from specific localities and Au-bearing ore samples were analyzed by EMP at UBC during this period. The initial database was augmented in two ways during the present study. Firstly, the inclusion suites present in each population of gold particles were established by visual examination on the scanning electron microscope (SEM; see below). The incidence of inclusions varies considerably [27,32], and in many cases, the number recorded in sample populations was insufficient to underpin rigorous classification. Secondly, the UBC collections contained additional particles from numerous localities, and these were mounted and analyzed in the present study to improve the quality of the final data set. The remit of the present project to generate a compositional template against which other gold samples may be compared requires gold samples whose deposit type provenance is unambiguous. Samples of placer and lode gold in the UoL collections relate to either locality-specific studies (Cariboo Gold District: [43]; Atlin, [55]) or deposit-type-specific studies (gold from alkalic porphyry systems, [26]). Lode samples are vital in this regard, but in many other cases, the source type of gold placer samples can be established with near certainty, particularly where the signature of the placer gold corresponds to that of proximal lode gold [26,43]. In other cases, the deposit type from which placer gold is derived remains unclear, and such sample populations cannot be used to generate compositional templates. Similarly, placer samples from some (commonly large) drainages may contain gold particles from two or more different deposit types. Around 30% of the gold particles in the UBC collections fall into this category (e.g., gold from the Fraser and Coquihalla river main valleys), because at the time of collection, the drivers for gold collection were to investigate variation in gold signatures between localities rather than to identify compositional signatures for gold from specific deposit types. For the purposes of the present study, the data set has been divided into sample populations where the source deposit type can be ascribed with confidence and others where, although the source deposit type is unclear, there is sufficient compositional information to establish a compositional signature. A full table showing details of the localities for which deposit types may be confidently ascribed is presented in Appendix A, and the data are summarized in Table 1. The data set comprises 11,840 gold particles from 160 localities.

Table 1. Overview of sample suites according to deposit type.

Deposit Type	No. Localities	Total No. Particles
Alkalic porphyry	10	897
Calc-alkalic porphyry	7	551
High sulphidation epithermal	2	40
Low sulphidation epithermal	8	1032
Orogenic	93	8724
Intrusion-related veins	7	38
Skarn	3	93
Ultramafic association	2	107
VMS	1	38

Gold from orogenic settings has the strongest representation in the data set, and this is an inevitable consequence of the amenability of orogenic gold to form placers. It is also clear that several deposit types (high-sulphidation epithermal, VMS, intrusion-related gold, and skarns) are poorly represented. In some cases, it has been possible to partially alleviate this issue by studying samples of polished blocks of ore, where the association of gold with coeval minerals can be used to predict elements of the inclusion signature. In addition, there is a bias in the whole data set according to previous studies in the Province that targeted gold from the Cariboo Gold District [43] and the sample suites describing gold from alkalic porphyry deposits [26].

The suite of samples for which provenance is unknown comprises a total of 2916 gold particles from 41 localities, and details are provided in Appendix B. However, only 8 of these yielded a sufficiently large inclusion suite to permit comparison with deposit-specific compositional templates (Table 2). In addition, sample populations from Bonaparte Mine, Granite Ck, Lilloet, Peers Ck, and Fairless Ck exhibited compositional characteristics that could be informative, and these are mentioned in the text.

Table 2. Details of sample populations where the source is unknown but there is sufficient data to characterize the signature. The 'inclusion tally' refers to the number of particles that contain a useful inclusion.

Location	No. of Particles	Inclusion Tally
Bridge R at Yalakom	88	13
Bridge R above Moha	45	16
Coquihalla R	83	24
Liard R	94	16
Tranquille Ck	164	25
Whipsaw Ck	328	21
Yalakom R	45	10
Ladner Ck	165	24

Polished blocks were inspected using the secondary electron (SE) and back scattered electron (BSE) facilities of a Quanta 650 FEG scanning electron microscope (SEM). Liberated or detrital gold particles are characterized through a combination of alloy profiles (determined by EMP) and inclusion assemblages (determined by visual inspection in both (SE) and (BSE) modes). Both approaches require particles to be sectioned and polished. Alloy analyses of most of the UBC sample suite were carried out at UBC, and all other analyses were carried out at UoL. The compatibility of results from the two analytical facilities was previously established by duplicate analyses of populations of gold particles from localities in Yukon [42]. All analysis regimes included Au, Ag, Cu, and Hg, but the early studies did not include Pd. An overview of the analytical conditions used for gold analysis for the full element range has been described previously [32]. All analyses quoted are mass%.

A summary of the workflow from gold collection to sample preparation is provided in Figure 2. The first stage in the sample characterization was a visual inspection of all gold particle sections. These studies were carried out at UoL using a SEM. Mineral inclusions were identified and chemical analyses generated using the energy dispersive spectrometer (EDS) facility. Mineral speciation was interpreted by comparing the spectra with those of reference minerals. In some cases, a small degree of substitution was observed (e.g., Cu in acanthite or Sb in galena). In these cases, a record was generated that influenced the scoring system used in the generation of radar diagrams, as described previously [28].

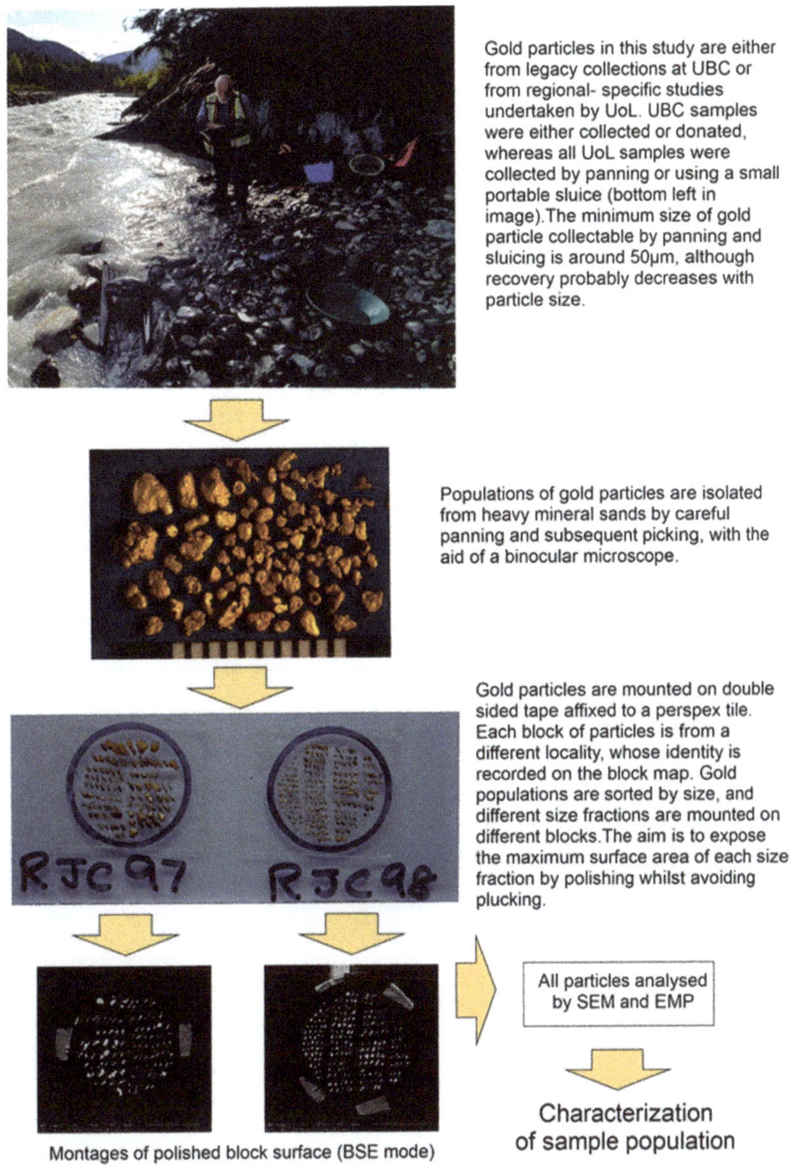

Figure 2. Schematic representation of the project workflow for the collection and characterization of gold particles.

3. Results

3.1. Variation in Mineral Assemblages between Gold-Bearing Deposit Types

The generic geological settings in which specific ore deposits form have a substantial influence on the overall deposit mineralogy, particularly regarding the assemblages coeval with gold. Examples of different mineral associations in gold-bearing deposit types are presented in Figure 3. Figure 3A–F shows a range of gold-mineral associations in samples from orogenic gold deposits. The simplest mineralogical associations are gold–quartz (Figure 3A) and gold–quartz–pyrite (Figure 3B). In some samples, abundant Fe oxides are the decomposition products of pyrite (Figure 3C). Common sulphides are important components at some localities, e.g., chalcopyrite (Figure 3C). Gold from Bralorne is an example of a gold ore associated with a range of accessory minerals such as arsenopyrite, galena, sulphosalts, and sphalerite, as illustrated in Figure 3D. Carbonate is an important component at many localities (Figure 3E), and in some cases, gold is associated with alteration products of the mineralizing event (Figure 3F).

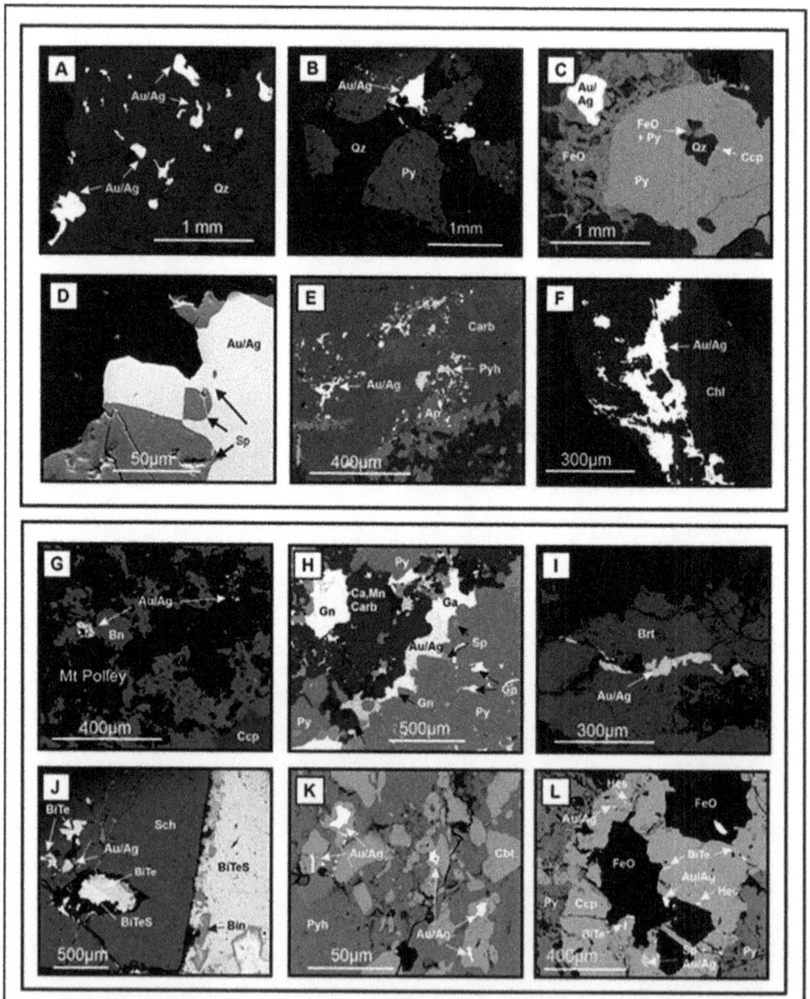

Figure 3. Mineral associations of gold in various deposit types. (**A–F**): Orogenic Au. (**A**): Murphy, (**B**): Erickson Eileen Vein, (**C**): Frasergold, (**D**): Bralorne, (**E**): Carolyn, (**F**): Aurum. (**G–L**): Gold mineral association in magmatic hydrothermal systems (**G**): Mt Polley Cu–Au alkalic porphyry, (**H**): Silback Premier low-sulphidation Au, (**I**): Albert's Hump High sulphidation Au, (**J**,**K**) Hedley Au skarn, (**L**): Bonaparte intrusion-related veins. Mineral abbreviations in all figures are approved [56], except where speciation is unknown; and in those cases, the elemental components are shown.

Gold from magmatic hydrothermal systems commonly exhibits intimate spatial relationships with a wider range of mineral types, and examples are provided in Figure 3G–L, although these by no means describe the full range of mineral associations for each deposit type. Gold particles associated with the potassic stage of Cu–Au porphyry formation are exolved from either chalcopyrite or bornite and are usually too small to be collected during field sampling using traditional panning techniques. An example of a relatively large Au particle associated with bornite is shown in Figure 3G. Variation between the mineralogy of high- and low-sulphidation epithermal deposits is the basis for their classification, and examples from each ore are depicted in Figure 3H,I. The Ag content of gold from the low-sulphidation ore at the Silback Premier Mine was determined by EDS rather than EMP and contained around 40 wt% Ag. The gold is associated with pyrite, galena, sphalerite, and Mn-bearing carbonates. In contrast, gold from Albert's Hump comprises barytes and a gold alloy containing only 0.5 wt% Ag. Samples of gold-bearing ore from the Hedley skarn deposit (Figure 3I,J) show two auriferous associations: one of around 10 wt% Ag with various minerals in the Bi–Te–S system and scheelite, and the other of around 5 wt% Ag with pyrrhotite and cobaltite. Finally, gold from the Bonaparte intrusion-related gold system is associated with Bi telluride, hessite, chalcopyrite, and pyrite.

3.2. Features of Natural Gold Particles That Permit Compositional Characterization

Gold particles may comprise homogeneous alloys (Figure 4A), but the variation in Ag contents of different homogenous particles may vary widely (Figure 4B). Placer gold particles commonly exhibit an Au-rich (equivalently Ag-depleted) rim typically 2–10 microns in thickness, and examples are visible in Figure 4C, where the particle core is highly heterogeneous, as indicated by the variation in grayscale when viewed in back-scattered electron (BSE) mode. The detailed images of different microfabrics presented in Figure 4C–G are interpreted to indicate modification of pre-existing Au–Ag alloy to Ag-rich alloy in the later stages of the mineralizing event [32]. Modification of the primary alloy by fluid ingress along grain boundaries yields Ag-rich films, which may or may not be associated with heterogeneity, as indicated by the variation in grayscale when viewed in back-scattered electron (BSE) mode. The detailed images of different microfabrics presented in Figure 4C–G are interpreted to indicate modification of pre-existing Au–Ag alloy to Ag-rich alloy in the later stages of the mineralizing event [32]. Modification of the primary alloy by fluid ingress along grain boundaries yields Ag-rich films, which may or may not be associated with grain boundary migration (Figure 4D,E). Where Cu contents in Au–Ag alloys are relatively high, Cu–Au intermetallic compounds exsolve on cooling (Figure 4F). Modifications to gold particles in the surficial environment comprise loss of Ag, sympathetic to grain boundaries in the interior of particles (Figure 4G) and also parallel to the particle surface, to form the Ag-depleted rims described above [39]. In summary, it is important to note that an individual gold particle may exhibit a chemical record of changes in the conditions of precipitation during the mineralization event, which may be subsequently partly altered during residence at or near the surficial environment. In these cases, it is not possible to derive a simple 'signature' from an individual particle without knowledge of the degree and nature of the heterogeneity. For example, spot analysis of the particle shown in Figure 4F could generate alloy compositions ranging from 85.8 wt% Au, 1.62% Ag, and 13.6 wt% Cu in the matrix to 75.7 wt% Au and 24.3 wt% Cu in the exolved laths, showing that in heterogeneous particles, compositional definition by a single measure is meaningless. Similarly, particles with varying Ag content (e.g., Figure 4C) are impossible to characterize with a single value. In these cases, the analysis value relates to the earliest paragenetic stage of the alloy that is identified through mutual spatial relationships within the section. The rationale is that this alloy is most useful in relating particle composition to the main episode of gold deposition. Where particles exhibit exsolution of intermetallic Au–Cu, the composition of the matrix is recorded.

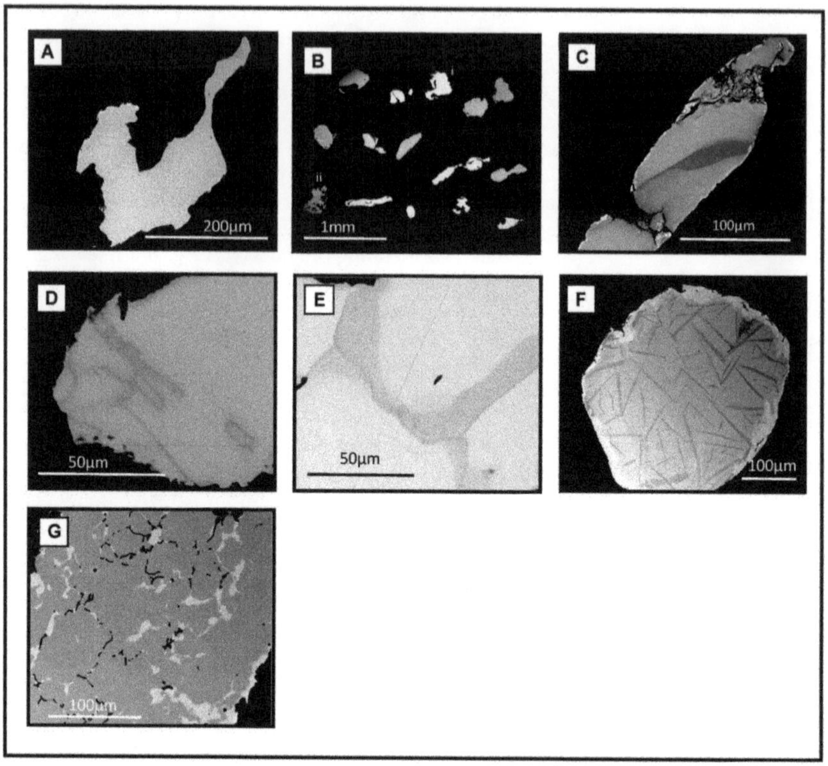

Figure 4. (**A**): Homogenous gold particle (Eight Mile Lake), (**B**): Example of variation in BSE response according to Ag content: range 4.6 wt% Ag (i) to 25.9 wt% Ag (ii) (Antler Creek), (**C**): Heterogeneity within a single particle (Granite Creek), (**D**,**E**): examples of late stage Ag-rich alloy emplaced sympathetic to grain boundaries (Mitchell Creek), (**F**): Exsolution of CuAu intermetallic from Au–Ag–Cu alloy (Coquihalla River), (**G**): Au-rich alloy (pale grey) resulting from Ag removal in the surficial environment in primary relatively Ag-rich gold (dark grey) (Tranquille Creek).

Mineral inclusions typically comprise 2–10 μm particles genetically related to the mineralization stage coeval with gold. They are recorded in gold particles from lode samples (e.g., Figure 3D) and are preserved by their encapsulation within the inert gold particles following erosion. Figure 5A–F shows examples of inclusions commonly found in gold from orogenic settings, where they typically comprise the entire inclusion suite. These minerals also occur in gold from magmatic hydrothermal settings but usually in association with minerals from other classes, e.g., tellurides or selenides. Figure 5E–K provides examples of inclusions observed in gold from magmatic hydrothermal systems. Many mineral species recorded in gold from magmatic hydrothermal deposits are apparently absent or extremely uncommon in gold from orogenic systems. Silver-bearing minerals such as proustite and aguilerite are confined to gold from these deposit types, and minerals in the Bi–Te–S system are very common, whereas they are extremely rare in gold from orogenic systems. Gold from alkalic porphyry systems exhibits a distinctive array of Pd and/or Hg-bearing minerals within the inclusion suite (e.g., Figure 5G). Various Cu sulphides (±Fe) are common in gold associated with ultramafic rocks (e.g., Figure 5L).

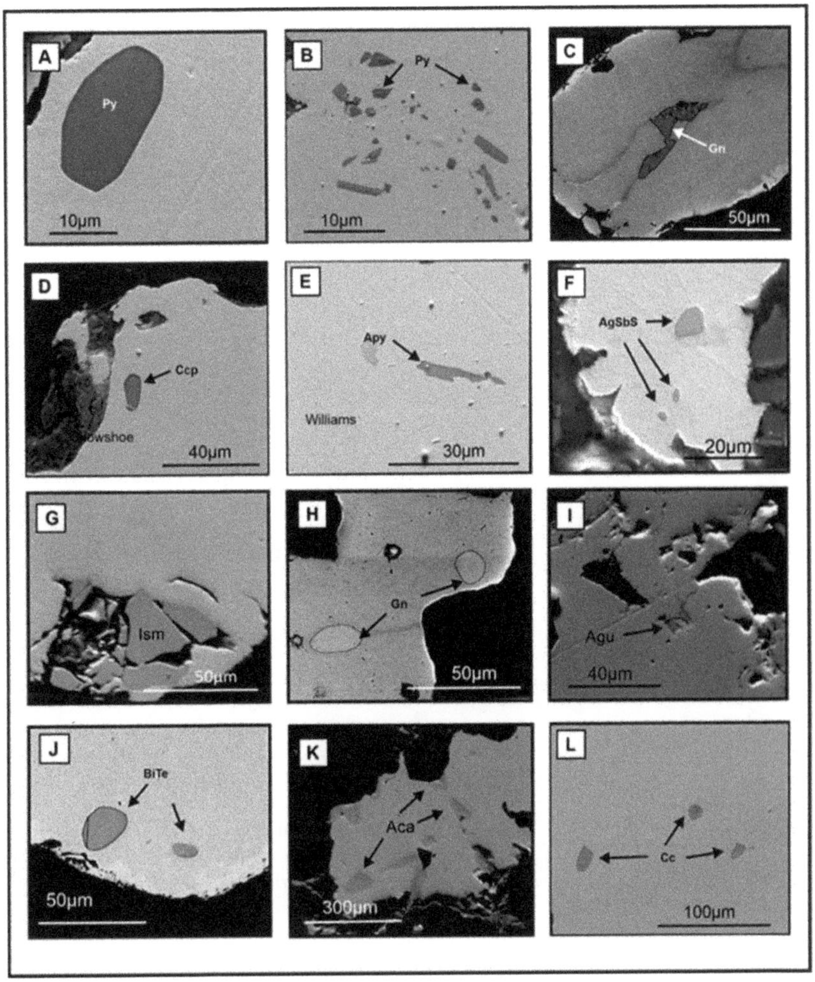

Figure 5. Examples of mineral inclusions. (**A–F**): Orogenic gold deposits (**A**), Spring Ck, Cassiar District, (**B**): Chisholm Ck, CGD, (**C**): Tertiary Mine, CGD, (**D**): Little Snowshoe Ck, CGD, (**E**): Williams Ck, CGD. (**F–K**): Examples of inclusions in gold from magmatic hydrothermal systems (**F**): Sulphurettes Ck, KSM deposit, (calc alkalic porphyry) (**G**): Friday Ck, Copper Mountain (alkalic porphyry) (**H–K**): gold from low-sulphidation epithermal deposits, (**H,I**): Blackdome, (**J**): Stirrup Ck, (**K**): Brucejack. (**L**): gold from an ultrabasic association: Wheaton Ck.

3.3. Characterization of Sample Populations

At the outset, it is useful to gain an overall impression of the data available to the study in order to get a sense of the broad differences between the compositional characteristics of gold from different deposit types. Differences in the signatures of gold within individual deposit types are considered subsequently. Figure 6 compares the Ag profiles of all populations whose genetic origins may be confidently ascribed.

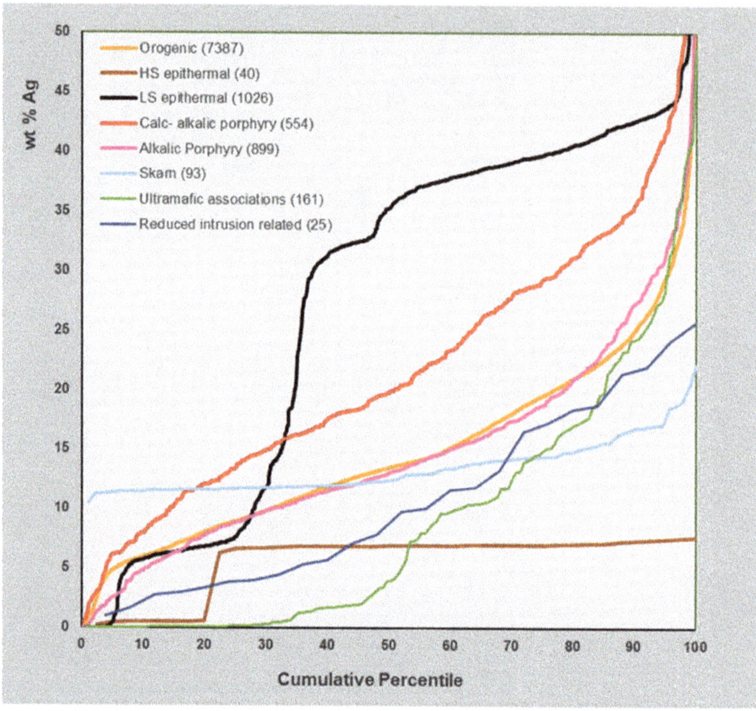

Figure 6. Ag profiles of samples for which the deposit type is known. Figures in parentheses refer to the number of particles within the population.

The majority of gold particles contain between 5 and 30 wt% Ag, irrespective of deposit type. The size of the populations available for study greatly influences confidence in ascribing generic characteristics to particular gold types. Gold from high-sulphidation epithermal, skarn, VMS, and ultramafic associations is relatively underrepresented in the current database. Gold derived from orogenic and porphyry sources generates plots with a continuous increase in Ag, whereas the Ag profile of gold from low-sulphidation epithermal systems typically shows a pronounced step, which is a consequence of the relatively large proportion of Ag-rich particles that all originate from the Blackdome deposit. The curve depicting gold from calc-alkalic porphyry systems is derived solely from sampling alluvial localities in the environs of the KSM porphyry deposit. The small sample set from the Britannia Mine is the only example of gold from a VMS system available for the study. The Ag profile is similar to that exhibited by the far larger sample suite from orogenic systems. Signatures of gold from VMS systems are not considered further in this study as a consequence of the small amount of data available.

The range of Ag contents of the deposit types in Figure 6 is a consequence of the variation in Ag profiles of the constituent populations. Figure 7 compares Ag plots for different individual localities according to deposit type. Individual Ag profiles of gold from orogenic settings (Figure 7A) show various characteristics in sub-populations, as evidenced by portions of the curve with markedly different gradients (e.g., Bralorne Mine and Lowhee Creek). In contrast, gold from the Eileen Lode of the Erickson Mine in the Cassiar area shows two mutually exclusive sub-populations, each displaying a narrow Ag range. Similarly, the gold from low-sulphidation epithermal systems shown in Figure 7B shows different plot forms; some are sub-horizontal, whereas others show profiles containing both shallow and steep gradients. Gold from alkalic porphyry deposits (Figure 7C) all show a continuum of compositions between 0 and 45 wt% Ag. The Ag profiles from some other deposit types are illustrated in Figure 7D–F, and although the data sets are relatively small, some useful observations can be made. Gold from the two skarn deposits (Figure 7D) exhibits curve shapes similar to those observed in gold from other deposit types, and Ag ranges are also comparable. The Ag profiles of gold from populations associated with ultramafic rocks (Figure 7E) exhibit a wide compositional range, including sub-populations where Ag is absent. Finally, the sample populations from intrusion-related systems show a continuum of Ag values from 0 to 25 wt% Ag.

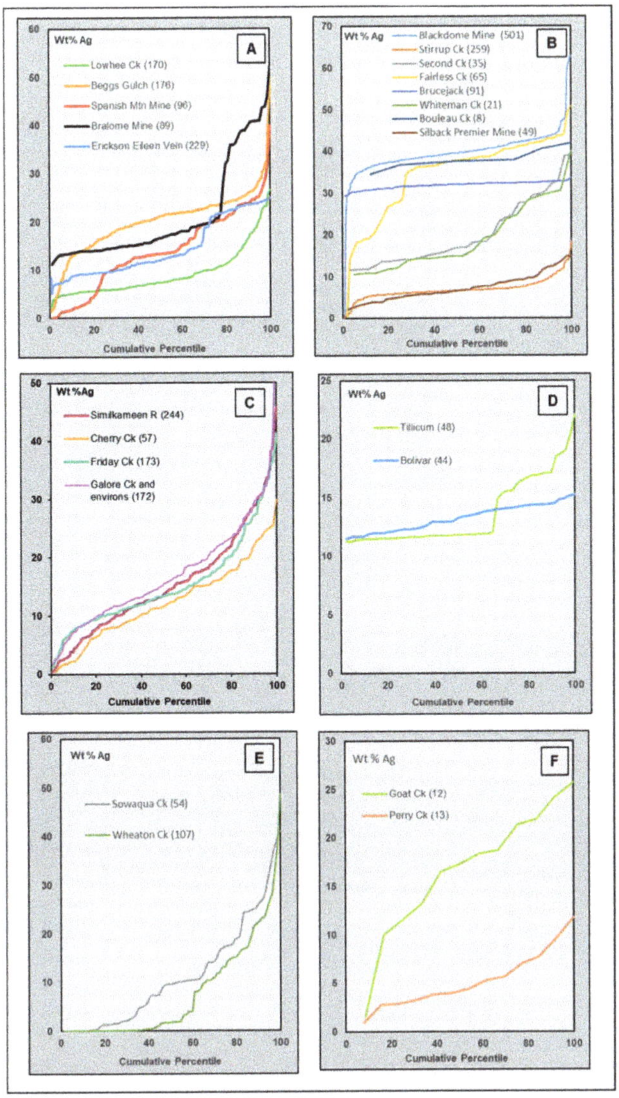

Figure 7. Intra-deposit type Ag comparisons. (**A**): selected Ag profiles from orogenic gold samples: (**B**): Low sulphidation epithermal systems, (**C**): Alkalic porphyries. (**D**): Skarn deposits, (**E**): Ultramafic associations, (**F**): Reduced intrusion associations.

The other minor components of the metal alloys detectable by EMP may, in some cases, be useful discriminants, and Figure 8 shows the concentration ranges of Cu, Hg, and Pd according to deposit type. A small proportion of gold particles from most deposit types may exhibit relatively high (>0.5 wt%) Cu values, but most are below the LOQ of 0.06 wt%. Concentrations of Cu in gold from ultramafic associations are frequently far higher, and compositions often conform to the mineral auricupride (AuCu). These particles are often highly heterogeneous with respect to Cu and Au (see Figure 4F), and the individual analyses that describe a particle are always derived from the low-Cu alloy matrix. The alloy composition and heterogeneous microfabrics observed in these particles are clearly distinctive. Some of the 40 gold particles from the Taylor Windfall high-sulphidation epithermal deposit show Cu values of nearly 1 wt% (Figure 9).

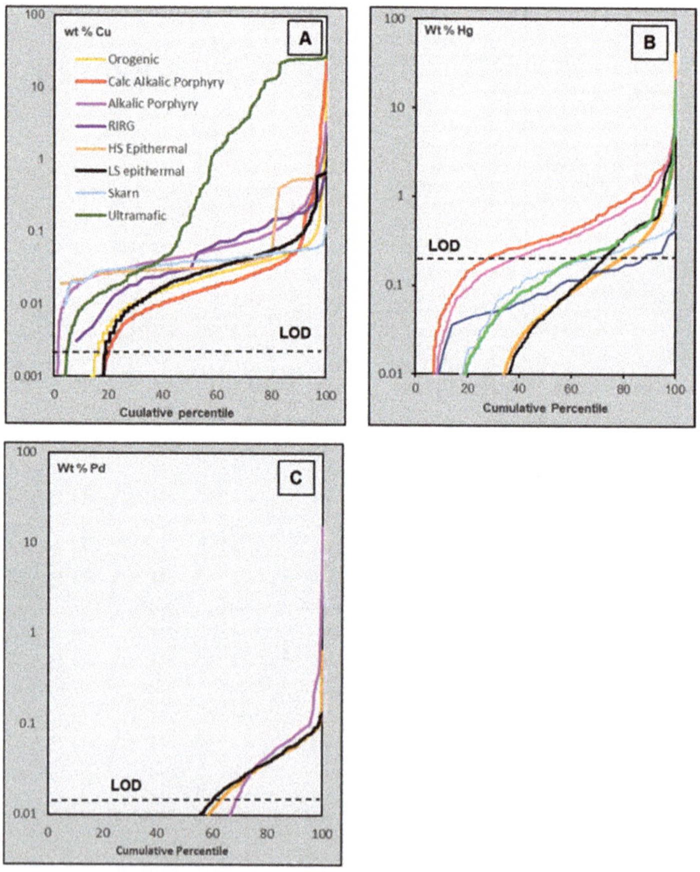

Figure 8. Minor alloy components of gold alloys according to deposit type. (**A**): Cu, (**B**): Hg, and (**C**): Pd. The number of particles in each of the populations is provided in Table 1.

The detection limit of Hg in Au–Ag alloys is relatively high (0.3 wt%), and therefore most of the data points depicted in Figure 8B are below the limit of quantification. The proportion of gold particles from porphyry environments exhibiting detectable Hg appears to be greater than for other deposits, but most deposit types yield some gold particles that exhibit Hg at percent levels.

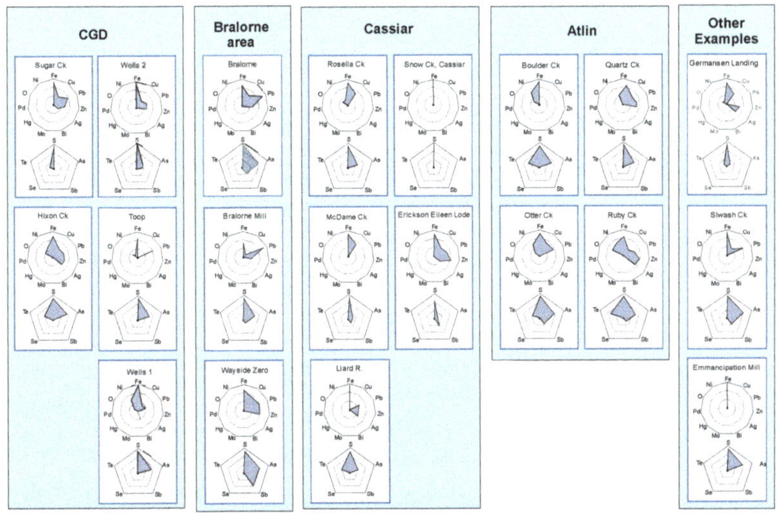

Figure 9. Radar diagrams for inclusion signatures of gold from orogenic settings.

Measurable palladium is confined to gold from alkalic porphyry systems, as previously reported [26], and although it is only detected in around 4% of the particles from each location (Figure 8C), where present, it comprises a clear discriminant.

A previous study of the signatures of gold from throughout the Canadian Cordillera established that inclusion suites were the most powerful tool in establishing deposit type [28]. Around 15 particles containing inclusions are normally required to confidently establish a robust signature, but in the present study, many sample populations have not revealed inclusion suites sufficiently large to permit characterization. Figure 9 shows the radar diagrams relating to samples of orogenic gold. Consideration of such data sets from orogenic gold localities globally showed that the non-metal component was usually most useful to classify gold of orogenic derivation [28]. Common associations are as follows: i. sulphides only; ii. sulphides and sulpharsenides $\pm$ minor sulphosalts; iii. sulphides, sulpharsenides, and tellurides; iv. sulphides and tellurides; and v. sulphides, sulpharsenides, tellurides, and sulphosalts. All associations have been observed in sample populations from BC. Figure 9 has grouped inclusion signatures from various localities in the same region, and it can be seen that gold from geographically close areas can exhibit different signatures. For example, two regional signatures in gold from the CGD have been identified [43] (Wells 1′ and Wells 2′, Figure 9), and the new sample from Toop conforms to one of these. Gold from Hixon Creek differs from the Wells 1 signature because it exhibits a Te component, as does gold from Sugar Ck, to the exclusion of As. Similarly, orogenic gold from localities in the Cassiar District conforms to either the S or (S + As $\pm$ Sb) signature. Gold from the Liard River could contain a distal component (because of the size of the drainage area), which might account for the presence of Te. The S–As–Sb signature is also observed in all samples from the Bralorne area. Most of the aforementioned inclusion suites comprise simple sulphides such as pyrite, arsenopyrite, galena, chalcopyrite, and sphalerite, but the gold from the four localities within the Atlin camp is the most mineralogically complex. There seem to be minor differences between inclusion suites of gold from different creeks in the Atlin area, although it is recognized that this may be a consequence of some relatively small data sets. In general, the mixed S–As–Te $\pm$ Sb signature is accompanied by a strong Ni–Co component (possibly reflecting the dominantly mafic-ultramafic host rocks for most gold occurrences in the Atlin area), and in 3 out of 4 cases, Ag. Unfortunately, no inclusions were observed in the sample population from Feather Ck, as other lines of investigation suggest that gold at this locality is fundamentally different and exhibits a strong association with cassiterite [57].

Radar diagrams describing gold from magmatic hydrothermal systems are presented in Figure 10, and it is immediately apparent that these signatures are more complicated than the signatures of gold from orogenic settings, as illustrated in Figure 9. The diagram includes some examples of gold signatures from calk-alkalic porphyry systems in Yukon (green tiles) because these emphasize generic deposit-type signatures and form useful comparators for the gold sample populations from the KSM drainage available from this study. Similarly, signatures of gold from the Clear Creek and Dublin Gulch intrusion-related systems in Yukon have been included to show their compatibility with the limited amount of mineralogical information available to infer inclusion suites of gold of this type from localities in BC (Figure 3L and Table 3).

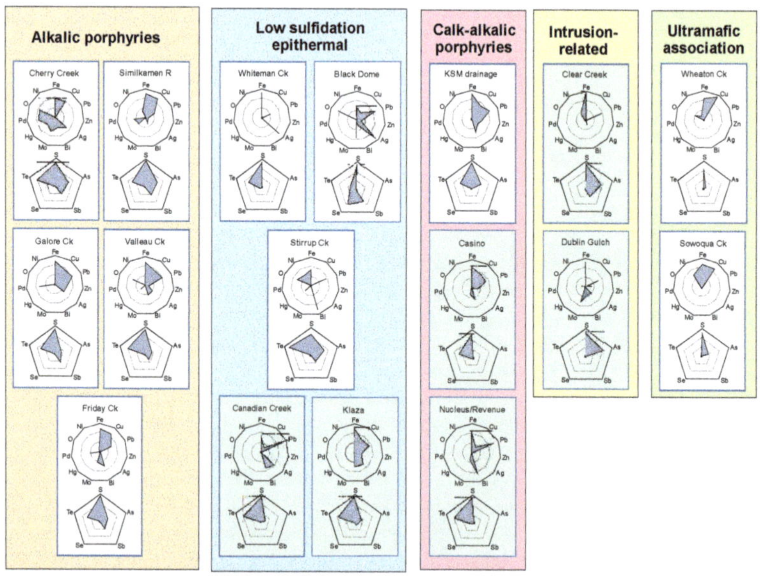

Figure 10. Characterization of inclusion suites in gold from magmatic-related systems. Diagrams in green relate to sample populations from Yukon: included here as further comparatives.

Table 3. Examples of small inclusion suites compatible with generic inclusion signatures.

Location	Inclusions	Deposit Type
Bonaparte	Matildite (AgBiS), undifferentiated Pb, Bi Telluride	IR veins
Brucejack	Acanthite (trace As)	LS epithermal
Silback Premier Mine	Acanthite	LS epithermal
Fairless Ck	Undifferentiated Au–Ag tellurosulphide	LS epithermal

The three inclusion suites derived from low-sulphidation epithermal mineralization are all relatively complex but show clear differences from those from the orogenic gold suite, for example in the relative importance of Ag and Bi. Tellurides are common to all, but gold from Blackdome shows a very strong Se signature. Gold from Stirrup Ck shows a strong similarity to gold from alkalic porphyry systems in the non-metallic components. A complex signature comprising Ag, Bi, and Te is also evident in gold from the Mitchell and Sulphurettes creek drainages at KSM.

Radar diagrams offer many advantages in depicting inclusion suites over other graphical approaches [28], but the elemental signatures provide no information on mineral speciation, which may itself be important. Differentiation between different mineral species can prove useful on a qualitative level; for example, the Cu signature of gold from orogenic settings is almost exclusively a consequence of chalcopyrite, whereas Cu mineral speciation in magmatic hydrothermal systems may also include bornite, chalcocite, and covellite. Gold associated with ultramafic rocks shows a complex Cu-bearing mineralogy, including all the species mentioned above and non-stoichiometric Cu–Fe sulphides. Tellurium-bearing species are encountered in gold from orogenic hydrothermal systems but almost exclusively as a consequence of hessite (Ag_2Te), whereas Bi-tellurides are the most common Te-bearing species in most magmatic hydrothermal systems.

In many cases, sample populations contain a few inclusions but are not sufficient to characterize the suite. However, where unusual inclusions are present, these may provide some useful information (Table 3). The inclusion suite of gold from the Bonaparte deposit (intrusion-related veins) shows an elemental signature suggestive of a magmatic hydrothermal system, although the presence of Te suggests a difference with the gold from intrusion-related systems in Yukon. Gold from the low-sulphidation epithermal occurrences at Brucejack contains inclusions of Ag-bearing minerals, and there is a close association of high (ca. 40 wt% Ag) gold with acanthite in an ore sample from Silback Premier Mine (Figure 3H). The small sample from Fairless Ck, near the Black Dome occurrence, contains Au–Ag sulpho-selenide inclusions, as does the gold from the Black Dome deposit itself.

4. Discussion

The discussion section has been divided into two sections: one reviewing the new data that contributes to refined compositional templates for specific deposit types, and the other applying these to gold from localities where the source deposit type is unknown.

4.1. Compositional Variation in Gold from Different Deposit Types

The data presented in Figure 6 shows that, when considered in isolation, the Ag content of gold alloy within a single particle is not diagnostic for source type. The ranges of Ag contents in gold alloys also vary between deposits of the same type (Figure 7), so it is not possible to make general statements relating Ag range to genesis, although values of >30 wt% appear more common in gold from magmatic hydrothermal systems. Gold from low-sulphidation deposits has been characterized as 'high Ag' (e.g., [57]), and the data depicted in Figure 7B does indeed show that some Ag ranges are notably higher than those in most gold from orogenic systems. However, high Ag contents of over 30 wt% are not a generic feature with gold from Whitman, Fairless, and Second creeks, where Ag ranges are lower and compatible with those of gold alloy from other deposit types. The profile of the Ag curves may either be horizontal/sub-horizontal or exhibit a gradient. These generic shapes of the Ag curve have been discussed in terms of the nature and evolution of the mineralizing hydrothermal system [42,43]. The Au/Ag ratio of the gold alloy is a function of (Au/Ag) (aq), temperature, pH, fS_2^-, and fCl^- [41], and consequently, a low Ag range is most likely indicative of stable conditions of alloy precipitation, whereas a curve with a steeper gradient indicates change in one of more parameters in an evolving system. Hydrothermal systems that comprise multiple episodes of fluid influx may generate sub-populations corresponding to either (or both) of these scenarios. The curves depicted in Figure 7A,B show that gold from orogenic and low-sulphidation epithermal mineralization may conform to either profile form, indicating that the plot shape is not diagnostic for deposit type. A comparison of the Ag ranges of gold from alkalic porphyry systems is shown in Figure 7C. In this case, all curves show Ag values mainly across a range of 0–30 wt%. The detrital gold collected from the environs of the calc-alkalic KSM porphyry exhibits a similar Ag profile (Figure 6). This result may be a consequence of the 'net effect' of sampling placer locations where gold is derived from the various mineralizing environments within the hydrothermal system as a whole. At Copper Mountain (alkalic porphyry), 22 distinct episodes of mineralization were previously reported [58]. The sample from Mitchell Ck (calc-alkalic porphyry) could contain gold particles from both the Mitchell porphyry and the adjacent Iron Cap epithermal deposit. A detailed paragenetic study of mineralization at Iron Cap revealed seven stages of mineralization, five of which contained gold [59], and multiple stages of veining have also been recorded in the Mitchell porphyry [60]. Consequently, the sample population of placer particles from Mitchell Creek almost certainly contains contributions from different mineralizing episodes, and the same is almost certainly true of gold in Sulphurets Creek, which drains the adjacent Kerr and Sulphurets porphyries. Data from gold sample populations from Mitchell and Sulphurrets creeks has been combined to generate a signature of the entire Kerr–Sulphurets–Mitchell (KSM) porphyry system. While this data set cannot identify the nuances between gold formed in individual hydrothermal episodes, it does provide an example of the signature obtained by sampling porphyry-epithermal systems of this type.

Gold particles from the skarn deposits depicted in Figure 7D suggest that conditions for gold precipitation may be variable within the same deposit. A limited amount of data describing gold compositions from the French Mine at Hedley also showed a range of Ag values from 10 to 20 wt% [61]. The very limited amount of data describing gold from high-sulphidation epithermal settings also shows that a compositional range with respect to Ag is possible. Gold particles from mineralization within reduced intrusion deposits are composites from several sampling sites, which may in part account for the range in Ag values. The sample populations from Wheaton and Sowaqua creeks are distinctive because of their very pronounced Cu signature (see below), but they also show a wide range of Ag values. This Ag profile could be explained either by a large variation in the signature of a single source or as a result of a mixture of gold types in a single placer locality. This subject is explored in more detail below.

Concentrations of minor metals have proved useful in some cases. The Cu contents of gold from high-sulphidation epithermal localities suggest that they may be a useful discriminant for gold of this deposit type more widely. Mercury contents of gold alloys are generally higher in gold from porphyry mineralization, but some individual localities in orogenic gold regions also yield gold with Hg at percent levels; for example, Dragon and California creeks in the CGD [43], so in isolation, Hg values are not a diagnostic discriminant. The previous assertion [26] that gold from alkalic porphyry systems exhibits both Hg and Pd-bearing inclusions is partly upheld, with Pd minerals observed in gold from Galore Creek and Hg-bearing minerals recorded in gold from Valleau.

These inclusion signatures recorded in gold from all deposit types are entirely consistent with the range of gold alloy-mineral associations observed in petrographic studies of ore assemblages. The dominance of mineralogically simple inclusion suites in gold from orogenic settings previously reported [28] has been confirmed, and the importance of Ag-bearing minerals in gold from low-sulphidation epithermal systems has been emphasized by the auriferous mineral assemblage of ore from Silback Premier Mine and the few inclusions recorded in gold from the Brucejack Mine and Fairless Creek. At present, there are insufficient data points to generate a robust inclusion template for gold from intrusion-related systems, but consideration of the inclusion suite in gold from the Bonaparte deposit and the minerals associated with gold in ore samples shows that minerals with a Bi–Te–Ag–Pb signature are likely to be important components. Samples of gold-bearing ore from the Hedley skarn deposit show two distinct mineral associations that would likely be represented in local detrital gold. While only a single example of high-sulphidation gold is available for this study, both the alloy and mineral association profiles are unlike any gold from low-sulphidation deposits and occurrences. The distinctive microfabrics in Cu-rich gold derived from ultramafic associations have previously been described [48]. These authors reported the presence of such gold particles in alluvial gold from the Coquihalla River drainage, and these were observed in the same sample set during the present study (Figure 11A). In addition, similar particles were recorded in gold from Bridge River (Figure 11B,C), Relay Ck (Figure 11D), and in one case from the Fraser R. above Williams

Lake (Figure 11E). In addition, around 10% of the sample population from the Bridge River, 1km upstream from Moha, exhibited Cu contents > 2%. In all these cases, many other gold particles in the sample populations contain no detectable Cu and may also exhibit inclusion species commonly associated with orogenic gold deposits. Nevertheless, the identification of specific alloy microfabrics and inclusion suites within gold of ultramafic association provides clear diagnostic markers for gold of this type.

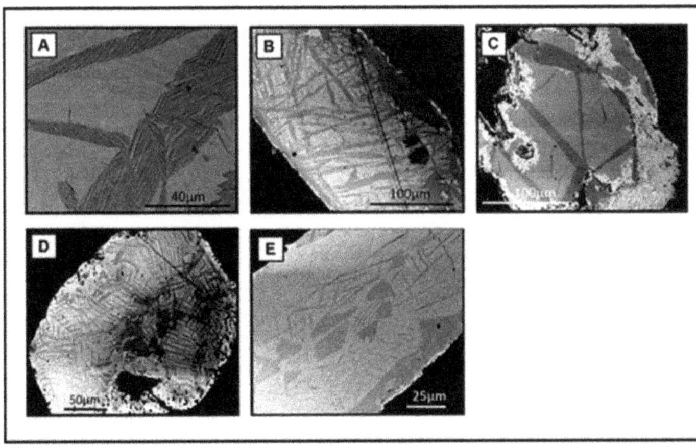

Figure 11. Gold from ultramafic sources identified through microfabrics in which auricupride has exsolved from a Au–Cu–Ag alloy. (**A**); Coquihalla R., (**B**): Bridge R. at Bralorne, (**C**): Bridge River at Yalakom confluence, (**D**): Relay Ck, (**E**): Fraser R. above Williams Lake.

While the focus of the study has been the gold metallogeny of British Columbia, it has been useful to refer to gold signatures derived from studies of deposit types elsewhere, particularly in the contiguous territory of Yukon (see green tiles in Figure 10). The range of signatures observed in orogenic gold in Yukon [28] has also been recorded in BC, and the generic characteristics of gold from porphyry and epithermal environments in Yukon resonate with signatures from those deposit types recorded in the present study. The Bi–Te signature in gold at the Hedley skarn deposit was also reported in gold from Ecuador [62]. Bismuth is a strong component of the inclusion signature in gold from the Wells area in the CGD, where it is present as cosalite ($Pb_2Bi_2S_5$) only, but in all magmatic hydrothermal systems, Bi minerals usually comprise various tellurides and sulphotellurides.

The methodology employed during the study has highlighted some shortcomings of gold characterization work where approaches to sample collection are not specifically designed to permit characterization of the source gold type. The recognition of the limitations of Ag content as a discriminant in this regard and the inability to utilize concentrations of minor metals for characterization as a consequence of their low concentrations and LOQ of the analytical technique hinder further insights into the origins of many of the sample populations studied. In some cases, there are neither sufficient gold particles nor sufficient sample populations (or both) to permit identification of robust compositional signatures (VMS, intrusion-related veins, high-sulphidation epithermal systems), but the data reported here forms a platform for future work. Similarly, the inclusion suites recorded are often too small to facilitate robust interpretation as a consequence of relatively small sample numbers and, most likely, their obliteration by morphological changes to gold particles that accompany fluvial transport. Nevertheless, the databases available to the study have contributed to a more complete understanding of the signatures of gold from deposit types from throughout British Columbia, and a summary of these characteristics is presented in Table 4.

Table 4. Summary of generic compositional characteristics of gold formed in various deposit types.

Deposit Type	Key Distinguishing Features			Comments
	Inclusion Suite Signature		Alloy Composition/Microfabrics	
	Non-Metallic	Metallic		
Orogenic: 1	S	Base metal sulphides, sulparsenides ± sulphosalts or tellurides. Dominated by pyrite	Variable Ag, Cu mostly below LOQ, Hg rarely detectable but in specific cases present to % levels, associated with relatively high Ag (>25 wt%). Where present, Sb and Te are usually minor components of the inclusion signature	Non-metallic signatures replicated at various localities, both within BC and worldwide. Signature type may differ between samples from within a gold camp
Orogenic: 2	S–As ± Sb			
Orogenic: 3	S–As–Te			
Orogenic: 4	S–Te			
Orogenic: 5	S–As–Sb–Te	Ni–Co very strong ± Ag	Wide range of Ag, Hg, and Cu < LOQ	Signature unique to (most) gold from Atlin camp.
Low sulphidation epithermal	S–As–Te ± Se	Ag and or Bi occurring as various minerals Base metal sulphides ± sulpharsenides		Ag ranges in gold from a specific locality may be small, but differences in Ag ranges between deposits can be large.
Alkalic porphyry	S–As–Te–Sb	Pd and or Hg-bearing inclusions plus a wide range of other species	Pd to wt% levels typically in c 4% of particles. Cu > LOQ in c 40% of particles	
Ultramafic association.	S>>As	Dominated by Cu ± minor Ni.	Diagnostic microfabrics of Cu–Au intermetallic exsolving form Au–Ag–Cu alloy	Speciation of Cu minerals distinct from Cu in OGDs: copper sulphides and bornite dominate.
Calc- alkali porphyry	S–As–Te	Base metals, Bi		Sample suite confined to environs of KSM deposit, and may be a mixture of gold from porphyry and epithermal environments. Inclusion signature compatible with that reported in gold of the same type in Yukon.
RIRG	Insufficient data to fully characterize	Bi and Te-bearing minerals identified in gold from Bonaparte	Wide range of Ag values in small sample set studied	Insufficient data to fully characterize, but Bi–Te association is clear
Skarn		BiTe–Au and Au-apy-po associations evident in polished block samples		BiTe–Au and Au-apy-po associations evident in polished block samples from Hedley.
VMS				Single sample population is insufficient to speculate on generic signature
High sulphidation epithermal			Some particles from Taylor–Windfall exhibit Cu values to c. 1 wt%: far higher than observed in any of the samples from low sulphidation epithermal systems	

4.2. *Application of Compositional Templates to Characterization of Placer Gold at Localities Where the Deposit Type(s) Are Unknown*

A summary of sample populations where the source type is unknown or where mixtures of gold from different gold types are considered likely is presented in Table 2, and details of localities form Appendix B. In this section, we have explored how some of the generic compositional characteristics of gold from specific deposit types have been identified in gold from other populations where the source deposit type is unknown. Given that Ag is not a particularly useful discriminant in this regard and that the main concentration of other alloying elements is below LOQ, there are in general limited characteristics of the population of unknown samples to examine. However, in some cases, inclusion suites and textural information can be applied to advantage, and several short case studies are described below.

Where populations of inclusions are sufficiently large, radar diagrams have been generated for comparison with those of inclusion suites in gold of known origin. 'Unkowns' (in pale red) have been paired with their closest analogues, and Figure 12A shows that the inclusion signature of gold from Lorne Mine is almost identical to others recorded in the Bralorne area. The strong Sb signature of gold from the Tertiary Mine in the Western CDG (Figure 12B) bears a striking resemblance to that previously observed in the Mooosehorn Range of Yukon and Alaska and also the environs of the Coffee Gold property, Yukon [28].

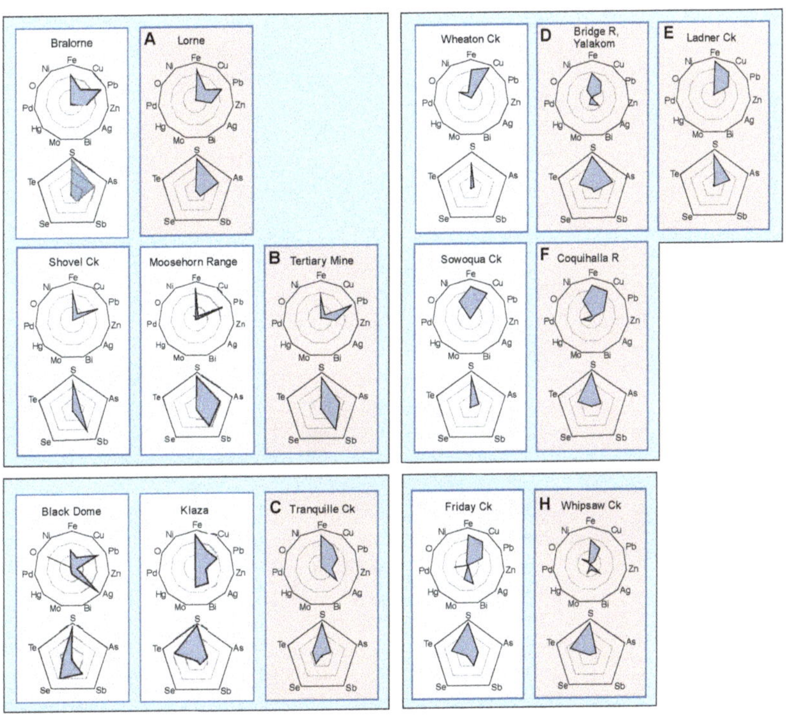

Figure 12. Inclusion assemblage comparisons of samples from 'known' and 'unknown' sources with relevant comparisons indicated by the horizontally adjacent groupings A-H 'Unknown' samples are depicted in pale red tiles and are horizontally adjacent to tiles showing signatures of gold with similar characteristics.

The placer deposit in Tranquille Creek near Kamloops was economically significant [33], although the source of the gold remains unclear. The inclusion assemblage depicted in Figure 12C is most similar to the signature of epithermal gold from Blackdome, which suggests a genetic association with local Eocene volcanic rocks. Gold from the Klaza intermediate sulphidation deposit in Southern Yukon shows a similar signature but contains Bi. Gold from Tranquille Ck contained chalcocite inclusions, which are more commonly associated with porphyry rather than epithermal systems, and consequently, this sample population may itself be a mixture of gold from different episodes within the evolution of a magmatic hydrothermal event.

Other examples of where the few inclusions available are potentially useful in determining the source type are Youbou, Vancouver Island (tsumoite (BiTe), HREE silicate, and loelingite (FeAs) suggesting a low-sulphidation environment), Granite Creek near Princeton (BiPdTe-indicative either of an alkalic porphyry source or perhaps related to the source of platinum in local placers), and the Fraser River at Quesnel Canyon, where a particle containing temagamite ($Pd3HgTe_4$) inclusions was recorded, again suggestive of an alkalic porphyry source.

Individual gold particles derived from ultramafic rocks are distinctive (Figure 11), but the sample populations that contain them often also contain gold particles where Cu is below LOD, and these may host either pyrite or arsenopyrite inclusions, suggesting a different source deposit type. Radar diagrams depicting inclusion suites for sample populations where the distinctive Cu–Au microfabrics were present in some particles are presented in Figure 12, where it can be seen that the strong Cu signature is accompanied by various contributions from As (a minor component of gold from Wheaton and Sowaqua creeks) and Pb (absent in gold from Wheton and Sowaqua creeks). The detailed relationship between inclusion species and host alloy composition is provided in Figure 13, which comprises a bivariate Ag–Co plot with the host alloys of specific inclusion types indicated. shows the relationship of different inclusion species to their alloy host and provides evidence for sample populations of alluvial gold that contain contributions from both gold of ultramafic origin (low Ag, high Cu) and gold from orogenic hydrothermal systems (low Cu, high Ag). There is no necessity for these two populations to be compositionally mutually exclusive, but overall, the compositional fields are clearly distinct. The incidence of the distinctive Cu-rich gold particles is focused around the Bridge River area (Bridge River, Relay Creek) and the Coquihalla and its tributaries (Sowoqua Creek and also Ladner Creek). Gold from Ladner Creek exhibits an inclusion signature that shows a strong resemblance to gold from Sowoqua Creek (Figure 12), but around 20% of the gold particles contain Cu to >2 wt%. Gold from Peers Ck (another tributary of the Coquihalla) also contained relatively Cu-rich particles, but no inclusions were observed. Similarly, gold from Thibert Ck exhibited some very high Cu values, and one of these particles contained a bornite inclusion. However, there were insufficient inclusions present to fully characterize a signature, but it was noted that most other inclusions were either galena or pyrite hosted in gold where Cu was below LOD. Finally, a single particle exhibiting Cu exsolution from a Cu-rich Au–Ag alloy was also recorded in the sample population from the Fraser River above Williams Lake. Variably altered ultramafic rocks occur as small to very extensive fault-bounded bodies within and adjacent to many of the major fault zones in Southwestern BC [63], and these are the likely sources of the distinctive gold particles reported here.

During this study, sample populations from localities that had previously been investigated were augmented to generate more robust data describing the inclusion suites. In the case of Whipsaw Creek near the Copper Mountain alkalic Cu–Au porphyry deposit, new data has permitted a re-evaluation of the previous classification of derivation from an alkalic porphyry source originally proposed because of some shared compositional characteristics with gold from the nearby Friday Creek and Similkameen River localities [26]. Examination of 124 additional gold particles provided a large data set in which neither Pd or Hg-bearing inclusions nor alloys containing detectable Pd were observed (Figure 12D). The inclusion suite differs from that observed in gold from Friday Creek and most likely represents a mixture of gold from different source types, possibly including that present in the drainages of Granite Creek and the Tulameen River to the north. Unfortunately, there is insufficient data to characterize gold from these two drainages to afford a robust comparison. Nevertheless, the availability of a significantly larger data set has permitted refinement of the previous classification.

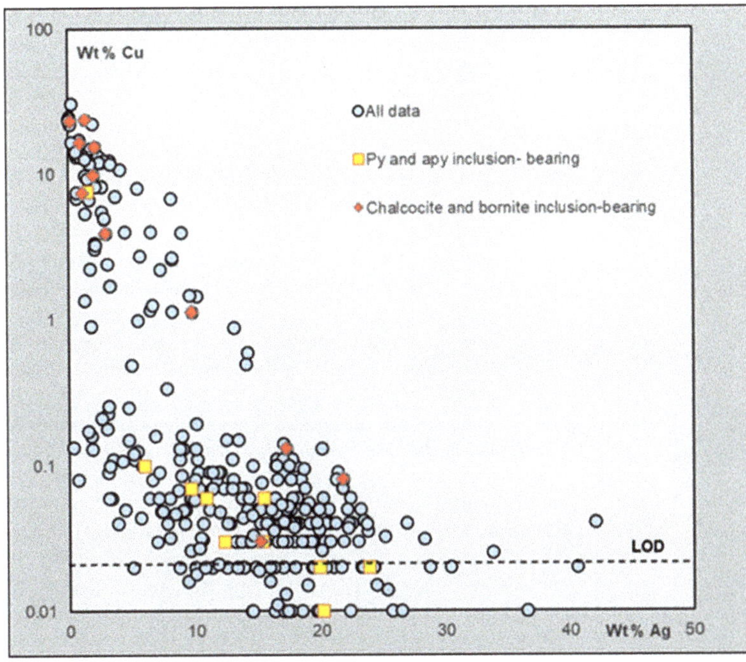

Figure 13. Distribution of inclusion species within different alloy compositional fields. 'All data' corresponds to sample populations from localities in which Cu–Au exsolution microfabrics were observed: Coquihalla R, Bridge River and Yalakom, Sowaqua Ck and Relay Ck.

The characterization of compositional ranges of gold from different deposit types in the Canadian Cordillera is an ongoing endeavor, with various targeted studies [26,27] feeding into a regional consideration [28] that has been expanded in the present study. From a global perspective, interpretations of large data sets such as this can be used to establish generic deposit-type signatures through integration with data sets describing gold from similar tectonic environments elsewhere [28]. The continuation of this work will without doubt involve the characterization of gold from deposit types where compositional data are relatively scarce. In addition, researchers interested in developing regional compositional templates should be aware of other gold signatures specific to specific deposit types, for example, those associated with relatively unusual hydrothermal systems [31] or those associated with a specific sub-class of a broader category [64].

5. Conclusions

This large-scale regional study has greatly increased our understanding of the range of compositional signatures of gold within the complex geological settings in British Columbia. Although the geochemistry and mineralogy of gold grains in the region are certainly far from simple, gold alloy compositions of populations of gold particles together with inclusion suites of opaque minerals have in many cases proven capable of discriminating between gold derived from different types of source deposits. In some cases, the project outcomes have confirmed those of previous work, but new insights have also been generated, principally by combining studies of in situ gold-bearing mineralization with compositional studies of detrital gold particles.

It is important to consider the full range of compositional data available when aiming to identify deposit-type signatures. In this study, the Ag content of the Au–Ag alloy has not proved useful in the majority of cases, as there is substantial overlap in the compositional ranges of gold from different deposit types and between samples from different localities of the same deposit type. Nevertheless, extreme values may prove informative. Where detectable, concentrations of minor metals (Cu, Hg, and Pd) can be useful as strong indicators of gold genesis, but in many cases they are below detection by EMP. The study of the relationship between the mineralogy of the inclusion suite in detrital gold particles and the ore mineralogy itself has proved far more illuminating.

Detailed petrographic studies of samples of gold-bearing ore from various deposit types have confirmed the relationship between ore mineralogy and the compositional signatures of gold particles in their erosional products. The range of well-constrained compositional templates for gold from orogenic and low-sulphidation epithermal systems shows strong similarities with mineralogical associations of gold with other coeval minerals within ore samples of those deposit types. This outcome supports two important claims about gold compositional studies. The first confirms that the inclusion suites of detrital gold from locations where the source is unknown can be used to elucidate the type of that source mineralization, and comparison of deposit-specific inclusion suites with those of gold from other localities in the Canadian Cordillera in neighboring Yukon confirms the generic nature of the signatures. The second has been brought into sharp focus in the present study and shows that an understanding of the mineral associations of gold in a specific deposit type may be used to predict the inclusion suite of any associated detrital gold liberated by erosional processes. Consequently, although there is limited inclusion data describing mineral suites in detrital gold from high-sulphidation, skarn, and intrusion-related deposits, their signature is predictable. Where small inclusion suites are available for comparison, they support this hypothesis.

For some localities in BC where the source of detrital gold is unknown, it has been possible to apply compositional templates to elucidate the source type. The success of this approach depends entirely on the quality of the sample population of the unknown sample in terms of the number of particles available and the degree to which they are representative of the whole population. Donated samples may not fulfill either of these criteria, and projects such as this generally demand dedicated sampling campaigns. This is especially important in regions such as BC, where gold from different source types may be present in a single drainage system, with the result that the detrital gold inventory may contain gold with different compositional signatures. An understanding of the various compositional ranges of gold from different deposit types permits discrimination between sub-populations, an accurate interpretation of local gold metallogeny, and an aid to the design of focused exploration campaigns.

The outcomes of the project can already underpin the examination of new sample suites where the source deposit types are unclear. At the very least, it is possible to clearly discriminate between gold from orogenic and magmatic hydrothermal systems as well as that associated with the ultramafic rocks present at several different localities. Further study is required to gain a better generic compositional template for gold from some deposit types and to identify the compositional nuances between gold from deposit types formed by magmatic hydrothermal systems.

Author Contributions: Conceptualization, R.C. and J.K.M.; methodology, R.C. and J.K.M.; formal analysis, R.C. and R.M.; investigation, R.C. and R.M.; resources, R.C. and J.K.M.; data curation, R.C.; writing—original draft preparation, R.C. and J.K.M.; writing—review and editing, R.C. and J.K.M.; visualization, R.C. and J.K.M. project administration, R.C.; funding acquisition, R.C. and J.K.M. All authors have read and agreed to the published version of the manuscript.

Funding: This research was funded by Geoscience BC, grant number 2018-013.

Data Availability Statement: The compositional data for all gold particles that contributed to this study can be accessed at https://www.geosciencebc.com/projects/2018-013/ (accessed on 9 August 2023).

Acknowledgments: The authors are indebted to Geoscience BC for funding this extensive project. UBC are thanked for making sample collections available to study, and Richard Walshaw at UoL provided constant support with all aspects of SEM and EMP analysis. Two anonymous reviewers are thanked for their constructive comments.

Conflicts of Interest: The authors declare no conflict of interest. The funders had no role in the design of the study; in the collection, analysis, or interpretation of data; in the writing of the manuscript; or in the decision to publish the results.

Appendix A. Summary of Sample Suites Where Source Deposit Type Is Known

Site	Region	Sample Type	No Particles	UTM	E	N
Alkalic porphyries						
Copper Mountain Magnetite Veins Virginia Pit	Similkameen	Hypogene	1	10	679873	5466650
Copper Mountain Pit 3 North Wall	Similkameen	Hypogene	2	10	680075	5466265
Copper Mtn Ingerbelle Pit	Similkameen	Hypogene	2	10	677637	5468093
Friday Ck	Similkameen	Detrital	173	10	674786	5463431
Galore Ck	Skeena	Detrital	55	9	351451	6334803
Galore, Scottsimpson Ck	Skeena	Detrital	95	9	347496	6327746
Mount Polley Springer Pit North West Face	Cariboo	Hypogene	1	10	591942	5822820
Mount Polley Underground Stockpile Wight Pit	Cariboo	Hypogene	1	10	591897	5822842
Mount Polley Wight Pit	Cariboo	hypogene	14	10	592861	5825510
Mount Polley Wight Pit	Cariboo	Detrital	1	10	592861	5825510
Mt Milligan King Richard Creek	Omineca	Detrital	39	10	432688	6108843
Mt Milligan MBX Pit	Omineca	Detrital	2	10	424363	6109380
Mt Milligan MBX Pit	Omineca	Hypogene	8	10	434572	6109300
Mt Milligan MBX Pit	Omineca	Hypogene	12	10	434698	6109460
Mt Milligan MBX Pit Blast hole 82-5522	Omineca	Hypogene	3	10	434669	6109040
New Afton	Kamloops	Hypogene	4	10	683647	5609456
New Afton	Kamloops	Hypogene	2	10	675628	5615087
New Afton Cherry I Ck	Kamloops	Detrital	57	10	672106	5615710
Similkameen River	Similkameen	Detrital	248	10	669117	5485510
Valleau Ck	Omineca	Detrital	177	10	386865	6133372
Calc-alkalic porphyries						
Iron Cap	Skeena	Hypogene	3	9	425500	6267000
Kerr Fan	Skeena	Detrital	5	9	421012	6261191
Mitchell Ck	Skeena	Detrital	164	9	416722	6262775
Relay	Lillooet	Detrital	64	10	509681	5664325
Sowchea	Omineca	Detrital	6	10	408890	6031602
Sulphurets Ck	Skeena	Detrital	304	9	419095	6261371
Tennyson Property	Skeena	Hypogene	5	9	427925	6236028
White Star Mine	Alberni	Hypogene	3	9	656988	5543526
High sulphidation epithermal						
Albert's Hump	Liard	Hypogene	8	9	594716	6371425
Taylor–Windfall	Clinton	Hypogene	32	10	475244	5661693
Low sulphidation epithermal						
Blackdome	Lillooet	Hypogene	501	10	531130	5653664
Bouleau Ck	Vernon	Detrital	8	11	318668	5568633
Brucejack	Skeena	Hypogene	90	9	426958	6258537
Fairless Ck	Lillooet	Detrital	65	10	530860	5653775
Mt Graves property	Omineca	Hypogene	6	9	621942	6361045
Second Ck	Clinton	Detrital	35	10	566452	5658769
Silbak–Premier	Skeena	Hypogene	47	9	500000	6209789
Stirrup Ck	Clinton	Detrital	259	10	557997	5659116
Whiteman–Boule Ck	Vernon	Detrital	21	11	315395	5566256

Site	Region	Sample Type	No Particles	UTM	E	N
Orogenic						
Amador Gulch	Cariboo	Detrital	75	10	588853	5876180
Antler Ck	Cariboo	Detrital	151	10	609056	5888223
Aurum Mine	New Westminster	Hypogene	52	10	623873	5485405
Baldhead Ck	Cariboo	Detrital	16	10	574196	5883884
Ballarat St Georges	Cariboo	Detrital	80	10	600450	5881650
Bassford Ck	Cariboo	Detrital	18	10	582546	5874825
BC Vein	Cariboo	Hypogene	25	10	596343	5883218
Beggs Gulch	Cariboo	Detrital	176	10	605454	5875266
Berube	Liard	Hypogene	29	9	462597	6570816
Boulder Ck	Atlin	Detrital	121	8	589229	6613761
Bralorne	Lillooet	Hypogene	173	10	515477	5623283
BRX Property	Lillooet	Hypogene	10	10	511246	5632230
Burns Ck	Cariboo	Detrital	132	10	590031	5881840
Burns Mountain (Perkins)	Cariboo	Hypogene	155	10	588790	5878034
California Ck	Cariboo	Detrital	101	10	606550	5873800
Cariboo Gold Qtz	Cariboo	Hypogene	20	10	596343	5883219
Cariboo R Bench	Cariboo	Detrital	142	10	602260	5835729
Carolin Mine	New Westminster	Hypogene	23	10	623873	5485405
Cayuse Ck	Lillooet	Detrital	75	10	567021	5610440
Cayuse Ck Balbernie	Lillooet	Detrital	55	10	572670	5613532
Chisholm Ck	Cariboo	Detrital	216	10	586791	5878197
Cottonwood Bar	Cariboo	Detrital	87	10	523229	5886799
Cottonwood R	Cariboo	Detrital	74	10	553205	5881895
Coulter Ck	Cariboo	Detrital	94	10	610550	5868650
Cow Mt	Cariboo	Hypogene	77	10	596343	5883219
Cunningham Ck	Cariboo	Detrital	164	10	606398	5870152
Dennis Ck	Liard	Detrital	39	9	476210	6579070
Devlin Bench	Cariboo	Detrital	35	10	597180	5888057
Dragon Ck	Cariboo	Detrital	193	10	582250	5885000
Eight Mile Lake	Cariboo	Detrital	279	10	597286	5666686
Emancipation_Mill	New Westminster	Hypogene	138	10	625900	5482980
Emancipation_	New Westminster	Hypogene	29	10	625900	5482980
Erickson Jennie	Liard	Hypogene	8	10	623873	5485405
Erickson Vollaug	Liard	Hypogene	2	9	461639	6564403
Erickson_Alison	Liard	Hypogene	18	9	461639	6564403
Erickson_Caitlin	Liard	Hypogene	8	9	461639	6564403
Erickson_Eileen	Liard	Hypogene	227	9	461639	6564403
Feather Ck	Atlin	Detrital	42	8	600785	6600303
Foster's Ledge	Cariboo	Hypogene	5	10	425565	5382210
Frasergold	Cariboo	Hypogene	15	10	579779	6650830
Frye Ck	Cariboo	Detrital	110	10	546145	5880934
Germansen	Omineca	Detrital	297	10	394405	6181362
Hibernia	Cariboo	Hypogene	23	10	586345	6011100
Hixon	Cariboo	Detrital	417	10	531790	5921604
Hurley R	Lillooet	Detrital	220	10	510797	5633030
Island Mt	Cariboo	Hypogene	7	10	594811	5884455
Jerry Ck Strathnaver	Cariboo	Detrital	80	10	558509	5937595
Keighley Ck	Cariboo	Detrital	95	10	604183	5849514
Lightening Ck	Cariboo	Detrital	146	10	561728	5874567
Likely	Cariboo	Detrital	26	10	604917	5828416
Lillooet	Lillooet	Detrital	62	10	576883	5618322
Little Snowshoe Ck	Cariboo	Detrital	35	10	604600	5856500
Lowhee Ck_1	Cariboo	Detrital	143	10	596500	5883750
Lowhee Ck_2	Cariboo	Detrital	28	10	597700	5881550
Manson Ck	Omineca	Detrital	26	10	406679	6170248
Maude Ck	Cariboo	Detrital	101	10	604300	5879450
McDame R	Liard	Detrital	124	9	475208	6570724
McKee Ck	Atlin	Detrital	1	8	581303	6592762

Site	Region	Sample Type	No Particles	UTM	E	N
Midas adit	Cariboo	Hypogene	247	10	606534	5856510
Montgomery Ck	Cariboo	Detrital	48	10	583950	5885450
Moorehead Ck	Cariboo	Detrital	102	10	580981	5833737
Mosquito Ck	Cariboo	Detrital	88	10	594500	5886100
Mosquito Mine	Cariboo	Hypogene	4	10	593903	5885241
Moustique Ck	Cariboo	Detrital	91	10	587500	5654570
Mt Calverly	Cariboo	Hypogene	2	10	604665	5827629
Mt Proserpine	Cariboo	Hypogene	30	10	600772	5877900
Murphy	New Westminster	Hypogene	59	10	613553	5473781
Myrtle	Cariboo	Hypogene	68	10	597414	5881788
Nelson Ck	Cariboo	Detrital	20	10	586700	5881900
Oregon Gulch	Cariboo	Detrital	74	10	586700	5879250
Otter Ck	Atlin	Detrital	306	8	590451	6610413
Perkins Gulch	Cariboo	Detrital	34	10	587655	5876715
Peter Ck	Cariboo	Detrital	42	10	611129	5863278
Pine Ck	Atlin	Detrital	27	8	575126	6603083
Pioneer	Lillooet	Hypogene	30	10	515477	5623283
Quartz Ck	Atlin	Detrital	100	8	600389	6613902
Quesnel Canyon	Cariboo	Detrital	106	10	538806	5872170
Rosella Ck	Liard	Detrital	146	9	478356	6590788
Ruby Ck	Atlin	Detrital	91	8	589293	6613778
Siwash Ck	New Westminster	Detrital	87	10	619497	5493434
Slough Bench	Cariboo	Detrital	96	10	587250	5883100
Snowy Ck placer	Liard	Hypogene	83	9	462875	6569894
Snowy_Crusher	Liard	Hypogene	12	9	463325	6569892
Snowy_Rich Vein	Liard	Hypogene	4	9	462300	6570388
Sooke R	Victoria	Detrital	86	10	447377	5371480
Sovereign Ck	Cariboo	Detrital	145	10	586700	5881900
Spanish Mt	Cariboo	Detrital	23	10	608007	5823814
Spanish Mt	Cariboo	lode	114	10	582763	6046650
Spring Ck	Liard	Detrital	17	9	476617	6582888
Spruce Ck	Atlin	Detrital	79	8	588902	6601480
Sugar Ck	Cariboo	Detrital	284	9	585695	5894707
Summit Ck	Cariboo	Detrital	50	10	589532	6029860
Tame Ck	Liard	Detrital	91	9	479889	6590234
Taurus Mine	Liard	Hypogene	9	9	460706	6570815
Tertiary	Cariboo	Hypogene	144	10	526980	5886565
Toop	Cariboo	Detrital	144	10	561974	5880404
Vancouver Mine	Nelson	Hypogene	6	11	490640	5443120
Warspite	Cariboo	Hypogene	46	10	601518	5876958
Wayside	Lillooet	Hypogene	102	10	512019	5636155
Wayside (Cariboo)	Cariboo	Hypogene	19	10	588155	5971100
Wells Adit	Cariboo	Hypogene	52	10	595950	5883250
Williams Ck	Cariboo	Detrital	54	10	599830	5881613
Wright Ck	Atlin	Detrital	92	8	593989	6607041
Yellowjacket Mine	Atlin	Hypogene	3	8	581908	6607172
Skarn						
Bolivar claim	Nanaimo	Hypogene	44	10	385757	5513482
Molly Claim	Nanaimo	Hypogene	1	10	390463	5509906
Tillicum	Slocan	Hypogene	48	11	449002	5537267
Ultramafic Intrusion related						
Sowaqua Ck	New Westminster	Detrital	54	10	625391	5478035
Thibert Ck	Liard	Detrital	106	9	421961	6522643
Wheaton Ck	Liard	Detrital	107	9	500300	6471824
VMS						
Britannia	Vancouver	Hypogene	38	10	489806	5495403

Site	Region	Sample Type	No Particles	UTM	E	N
Intrusion-related veins						
Bonaparte	Kamloops	Hypogene	4	10	679829	5653693
Bohan	Nelson	Detrital	1	11	546550	5459630
Cranbrook	Fort Steel	Hypogene	5	11	600110	5472550
Goat	Nelson	Detrital	12	11	545436	5465307
Kithchener	Fort Steel	Detrital	1	11	548995	5445300
Lamb	Fort Steel	Detrital	2	11	581529	5464916
Lewis Ck	Fort Steel	Detrital	1	11	568159	5467402
Moyie Lake	Fort Steel	Detrital	2	11	582540	5471680
Perry	Fort Steel	Detrital	14	11	565250	5480050

Appendix B. Summary of Sample Suites for Which Either the Source Deposit Type Is Unknown or Which Are Most Likely Mixtures of Gold Particles Form Different Source Types

Site	Region	No Particles	UTM	E	N
15 Mile	New Westminster	193	10	627488	5483762
Big Bar	Clinton	41	10	561069	5670320
Black Ck	Cariboo	4	10	629950	5797000
Bridge River	Lillooet	20	10	569676	5627361
Bridge–Yalakom	Lillooet	35	10	558221	5634927
Bridge River	Lillooet	119	10	510540	5632938
Canal flats	Golden	2	11	580880	5561320
Chilliwack	New Westminster	81	10	576159	5438322
Chimney Ck	New Westminster	13	10	614722	5506237
Coquihalla R	New Westminster	83	10	625304	5478308
Coquihalla–Dewdeney	New Westminster	3	10	626389	5481613
Elizabeth–Yalakom	Lillooet	4	10	531497	5653411
Findlay Ck	Golden	2	11	568380	5552920
Fountain Bar	Lillooet	143	10	579673	5622274
Fraser R	Clinton	8	10	576135	5650008
Gold Pan	Kamloops	56	10	614002	5579526
Gold Ck	Cariboo	100	10	559062	5910736
Gordon River	Victoria	34	10	402046	5404036
Granite Ck	Similkameen	86	10	667000	5485308
Haney Pit	Vancouver	50	10	547292	5447988
Kanaka Bar	Kamloops	2	10	602627	5552726
Ladner Creek	New Westminster	166	10	627367	5483766
Ladner Creek	New Westminster	40	10	626625	5490192
Liard R	Liard	95	9	497300	6672000
Lorne	Lilloet	90	10	512632	5624910
Lytton	Kamloops	74	10	596648	5572573
McConnel R	Omineca	19	9	655440	6305441
Peers Ck	New Westminster	50	10	622486	5471341
Pend d'Oreille R	Nelson	18	11	454747	5428220
Relay	Lilloet	64	10	509681	5664325
Scuzzy	New Westminster	35	10	607605	5521594
Slate Ck	Omineca	42	10	401829	6170664
Thibert Ck	Liard	106	9	421961	6522643
Tranquille R	Kamloops	164	10	687634	5617654
Tulameen	Similkameen	91	10	660813	5489829
Upper Fraser	Clinton	126	10	549640	5759834
Vedder (Chilliwack) R	New Westminster	23	10	584673	5435977
Whipsaw Creek	Similkameen	328	10	676644	5471200
Yalakom	Lillooet	45	10	535584	5657087
Yale	New Westminster	230	10	613877	5488020
Youbou	Victoria	31	10	412348	5414107

References

1. Averill, S.A. Viable indicator minerals in surficial sediments for two major base metal deposit types: Ni-Cu-PGE and porphyry Cu. *Geochem. Explor. Environ. Anal.* **2011**, *11*, 279–291. [CrossRef]
2. McClenaghan, M.B.; Paulen, R.C. Application of till mineralogy and geochemistry to mineral exploration. In *Past Glacial Environments*; Elsevier: Amsterdam, The Netherlands, 2018; pp. 689–751.
3. Layton-Matthews, D.; McClenaghan, M.B. Current Techniques and Applications of Mineral Chemistry to Mineral Exploration; Examples from Glaciated Terrain: A Review. *Minerals* **2022**, *12*, 59. [CrossRef]
4. McClenaghan, M.B.; Kjarsgaard, B.A. Indicator mineral and surficial geochemical exploration methods for kimberlite in glaciated terrain: Examples from Canada. In *Mineral Deposits of Canada: A Synthesis of Major Deposit-Types, District Metallogeny, the Evolution of Geological Provinces, and Exploration Methods*; Goodfellow, W.D., Ed.; Geological Association of Canada, Minnesota Department of Public Safety, Special Publication: Amsterdam, The Netherlands, 2007; Volume 5, pp. 983–1006.
5. Barnett, P.J.; Averill, S. Heavy mineral dispersal trains in till in the area of the Lac des Iles PGE deposit, northwestern Ontario, Canada. *Geochem. Explor. Environ. Anal.* **2010**, *10*, 391–399. [CrossRef]
6. Averill, S.A. The Blackwater gold-spessartine-pyrolusite glacial dispersal train, British Columbia, Canada; influence of sampling depth on indicator mineralogy and geochemistry. *Geochem. Explor. Environ. Anal.* **2011**, *17*, 43–60. [CrossRef]
7. Manéglia, N.; Beaudoin, G.; Simard, M. Indicator Minerals of the Meliadine Orogenic Gold Deposits, Nunavut (Canada), And Application to Till Surveys. *Geochem. Explor. Environ. Anal.* **2018**, *18*, 241–251. [CrossRef]
8. Celis, M.A.; Bouzari, F.; Bissig, T.; Hart, C.J.R.; Ferbey, T. Petrographic characteristics of porphyry indicator minerals from alkalic porphyry copper-gold deposits in south-central British Columbia (NTS 092, 093). In *Geoscience BC Summary of Activities 2013*; Report 2014-1; Geoscience BC: Vancouver, BC, Canada, 2013.
9. Pisiak, L.K.; Canil, D.; Grondahl, C.; Plouffe, A.; Ferbey, T.; Anderson, R.G. Magnetite as a porphyry copper indicator mineral in till: A test using the Mount Polley porphyry copper-gold deposit, south-central British Columbia (NTS 093A). In *Geoscience BC Summary of Activities 2014*; Report 2015-1; Geoscience BC: Vancouver, BC, Canada, 2015; pp. 141–150.
10. Bouzari, F.; Hart, C.J.R.; Barker, S.; Bissig, T. Porphyry indicator minerals (PIMs): Exploration for concealed deposits in south central British Columbia (NTS 092I/06, 093A/12, 093N/01, /14). In *Geoscience BC Summary of Activities 2009*; Report 2010-1; Geoscience BC: Vancouver, BC, Canada, 2010; pp. 25–32.
11. Bouzari, F.; Hart, C.J.R.; Barker, S. Hydrothermal alteration revealed by apatite luminescence and chemistry: A potential indicator mineral for exploring covered porphyry copper deposits. *Econ. Geol.* **2016**, *111*, 1397–1410. [CrossRef]
12. Mao, M.; Rukhlov, A.S.; Rowins, S.M.; Spence, J.; Coogan, L.A. Apatite trace element compositions: A robust new tool for mineral exploration. *Econ. Geol.* **2016**, *111*, 1187–1222. [CrossRef]
13. McClenaghan, M.B.; Beckett-Brown, C.E.; McCurdy, M.W.; McDonald, A.M.; Leybourne, M.I.; Chapman, J.B.; Plouffe, A.; Ferbey, T. Mineral markers of porphyry copper mineralization: Progress report on the evaluation of tourmaline as an indicator mineral. *Target. Geosci.* **2017**, *8358*, 69–77. [CrossRef]
14. McClenaghan, M.B.; Cabri, L.J. Review of gold and platinum group element (PGE) indicator minerals methods for surficial sediment sampling. *Geochem. Expl. Environ. Anal.* **2011**, *2011 11*, 251–264. [CrossRef]
15. Girard, R.; Tremblay, J.; Néron, A.; Longuépée, H. Automated gold grain counting. Part 1: Why counts matter! *Minerals* **2021**, *11*, 337. [CrossRef]
16. Nevolko, P.A.; Kolpakov, V.V.; Nesterenko, G.G.; Fominykh, P.A. Alluvial placer gold of the Egor'evsk district (northern-Western Salair): Composition characteristics, types and mineral microinclusions. *Russ. Geol. Geophys.* **2019**, *60*, 67–85. [CrossRef]
17. Svetlitskaya, T.V.; Nevolko, P.A.; Kolpakov, V.V.; Tolstykh, N.D. Native gold from the Inagli Pt–Au placer deposit (the Aldan Shield, Russia): Geochemical characteristics and implications for possible bedrock sources. *Miner Depos.* **2018**, *53*, 323–338. [CrossRef]
18. Lalomov, A.V.; Chefranov, R.M.; Naumov, V.A.; Naumova, O.B.; Lebarge, W.; Dilly, R.A. Typomorphic features of placer gold of Vagran cluster (the Northern Urals) and search indicators for primary bedrock gold deposits. *Ore Geol. Rev.* **2017**, *85*, 321–335. [CrossRef]
19. Zaykov, V.V.; Melekestseva, I.Y.; Zaykova, E.V.; Kotlyarov, V.A.; Kraynev, Y.D. Gold and platinum group minerals in placers of the South Urals: Composition, microinclusions of ore minerals and primary sources. *Ore Geol. Rev.* **2017**, *85*, 299–320. [CrossRef]
20. Fominykh, P.A.; Nevolko, P.A.; Svetlitskaya, T.V.; Kolpakov, V.V. Native gold from the Kamenka-Barabanovsky and Kharuzovka alluvial placers (Northwest Salair Ridge, Western Siberia, Russia): Typomorphic features and possible bedrock sources. *Ore Geol. Rev.* **2020**, *126*, 103781. [CrossRef]
21. Nikiforova, Z.S.; Tolstov, A.V. Gold-bearing placer assemblages in the east of the siberian platform: Origin and prospects. *Geol. Ore Dep.* **2022**, *64*, 1–25. [CrossRef]
22. Becerra, P.; Sanchez-Alfaro, P.; Piquer, J.; Plissart, G.; Garroz, B.; Kunstmann, D. Gold Provenance in placers from Pureo area, southern Chile coastal cordillera, and their relationship with paleozoic metamorphic rocks. *Minerals* **2022**, *12*, 1147. [CrossRef]
23. Alam, M.; Li, S.R.; Santosh, M.; Yuan, M.W. Morphology and chemistry of placer gold in the Bagrote and Dainter streams, northern Pakistan: Implications for provenance and exploration. *Geol. J.* **2018**, *54*, 1672–1687. [CrossRef]
24. Ali, L.; Chapman, R.; Farhan, M.; Shah, M.T.; Khattak, S.A.; Ali, A. Exploration methodology using morphology and alloy composition of alluvial gold: A case study from quaternary deposits of the Nowshera District, Khyber Pakhtunkhwa, Pakistan. *Min. Metall. Explor.* **2021**, *2021 38*, 367–377.

25. Gas'kov, I.V. Major impurity elements in native gold and their association with gold mineralization settings in deposits of Asian folded areas. *Russ. Geol. Geophys.* **2017**, *58*, 1080–1092. [CrossRef]
26. Chapman, R.J.; Mileham, T.J.; Allan, M.M.; Mortensen, J.K. A distinctive Pd-Hg signature in detrital gold derived from alkalic Cu-Au porphyry systems. *Ore Geol. Rev.* **2017**, *83*, 84–102. [CrossRef]
27. Chapman, R.J.; Allan, M.M.; Mortensen, J.K.; Wrighton, T.M.; Grimshaw, M.R. A new indicator mineral methodology based on a generic Bi-Pb-Te-S mineral inclusion signature in detrital gold from porphyry and low/intermediate sulfidation epithermal environments in Yukon Territory, Canada. *Miner. Depos.* **2018**, *53*, 815–834. [CrossRef]
28. Chapman, R.J.; Mortensen, J.K.; Allan, M.M.; Walshaw, R.D.; Bond, J.; MacWilliam, K. A New Approach to Characterizing Deposit Type Using Mineral Inclusion Assemblages in Gold Particles. *Econ. Geol.* **2022**, *117*, 361–381. [CrossRef]
29. Townley, B.K.; Hérail, G.; Maksaev, V.; Palacios, C.; De Parseval, P.; Sepulveda, F.; Orellana, R.; Rivas, P.; Ulloa, C. Gold grain morphology and composition as an exploration tool: Application to gold exploration in covered areas. *Geochem. Explor. Environ. Anal.* **2003**, *3*, 29–38. [CrossRef]
30. Moles, N.R.; Chapman, R.J.; Warner, R.B. The significance of copper concentrations in natural gold alloy for reconnaissance exploration and understanding gold-depositing hydrothermal systems. *Geochem. Explor. Environ. Anal.* **2013**, *13*, 115–130. [CrossRef]
31. Chapman, R.J.; Leake, R.C.; Bond, D.P.; Stedra, V.; Fairgrieve, B. Chemical and mineralogical signatures of gold formed in oxidizing chloride hydrothermal systems and their significance within populations of placer gold grains collected during reconnaissance. *Econ. Geol.* **2009**, *104*, 563–585. [CrossRef]
32. Chapman, R.J.; Banks, D.A.; Styles, M.T.; Walshaw, R.D.; Piazolo, S.; Morgan, D.J.; Grimshaw, M.R.; Spence-Jones, C.P.; Matthews, T.J.; Borovinskaya, O. Chemical and physical heterogeneity within native gold: Implications for the design of gold particle studies. *Miner. Depos.* **2021**, *56*, 1563–1588. [CrossRef]
33. Holland, S.S. *Placer Gold Production of British Columbia*; BC Ministry of Energy, Mines and Petroleum Resources, Bull: Victoria, BC, Canada, 1950; **Bull**. 28.
34. Plouffe, A.; Ferby, T. Porphyry Cu indicator minerals in till: A method to discover buried mineralization. In *Indicator Minerals in Till and Stream Sediments of the Canadian Cordillera*; Ferbey, T., Plouffe, A., Hickin, A., Eds.; Topics in Mineral Sciences; Mineralogical Association of Canada; The Geological Association of Canada: Newfoundland, NL, Canada, Special Paper 50; 2017; Volume 47, pp. 129–159, ISBN 978-0-921294-60-3.
35. Wrighton, T.M. Placer Gold Microchemical Characterization and Shape Analysis Applied as an Exploration Tool in Western Yukon. MSC Thesis, University of British Columbia, Vancouver, BC, Canada, 2013.
36. Nelson, J.L.; Colpron, M.; Israel, S. The cordillera of British Columbia, Yukon, and Alaska: Tectonics and metallogeny. In *Tectonics, Metallogeny, and Discovery: The North American Cordillera and Similar Accretionary Settings*; SEG Special Publication: Amsterdam, The Netherlands, 2013; Volume 17, pp. 53–109.
37. Nelson, J.O.; Colpron, M.A.; Goodfellow, W.D. Tectonics and metallogeny of the British Columbia, Yukon and Alaskan Cordillera, 1.8 Ga to the present. In *Mineral Deposits of Canada: A Synthesis of Major Deposit-Types, District Metallogeny, the Evolution of Geological Provinces, and Exploration Methods*; The Geological Association of Canada, Mineral Deposits Division, Special Publication: Newfoundland, NL, Canada, 2007; Volume 5, pp. 755–791.
38. Mortensen, J.K.; Craw, D.; MacKenzie, D.J. Concepts and revised models for Phanerozoic orogenic gold deposits. In *Recent Advances in Understanding Gold Deposits: From Orogeny to Alluvium*; Torvela, T.M., Chapman, R.J., Lambert-Smith, J., Eds.; Special Publication: London, UK, 2022; Volume 516, pp. 15–46.
39. Chapman, R.J.; Craw, D.; Moles, N.R.; Banks, D.A.; Grimshaw, M.R. Evaluation of the contributions of gold derived from hypogene, supergene and surficial processes in the formation of placer gold deposits. In *Recent Advances in Understanding Gold Deposits: From Orogeny to Alluvium*; Torvela, T.M., Chapman, R.J., Lambert-Smith, J., Eds.; The Geological Society, Special Publication: London, UK, 2022; Volume 516, pp. 337–352.
40. Chapman, R.J.; Moles, N.R.; Bluemel, B.; Walshaw, R.D. Detrital gold as an indicator mineral. In *Recent Advances in Understanding Gold Deposits: From Orogeny to Alluvium*; Torvela, T.M., Chapman, R.J., Lambert-Smith, J., Eds.; The Geological Society, Special Publication: London, UK, 2022; Volume 516, pp. 291–312.
41. Gammons, C.H.; Williams-Jones, A.E. Hydrothermal geochemistry of electrum; thermodynamic constraints. *Econ. Geol.* **1995**, *90*, 420–432. [CrossRef]
42. Chapman, R.J.; Mortensen, J.K.; Crawford, E.C.; Lebarge, W. Microchemical studies of placer and lode gold in the Klondike District, Yukon, Canada: 1. Evidence for a small, gold-rich, orogenic hydrothermal system in the Bonanza and Eldorado Creek area. *Econ. Geol.* **2010**, *105*, 1369–1392. [CrossRef]
43. Chapman, R.J.; Mortensen, J.K. Characterization of gold mineralization in the northern Cariboo gold district, British Columbia, Canada, through integration of compositional studies of lode and detrital gold with historical placer production: A template for evaluation of orogenic gold districts. *Econ. Geol.* **2016**, *111*, 1321–1345.
44. Knight, J.B. A Microprobe Study of Placer Gold and Its Origin in the Lower Fraser River Drainage Basin. MSc Thesis, The University of British Columbia, Vancouver, BC, Canada, 1985.
45. Knight, J.; McTaggart, K.C. The composition of placer and lode gold from the lower Fraser River drainage area, south western British Columbia. *Geol. J. Can. Inst. Min. Metall.* **1986**, *1*, 21–30.

46. Knight, J.; McTaggart, K.C. Composition of gold from southwestern British Columbia: A progress report. In *Geological Fieldwork 1988*; BC Min. Energy, Min Pet. Resources, Paper 1989-1; Geoscience BC: Vancouver, BC, Canada, 1989; pp. 387–394.
47. Knight, J.; McTaggart, K.C. Lode and placer gold of the Coquihalla and Wells area, British Columbia (92H93H). In *Exploration in British Columbia 1989*; Min. Energy, Min. Pet Resources; Geoscience BC: Vancouver, BC, Canada, 1990; pp. 105–118.
48. McTaggart, K.C.; Knight, J. *Geochemistry of Lode and Placer Gold of the Cariboo District, BC*; BC Ministry of Energy, Mines and Petroleum Resources, Open File: Vancouver, BC, Canada, 1993; Open File 1993-30; 25p.
49. Knight, J.; Leitch, C.H. Phase relations in the system Au–Cu–Ag at low temperatures, based on natural assemblages. *Can. Min.* **2001**, *39*, 889–905. [CrossRef]
50. Banks, D.A.; Chapman, R.J.; Spence-Jones, C. *Detrital Gold as a Deposit-Specific Indicator Mineral by LAICP-MS Analysis*; Geoscience BC: Vancouver, BC, Canada, 2018; Report 2018-21.
51. Johnston, W.A.; Uglow, W.L. *Placer and Vein Deposits of Barkerville, Cariboo District, British Columbia*; Geol. Survey of Canada, Memoir 149; Geoscience BC: Vancouver, BC, Canada, 1926; 246p.
52. Eyles, N. Characteristics and origin of coarse gold in Late Pleistocene sediments of the Cariboo placer mining district, British Columbia, Canada. *Sed. Geol.* **1995**, *95*, 69–95. [CrossRef]
53. Eyles, N.; Kocsis, S.P. Sedimentological controls on gold in a late Pleistocene glacial placer deposit, Cariboo Mining District, British Columbia, Canada. *Sed. Geol.* **1989**, *65*, 45–68. [CrossRef]
54. Reith, F.; Fairbrother, L.; Nolze, G.; Wilhelmi, O.; Clode, P.L.; Gregg, A.; Parsons, J.E.; Wakelin, S.A.; Pring, A.; Hough, R.; et al. Nanoparticle factories: Biofilms hold the key to gold dispersion and nugget formation. *Geology* **2010**, *38*, 843–846. [CrossRef]
55. Spence-Jones, C.P. Geochemical Signatures of Native Gold Alloys as a Tool for Understanding Auriferous Ore Deposits. Ph.D. Thesis, University of Leeds, Leeds, UK, 2022.
56. Warr, L.N. IMA–CNMNC approved mineral symbols. *Mineral Mag.* **2021**, *85*, 291–320. [CrossRef]
57. Sack, P.J.; Mihalynuk, M.G. *Proximal Gold Cassiterite Nuggets and Composition of the Feather Creek Placer Gravels: Clues to a Lode Source near Atlin, BC*; Geological Fieldwork: Vancouver, BC, Canada, 2003; 2004-1; pp. 147–161.
58. Stanley, C.R.; Holbek, P.M.; Huyck, H.L.O.; Lang, J.R.; Preto, V.A.G.; Blower, S.J.; Bottaro, J.C. Geology of the Copper Mountain alkalic copper-gold porphyry deposits, Princeton, British Columbia. In *Porphyry Deposits of the Northwestern Cordillera of North America*; Schroeter, T.G., Quebec, C.I., Eds.; Min, Metal, Pet Special; Geological Fieldwork: Vancouver, BC, Canada, 1995; CIM Special Volume 46; pp. 537–564.
59. Graham, H.C. Trace Elements in Ore Minerals as Indicators of Hydrothermal Fluid Evolution in the Au-Rich Porphyry System of Iron Cap, British Columbia, Canada. Ph.D. Thesis, University of Leeds, Leeds, UK, 2022.
60. Campbell, M.E.; Dilles, J.H. *Magmatic History of the Kerr-Sulphurets-Mitchell Copper-Gold Porphyry District, Northwestern British Columbia (NTS 104B)*; Geoscience BC Summary of Activities: Vancouver, BC, Canada, 2016; 2017-1.
61. Dawson, G.L. Geological Setting of the Hedley Gold Skarn Camp with Specific Reference to the French Mine, South-Central British Columbia. MSC Thesis, University of British Columbia, Vancouver, BC, Canada, 1994.
62. Potter, M.; Styles, M.T. Gold characterization as a guide to bedrock sources for the Estero Hondo alluvial gold mine, western Ecuador. *Trans. Inst. Min. Metall. (Sect. B Appl. Earth Sci.)* **2003**, *112*, 297–304.
63. Schiarizza, P.O.; Gaba, R.G.; Coleman, M.; Garver, G.I.; Glover, M. *Geology and Mineral Occurrences of the Yalakom River Area (920/1,2, 92J/15,16)*; BC Geological Survey, Geological Fieldwork: Vancouver, BC, Canada, 1989.
64. Lalomov, A.; Grigorieva, A.; Kotov, A.; Ivanova, L. Typomorphic Features and Source of Native Gold from the Sykhoi Log Area Placer Deposits, Bodaibo Gold-Bearing District, Siberia, Russia. *Minerals* **2023**, *13*, 707. [CrossRef]

MDPI
St. Alban-Anlage 66
4052 Basel
Switzerland
www.mdpi.com

Minerals Editorial Office
E-mail: minerals@mdpi.com
www.mdpi.com/journal/minerals

www.ingramcontent.com/pod-product-compliance
Lightning Source LLC
Chambersburg PA
CBHW061546220926
51850CB00005B/262
* 9 7 8 3 0 3 6 5 9 3 8 1 4 *